Birkhäuser

Progress in Nonlinear Differential Equations and Their Applications

Volume 96

More information about this series at https://link.springer.com/bookseries/4889

Haïm Brezis · Petru Mironescu

Sobolev Maps to the Circle

From the Perspective of Analysis, Geometry, and Topology

Birkhäuser

Haïm Brezis
Visiting Distinguished Professor,
Department of Mathematics
Rutgers University
Piscataway, NJ 08854, USA

and

Visiting Distinguished Professor,
Departments of Mathematics
and Computer Science, Technion
32.000 Haifa, Israel

and

Professeur émérite,
Laboratoire J.L. Lions
Sorbonne Université
Paris Cedex 05 75252, France

Petru Mironescu
Institut Camille Jordan
Université Claude Bernard Lyon 1
Villeurbanne, France

ISSN 1421-1750 ISSN 2374-0280 (electronic)
Progress in Nonlinear Differential Equations and Their Applications
ISBN 978-1-0716-1510-2 ISBN 978-1-0716-1512-6 (eBook)
https://doi.org/10.1007/978-1-0716-1512-6

Mathematics Subject Classification: 26A45, 28A75, 35J25, 46A55, 46E35, 49Q20, 55M25, 58B05, 58E20, 58E50, 90C46

This book is published under the imprint Birkhäuser, www.birkhauser-science.com by the registered company Springer Science+Business Media, LLC
The registered company address is: 1 New York Plaza, New York, NY 10004, U.S.A.

Dedicated to the memory of Jean Bourgain,
with immense admiration and gratitude

Preface

The theory of Sobolev functions from a domain $\Omega \subset \mathbb{R}^N$ with values in $\mathbb{R}$ (or $\mathbb{R}^k$) is by now a classical part of analysis, and has a wide range of applications in pure and applied mathematics. By contrast, the study of Sobolev maps from Ω to $\mathscr{N}$, where $\mathscr{N}$ is a compact manifold embedded in some $\mathbb{R}^\ell$, e.g., $\mathscr{N} = \mathbb{S}^k$, with $k \geq 1$, is relatively new, not yet fully developed, and, up to now, reserved for a small "club" of researchers. The incentive to explore these spaces came, in recent years, from two directions: geometry and physics. In geometry, they appear in the study of harmonic maps. In physics, they arise, for example, in the Oseen–Frank theory of nematic liquid crystals, in the Ginzburg–Landau theory of superconductors, in superfluids and in micromagnetics. The reason one is led to work with Sobolev maps rather than smooth maps is to allow *singularities* such as $\dfrac{x}{|x|}$ in 2D, or line singularities in 3D, which occur naturally both in geometry and physics.

In this monograph, we only briefly mention the connections to harmonic maps and to physics, for which we refer the reader to specialized works such as Schoen and Uhlenbeck [317, 318], Bethuel, Brezis, and Hélein [31] and Sandier and Serfaty [315]. Our main goal here is the study of the *intrinsic properties* of the Sobolev spaces

$$W^{1,p}(\Omega; \mathbb{S}^1) := \{u \in W^{1,p}(\Omega; \mathbb{R}^2);\ |u(x)| = 1 \text{ a.e. on } \Omega\},\ 1 \leq p < \infty,$$

and, more generally, the fractional Sobolev spaces $W^{s,p}(\Omega; \mathbb{S}^1)$, $0 < s < \infty$, $1 \leq p < \infty$.

In order to simplify the presentation, we concentrate on $\mathbb{S}^1$, rather than $\mathbb{S}^k$, $k \geq 1$, since the $\mathbb{S}^1$ theory is fairly polished. However, we mention here and there, in Chapter 4 and in sections entitled "Complements and Open Problems," what needs to be modified – or what remains to be proved – when the target space is $\mathbb{S}^k$.

Our decision (a long time ago) to write a book that would include "everything you wanted to know" about Sobolev maps to the circle, prompted us to address many unsettled questions. In turn, this led to a series of publications, whose results have entered the current book. We traveled part of the road jointly with our dearly missed friend Jean Bourgain, a luminous and inspiring collaborator. As a matter of fact, Jean became attracted by the subject when one of us (HB) asked him the seemingly innocuous question whether a map $u \in H^s((0,1); \mathbb{S}^1)$ admits a lifting (i.e., a phase) $\varphi \in H^s((0,1); \mathbb{R})$ when $0 < s < 1/2$. Jean proved that the answer is positive, but the question turned out to be tougher – and the proof much more involved – than expected. This result became the heart of our first joint paper [50]. Incidentally, we still do not have a simple proof of this fact (see Open Problem 7).

As the reader will discover, the classes $W^{s,p}(\Omega; \mathbb{S}^1)$ have an *amazingly rich structure*. Topological and geometrical effects are already visible, even in this simple framework, since $\mathbb{S}^1$ has non-trivial topology. Moreover, the fact that the target space is $\mathbb{S}^1$ (as opposed to $\mathbb{S}^k$) offers the option to consider liftings of maps $u : \Omega \to \mathbb{S}^1$. The search for "optimal liftings" leads to an interesting connection with minimal surfaces spanned by a curve.

The reader will also find throughout this book unexpected connections with a range of topics such as the Monge–Kantorovich theory in optimal transport, items in geometric measure theory, such as the fine theory of BV functions, rectifiable currents, minimal surfaces, etc., Fourier series, non-local functionals occurring, for example, as denoising filters in image processing, and more.

A final word of warning. Anyone familiar with standard properties of the space $W^{1,p}(\Omega; \mathbb{R})$ should expect surprises. Here are, for example, three striking differences, that we discuss in great length in this text:

a) $C^\infty(\overline{\Omega}; \mathbb{R})$ is dense in $W^{1,p}(\Omega; \mathbb{R})$; by contrast, $C^\infty(\overline{\Omega}; \mathbb{S}^1)$ need *not* be dense in $W^{1,p}(\Omega; \mathbb{S}^1)$.
b) The trace space for $W^{1,p}(\Omega; \mathbb{R})$ is precisely $W^{1-1/p,p}(\partial\Omega; \mathbb{R})$, $1 < p < \infty$. By contrast, the trace space for $W^{1,p}(\Omega; \mathbb{S}^1)$ is contained in $W^{1-1/p,p}(\partial\Omega; \mathbb{S}^1)$, but the inclusion can be *strict*.
c) If $\varphi \in W^{1,p}(\Omega; \mathbb{R})$, then $u := e^{\imath\varphi} \in W^{1,p}(\Omega; \mathbb{S}^1)$ and $|\nabla u| = |\nabla \varphi|$ a.e. on Ω. One might expect (by analogy with the C^k case) that, conversely, any $u \in W^{1,p}(\Omega; \mathbb{S}^1)$ comes from a $\varphi \in W^{1,p}(\Omega; \mathbb{R})$. This is *not* true!

We hope that intrigued readers will "join the club" and tackle some of the open problems scattered throughout the book.

Acknowledgments

HB is indebted to Jerald Ericksen and David Kinderlehrer for introducing him to the enchanted world of sphere-valued maps during the special program on liquid crystals they organized at IMA (the University of Minnesota) in 1984–1985.

The earliest papers of HB on this topic were written (in the 1980s and 1990s) jointly with Jean-Michel Coron, Elliott Lieb, and later with Fabrice Bethuel and Frédéric Hélein. They were followed by several contributions with the late Louis Nirenberg and with Yanyan Li. He thanks his coauthors for these fruitful collaborations that paved the way for numerous subsequent developments.

More recently, we enjoyed working with Luigi Ambrosio, Alessio Figalli, Emmanuel Russ, Itai Shafrir, and Yannick Sire on some topics that are presented in this book.

In the early 2000s, several talented young mathematicians shared our gusto for Sobolev maps to manifolds: Pierre Bousquet, Juan Dávila, Radu Ignat, Benoît Merlet, Vincent Millot, Hoai-Minh Nguyen, Adriano Pisante, Arkady Poliakovsky, Augusto Ponce, and Jean Van Schaftingen. Lately, Eduard Curcă, Xavier Lamy, and Ioana Molnar "joined the club."

Over the years, HB taught graduate courses at the Université Paris VI (now called Sorbonne Université) and at Rutgers, based on sections of this monograph. He received useful feedback from the attendees, in particular from Vicenţiu Rădulescu and David Herrera.

The *Rencontres d'analyse*, co-organized by PM at the Camille Jordan Institute in Lyon, have been a great environment for discussing some of the results in this book, and for fruitful collaborations.

The presentation of the first chapter owes much to Itai Shafrir, who shared with us an illuminating duality approach. We warmly thank him for his precious input on various other sections, and also for the constructive reading of preliminary versions of parts of this book.

Throughout the years of work on this book, we have received help and advice from many colleagues: Eli Aljadeff, Eric Baer, John Ball, Vincent Borrelli (with special thanks from PM), George Dinca, Lawrence Craig Evans, Michael Frazier, Mariano Giaquinta, Robert Hardt, Bernd Kawohl, Fanghua Lin, Jean Mawhin, Vladimir Maz'ya, Élisabeth Mironescu, Tristan Rivière, Jacob Rubinstein, Étienne Sandier, Ovidiu Savin, Peter Sternberg, and Cédric Villani, to whom we express our gratitude.

During part of the preparation of this monograph, HB was partially supported by NSF grants DMS 0802958 and 1207793, and by grant number 238702 of the European Commission (ITN, project FIRST). PM acknowledges the support of the Labex MILYON/ANR-10-LABX-0070. A significant part of this work has been written during several visits of PM to the Mathematics Departments of Rutgers University and the Technion, and a sabbatical year at the Simion Stoilow Institute of Mathematics of the Romanian Academy in Bucharest. He is very grateful to these institutions for their hospitality.

We warmly thank Barbara Mastrian for her kind dedication and professional help during the preparation of this text.

And, finally, we are extremely grateful to our spouses Élisabeth and Michal for their patience and encouragement for all these years; this book would not have been possible without their constant support.

Paris, New Brunswick, Jerusalem, Haifa
Lyon, Bucharest
September 2020

Haïm Brezis
Petru Mironescu

Overview

Our text is structured as follows. Each chapter deals with a single topic. They start with an Introduction, which consists of a rather detailed overview. The first sections of each chapter are written in a textbook style, focusing on results and giving rather complete proofs. The Complements and Open problems sections provide short introductions to various subsequent developments or related topics, and suggest new directions of research. The Comments sections present some historical perspectives on each topic, together with a rather comprehensive list of references.

For the sake of simplicity, we leave aside throughout the monograph (except in Chapter 14) the topological effects that may possibly arise when Ω has non-trivial topology. In order to further simplify the presentation, we assume in the overview of Chapters 1–13 that Ω is a ball.

Chapter 1. We investigate here the question of lifting within the framework of Sobolev spaces $W^{1,p}$. In the process, we are led to the introduction of two *fundamental* tools, Ju and $\Sigma(u)$, which are ubiquitous throughout the entire book.

Given a map $u : \Omega \to \mathbb{S}^1$, we may always write $u = e^{\imath\varphi}$, for some $\varphi : \Omega \to \mathbb{R}$. The function φ is called a *lifting* or a *phase* of u. Some liftings φ can be very rough; for example, $u \equiv 1$ admits non-measurable liftings. However, when u is continuous (or better), we may find a lifting φ as smooth as u: for example, if $u \in C^k$, then we may find $\varphi \in C^k$.

This need not be the case in the context of Sobolev spaces. More specifically, consider, for $1 \le p < \infty$, the space

$$\begin{aligned} W^{1,p}(\Omega; \mathbb{S}^1) :=& \{u \in W^{1,p}(\Omega; \mathbb{C});\ |u(x)| = 1 \text{ for a.e. } x \in \Omega\} \\ \simeq& \{u \in W^{1,p}(\Omega; \mathbb{R}^2);\ |u(x)| = 1 \text{ for a.e. } x \in \Omega\}. \end{aligned}$$

When $N = 1$ and $u \in W^{1,p}(\Omega; \mathbb{S}^1)$, then u is continuous, and any continuous lifting φ of u belongs to $W^{1,p}$. However, when $N \ge 2$, a map $u \in W^{1,p}(\Omega; \mathbb{S}^1)$ *need not have a* $W^{1,p}(\Omega; \mathbb{R})$ *phase*. For example, if $N = 2$, $0 \in \Omega$, and

$$u(x) := \frac{x}{|x|}, \tag{1}$$

then $u \in W^{1,p}$, $\forall\, 1 \le p < 2$, but u has no lifting in $W^{1,p}$, and not even in $W^{1,1}_{loc}$.

This is a "typical" example of non-lifting. By contrast, when $N \ge 2$ and $p \ge 2$, $W^{1,p}(\Omega; \mathbb{S}^1)$ does have the *lifting property*, i.e., *every* $u \in W^{1,p}(\Omega; \mathbb{S}^1)$ can be written as $u = e^{\imath\varphi}$ for some $\varphi \in W^{1,p}(\Omega; \mathbb{R})$.

In the "nasty" range $N \ge 2$, $1 \le p < 2$, where the lifting property fails, there are two natural directions to investigate.

Direction 1. *Characterize* the u's in $W^{1,p}(\Omega; \mathbb{S}^1)$ that can be written as $u = e^{\imath\varphi}$ for some $\varphi \in W^{1,p}(\Omega; \mathbb{R})$.

Direction 2. If $u \in W^{1,p}(\Omega; \mathbb{S}^1)$ admits no lifting in $W^{1,p}$, what are the "best possible" alternative classes in which a lifting exists?

In the first direction, a major role is played by the "distributional Jacobian" Ju. To motivate its introduction in our context, we start with the following elementary calculation. Let $u = (u_1, u_2) \in W^{1,p}(\Omega; \mathbb{S}^1)$ and assume that

$$u = e^{\imath\varphi} \text{ for some } \varphi \in W^{1,p}(\Omega; \mathbb{R}). \tag{2}$$

Differentiating the equalities

$$u_1 = \cos\varphi, \quad u_2 = \sin\varphi,$$

we find that

$$\nabla u_1 = -\sin\varphi\, \nabla\varphi, \quad \nabla u_2 = \cos\varphi\, \nabla\varphi,$$

and therefore

$$\nabla\varphi = \cos\varphi\, \nabla u_2 - \sin\varphi\, \nabla u_1 = u_1 \nabla u_2 - u_2 \nabla u_1 := u \wedge \nabla u. \tag{3}$$

Since the left-hand side of (3) is a gradient, it follows that the curl of the right-hand side (3) (considered in the sense of distributions) vanishes. We can express this fact as $Ju = 0$ in $\mathscr{D}'(\Omega)$, where Ju is the $N \times N$ antisymmetric matrix with entries in $\mathscr{D}'(\Omega; \mathbb{R})$, given, $\forall\, 1 \le i, j \le N$, by

$$\begin{aligned}(Ju)_{ij} :=& \frac{1}{2}\left[\frac{\partial}{\partial x_i}(u \wedge \nabla u)_j - \frac{\partial}{\partial x_j}(u \wedge \nabla u)_i\right] \\ =& \frac{1}{2}\left[\frac{\partial}{\partial x_i}\left(u \wedge \frac{\partial u}{\partial x_j}\right) - \frac{\partial}{\partial x_j}\left(u \wedge \nabla \frac{\partial u}{\partial x_i}\right)\right].\end{aligned} \tag{4}$$

For reasons that will be explained in Chapter 1, this object is called the distributional Jacobian of u.

Thus, the condition $Ju = 0$ is a *necessary condition* for the existence of $\varphi \in W^{1,p}$ satisfying (2). Remarkably, this condition is also *sufficient*, and this gives a complete and useful answer to Direction 1.

In particular, this implies that, for the u given by (1), we have $Ju \neq 0$. More specifically, we have the following formula (which will be generalized considerably in Chapters 2 and 3):

$$Ju = \begin{pmatrix} 0 & \pi\,\delta_0 \\ -\pi\,\delta_0 & 0 \end{pmatrix}. \tag{5}$$

As we are going to see, there is a way to "quantify" the failure of the condition $Ju = 0$. The quantity $\Sigma(u)$ defined below measures the "deviation" of $u \wedge \nabla u$ from the space of gradients. Given $u \in W^{1,1}(\Omega; \mathbb{S}^1)$, we set

$$\Sigma(u) := \inf\{\|u \wedge \nabla u - \nabla\eta\|_{L^1};\ \eta \in W^{1,1}(\Omega; \mathbb{R})\}. \tag{6}$$

In Section 1.5, we will prove (via duality) that $\Sigma(u)$ coincides with $\|Ju\|$ for some appropriate seminorm $\|\ \|$ on the space of distributions $(W^{-1,1})^{N\times N}$ (to which Ju belongs).

In summary, we have, $\forall\, u \in W^{1,1}(\Omega; \mathbb{S}^1)$,

$$\Sigma(u) = 0 \iff Ju = 0 \iff u = e^{\imath\varphi} \text{ for some } \varphi \in W^{1,1}(\Omega; \mathbb{R}). \tag{7}$$

Here are three important occurrences of $\Sigma(u)$.

A) From the above discussion, we know that, if $u \in W^{1,1}(\Omega; \mathbb{S}^1)$ is such that $Ju \neq 0$, then there exists no lifting φ of u with $\varphi \in W^{1,1}(\Omega; \mathbb{R})$. However, *any* $u \in W^{1,1}(\Omega; \mathbb{R})$ may be written as

$$u = e^{\imath\varphi} \text{ for some } \varphi \in BV(\Omega; \mathbb{R}), \tag{8}$$

where BV denotes the spaces of functions of bounded variation – a space slightly larger than $W^{1,1}$. For example, the map u in (1) can be written as $u = e^{\imath\varphi}$, where $\varphi := \arg u \in [0, 2\pi)$ is the principal branch of the argument of u. This φ belongs to BV, and jumps by 2π on $\Omega \cap ([0, \infty) \times \{0\})$.

As the above example shows, a major difference between BV and $W^{1,1}$ is that BV allows discontinuities along hypersurfaces (e.g., line discontinuities when $N = 2$).

Another major difference between lifting in the space of continuous functions (or, as we are going to see in Chapter 6, in $W^{1,1}$) and lifting in BV is that φ in (8) is not unique, even (mod 2π): for example, if B is a ball such that $B \subset \Omega$ and φ satisfies (8), then $\varphi + 2\pi\,\chi_B$ still satisfies (8).

We may thus ask whether some BV liftings are "better" than others. It turns out that a natural way of selecting "best" BV liftings involves the "energy"

$$E(u) := \min\{|\varphi|_{BV};\ \varphi \in BV(\Omega;\mathbb{R}) \text{ and } u = e^{\imath\varphi} \text{ in } \Omega\}, \tag{9}$$

where $|\varphi|_{BV}$ is the mass of the measure $|D\varphi|$.

One of the main results in this chapter is that

$$E(u) = \int_\Omega |\nabla u| + \Sigma(u),\ \forall\, u \in W^{1,1}(\Omega;\mathbb{S}^1).$$

B) As we are going to see in Chapters 2 and 3, the distribution $(1/\pi)\,Ju$ is supported by the "singular set" of u and carries information about the location and strength of the topological singularities of u. With this in mind, the quantity

$$S_p(u) := \inf\left\{\int_\Omega |\nabla v|^p;\ v \in W^{1,p}(\Omega;\mathbb{S}^1),\ Jv = Ju\right\},$$

defined, for every $u \in W^{1,p}(\Omega;\mathbb{S}^1)$, represents the *least* energy required to "produce" a map $v \in W^{1,p}(\Omega;\mathbb{S}^1)$ having the same topological singularities as u.

Here again $\Sigma(u)$ pops up:

$$S_1(u) = 2\pi\ \Sigma(u),\ \forall\, u \in W^{1,1}(\Omega;\mathbb{S}^1). \tag{10}$$

C) The *relaxed energy* is defined for every $u \in W^{1,p}(\Omega;\mathbb{S}^1)$ by

$$R_p(u) := \inf\left\{\liminf_{n\to\infty}\int_\Omega |\nabla u_n|^p;\ u_n \in C^\infty(\overline{\Omega};\mathbb{S}^1),\ u_n \to u \text{ a.e. on } \Omega\right\},$$

where the first inf is taken over all sequences (u_n) in $C^\infty(\overline{\Omega};\mathbb{S}^1)$ such that $u_n \to u$ a.e. on Ω. We will prove that

$$R_1(u) = \int_\Omega |\nabla u| + \Sigma(u),\ \forall\, u \in W^{1,1}(\Omega;\mathbb{S}^1). \tag{11}$$

Chapter 2. In this chapter, we give illuminating geometric interpretations for Ju and $\Sigma(u)$ when $N = 2$. In 2D, we naturally identify Ju, given by (4), with the *scalar* distribution $(Ju)_{12}$. Assuming, for example, that $u \in W^{1,1}(\Omega;\mathbb{S}^1)$ is smooth except at some point $a \in \Omega$, and that the winding number, $\deg(u,a)$, of u on small circles around a, equals one, we have the following generalization of (5):

$$Ju = \pi\delta_a \text{ in } \mathscr{D}'(\Omega). \tag{12}$$

More generally, if $u \in W^{1,1}(\Omega;\mathbb{S}^1)$ is a "nice" map, i.e., continuous on Ω except at a finite number of distinct points $a_1,\ldots,a_k$, we have

$$Ju = \pi\sum_{j=1}^{k}\deg(u,a_j)\,\delta_{a_j} \text{ in } \mathscr{D}'(\Omega). \tag{13}$$

Nice maps u play an important role because they are "generic" in $W^{1,1}(\Omega; \mathbb{S}^1)$ (as explained in Chapter 10). This allows results valid for nice maps to be transferred to general u's in $W^{1,1}(\Omega; \mathbb{S}^1)$. A typical result is the following extension of (13): for every $u \in W^{1,1}(\Omega; \mathbb{S}^1)$, there exist points $P_j, N_j \in \overline{\Omega}$ such that

$$\sum_{j=1}^{\infty} |P_j - N_j| < \infty, \tag{14}$$

$$Ju = \pi \sum_{j=1}^{\infty} (\delta_{P_j} - \delta_{N_j}) \text{ in } \mathscr{D}'(\Omega). \tag{15}$$

Conversely, given any points $P_j, N_j \in \overline{\Omega}$ satisfying (14), there exists some $u \in W^{1,1}(\Omega; \mathbb{S}^1)$ such that (15) holds.

Next, we turn to the geometric interpretation of $\Sigma(u)$ as a "minimal length" required to "connect the singularities."

For simplicity, assume again that u is a nice map, continuous except at the distinct points $P_j, N_j, j = 1, \ldots, m$, satisfying $\deg(u, P_j) = 1$, $\deg(u, N_j) = -1$, $\forall j$. Then we have

$$\Sigma(u) = \min_{\sigma \in S_m} 2\pi \sum_{j=1}^{m} d(P_j, N_{\sigma(j)}), \tag{16}$$

where S_m denotes the group of permutations of the integers $\{1, \ldots, m\}$ and $d(P, N)$ is the pseudometric defined on $\overline{\Omega}$ by

$$d(P, N) := \min\{\operatorname{dist}(P, \partial\Omega) + \operatorname{dist}(N, \partial\Omega),\ |P - N|\}.$$

Obviously, the right-hand side of (16) has a flavor of optimal transport, and indeed the Monge–Kantorovich formula plays a role in the proof of (16).

Again, we can adapt formula (16) to general u's in $W^{1,1}(\Omega; \mathbb{S}^1)$; this involves the minimization of the length of 1-rectifiable currents $\mathscr{C}$ (in the sense of geometric measure theory) such that $\partial\mathscr{C} = (1/\pi) Ju$.

Chapter 3. This chapter is the 3D (and higher) counterpart of the previous one. Here, geometry enters decisively, the basic objects being curves of singularities, which play in 3D the same role as points in 2D.

We illustrate this analogy when $N = 3$ and u has the simplest possible singular set: a closed curve Γ.

In 3D, we naturally identify Ju given by (4) with a vector field whose entries are scalar distributions:

$$Ju \simeq ((Ju)_{23}, (Ju)_{31}, (Ju)_{12}).$$

Assume that $u \in W^{1,1}(\Omega; \mathbb{S}^1) \cap C(\Omega \setminus \Gamma)$, where $\Gamma \subset \Omega$ is a smooth simple closed oriented curve. If the winding number of u around Γ equals one, then $(1/\pi)\, Ju$ can be identified with the integration along Γ, i.e.,

$$\langle Ju, \zeta \rangle = \pi \int_\Gamma \tau \cdot \zeta, \ \forall \zeta \in C_c^\infty(\Omega; \mathbb{R}^3),$$

where τ denotes the unit tangent to Γ; this is the analogue of (12).

For such maps, we have the following counterpart of (16):

$$\Sigma(u) = 2\pi \ \inf\{|M|; M \subset \Omega \text{ is a smooth oriented surface with boundary } \Gamma\}, \tag{17}$$

where $|M|$ is the area of M and the inf in (17) is the quantity that arises in the celebrated Plateau problem.

Incidentally, as a byproduct of our analysis, we obtain the following formula for the least area spanned by a smooth simple closed oriented curve $\Gamma \subset \mathbb{R}^3$:

$$\begin{aligned} &\inf\{|M|;\ M \subset \mathbb{R}^3 \text{ is a smooth oriented surface with boundary } \Gamma\} \\ &= \sup\left\{\int_\Gamma \tau \cdot \zeta;\ \zeta \in C_c^\infty(\mathbb{R}^3; \mathbb{R}^3),\ \|\operatorname{curl} \zeta\|_{L^\infty} \le 1\right\}. \end{aligned} \tag{18}$$

Note that assertion (18) does not involve $\mathbb{S}^1$-valued maps. Instead, it provides a new approach to minimal surfaces via a kind of "Monge–Kantorovich formula." Interestingly, the proof of (18) that we present transits via $\mathbb{S}^1$-valued maps.

Chapter 4. The purpose of this brief introduction to the world of sphere-valued maps is to explain how some of the tools and results presented in the previous chapters extend to maps $u : \Omega \to \mathbb{S}^k$, with $1 \le k \le N-1$, and where $W^{1,k}(\Omega; \mathbb{S}^k)$ plays the role of $W^{1,1}(\Omega; \mathbb{S}^1)$.

When $k \ge 2$, lifting is not available, but there are natural counterparts for Ju, $\Sigma(u)$, $S_p(u)$, and $R_p(u)$.

The "historical" case $k = 2$, $N = 3$, which was originally motivated by liquid crystals, is very similar to the case $k = 1$, $N = 2$ discussed in Chapter 2. When u is nice, there are formulas similar to (13) and (16). The counterparts of (10) and (11) are

$$S_2(u) = 2\,\Sigma(u),\ \forall u \in H^1(\Omega; \mathbb{S}^2), \tag{19}$$

and

$$R_2(u) = \int_\Omega |\nabla u|^2 + 2\,\Sigma(u),\ \forall u \in H^1(\Omega; \mathbb{S}^2). \tag{20}$$

At the next levels of the extreme case $k = N - 1$, i.e., $k = 3$, $N = 4$, etc., the situation is still rather satisfactory. Formula (19) has to be replaced by

$$S_k(u) = k^{k/2}\, \Sigma(u),\ \forall\, u \in W^{1,k}(\Omega; \mathbb{S}^k). \tag{21}$$

The analogue of (20) has not been established, but is highly plausible.

However, in the "intermediate" range, $2 \le k \le N - 2$ (starting with $N = 4$, $k = 2$), a new phenomenon occurs. One can still assert that $S_k(u) \ge k^{k/2}\, \Sigma(u)$, but in general one may have strict inequality. However, the geometric flavor of $S_k(u)$ persists. Equality (21) has to be replaced by

$$S_k(u) = k^{k/2}\, \Sigma^*(u),\ \forall\, u \in W^{1,k}(\Omega; \mathbb{S}^k), \tag{22}$$

where $\Sigma^*(u)$ corresponds to the least mass among all $(N - k)$-rectifiable currents spanned by the singular set of u.

Here, it is also expected (but not proved when $3 \le k \le N - 1$) that

$$R_k(u) = \int_\Omega |\nabla u|^k + k^{k/2}\, \Sigma^*(u),\ \forall\, u \in W^{1,k}(\Omega; \mathbb{S}^k).$$

Chapter 5. Starting from this chapter, we introduce the reader into the realm of fractional Sobolev spaces. The first question we ask, in the spirit of Chapter 1, is: does *any* $u \in W^{s,p}(\Omega; \mathbb{S}^1)$ admit a lifting $\varphi \in W^{s,p}(\Omega; \mathbb{R})$? Here, for $0 < s < \infty$ and $1 \le p < \infty$, we set

$$\begin{aligned} W^{s,p}(\Omega; \mathbb{S}^1) :=& \{u \in W^{s,p}(\Omega; \mathbb{C});\ |u(x)| = 1 \text{ for a.e. } x \in \Omega\} \\ \simeq& \{u \in W^{s,p}(\Omega; \mathbb{R}^2);\ |u(x)| = 1 \text{ for a.e. } x \in \Omega\}. \end{aligned}$$

We give a full answer to this question.

When $N = 1$, the answer is positive, $\forall\, s$, $\forall\, p$, but the proof is surprisingly difficult if $sp < 1$. When $N \ge 2$, a significant "phase transition" occurs at $s = 1$. In the case $s < 1$, the answer is positive if and only if $sp < 1$ or $sp \ge N$. In the case $s \ge 1$, the answer is positive if and only if $sp \ge 2$ – a fact that is consistent with the result in Chapter 1 concerning $s = 1$.

Chapter 6. We investigate here the question of *uniqueness* (mod 2π) of a lifting. More precisely, assume that we have two liftings φ_1, φ_2 for the same u. Then, $\varphi_1(x) - \varphi_2(x) = 2\pi\, \psi(x)$ for some $\psi : \Omega \to \mathbb{Z}$. Therefore, we are led to the question of finding minimal assumptions on a measurable function $\psi : \Omega \to \mathbb{Z}$ implying that ψ is constant a.e. Clearly, continuity is sufficient, but, as we are going to see, constancy holds for a surprisingly large class of possibly *discontinuous* functions.

It is not difficult to see that any $\psi \in VMO\,(\Omega; \mathbb{Z})$ must be constant a.e., where *VMO* denotes the space of functions of vanishing mean oscillation. This is a small step beyond continuity.

It is also well-known that any $\psi \in W^{1,1}(\Omega; \mathbb{Z})$ must be constant a.e., simply because $\nabla\psi = 0$ a.e. on Ω.

A more surprising (and not so well-known) fact is that any $\psi \in W^{1/p,p}(\Omega; \mathbb{Z})$, $1 < p < \infty$, must be constant a.e. For the enjoyment of the reader, we present several proofs of this result.

Our favorite one (but not the shortest one) goes as follows. If $\psi \in W^{1/p,p}$, then (by definition)

$$\int_\Omega \int_\Omega \frac{|\psi(x) - \psi(y)|^p}{|x - y|^{N+1}}\, dxdy < \infty,$$

and thus (using the fact that $\psi(x) - \psi(y) \in \mathbb{Z}$)

$$\int_\Omega \int_\Omega \frac{|\psi(x) - \psi(y)|}{|x - y|^{N+1}}\, dxdy < \infty. \tag{23}$$

At this stage, one can discard the assumption that ψ is integer-valued and prove that any measurable function $\psi : \Omega \to \mathbb{R}$ satisfying (23) must be constant a.e.

Here enters the "BBM formula," which asserts that, for any measurable function $f : \Omega \to \mathbb{R}$,

$$\lim_{\varepsilon \to 0} \int_\Omega \int_\Omega \frac{|f(x) - f(y)|}{|x - y|} \rho_\varepsilon(|x - y|)\, dxdy = K_N \int_\Omega |Df|, \tag{24}$$

where (ρ_ε) is a family of radial mollifiers, K_N is a constant depending only on N, and $\int_\Omega |Df|$ is understood to be $+\infty$ if f is not a BV function. Applying (24) to $\rho_\varepsilon(r) := \dfrac{C\varepsilon}{r^{N-\varepsilon}}\, \chi(r)$, where χ is a cutoff function, and using (23), yields $\psi \in BV$ and $D\psi = 0$ in $\mathscr{D}'$, and thus ψ is constant a.e.

We seize this opportunity to "advertise" some of the many avatars of the BBM formula, and its potential applications in a variety of directions, including "non-local minimal surfaces," "s-perimeters," and image processing.

Chapter 7. As explained above, a map $u \in W^{s,p}(\Omega; \mathbb{S}^1)$, where $N \geq 2$, $0 < s < 1$, $1 \leq p < \infty$ and $1 \leq sp < N$, need not have a phase in $W^{s,p}$.

The main (and highly non-trivial) result in this chapter asserts, in particular, that any $u \in W^{s,p}(\Omega; \mathbb{S}^1)$, where $N \geq 1, 0 < s < 1, 1 \leq p < \infty$ and $sp \geq 1$, can be "factorized" as

$$u = e^{\imath\varphi}\, v, \;\text{ with } \varphi \in W^{s,p}(\Omega; \mathbb{R}) \text{ and } v \in W^{1,sp}(\Omega; \mathbb{S}^1),$$

with corresponding estimates.

Note that v has much better regularity than u – it admits a "full" gradient, as opposed to u, which has only "fractional" derivatives – moreover, $W^{1,sp}(\Omega; \mathbb{S}^1) \hookrightarrow$

$W^{s,p}(\Omega; \mathbb{S}^1)$ when $sp > 1$. Consequently, we can apply to v all the results in Chapters 1–3, including the analysis of its topological singularities. As we are going to see in Chapter 8, this basic tool allows us to give a proper meaning to the topological singularities of any $u \in W^{s,p}(\Omega; \mathbb{S}^1)$ with $N \geq 2$, $0 < s < 1$, $1 \leq p < \infty$, and $sp \geq 1$ – for example $u \in H^{1/2}(\Omega; \mathbb{S}^1)$.

Chapter 8. We present various applications of the factorization. Arguably the most spectacular one is the possibility of giving a "robust" definition to $u \wedge \nabla u$ when $u \in W^{1/p,p}(\Omega; \mathbb{S}^1)$, with $1 < p < \infty$. As a byproduct, we also obtain a robust definition of the distribution Ju for such u. For the sake of completeness, we mention that there exists a totally different technique for defining Ju when $u \in W^{1/p,p}$, that does not rely on factorization and can be adapted to $\mathbb{S}^k$-valued maps.

We explain how the factorization allows us to complement the results on lifting obtained in Chapter 3, by providing alternative lifting spaces in the cases where the answer to the lifting problem is negative.

Chapter 9. While the previous chapters were concerned with the *existence* and *uniqueness* of a lifting, this chapter investigates the matter of *estimates* for the phase φ of u. We exhibit two unusual aspects: "existence without estimates," and "estimates without existence."

Chapter 10. We investigate here density questions. For *real-valued* Sobolev spaces, $C^\infty(\overline{\Omega}; \mathbb{R})$ is dense in $W^{s,p}(\Omega; \mathbb{R})$, for any $s > 0$ and $1 \leq p < \infty$. This need not be true for the Sobolev spaces $W^{s,p}(\Omega; \mathscr{N})$, where $\mathscr{N}$ is a manifold. In particular, this is not always the case when $\mathscr{N} = \mathbb{S}^1$.

We present the optimal conditions on s and p so that $C^\infty(\overline{\Omega}; \mathbb{S}^1)$ is dense in $W^{s,p}(\Omega; \mathbb{S}^1)$. More precisely, we prove that density holds except in the "nasty" range $N \geq 2$, $s > 0$, $1 \leq p < \infty$, and $1 \leq sp < 2$.

We prove that, in the nasty range, maps in $W^{s,p}(\Omega; \mathbb{S}^1)$ that are smooth except on "small sets" $\Sigma \subset \Omega$ are dense in $W^{s,p}(\Omega; \mathbb{S}^1)$. In particular, when $N = 2$, Σ can be a finite collection of points.

In the same range, we characterize the closure of $C^\infty(\overline{\Omega}; \mathbb{S}^1)$ in $W^{s,p}(\Omega; \mathbb{S}^1)$. More specifically, we prove that

$$\overline{C^\infty(\overline{\Omega}; \mathbb{S}^1)}^{W^{s,p}} = \{u \in W^{s,p}(\Omega; \mathbb{S}^1);\ Ju = 0\}.$$

When $0 < s < 1$, we use here the definition of Ju introduced in Chapter 8.

Chapter 11. We present a complete trace theory for $\mathbb{S}^1$-valued maps. When $s > 0$ is not an integer and $1 \leq p < \infty$, standard trace theory for *real-valued* maps asserts that

$$\operatorname{tr}_\Omega W^{s+1/p,p}(\Omega \times (0,1); \mathbb{R}) = W^{s,p}(\Omega; \mathbb{R}),$$

where we identify Ω with $\Omega \times \{0\}$. When $\mathbb{R}$ is replaced by a manifold $\mathscr{N}$, in general we only have the inclusion

$$\operatorname{tr}_\Omega W^{s+1/p,p}(\Omega \times (0,1); \mathcal{N}) \subset W^{s,p}(\Omega; \mathcal{N}), \tag{25}$$

and equality may fail.

In the special case where $\mathcal{N} = \mathbb{S}^1$, we prove that equality occurs in (25) if and only if $sp < 1$ or $sp \geq N$.

In the remaining cases, where $1 \leq sp < N$, we characterize the range of the mapping $v \mapsto \operatorname{tr}_\Omega v$, with $v \in W^{s+1/p,p}(\Omega \times (0,1); \mathbb{S}^1)$.

Chapter 12. We revisit the notion of topological degree $\deg f$ (aka index or winding number) for maps $f : \mathbb{S}^1 \to \mathbb{S}^1$. This is a classical concept when f is continuous: $\deg f$ counts "how many times $f(\mathbb{S}^1)$ covers $\mathbb{S}^1$, taking into account algebraic multiplicity." One can still give a robust definition for $\deg f$ when f belongs merely to $VMO(\mathbb{S}^1; \mathbb{S}^1)$, and thus, by the Sobolev embeddings, for maps in the critical spaces $W^{1/p,p}(\mathbb{S}^1; \mathbb{S}^1)$, with $1 < p < \infty$. We establish some basic properties of this degree.

We also derive an integral formula for the degree of VMO maps, generalizing the Kronecker formula

$$\deg f = \frac{1}{\pi} \int_{\mathbb{D}} \operatorname{Jac} u, \ \forall f \in C^1(\mathbb{S}^1; \mathbb{S}^1), \ \forall u \in C^1(\overline{\mathbb{D}}; \mathbb{R}^2) \text{ with } u_{|\partial\mathbb{D}} = f.$$

Note that, when $f \in VMO(\mathbb{S}^1; \mathbb{S}^1)$ – or even $f \in C(\mathbb{S}^1; \mathbb{S}^1)$ – standard trace theory does not yield the existence of a map $u : \mathbb{D} \to \mathbb{R}^2$ such that $u_{|\partial\mathbb{D}} = f$ and $u \in H^1(\mathbb{D})$ – and thereby $\operatorname{Jac} u \in L^1(\mathbb{D})$. We will explain how to bypass this difficulty.

We investigate bounds of $|\deg f|$ in terms of some natural norms; a typical estimate is

$$|\deg f| \leq C_p |f|^p_{W^{1/p,p}}, \ \forall f \in W^{1/p,p}(\mathbb{S}^1; \mathbb{S}^1), \ \forall 1 < p < \infty.$$

A full section is dedicated to the formula

$$\deg f = \sum_{n \in \mathbb{Z}} n\, |a_n|^2, \ \forall f \in H^{1/2}(\mathbb{S}^1; \mathbb{S}^1),$$

where the a_n's are the Fourier coefficients of f.

This striking formula is the starting point of the intriguing saga "Can one hear the degree?," which involves deep interactions with Fourier analysis and still raises challenging open problems.

Chapter 13. We investigate minimization problems of the form

$$\min \left\{ \int_\Omega |\nabla u|^p; \ u \in W^{1,p}_g(\Omega; \mathbb{S}^1) \right\}, \tag{26}$$

where $1 \leq p < \infty$ and $g \in C^\infty(\partial\Omega; \mathbb{S}^1)$ is a given boundary condition (satisfying also $\deg g = 0$ when $N = 2$).

We prove that a "phase transition" appears at $p = 2$. When $p \geq 2$, the minimizer in (26) is unique and smooth; lifting theory from Chapter 1 enters in the proof. By contrast, when $1 < p < 2$, minimizers may have singularities. This is a consequence of the following "gap phenomenon," which occurs for some g's:

$$\min\left\{\int_\Omega |\nabla u|^p;\ u \in W^{1,p}_g(\Omega;\mathbb{S}^1)\right\} < \inf\left\{\int_\Omega |\nabla u|^p;\ u \in C^1_g(\overline{\Omega};\mathbb{S}^1)\right\} < \infty.$$

We discuss the nature of these singularities when $N = 2$.

We also survey what happens to (26) when the class $W^{1,p}_g(\Omega;\mathbb{S}^1)$ is empty. This occurs, for example, when $N = 2$, $p \geq 2$, $g \in C^\infty(\partial\Omega;\mathbb{S}^1)$, and $\deg g \neq 0$. Alternatively, this may occur when $N \geq 3$ and $2 \leq p < N$, for some g's such that $g \in W^{1-1/p,p}(\partial\Omega;\mathbb{S}^1)$, whereas $W^{1,p}_g(\Omega;\mathbb{S}^1) = \emptyset$ (see Chapter 11). These are typical examples of "infinite energy" minimization problems, and they can be tackled via a Ginzburg–Landau-type approximation

$$\min\left\{\int_\Omega |\nabla u|^p + \frac{1}{\varepsilon^2}\int_\Omega (|u|^2 - 1)^2;\ u \in W^{1,p}_g(\Omega;\mathbb{R}^2)\right\},$$

analyzing the asymptotic behavior of minimizers as $\varepsilon \to 0$. This analysis can be extremely delicate, and we only sketch some of the highlights, whose details are beyond the scope of this book.

Chapter 14. We investigate the impact of the topology of Ω on the main topics presented above (existence of lifting, relaxed energy, density of smooth maps, etc.).

Appendices. In the final appendices, we recall standard properties of Sobolev spaces used throughout the book, and we gather the most technical parts of some of the proofs presented in the main text.

Notation

N.1 We denote $\mathbb{N} = \{0, 1, \ldots\}$ and $\mathbb{N}^* = \{1, 2, \ldots\}$
N.2 N is the space dimension
N.3 The standard Euclidean norm in $\mathbb{R}^N$ is denoted by $|\ |$
N.4 $B_R(x) = B(x, R) := \{y \in \mathbb{R}^N;\ |x - y| < R\}$
N.5 $S(x, R) := \{y \in \mathbb{R}^N;\ |x - y| = R\}$
N.6 $C(x, R) := \{y \in \mathbb{R}^2;\ |x - y| = R\}$
N.7 $\mathbb{D}$ is the (Euclidean) unit disk in $\mathbb{R}^2$, and $\mathbb{D}_r$ is the disk of radius r centered at the origin
N.8 $\mathbb{S}^k$ is the (Euclidean) unit sphere in $\mathbb{R}^{k+1}$. In particular, $\mathbb{S}^1$ is the unit circle. $\mathbb{B}^N$ is the (Euclidean) unit ball in $\mathbb{R}^N$
N.9 The Lebesgue measure of a set $A \subset \mathbb{R}^N$ is denoted by $|A|$. Occasionally, A can be a subset of a k-dimensional manifold, and then $|A|$ stands for the k-dimensional Hausdorff measure of A: $|A| := \mathscr{H}^k(A)$
N.10 $\omega_N := |\mathbb{B}^N|$ is the volume of the unit Euclidean ball $\mathbb{B}^N$ of $\mathbb{R}^N$
N.11 $\fint$ stands for the average integral: $\fint_A f := \frac{1}{|A|}\int_A f$
N.12 The characteristic function of A is denoted by χ_A
N.13 Ω stands for a domain in $\mathbb{R}^N$. A "standard" domain is either a smooth bounded domain, a half space, or $\mathbb{R}^N$ itself
N.14 The complement of A is denoted by A^c
N.15 The set of points x having some property is denoted within [ ]. For example, the set $\{x;\ u(x) = v(x)\}$ will be denoted by $[u = v]$
N.16 If $f : X \to Y$, where (X, d) and (Y, δ) are metric spaces, then $|f|_{\text{Lip}}$ denotes the Lipschitz constant of f:

$$|f|_{\text{Lip}} := \left\{\sup \frac{\delta(f(x), f(y))}{d(x, y)};\ x, y \in X, x \neq y\right\}$$

N.17 When s is an integer and $1 \leq p \leq \infty$, we define the Sobolev space $W^{s,p}(\Omega; \mathbb{R}^m) := \{u : \Omega \to \mathbb{R}^m;\ \|u\|_{W^{s,p}} < \infty\}$, where $\|u\|_{W^{s,p}} := \sum_{j \leq s}$

$\|D^j u\|_{L^p}$. Here, $\|u\|_{L^p}^p := \int_\Omega |u(x)|^p \, dx$ when $1 \le p < \infty$, resp. $\|u\|_{L^\infty} :=$ $\operatorname{esssup}_\Omega |u(x)|$ when $p = \infty$, where $|u(x)|$ is the Euclidean norm of $u(x)$ in $\mathbb{R}^m$

N.18 When $s > 0$ is not an integer and $1 \le p < \infty$, $W^{s,p}$ is the Slobodeskii space (referred to in the book as the fractional Sobolev space), $W^{s,p}(\Omega; \mathbb{R}^m) := \{u : \Omega \to \mathbb{R}^m; \|u\|_{W^{s,p}} < \infty\}$. Here, we write $s = n + \sigma$, with $n \in \mathbb{N}$ and $0 < \sigma < 1$, and we let

$$\|u\|_{W^{s,p}} := \sum_{j \le n} \|D^j u\|_{L^p} + \left(\iint_{\Omega \times \Omega} \frac{|D^n u(x) - D^n u(y)|^p}{|x - y|^{N + \sigma p}} \, dxdy \right)^{1/p}$$

N.19 When $0 < s < 1$ and $1 \le p < \infty$, $|u|_{W^{s,p}}$ denotes the seminorm

$$|u|_{W^{s,p}} := \left(\iint_{\Omega \times \Omega} \frac{|u(x) - u(y)|^p}{|x - y|^{N + \sigma p}} \, dxdy \right)^{1/p}$$

N.20 When $s \ge 1$, $1 \le p < \infty$ and Ω is smooth bounded connected, we take as seminorm in $W^{s,p}(\Omega)$ one of the two equivalent quantities

1. $|u|_{W^{s,p}} := \left\| u - \fint_\Omega u \right\|_{W^{s,p}}$

or

2. $|u|_{W^{s,p}} := \|\nabla u\|_{W^{s-1,p}}$

N.21 $\mathscr{M}(\Omega; \mathbb{R}^k)$ denotes the set of $\mathbb{R}^k$-valued finite Borel measures. Often, we simply denote this set by $\mathscr{M}$

N.22 We denote by $\|\mu\|_{\mathscr{M}}$ the mass of an $\mathbb{R}^k$-valued Borel measure $\mu = (\mu_1, \dots, \mu_k)$ in Ω, that is:

$$\|\mu\|_{\mathscr{M}} := \sup \left\{ \int_\Omega \sum_{j=1}^k \varphi_j \, d\mu_j; \; \varphi_j \in C_c(\Omega), j = 1, \dots, k, \sum_{j=1}^k \varphi_j^2 \le 1 \right\}.$$

Alternatively, we have $\|\mu\|_{\mathscr{M}} = |\mu|(\Omega)$, where $|\mu|$ is the total variation of the vector-valued measure μ

N.23 $BV(\Omega)$ is the space of functions of bounded variation:

$$BV(\Omega) := \{u \in L^1(\Omega); Du \in \mathscr{M}\}.$$

Some of the main properties of BV functions (properties of the jump set, structure of the differential, the chain rule) are recalled in Appendix 15.7 (see also Section 1.4). We also refer to these sections for the notation $|\ |_{BV}$, J_u, $D^a u$, $D^j u$ and ∇u

N.24 When $X(\Omega)$ is a function space on Ω and f is a function defined on $\partial\Omega$, we let, whenever this makes sense, $X_f(\Omega)$ denote the space of functions in $X(\Omega)$ with trace f. Examples: $W_f^{1,p}(\Omega)$, $BV_f(\Omega)$

N.25 *BMO* (resp. *VMO*) denotes the space of functions with bounded (resp. vanishing) mean oscillation. The definition and some properties of these spaces are recalled in Section 5.2

N.26 When $X(\Omega)$ is a function space on Ω,

$$X(\Omega; \mathbb{S}^k) := \{u \in X(\Omega; \mathbb{R}^{k+1});\ |u(x)| = 1 \text{ a.e. in } \Omega\}.$$

In particular,

$$W^{s,p}(\Omega; \mathbb{S}^1) := \{u \in W^{s,p}(\Omega; \mathbb{R}^2);\ |u(x)| = 1 \text{ a.e. in } \Omega\}$$

N.27 All the above function spaces have "local" versions. For example, we set

$$W^{s,p}_{loc}(\Omega) := \{u : \Omega \to \mathbb{R};\ u_{|B} \in W^{s,p}(B) \text{ for each ball } B \text{ such that } \overline{B} \subset \Omega\},$$

endowed with its usual metric (Fréchet) structure, for which

$$[u_n \to u \in W^{s,p}_{loc}(\Omega)] \iff [u_{n|B} \to u_{|B} \text{ in } W^{s,p}(B) \text{ for each } B \text{ as above }]$$

One defines similarly $W^{s,p}_{loc}(\Omega; \mathbb{S}^1)$, VMO_{loc}, etc.

N.28 If $a = a_1 + \imath a_2$, $b = b_1 + \imath b_2 \in \mathbb{C}$, then

1. $a \wedge b := a_1 b_2 - a_2 b_1 = \det \begin{pmatrix} a_1 & b_1 \\ a_2 & b_2 \end{pmatrix}$
2. ab is the usual complex multiplication
3. $a \cdot b := a_1 b_1 + a_2 b_2$

N.29 We note the following identities, valid for $a, b, c \in \mathbb{C}$

1. $\overline{a}b = a \cdot b + \imath(a \wedge b)$
2. $|a|^2|b|^2 = (a \cdot b)^2 + (a \wedge b)^2$
3. $(ac) \wedge (bc) = |c|^2(a \wedge b)$
   In particular, if $c \in \mathbb{S}^1$, then $(ac) \wedge (bc) = a \wedge b$

N.30 If $a \in \mathbb{C}$, $\overrightarrow{B} \in \mathbb{C}^N$, then

1. $a\overrightarrow{B} := (aB_j)_{j=1,\ldots,N}$
2. $a \wedge \overrightarrow{B} := (a \wedge B_j)_{j=1,\ldots,N}$

N.31 If $u, v : \Omega \to \mathbb{C}$, then

1. $\nabla u \cdot \nabla v := \nabla u_1 \cdot \nabla v_1 + \nabla u_2 \cdot \nabla v_2 \in \mathbb{R}$
2. $|\nabla u|^2 = \nabla u \cdot \nabla u = |\nabla u_1|^2 + |\nabla u_2|^2$
3. $u \wedge \nabla v := u_1 \nabla v_2 - u_2 \nabla v_1$

4. $\partial_j u = \dfrac{\partial u}{\partial x_j}$

N.32 $*$ denotes the Hodge star operator mapping k-forms onto $(N-k)$-forms

N.33 In two dimensions, $\perp$ denotes the rotation by $\dfrac{\pi}{2}$. Thus, $(a+\imath b)^\perp = -b+\imath a$ and $\nabla^\perp u(x,y) = (-\partial_y u, \partial_x u)(x,y)$

N.34 $\mathscr{H}^\ell$ denotes the ℓ-dimensional Hausdorff measure; in what follows, ℓ is an integer. When $S \subset \mathbb{R}^N$ is an ℓ-rectifiable set (typically, a Borel subset of a C^1-oriented embedded ℓ-submanifold $\Sigma \subset \mathbb{R}^N$), we let $\mathscr{H}^\ell \llcorner S$ denote the restriction of $\mathscr{H}^\ell$ to S

N.35 If A is a subset of a (real) vector space, $\operatorname{Ext}(A)$ denotes the set of the extreme points of A, and $\operatorname{conv}(A)$ its convex envelope

N.36 $\mathscr{P}(X)$ is the set of probability measures on X. In practice, X will be a metric space, and it is understood that measures are Borel measures

N.37 $f^\sharp\omega$ denotes the pullback of the form ω by the map f

N.38 A normalized bump function in $\mathbb{R}^N$ is a function $\rho \in C_c^\infty(\mathbb{R}^N)$ such that: $\operatorname{supp}\rho \subset B(0,1)$, $\rho \geq 0$, $\displaystyle\int \rho = 1$

N.39 If ρ is a normalized bump function (see the previous item), then

$$\rho_\varepsilon(x) := \frac{1}{\varepsilon^N}\rho(x/\varepsilon),\ \forall x \in \mathbb{R}^N, \forall \varepsilon > 0$$

is a family of standard mollifiers. If, in addition, ρ is radial, (ρ_ε) is a family of standard radial mollifiers

N.40 A family $(\rho_\varepsilon)_{\varepsilon>0}$ is a family of mollifiers if $\rho_\varepsilon \geq 0$ and $\displaystyle\int \rho_\varepsilon = 1, \forall \varepsilon > 0$, and

$$\lim_{\varepsilon\to 0}\int_{|x|>\delta} \rho_\varepsilon(x)\,dx = 0,\ \forall \delta > 0.$$

N.41 Λ^k is the space of k-forms (in $\mathbb{R}^N$).
If X is a function or distribution space on Ω, then $X(\Omega;\Lambda^k)$ is the space of k-forms with coefficients in X. Examples: $\mathscr{D}'(\Omega;\Lambda^k), \mathscr{M}(\Omega;\Lambda^k), L^1(\Omega;\Lambda^k)$

N.42 $\#A$ is the number of elements of the set A

Contents

Preface vii

Overview xi

Notation xxiii

Contents xxvii

1 Lifting in $W^{1,p}$ 1
- 1.0 Introduction 1
- 1.1 Warm up. Lifting of continuous maps: local and global aspects 5
- 1.2 Lifting of $\mathbb{S}^1$-valued maps in $W^{1,p}$. Definition of Ju 11
- 1.3 Square root 22
- 1.4 Lifting of $\mathbb{S}^1$-valued maps in $W^{1,1}$ and BV. Definition of $\Sigma(u)$ 23
- 1.5 $\Sigma(u)$ computed via duality: the basic relation between $\Sigma(u)$ and Ju 37
- 1.6 An excursion into Monge–Kantorovich (=MK) territory 44
- 1.7 Relaxed energy 49
- 1.8 Least energy with prescribed Jacobian 51
- 1.9 Complements and open problems 52
- 1.10 Comments 56

2 The geometry of Ju and $\Sigma(u)$ in 2D; point singularities and minimal connections 65
- 2.0 Introduction 65
- 2.1 Ju as a sum of Dirac masses 71
- 2.2 Where optimal transport (=OT) enters: the quantity $L(a, d)$ and minimal configurations 76
- 2.3 Returning to u: $\Sigma(u) = 2\pi L(a, d)$ 88
- 2.4 $\Sigma(u) = S_1(u) \geq 2\pi\, L(a, d)$ via the coarea formula 91
- 2.5 $\Sigma(u) = S_1(u) \leq 2\pi L(a, d)$ via the dipole construction 94

2.6 $E(u) \leq \int_\Omega |\nabla u| + 2\pi L(a, d)$ via minimal configurations 96
2.7 Connections associated with (a, d). Minimal connections 101
2.8 Returning to u: a bijective correspondence between BV liftings and connections 105
2.9 Describing Ju and $\Sigma(u)$ for a general $u \in W^{1,1}$ 107
2.10 An integral representation of the distribution Ju 119
2.11 Complements and open problems 123
2.12 Comments 128

3 The geometry of Ju and $\Sigma(u)$ in 3D (and higher); line singularities and minimal surfaces 133
3.0 Introduction 133
3.1 Examples. Ju as path integration 135
3.2 $\Sigma(u)$ as a least area: $\Sigma(u) = 2\pi A_0(\Gamma)$ 142
3.3 $\Sigma(u) \geq 2\pi A_0(\Gamma)$ via the coarea formula 145
3.4 $\Sigma(u) \leq 2\pi A_0(\Gamma)$ via the dipole construction 148
3.5 Least area spanned by a contour from the perspective of MK 149
3.6 The structure of Ju for a general $u \in W^{1,1}$. Where Federer encounters Kantorovich 150
3.7 Further properties when Ju is a measure 157
3.8 Complements and open problems 160
3.9 Comments 170

4 A digression: sphere-valued maps 173
4.0 Introduction 173
4.1 The "historical" case: $N = 3$ and $k = 2$, where everything fits into place! 177
4.2 A distinguished class of currents. Definition of $\mathscr{F}^\ell$ 183
4.3 The case $N = 4$ and $k = 2$; where complications appear 186
4.4 The general case: $N \geq 2$ and $1 \leq k \leq N - 1$ 190
4.5 Complements and open problems 196
4.6 Comments 201

5 Lifting in fractional Sobolev spaces and in *VMO* 205
5.0 Introduction 205
5.1 Lifting of $\mathbb{S}^1$-valued maps in fractional Sobolev spaces 206
5.2 Lifting of $\mathbb{S}^1$-valued maps in *VMO* and *BMO* 216
5.3 Lifting in $W^{s,p}$, $sp < 1$, upgraded 223
5.4 Complements and open problems 224
5.5 Comments 225

6 Uniqueness of lifting and beyond 227
6.0 Introduction 227
6.1 Constancy in $VMO\,(\Omega; \mathbb{Z})$ 228
6.2 Constancy in $W^{1,1}(\Omega; \mathbb{Z})$ 229

6.3 Constancy in $W^{1/p,p}(\Omega; \mathbb{Z})$, $1 < p < \infty$ 229
6.4 Connectedness of the essential range 234
6.5 A new function space. Applications to sums 235
6.6 Proof of Theorem 6.2 (the BBM formula) 240
6.7 Complements and open problems 242
6.8 Comments 250

7 Factorization 253
7.0 Introduction 253
7.1 Proof of Theorem 7.2 255
7.2 Outline of the proof of Theorem 7.1 258
7.3 A glimpse of the theory of weighted Sobolev spaces 259
7.4 Proof of Theorem 7.1 260
7.5 Complements and open problems 274
7.6 Comments 277

8 Applications of the factorization 279
8.0 Introduction 279
8.1 Existence of Ju for u in $W^{1/p,p}$ 280
8.2 Lifting revisited 285
8.3 Least energy with prescribed Jacobian 288
8.4 Relaxed energy 289
8.5 Square root 291
8.6 Minimizing the BV part of the phase 292
8.7 Complements and open problems 294
8.8 Comments 297

9 Estimates of phases: positive and negative results 299
9.0 Introduction 299
9.1 $N \geq 1$ and $s \geq 1$ 300
9.2 $N \geq 1$ and $sp < 1$ 300
9.3 $N = 1$, $s = 1/p$, and $p > 1$ 302
9.4 $N = 1$, $0 < s < 1$, and $sp > 1$ 303
9.5 $N \geq 2$, $0 < s < 1$, and $sp \geq N$ 304
9.6 $N \geq 2$, $0 < s < 1$, and $1 \leq sp < N$ 304
9.7 Complements and open problems 306
9.8 Comments 308

10 Density 311
10.0 Introduction 311
10.1 When are smooth maps dense? 312
10.2 Density of $\mathscr{R}$. Answer to Question 1 313
10.3 Characterization of $\overline{C^\infty(\overline{\Omega}; \mathbb{S}^1)}^{W^{s,p}}$. Answer to Question 2 318
10.4 Weak sequential density. Answer to Question 3 320

10.5 Distance to smooth maps ... 320
10.6 Complements and open problems ... 324
10.7 Comments ... 327

11 Traces ... 331
11.0 Introduction ... 331
11.1 Proof of Theorem 11.1 ... 332
11.2 A sharp form of the extension problem ... 334
11.3 Complements and open problems ... 334
11.4 Comments ... 336

12 Degree ... 339
12.0 Introduction ... 339
12.1 Degree and *VMO* ... 340
12.2 Degree, lifting, and traces in $W^{s,p}(\mathbb{S}^1; \mathbb{S}^1)$ ... 344
12.3 Integral representations for the degree ... 345
12.4 A "distributional degree" ... 350
12.5 Estimates for the degree ... 352
12.6 Homotopy classes ... 358
12.7 Degree and Fourier coefficients ... 358
12.8 Complements and open problems ... 364
12.9 Comments ... 379

13 Dirichlet problems. Gaps. Infinite energies ... 381
13.0 Introduction ... 381
13.1 Minimizing the $W^{1,p}$ energy when $p \geq 2$ ... 382
13.2 Minimizing the $W^{1,p}$ energy when $1 < p < 2$ ... 383
13.3 Gaps ... 386
13.4 More about minimizers when $\Omega = \mathbb{D}$ and $1 < p < 2$ ... 388
13.5 Further regularity and no gap for $p < 2$, near $p = 2$, when $N = 2$ ... 390
13.6 Complements and open problems ... 394
13.7 Comments ... 401

14 Domains with topology ... 403
14.0 Introduction ... 403
14.1 Lifting in $W^{1,p}(\Omega; \mathbb{S}^1)$ revisited ... 403
14.2 Relaxed energy revisited ... 409
14.3 Lifting in $W^{s,p}(\Omega; \mathbb{S}^1)$ revisited ... 417
14.4 Density in $W^{s,p}(\Omega; \mathbb{S}^1)$ revisited ... 421
14.5 Homotopy classes ... 423
14.6 Complements and open problems ... 425
14.7 Comments ... 430

15 Appendices ... 431
15.1 Sobolev spaces ... 431
15.2 Sobolev embeddings and Gagliardo–Nirenberg inequalities ... 435
15.3 Composition in Sobolev spaces ... 441
15.4 Standard (and non-standard) examples of maps in Sobolev spaces ... 443
15.5 Further results on *BMO* and *VMO* ... 451
15.6 Enlarging Ω ... 461
15.7 Fine theory of *BV* maps ... 467
15.8 Description of (minimal) connections associated with (a, d) ... 475
15.9 A pathological case ... 489
15.10 Jacobians of $W^{1,p}(\Omega; \mathbb{S}^1)$ maps ... 493
15.11 When Ju is a measure ... 496
15.12 Products in fractional Sobolev spaces ... 497
15.13 Maximal estimates ... 499
15.14 Smoothing ... 500

References ... 507

Symbol Index ... 523

Subject Index ... 525

Author Index ... 527

Chapter 1
Lifting in $W^{1,p}$

1.0 Introduction

Given a map $u : \Omega \to \mathbb{S}^1$, we may always write $u = e^{\imath\varphi}$, for some $\varphi : \Omega \to \mathbb{R}$. The function φ is called a *lifting* or a *phase* of u. Some liftings φ can be very rough; for example, $u \equiv 1$ admits non-measurable liftings. Our goal is to find a lifting φ "as smooth as u permits." Even when u is smooth, the topology of Ω may be an obstruction to the existence of a smooth φ. As mentioned in the Overview, we assume throughout most of this book that *Ω is a smooth bounded open simply connected set in $\mathbb{R}^N$*, $N \geq 1$, so that the topology of Ω does not interfere with lifting. [The case of multiply connected domains will be discussed in Chapter 14.] Most of the results in this chapter, if not all, extend to the case where Ω is contained in a manifold M of dimension N; for example, we could take $\Omega = \mathbb{S}^N$ with $N \geq 2$ (see also Section 1.9.3).

In Section 1.1, we recall some classical facts about lifting, e.g., the fact that C^k, $k \geq 0$, maps admit C^k liftings. In Section 1.2, we investigate the same question in the context of the Sobolev space $W^{1,p}(\Omega; \mathbb{S}^1)$. In view of the above, it is tempting to expect every map $u \in W^{1,p}(\Omega; \mathbb{S}^1)$ with $p \geq 1$ to have a $W^{1,p}$ phase. This is indeed true when $N = 1$. But, surprisingly, this is not the case when $N \geq 2$. Therefore, one is led to three natural questions.

Question 1. Characterize the p's such that $W^{1,p}(\Omega; \mathbb{S}^1)$ has the *lifting property*, i.e., *every* $u \in W^{1,p}(\Omega; \mathbb{S}^1)$ can be written as $u = e^{\imath\varphi}$ for some $\varphi \in W^{1,p}$ $(\Omega; \mathbb{R})$.

Assuming that $W^{1,p}(\Omega; \mathbb{S}^1)$ does *not* have the lifting property, we may then ask

Question 2. *Characterize* the u's in $W^{1,p}(\Omega; \mathbb{S}^1)$ that can be written as $u = e^{\imath\varphi}$ for some $\varphi \in W^{1,p}(\Omega; \mathbb{R})$.

Question 3. Assume that $u \in W^{1,p}(\Omega; \mathbb{S}^1)$ admits no lifting in $W^{1,p}$. Which are the "best possible" liftings of u (in a sense to be clarified below)?

As we are going to see, the lifting property for $W^{1,p}(\Omega; \mathbb{S}^1)$ holds precisely when $N = 1$ and $p \in [1, \infty)$, and when $N \geq 2$ and $p \geq 2$ (Theorem 1.1).

H. Brezis and P. Mironescu, *Sobolev Maps to the Circle*, Progress in Nonlinear Differential Equations and Their Applications 96,
https://doi.org/10.1007/978-1-0716-1512-6_1

We next discuss the answer to Question 2. Let us start with a smooth map $u : \overline{\Omega} \to \mathbb{S}^1$ and let φ be a smooth phase of u. Write

$$u = u_1 + \imath\, u_2 \text{ with } u_1 = \cos\varphi \text{ and } u_2 = \sin\varphi. \tag{1.1}$$

Differentiating with respect to x_i yields

$$\frac{\partial\varphi}{\partial x_i} = \cos\varphi\,\frac{\partial u_2}{\partial x_i} - \sin\varphi\,\frac{\partial u_1}{\partial x_i} = u_1\,\frac{\partial u_2}{\partial x_i} - u_2\,\frac{\partial u_1}{\partial x_i} = u \wedge \frac{\partial u}{\partial x_i} = (u \wedge \nabla u)_i. \tag{1.2}$$

Therefore, we have, $\forall\, i, j = 1, \ldots, N$,

$$(Ju)_{ij} := \frac{1}{2}\left[\frac{\partial}{\partial x_i}\,(u \wedge \nabla u)_j - \frac{\partial}{\partial x_j}\,(u \wedge \nabla u)_i\right] = 0. \tag{1.3}$$

Suppose now that u belongs merely to $W^{1,1}_{loc}(\Omega; \mathbb{S}^1)$ and *assume* that u admits a phase $\varphi \in W^{1,1}_{loc}(\Omega; \mathbb{R})$. Equality (1.2) still holds (in the L^1 sense). Moreover, Ju defined by (1.3) is an $N \times N$ antisymmetric matrix with entries in $\mathscr{D}'(\Omega; \mathbb{R})$ and

$$Ju = 0 \text{ in the sense of } \mathscr{D}'. \tag{1.4}$$

Finally, suppose that $u \in W^{1,1}(\Omega; \mathbb{R})$, but do *not* assume that u admits a phase in $W^{1,1}(\Omega; \mathbb{R})$. Then, Ju *still makes sense* in $(\mathscr{D}')^{N\times N}$. *However,* Ju need *not* be zero as above. Here is a typical example in a domain $\Omega \subset \mathbb{R}^2$ containing 0. Let $u(x) := \dfrac{x}{|x|}$; as we are going to see in Remark 1.9, we have $u \in W^{1,p}(\Omega; \mathbb{S}^1)$ for every $p < 2$ and

$$Ju = \begin{pmatrix} 0 & \pi\,\delta_0 \\ -\pi\,\delta_0 & 0 \end{pmatrix}.$$

The answer to Question 2 is given by Theorem 1.2: for $u \in W^{1,p}(\Omega; \mathbb{S}^1)$ with $1 \le p < 2$, we have

$$[u = e^{\imath\varphi} \text{ for some } \varphi \in W^{1,p}(\Omega; \mathbb{R})] \iff [Ju = 0 \text{ in } \mathscr{D}']. \tag{1.5}$$

This result is consistent with the answer to Question 1, since we have $Ju = 0$, $\forall\, u \in W^{1,p}(\Omega; \mathbb{S}^1)$ with $p \ge 2$ (see (1.61)).

The object Ju, called the *Jacobian* of u, defined by (1.3) for every $u \in W^{1,1}_{loc}(\Omega; \mathbb{S}^1)$, plays a central role throughout the book, much beyond its application to questions of lifting. As we are going to see in Chapters 2 and 3, Ju carries information about the location and the “strength” of the *topological singularities of* u.

We now turn to Question 3, again in the relevant range $1 \le p < 2$, with $N \ge 2$. Let $u \in W^{1,p}(\Omega; \mathbb{S}^1)$ be such that $Ju \ne 0$. We know, by (1.5), that u cannot be written as $u = e^{\imath\varphi}$ with $\varphi \in W^{1,p}(\Omega; \mathbb{R})$. In fact, the non-existence assertion is even more acute, since u cannot be written as $u = e^{\imath\varphi}$ with $\varphi \in W^{1,1}(\Omega; \mathbb{R})$. In some sense the "next best" we could hope is to write $u = e^{\imath\varphi}$ with $\varphi \in BV(\Omega; \mathbb{R})$. This is indeed true, by Theorem 1.4. However, formula (1.2) fails, otherwise we would deduce that $\varphi \in W^{1,1}$, which contradicts the assumption $Ju \neq 0$. This failure is also related to the fact that the standard chain rule formula does not hold for BV functions and must be replaced by the more delicate Volpert's chain rule (see (1.109) and Appendix 15.7). Of course, any given u has plenty of BV phases; for example, one may add $2\pi\, \chi_\omega$ (where χ_ω is the characteristic function of a smooth domain $\omega \subset \Omega$) to any BV phase φ. Therefore, it is natural to ask whether some phases are "better" than others. Such "minimal liftings" are defined to be the minimizers of

$$E(u) := \inf\{|\varphi|_{BV};\ \varphi \in BV(\Omega; \mathbb{R}) \text{ and } u = e^{\imath\varphi} \text{ in } \Omega\}. \tag{1.6}$$

It turns out that there is a neat formula for $E(u)$, given by Theorem 1.5: for every $u \in W^{1,1}(\Omega; \mathbb{S}^1)$,

$$E(u) = \int_\Omega |\nabla u| + \Sigma(u), \tag{1.7}$$

where

$$\Sigma(u) := \inf\{\|u \wedge \nabla u - \nabla\eta\|_{L^1};\ \eta \in W^{1,1}(\Omega; \mathbb{R})\}. \tag{1.8}$$

The proof of (1.7) is somewhat tricky and relies on fine properties of BV functions. The quantity $\Sigma(u)$ also plays a central role throughout the book. Here is a brief list of seemingly unrelated results involving $\Sigma(u)$ from this chapter alone.

1. We have, for every $u \in W^{1,1}(\Omega; \mathbb{S}^1)$,

$$\Sigma(u) = \inf\left\{\int_\Omega |\nabla(e^{-\imath\eta}\, u)|;\ \eta \in W^{1,1}(\Omega; \mathbb{R})\right\}. \tag{1.9}$$

Indeed, we may write

$$\nabla(e^{-\imath\eta}\, u) = e^{-\imath\eta}\, \nabla u - \imath e^{-\imath\eta}\, u\, \nabla\eta = \imath e^{-\imath\eta}\, u\, (-\imath\, \overline{u}\, \nabla u - \nabla\eta).$$

On the other hand,

$$\begin{aligned}\overline{u}\, \nabla u &= (u_1 - \imath\, u_2)\, (\nabla u_1 + \imath\, \nabla u_2) = (u_1\, \nabla u_1 + u_2\, \nabla u_2) + \imath\, u \wedge \nabla u \\ &= \frac{1}{2}\, \nabla(u_1^2 + u_2^2) + \imath\, u \wedge \nabla u = \imath\, u \wedge \nabla u,\end{aligned}$$

and thus

$$\nabla(e^{-\imath\eta}\, u) = \imath e^{-\imath\eta}\, u\, (u \wedge \nabla u - \nabla\eta),$$

which yields (1.9). In view of (1.9), we may say that $\Sigma(u)$ measures how much a given $u \in W^{1,1}(\Omega; \mathbb{S}^1)$ "deviates" from maps of the form $e^{\imath\eta}$, with

$\eta \in W^{1,1}(\Omega; \mathbb{R})$. On a related topic, see Theorem 10.5 and Brezis, Mironescu, and Shafrir [94].

2. We have, for every $u \in W^{1,1}(\Omega; \mathbb{S}^1)$,

$$\Sigma(u) = \inf\left\{\int_\Omega |\nabla v|;\ v \in W^{1,1}(\Omega; \mathbb{S}^1),\ Jv = Ju \text{ in } \mathscr{D}'\right\}. \tag{1.10}$$

Indeed (see (1.57) and Lemma 1.8), $J(u\overline{v}) = Ju - Jv = 0$ corresponds to $u\overline{v} = e^{\imath\eta}$ for some $\eta \in W^{1,1}(\Omega; \mathbb{R})$ and we are reduced to (1.9). In view of (1.10), we may say that $\Sigma(u)$ represents the least $W^{1,1}$ energy of maps having the "same singularities" as u. See Section 1.8 for further developments.

3. As a direct consequence of (1.10), we find that $\Sigma(u)$ depends only on Ju; more precisely,

$$[Ju = Jv] \implies [\Sigma(u) = \Sigma(v)]. \tag{1.11}$$

In fact, (1.11) could also be derived from a deeper result: $\Sigma(u)$ is the norm of Ju in a suitable dual space. For simplicity, we state the result only when $N = 2, 3$, and refer to Section 1.5 for a general N. When $N = 2$, we identify Ju with the scalar distribution given by

$$Ju = \frac{1}{2}\left[\frac{\partial}{\partial x_1}\left(u \wedge \frac{\partial u}{\partial x_2}\right) - \frac{\partial}{\partial x_2}\left(u \wedge \frac{\partial u}{\partial x_1}\right)\right],$$

and then we have

$$\Sigma(u) = 2\sup\{\langle Ju, \zeta\rangle;\ \zeta \in C_c^\infty(\Omega; \mathbb{R}) \text{ with } \|\nabla\zeta\|_{L^\infty} \le 1\}.$$

When $N = 3$, we identify Ju with the distributional vector field given by

$$Ju = \frac{1}{2}\operatorname{curl}(u \wedge \nabla u),$$

and then we have

$$\Sigma(u) = 2\sup\{\langle Ju, \zeta\rangle;\ \zeta \in C_c^\infty(\Omega; \mathbb{R}^3) \text{ with } \|\operatorname{curl}\zeta\|_{L^\infty} \le 1\}.$$

4. The *relaxed energy* is defined for every $u \in W^{1,1}(\Omega; \mathbb{S}^1)$ by

$$R_1(u) := \inf\left\{\liminf_{n\to\infty}\int_\Omega |\nabla u_n|;\ u_n \in C^\infty(\overline{\Omega}; \mathbb{S}^1),\ u_n \to u \text{ a.e. on } \Omega\right\},$$

where the first inf is taken over all sequences (u_n) in $C^\infty(\overline{\Omega}; \mathbb{S}^1)$ such that $u_n \to u$ a.e. on Ω. [In general, there is no sequence (u_n) in $C^\infty(\overline{\Omega}; \mathbb{S}^1)$ such that $u_n \to u$ in $W^{1,1}$, unless $Ju = 0$. However, it is always possible to find a sequence (u_n) in $C^\infty(\overline{\Omega}; \mathbb{S}^1)$ such that $u_n \to u$ a.e. on Ω.] Theorem 1.11 asserts that

$$R_1(u) = \int_\Omega |\nabla u| + \Sigma(u).$$

In Section 1.6, we show that some of the various ways of computing $\Sigma(u)$ are surprisingly analogous to various formulas giving the least cost in the Monge–Kantorovich theory of optimal transport.

1.1 Warm up. Lifting of continuous maps: local and global aspects

We first recall some well-known simple facts about lifting of continuous (or smooth) $\mathbb{S}^1$-valued maps.

Let $u = u_1 + \imath u_2 \in C^1(\Omega; \mathbb{S}^1)$. By differentiating the identity $|u|^2 \equiv 1$, we obtain

$$u \cdot \frac{\partial u}{\partial x_j} = u_1 \frac{\partial u_1}{\partial x_j} + u_2 \frac{\partial u_2}{\partial x_j} = 0. \tag{1.12}$$

If $u = e^{\imath\varphi}$, with $\varphi \in C^1$, then $\nabla u = \imath \nabla\varphi\, u$, and thus $\nabla\varphi = -\imath\overline{u}\,\nabla u$. Combining this identity with (1.12), we obtain

$$\nabla\varphi = -\imath\overline{u}\,\nabla u = u_1\nabla u_2 - u_2\nabla u_1. \tag{1.13}$$

The right-hand side in (1.13) plays an important role and will be denoted by $u \wedge \nabla u$. Clearly,

$$|\nabla\varphi| = |u \wedge \nabla u| = |\nabla u|. \tag{1.14}$$

Lemma 1.1. *Let $u \in C(\overline{\Omega}; \mathbb{S}^1)$. Then there exists some $\varphi \in C(\overline{\Omega}; \mathbb{R})$ such that $u = e^{\imath\varphi}$. Moreover, φ is unique (mod 2π).*

If, in addition, $u \in C^k$ for some $k \geq 1$, then $\varphi \in C^k$.

Similarly, if, in addition, $u \in W^{1,p}(\Omega)$ for some $p \in [1, \infty]$, then $\varphi \in W^{1,p}(\Omega)$.

Proof. We present a non-standard proof of this standard result. The same approach will be useful later within the framework of Sobolev spaces (even in cases where Sobolev maps need not be continuous; see the proof of Lemma 1.8).

Step 1. Proof when $u \in C^2(\overline{\Omega}; \mathbb{S}^1)$

Let $f := u_1\nabla u_2 - u_2\nabla u_1 \in C^1(\overline{\Omega}; \mathbb{R}^N)$. In order to find φ, we first solve (1.13). By Poincaré's lemma, existence of a $\varphi \in C^2$ satisfying (1.13) is equivalent to

$$\frac{\partial f_i}{\partial x_j} = \frac{\partial f_j}{\partial x_i}, \ \forall\, i, j \in \{1, \dots, N\}. \tag{1.15}$$

Since $u \in C^2$, we have

$$\frac{\partial f_j}{\partial x_i} - \frac{\partial f_i}{\partial x_j} = 2\left(\frac{\partial u_1}{\partial x_i}\frac{\partial u_2}{\partial x_j} - \frac{\partial u_1}{\partial x_j}\frac{\partial u_2}{\partial x_i}\right),$$

so that (1.15) is equivalent to

$$\det\begin{pmatrix} \dfrac{\partial u_1}{\partial x_i} & \dfrac{\partial u_1}{\partial x_j} \\ \dfrac{\partial u_2}{\partial x_i} & \dfrac{\partial u_2}{\partial x_j} \end{pmatrix} = 0, \ \forall\, i, j \in \{1, \ldots, N\}. \tag{1.16}$$

Property (1.16) is established as follows: (1.12) implies that, for every $j=1, 2, \ldots, N$, the vector $\left(\dfrac{\partial u_1}{\partial x_j}, \dfrac{\partial u_2}{\partial x_j}\right)$ is orthogonal in $\mathbb{R}^2$ to the unit vector (u_1, u_2). It follows that the vectors $\left(\dfrac{\partial u_1}{\partial x_i}, \dfrac{\partial u_2}{\partial x_i}\right)$ and $\left(\dfrac{\partial u_1}{\partial x_j}, \dfrac{\partial u_2}{\partial x_j}\right)$ are collinear, and thus (1.16) holds.

We complete now the existence of a lifting $\varphi \in C^2(\overline{\Omega}; \mathbb{R})$ when $u \in C^2(\overline{\Omega}; \mathbb{S}^1)$. Let $\varphi \in C^2(\overline{\Omega}; \mathbb{R})$ solve (1.13), i.e., $\nabla\varphi = -\imath\overline{u}\nabla u$. Then,

$$\nabla(u\, e^{-\imath\varphi}) = e^{-\imath\varphi}(\nabla u - \imath u \nabla\varphi) = e^{-\imath\varphi}[\nabla u + \imath^2 u\overline{u}\nabla u] = 0. \tag{1.17}$$

Consequently, $u = e^{\imath(\varphi + C)}$ for some appropriate $C \in \mathbb{R}$.

Step 2. Proof when $u \in C(\overline{\Omega}; \mathbb{S}^1)$ is close to a constant
Assume that $\|u - 1\|_{L^\infty} < 2$. Then, $u(\overline{\Omega}) \subset \mathbb{C} \setminus \mathbb{R}_-$, and thus $u = e^{\imath\varphi}$, where $\varphi := -\imath \ln u$, and ln is the principal branch of the logarithm.

Step 3. Proof for a general $u \in C(\overline{\Omega}; \mathbb{S}^1)$
We will prove that we may write $u = v_1 v_2$, with $v_1 \in C^\infty(\overline{\Omega}; \mathbb{S}^1)$ and $v_2 \in C(\overline{\Omega}; \mathbb{S}^1)$ satisfies $\|v_2 - 1\|_{L^\infty} < 2$. Assuming the claim proved for the moment, we write $v_j = e^{\imath\varphi_j}$, with φ_j continuous, $j = 1, 2$, and let $\varphi := \varphi_1 + \varphi_2$.

We now turn to the proof of the claim. Given $u \in C(\overline{\Omega}; \mathbb{S}^1)$, there exists a sequence $(u_n) \subset C^\infty(\overline{\Omega}; \mathbb{S}^1)$ such that $u_n \to u$ uniformly on $\overline{\Omega}$. Indeed, it is standard that there exists a sequence $(U_n) \subset C^\infty(\overline{\Omega}; \mathbb{R}^2)$ such that $U_n \to u$ uniformly on $\overline{\Omega}$. We may then choose $u_n := \dfrac{U_n}{|U_n|}$ for large n. Next, set $v_1 := u_n$ and $v_2 := u/v_1$. Then, $\|v_2 - 1\|_{L^\infty} < 2$ provided n is sufficiently large.

Uniqueness of φ is straightforward: if $u{=}e^{\imath\varphi_1}{=}e^{\imath\varphi_2}$, then $\varphi_1 - \varphi_2 \in C(\overline{\Omega}; 2\pi\mathbb{Z})$, and thus $\varphi_1 - \varphi_2$ is a constant multiple of 2π.

Step 4. Proof when u belongs, in addition, to $C^k(\Omega)$ or $W^{1,p}(\Omega)$
Locally, we have $\varphi = -\imath \ln u$, where ln is an appropriate branch of the logarithm. We find that, if $u \in C^k$, then $\varphi \in C^k$. The case where $u \in W^{1,p}$ is treated similarly, using the chain rule for $W^{1,p}$ maps [75, Proposition 9.5]; we will come back to this in Section 1.2. □

The situation is more complicated in multiply connected domains. We will discuss this in detail in Chapter 14. Roughly speaking, the existence of a *continuous* lifting requires *additional topological* assumptions on u. Here is a model case.

Given $u \in C(\mathbb{S}^1; \mathbb{S}^1)$, we address the question whether u admits a lifting $\varphi \in C(\mathbb{S}^1; \mathbb{R})$. By Lemma 1.1, the continuous map

$$h : [0, 2\pi] \to \mathbb{S}^1, \ h(\theta) := u(e^{\imath\theta}), \tag{1.18}$$

admits a lifting $\psi \in C([0, 2\pi]; \mathbb{R})$. Define the *degree* of u by

$$\deg u := \frac{\psi(2\pi) - \psi(0)}{2\pi} \in \mathbb{Z}. \tag{1.19}$$

The definition does not depend on the choice of ψ, by uniqueness of ψ (mod 2π).

Here are some straightforward properties of the degree.

Lemma 1.2. *The degree is continuous with respect to uniform convergence, or, equivalently, locally constant in $C(\mathbb{S}^1; \mathbb{S}^1)$.*

Moreover,

$$\deg \overline{u} = -\deg u, \ \forall\, u \in C(\mathbb{S}^1; \mathbb{S}^1) \tag{1.20}$$

and

$$\deg(uv) = \deg u + \deg v, \ \forall\, u, v \in C(\mathbb{S}^1; \mathbb{S}^1). \tag{1.21}$$

Proof. Properties (1.20) and (1.21) are clear, by (1.18)–(1.19).

We next prove that deg is locally constant. If $\|v - u\|_{L^\infty} < \sqrt{2}$, then $\|w - 1\|_{L^\infty} < \sqrt{2}$, where $w := u/v$. As in Step 3 in the proof of Lemma 1.1, we may write $w(e^{\imath\theta}) = e^{\imath\psi(\theta)}$, for some continuous ψ such that $|\psi(\theta)| < \pi/2$, $\forall\, \theta$. By (1.19), we have $|\deg w| < 1/2$, and thus $\deg w = 0$. Finally, (1.21) implies that

$$\deg v = \deg\,(uw) = \deg u + \deg w = \deg u.$$

We will return to the properties of the degree in Chapter 12. For the time being, we state a result that relates the degree to lifting.

Lemma 1.3. *For $u \in C(\mathbb{S}^1; \mathbb{S}^1)$, the following assertions are equivalent:*

1. $\deg u = 0$.
2. $u = e^{\imath\varphi}$ *for some* $\varphi \in C(\mathbb{S}^1; \mathbb{R})$.

If in addition $u \in W^{1,p}(\mathbb{S}^1)$ for some $p \in [1, \infty]$, then $\varphi \in W^{1,p}(\mathbb{S}^1)$.

Proof. If $\deg u = 0$, then $\psi(0) = \psi(2\pi)$. Thus, we may extend ψ as a 2π-periodic continuous function on $\mathbb{R}$. Let $z \in \mathbb{S}^1$ and set $\varphi(z) := \psi(-\imath \ln z)$. Here, ln is a branch of the logarithm, smooth near z. Since ψ is 2π-periodic, the definition of φ does not depend on the choice of the branch, and yields a continuous phase φ of u.

Conversely, if $u = e^{\imath\varphi}$, let $\psi(\theta) := \varphi(e^{\imath\theta})$. Then, $h = e^{\imath\psi}$ and $\psi(2\pi) = \psi(0)$, so that $\deg u = 0$. □

By Lemma 1.3, a map $u \in C^\infty(\mathbb{S}^1; \mathbb{S}^1)$ such that $\deg u \neq 0$ does not admit a continuous lifting. On the other hand, u has measurable liftings φ; e.g., one may take $\varphi := \arg u = -\imath \ln u$. Here, arg and ln stand resp. for the principal branches of the argument and of the logarithm of complex numbers. It is natural to search for liftings φ "as smooth as possible." By the proof of Lemma 1.3, it is possible to choose φ smooth on $\mathbb{S}^1$ except at $z = 1$. When $\mathbb{S}^1$ is equipped with the standard (counterclockwise) orientation, this φ jumps by $-2\pi \deg u$ at $z = 1$. Such a φ belongs to BV. The space BV plays a natural role through the concept of minimal lifting. Set

$$E(u) := \inf\{|\varphi|_{BV};\ \varphi \in BV(\mathbb{S}^1; \mathbb{R}) \text{ and } u = e^{\imath\varphi}\}. \tag{1.22}$$

Here, $|\varphi|_{BV}$ denotes the total variation of the measure $\dfrac{\partial \varphi}{\partial \tau}$, where $\dfrac{\partial}{\partial \tau}$ stands for tangential differentiation on $\mathbb{S}^1$ with the standard orientation. By definition, a *minimal lifting* is a minimizer in (1.22). Note that inf is attained in (1.22). Indeed, since we may add a multiple of 2π to any lifting, it suffices to consider liftings satisfying $0 \le \int_{\mathbb{S}^1} \varphi \le 4\pi^2$. The class of such liftings contains a minimal one, by the compactness of the embedding $BV \hookrightarrow L^1$.

Lemma 1.4. *Let $u \in W^{1,1}(\mathbb{S}^1; \mathbb{S}^1)$. Then we have the Cauchy formula*

$$\deg u = \frac{1}{2\pi} \int_{\mathbb{S}^1} u \wedge \frac{\partial u}{\partial \tau}. \tag{1.23}$$

Moreover,

$$E(u) = \int_{\mathbb{S}^1} \left| \frac{\partial u}{\partial \tau} \right| + 2\pi |\deg u|. \tag{1.24}$$

Remark 1.1. Equation (1.23) is just another way of writing the classical *Cauchy formula*

$$\deg u = \frac{1}{2\imath\pi} \int_{\mathbb{S}^1} \frac{1}{u} \left(\frac{\partial u}{\partial \tau} \right)$$

when $u \in W^{1,1}(\mathbb{S}^1; \mathbb{S}^1)$. Indeed, we have (see (1.13))

$$\frac{1}{\imath u} \left(\frac{\partial u}{\partial \tau} \right) = -\imath \bar{u} \left(\frac{\partial u}{\partial \tau} \right) = u \wedge \frac{\partial u}{\partial \tau}.$$

Proof. When $u \in W^{1,1}(\mathbb{S}^1; \mathbb{S}^1)$, the map h, defined by (1.18), belongs to $W^{1,1}$ $([0, 2\pi]; \mathbb{S}^1)$. By Lemma 1.1, so does its continuous lifting ψ. By the chain rule, we have (as in (1.13))

$$\psi'(\theta) = h(\theta) \wedge h'(\theta) = u(e^{\imath\theta}) \wedge \frac{\partial u}{\partial \tau}(e^{\imath\theta}), \tag{1.25}$$

and thus

$$\deg u = \frac{1}{2\pi}(\psi(2\pi) - \psi(0)) = \frac{1}{2\pi}\int_0^{2\pi} \psi'(\theta)\, d\theta = \frac{1}{2\pi}\int_{\mathbb{S}^1} u \wedge \frac{\partial u}{\partial \tau}.$$

On the other hand, the function

$$\varphi_0(e^{\imath\theta}) := \psi(\theta), \ \forall \theta \in (0, 2\pi), \tag{1.26}$$

belongs to $W^{1,1}(\mathbb{S}^1 \setminus \{1\}; \mathbb{R})$. At $z = 1$, φ_0 jumps by $-2\pi \ \deg u$. Hence, φ_0 is a lifting of u that belongs to $BV(\mathbb{S}^1; \mathbb{R})$. By (1.25), we have

$$\frac{\partial \varphi_0}{\partial \tau} = u \wedge \frac{\partial u}{\partial \tau} \text{ in } \mathbb{S}^1 \setminus \{1\}, \tag{1.27}$$

and thus

$$\frac{\partial \varphi_0}{\partial \tau} = u \wedge \frac{\partial u}{\partial \tau} - 2\pi \ \deg u\, \delta_1 \text{ in } \mathscr{D}'(\mathbb{S}^1). \tag{1.28}$$

Combining (1.28) with (1.14), we find that

$$E(u) \le |\varphi_0|_{BV} = \int_{\mathbb{S}^1} \left|\frac{\partial u}{\partial \tau}\right| + 2\pi |\deg u|.$$

For the reverse inequality, we start by characterizing all the BV liftings of u.

Lemma 1.5. *Let $\varphi \in BV(\mathbb{S}^1; \mathbb{R})$ be any lifting of $u \in W^{1,1}(\mathbb{S}^1; \mathbb{S}^1)$. Then,*

$$\frac{\partial \varphi}{\partial \tau} = u \wedge \frac{\partial u}{\partial \tau} - 2\pi \sum \alpha_j\, \delta_{a_j}, \tag{1.29}$$

for a finite collection of points $a_j \in \mathbb{S}^1$ and some integers $\alpha_j \in \mathbb{Z}$ such that $\sum \alpha_j = \deg u$.

And conversely, any $\varphi \in BV(\mathbb{S}^1; \mathbb{R})$ satisfying (1.29) *is a lifting of u (modulo a constant).*

Proof of Lemma 1.5. Let φ_0 be as in (1.26), and set $\varphi_1 := \varphi - \varphi_0$. Clearly, $e^{\imath\varphi_1} = u\,\overline{u} = 1$, so that $\varphi_1 \in BV(\mathbb{S}^1; 2\pi\mathbb{Z})$. Hence, we may write

$$\varphi_1 = 2\pi \sum_{j=1}^{n} k_j\, \chi_{(a_j, a_{j+1})},$$

for some sequence (k_j) in $\mathbb{Z}$, and some increasing sequence of points (a_j), $j = 1, \dots, n$, on $\mathbb{S}^1$, where $a_{n+1} := a_1$ and $\chi_{(a_j,a_{j+1})}$ denotes the characteristic function of the arc (a_j, a_{j+1}) from a_j to a_{j+1} (with the direct orientation on $\mathbb{S}^1$). Thus,

$$\frac{\partial \varphi_1}{\partial \tau} = 2\pi \sum_{j=1}^{n} (k_j - k_{j-1})\, \delta_{a_j},$$

where $k_0 := k_n$. Combining this with (1.28) yields

$$\frac{\partial\varphi}{\partial\tau}=\frac{\partial\varphi_0}{\partial\tau}+\frac{\partial\varphi_1}{\partial\tau}=u\wedge\frac{\partial u}{\partial\tau}-2\pi\ \deg u\,\delta_1+2\pi\sum_{j=1}^{n}(k_j-k_{j-1})\,\delta_{a_j}.$$

We deduce that (1.29) holds, where the summation in (1.29) runs from $j=0$ to $j=n$, with $\alpha_0:=\deg u$, $\alpha_j:=-(k_j-k_{j-1}), j=1,\ldots,n$, $a_0:=1$, and a_j as above for $j=1,\ldots,n$.

Conversely, if (1.29) holds, set

$$\nu:=u\wedge\frac{\partial u}{\partial\tau}-2\pi\sum\alpha_j\delta_{a_j},$$

and consider $\varphi\in BV(\mathbb{S}^1)$ satisfying $\dfrac{\partial\varphi}{\partial\tau}=\nu$. Such a φ exists since the compatibility condition $\displaystyle\int_{\mathbb{S}^1}\nu=0$ is satisfied. Then, $e^{\imath\varphi}$ is continuous at the a_j's and $\dfrac{\partial(u\,e^{-\imath\varphi})}{\partial\tau}=0$ in $\mathbb{S}^1\setminus\cup\{a_j\}$. It follows that $u\,e^{-\imath\varphi}$ is constant, and thus $e^{\imath(\varphi+C)}=u$. □

Proof of Lemma 1.4 *completed.* If φ is any BV lifting of u and α_j, a_j are as in (1.29), then $\sum|\alpha_j|\geq\left|\sum\alpha_j\right|=|\deg u|$, and thus

$$|\varphi|_{BV}=\int_{\mathbb{S}^1}\left|\frac{\partial u}{\partial\tau}\right|+2\pi\sum|\alpha_j|\geq\int_{\mathbb{S}^1}\left|\frac{\partial u}{\partial\tau}\right|+2\pi|\deg u|.$$

Therefore, we have established (1.24), and we have also shown that the lifting φ_0 in (1.26) is a minimal lifting of u. □

Remark 1.2. One may extend the above results to the case of maps $u:\Gamma\to\mathbb{S}^1$, where $\Gamma\subset\mathbb{R}^n$ is a simple closed oriented curve or a finite (disjoint) union of such curves. For example, if $u\in W^{1,1}(\Gamma;\mathbb{S}^1)$, then

$$\deg(u,\Gamma)=\frac{1}{2\pi}\int_\Gamma u\wedge\frac{\partial u}{\partial\tau}. \tag{1.30}$$

Here is a final elementary result, which will be used later.

Lemma 1.6. *Let* $u\in C(\overline{\mathbb{D}}\setminus\{0\};\mathbb{S}^1)$ *be such that* $\deg(u,\mathbb{S}^1)=0$. *Then there exists some* $\varphi\in C(\overline{\mathbb{D}}\setminus\{0\};\mathbb{R})$ *such that* $u=e^{\imath\varphi}$.

Here, $\mathbb{D}$ is the unit disk in $\mathbb{R}^2$.

Proof. By Lemma 1.2, the degree of the map $\mathbb{S}^1\ni x\mapsto u(rx)\in\mathbb{S}^1$ does not depend on $r\in(0,1]$, and thus is always 0. By Lemma 1.3, we may write, for each r, $u=e^{\imath\eta_r}$ on $C(0,r)$, where η_r is continuous on $C(0,r)$. In particular, for each r, the restriction of u to $C(0,r)$ has a continuous $\mathbb{S}^1$-valued extension u_r to $\overline{\mathbb{D}}_r:=\{x;\,|x|\leq r\}$; it suffices to take a continuous extension ψ_r of η_r and let $u_r:=e^{\imath\psi_r}$. Let $v_r\in C(\overline{\mathbb{D}};\mathbb{S}^1)$, $v_r:=\begin{cases}u, & \text{in } \overline{\mathbb{D}}\setminus\overline{\mathbb{D}}_r\\ u_r, & \text{in } \overline{\mathbb{D}}_r\end{cases}$. By Lemma 1.1, we may write, in $\overline{\mathbb{D}}$, $v_r=e^{\imath\varphi_r}$, with φ_r continuous. In addition, we may take $\varphi_r(1)$ independent of r. Then, for $0<s<r<1$, we have $\varphi_s=\varphi_r$ in $\overline{\mathbb{D}}\setminus\mathbb{D}_r$. We find that the map

$$\varphi(x) := \varphi_r(x) \text{ if } r < |x| \leq 1$$

is a well-defined continuous lifting of u in $\overline{\mathbb{D}} \setminus \{0\}$. □

1.2 Lifting of $\mathbb{S}^1$-valued maps in $W^{1,p}$. Definition of *Ju*

Recall the chain rule for $W^{1,p}$ maps (See, for example, Gilbarg and Trudinger [164, Theorem 7.8] or Brezis [75, Proposition 9.5]): if $f : \mathbb{R} \to \mathbb{R}$ is a C^1 function with $f' \in L^\infty(\mathbb{R})$ and if $\varphi \in W^{1,p}(\Omega; \mathbb{R})$ for some $1 \leq p < \infty$, then $f \circ \varphi \in W^{1,p}(\Omega; \mathbb{R})$. Moreover, we have

$$\frac{\partial}{\partial x_i}(f \circ \varphi) = f'(\varphi)\frac{\partial \varphi}{\partial x_i}. \tag{1.31}$$

In particular, if $\varphi \in W^{1,p}(\Omega; \mathbb{R})$, then $u := e^{\iota\varphi} \in W^{1,p}(\Omega; \mathbb{C})$ and, in addition, $|u| = 1$, that is, $u(x) \in \mathbb{S}^1$ for a.e. $x \in \Omega$. [When there is no confusion, we often omit "a.e."]

We are concerned with the converse. For this purpose, we set

$$W^{1,p}(\Omega; \mathbb{S}^1) := \{u \in W^{1,p}(\Omega; \mathbb{C}); |u| = 1 \text{ in } \Omega\}.$$

[We call the attention of the reader to the fact that there is another natural concept, namely the closure in $W^{1,p}(\Omega; \mathbb{R}^2)$ of $C^\infty(\overline{\Omega}; \mathbb{S}^1)$. In general, this a strict subset of $W^{1,p}(\Omega; \mathbb{S}^1)$; see Chapter 10.]

As mentioned in the Introduction, we may raise two different questions.

Question 1. Fix p, and ask whether *any* $u \in W^{1,p}(\Omega; \mathbb{S}^1)$ can be written as $u = e^{\iota\varphi}$ for some $\varphi \in W^{1,p}(\Omega; \mathbb{R})$.

Assuming that, for some p, the answer to the first question is negative, we may then ask the following

Question 2. *Characterize* the u's in $W^{1,p}$ that can be written as $u = e^{\iota\varphi}$ for some $\varphi \in W^{1,p}(\Omega; \mathbb{R})$.

We are going to prove (Theorem 1.1 and Remark 1.6) that the answer to Question 1 is positive if $p \geq 2$ and $N \geq 2$ (resp. $p \geq 1$ when $N = 1$) and negative if $1 \leq p < 2$ and $N \geq 2$. This is a remarkable result of Bethuel and Zheng [37].

Motivated by (1.13) and by the proof of Lemma 1.1, we introduce two quantities that will play a *fundamental* role in the analysis of Questions 1 and 2, and throughout many parts of the book.

Given $u \in W^{1,1}_{loc}(\Omega; \mathbb{R}^2) \cap L^\infty_{loc}(\Omega; \mathbb{R}^2)$, set

$$u \wedge \nabla u := u_1 \nabla u_2 - u_2 \nabla u_1 \in L^1_{loc}(\Omega; \mathbb{R}^N), \tag{1.32}$$

i.e., in components,

$$(u \wedge \nabla u)_j = u_1 \frac{\partial u_2}{\partial x_j} - u_2 \frac{\partial u_1}{\partial x_j}, \ \forall j = 1, 2, \ldots, N.$$

When $N \geq 2$, consider the vector-valued distribution defined by:

$$Ju := \frac{1}{2} \operatorname{curl} (u \wedge \nabla u), \tag{1.33}$$

which will be called the *Jacobian* of u. This terminology is motivated by formula (1.36) below.

In components, we have

$$\begin{aligned} (Ju)_{ij} &= \frac{1}{2} \left[\frac{\partial}{\partial x_i} (u \wedge \nabla u)_j - \frac{\partial}{\partial x_j} (u \wedge \nabla u)_i \right] \\ &= \frac{1}{2} \left[\frac{\partial}{\partial x_i} \left(u_1 \frac{\partial u_2}{\partial x_j} - u_2 \frac{\partial u_1}{\partial x_j} \right) - \frac{\partial}{\partial x_j} \left(u_1 \frac{\partial u_2}{\partial x_i} - u_2 \frac{\partial u_1}{\partial x_i} \right) \right], \end{aligned} \tag{1.34}$$

for $1 \leq i, j \leq N$. Note that Ju is an $N \times N$ antisymmetric matrix with entries in $\mathscr{D}'(\Omega)$.

When $N = 2$, Ju is identified with the *scalar* distribution

$$Ju = \frac{1}{2} \left[\frac{\partial}{\partial x_1} \left(u_1 \frac{\partial u_2}{\partial x_2} - u_2 \frac{\partial u_1}{\partial x_2} \right) - \frac{\partial}{\partial x_2} \left(u_1 \frac{\partial u_2}{\partial x_1} - u_2 \frac{\partial u_1}{\partial x_1} \right) \right]. \tag{1.35}$$

Note that, if $u = (u_1, u_2)$ is a smooth map from $\Omega \subset \mathbb{R}^2$ to $\mathbb{R}^2$, then

$$Ju = \frac{\partial u_1}{\partial x_1} \frac{\partial u_2}{\partial x_2} - \frac{\partial u_1}{\partial x_2} \frac{\partial u_2}{\partial x_1} \tag{1.36}$$

is the Jacobian of u. When u is merely in $W^{1,1}_{loc}(\Omega; \mathbb{R}^2) \cap L^\infty_{loc}(\Omega; \mathbb{R}^2)$, Ju coincides with the "distributional Jacobian" in the sense of Ball [20, Section 6] (see also Brezis and Nguyen [96] and the references therein).

When $N = 3$, Ju is identified with the 3-dimensional vector distribution

$$Ju = (Ju)_{23} e_1 + (Ju)_{31} e_2 + (Ju)_{12} e_3, \tag{1.37}$$

where

$$\begin{aligned} (Ju)_{23} &= \frac{1}{2} \left[\frac{\partial}{\partial x_2} \left(u_1 \frac{\partial u_2}{\partial x_3} - u_2 \frac{\partial u_1}{\partial x_3} \right) - \frac{\partial}{\partial x_3} \left(u_1 \frac{\partial u_2}{\partial x_2} - u_2 \frac{\partial u_1}{\partial x_2} \right) \right], \\ (Ju)_{31} &= \frac{1}{2} \left[\frac{\partial}{\partial x_3} \left(u_1 \frac{\partial u_2}{\partial x_1} - u_2 \frac{\partial u_1}{\partial x_1} \right) - \frac{\partial}{\partial x_1} \left(u_1 \frac{\partial u_2}{\partial x_3} - u_2 \frac{\partial u_1}{\partial x_3} \right) \right], \\ (Ju)_{12} &= \frac{1}{2} \left[\frac{\partial}{\partial x_1} \left(u_1 \frac{\partial u_2}{\partial x_2} - u_2 \frac{\partial u_1}{\partial x_2} \right) - \frac{\partial}{\partial x_2} \left(u_1 \frac{\partial u_2}{\partial x_1} - u_2 \frac{\partial u_1}{\partial x_1} \right) \right]. \end{aligned} \tag{1.38}$$

Remark 1.3. It will sometimes be convenient to use the language of differential forms (or currents) (see, for example, Section 1.5). More precisely, if $u \in W^{1,1}(\Omega; \mathbb{R}^2) \cap L^\infty(\Omega; \mathbb{R}^2)$, we identify $u \wedge \nabla u$ with the 1-form in $L^1(\Omega; \Lambda^1)$ defined in coordinates by

$$u \wedge \nabla u = \sum_{j=1}^{N} (u \wedge \nabla u)_j \, dx_j, \tag{1.39}$$

and Ju with the 2-form in $\mathscr{D}'(\Omega; \Lambda^2)$ defined by

$$Ju = \frac{1}{2} d(u \wedge \nabla u), \tag{1.40}$$

where d denotes the exterior differential of forms. In coordinates, we have

$$Ju = \sum_{i<j} (Ju)_{ij} \, dx_i \wedge dx_j,$$

where $(Ju)_{ij}$ is defined by (1.34).

For example, when $N = 2$, Ju is the 2-form $Ju = (Ju)_{12} \, dx_1 \wedge dx_2$. The corresponding 0-form $*Ju$ is given by $*Ju = (Ju)_{12}$, where $*$ denotes the Hodge star operator mapping k-forms onto $(N-k)$-forms (see, for example, Federerer [147, 1.7.8] or Giaquinta, Modica, and Souček [156, 5.2.2]). As usual, we identify 0-forms with scalar functions (or distributions). In our case, $*Ju$ is identified with the right-hand side of (1.35).

When $N = 3$, Ju is the 2-form given in coordinates by

$$Ju = (Ju)_{12} \, dx_1 \wedge dx_2 + (Ju)_{13} \, dx_1 \wedge dx_3 + (Ju)_{23} \, dx_2 \wedge dx_3, \tag{1.41}$$

whereas $*Ju$ is the 1-form given by

$$*Ju = (Ju)_{23} \, dx_1 + (Ju)_{31} \, dx_2 + (Ju)_{12} \, dx_3. \tag{1.42}$$

As usual, we identify 1-forms with vector fields and then $*Ju$ is identified with the right-hand side of (1.37).

Remark 1.4. For $u \in W^{1,1}(\Omega; \mathbb{R}^2) \cap L^\infty(\Omega; \mathbb{R}^2)$, we have defined $Ju \in [\mathscr{D}'(\Omega)]^{N \times N}$ via (1.34) and (1.35). But, in fact, Ju belongs to the smaller space $[(W_0^{1,\infty}(\Omega))^{N \times N}]^*$, in view of the formula

$$\langle Ju, \zeta \rangle = \frac{1}{2} \sum_{i,j} \int_\Omega \left[-\left(u \wedge \frac{\partial u}{\partial x_j} \right) \frac{\partial \zeta_{ij}}{\partial x_i} + \left(u \wedge \frac{\partial u}{\partial x_i} \right) \frac{\partial \zeta_{ij}}{\partial x_j} \right], \ \forall \zeta \in C_c^\infty(\Omega; \mathbb{R}^{N \times N}).$$

Note that, if $u_n \to u$ in $W^{1,1}(\Omega; \mathbb{R}^2)$ with $\|u_n\|_{L^\infty} \le C$, then $Ju_n \to Ju$ in $[W_0^{1,\infty}(\Omega; \mathbb{R})]^*$ (and in particular in $\mathscr{D}'$), since

$$\|Ju_n - Ju\|_{[W_0^{1,\infty}]^*} \le C \int_\Omega |u_n \wedge \nabla u_n - u \wedge \nabla u| \to 0.$$

When $N \ge 2$ and $u \in W^{1,2}(\Omega; \mathbb{R}^2)$, we may also consider $\widetilde{J}u$, whose components are given by

$$(\widetilde{J}u)_{ij} := \frac{\partial u_1}{\partial x_i}\frac{\partial u_2}{\partial x_j} - \frac{\partial u_1}{\partial x_j}\frac{\partial u_2}{\partial x_i}, \ \forall\, 1 \le i, j \le N, \tag{1.43}$$

which is an $N \times N$ antisymmetric matrix with coefficients in $L^1(\Omega; \mathbb{R})$.

In view of (1.36), it is natural to compare Ju with $\widetilde{J}u$ in the case where both are defined. We have

Lemma 1.7. *Assume that $N \ge 2$ and $u \in W^{1,2}(\Omega; \mathbb{R}^2)$. Then,*

$$Ju = \widetilde{J}u \text{ in } \mathscr{D}'. \tag{1.44}$$

Proof. By standard density arguments (see, for example, Adams [4, Chapter III] or Brezis [75, Chapter 9]) there exist sequences (u_{1n}) and (u_{2n}) in $C^\infty(\overline{\Omega}; \mathbb{R})$ such that $u_{1n} \to u_1$, $u_{2n} \to u_2$ in $W^{1,2}$ and a.e. in Ω, and moreover $\|u_{1n}\|_{L^\infty} \le \|u\|_{L^\infty}$, $\|u_{2n}\|_{L^\infty} \le \|u\|_{L^\infty}$.

Set $u_n := (u_{1n}, u_{2n})$. Clearly, we have

$$Ju_n = \widetilde{J}u_n, \ \forall n, \tag{1.45}$$

and $u_n \wedge \nabla u_n \to u \wedge \nabla u$ in L^1, whereas $\widetilde{J}u_n \to \widetilde{J}u$ in L^1. Passing to the limit in (1.45), we obtain (1.44). □

We now turn to the special case where u takes its values in $\mathbb{S}^1$. Assume that $u \in W^{1,1}(\Omega; \mathbb{S}^1)$. Here is a list of useful elementary properties of $u \wedge \nabla u$:

$$u \wedge \nabla u = -\imath \overline{u}\, \nabla u \in L^1(\Omega; \mathbb{R}^N) \tag{1.46}$$

(where $\imath \overline{u}\, \nabla u$ denotes the vector $\nabla u \in \mathbb{C}^N$ multiplied by the complex scalar $\imath \overline{u}$),

$$|u \wedge \nabla u| = |\nabla u|, \tag{1.47}$$

and therefore

$$u \wedge \nabla u = 0 \text{ iff } u \text{ is a constant}, \tag{1.48}$$

$$\overline{u} \wedge \nabla \overline{u} = -u \wedge \nabla u, \tag{1.49}$$

$$(uv) \wedge \nabla(uv) = (u \wedge \nabla u) + (v \wedge \nabla v), \forall u, v \in W^{1,1}(\Omega; \mathbb{S}^1), \tag{1.50}$$

and, in particular,

$$(\alpha u) \wedge \nabla(\alpha u) = u \wedge \nabla u, \ \forall \alpha \in \mathbb{S}^1, \ \forall u \in W^{1,1}(\Omega; \mathbb{S}^1), \tag{1.51}$$

$$\text{if } \varphi \in W^{1,1}(\Omega; \mathbb{R}) \text{ and } u = e^{\imath\varphi}, \text{ then } u \wedge \nabla u = \nabla\varphi, \tag{1.52}$$

and, conversely,

$$[u \in W^{1,1}(\Omega; \mathbb{S}^1), \varphi \in W^{1,1}(\Omega; \mathbb{R}), u \wedge \nabla u = \nabla \varphi] \Longrightarrow u = e^{\imath(\varphi + C)}. \tag{1.53}$$

We prove (1.46), which is the generalization of (1.13) to $W^{1,1}$ maps. By the proof of (1.13), (1.46) amounts to

$$u \cdot \frac{\partial u}{\partial x_j} = 0 \text{ a.e. in } \Omega. \tag{1.54}$$

This is obtained by differentiating the identity $|u|^2 \equiv 1$ and using the following fact [75, Chapter 9]: if $g, h \in W^{1,p}(\Omega) \cap L^\infty(\Omega)$, with $1 \le p < \infty$, then $gh \in W^{1,p}(\Omega)$ and

$$\frac{\partial}{\partial x_j}(gh) = g \frac{\partial h}{\partial x_j} + h \frac{\partial g}{\partial x_j}.$$

[Alternatively, (1.54) can be obtained by noting that the chain rule (1.31) holds when f is supposed only C^1 (but not necessarily Lipschitz) provided $\varphi \in W^{1,p} \cap L^\infty$.]

Properties (1.47)–(1.50) are straightforward consequences of (1.46). To establish (1.52), note that, if $u = e^{\imath\varphi}$, then $u = u_1 + \imath u_2$ with $u_1 = \cos\varphi$ and $u_2 = \sin\varphi$. Thus, $\nabla u_1 = -(\sin\varphi)\nabla\varphi$ and $\nabla u_2 = (\cos\varphi)\nabla\varphi$, so that $u \wedge \nabla u = u_1 \nabla u_2 - u_2 \nabla u_1 = \nabla\varphi$.

For (1.53), set $w := ue^{-\imath\varphi}$ and write, using successively (1.50), (1.49), and (1.52),

$$w \wedge \nabla w = u \wedge \nabla u + (e^{-\imath\varphi}) \wedge \nabla(e^{-\imath\varphi}) = u \wedge \nabla u - \nabla\varphi = 0.$$

Therefore, w is constant, by (1.48).

From the above, we deduce immediately some properties of J:

$$J(\alpha u) = Ju, \ \forall\, u \in W^{1,1}(\Omega; \mathbb{S}^1), \ \forall\, \alpha \in \mathbb{S}^1, \tag{1.55}$$

$$J(uv) = Ju + Jv, \ \forall\, u, v \in W^{1,1}(\Omega; \mathbb{S}^1), \tag{1.56}$$

$$J(u\overline{v}) = Ju - Jv, \ \forall\, u, v \in W^{1,1}(\Omega; \mathbb{S}^1), \tag{1.57}$$

$$\text{if } \varphi \in W^{1,1}(\Omega; \mathbb{R}) \text{ and } u = e^{\imath\varphi}, \text{ then } Ju = 0. \tag{1.58}$$

The converse of (1.58) is also true, but requires more work.

Lemma 1.8. *Assume that $u \in W^{1,1}(\Omega; \mathbb{S}^1)$ satisfies $Ju = 0$. Then there exists $\varphi \in W^{1,1}(\Omega; \mathbb{R})$ such that $u = e^{\imath\varphi}$.*

Proof. We proceed as in the proof of Lemma 1.1: we solve the system

$$\nabla\varphi = u \wedge \nabla u \tag{1.59}$$

with the help of a generalized form of Poincaré's lemma.

Lemma 1.9. *Let $1 \le p < \infty$ and let $f \in L^p(\Omega; \mathbb{R}^N)$. Then, the following properties are equivalent:*

1. *There exists some $\varphi \in W^{1,p}(\Omega; \mathbb{R})$ such that $\nabla\varphi = f$.*
2. *One has*

$$\frac{\partial f_i}{\partial x_j} = \frac{\partial f_j}{\partial x_i}, \ \forall i, j \in \{1, \ldots, N\}, \tag{1.60}$$

in the sense of distributions, i.e.,

$$\int f_i \frac{\partial \psi}{\partial x_j} = \int f_j \frac{\partial \psi}{\partial x_i}, \ \forall \psi \in C_c^\infty(\Omega).$$

We emphasize that, as in the case of the classical Poincaré lemma, the assumption that Ω is simply connected is essential in this lemma.

Proof of Lemma 1.9. The implication "1 $\implies$ 2" is obvious. To prove the converse, let $\widetilde{f}$ be the extension of f by 0 outside Ω, and let $\widetilde{f}_\varepsilon := \widetilde{f} * \rho_\varepsilon$, where (ρ_ε) is a family of standard mollifiers. The $\widetilde{f}_\varepsilon$'s satisfy (1.60) on every compact subset of Ω (for ε sufficiently small). In particular, on every smooth simply connected domain $\omega \subset \Omega$ with compact closure in Ω, there exists a function ψ_ε such that $\nabla\psi_\varepsilon = \widetilde{f}_\varepsilon$ in ω. [Here we use the standard Poincaré lemma.] Assuming, in addition, that ψ_ε has zero average in ω, we may pass to the limit $\varepsilon \to 0$ and we obtain some $\psi \in W^{1,p}(\omega)$ such that $\nabla\psi = f$ in ω (the details are left to the reader). Finally, we write Ω as an increasing union of ω_n as above and obtain a corresponding sequence (ψ_n). In the limit, we find some $\varphi \in L^1_{\mathrm{loc}}(\Omega)$ with $\nabla\varphi = f$ in Ω. Using the regularity of Ω and a standard property of Sobolev spaces (see, for example, Maz'ya [240, Section 1.1.11, Corollary]), we conclude that $\varphi \in W^{1,p}(\Omega)$. □

Remark 1.5. For later reference, we mention that a similar conclusion holds when f is a measure. More precisely, let $f \in \mathcal{M}(\Omega; \mathbb{R}^N)$. Then there exists some $\varphi \in BV(\Omega; \mathbb{R})$ such that $D\varphi = f$ if and only if (1.60) holds. The proof is basically unchanged.

Proof of Lemma 1.8 *completed.* Set $f := u \wedge \nabla u$. From (1.34), we see that condition (1.60) is equivalent to $Ju = 0$. By Lemma 1.9, we know that there exists $\varphi_0 \in W^{1,1}(\Omega; \mathbb{R})$ such that $\nabla\varphi_0 = f = u \wedge \nabla u$. Applying (1.53), we conclude that $u = e^{\imath(\varphi_0 + C)}$. Hence, $\varphi := \varphi_0 + C$ has the desired properties. □

We are now ready to answer Questions 1 and 2.

Answer to Question 1

If $N = 1$, i.e., Ω is a bounded interval, the answer is positive for any p, $1 \le p < \infty$. Indeed, since u is continuous, it may be written as $u = e^{\imath\varphi}$ for some continuous function φ. Locally, we have $\varphi = -\imath \ln u$, where ln is a branch of the logarithm, and thus $\varphi \in W^{1,p}$, by the chain rule.

Theorem 1.1. *Assume that $N \ge 2$, and let $u \in W^{1,p}(\Omega; \mathbb{S}^1)$, with $p \ge 2$. Then there exists some $\varphi \in W^{1,p}(\Omega; \mathbb{R})$ such that $u = e^{\imath\varphi}$. Moreover, φ is unique (mod 2π).*

Proof. Recall that, by Lemma 1.7, we have $Ju = \widetilde{J}u$ in $\mathscr{D}'$, where $\widetilde{J}u$ denotes the vector-valued L^1 function whose components are given by (1.43). Here, the assumption $p \geq 2$ plays an important role.

As in the proof of (1.16) (but using (1.54) instead of (1.12)), we find that $\widetilde{J}u = 0$; that is,

$$Ju = 0, \ \forall\, u \in W^{1,p}(\Omega; \mathbb{S}^1), \ p \geq 2. \tag{1.61}$$

We conclude, with the help of Lemma 1.8, that there exists some $\varphi \in W^{1,1}(\Omega; \mathbb{R})$ such that $u = e^{\imath\varphi}$. Moreover, $\varphi \in W^{1,p}$, since $\nabla\varphi = u \wedge \nabla u \in L^p$.

Uniqueness follows from the fact that, if $u = e^{\imath\varphi_1} = e^{\imath\varphi_2}$, then $\nabla\varphi_1 = \nabla\varphi_2$. [We will return to the question of uniqueness in Chapter 6.] □

Remark 1.6. The conclusion of Theorem 1.1 fails when $p < 2$ (in any dimension $N \geq 2$), i.e.,

$$\begin{aligned} &\text{for every } p \in [1,2), \text{ there exists some } u \in W^{1,p}(\Omega; \mathbb{S}^1) \\ &\text{that } \textit{cannot} \text{ be written as } u = e^{\imath\varphi} \text{ for some } \varphi \in W^{1,p}. \end{aligned} \tag{1.62}$$

To see this, assume first that $N = 2$ and, for simplicity, that $0 \in \Omega$. Set $u(x) := \dfrac{x}{|x|}$. We start by proving that

$$u \in W^{1,p}(\Omega; \mathbb{S}^1) \text{ for every } 1 \leq p < 2 \tag{1.63}$$

(but $u \notin W^{1,p}$ for $p \geq 2$). The proof of (1.63) relies on (1.65) and Lemma 1.10 below, and goes as follows. First, in $\Omega \setminus \{0\}$, we have

$$\frac{\partial u_1}{\partial x_1} = \frac{x_2^2}{|x|^3}, \ \frac{\partial u_1}{\partial x_2} = -\frac{x_1 x_2}{|x|^3}, \ \frac{\partial u_2}{\partial x_1} = -\frac{x_1 x_2}{|x|^3}, \ \frac{\partial u_2}{\partial x_2} = \frac{x_1^2}{|x|^3}, \tag{1.64}$$

and thus

$$|\nabla u(x)| = \frac{1}{|x|} \in L^p(\Omega \setminus \{0\}), \ \forall\, p \in [1,2). \tag{1.65}$$

By itself, (1.65) does not imply that $u \in W^{1,p}(\Omega)$. [Think of the Heaviside function on $(-1,1)$.] However, the desired conclusion follows from the following lemma (stated in greater generality than needed here, and applied with $K := \{0\}$).

Lemma 1.10. *Assume that $N \geq 2$. Let K be a compact subset of Ω such that $\mathscr{H}^{N-1}(K) = 0$. Let $u \in W^{1,1}_{loc}(\Omega \setminus K)$ be such that $\int_{\Omega\setminus K} |\nabla u| < \infty$. Then, $u \in W^{1,1}_{loc}(\Omega)$.*

As a consequence, if, in addition, $\int_{\Omega\setminus K} |\nabla u|^p < \infty$ for some $p > 1$, then $u \in W^{1,p}_{loc}(\Omega)$.

Here, $\mathscr{H}^{N-1}$ stands for the $(N-1)$-dimensional Hausdorff measure.

Proof of Lemma 1.10. After multiplying by a cutoff, we may assume, with no loss of generality, that $u \in W^{1,1}_{loc}(\mathbb{R}^N \setminus K)$ and $\operatorname{supp} u \subset (0,1)^N$. In this special case, the conclusion of the lemma is equivalent to $u \in L^1(\mathbb{R}^N)$ and

$$\int_{\mathbb{R}^N} u\frac{\partial \varphi}{\partial x_j} = -\int_{\mathbb{R}^N} \frac{\partial u}{\partial x_j}\varphi, \ \forall \varphi \in C_c^\infty(\mathbb{R}^N).$$

[The fact that u and ∇u are not defined in K is irrelevant, since $|K| = 0$.] Let $\varepsilon > 0$. Let (C_j) be a finite collection of open cubes with sides parallel to the coordinate axes and such that $K \subset \cup_j C_j$ and $\sum_j \ell_j^{N-1} < \varepsilon$, where ℓ_j is the size of C_j. We identify $\mathbb{R}^{N-1} \times \{0\}$ with $\mathbb{R}^{N-1}$ and let D_j be the projection of C_j on $\mathbb{R}^{N-1}$. Let $V_\varepsilon := \cup_j D_j$ and $U_\varepsilon := V_\varepsilon \times [0,1]$. Since $K \subset U_\varepsilon$ (and thus $u \in W^{1,1}(\mathbb{R}^N \setminus U_\varepsilon)$), we find that, for a.e. $x' \notin V_\varepsilon$, the map $[0,1] \ni t \mapsto u(x',t)$ belongs to $W_0^{1,1}((0,1))$. We obtain

$$\int_{\mathbb{R}^N\setminus U_\varepsilon} |u| = \int_{\mathbb{R}^{N-1}\setminus V_\varepsilon} \left(\int_0^1 |u(x',t)|\, dt\right) dx' \le C \int_{\mathbb{R}^N} \left|\frac{\partial u}{\partial x_1}\right|. \tag{1.66}$$

Using (1.66) and the fact that $|U_\varepsilon| < \varepsilon$, we easily find (by letting $\varepsilon \to 0$) that

$$\int_{\mathbb{R}^N} |u| \le C \int_{\mathbb{R}^N} \left|\frac{\partial u}{\partial x_1}\right| < \infty. \tag{1.67}$$

On the other hand, (1.67) combined with the identity

$$\int_{\mathbb{R}^N\setminus U_\varepsilon} u\frac{\partial \varphi}{\partial x_1} = -\int_{\mathbb{R}^N\setminus U_\varepsilon} \frac{\partial u}{\partial x_1}\varphi, \ \forall \varphi \in C_c^\infty(\mathbb{R}^N),$$

(which follows by Fubini), yields

$$\int_{\mathbb{R}^N} u\frac{\partial \varphi}{\partial x_1} = -\int_{\mathbb{R}^N} \frac{\partial u}{\partial x_1}\varphi, \ \forall \varphi \in C_c^\infty(\mathbb{R}^N).$$

A similar identity holds for the other first order derivatives. □

Remark 1.7. The lemma applies in particular to the case where K reduces to a point. [The reader is encouraged to find his/her own proof in this special case!] More generally, we may let K be a k-dimensional compact Lipschitz manifold with $k \le N-2$. We may actually replace compact manifolds by manifolds that are closed in Ω. The latter case, which does not follow directly from Lemma 1.10, can be established by copying its proof.

Remark 1.8. Results related to Lemma 1.10 may be found, for example, in Koskela [215, Introduction], Dupaigne and Ponce [142, Lemma 3], and the references therein. Let us also recall that our condition $\mathscr{H}^{N-1}(K) = 0$ is equivalent to the fact that K has null $W^{1,1}$-capacity [146, Theorem 3, p. 193].

Proof of (1.62). We claim that there *does not* exist any $\varphi \in W^{1,1}(\Omega; \mathbb{R})$ (thus, there exists no $\varphi \in W^{1,p}$) such that $u = e^{\imath\varphi}$, where $u(x) := \dfrac{x}{|x|}$ as above. Argue by contradiction, and suppose that such φ exists. By (1.52), and since $u \in C^\infty(\Omega\backslash\{0\})$, we see that $\varphi \in C^\infty(\Omega\backslash\{0\})$.

Fix $r > 0$ so small that $C_r := \{x;\, |x| = r\} \subset \Omega$. On C_r, we have $e^{\imath\varphi(r,\theta)} = e^{\imath\theta}$, and thus $\varphi(re^{\imath\theta}) = \theta + 2\pi k(\theta)$, $\forall\, \theta \in \mathbb{R}$, for some $k(\theta) \in \mathbb{Z}$. By continuity of φ, k is constant on $[0, 2\pi]$. Hence,

$$2\pi k = \varphi(r\, e^{\imath 0}) = \varphi(r\, e^{\imath 2\pi}) = 2\pi + 2\pi k,$$

a contradiction.

The same conclusion can also be reached using degree theory (see (1.19) and also Chapter 12). [For a general introduction to degree theory, see, for example, Dinca and Mawhin [141, Chapter 1].] The map $u(x) := \dfrac{x}{|x|}$ provides a canonical example of *topological obstruction* to lifting.

When $N \geq 3$, we assume, as above, that $0 \in \Omega$, and consider

$$u(x) := \frac{(x_1, x_2)}{(x_1^2 + x_2^2)^{1/2}}.$$

Clearly, $u \in W^{1,p}(\Omega; \mathbb{S}^1)$ for every p with $1 \leq p < 2$, and moreover $u \in C^\infty(\omega)$, where $\omega := \Omega \setminus \{x \in \mathbb{R}^N;\, x_1 = x_2 = 0\}$.

If $\varphi \in W^{1,1}(\Omega; \mathbb{R})$ satisfying $u = e^{\imath\varphi}$ exists, then we see, using (1.52), that $\varphi \in C^\infty(\omega)$. Since $u = e^{\imath\varphi}$ a.e. in Ω, we have, by continuity, $u = e^{\imath\varphi}$ *everywhere* in ω. Let

$$C_r := \{x \in \mathbb{R}^N;\, x_1^2 + x_2^2 = r^2 \text{ and } x_3 = \cdots = x_N = 0\},$$

and fix $r > 0$ so small that $C_r \subset \omega$. Since $u = e^{\imath\varphi}$ on C_r, we reach the same contradiction as in the case $N = 2$. □

Answer to Question 2

We may now summarize part of the preceding discussion in the following

Theorem 1.2. *Let $N \geq 2$ and let $1 \leq p < 2$. Given $u \in W^{1,p}(\Omega; \mathbb{S}^1)$, the following assertions are equivalent:*

a) u can be written as $u = e^{\imath\varphi}$ for some $\varphi \in W^{1,p}(\Omega; \mathbb{R})$
b) $Ju = 0$,

and, if φ exists, it is unique (mod 2π).

In particular, if $u, v \in W^{1,p}(\Omega; \mathbb{S}^1)$, the following are equivalent:

(i) $u = e^{\imath\varphi} v$ for some $\varphi \in W^{1,p}(\Omega; \mathbb{R})$
(ii) $Ju = Jv$.

If $Ju = Jv$, we say that u and v have the *same topological singularities*. This terminology will become transparent in Chapters 2 and 3.

Combining Theorem 1.2 with Lemma 1.1, we obtain the following

Corollary 1.1. *Let* $u \in W^{1,1}(\Omega; \mathbb{S}^1) \cap C(\overline{\Omega}; \mathbb{S}^1)$. *Then there exists some* $\varphi \in W^{1,1}(\Omega; \mathbb{R}) \cap C(\overline{\Omega}; \mathbb{R})$ *such that* $u = e^{\imath\varphi}$. *In particular,* $Ju = 0$.

Remark 1.9. Let us now return to the map $u(x) = \dfrac{x}{|x|}$ considered in Remark 1.6 for $N = 2$. Since $u \in W^{1,1}(\Omega; \mathbb{S}^1)$ and u admits no lifting in $W^{1,1}(\Omega; \mathbb{R})$, we know, by Theorem 1.2, that $Ju \neq 0$. The reader may wonder what is the value of Ju. We claim that $Ju = \pi\delta_0$. Indeed, let

$$u^\varepsilon(x) := \begin{cases} u(x), & \text{if } |x| \geq \varepsilon \\ x/\varepsilon, & \text{if } |x| < \varepsilon \end{cases}.$$

Note that $u^\varepsilon \in W^{1,2}(\Omega; \mathbb{R}^2)$ (u^ε is even Lipschitz) and $u^\varepsilon \to u$ in $W^{1,p}$ for every $p < 2$. Clearly, $u^\varepsilon \wedge \nabla u^\varepsilon \to u \wedge \nabla u$ in L^p for every $p < 2$, and thus

$$Ju^\varepsilon = \frac{1}{2}\operatorname{curl}(u^\varepsilon \wedge \nabla u^\varepsilon) \to Ju \text{ in } \mathscr{D}'.$$

On the other hand, since $u^\varepsilon \in W^{1,2} \cap L^\infty$, we know from Lemma 1.7 that

$$Ju^\varepsilon = \widetilde{J}u^\varepsilon = \frac{\partial u_1^\varepsilon}{\partial x_1}\frac{\partial u_2^\varepsilon}{\partial x_2} - \frac{\partial u_1^\varepsilon}{\partial x_2}\frac{\partial u_2^\varepsilon}{\partial x_1}.$$

By a direct computation, we find that $\widetilde{J}u^\varepsilon = \dfrac{1}{\varepsilon^2}\chi_{B_\varepsilon(0)}$. Indeed, $\widetilde{J}u^\varepsilon$ vanishes in the set $\{|x| > \varepsilon\}$ (since u^ε is smooth and $\mathbb{S}^1$-valued in that set), while it is clear that $\widetilde{J}u^\varepsilon = 1/\varepsilon^2$ in the set $\{|x| < \varepsilon\}$. Finally, we note that $\lim\limits_{\varepsilon\to 0} \dfrac{1}{\varepsilon^2}\chi_{B_\varepsilon(0)} = \pi\delta_0$, and therefore

$$Ju = \pi\delta_0. \tag{1.68}$$

Alternatively, the reader may easily check that

$$(u \wedge \nabla u)(x) = \frac{1}{|x|^2}(-x_2, x_1)$$

and $\operatorname{curl}(u \wedge \nabla u) = 2\pi\delta_0$.

In Chapter 2, we will present far-reaching generalizations of (1.68). For example, we will see that, if $u \in W^{1,1}(\Omega; \mathbb{S}^1) \cap C(\Omega \setminus \cup\{a_j\})$ for a finite collection of points (a_j) in Ω, then $Ju = \pi\sum_j \deg(u, a_j)\delta_{a_j}$, where $\deg(u, a_j)$ denotes the topological degree of u at a_j (see Chapter 2 for a precise definition). A striking fact discussed in Section 2.9 asserts that, for *every* $u \in W^{1,1}(\Omega; \mathbb{S}^1)$, the distribution $\frac{1}{\pi} Ju$ is a (possibly infinite) sum of $\pm$ Dirac masses located at the topological singularities of u. In Chapter 3, we will discuss representation formulas for Ju when $N \geq 3$.

Using Theorem 1.1, it is easy to answer Question 1 for $W^{m,p}$ maps, when $m = 2, 3, \ldots$ More specifically, we have the following

Corollary 1.2. *Let $m \in \{2, 3, \ldots\}$ and $1 \leq p < \infty$. Then, each map $u \in W^{m,p}(\Omega; \mathbb{S}^1)$ can be written as $u = e^{\imath\varphi}$ for some $\varphi \in W^{m,p} \cap W^{1,mp}$.*

Conversely, for every $\varphi \in W^{m,p} \cap W^{1,mp}$, the map $u = e^{\imath\varphi}$ belongs to $W^{m,p}(\Omega; \mathbb{S}^1)$. In addition, φ satisfies the estimates

$$\sum_{k=1}^{m} \|D^k \varphi\|_{L^p} \leq C|u|_{W^{m,p}}. \tag{1.69}$$

Proof. We first note that $W^{m,p} \cap L^\infty \subset W^{1,mp}$ (by the Gagliardo–Nirenberg embeddings; see Nirenberg [283] and Appendix 15.2). Since $mp \geq 2$, Theorem 1.1 implies that each $u \in W^{m,p}(\Omega; \mathbb{S}^1)$ can be written as $u = e^{\imath\varphi}$ for some $\varphi \in W^{1,mp}$. Equality (1.52) implies that $\nabla\varphi$ is actually in $W^{m-1,p}$. [This is a straightforward consequence of the Gagliardo–Nirenberg embeddings $W^{m,p} \cap L^\infty \subset W^{j,mp/j}$, $j = 1, \ldots, m-1$; see also Lemma 5.2.] It follows that $\varphi \in W^{m,p}$.

The converse follows easily from Theorem 15.8 in Appendix 15.3.

Estimate (1.69) is a consequence of (1.52) and of Gagliardo–Nirenberg. □

Assume that $u \in W^{1,1}(\Omega; \mathbb{S}^1)$ and that $f \in \mathrm{Lip}(\mathbb{S}^1; \mathbb{S}^1)$. Then, $v := f \circ u \in W^{1,1}(\Omega; \mathbb{S}^1)$ (see Marcus and Mizel [235]) and one may consider the expressions $v \wedge \nabla v$ and Jv. The next result relates these expressions to $u \wedge \nabla u$ and Ju.

Theorem 1.3. *Under the above assumptions,*

$$v \wedge \nabla v = \left(f(u) \wedge \frac{\partial f}{\partial \tau}(u)\right) (u \wedge \nabla u) \tag{1.70}$$

and

$$Jv = (\deg f)\, Ju. \tag{1.71}$$

Note that the meaning of (1.70) is clear when $f \in C^1$. However, when f is merely Lipschitz, the quantity $(\partial f/\partial\tau)(u)$ is not well-defined, e.g., when u is a constant, $u \equiv C$, and $(\partial f/\partial\tau)(C)$ is not well-defined. Nevertheless, the right-hand side of (1.70) is well-defined, as a consequence of the fact that $(\partial f/\partial\tau)(a)$ is defined for $a \in \mathbb{S}^1 \setminus E$, with E a null subset of $\mathbb{S}^1$, and of the following property: if $w \in W^{1,1}(\Omega; \mathbb{R})$ and $F \subset \mathbb{R}$ is a null set, then $\nabla w = 0$ a.e. on $\{x \in \Omega;\ w(x) \in F\}$ (see Serrin and Varberg [320], Marcus and Mizel [235], Lieb and Loss [225, Theorem 6.19]).

Proof of Theorem 1.3. Set $\dot{f} := \partial f/\partial\tau$ and assume for simplicity that $f \in C^1$. [The general case is similar, relying on the appropriate chain rule [320].] We claim that

$$\frac{\partial v}{\partial x_j} = \left(u \wedge \frac{\partial u}{\partial x_j}\right) \dot{f}(u). \tag{1.72}$$

Indeed, it clearly suffices to prove that (1.72) holds in 1D. If $u \in W^{1,1}(I; \mathbb{S}^1)$, with I an interval, then we may write $u = e^{\imath\varphi}$, with $\varphi \in W^{1,1}(I; \mathbb{R})$ (Lemma 1.1). If $y \in I$, then we may write, locally near $u(y)$, $f(e^{\imath\theta}) = e^{\imath\psi(\theta)}$, with $\psi \in C^1$, and then locally near y we have $v(x) = e^{\imath\psi(\varphi(x))}$. Therefore,

$$v'(x) = \imath\, e^{\imath\psi(\varphi(x))}\, \psi'(\varphi(x))\, \varphi'(x), \;\; u(x) \wedge u'(x) = \varphi'(x),$$
$$\dot{f}(u(x)) = \imath\, e^{\imath\psi(\varphi(x))}\, \psi'(\varphi(x)),$$

and (1.72) is clear.

By (1.72), we have

$$v \wedge \nabla v = (f(u) \wedge \dot{f}(u))\, (u \wedge \nabla u).$$

For the proof of (1.71), we write (using Lemmas 1.2 and 1.3) $f(z) = z^k\, e^{\imath\varphi(z)}$, $z \in \mathbb{S}^1$, for some $\varphi \in \mathrm{Lip}(\mathbb{S}^1; \mathbb{R})$, where $k := \deg f$. Then, $v = u^k\, e^{\imath\varphi(u)}$, so that (by (1.56), (1.57), and Lemma 1.8),

$$Jv = Ju^k + Je^{\imath\varphi(u)} = Ju^k = k\, Ju = (\deg f)\, Ju$$

(since $\varphi(u) \in W^{1,1}(\Omega; \mathbb{R})$). □

1.3 Square root

Here is a natural question: given $u \in W^{1,p}(\Omega; \mathbb{S}^1)$, does there exist $v \in W^{1,p}(\Omega; \mathbb{S}^1)$ such that $v^2 = u$? Or, more generally, $v^m = u$? Theorem 1.1 gives a positive answer if $p \geq 2$. When $1 \leq p < 2$, the answer is, in general, negative. To see this, we argue as in Remark 1.6. Let $u(x) := \dfrac{x}{|x|}$ in the unit disk Ω of $\mathbb{R}^2$. If $u = v^2$ for some v, then $v \in C^\infty(\Omega \setminus \{0\})$. Indeed, we may differentiate the equation $u = v^2$ and find that

$$\nabla v = \frac{1}{2} \bar{v}\, \nabla u \text{ in } \Omega \setminus \{0\}. \tag{1.73}$$

Thus, $\nabla v \in W^{1,p}_{loc}(\Omega \setminus \{0\})$, and therefore $v \in W^{2,p}_{loc}(\Omega \setminus \{0\})$. Iterating this argument, we end up with $v \in C^\infty(\Omega \setminus \{0\})$.

We find that, on a circle $C(0, r)$, with $0 < r < 1$,

$$1 = \deg(u, C(0, r)) = \deg(v^2, C(0, r)) = 2 \deg(v, C(0, r)),$$

which is impossible.

We now give a necessary and sufficient condition for the existence of the square root.

Corollary 1.3. *Let $1 \le p < 2$. Let $u \in W^{1,p}(\Omega; \mathbb{S}^1)$. Then, the following conditions are equivalent:*

a) u can be written as $u = v^2$ for some $v \in W^{1,p}(\Omega; \mathbb{S}^1)$
b) There exists some $\widetilde{v} \in W^{1,p}(\Omega; \mathbb{S}^1)$ such that $Ju = 2J\widetilde{v}$
c) There exists some $w \in W^{1,1}(\Omega; \mathbb{S}^1)$ such that $Ju = 2Jw$.

Proof. In view of (1.56), it suffices to prove that *c)* implies *a)*. The map u/w^2 belongs to $W^{1,1}(\Omega; \mathbb{S}^1)$ and satisfies $J(u/w^2) = 0$ (by (1.56), (1.57)). By Theorem 1.2, there exists some $\eta \in W^{1,1}(\Omega; \mathbb{S}^1)$ such that $u/w^2 = \eta^2$. Set $v := \eta w \in W^{1,1}(\Omega; \mathbb{S}^1)$. Using (1.73), we find that $\nabla v \in L^p$, and thus $v \in W^{1,p}(\Omega; \mathbb{S}^1)$. □

In Section 8.5, we will return to the square root problem within the framework of fractional Sobolev spaces.

1.4 Lifting of $\mathbb{S}^1$-valued maps in $W^{1,1}$ and BV. Definition of $\Sigma(u)$

1.4.1 BV maps

Let Ω be a smooth bounded simply connected domain in $\mathbb{R}^N$. A map $u = (u_1, \dots, u_k) \in L^1(\Omega; \mathbb{R}^k)$ is said to be of bounded variation, and we write $u \in BV(\Omega; \mathbb{R}^k)$, if

$$\sup \left\{ \sum_{i=1}^{k} \int u_i \operatorname{div} g_i \, dx;\ g_i \in C_c^\infty(\Omega; \mathbb{R}^N),\ \sum_{i=1}^{N} |g_i|^2 \le 1 \text{ in } \Omega \right\} < \infty .$$

We denote the above quantity by $|u|_{BV}$, the BV seminorm of u. If u belongs to $W^{1,1}$, then $|u|_{BV}$ coincides with $\| |\nabla u| \|_{L^1}$. Here, $|\nabla u|$ stands for the Euclidean norm of ∇u. It is clear from the definition that, if $u^n, u \in BV$ and $u^n \to u$ in L^1, then $|u|_{BV} \le \liminf |u^n|_{BV}$.

Our starting point is the following standard

Lemma 1.11. *Let $\varphi \in BV(\Omega; \mathbb{R}^k)$. Let $L : \mathbb{R}^k \to \mathbb{R}^l$ be a Lipschitz map. Then, $L \circ \varphi \in BV(\Omega; \mathbb{R}^l)$ and $|L \circ \varphi|_{BV} \le |L|_{\mathrm{Lip}} |\varphi|_{BV}$.*

Here, $|L|_{\mathrm{Lip}}$ stands for the Lipschitz constant of L.

Sketch of proof. One proceeds by regularization; there are two possible methods.

Method 1. One regularizes both L and φ. More precisely, there exists a sequence (L^n) of smooth maps such that $L^n \to L$ uniformly on compact sets and $|L^n|_{\mathrm{Lip}} \le |L|_{\mathrm{Lip}}$. On the other hand, there exists a sequence $(\varphi^n) \subset C^\infty(\Omega)$ such that $\varphi^n \to \varphi$ in L^1 and a.e., and $\| |\nabla \varphi^n| \|_{L^1} \to |\varphi|_{BV}$ (see, for example, Giusti [167, Theorem 1.17] or

Ambrosio, Fusco, and Pallara [13, Theorem 3.9]). The conclusion of Lemma 1.11 is clear for $L^n \circ \varphi^n$. The case of a general L and u follows by letting $n \to \infty$.

Method 2. One obtains Lemma 1.11 as a straightforward consequence of the "discretization" of the BV seminorm described in Theorem 6.2. □

As a consequence of Lemma 1.11, if $\varphi \in BV(\Omega;\mathbb{R})$, then $u := e^{\imath\varphi}$ belongs to $BV(\Omega;\mathbb{S}^1)$, i.e., $u \in BV(\Omega;\mathbb{R}^2)$ and $|u| = 1$ in Ω. We address the converse question, i.e., whether every $u \in BV(\Omega;\mathbb{S}^1)$ is of the form $u = e^{\imath\varphi}$ for some $\varphi \in BV(\Omega;\mathbb{R})$. The answer is positive, as proved by Giaquinta, Modica, and Souček [157, Section 6.2]. We now state a sharper result.

Theorem 1.4. *Let $u \in BV(\Omega;\mathbb{S}^1)$. Then there exists a lifting $\varphi \in BV(\Omega;\mathbb{R})$ of u such that $\|\varphi\|_{L^\infty} \leq 2\pi$ and*

$$|\varphi|_{BV} \leq 2|u|_{BV}. \tag{1.74}$$

Remark 1.10. For $N \geq 2$, the constant 2 in (1.74) is minimal [129]. In fact, let $u(x) := \dfrac{x}{|x|}$, defined in the unit disk Ω in $\mathbb{R}^2$. Then, for every lifting $\varphi \in BV(\Omega;\mathbb{R})$ of u we have $|\varphi|_{BV} \geq E(u) = 4\pi = 2|u|_{BV}$ (see Example 2.4 and Theorem 2.5 in Chapter 2).

Proof of Theorem 1.4. The construction of φ we present here is taken from Dávila and Ignat [129]. For the proof of the estimate (1.74), we follow Merlet [247].

Let, for $0 \leq \alpha < 2\pi$ and $z \in \mathbb{S}^1$, $\Theta_\alpha(z)$ be the unique argument of z in the interval $(\alpha - 2\pi, \alpha]$. The explicit formula giving Θ_α is

$$\Theta_\alpha(e^{\imath\beta}) = \begin{cases} \beta - 2\pi, & \text{if } \alpha < \beta < 2\pi \\ \beta, & \text{if } 0 \leq \beta \leq \alpha \end{cases}. \tag{1.75}$$

Using (1.75), it is simple to establish the following result, whose proof will be omitted.

Lemma 1.12. *Let $\omega := e^{\imath\beta}$, $\sigma := e^{\imath\gamma}$, where $\beta, \gamma \in [0, 2\pi)$. Then*

$$\int_0^{2\pi} |\Theta_\alpha(\omega) - \Theta_\alpha(\sigma)|\,d\alpha = 2|\beta - \gamma|(2\pi - |\beta - \gamma|) \leq 4\pi|\omega - \sigma|. \tag{1.76}$$

Let now $u \in BV(\Omega;\mathbb{S}^1)$. By (1.76), we have, for every family (ρ_ε) of non-negative radial mollifiers,

$$\begin{aligned} \int_0^{2\pi}\int_\Omega\int_\Omega \frac{|\Theta_\alpha(u(x)) - \Theta_\alpha(u(y))|}{|x-y|}\rho_\varepsilon(x-y)\,dxdy\,d\alpha \leq \\ 4\pi\int_\Omega\int_\Omega \frac{|u(x)-u(y)|}{|x-y|}\rho_\varepsilon(x-y)\,dxdy. \end{aligned} \tag{1.77}$$

By (6.5) and (6.7) in Chapter 6, (1.77), and Fatou's lemma, we find that, for almost every α, we have $\Theta_\alpha(u) \in BV$, $\alpha \mapsto |\Theta_\alpha(u)|_{BV}$ is measurable and

$$\int_0^{2\pi} |\Theta_\alpha(u)|_{BV}\, d\alpha \leq 4\pi\, |u|_{BV}.$$

Thus, for an appropriate α, the phase $\varphi := \Theta_\alpha(u)$ belongs to BV and satisfies $|\varphi|_{BV} \leq 2|u|_{BV}$. □

1.4.2 $W^{1,1}$ maps. Minimal liftings. Definition of $\Sigma(u)$

We start with some general considerations about maps in $BV(\Omega;\mathbb{S}^1)$. From Theorem 1.4, we obtain in particular that every $u \in BV(\Omega;\mathbb{S}^1)$ has a lifting in BV. But there are, of course, many BV liftings (in contrast with the $W^{1,1}$ situation, see the uniqueness part in the proof of Theorem 1.1). Therefore, it is natural to introduce the quantity

$$E(u) := \inf\{|\varphi|_{BV};\ \varphi \in BV(\Omega;\mathbb{R}) \text{ is such that } u = e^{\imath\varphi} \text{ in } \Omega\}. \tag{1.78}$$

By Theorem 1.4 and Lemma 1.11, we have

$$|u|_{BV} \leq E(u) \leq 2|u|_{BV}. \tag{1.79}$$

Note that inf in (1.78) is attained. Indeed, since we may add an integer multiple of 2π to any lifting, it suffices to consider liftings φ satisfying $0 \leq \int_\Omega \varphi \leq 2\pi|\Omega|$. By Theorem 1.4 and Poincaré's inequality in BV, it follows that the set

$$B := \{\varphi \in BV(\Omega) \text{ is a lifting of } u,\ |\varphi|_{BV} \leq 2|u|_{BV},\ 0 \leq \int_\Omega \varphi \leq 2\pi\, |\Omega|\}$$

is non-empty and bounded in $BV(\Omega)$ (and thus relatively compact in L^1). Since the BV norm is lower semicontinuous with respect to L^1 convergence, it follows that B contains a minimizer in (1.78).

Remark 1.11. By the above argument, $u \mapsto E(u)$ is lower semicontinuous in the following sense: if $(u_n) \subset BV(\Omega;\mathbb{S}^1)$ is bounded and $u_n \to u$ a.e., then $E(u) \leq \liminf E(u_n)$.

Any minimizer of $E(u)$ in (1.78) is called a *minimal lifting* of u. Minimal liftings have very interesting properties. We will return to this topic in Chapters 2 and 3. As we are going to see in Theorems 2.12 and 3.9, there is a one-to-one correspondence between minimal liftings and minimal connections (resp. minimal surfaces) spanned by the singularities of u.

In what follows, we assume that $u \in W^{1,1}(\Omega;\mathbb{S}^1)$. Our goal is to obtain a formula for $E(u)$, or more precisely for the (non-negative) "gap" $E(u) - \int_\Omega |\nabla u|$. It involves the quantity $\Sigma(u)$ defined by

$$\Sigma(u) := \inf_{\eta\in W^{1,1}(\Omega;\mathbb{R})} \|u\wedge\nabla u - \nabla\eta\|_{L^1}. \tag{1.80}$$

In particular,

$$\Sigma(u) \le \int_\Omega |\nabla u|. \tag{1.81}$$

The infimum in (1.80) is not achieved in general (see Proposition 1.2 and Section 2.11.1). However, if $W^{1,1}$ is replaced by BV in (1.80), then the value of the infimum is unchanged and it is achieved:

Lemma 1.13. *We have*

$$\Sigma(u) = \min_{\eta\in BV(\Omega;\mathbb{R})} \|u\wedge\nabla u - D\eta\|_{\mathscr{M}}. \tag{1.82}$$

A word about notation. Throughout the book, if $\eta\in BV$, then $D\eta$ denotes the distributional derivative of η, which is a vector-valued finite measure. We denote by $\nabla\eta$ the absolutely continuous part of $D\eta$. In particular, if $\eta\in W^{1,1}$, then $D\eta=\nabla\eta$. We will return to this subject in the proof of Theorem 1.6.

Proof. We start by noting that inf over $\eta\in BV$ is indeed a minimum. This uses the fact that each minimizing sequence (η_n), normalized by the condition $\displaystyle\int \eta_n = 0$, is bounded in BV. Passing to a subsequence, we find some $\eta\in BV$ such that $\eta_n\to\eta$ in L^1, and this η minimizes $\Sigma(u)$.

The fact that inf over $W^{1,1}$ coincides with the minimum over BV follows from a standard property of BV: for $F\in L^1(\Omega;\mathbb{R}^N)$ and $\eta\in BV(\Omega)$, there exists a sequence $(\eta_n)\subset C^\infty(\Omega)\cap W^{1,1}(\Omega)$ satisfying

$$\eta_n\to\eta \text{ in } L^1(\Omega) \text{ and } \|F-D\eta\|_{\mathscr{M}} = \lim_{n\to\infty}\int_\Omega |F-\nabla\eta_n|. \tag{1.83}$$

Assertion (1.83) is usually proved in the special case where $F=0$, but the same proof yields (1.83) in the general case (see the proof of Theorem 1.17 in Giusti [167] or the proof of Theorem 3.9 in Ambrosio, Fusco, and Pallara [13]; see also Lemma 15.32 in Appendix 15.7). □

$\Sigma(u)$ plays a *very important role* throughout the book (see, for example, Theorems 1.5, 1.8, 1.9, and 1.11, Corollary 1.6, formula (1.84), and Chapters 2, 3, 8, and 10). We point out that $\Sigma(u)$ depends only on Ju; this is a consequence of the following

Lemma 1.14. *We have*

$$\Sigma(u) = \inf\left\{\int_\Omega |\nabla v|;\ v\in W^{1,1}(\Omega;\mathbb{S}^1),\ Jv = Ju \textit{ in } \mathscr{D}'\right\} \tag{1.84}$$

and

$$\Sigma(u) = \min\{\|\mu\|_{\mathscr{M}};\ \mu\in\mathscr{M}(\Omega;\mathbb{R}^N),\ \operatorname{curl}\mu = 2Ju \textit{ in } \mathscr{D}'\}. \tag{1.85}$$

Proof. From the properties of J ((1.57), (1.58), and Lemma 1.8) we know that

$$Ju = Jv \iff [v \wedge \nabla v = u \wedge \nabla u - \nabla\varphi \text{ for some } \varphi \in W^{1,1}(\Omega; \mathbb{R})].$$

Moreover (by (1.47)), $|v \wedge \nabla v| = |\nabla v|$, and thus (1.84) holds.

Next we establish (1.85). Note that, if $\eta \in BV(\Omega; \mathbb{R})$, then $\mu := u \wedge \nabla u - D\eta$ satisfies

$$\operatorname{curl} \mu = \operatorname{curl} (u \wedge \nabla u) = 2Ju$$

(by definition of Ju). Conversely, if $\mu \in \mathcal{M}(\Omega; \mathbb{R}^N)$ satisfies $\operatorname{curl} \mu = 2Ju$, then $\operatorname{curl} (u \wedge \nabla u - \mu) = 0$, and therefore there exists some $\eta \in BV(\Omega; \mathbb{R})$ such that $u \wedge \nabla u - \mu = D\eta$ (this comes from the variant of Poincaré's lemma mentioned in Remark 1.5). □

We deduce from Theorem 1.2, Lemma 1.13, and (1.52)–(1.53) that, given $u \in W^{1,1}(\Omega; \mathbb{S}^1)$, we have

$$\Sigma(u) = 0 \iff [u = e^{\imath\varphi} \text{ for some } \varphi \in W^{1,1}] \iff Ju = 0. \tag{1.86}$$

Here are two fundamental properties.

Theorem 1.5. *For every $u \in W^{1,1}(\Omega; \mathbb{S}^1)$, we have*

$$E(u) = \int_\Omega |\nabla u| + \Sigma(u). \tag{1.87}$$

Moreover, every minimal lifting $\varphi \in BV(\Omega; \mathbb{R})$ of u satisfies

$$\Sigma(u) = \|u \wedge \nabla u - D\varphi\|_{\mathcal{M}}, \tag{1.88}$$

i.e., φ is a minimizer in (1.82).

Remark 1.12. The converse to the second part of Theorem 1.5 is not true: a minimizer η for (1.82) need not be a minimal lifting of u. Actually, a minimizer η for (1.82) need not even be a lifting of u. Take, for example, $\Omega := \mathbb{D}$ and $u(x) := x/|x|$. We know (by (1.47) and (1.65)) that $\|u \wedge \nabla u\|_{L^1} = 2\pi$. On the other hand (see Example 2.4 and Theorem 2.5 in Chapter 2), we have $\Sigma(u) = 2\pi$, and therefore $\eta = 0$, which is not a lifting of u, is a minimizer in (1.82).

Remark 1.13. Theorem 1.5 was generalized by Ignat [193] (see also Giaquinta, Modica, and Souček [157]) to the case where u is merely BV. One of the difficulties consists in giving a meaning to the quantity $u \wedge Du$ when u is in BV. We will explain in Section 1.9.1 how Ignat proceeds.

The proof of Theorem 1.5 is subtle (especially the proof of "$\geq$" in (1.87)) and is divided into four steps.

Step 1. We establish the following

Lemma 1.15. *Let $\delta > 0$ and let $v \in W^{1,1}(\Omega; \mathbb{S}^1)$. Then there exists some $\eta \in W^{1,1}(\Omega; \mathbb{R})$ such that*

$$\int_\Omega |v \wedge \nabla v - \nabla \eta| \leq \int_\Omega |\nabla v| + \delta, \tag{1.89}$$

$$\text{meas}\{x \in \Omega;\ v(x) \neq e^{\imath\eta(x)}\} := |[v \neq e^{\imath\eta}]| \leq \delta, \tag{1.90}$$

and in particular

$$\int_\Omega \left|v - e^{\imath\eta}\right| \leq 2\delta. \tag{1.91}$$

Here is a simple consequence of Lemma 1.15.

Proposition 1.1. *Let $u \in W^{1,1}(\Omega; \mathbb{S}^1)$. Then there exists a sequence $(u_n) \subset W^{1,1}(\Omega; \mathbb{S}^1)$ such that*

$$Ju_n = Ju,\ \forall\, n, \tag{1.92}$$

$$u_n \to 1 \text{ a.e.}, \tag{1.93}$$

and

$$\lim_{n\to\infty} \int_\Omega |\nabla u_n| = \Sigma(u). \tag{1.94}$$

The above result is sometimes called a "dipole construction." The terminology becomes clear in the next chapter (see in particular Section 2.5).

Proof of Proposition 1.1. (1.92) implies that

$$\int_\Omega |\nabla u_n| = \int_\Omega |u_n \wedge \nabla u_n| \geq \Sigma(u_n) = \Sigma(u),\ \forall\, n.$$

Therefore, it suffices to construct u_n such that (1.92), (1.93), and "$\leq$" in (1.94) hold.

Consider a sequence $(\psi_n) \subset W^{1,1}(\Omega; \mathbb{R})$ such that

$$\int_\Omega |u \wedge \nabla u - \nabla \psi_n| \leq \Sigma(u) + \frac{1}{n}. \tag{1.95}$$

Applying Lemma 1.15 to $v := u\, e^{-\imath\psi_n}$, we obtain a sequence (η_n) such that

$$\int_\Omega \left|\nabla\left(u\, e^{-\imath(\psi_n+\eta_n)}\right)\right| = \int_\Omega |u \wedge \nabla u - \nabla\psi_n - \nabla\eta_n| \leq \Sigma(u) + \frac{2}{n}$$

and

$$\int_\Omega |u\, e^{-\imath(\psi_n+\eta_n)} - 1| = \int_\Omega |u\, e^{-\imath\psi_n} - e^{\imath\eta_n}| \leq \frac{2}{n}.$$

Possibly after passing to a subsequence, $u_n := u\, e^{-\imath(\psi_n+\eta_n)}$ has all the required properties. □

Step 2. We establish the estimate

$$E(u) \leq \int_\Omega |\nabla u| + \Sigma(u). \tag{1.96}$$

Step 3. We prove

Theorem 1.6. *Let $\varphi \in BV(\Omega; \mathbb{R})$ be such that $u := e^{\imath\varphi} \in W^{1,1}(\Omega; \mathbb{S}^1)$. Then*

$$|D\varphi| = |D\varphi - u \wedge \nabla u| + |u \wedge \nabla u| = |D\varphi - u \wedge \nabla u| + |\nabla u|. \tag{1.97}$$

Here, $|\ |$ denotes the total variation of vector-valued measures. The proof of Theorem 1.6 involves fine properties of BV functions, and in particular Volpert's chain rule.

Step 4. Conclusion: proof of Theorem 1.5
This step is easy. Indeed, equality (1.87) follows from Lemma 1.13, (1.96), and (1.97).

On the other hand, let $\varphi \in BV(\Omega; \mathbb{R})$ be a minimal lifting of u. By (1.87), we know that

$$\|D\varphi\|_{\mathscr{M}} = E(u) = \int_\Omega |\nabla u| + \Sigma(u). \tag{1.98}$$

If we compare (1.98) with (1.97), we obtain (1.88). □

An immediate consequence of Theorems 1.5 and 1.6 is the following

Corollary 1.4. *Let $u \in W^{1,1}(\Omega; \mathbb{S}^1)$. Then $\varphi \in BV(\Omega; \mathbb{R})$ is a minimal lifting of u if and only if*

$$u = e^{\imath\varphi} \text{ and } \|u \wedge \nabla u - D\varphi\|_{\mathscr{M}} = \Sigma(u). \tag{1.99}$$

We now present the proofs of Steps 1, 2, and 3.

Step 1. Proof of Lemma 1.15. Let $0 < \varepsilon < 2\pi$. For $\theta \in [0, 2\pi)$, let $I_{\varepsilon,\theta}$ be the open arc of $\mathbb{S}^1$ centered at $e^{\imath\theta}$ and of length ε. Define $\Phi_{\varepsilon,\theta} : \mathbb{S}^1 \to \mathbb{S}^1$ by

$$\Phi_{\varepsilon,\theta}(z) := \begin{cases} 1, & \text{if } z \notin I_{\varepsilon,\theta} \\ \exp\left(\imath\pi\left(\dfrac{2(\gamma-\theta)}{\varepsilon}+1\right)\right), & \text{if } z = e^{\imath\gamma} \text{ with } |\gamma-\theta| < \dfrac{\varepsilon}{2} \end{cases}.$$

This map satisfies $\left|\dfrac{\partial \Phi_{\varepsilon,\theta}}{\partial \tau}(z)\right| = \begin{cases} 0, & \text{if } z \notin I_{\varepsilon,\theta} \\ \dfrac{2\pi}{\varepsilon}, & \text{if } z \in I_{\varepsilon,\theta} \end{cases}.$

Set $w_{\varepsilon,\theta} := \Phi_{\varepsilon,\theta} \circ v$. Then

$$\int_0^{2\pi}\int_\Omega |\nabla w_{\varepsilon,\theta}|\,dxd\theta = \frac{2\pi}{\varepsilon}\int_\Omega |\nabla v(x)||\{\theta \in [0,2\pi);\ v(x)\in I_{\varepsilon,\theta}\}|\,dx \\ = 2\pi\,\|\nabla v\|_{L^1}. \tag{1.100}$$

On the other hand, we have

$$\int_0^{2\pi} |[w_{\varepsilon,\theta}\neq 1]|\,d\theta = \int_\Omega |\{\theta\in[0,2\pi);\ v(x)\in I_{\varepsilon,\theta}\}|\,dx = \varepsilon|\Omega|. \tag{1.101}$$

We next rely on the following: if f, g are integrable functions on $(0, L)$ such that $\int_0^L f \le A$ and $\int_0^L g \le B$, then there is some $t \in (0, L)$ such that $f(t) + g(t)/\sqrt{B} \le A/L + \sqrt{B}/L$. [This follows from a mean value argument applied to $f + g/\sqrt{B}$.] This fact, combined with (1.100) and (1.101), implies that, for small ε, there exists some $\theta = \theta_\varepsilon \in [0, 2\pi)$ such that

$$\int_\Omega |\nabla w_{\varepsilon,\theta}|\,dx \le \|\nabla v\|_{L^1} + \delta \tag{1.102}$$

and

$$|[w_{\varepsilon,\theta}\neq 1]| \le \delta. \tag{1.103}$$

In order to conclude, it suffices to check that there exists some $\eta = \eta_{\varepsilon,\theta} \in W^{1,1}(\Omega;\mathbb{R})$ such that $w_{\varepsilon,\theta} = v\,e^{-\imath\eta}$. Indeed, $v\,\overline{w_{\varepsilon,\theta}}$ takes its values in $\mathbb{S}^1\setminus I_{\varepsilon,\theta}$ (this amounts to the fact that $z\,\overline{\Phi_{\varepsilon,\theta}(z)} \notin I_{\varepsilon,\theta}$ when $z\in\mathbb{S}^1$). Thus, if we set $\eta := \Theta_\theta\left(v\,\overline{w_{\varepsilon,\theta}}\right)$ (where Θ_θ is the argument introduced in (1.75)), then this η belongs to $W^{1,1}$, and $w_{\varepsilon,\theta} = v\,e^{-\imath\eta}$. □

Step 2. Proof of (1.96). Consider the sequence (u_n) given by Proposition 1.1. Since $J(u\,\overline{u_n}) = 0$, we may write $u\,\overline{u_n} = e^{\imath\varphi_n}$ for some $\varphi_n \in W^{1,1}(\Omega;\mathbb{R})$. Thus

$$\int_\Omega |\nabla\varphi_n| \le \int_\Omega |\nabla u| + \int_\Omega |\nabla u_n| \le C,\ \text{ by (1.94).}$$

Passing to a subsequence and adding suitable multiples of 2π, we find that φ_n converges a.e. to some $\varphi\in BV$ satisfying $u = e^{\imath\varphi}$ and $\int_\Omega |\nabla\varphi| \le \int_\Omega |\nabla u| + \Sigma(u)$. Hence, $E(u) \le \int_\Omega |\nabla u| + \Sigma(u)$. □

Step 3. Proof of Theorem 1.6. Recall that, if $\varphi \in BV(\Omega;\mathbb{R})$, then the measure $D\varphi$ can be (uniquely) written as a sum of an absolutely continuous part with respect to the Lebesgue measure, $D^a\varphi$, whose density is denoted by $\nabla\varphi$, a Cantor part $D^c\varphi$ and a jump part $D^j\varphi$. With an abuse of notation, we write this decomposition as:

$$D\varphi = \nabla\varphi + D^c\varphi + D^j\varphi. \tag{1.104}$$

For this decomposition, for the concept of precise representative and for the chain rule (1.109), we refer the reader, for example, to Ambrosio, Fusco, and Pallara [13,

Chapter 3]. For the convenience of the reader, we have collected some basic facts about BV functions in Appendix 15.7.

Lemma 1.16. *Let $\varphi \in BV(\Omega; \mathbb{R})$ be such that $u := e^{\imath\varphi} \in W^{1,1}$. Then, we have*

$$\nabla\varphi = u \wedge \nabla u \tag{1.105}$$

$$D^c\varphi = 0, \tag{1.106}$$

$$D^j\varphi = 2\pi \sum_{i=0}^{\infty} k_i\, \nu_i\, \mathscr{H}^{N-1} \llcorner S_i. \tag{1.107}$$

Here,

1. *$k_i \in \mathbb{Z}$;*
2. *S_i are Borel subsets of C^1 hypersurfaces Σ_i;*
3. *ν_i is a normal vector orienting Σ_i such that $\Sigma_i \ni x \mapsto \nu_i(x)$ is a Borel function;*
4. *$\mathscr{H}^{N-1} \llcorner S_i$ is the restriction to S_i of the $(N-1)$-dimensional Hausdorff measure $\mathscr{H}^{N-1}$;*
5. *$\sum_{i=0}^{\infty} |k_i|\mathscr{H}^{N-1}(S_i) < \infty$.*

   In particular, we have

$$D\varphi = u \wedge \nabla u + D^j\varphi. \tag{1.108}$$

Remark 1.14. We return here briefly to Question 3 mentioned in the Introduction. Assume that $u \in W^{1,p}(\Omega; \mathbb{S}^1)$, with $1 \le p < 2$. In view of the above results, we know that any minimal lifting $\varphi \in BV$ of u satisfies $\nabla\varphi = u \wedge \nabla u \in L^p$, $\|D^j\varphi\|_{\mathscr{M}} = \Sigma(u)$, and $D^c\varphi = 0$.

Remark 1.15. It follows from (1.105)–(1.107) that, if $\varphi \in BV(\Omega; \mathbb{R})$ is such that $u := e^{\imath\varphi} \in W^{1,1}(\Omega; \mathbb{S}^1)$, then φ belongs to $SBV(\Omega; \mathbb{R})$, where

$$SBV(\Omega; \mathbb{R}) := \{\varphi \in BV(\Omega; \mathbb{R});\ D^c\varphi = 0\}.$$

Proof of Lemma 1.16. Volpert's chain rule asserts that, when f is C^1 and Lipschitz and φ is BV, we have

$$D(f \circ \varphi) = f'(\varphi)\nabla\varphi + f'(\varphi)D^c\varphi + \frac{f(\varphi^+) - f(\varphi^-)}{\varphi^+ - \varphi^-} D^j\varphi. \tag{1.109}$$

The meaning of this identity needs to be clarified. Let C_φ be the set of points where φ has approximate limits, i.e., $x \in C_\varphi$ if and only if there exists some $\ell(x) \in \mathbb{R}$ such that

$$\lim_{r\to 0} \fint_{B_r(x)} |\varphi(y) - \ell(x)|\, dy = 0.$$

For $x \in C_\varphi$, one may define the precise representative of φ at x as the approximate limit $\ell(x)$ of φ at x. In what follows, we systematically replace, in C_φ, φ by its precise representative. By Lebesgue's differentiation theorem, C_φ is of full measure and φ coincides a.e. with its precise representative in C_φ.

Let S_φ be the complement of C_φ (S_φ is the set of points of approximate discontinuity of φ). Recall that, for every function $\varphi \in BV(\Omega)$, S_φ is a Borel set contained, up to an $\mathscr{H}^{N-1}$-negligible set, in a countable union of disjoint C^1-hypersurfaces Σ_i.

Let $J_\varphi \subset S_\varphi \cap (\cup \Sigma_i)$ denote the set of jump points of φ. This is a countable $\mathscr{H}^{N-1}$-rectifiable set, whose points are characterized by the following property: if $x \in J_\varphi \cap \Sigma_i$ for some i, then φ has (different) approximate one-sided limits at x along the direction $\nu_i(x)$ normal to Σ_i at x. Equivalently, for such x the quantities $\varphi^+(x)$ and $\varphi^-(x)$ satisfying

$$\lim_{r\to 0} \fint_{B_r(x)\cap[\pm(y-x)\cdot\nu_i(x)>0]} |\varphi(y) - \varphi^\pm(x)|\, dy = 0$$

exist (and are different). The numbers $\varphi^\pm(x)$ are the two one-sided approximate limits of φ at x. The set J_φ covers most of S_φ, in the sense that $S_\varphi \setminus J_\varphi$ is an $\mathscr{H}^{N-1}$-null set.

The jump part of $D\varphi$ is given by the formula

$$D^j\varphi = \sum_i (\varphi^+ - \varphi^-)\, \nu_i\, \mathscr{H}^{N-1} \llcorner S_i, \quad \text{with } S_i := S_\varphi \cap \Sigma_i. \tag{1.110}$$

We are now in a position to explain the meaning of (1.109). This identity is obtained by superposing the three following formulas:

$$\begin{aligned}
&\nabla f(\varphi) = f'(\varphi)\nabla\varphi;\\
&D^c f(\varphi) = f'(\varphi) D^c\varphi;\\
&D^j f(\varphi) = \sum_i (f(\varphi^+) - f(\varphi^-))\, \nu_i\, \mathscr{H}^{N-1} \llcorner S_i = \frac{f(\varphi^+) - f(\varphi^-)}{\varphi^+ - \varphi^-} D^j\varphi.
\end{aligned}$$

In the second identity above, it is crucial to consider the precise representative of φ. The fact that the product $f'(\varphi)D^c\varphi$ is meaningful follows from the inclusion $\operatorname{supp} D^c\varphi \subset C_\varphi$.

For further use, let us note that, for $\varphi \in BV$, we have

$$\|D\varphi\|_{\mathscr{M}} = \int_\Omega |\nabla\varphi| + \|D^c\varphi\|_{\mathscr{M}} + \int_{J_\varphi} |\varphi^+ - \varphi^-|\, d\mathscr{H}^{N-1}; \tag{1.111}$$

this follows from the fact that $D^a\varphi$, $D^c\varphi$ and $D^j\varphi$ are mutually orthogonal.

We now proceed to the proof of the lemma. We apply the chain rule to the map $f(t) := e^{\imath t}$. Since $u \in W^{1,1}$, it follows that $D^c u = D^j u = 0$, so that $D^c\varphi = 0$ and

$$\nabla u = f'(\varphi)\nabla\varphi = \imath u\, \nabla\varphi. \tag{1.112}$$

From (1.112), we obtain that $u \wedge \nabla u = -\imath \bar{u}\, \nabla u = \nabla\varphi$.

Finally, (1.109) implies that $\dfrac{f(\varphi^+)-f(\varphi^-)}{\varphi^+-\varphi^-}D^j\varphi = 0$, i.e.,

$$\varphi^+ - \varphi^- \text{ is } \mathscr{H}^{N-1}\text{-a.e. on } J_\varphi \text{ a multiple of } 2\pi. \tag{1.113}$$

Using (1.110), we find that (1.107) holds (possibly after changing the S_i's). □

Proof of Theorem 1.6 *completed.* Since the decomposition (1.104) consists of mutually orthogonal measures, we have

$$\begin{aligned}|D\varphi| &= |\nabla\varphi| + |D^c\varphi| + |D^j\varphi| = |\nabla\varphi| + |D^j\varphi|\\ &= |u\wedge\nabla u| + |D^j\varphi| = |u\wedge\nabla u| + |D\varphi - u\wedge\nabla u|\\ &= |\nabla u| + |D\varphi - u\wedge\nabla u|,\end{aligned} \tag{1.114}$$

by (1.105), (1.106), and (1.108). □

In the spirit of Corollary 1.4, we have the following

Corollary 1.5. *For every $u\in W^{1,1}(\Omega;\mathbb{S}^1)$, we have*

$$\Sigma(u) = \min\left\{\|D^j\varphi\|_{\mathscr{M}};\ \varphi\in BV(\Omega;\mathbb{R}) \text{ and } u = e^{\imath\varphi}\right\}; \tag{1.115}$$

moreover, φ achieves the min *in* (1.115) *if and only if it is a minimal lifting of u.*

Proof. If $\varphi\in BV(\Omega;\mathbb{R})$ is a lifting of u, then, by (1.114), we have

$$\|D\varphi\|_{\mathscr{M}} = \int_\Omega |\nabla u| + \|D^j\varphi\|_{\mathscr{M}}. \tag{1.116}$$

Combining (1.116) and Theorem 1.5, we deduce that

$$\|D^j\varphi\|_{\mathscr{M}} \geq \Sigma(u). \tag{1.117}$$

Finally, we note that φ is (by definition) a minimal lifting of u if and only if $\|D\varphi\|_{\mathscr{M}} = E(u)$, or, equivalently, $\|D^j\varphi\|_{\mathscr{M}} = \Sigma(u)$. □

Remark 1.16. Instead of proving the inequality

$$E(u) \geq \int_\Omega |\nabla u| + \Sigma(u) \tag{1.118}$$

via Theorem 1.6, a natural alternative strategy is the following. Let $\psi\in BV$ be a lifting of u. We need to show that $\|D\psi\|_{\mathscr{M}} \geq \int_\Omega |\nabla u| + \Sigma(u)$. Let (ψ_n) be a sequence of smooth functions such that $\int_\Omega |\nabla\psi_n| \to \|D\psi\|_{\mathscr{M}}$ and $\psi_n\to\psi$ a.e. Set $u_n := e^{\imath\psi_n}$. Then, $u_n\in C^\infty(\overline{\Omega};\mathbb{S}^1)$ and $u_n\to u$ a.e. Set $v_n := u_n/u$. Note that

$$\int_\Omega |\nabla v_n| = \int_\Omega |u\wedge\nabla u - \nabla\varphi_n| \geq \Sigma(u), \tag{1.119}$$

by definition of $\Sigma(u)$. (1.118) is equivalent to

$$\lim \int_{\Omega} |\nabla(uv_n)| \geq \int_{\Omega} |\nabla u| + \Sigma(u). \tag{1.120}$$

In view of (1.119), it is tempting to establish the following:

$$\liminf \int_{\Omega} |\nabla(uv_n)| = \int_{\Omega} |\nabla u| + \liminf \int_{\Omega} |\nabla v_n| \text{ if } v_n \to 1 \text{ a.e.} \tag{1.121}$$

This would be similar to other well-known "orthogonal decompositions," successfully used in Brezis and Lieb [82], Bethuel, Brezis, and Coron [30, Proof of Theorem 3], and Bourgain, Brezis, and Mironescu [54, Lemma 14'] . However, we warn the reader that (1.121) is wrong (see Lemma 1.17 below). Instead, we had to rely on the fine structure of BV functions, as explained in the proof of Theorem 1.6.

Lemma 1.17. *There exist a map* $u : W^{1,1}((0,1); \mathbb{S}^1)$ *and a sequence* $(v_n) \subset W^{1,1}((0,1); \mathbb{S}^1)$ *such that* $v_n \to 1$ *a.e. and*

$$\liminf \int_0^1 |(u\, v_n)'| < \int_0^1 |u'| + \liminf \int_0^1 |v_n'|.$$

Proof. Let $\varphi(x) := x/2$ and define the "zigzag" φ_n by $\varphi_n(0) = 0$ and

$$\varphi_n' := \begin{cases} 1, & \text{if } x \in (0, 1/2n) \cup (2/2n, 3/2n) \cup \ldots \cup ((2n-2)/2n, (2n-1)/2n) \\ -1, & \text{if } x \in (1/2n, 2/2n) \cup (3/2n, 4/2n) \cup \ldots \cup ((2n-1)/2n, 1) \end{cases}.$$

It is easy to see that $\varphi_n \to 0$ uniformly. Set $u := e^{\imath\varphi}$ and $v_n := e^{\imath\varphi_n}$. Then, $v_n \to 1$ uniformly and

$$\int |(uv_n)'| = \int |\varphi' + \varphi_n'| = 1,\ \int |u'| = \int |\varphi'| = \frac{1}{2},\ \int |v_n'| = \int |\varphi_n'| = 1.$$

□

Remark 1.17. Let us point out that, if φ is a minimal lifting of u, then $\mu := -D^j\varphi$ is a minimizer in (1.85), since $D\varphi = u \wedge \nabla u + D^j\varphi$ (by (1.108)), and thus

$$\operatorname{curl}(-D^j\varphi) = \operatorname{curl}(u \wedge \nabla u - D\varphi) = \operatorname{curl}(u \wedge \nabla u) = 2Ju.$$

On the other hand, by (1.88) and (1.97), we have $\|D^j\varphi\|_{\mathcal{M}} = \Sigma(u)$. We warn the reader that there are minimizing measures μ for (1.85) that are *not* of the form $-D^j\varphi$ for some minimal lifting φ. For example, if $u(x) = x/|x|$ in the unit disk $\mathbb{D}$ of $\mathbb{R}^2$, then $\mu := u \wedge \nabla u$ is a minimizer in (1.85). Indeed, we have $\|\mu\|_{\mathcal{M}} = 2\pi$, whereas $\Sigma(u) = 2\pi$ (as in Remark 1.12).

We next present an interesting refinement of Theorem 1.5. Fix $u \in W^{1,1}(\Omega; \mathbb{S}^1)$; for every $\eta \in BV(\Omega; \mathbb{R})$ define $V(\eta) := \|u \wedge \nabla u - D\eta\|_{\mathcal{M}}$. Let

$$A := \left\{ \psi_0 \in BV(\Omega; \mathbb{R});\ V(\psi_0) = \Sigma(u) \text{ and } \int_{\Omega} \psi_0 = 0 \right\}, \tag{1.122}$$

where we recall that $\Sigma(u) = \min_{\eta \in BV} V(\eta)$, by (1.82). By Poincaré's inequality, the set A is a non-empty convex and compact set in $L^1(\Omega)$. By Theorem 1.5, we know that minimal liftings of u are elements of A (modulo constants). We also know that every lifting of u (modulo constants) that belongs to A is a minimal lifting (Corollary 1.4). We warn the reader that some (even "most") elements of A are not liftings (see Remark 1.12). However, we have the following remarkable result.

Theorem 1.7. *Let ψ be an extreme point of A. Then there exists some $C \in \mathbb{R}$ such that $\psi + C$ is a minimal lifting of u.*

Remark 1.18. Let $\Omega := \mathbb{D}$ and $u(z) := \left(\dfrac{z}{|z|}\right)^2$. Then, u has minimal liftings that are not (modulo constants) extreme points of the set A defined in (1.122). In other words, the converse to Theorem 1.7 fails. Indeed, let $0 \le \alpha_1 < \alpha_2 < 2\pi$ and set $\varphi := \Theta_{\alpha_1} + \Theta_{\alpha_2}$; here, Θ_α is as in (1.75). We claim that φ is a minimal lifting of u. Indeed, by direct calculations we have $|u|_{W^{1,1}} = 4\pi$ and $|\varphi|_{BV} = 8\pi$. On the other hand, we have $\Sigma(u) = 4\pi$ (by Example 2.5 and Theorem 2.5 in Chapter 2) and thus $E(u) = 8\pi = |\varphi|_{BV}$. Therefore, φ is a minimal lifting of u.

However, we have $\varphi = \dfrac{1}{2}(\varphi_1 + \varphi_2)$, with $\varphi_j := 2\Theta_{\alpha_j}, j = 1, 2$. Arguing as above, each φ_j is a minimal lifting of u, and is clearly different (modulo constants) from φ. Thus, φ is a minimal lifting of u that is not an extreme point of A.

Remark 1.19. Combined with the Krein–Milman theorem, Theorem 1.7 implies that, modulo constants, the minimizers of V coincide with the closed convex hull of minimal liftings of u.

Proof of Theorem 1.7. The theorem follows at once from the following construction: we let $\psi \in BV(\Omega)$ be such that the function $e^{-\imath\psi}u$ is not constant and we exhibit two maps, $\psi_\pm \in BV(\Omega)$, such that $\psi_+ - \psi_-$ is not a constant, $\psi = \dfrac{\psi_+ + \psi_-}{2}$, and $V(\psi) = \dfrac{1}{2}V(\psi_+) + \dfrac{1}{2}V(\psi_-)$.

The functions ψ_+, ψ_- are constructed as follows. Consider a BV lifting φ of u. The fact that $e^{-\imath\psi}u$ is not constant is equivalent to the fact that at least one of the functions $\cos(\psi - \varphi)$ or $\sin(\psi - \varphi)$ is not constant. We assume, for example, that $\sin(\psi - \varphi)$ is not constant. We then let $\psi_\pm := \psi \pm \sin(\psi - \varphi)$ and check that these maps have all the required properties.

To start with, $\psi_+ - \psi_-$ is not constant and $\psi = \dfrac{\psi_+ + \psi_-}{2}$. Next, by Theorem 1.6, we have $\nabla\varphi = u \wedge \nabla u$ and the Cantor part $D^c\varphi$ of $D\varphi$ vanishes. Combining this with Volpert's chain rule, we find that

$$\begin{aligned} D\psi_\pm = & \underbrace{\nabla\psi \pm \cos(\psi - \varphi)(\nabla\psi - u \wedge \nabla u)}_{\nabla\psi_\pm} + \underbrace{(1 \pm \cos(\psi - \varphi))D^c\psi}_{D^c\psi_\pm} \\ & \underbrace{+ D^j\psi \pm \frac{\sin(\psi^+ - \varphi^+) - \sin(\psi^- - \varphi^-)}{(\psi^+ - \varphi^+) - (\psi^- - \varphi^-)} D^j(\psi - \varphi)}_{D^j\psi_\pm}, \end{aligned} \tag{1.123}$$

which implies

$$u \wedge \nabla u - D\psi_{\pm} = (1 \pm \cos(\psi - \varphi))(u \wedge \nabla u - \nabla\psi) - D^c\psi_{\pm} - D^j\psi_{\pm}. \quad (1.124)$$

[Recall that $\psi^{\pm}$ (resp. $\varphi^{\pm}$) represent the one-sided limits of ψ on the jump set J_{ψ} (resp. of φ on J_{φ}).]

We next obtain a more informative formula for the jump part of $D\psi_{\pm}$. Let us first note that the jump set $J_{\psi_{\pm}}$ of $\psi_{\pm}$ is contained in $J_{\psi} \cup J_{\varphi}$. We claim that the jump $\mathscr{J}_{\psi_{\pm}} := (\psi_{\pm})^{+} - (\psi_{\pm})^{-}$ of $\psi_{\pm}$ on $J_{\psi} \cup J_{\varphi}$ satisfies, $\mathscr{H}^{N-1}$-a.e. on $J_{\psi} \cup J_{\varphi}$,

$$\mathscr{J}_{\psi_{\pm}} = \begin{cases} [\psi^{+} - \psi^{-}] \pm [\sin(\psi^{+} - \varphi^{+}) - \sin(\psi^{-} - \varphi^{+})], & \text{on } J_{\psi} \\ 0, & \text{on } J_{\varphi} \setminus J_{\psi} \end{cases}. \quad (1.125)$$

Indeed, by (1.113), the jumps of φ are $\mathscr{H}^{N-1}$-a.e. multiples of 2π. Therefore, we have, $\mathscr{H}^{N-1}$-a.e. on $J_{\psi} \cup J_{\varphi}$,

$$\sin(\psi^{+} - \varphi^{+}) - \sin(\psi^{-} - \varphi^{-}) = \sin(\psi^{+} - \varphi^{+}) - \sin(\psi^{-} - \varphi^{+}),$$

which easily implies (1.125).

Recalling the identity (1.111) and using (1.123)–(1.125), we find that

$$\begin{aligned} V(\psi_{\pm}) = \int & |1 \pm \cos(\psi - \varphi)||u \wedge \nabla u - \nabla\psi| \\ & + \|(1 \pm \cos(\psi - \varphi))D^c\psi\|_{\mathscr{M}} + \int_{J_{\psi}} |\mathscr{J}_{\psi_{\pm}}|\, d\mathscr{H}^{N-1}. \end{aligned} \quad (1.126)$$

We now make two observations. First, we have

$$1 \pm \cos(\psi - \varphi) \geq 0. \quad (1.127)$$

Second, we have

$$1 \pm \frac{\sin(\psi^{+} - \varphi^{+}) - \sin(\psi^{-} - \varphi^{+})}{|\psi^{+} - \psi^{-}|} \geq 0, \quad (1.128)$$

and thus, $\mathscr{H}^{N-1}$-a.e. on J_{ψ},

$$|\mathscr{J}_{\psi_{\pm}}| = |\psi^{+} - \psi^{-}| \times \left(1 \pm \frac{\sin(\psi^{+} - \varphi^{+}) - \sin(\psi^{-} - \varphi^{+})}{|\psi^{+} - \psi^{-}|}\right). \quad (1.129)$$

Combining (1.127)–(1.129) with (1.126), equality $V(\psi) = \dfrac{V(\psi_{+}) + V(\psi_{-})}{2}$ becomes clear. □

Another natural question related to lifting is the following. Assume that $u \in W^{1,1}(\Omega; \mathbb{S}^1)$ and let $\varphi \in BV(\Omega; \mathbb{R})$ be such that $u = e^{\imath\varphi}$. Write $\varphi = \varphi_1 + \varphi_2$, with

$\varphi_1 \in W^{1,1}$ and $\varphi_2 \in BV$. We now attempt to minimize the BV seminorm of the "parasite" term φ_2. Here is the answer.

Theorem 1.8. *Let $u \in W^{1,1}(\Omega; \mathbb{S}^1)$. Then we have*

$$\inf\{\|D\psi\|_{\mathcal{M}};\ u = e^{\imath(\varphi+\psi)} \text{ with } \varphi \in W^{1,1} \text{ and } \psi \in BV\} = 2\Sigma(u). \tag{1.130}$$

Proof. Let $\varphi_0 \in W^{1,1}$. By Theorem 1.4, there exists some $\psi_0 \in BV$ such that

$$\|D\psi_0\|_{\mathcal{M}} \le 2\|\nabla(ue^{-\imath\varphi_0})\|_{L^1} \text{ and } ue^{-\imath\varphi_0} = e^{\imath\psi_0}. \tag{1.131}$$

By (1.47), (1.49), (1.50), and (1.52), we have $|\nabla(ue^{-\imath\varphi_0})| = |u \wedge \nabla u - \nabla\varphi_0|$. Since φ_0 is arbitrary in $W^{1,1}$, we find that the left-hand side of (1.130) is at most

$$2 \inf_{\varphi \in W^{1,1}} \|u \wedge \nabla u - \nabla\varphi\|_{L^1} = 2\Sigma(u),$$

by definition of $\Sigma(u)$.

For the reverse inequality, we rely on (1.97) applied to $ue^{-\imath\varphi} = e^{\imath\psi}$, which implies that

$$\begin{aligned}\|D\psi\|_{\mathcal{M}} &= \|D\psi - (ue^{-\imath\varphi}) \wedge \nabla(ue^{-\imath\varphi})\|_{\mathcal{M}} + \|\nabla(ue^{-\imath\varphi})\|_{L^1} \\ &= \|D(\psi + \varphi) - u \wedge \nabla u\|_{\mathcal{M}} + \|u \wedge \nabla u - \nabla\varphi\|_{L^1} \ge 2\Sigma(u),\end{aligned} \tag{1.132}$$

by (1.82). □

Remark 1.20. As we are going to see later, inf in (1.130) need not be achieved (see Proposition 1.2 and Section 2.11.1).

1.5 $\Sigma(u)$ computed via duality: the basic relation between $\Sigma(u)$ and Ju

In this section, we present a formula expressing $\Sigma(u)$ as the norm of Ju in a suitable dual space. This highlights once more the important fact that for every $u \in W^{1,1}(\Omega; \mathbb{S}^1)$, $\Sigma(u)$ depends only on Ju (see Lemma 1.14). For the convenience of the reader, we will state the next result first when $N = 2, 3$, and then for a general N.

Theorem 1.9. *Let $u \in W^{1,1}(\Omega; \mathbb{S}^1)$.*
When $N = 2$, we have

$$\Sigma(u) = 2 \sup\left\{\langle Ju, \zeta\rangle;\ \zeta \in C_c^\infty(\Omega; \mathbb{R}),\ \|\nabla\zeta\|_{L^\infty} \le 1\right\}. \tag{1.133}$$

When $N = 3$, we have

$$\Sigma(u) = 2 \sup\left\{\langle Ju, \zeta\rangle;\ \zeta \in C_c^\infty(\Omega; \mathbb{R}^3),\ \|\operatorname{curl}\zeta\|_{L^\infty} \le 1\right\}. \tag{1.134}$$

For a general N, we have

$$\Sigma(u) = 2\sup\left\{\langle Ju, *\zeta\rangle;\ \zeta \in C_c^\infty(\Omega; \Lambda^{N-2}),\ \|d\zeta\|_{L^\infty} \le 1\right\}. \tag{1.135}$$

The above requires some explanation. When $N = 2$, the meaning of Ju in (1.133) is the scalar distribution given by (1.35), so that

$$\langle Ju, \zeta\rangle = -\frac{1}{2}\int_\Omega (u \wedge \nabla u)\cdot \nabla^\perp \zeta, \tag{1.136}$$

where

$$\nabla^\perp\zeta := (-\partial\zeta/\partial x_2, \partial\zeta/\partial x_1). \tag{1.137}$$

When $N = 3$, the meaning of Ju in (1.134) is the distributional vector field $Ju \in \mathscr{D}'(\Omega; \mathbb{R}^3)$ given by (1.37)–(1.38), so that

$$\langle Ju, \zeta\rangle = \frac{1}{2}\int_\Omega (u \wedge \nabla u)\cdot \operatorname{curl}\zeta. \tag{1.138}$$

For a general N, the meaning of Ju in (1.135) is the distributional 2-form

$$Ju = \frac{1}{2} d(u \wedge \nabla u) \in \mathscr{D}'(\Omega; \Lambda^2), \tag{1.139}$$

so that

$$\langle Ju, *\zeta\rangle = \frac{1}{2}\langle u \wedge \nabla u, d^*(*\zeta)\rangle. \tag{1.140}$$

Here, $*$ is the Hodge operator, transforming k-forms into $(N-k)$-forms, and d^* is the (formal) adjoint of d, given by

$$d^*\omega = (-1)^{Nk+N+1} * d * \omega,\ \forall\, \omega \in C^1(\overline{\Omega}; \Lambda^k). \tag{1.141}$$

In (1.140), we let $*$ act on $(N-2)$-forms, and d^* acts on 2-forms.
For further use, we also note the identity

$$** = (-1)^{k(N-k)}\,\mathrm{Id}_k \text{ on } k\text{-forms}. \tag{1.142}$$

[For the above formulas, see, for example, Giaquinta, Modica, and Souček [156, Sections 5.2.2–5.2.4].] Noting that

$$d^*(*\zeta) = (-1)^{N+1} * d\zeta,$$

we obtain

$$\langle Ju, *\zeta\rangle = \frac{(-1)^{N+1}}{2}\int_\Omega (u \wedge \nabla u)\cdot(*d\zeta). \tag{1.143}$$

This is consistent with (1.136) and (1.138), since $*d\zeta = \nabla^{\perp}\zeta$ when $N = 2$, and $*d\zeta = \operatorname{curl}\zeta$ when $N = 3$ (once we identify 1-forms with vector fields).

Proof of Theorem 1.9. For simplicity, we will present only the proof of (1.134), corresponding to $N = 3$; the general case is similar. Set $F := u \wedge \nabla u \in L^1(\Omega)$.
Proof of "$\geq$." Recall that (see Lemma 1.13)

$$\Sigma(u) = \min\{\|F - D\varphi\|_{\mathcal{M}};\ \varphi \in BV(\Omega;\mathbb{R})\}.$$

Let $\varphi \in BV(\Omega;\mathbb{R})$ and let $\zeta \in C_c^\infty(\Omega;\mathbb{R}^3)$ be such that $|\operatorname{curl}\zeta| \leq 1$ in Ω. Then

$$2\langle Ju, \zeta\rangle = \langle \operatorname{curl} F, \zeta\rangle = \int_\Omega F\cdot(\operatorname{curl}\zeta) = \int_\Omega (F - D\varphi)\cdot(\operatorname{curl}\zeta) \leq \|F - D\varphi\|_{\mathcal{M}}.$$

We conclude by taking in the above sup over ζ and inf over φ.
Proof of "$\leq$." Let

$$G : \{\operatorname{curl}\zeta;\ \zeta \in C_c^\infty(\Omega;\mathbb{R}^3)\} \to \mathbb{R},\ G(\operatorname{curl}\zeta) := 2\langle Ju, \zeta\rangle = \int_\Omega F\cdot(\operatorname{curl}\zeta).$$

Then, G is linear and

$$|G(\operatorname{curl}\zeta)| \leq m\|\operatorname{curl}\zeta\|_{L^\infty},$$

where m is the right-hand side of (1.134). By the Hahn–Banach theorem, G can be extended as a linear functional of norm m to the space

$$C_0(\overline{\Omega};\mathbb{R}^3) := \{f \in C(\overline{\Omega};\mathbb{R}^3);\ f = 0 \text{ on } \partial\Omega\}.$$

Therefore, we obtain some $\mu \in \mathcal{M}(\Omega;\mathbb{R}^3)$ such that $\|\mu\|_{\mathcal{M}} = m$,

$$2\langle Ju, \zeta\rangle = G(\operatorname{curl}\zeta) = \int_\Omega (\operatorname{curl}\zeta)\cdot\mu = \langle \operatorname{curl}\mu, \zeta\rangle,\ \forall\zeta \in C_c^\infty(\Omega;\mathbb{R}^3).$$

We deduce that $\operatorname{curl}\mu = 2\,Ju$. Since $\|\mu\|_{\mathcal{M}} = m$, we conclude via Lemma 1.14 that $\Sigma(u) \leq m$. □

Remark 1.21. The fact that $F = u \wedge \nabla u$ plays no role in this proof. We have actually proved the following assertion: if $F \in \mathcal{M}(\Omega;\mathbb{R}^3)$, then

$$\begin{aligned}&\min\{\|F - D\varphi\|_{\mathcal{M}};\ \varphi \in BV(\Omega;\mathbb{R})\}\\ &= \sup\{\langle F, \operatorname{curl}\zeta\rangle;\ \zeta \in C_c^\infty(\Omega;\mathbb{R}^3),\ \|\operatorname{curl}\zeta\|_{L^\infty} \leq 1\}.\end{aligned}$$

The N-dimensional version of this equality is the following: if $F \in \mathcal{M}(\Omega;\Lambda^1)$ (i.e., F is a 1-form with coefficients measures), then

$$\begin{aligned}&\min\{\|F - D\varphi\|_{\mathcal{M}};\ \varphi \in BV(\Omega;\mathbb{R})\}\\ &= \sup\{\langle F, *d\zeta\rangle;\ \zeta \in C_c^\infty(\Omega;\Lambda^{N-2}),\ \|d\zeta\|_{L^\infty} \leq 1\}.\end{aligned}$$

The next result is similar to Theorem 1.9, except that we consider a wider class of test functions.

Corollary 1.6. *Let* $u \in W^{1,1}(\Omega; \mathbb{S}^1)$.
When $N = 2$, *we have*

$$\Sigma(u) = \max\left\{\int_\Omega (u \wedge \nabla u) \cdot \nabla^\perp \zeta;\ \zeta \in W_0^{1,\infty}(\Omega; \mathbb{R}),\ \|\nabla\zeta\|_{L^\infty} \le 1\right\}. \tag{1.144}$$

When $N = 3$, *we have*

$$\Sigma(u) = \sup\left\{\int_\Omega (u \wedge \nabla u) \cdot (\operatorname{curl}\zeta);\ \zeta \in W^{1,\infty}(\Omega; \mathbb{R}^3),\right.$$
$$\left. \zeta \parallel \nu \text{ on } \partial\Omega,\ \|\operatorname{curl}\zeta\|_{L^\infty} \le 1\right\} \tag{1.145}$$

and

$$\Sigma(u) = \max\left\{\int_\Omega (u \wedge \nabla u) \cdot \xi;\ \xi \in L^\infty(\Omega; \mathbb{R}^3),\ \|\xi\|_{L^\infty} \le 1,\right.$$
$$\left. \operatorname{div}\xi = 0 \text{ in } \Omega \text{ and } \xi \cdot \nu = 0 \text{ on } \partial\Omega\right\}. \tag{1.146}$$

For a general N, *we have*

$$\Sigma(u) = \sup\left\{\int_\Omega (u \wedge \nabla u) \cdot (*d\zeta);\ \zeta \in W^{1,\infty}(\Omega; \Lambda^{N-2}),\right.$$
$$\left. \zeta_T = 0 \text{ on } \partial\Omega,\ \|d\zeta\|_{L^\infty} \le 1\right\}, \tag{1.147}$$

and

$$\Sigma(u) = \max\left\{\int_\Omega (u \wedge \nabla u) \cdot \xi;\ \xi \in L^\infty(\Omega; \mathbb{R}^N),\ \|\xi\|_{L^\infty} \le 1,\right.$$
$$\left. \operatorname{div}\xi = 0 \text{ in } \Omega \text{ and } \xi \cdot \nu = 0 \text{ on } \partial\Omega\right\}. \tag{1.148}$$

Some explanations. In (1.145), (1.146), and (1.148), ν is the normal to $\partial\Omega$. In (1.146) and (1.148), the conditions $\operatorname{div}\xi = 0$ and $\xi \cdot \nu = 0$ are to be understood in the weak sense, i.e.,

$$\int_\Omega \xi \cdot \nabla\varphi = 0,\ \forall \varphi \in W^{1,1}(\Omega; \mathbb{R}); \tag{1.149}$$

equivalently, (1.149) holds for every $\varphi \in C^\infty(\overline{\Omega}; \mathbb{R})$.

In (1.147), $\cdot$ is the inner (or scalar) product of 1-forms (identified with vectors in $\mathbb{R}^N$); an alternative standard notation for $(u \wedge \nabla u) \cdot (*d\zeta)$ is $\langle u \wedge \nabla u, *d\zeta\rangle$.

Also in (1.147), ζ_T is the tangential part of ζ. The form ζ_T is easily computed in coordinates: if $e_1, \ldots, e_{N-1}$ is a direct orthonormal basis of $T_x(\partial\Omega)$, completed with the outward normal vector $e_N = \nu$, if $e^1, \ldots, e^N$ is the dual basis, and if we write

$$\zeta = \sum_{j<k} \zeta_{j,k}\, e^1 \wedge \ldots \wedge \widehat{e^j} \ldots \wedge \widehat{e^k} \wedge \ldots \wedge e^N,$$

then ζ_T is the part of ζ not containing e^N, i.e.,

$$\zeta_T := \sum_{j<N} \zeta_{j,N}\, e^1 \wedge \ldots \wedge \widehat{e^j} \ldots \wedge \widehat{e^N};$$

here, we use the usual convention that $\widehat{\ }$ denotes a missing term.

The definition of ζ_T when ζ is a k-form is similar. Of interest to us is the following integration by parts formula:

$$\int_\Omega \langle d\alpha, \beta\rangle = \int_{\partial\Omega} \alpha_T \wedge (*\beta)_T + (-1)^{Nk+N+1} \int_\Omega \langle \alpha, *d*\beta\rangle, \quad \forall \alpha \in W^{1,\infty}(\Omega; \Lambda^k),\ \forall \beta \in C^\infty(\overline{\Omega}; \Lambda^{k+1}); \tag{1.150}$$

here, $\langle\cdot,\cdot\rangle$ is the scalar product of forms of the same order.

For more details concerning the tangential part ζ_T, see, for example, Iwaniec [201, Sections 1.8 and 2.1] and Giaquinta, Modica, and Souček [156, Section 5.2.6].

Remark 1.22. When $N = 2$, formula (1.144) asserts that

$$\Sigma(u) = 2\|Ju\|_W, \tag{1.151}$$

where $\|T\|_W$ denotes the norm of T in the dual of $W_0^{1,\infty}(\Omega)$ equipped with the norm $\zeta \mapsto \|\nabla\zeta\|_{L^\infty}$, i.e.,

$$\|T\|_W := \max\{\langle T, \zeta\rangle;\ \zeta \in W_0^{1,\infty}(\Omega),\ \|\nabla\zeta\|_{L^\infty} \le 1\}. \tag{1.152}$$

When $T = \mu_1 - \mu_2$, with μ_1, μ_2 probability measures, $\|T\|_W$ is the Wasserstein distance $W_1(\mu_1, \mu_2)$.

Remark 1.23. When $N = 2$, the class of test functions in (1.144), namely

$$\{\nabla^\perp\zeta;\ \zeta \in W_0^{1,\infty}(\Omega; \mathbb{R}),\ \|\nabla\zeta\|_{L^\infty} \le 1\},$$

coincides precisely with

$$\{\xi \in L^\infty(\Omega; \mathbb{R}^2);\ \|\xi\|_{L^\infty} \le 1,\ \operatorname{div}\xi = 0 \text{ in } \Omega \text{ and } \xi\cdot\nu = 0 \text{ on } \partial\Omega\}.$$

This is a standard equality. The reader may find a proof of a more general result in Chapter 14 (see the proof of Corollary 1.6″).

However, when $N \geq 3$, there is *strict inclusion* of the class

$$\{*d\zeta;\ \zeta \in W^{1,\infty}(\Omega; \Lambda^{N-2}),\ \zeta_T = 0 \text{ on } \partial\Omega \text{ and } \|d\zeta\|_{L^\infty} \leq 1\}$$

of test functions in (1.147) in the class

$$\{\xi \in L^\infty(\Omega; \mathbb{R}^N);\ |\xi| \leq 1 \text{ in } \Omega,\ \operatorname{div} \xi = 0 \text{ in } \Omega \text{ and } \xi \cdot \nu = 0 \text{ on } \partial\Omega\},$$

which appears in (1.148). The inclusion of the first class in the second one is clear, since

$$\begin{aligned}\int_\Omega (*d\zeta) \cdot d\varphi &= \pm \int_\Omega \langle *d\varphi, d\zeta \rangle \\ &= \pm \int_{\partial\Omega} \zeta_T \wedge (d\varphi)_T \pm \int_\Omega \langle \zeta, *d^2\varphi \rangle = 0,\end{aligned}$$

by the assumption $\zeta_T = 0$ on $\partial\Omega$ and by (1.150) (applied with $k := N - 2$, $\alpha := \zeta$, and $\beta := *d\varphi$, where $\varphi \in C^\infty(\overline{\Omega}; \mathbb{R})$).

The strict inclusion follows from the fact that there exists some $\xi \in L^\infty_c(\mathbb{R}^N; \mathbb{R}^N)$ such that $\operatorname{div} \xi = 0$, but ξ cannot be written as $\xi = *d\zeta$ with $\zeta \in W^{1,\infty}_{loc}(\mathbb{R}^N; \Lambda^{N-2})$ (see Curcă [123]).

Proof of Corollary 1.6. Equations (1.136), (1.138), and (1.143) imply that all the inequalities "$\Sigma(u) \leq$" in the corollary hold, since the class of test functions here is wider than the one in Theorem 1.9. For the reverse inequality, we consider directly the case of a general $N \geq 2$. Let ξ be a test function as in (1.148), i.e., $\xi \in L^\infty(\Omega; \mathbb{R}^N)$, $\|\xi\|_{L^\infty} \leq 1$, and (1.149) holds. Then, for every $\varphi \in W^{1,1}(\Omega; \mathbb{R})$, we have

$$\int_\Omega (u \wedge \nabla u) \cdot \xi = \int_\Omega [u \wedge \nabla u - \nabla\varphi] \cdot \xi \leq \|u \wedge \nabla u - \nabla\varphi\|_{L^1}.$$

Taking inf over φ yields $\int_\Omega (u \wedge \nabla u) \cdot \xi \leq \Sigma(u)$, which is the desired inequality. □

Finally, we present some stability results for $\Sigma(u)$.

Corollary 1.7. *For every $u, v \in W^{1,1}(\Omega; \mathbb{S}^1)$, we have*

$$|\Sigma(u) - \Sigma(v)| \leq 2\|\nabla(u - v)\|_{L^1} \tag{1.153}$$

and

$$|\Sigma(u) - \Sigma(v)| \leq \|u - v\|_{L^{p'}} (\|\nabla u\|_{L^p} + \|\nabla v\|_{L^p}),\ \forall\, 1 \leq p \leq \infty. \tag{1.154}$$

Remark 1.24. Inequality (1.154) implies in particular that, if $u_n, u \in W^{1,p}(\Omega; \mathbb{S}^1)$ satisfy $\|\nabla u_n\|_{L^p} \leq C$ for some $p > 1$ and $u_n \to u$ a.e., then $\Sigma(u_n) \to \Sigma(u)$. [Indeed, in this case we have $\|u_n - u\|_{L^{p'}} \to 0$ by dominated convergence.] Note that the conclusion fails if we assume only that $\|\nabla u_n\|_{L^1} \leq C$ and $u_n \to u$ a.e. Indeed, take any $u \in W^{1,1}(\Omega; \mathbb{S}^1)$ such that $Ju \neq 0$ and let $\varphi \in BV(\Omega; \mathbb{R})$ be a phase of u. Let

$(\varphi_n) \subset C^\infty(\overline{\Omega}; \mathbb{R})$ be such that $\int_\Omega |\nabla \varphi_n| \to |\varphi|_{BV}$ and $\varphi_n \to \varphi$ a.e. If we set $u_n := e^{\imath \varphi_n}$, then $\int_\Omega |\nabla u_n| \to |\varphi|_{BV}$ and $u_n \to u$ a.e. However, we have $\Sigma(u_n) = 0$, whereas $\Sigma(u) > 0$.

Let us finally note that the argument above shows that, when $p > 1$, the Jacobian map

$$J : W^{1,p}(\Omega; \mathbb{S}^1) \to \mathscr{D}'(\Omega; \Lambda^2),\ u \mapsto Ju, \tag{1.155}$$

is continuous with respect to weak sequential $W^{1,p}$ convergence.

Proof of Corollary 1.7. We start by establishing a useful identity. Let $U, V \in W^{1,1}(\Omega; \mathbb{R}^2) \cap L^\infty(\Omega; \mathbb{R}^2)$. If we identify the gradient with a 1-form, then we have

$$0 = d^2(U \wedge V) = d(\nabla U \wedge V + U \wedge \nabla V),$$

so that

$$d(U \wedge \nabla V) = -d(\nabla U \wedge V),\ \forall U, V \in W^{1,1}(\Omega; \mathbb{R}^2) \cap L^\infty(\Omega; \mathbb{R}^2). \tag{1.156}$$

Applying (1.156) to the identity

$$Ju - Jv = \frac{1}{2} d((u - v) \wedge \nabla u) + \frac{1}{2} d(v \wedge \nabla(u - v)),$$

we find that

$$Ju - Jv = \frac{1}{2} d((u - v) \wedge \nabla(u + v)) = \frac{1}{2} d((u + v) \wedge \nabla(u - v)).$$

When $\zeta \in C_c^\infty(\Omega; \Lambda^{N-2})$ is such that $\|d\zeta\|_{L^\infty} \le 1$, we obtain

$$\begin{aligned} 2\langle Ju, *\zeta\rangle &= 2\langle Jv, *\zeta\rangle + \langle (u + v) \wedge \nabla(u - v), d^* * \zeta\rangle \\ &\le 2\langle Jv, *\zeta\rangle + 2\|\nabla(u - v)\|_{L^1}, \end{aligned} \tag{1.157}$$

and similarly

$$\begin{aligned} 2\langle Ju, *\zeta\rangle &= 2\langle Jv, *\zeta\rangle + \langle (u - v) \wedge \nabla(u + v), d^* * \zeta\rangle \\ &\le 2\langle Jv, *\zeta\rangle + \|u - v\|_{L^{p'}}(\|\nabla u\|_{L^p} + \|\nabla v\|_{L^p}). \end{aligned} \tag{1.158}$$

We obtain (1.153) (resp. (1.154)) from (1.157) (resp. (1.158)). □

Remark 1.25. The above argument also yields

$$\|Ju - Jv\|_{[W_0^{1,\infty}]^*} \le \|\nabla(u - v)\|_{L^1},\ \forall u, v \in W^{1,1}(\Omega; \mathbb{S}^1), \tag{1.159}$$

and

$$\|Ju - Jv\|_{[W_0^{1,\infty}]^*} \leq \frac{1}{2}\|u - v\|_{L^{p'}} (\|\nabla u\|_{L^p} + \|\nabla v\|_{L^p}), \quad \forall\, u, v \in W^{1,p}(\Omega; \mathbb{S}^1) \text{ with } 1 \leq p \leq \infty. \tag{1.160}$$

1.6 An excursion into Monge–Kantorovich (=MK) territory

Let us pause briefly to compare some quantities introduced in the previous sections, and that are associated with a given $u \in W^{1,1}(\Omega; \mathbb{S}^1)$. Set

$$A(u) := \inf \left\{\|D^j\varphi\|_{\mathcal{M}};\ \varphi \in BV(\Omega; \mathbb{R}) \text{ and } u = e^{\imath\varphi}\right\}, \tag{1.161}$$

$$B(u) := \min \left\{\|\mu\|_{\mathcal{M}};\ \mu \in \mathcal{M}(\Omega; \mathbb{R}^N) \text{ and } \operatorname{curl} \mu = 2Ju\right\}, \tag{1.162}$$

and

$$C(u) := \sup \left\{\langle 2Ju, *\zeta\rangle;\ \zeta \in C_c^\infty(\Omega; \Lambda^{N-2}) \text{ and } \|d\zeta\|_{L^\infty} \leq 1\right\}. \tag{1.163}$$

Equivalently, in (1.162), we may view μ as an element of $\mathcal{M}(\Omega; \Lambda^1)$, and then the requirement on μ reads $d\mu = 2Ju = d(u \wedge du)$ in the sense of 2-forms.

Corollary 1.8. *For every $u \in W^{1,1}(\Omega; \mathbb{S}^1)$ we have $A(u) = B(u) = C(u) = \Sigma(u)$. Moreover, the* inf *in (1.161) is achieved.*

Proof. Apply successively Corollary 1.5, (1.85), and (1.135). □

Alternative proof of Corollary 1.8. The goal is to present a proof "from scratch" of the equality

$$A(u) = B(u) = C(u). \tag{1.164}$$

It does not rely on any of the results involving $\Sigma(u)$ established in Section 1.4; in particular, it does not make use of Theorem 1.5 (which entered the previous proof via Corollary 1.5). The only ingredients are Hahn–Banach, Volpert's chain rule (more precisely, (1.108)), the Krein–Milman theorem (in fact, just the existence of an extreme point in any non-empty convex compact set), and Theorem 1.7, whose proof is self-contained – it uses only the definition of $\Sigma(u)$, or rather Lemma 1.13.

Since $\operatorname{curl}(-D^j\varphi) = 2Ju$ for any $\varphi \in BV(\Omega; \mathbb{R})$ such that $u = e^{\imath\varphi}$, we deduce that

$$B(u) \leq A(u). \tag{1.165}$$

Inequality $C(u) \leq B(u)$ is trivial. As explained in Section 1.5 (see also Section 1.9.2 and Comment 12 in Section 1.10), equality $C(u) = B(u)$ is a consequence of standard duality theory.

It remains to prove that

$$A(u) \leq B(u). \tag{1.166}$$

Set

$$\mathscr{Q}(u) := \left\{\mu \in \mathscr{M}(\Omega; \mathbb{R}^N);\ \operatorname{curl} \mu = 2Ju \text{ and } \|\mu\|_{\mathscr{M}} \leq B(u)\right\}. \tag{1.167}$$

Clearly, $\mathscr{Q}(u)$ is a non-empty convex, weak* compact subset of $\mathscr{M}(\Omega; \mathbb{R}^N)$. By the Krein-Milman theorem, $\mathscr{Q}(u)$ admits an extreme point μ_0. The next lemma asserts that $\mu_0 = -D^j\psi_0$ for some $\psi_0 \in BV(\Omega; \mathbb{R})$ such that $u = e^{\imath\psi_0}$. Thus

$$A(u) \leq \|D^j\psi_0\|_{\mathscr{M}} = \|\mu_0\|_{\mathscr{M}} \leq B(u). \qquad \square$$

Lemma 1.18. *Let μ_0 be an extreme point of $\mathscr{Q}(u)$. Then*

$$\mu_0 = -D^j\psi_0 \textit{ for some } \psi_0 \in BV(\Omega; \mathbb{R}) \textit{ such that } u = e^{\imath\psi_0}. \tag{1.168}$$

Proof. Any $\mu \in \mathscr{M}(\Omega; \mathbb{R}^N)$ such that $\operatorname{curl} \mu = 2Ju = \operatorname{curl}(u \wedge \nabla u)$ may be written as $\mu = u \wedge \nabla u - D\psi$ for some $\psi \in BV(\Omega; \mathbb{R})$ with $\int_\Omega \psi = 0$, and conversely. Thus,

$$\begin{aligned}&\{\mu \in \mathscr{M}(\Omega; \mathbb{R}^N);\ \operatorname{curl} \mu = 2Ju\}\\ &= \left\{u \wedge \nabla u - D\psi;\ \psi \in BV(\Omega; \mathbb{R}),\ \int_\Omega \psi = 0\right\}\end{aligned} \tag{1.169}$$

and

$$B(u) = \min\left\{\|u \wedge \nabla u - D\psi\|_{\mathscr{M}};\ \psi \in BV(\Omega; \mathbb{R}),\ \int_\Omega \psi = 0\right\} = \Sigma(u),$$

by Lemma 1.13. Hence, we may write

$$\mu_0 = u \wedge \nabla u - D\widetilde{\psi}_0 \tag{1.170}$$

for some $\widetilde{\psi}_0 \in BV(\Omega; \mathbb{R})$ such that $\int_\Omega \widetilde{\psi}_0 = 0$ and

$$\|u \wedge \nabla u - D\widetilde{\psi}_0\|_{\mathscr{M}} = \|\mu_0\|_{\mathscr{M}} = B(u) = \Sigma(u).$$

Thus, $\widetilde{\psi}_0 \in A$, where A is the convex set defined in (1.122). We claim that

$$\widetilde{\psi}_0 \text{ is an extreme point of } A. \tag{1.171}$$

Indeed, suppose that

$$\widetilde{\psi}_0 = t\widetilde{\psi}_1 + (1-t)\widetilde{\psi}_2, \tag{1.172}$$

with $0 < t < 1$ and $\widetilde{\psi}_1, \widetilde{\psi}_2 \in A$.

From (1.170) and (1.172), we see that

$$\mu_0 = t(u \wedge \nabla u - D\widetilde{\psi}_1) + (1-t)(u \wedge \nabla u - D\widetilde{\psi}_2) := t\mu_1 + (1-t)\mu_2.$$

Note that $\mu_1, \mu_2 \in \mathscr{Q}(u)$, since

$$\|\mu_1\|_{\mathscr{M}} = \|u \wedge \nabla u - D\widetilde{\psi}_1\|_{\mathscr{M}} = B(u),$$
$$\|\mu_2\|_{\mathscr{M}} = \|u \wedge \nabla u - D\widetilde{\psi}_2\|_{\mathscr{M}} = B(u),$$

and

$$\text{curl}\,\mu_1 = \text{curl}\,\mu_2 = \text{curl}\,(u \wedge \nabla u).$$

Since μ_0 is an extreme point of $\mathscr{Q}(u)$ (by assumption), we deduce that $\mu = \mu_1 = \mu_2$, i.e., $D\widetilde{\psi}_1 = D\widetilde{\psi}_2$, and thus $\widetilde{\psi}_1 = \widetilde{\psi}_2$ (since $\int_\Omega \widetilde{\psi}_1 = \int_\Omega \widetilde{\psi}_2 = 0$). This completes the proof of (1.171). Applying Theorem 1.7, we infer that, for some constant C, $\psi_0 := \widetilde{\psi}_0 + C$ is a minimal lifting of u.

Therefore (by (1.108)), $D\psi_0 = u \wedge \nabla u + D^j\psi_0$ and thus (by (1.170))

$$\mu_0 = u \wedge \nabla u - D\psi_0 = -D^j\psi_0,$$

i.e., (1.168) is established. □

We now investigate the structure of the minimizers of (1.161) and (1.162). Given $u \in W^{1,1}(\Omega; \mathbb{S}^1)$, set

$$\mathscr{R}(u) := \left\{-D^j\varphi;\ \varphi \in BV(\Omega; \mathbb{R}),\ u = e^{\imath\varphi} \text{ and } \|D^j\varphi\|_{\mathscr{M}} = \Sigma(u)\right\} \tag{1.173}$$

and (as above)

$$\mathscr{Q}(u) := \left\{\mu \in \mathscr{M}(\Omega; \mathbb{R}^N);\ \text{curl}\,\mu = 2Ju \text{ and } \|\mu\|_{\mathscr{M}} = \Sigma(u)\right\}.$$

From Corollary 1.5, we know that

$$\mathscr{R}(u) = \{-D^j\varphi;\ \varphi \text{ is a minimal lifting of } u\}. \tag{1.174}$$

On the other hand, recall that, if $\varphi \in BV$ is a lifting of u, we have, by (1.108),

$$0 = \text{curl}\ D\varphi = \text{curl}\,(u \wedge \nabla u) + \text{curl}\ D^j\varphi = 2Ju + \text{curl}\ D^j\varphi. \tag{1.175}$$

In particular, if φ is a minimal lifting of u then, by (1.175) and Corollary 1.5, we have $-D^j\varphi \in \mathscr{Q}(u)$, i.e.,

$$\mathscr{R}(u) \subset \mathscr{Q}(u). \tag{1.176}$$

Inclusion (1.176) may be strict. Consider, for example, $\Omega := \mathbb{D} \subset \mathbb{R}^2$ and $u(x) := x/|x|$. Since (by definition) curl $(u \wedge \nabla u) = 2Ju$, we see that $\mu_0 = u \wedge \nabla u$ is a competitor in (1.162). From (1.65), we have $|\nabla u(x)| = 1/|x|$ and thus

$$\|\mu_0\|_{\mathscr{M}} = \|u \wedge \nabla u\|_{\mathscr{M}} = \|\nabla u\|_{L^1} = 2\pi.$$

On the other hand, from (1.68) we know that $Ju = \pi\delta_0$. This implies (by (2.87) and Example 2.4) that $\Sigma(u) = 2\pi$. Hence, μ_0 is a minimizer in (1.162); however, $\mu_0 \notin \mathscr{R}(u)$, since $\mu_0 \in L^1$, whereas measures in $\mathscr{R}(u)$ are singular.

The next result, which is an immediate consequence of Lemma 1.18, is a far-reaching refinement of (1.176). Clearly, $\mathscr{Q}(u)$ is a convex, weak* compact subset of $\mathscr{M}(\Omega;\mathbb{R}^N)$, but in general $\mathscr{R}(u)$ is not convex. Denote by $\mathrm{Ext}(\mathscr{Q}(u))$ the extreme points of $\mathscr{Q}(u)$.

Corollary 1.9. *We have*

$$\mathrm{Ext}(\mathscr{Q}(u)) \subset \mathscr{R}(u), \tag{1.177}$$

and consequently

$$\overline{\mathrm{conv}(\mathscr{R}(u))}^{\text{weak}^*} = \mathscr{Q}(u). \tag{1.178}$$

Proof. Assertion (1.178) is a consequence of (1.177) and the Krein–Milman theorem, which says that

$$\overline{\mathrm{conv}(\mathrm{Ext}(\mathscr{Q}(u)))}^{\text{weak}^*} = \mathscr{Q}(u).$$

□

Remark 1.26. The inclusion (1.177) may be strict (see Remark 1.18 and Remark 2.14 in Chapter 2).

Remark 1.27. The quantities $A(u)$, $B(u)$ $C(u)$, as well as the sets $\mathscr{R}(u)$ and $\mathscr{Q}(u)$, depend *only* on Ju (and not on the specific u). This is clear for $A(u){=}B(u) = C(u) = \Sigma(u)$, since, by (1.84), $\Sigma(u)$ depends only on Ju. This is also clear for $\mathscr{Q}(u)$ (from the definition (1.167)). We now establish that, if $u_1, u_2{\in}W^{1,1}(\Omega;\mathbb{S}^1)$, then

$$[Ju_1 = Ju_2] \implies [\mathscr{R}(u_1) = \mathscr{R}(u_2)]. \tag{1.179}$$

Since $Ju_1 = Ju_2$, we may write $u_2 = u_1 e^{\imath\varphi_0}$ for some $\varphi_0 \in W^{1,1}(\Omega;\mathbb{R})$ (see Theorem 1.2). Let $\varphi_1 \in BV(\Omega;\mathbb{R})$ be a minimal lifting for u_1. Then, $u_2 = e^{\imath(\varphi_0+\varphi_1)}$, and thus $\psi = \varphi_0 + \varphi_1$ is a lifting of u_2. Moreover, $D^j\psi = D^j\varphi_1$, and therefore $\|D^j\psi\|_{\mathscr{M}} = \|D^j\varphi_1\|_{\mathscr{M}} = \Sigma(u_1)$, by Corollary 1.5. On the other hand, $\Sigma(u_1) = \Sigma(u_2)$ by (1.84). Hence, $\|D^j\psi\|_{\mathscr{M}} = \Sigma(u_2)$. Applying Corollary 1.5 once more, we deduce that ψ is a minimal lifting of u_2. Thus, $-D^j\psi \in \mathscr{R}(u_2)$, i.e., $-D^j\varphi_1 \in \mathscr{R}(u_2)$, and we have established that $\mathscr{R}(u_1) \subset \mathscr{R}(u_2)$. Reversing the roles of u_1 and u_2 yields $\mathscr{R}(u_1) = \mathscr{R}(u_2)$. □

There are striking analogies between Corollary 1.8 and Corollary 1.9 on one side, and the Monge–Kantorovich (MK) optimal transport (OT) theory on the other side.

Let us consider, for example, the MK theory in the discrete case, with a cost function being a distance $d(x,y)$ on a metric space Z. Given $2m$ points $P_1,\ldots,P_m,N_1,\ldots,N_m$ in Z, we introduce, following the presentation in Brezis [77], three quantities. The first one, denoted by M (for Monge), is defined by

$$M := \min_{\sigma\in S_m}\sum_{\ell=1}^{m} d(P_\ell, N_{\sigma(\ell)}), \tag{1.180}$$

where the minimum is taken over the set S_m of all permutations of the integers $\{1,\dots,m\}$.

The second one, denoted by K (for Kantorovich), is defined by

$$K := \min_{\mathrm{DS}}\left\{\sum_{i,j=1}^{m} a_{ij}d(P_i,N_j);\ \mathrm{DS}=(a_{ij}) \text{ is doubly stochastic}\right\}. \tag{1.181}$$

Recall that a matrix $\mathrm{DS}=(a_{ij})$ is *doubly stochastic* if

$$a_{ij}\geq 0,\ \forall\, i,j,\ \sum_{i=1}^{m} a_{ij}=1,\ \forall j,\ \text{and}\ \sum_{j=1}^{m} a_{ij}=1,\ \forall\, i.$$

Finally, define D (for duality) by

$$D := \sup\left\{\sum_{j=1}^{m}(\zeta(P_j)-\zeta(N_j));\ \zeta: Z\to\mathbb{R},\ |\zeta|_{\mathrm{Lip}}\leq 1\right\}. \tag{1.182}$$

Note that (1.180) is *not* a convex problem, and (1.181) is a kind of *convex relaxation* of (1.180). Problem (1.182) is also a convex problem, which can be viewed as *dual* to (1.181), either in the sense of linear programming or in the sense of conjugate convex functions.

A central result in the MK theory is the following

Theorem 1.10. *We have $M = K = D$.*

It is obvious that $K\leq M$, since permutation matrices are doubly stochastic. On the other hand, $D\leq K$, since

$$\sum_{j=1}^{m}(\zeta(P_j)-\zeta(N_j)) = \sum_{i,j=1}^{m} a_{ij}(\zeta(P_i)-\zeta(N_j)),\ \forall\,(a_{ij}) \text{ doubly stochastic}.$$

Equality $D=M$ is restated and proved in Chapter 2 (see Theorem 2.4).

Recall that Birkhoff's theorem on doubly stochastic matrices (also called the Birkhoff–von Neumann theorem) asserts that the extreme points of the convex set of the doubly stochastic matrices are precisely the permutation matrices.

Denote by $\sigma_1,\sigma_2,\dots,\sigma_k$ the minimizing permutations in (1.180), and by $\Lambda_1,\dots,\Lambda_k$ the corresponding permutation matrices. Consider the convex set Q of minimizers in (1.181), i.e.,

$$Q := \left\{(a_{ij});\ (a_{ij}) \text{ is doubly stochastic and } \sum_{i,j=1}^{m} a_{ij}\, d(P_i,N_j)\leq K\right\}.$$

Here is an immediate consequence of Birkhoff's theorem.

Corollary 1.10. *We have*

$$Q = \text{conv}\{\Lambda_1, \Lambda_2 \ldots, \Lambda_k\}$$

and

$$\text{Ext}(Q) = \{\Lambda_1, \Lambda_2 \ldots, \Lambda_k\}.$$

Note the following striking correspondence:

$W^{1,1}(\Omega; \mathbb{S}^1)$ setting	MK setting
Ju	$\sum_{j=1}^{m}(\delta_{P_j} - \delta_{N_j})$
$A(u)$	M
$B(u)$	K
$C(u)$	D
Corollary 1.8	Theorem 1.10
Corollary 1.9	Corollary 1.10

As we are going to see in Chapters 2 (Section 2.9.4) and 3 (Sections 3.5, 3.6 and 3.8), this correspondence can also be interpreted within the framework of area minimizing $(N-1)$-rectifiable currents with prescribed boundary.

1.7 Relaxed energy

A quantity that is commonly studied within the framework of Sobolev maps with values in manifolds (see Bethuel, Brezis, and Coron [30], and also Giaquinta, Modica, and Souček [157]) is the *relaxed energy*. For a given $u \in W^{1,p}(\Omega; \mathbb{S}^1)$, $1 \le p < \infty$, the relaxed energy is defined by:

$$R_p(u) := \inf\left\{\liminf_{n\to\infty} \int_\Omega |\nabla u_n|^p;\ u_n \in C^\infty(\overline{\Omega}; \mathbb{S}^1),\ u_n \to u \text{ a.e.}\right\} \le \infty, \tag{1.183}$$

where the first inf is taken over all sequences (u_n) such that $u_n \in C^\infty(\overline{\Omega}; \mathbb{S}^1)$ and $u_n \to u$ a.e.

We first observe the existence of competitors in (1.183).

Lemma 1.19. *Let $u : \Omega \to \mathbb{S}^1$ be any measurable map. Then there exists a sequence $(u_n) \subset C^\infty(\overline{\Omega}; \mathbb{S}^1)$ such that $u_n \to u$ a. e.*

Proof. Write $u = e^{\imath\varphi}$ with $\varphi \in L^\infty(\Omega; \mathbb{R})$, and let $u_n := e^{\imath\varphi_n}$, where $\varphi_n \in C^\infty(\overline{\Omega}; \mathbb{R})$ and $\varphi_n \to \varphi$ a.e. □

The next result identifies the relaxed energy $R_1(u)$ with the quantity $E(u)$ defined by (1.78). Recall that $E(u) = \int_\Omega |\nabla u| + \Sigma(u)$, by Theorem 1.5.

Theorem 1.11. *For every $u \in W^{1,1}(\Omega; \mathbb{S}^1)$, we have*

$$R_1(u) = E(u) = \int_\Omega |\nabla u| + \Sigma(u). \tag{1.184}$$

Proof. "$\leq$" Let $\varphi \in BV(\Omega; \mathbb{R})$ be a minimal lifting of u. Let $(\varphi_n) \subset C^\infty(\overline{\Omega}; \mathbb{R})$ be such that $\varphi_n \to \varphi$ a.e. and $\int_\Omega |\nabla \varphi_n| \to |\varphi|_{BV}$ (see, for example, Lemma 15.32). We define $u_n := e^{\imath\varphi_n} \in C^\infty(\overline{\Omega}; \mathbb{S}^1)$. Then, $u_n \to u$ a.e. and $\int_\Omega |\nabla u_n| = \int_\Omega |\nabla \varphi_n| \to |\varphi|_{BV} = E(u)$, so that "$\leq$" follows.

"$\geq$" Let $(u_n) \subset C^\infty(\overline{\Omega}; \mathbb{S}^1)$ be such that $u_n \to u$ a.e. and $\int_\Omega |\nabla u_n| \to R_1(u)$. Since Ω is simply connected, we may write $u_n = e^{\imath\varphi_n}$, with $\varphi_n \in C^\infty(\overline{\Omega}; \mathbb{R})$. We have $\int_\Omega |\nabla \varphi_n| = \int_\Omega |\nabla u_n| \to R_1(u)$. Therefore, there exists some $\varphi \in BV(\Omega; \mathbb{R})$ such that, up to a subsequence and after subtracting a suitable integer multiple of 2π from φ_n, we have $\varphi_n \to \varphi$ a.e. We then conclude that $|\varphi|_{BV} \leq \liminf_n \int_\Omega |\nabla \varphi_n| = R_1(u)$. □

Remark 1.28. The conclusion of Theorem 1.11 may fail when Ω is a multiply connected domain. This is discussed in great detail in Section 14.2.

Remark 1.29. There is a variant of Theorem 1.11 for maps with a prescribed boundary condition (see Comment 9 in Section 1.10).

The situation is completely different when $p > 1$.

Theorem 1.12. *Assume that either $N = 1$ or $N \geq 2$ and $p \geq 2$. Then*

$$R_p(u) = \int_\Omega |\nabla u|^p, \ \forall u \in W^{1,p}(\Omega; \mathbb{S}^1).$$

Proof. In these cases we may write $u = e^{\imath\varphi}$, with $\varphi \in W^{1,p}$. We consider a sequence $(\varphi_n) \subset C^\infty(\overline{\Omega}; \mathbb{R})$ such that $\varphi_n \to \varphi$ in $W^{1,p}$ and let $u_n := e^{\imath\varphi_n}$. Then $\int_\Omega |\nabla u_n|^p \to \int_\Omega |\nabla u|^p$, and thus $R_p(u) \leq \int_\Omega |\nabla u|^p$. On the other hand, we always have "$\geq$." □

Theorem 1.13. *Assume that $N \geq 2$ and $1 < p < 2$. Then, for every $u \in W^{1,p}(\Omega; \mathbb{S}^1)$, we have*

$$R_p(u) = \begin{cases} \int_\Omega |\nabla u|^p, & \text{if } Ju = 0 \\ \infty, & \text{if } Ju \neq 0 \end{cases}.$$

Proof. If $Ju = 0$, then we may write $u = e^{\imath\varphi}$, with $\varphi \in W^{1,p}$, and then we argue as above.

We next consider the case where $Ju \neq 0$. Then there is no sequence $(u_n) \subset C^\infty(\overline{\Omega}; \mathbb{S}^1)$ such that $u_n \to u$ a.e. and (u_n) is bounded in $W^{1,p}$. Argue by contradiction. Passing to a subsequence, we have $u_n \rightharpoonup u$ in $W^{1,p}$, and thus $Ju_n \to Ju$ in $\mathscr{D}'$, by Remark 1.24. This is impossible, since $Ju_n = 0$. Thus, $R_p(u) = \infty$. □

We will return in Section 8.4 to the study of the relaxed energy in the setting of fractional Sobolev spaces.

1.8 Least energy with prescribed Jacobian

In this section, we establish a connection between $\Sigma(u)$ and the least energy of maps having "prescribed singularities." More specifically, we define, for a given $u \in W^{1,p}(\Omega; \mathbb{S}^1)$ with $1 \leq p < 2$,

$$S_p(u) := \inf \left\{ \int_\Omega |\nabla v|^p;\ v \in W^{1,p}(\Omega; \mathbb{S}^1),\ Jv = Ju \text{ in } \mathscr{D}' \right\}. \tag{1.185}$$

We do not consider the case where $p \geq 2$, since in that case we have $Ju = 0$ for every $u \in W^{1,p}(\Omega; \mathbb{S}^1)$, by (1.61).

As we are going to see in Chapters 2 and 3, the Jacobian Ju describes precisely the topological singularities of u. Thus, $S_p(u)$ can be viewed as the least $W^{1,p}$ energy among all maps having the same topological singularities as u.

We examine first the case $p{=}1$. Recall (see Lemma 1.14) that, for $u{\in}W^{1,1}(\Omega; \mathbb{S}^1)$, we have

$$S_1(u) = \Sigma(u). \tag{1.186}$$

The existence of a minimizer in (1.185) when $p = 1$ is a delicate question that is still widely open even when $N = 2$. The bottom line is that minimizers exist only for an "exceptional" class of u's (see Section 2.11.1 and Brezis and Mironescu [87]). Surprisingly, this question turns out to be *equivalent* to the existence of a minimizer of various other minimization problems we have encountered throughout this chapter.

Proposition 1.2. *Given $u \in W^{1,1}(\Omega; \mathbb{S}^1)$, the following properties are equivalent:*

1. *There exists some $v \in W^{1,1}(\Omega; \mathbb{S}^1)$ satisfying*

$$Jv = Ju \text{ and } \int_\Omega |\nabla v| = S_1(u). \tag{1.187}$$

2. *There exists a minimizer $\varphi \in W^{1,1}(\Omega; \mathbb{S}^1)$ in* (1.80).
3. *There exists a minimizer $\mu \in L^1$ in* (1.85).
4. *There exists a minimizer ψ in* (1.130).

Proof. The equivalences of properties 1. to 3. is clear, by inspection of (1.80) and the proofs of Lemmas 1.13 and 1.14.

"4. $\Longrightarrow$ 1." Let ψ be a minimizer in (1.130). By (1.132), the corresponding φ must satisfy $\|u \wedge \nabla u - \nabla\varphi\|_{L^1} = \Sigma(u)$. Let $v := ue^{-\imath\varphi}$, so that $v \in W^{1,1}(\Omega; \mathbb{S}^1)$, $|u \wedge \nabla u - \nabla\varphi| = |\nabla v|$ and $\int_\Omega |\nabla v| = \Sigma(u)$. By Theorem 1.2 and (1.84), we have $Jv = Ju$ and v satisfies (1.187).

"1. $\Longrightarrow$ 4." Assume that v satisfies (1.187). Then (again by Theorem 1.2 and (1.84)) we may write $v = ue^{-\imath\varphi}$, where $\varphi \in W^{1,1}(\Omega;\mathbb{R})$ and

$$\|u \wedge \nabla u - \nabla\varphi\|_{L^1} = \Sigma(u). \tag{1.188}$$

Let $\eta \in BV(\Omega;\mathbb{R})$ be a minimal lifting of u and set $\psi := \eta - \varphi$. By Theorem 1.5, we have

$$\|D(\psi + \varphi) - u \wedge \nabla u\|_{\mathscr{M}} = \Sigma(u). \tag{1.189}$$

Combining (1.132), (1.188), and (1.189), we find that ψ is a minimizer in (1.130). □

When $1 < p < 2$, nothing is known about the exact value of $S_p(u)$. However, by contrast with the case $p = 1$, minimizers of (1.185) exist.

Proposition 1.3. *Let $1 < p < 2$ and $u \in W^{1,p}(\Omega;\mathbb{S}^1)$. Then there exists some $v \in W^{1,p}(\Omega;\mathbb{S}^1)$ such that $Jv = Ju$ and $S_p(u) = \int_\Omega |\nabla v|^p$.*

Proof. By Remark 1.24, $v \mapsto Jv$ is weakly sequentially continuous in $W^{1,p}$. Therefore, it suffices to consider a minimizing sequence in the definition of $S_p(u)$ and pass to the weak limit. □

We will return to this topic in Chapter 13.

1.9 Complements and open problems

1.9.1 Minimal liftings of BV maps (following Ignat)

We give here a brief account (without proofs) of the results of Ignat [193] concerning minimal lifting for maps in $BV(\mathbb{S}^2;\mathbb{S}^1)$. [The results of Ignat were subsequently extended by Mucci [274] to $BV(\mathbb{S}^N;\mathbb{S}^1)$. Further extensions to maps in $BV(\Omega;\mathscr{N})$, where $\mathscr{N}$ is a compact manifold without boundary, were obtained by Giaquinta and Mucci [160, 162].]

We start with the definition of $u \wedge Du$ for $u \in BV(\mathbb{S}^2;\mathbb{S}^1)$. Let $\rho : \mathbb{S}^1 \times \mathbb{S}^1 \to [-\pi,\pi]$ be the signed geodesical distance on $\mathbb{S}^1$, defined by the formula

$$\rho(\omega_1,\omega_2) := \begin{cases} \Theta_\pi(\omega_1/\omega_2), & \text{if } \omega_1 \neq -\omega_2 \\ \Theta_\pi(\omega_1) - \Theta_\pi(\omega_2), & \text{if } \omega_1 = -\omega_2 \end{cases}.$$

Here, Θ_π is as in (1.75). We then define, following [193],

$$u \wedge Du := u \wedge \nabla u + u \wedge D^c u + \rho(u^+,u^-)\nu \mathscr{H}^1 \llcorner J_u \in \mathscr{M},$$

where ν stands for the normal to J_u. Note that, when $u \in W^{1,1}$, we obtain the usual definition of $u \wedge \nabla u$.

We set, for $\psi \in BV(\mathbb{S}^2;\mathbb{R})$, $V(\psi) := \|u \wedge Du - D\psi\|_{\mathscr{M}}$ and we let $\Sigma(u) := \inf_{\psi \in BV} V(\psi)$.

The paper Ignat [193] contains two main results. The first one is a formula for $E(u)$ when $u \in BV(\mathbb{S}^2;\mathbb{S}^1)$. This rather complicated formula, generalizing (1.87), will not be given here; we refer the reader to [193, Theorem 2] (see also Giaquinta, Modica, and Souček [157, Chapter 6]). The second one connects $\Sigma(u)$ to a lifting of u. Theorem 3 in [193] asserts that, among the minimizers of $\Sigma(u)$ in $BV(\mathbb{S}^2;\mathbb{R})$, there is at least a lifting of u. One should compare this result with Theorem 1.5.

A subsequent work of Ignat and Lamy [196] extends this approach to BV maps into the projective space $\mathbb{RP}^k$ with $k \geq 2$.

1.9.2 Duality and density revisited

We now return to Remark 1.21 and expand it. As we already pointed out, the specific form $F = u \wedge \nabla u$ of F plays no role. What we actually proved is the following

Theorem 1.14. *Let $F \in L^1(\Omega;\mathbb{R}^N) \simeq L^1(\Omega;\Lambda^1)$. Then, $A = B = C = D$, where*

$$A := \inf\{\|F - \nabla\varphi\|_{L^1};\ \varphi \in W^{1,1}(\Omega;\mathbb{R})\},$$

$$B := \min\{\|\mu\|_{\mathscr{M}};\ \mu \in \mathscr{M}(\Omega;\Lambda^1),\ d\mu = dF\},$$

$$C := \sup\left\{\int_\Omega \langle F, *d\zeta\rangle;\ \zeta \in C_c^\infty(\Omega;\Lambda^{N-2}),\ \|d\zeta\|_{L^\infty} \leq 1\right\},$$

and

$$D := \max\left\{\int_\Omega F\cdot\xi;\ \xi \in L^\infty(\Omega;\mathbb{R}^N),\ \|\xi\|_{L^\infty} \leq 1,\ \operatorname{div}\xi = 0 \text{ in } \Omega \text{ and } \xi\cdot\nu = 0 \text{ on } \partial\Omega\right\}.$$

The next density result is equivalent to the equality $C = D$.

Corollary 1.11. *Let $\xi_0 \in L^\infty(\Omega;\mathbb{R}^N) \simeq L^\infty(\Omega;\Lambda^1)$ be such that $\|\xi_0\|_{L^\infty} \leq 1$, $\operatorname{div}\xi_0 = 0$ in Ω and $\xi_0\cdot\nu = 0$ on $\partial\Omega$. Then there exists a sequence $(\zeta_j) \subset C_c^\infty(\Omega;\Lambda^{N-2})$ such that $\xi_j := *d\zeta_j \in C_c^\infty(\Omega;\Lambda^1) \simeq C_c^\infty(\Omega;\mathbb{R}^N)$ satisfies $\|\xi_j\|_{L^\infty} \leq 1$, $\operatorname{div}\xi_j = 0$ in Ω and $\xi_j \to \xi_0$ a.e.*

Remark 1.30. When $N = 2$, Corollary 1.11 asserts that for every function $\zeta \in W_0^{1,\infty}(\Omega;\mathbb{R})$ such that $\|\nabla\zeta\|_{L^\infty} \leq 1$, there exists a sequence of functions $(\zeta_j) \subset C_c^\infty(\Omega;\mathbb{R})$ such that $\|\nabla\zeta_j\|_{L^\infty} \leq 1$ and $\nabla\zeta_j \to \nabla\zeta$ a.e. in Ω. When $N = 3$, Corollary 1.11 implies that for every $\zeta \in W^{1,\infty}(\Omega;\mathbb{R}^3)$ such that $\|\operatorname{curl}\zeta\|_{L^\infty} \leq 1$ and $\zeta \parallel \nu$ on $\partial\Omega$, there exists a sequence $(\zeta_j) \subset C_c^\infty(\Omega;\mathbb{R}^3)$ such that $\|\operatorname{curl}\zeta_j\|_{L^\infty} \leq 1$ and $\operatorname{curl}\zeta_j \to \operatorname{curl}\zeta$ a.e. in Ω.

Proof of Corollary 1.11. Let Y be the closure in $L^1(\Omega; \mathbb{R}^N) \simeq L^1(\Omega; \Lambda^1)$ of the convex set

$$Z := \{*d\zeta;\ \zeta \in C_c^\infty(\Omega; \Lambda^{N-2}),\ \|d\zeta\|_{L^\infty} \le 1\}.$$

It is easy to see that the conclusion of the corollary is equivalent to the fact that $\xi_0 \in Y$. In order to prove that $\xi_0 \in Y$, we argue by contradiction. Otherwise, by the Hahn–Banach theorem, there exists some $F \in L^\infty(\Omega; \mathbb{R}^N)$ such that

$$\sup\left\{\int_\Omega F \cdot \xi;\ \xi \in Z\right\} < \int_\Omega F \cdot \xi_0.$$

The above contradicts the equality $C = D$ in Theorem 1.14. □

Similar results hold if we replace, in Theorem 1.14, the assumption $F \in L^1(\Omega; \Lambda^1)$ by $F \in L^p(\Omega; \Lambda^k)$, where $1 \le p < \infty$ and $1 \le k \le N-1$. When $k \ne 1$, the assumption that Ω is simply connected has to be replaced by the assumption that an appropriate de Rham cohomology group is trivial. In what follows, we assume for simplicity that Ω is *contractible*; this implies that all these groups are trivial, i.e., all (smooth) closed forms are exact. We present only the case where $p = 1$, which is the most difficult.

Theorem 1.15. *Assume that Ω is a smooth bounded contractible domain. Let $F \in L^1(\Omega; \Lambda^k)$, $1 \le k \le N-1$. Then $A = B = C = D$, where*

$$A := \inf\{\|F - d\varphi\|_{L^1};\ \varphi \in W^{1,1}(\Omega; \Lambda^{k-1})\},$$
$$B := \min\{\|\mu\|_{\mathcal{M}};\ \mu \in \mathcal{M}(\Omega; \Lambda^k),\ d\mu = dF\},$$
$$C := \sup\left\{\int_\Omega \langle F, *d\zeta\rangle;\ \zeta \in C_c^\infty(\Omega; \Lambda^{N-k-1}),\ \|d\zeta\|_{L^\infty} \le 1\right\},$$

and

$$D := \max\left\{\int_\Omega \langle F, \xi\rangle;\ \xi \in L^\infty(\Omega; \Lambda^k),\ \|\xi\|_{L^\infty} \le 1,\ d(*\xi) = 0 \text{ in } \Omega \text{ and } (*\xi)_T = 0 \text{ on } \partial\Omega\right\}.$$

Here, the conditions $d(*\xi) = 0$ and $(*\xi)_T = 0$ are to be interpreted in the weak sense, i.e.,

$$\int_\Omega \langle \xi, d\varphi\rangle = 0,\ \forall \varphi \in W^{1,1}(\Omega; \Lambda^{k-1}) \text{ (resp. } \forall \varphi \in C^\infty(\overline{\Omega}; \Lambda^{k-1})). \tag{1.190}$$

Proof. Step 1. We have $B \le C \le D \le A$
Inequality $B \le C$ relies on the Hahn–Banach theorem, as in the proof of "$\le$" in Theorem 1.9. More precisely, there exists a measure $\mu_0 \in \mathcal{M}(\Omega; \Lambda^k)$ such that $\|\mu_0\|_{\mathcal{M}} \le C$ and $d\mu_0 = dF$. Thus, $B \le C$.

Inequality $C \le D$ is obvious, since the class of test functions is wider in D. Inequality $D \le A$ is also obvious, since for every test function ξ in D we have

$$\int_\Omega \langle F, \xi\rangle = \int_\Omega \langle F - d\varphi, \xi\rangle \le \|F - d\varphi\|_{L^1}, \ \forall \varphi \in W^{1,1}(\Omega; \Lambda^{k-1}).$$

Step 2. Proof of the inequality $A \le B$

This inequality is a consequence of the following cousin of (1.83) applied with $\nu := F - \mu$.

Lemma 1.20. *Let $\Omega \subset \mathbb{R}^N$ be a smooth bounded contractible domain. Let $1 \le k \le N-1$. Let $F \in L^1(\Omega; \Lambda^k)$ and let $\nu \in \mathscr{M}(\Omega; \Lambda^k)$ be such that $d\nu = 0$. Then there exists a sequence $(\varphi_j) \subset C^\infty(\overline{\Omega}; \Lambda^{k-1})$ such that*

$$\lim_{j\to\infty} \int_\Omega |F - d\varphi_j| = \|F - \nu\|_{\mathscr{M}}. \tag{1.191}$$

The proof of Lemma 1.20 is presented in Appendix 15.7 (see Lemma 15.33). □

From Theorem 1.15, we easily derive the following density result (with the same proof as that of Corollary 1.11).

Corollary 1.12. *Let $\eta_0 \in L^\infty(\Omega; \Lambda^\ell)$, $1 \le \ell \le N-1$, be such that $\|\eta_0\|_{L^\infty} \le 1$, $d\eta_0 = 0$ in Ω and $(\eta_0)_T = 0$ on $\partial\Omega$. Then there exists a sequence $(\zeta_j) \subset C_c^\infty(\Omega; \Lambda^{\ell-1})$ such that $|d\zeta_j| \le 1$ in Ω and $d\zeta_j \to \eta_0$ a.e.*

Remark 1.31. Applying in particular Corollary 1.12 with $\ell = 1$, we find that, given any $\varphi_0 \in W_0^{1,\infty}(\Omega; \mathbb{R})$ such that $|\nabla\varphi_0| \le 1$ in Ω, there exists a sequence $(\varphi_j) \subset C_c^\infty(\Omega; \mathbb{R})$ such that $|\nabla\varphi_j| \le 1, \forall j$, and $\nabla\varphi_j \to \nabla\varphi_0$ a.e.

Finally, we state the following L^p version of Theorem 1.15, whose proof, somewhat simpler than that of Theorem 1.15, is left to the reader.

Theorem 1.16. *Assume that Ω is a smooth bounded contractible domain. Let $F \in L^p(\Omega; \Lambda^k)$, $1 \le k \le N-1$, $1 < p < \infty$. Then, $A = B = C = D$, where*

$$A := \min\{\|F - d\varphi\|_{L^p};\ \varphi \in W^{1,p}(\Omega; \Lambda^{k-1})\},$$

$$B := \min\{\|f\|_{L^p};\ f \in L^p(\Omega; \Lambda^k),\ df = dF\},$$

$$C := \sup\left\{\int_\Omega \langle F, *d\zeta\rangle;\ \zeta \in C_c^\infty(\Omega; \Lambda^{N-k-1}),\ \|d\zeta\|_{L^{p'}} \le 1\right\},$$

and

$$D := \max\left\{\int_\Omega \langle F, \xi\rangle; \xi \in L^{p'}(\Omega; \Lambda^k),\ \|\xi\|_{L^{p'}} \le 1,\ d(*\xi) = 0 \text{ in } \Omega \right.$$
$$\left. \text{and } (*\xi)_T = 0 \text{ on } \partial\Omega\right\}.$$

1.9.3 Maps on $\partial\Omega$

Most of the analysis in this chapter extends to maps defined on a boundary, but the case where $N = 2$ requires a special treatment.

We start with $\Omega \subset \mathbb{R}^N$, with $N \geq 3$. We assume that $\partial\Omega$ is simply connected. For maps $u \in W^{1,1}(\partial\Omega; \mathbb{S}^1)$, we may define BV phases, minimal liftings, $Ju \in \mathscr{D}'(\partial\Omega; \Lambda^2)$, $\Sigma(u)$, and the relaxed energy. In particular, u has a $W^{1,1}$ phase if and only if $Ju = 0$. The proofs are basically unchanged, except for Theorem 1.4. On a manifold, that result can be obtained following, for example, the original argument in Dávila and Ignat [129]. Alternatively, it is possible to extend the arguments presented in Section 1.4; this requires replacing the Euclidean distance by the geodesic distance d on $\partial\Omega$, and $\rho_\varepsilon(x - y)$ by $\rho_\varepsilon(d(x, y))$.

When $N = 2$, the role of the Jacobian J is played by the degree, as explained in Section 1.1.

1.9.4 Open problems

Open Problem 1. Find proofs of Theorems 1.5 and 1.7 that do not make use of the fine theory of BV functions. See also a more precise question when $N = 2$, described in Open Problem 3 at the end of Chapter 2.

Open Problem 2. Assume $N \geq 3$. Consider the set

$$U := \{u \in W^{1,1}(\Omega; \mathbb{S}^1);\ u \text{ admits a unique minimal lifting } (\text{mod} 2\pi)\}.$$

Is U dense in $W^{1,1}(\Omega; \mathbb{S}^1)$? Furthermore, is U a residual set? [i.e., is the property "u admits a unique minimal lifting" generic?]

The answer is positive when $N = 2$ (see Curcă [124]).

1.10 Comments

1. Formula (1.24) in Lemma 1.4, as well as Lemma 1.5 are taken from Brezis, Mironescu, and Ponce [92].
2. As already mentioned, Theorem 1.1 and Remark 1.6 are due to Bethuel and Zheng [37]. They were originally concerned with the question whether $C^\infty(\overline{\Omega}; \mathbb{S}^1)$ is dense in $W^{1,p}(\Omega; \mathbb{S}^1)$. As we are going to see in Chapter 10, that question is closely related to Question 1 about lifting, discussed in Section 1.2.
3. Theorem 1.2 is due to Demengel [132]. Her goal was to characterize the closure of $C^\infty(\overline{\Omega}; \mathbb{S}^1)$ in $W^{1,p}(\Omega; \mathbb{S}^1)$ (a topic closely related to Question 2; see Chapter

10). She was motivated by an earlier work by Bethuel [25], who characterized the closure of $C^\infty(\overline{\mathbb{B}^3}; \mathbb{S}^2)$ in $H^1(\mathbb{B}^3; \mathbb{S}^2)$ and solved thereby a conjecture of Brezis. The proofs of Theorems 1.1 and 1.2 presented in Section 1.2 are due to Bourgain, Brezis, and Mironescu [50], and they are simpler than the original ones (see also Carbou [115]).

4. In his seminal contribution to the theory of non-linear elasticity, Ball [20, Section 6] made extensive use of the divergence structure of the Jacobian, which allows in particular to define the Jacobian in the sense of distributions for maps $u \in W^{1,\ell-1}(\Omega; \mathbb{R}^\ell) \cap L^\infty(\Omega; \mathbb{R}^\ell)$, with $\Omega \subset \mathbb{R}^\ell$; Ball coined the term "distributional Jacobian," which he denoted by $\operatorname{Det}(\nabla u)$. When $\ell = 2$, this corresponds to formula (1.35). Ball considered in particular u's of the form $u(x) = R(x)\dfrac{x}{|x|}$, where $R(x)$ is a smooth scalar function, and he noticed that $\operatorname{Det}(\nabla u)$ "has an atom at $x = 0$" [20, Section 6]. This can be viewed as a precursor to formula (1.68).
5. In Theorem 1.3, formula (1.70) is from Brezis, Mironescu, and Shafrir [93]. Formula (1.71) appears in Alberti, Baldo, and Orlandi [6, Corollary 3.9].
6. The study of the square root problem within the framework of Sobolev spaces was proposed by Bethuel and Chiron [32].
7. Theorem 1.4 is due to Dávila and Ignat [129]. The proof we present is due to Merlet [247] and simplifies the original one in [129]. The approach of Merlet is based on Theorem 6.2; it avoids the use of Volpert's chain rule for BV maps (see (1.109) and Appendix 15.7), which was an essential ingredient in [129].
8. The main result in Section 1.4, Theorem 1.5, is due to Brezis, Mironescu, and Ponce [92] when $N = 2$ and $\Omega = \mathbb{S}^2$ (see also [91] for the case where $\Omega \subset \mathbb{R}^2$) and to Poliakovsky [293] when $N \geq 2$ (see below). Theorem 1.5 is inspired by a result of Bethuel, Brezis, and Coron [30], who obtained an explicit formula (similar to (1.184)) for the relaxed Dirichlet energy of maps $u : \Omega \subset \mathbb{R}^3 \to \mathbb{S}^2$ (see Theorem 4.4).

   The technique used in [30] relies heavily on tools developed by Brezis, Coron, and Lieb [79]; in particular, the "length of a minimal connection" and the "dipole construction" (see Sections 2.3 and 2.5).

   The proof presented here of Theorem 1.5 consists of two independent parts:

   a) $E(u) \leq \int_\Omega |\nabla u| + \Sigma(u)$.
   The main novelty is Lemma 1.15, taken from Bousquet and Mironescu [63], whose proof borrows an idea suggested by Brezis to Escobedo, who incorporated it in Escobedo [145]. The use of Lemma 1.15 – which is a kind of quick (and efficient!) dipole construction – shortens noticeably the argument used in [92], valid for $N = 2$ and whose extension to $N \geq 3$ would require much effort.
   b) $E(u) \geq \int_\Omega |\nabla u| + \Sigma(u)$.
   The main ingredient, Theorem 1.6, is taken from [92] (the proof there is valid in any dimension $N \geq 2$). By contrast, the proof of the lower bound for the relaxed energy given in [30] relies on a cheap decomposition à la Brezis and

Lieb [82], which does not seem to be suitable in our context (see Lemma 1.17).

9. Theorem 1.7 is a remarkable result of Poliakovsky [293], who solved a conjecture stated in Brezis, Mironescu, and Ponce [92]. We follow his proof (which does not make use of Theorem 1.5). Note that Theorem 1.5 can easily be deduced from Theorems 1.6 and 1.7 as follows. Let ψ be an extreme point of A (which exists, by the Krein–Milman theorem). By Theorem 1.7, $\widetilde{\psi} := \psi + C$ is a minimal lifting of u for an appropriate C. Thus, $u = e^{\imath\widetilde{\psi}}$ and $\|u \wedge \nabla u - D\widetilde{\psi}\|_{\mathscr{M}} = \Sigma(u)$. By Theorem 1.6, we have $|D\widetilde{\psi}| = |u \wedge \nabla u - D\widetilde{\psi}| + |\nabla u|$, so that $\int |\nabla u| + \Sigma(u) = \|D\widetilde{\psi}\|_{\mathscr{M}}$. Consequently, $E(u) \leq \int |\nabla u| + \Sigma(u)$. On the other hand, $E(u) \geq \int |\nabla u| + \Sigma(u)$ (by Theorem 1.6), and thus (1.87) holds. Next, let φ be a minimal lifting. Then, $u = e^{\imath\varphi}$ and $\|D\varphi\|_{\mathscr{M}} = E(u) = \int |\nabla u| + \Sigma(u)$. Applying Theorem 1.6 once more, we obtain $\|D\varphi\|_{\mathscr{M}} = \int |\nabla u| + \|u \wedge \nabla u - D\varphi\|_{\mathscr{M}}$. Thus, $\|u \wedge \nabla u - D\varphi\|_{\mathscr{M}} = \Sigma(u)$, i.e., (1.88) holds.
10. In [161, Section 7.9], [163], Giaquinta and Mucci revisited Theorems 1.5, 1.9, and 1.11 within the framework of Cartesian currents as developed in Giaquinta, Modica, and Souček [156], [157]. In particular, they study the following quantities, which we present for simplicity only when $N = 2$.
Let $\Omega = \mathbb{D}$ be the unit disk. Fix some smooth boundary condition $g : \partial\Omega \to \mathbb{S}^1$ such that

$$\deg g = 0. \tag{1.192}$$

Set, for $u \in W^{1,1}_g(\Omega; \mathbb{S}^1)$,

$$A := \inf \left\{ |\varphi|_{BV};\ \varphi \in BV(\Omega; \mathbb{R}) \text{ is such that } u = e^{\imath\varphi} \text{ in } \Omega \right\};$$

$$B := \inf \left\{ \liminf_{n\to\infty} \int_\Omega |\nabla u_n|;\ u_n \in C^1_g(\overline{\Omega}; \mathbb{S}^1),\ u_n \to u \text{ a.e.} \right\};$$

$$C := \min_{\psi \in BV(\Omega;\mathbb{R})} \|u \wedge \nabla u - D\psi\|_{\mathscr{M}};$$

$$D := \sup \left\{ -\int_\Omega (u \wedge \nabla u) \cdot \nabla^\perp \zeta + \int_{\partial\Omega} (g \wedge \partial_\tau g)\, \zeta;\ \zeta \in \mathrm{Lip}(\overline{\Omega}), \right.$$
$$\left. \|\nabla \zeta\|_{L^\infty} \leq 1 \right\}.$$

Here, ∂_τ stands for the tangential derivative with the standard (counterclockwise) orientation on $\mathbb{S}^1$.
In Giaquinta and Mucci [161, Corollary 7.64, Theorem 7.66, top of p. 375], it is claimed that

$$A = B = \int_\Omega |\nabla u| + C = \int_\Omega |\nabla u| + D. \tag{1.193}$$

Part of (1.193) is not quite correct. More specifically (as a consequence of the results in this chapter), we always have

$$A = \int_\Omega |\nabla u| + C,$$

while clearly

$$A \le B$$

and

$$C \le D.$$

The assertion

$$B = \int_\Omega |\nabla u| + D \tag{1.194}$$

is also correct, as explained in Brezis and Mironescu [90].
However, in general we have $C < D$ (and thus $A < B$). Here is an example. Take

$$u(z) := \frac{z-a}{|z-a|}\frac{\overline{z-b}}{|z-b|}, \quad \text{with } a, b \text{ distinct points in } \Omega.$$

Note that $g := u_{|\partial\Omega}$ satisfies condition (1.192). By Theorem 2.5, we know that

$$C = \Sigma(u) = 2\pi \min\{\mathrm{dist}(a, \partial\Omega) + \mathrm{dist}(b, \partial\Omega), |a-b|\}. \tag{1.195}$$

On the other hand, the first proof of Theorem 2.1 shows that

$$-\int_\Omega (u \wedge \nabla u) \cdot \nabla^\perp \zeta + \int_{\partial\Omega} (g \wedge \partial_\tau g)\, \zeta = 2\pi(\zeta(a) - \zeta(b)). \tag{1.196}$$

This implies that

$$D = 2\pi\, |b-a|; \tag{1.197}$$

indeed, using (1.196), we find at once that $D \le 2\pi\, |a-b|$. With the choice $\zeta(z) = |z-b|$, we see that $D \ge 2\pi\, |a-b|$.
Let now a be close to -1 and b be close to 1. By (1.195) and (1.197), for such a and b we have $C < D$.
As explained in [90], formula (1.193) has to be amended as follows. Let $\eta \in C^\infty(\partial\Omega; \mathbb{R})$ be such that $g = e^{\imath\eta}$. Such an η exists, by (1.192). Then we have

$$A' = B = \int_\Omega |\nabla u| + C' = \int_\Omega |\nabla u| + D, \tag{1.198}$$

where

$$A' := \inf \left\{ |\varphi|_{BV};\ \varphi \in BV_\eta(\Omega; \mathbb{R}) \text{ is such that } u = e^{\imath\varphi} \text{ in } \Omega \right\};$$

and

$$\begin{aligned} C' :&= \inf_{\psi \in BV(\Omega;\mathbb{R})} \left\{ \|u \wedge \nabla u - D\psi\|_{\mathscr{M}} + \int_{\partial\Omega} |\operatorname{tr} \psi - \eta| \right\} \\ &= \inf_{\psi \in W^{1,1}_\eta(\Omega;\mathbb{R})} \|u \wedge \nabla u - \nabla\psi\|_{L^1}. \end{aligned}$$

11. Theorem 1.8 with $N = 2$ is from Brezis, Mironescu, and Ponce [92]. The first result of this type was obtained, for $H^{1/2}(\mathbb{D}; \mathbb{S}^1)$ maps, in Bourgain, Brezis, and Mironescu [54].
In Section 1.4, we have not discussed any constructive approach for generating minimal liftings. This will be done in Sections 2.7 and 2.8 (for nice u's in two dimensions), in Section 2.9.3 (for general u's in two dimensions), and in Sections 3.6, 3.8.1, and 3.8.2 (in three and higher dimensions). We discuss here another construction, which relies on approximation. A possible way to approximate φ is to consider, for a fixed parameter $1 \le p < \infty$, the family of energy functionals defined on $H^1(\Omega)$ by

$$\mathscr{J}_\varepsilon(\varphi) = \int_\Omega \left(\varepsilon|\nabla\varphi|^2 + \frac{1}{\varepsilon}|u - e^{\imath\varphi}|^{2p} \right) dx,\ \forall\, \varepsilon > 0. \tag{1.199}$$

Denoting by φ_ε a minimizer of $\mathscr{J}_\varepsilon$, one may hope that, owing to the presence of the penalizing term, the family (φ_ε) will converge, as ε goes to 0 and up to a subsequence, to a minimal lifting of u. Poliakovsky [294] studied this problem within the framework of Γ-convergence. He proved that in the case of $u \in W^{1,1}(\Omega; \mathbb{S}^1)$, for any sequence $\varepsilon_j \to 0$ there exists a subsequence satisfying $\varphi_{\varepsilon_{j_k}} \to \varphi_0$ in $L^1(\Omega)$, where φ_0 is a minimal lifting of u. The same result holds for $u \in BV(\Omega; \mathbb{S}^1)$, provided that $p = 2$. In the case of a general $u \in BV(\Omega; \mathbb{S}^1)$ and $p \ne 2$, a subsequence $(\varphi_{\varepsilon_{j_k}})$ still converges to a limit φ_0 in $L^1(\Omega)$, and the limit φ_0 is characterized as a minimizer of a certain functional [294]. But an example due to Ignat and Poliakovsky [199] shows that the limit φ_0 is not necessarily a minimal lifting.
12. Corollary 1.6 and the related developments presented in Section 1.9.2 are new. Equality (1.148) in Corollary 1.6 is originally due to Shafrir [321], and his proof relies on a different version of the Hahn–Banach theorem. More precisely, we start from the equality

$$\operatorname{dist}(p, M) = \inf_{m \in M} \|p - m\| = \max\{\langle \xi, p\rangle;\ \xi \in M^\perp,\ \|\xi\| \le 1\},$$

where E is a Banach space, $p \in E$, and M is a linear subspace of E (see, for example, Brezis [75, Section 1.4, Example 3]). If we take $E := L^1(\Omega; \mathbb{R}^N)$, $p := u \wedge \nabla u$, $M := \{\nabla\varphi;\ \varphi \in W^{1,1}(\Omega; \mathbb{R})\}$, then we have

$$M^\perp = \{\xi \in L^\infty(\Omega; \mathbb{R}^N);\ \operatorname{div} \xi = 0 \text{ in } \Omega \text{ and } \xi \cdot \nu = 0 \text{ on } \partial\Omega\},$$

where ν is the outward normal to $\partial\Omega$, and we obtain immediately (1.148). This simple and illuminating observation of Shafrir (with roots in Brezis, Coron, and Lieb [79]) paved the way for the sharper results described in Sections 1.5 and 1.9.2.

13. Corollary 1.7 is inspired by estimates of Brezis and Nguyen [96, Theorem 1] concerning Jacobian determinants.
14. The results in Section 1.6 are new. The usefulness of the MK theory as a *tool* in the study of energy-minimizing problems for $\mathbb{S}^k$-valued maps was originally demonstrated in Brezis, Coron, and Lieb [79] (see also Sections 2.2–2.3 and Comments 2–3 in Chapter 2). Here, the goal is different – we do not apply any result from the MK theory. Instead, we show that there are striking "structural" analogies between some aspects of the MK theory and some topics arising in the analysis of energy-minimizing $\mathbb{S}^1$-valued maps. In Chapters 2 and 3, we will present similar analogies with some area-minimizing issues in geometric measure theory.
15. The concept of relaxed energy occurs in a variety of distinct situations in the calculus of variations. Here are some of the most typical ones.

   a) Let $\Phi = \Phi(x, s, z) : \Omega \times \mathbb{R} \times \mathbb{R}^N \to \mathbb{R}$ be a (smooth) function such that

$$\alpha\, |z|^p - c \leq \Phi(x, s, z) \leq \beta\, |z|^p + c, \ \forall x,\ s,\ z, \tag{1.200}$$

for some constants $\alpha > 0$, $\beta > 0$, $c > 0$ and $p > 1$. Set

$$F(u) := \int_\Omega \Phi(x, u(x), \nabla u(x)), \ \forall\, u \in W^{1,p}(\Omega). \tag{1.201}$$

The minimization problem

$$\inf_{u \in W^{1,p}(\Omega)} F(u) \tag{1.202}$$

does not necessarily admit a minimizer, since F need not be lower semicontinuous for the weak sequential convergence in $W^{1,p}$. It is customary (see, for example, Buttazzo [110]) to introduce the *relaxed energy* defined on $W^{1,p}(\Omega)$ by

$$\overline{F}(u) := \inf\{\liminf_{n\to\infty} F(u_n);\ u_n \in W^{1,p}(\Omega),\ u_n \to u \text{ a.e.}\}.$$

Clearly,

$$\inf_{u \in W^{1,p}(\Omega)} F(u) = \min_{u \in W^{1,p}(\Omega)} \overline{F}(u), \tag{1.203}$$

and also

$$\inf_{u \in W^{1,p}(\Omega)} F(u) = \inf_{u \in C^\infty(\overline{\Omega})} F(u). \tag{1.204}$$

A very natural question consists in finding an integral representation formula, i.e.,

$$\overline{F}(u) = \int_{\Omega} \overline{\Phi}(x, u(x), \nabla u(x))\, dx$$

for some $\overline{\Phi}$. There is a vast body of literature on this subject. Similarly when $W^{1,p}$ is replaced by $W_0^{1,p}$.

b) Assume now that (1.200) holds with $p = 1$. Clearly, F is well-defined on $W^{1,1}$ (but not necessarily on BV). Here, it is customary to introduce the relaxed energy defined on BV by

$$\overline{F}(u) := \inf\{\liminf_{n\to\infty} F(u_n);\ u_n \in W^{1,1}(\Omega),\ u_n \to u \text{ a.e.}\}.$$

In this context, formula (1.203) becomes

$$\inf_{u\in W^{1,1}(\Omega)} F(u) = \min_{u\in BV(\Omega)} \overline{F}(u),$$

and (1.204) still holds with $p = 1$.
Again, one may try to find an integral representation for $\overline{F}$. A famous example related to minimal surfaces is $\Phi(x, s, z) := \sqrt{1 + |z|^2}$. It was originally investigated by Goffman and Serrin [169], who established that

$$\overline{F}(u) = \int_{\Omega} \sqrt{1 + |\nabla u|^2} + \|D^j u + D^c u\|_{\mathscr{M}},\ \forall\, u \in BV;$$

here, ∇u, $D^j u$, and $D^c u$ are as in (1.104).

c) A third example is related to the celebrated Lavrentiev gap, which corresponds to the following phenomenon: there exist some Φ's such that

$$\inf_{u\in W_0^{1,1}} F(u) < \inf_{u\in W_0^{1,\infty}} F(u) = \inf_{u\in C_c^{\infty}(\Omega)} F(u).$$

Such a phenomenon may occur even when Φ is convex in z (see Buttazzo and Belloni [111]).
In the process of investigating this phenomenon, one is led to a relaxed functional $\overline{F}$, defined for $u \in W_0^{1,1}(\Omega)$ by

$$\overline{F}(u) := \inf\{\liminf_{n\to\infty} F(u_n);\ u_n \in W_0^{1,\infty},\ u_n \to u \text{ in } \mathscr{T}\},$$

where the topology $\mathscr{T}$ might be weak $W^{1,1}$, weak $W^{1,q}$ for some appropriate $1 \le q < \infty$, strong L^1, etc. In this direction, see, for example, Acerbi, Bouchitté, and Fonseca [1], Buttazzo and Belloni [111], Buttazzo and Mizel [112], Mingione and Mucci [251], and the abundant literature cited therein.

d) Another type of relaxed energy that has attracted much attention in recent years corresponds to manifold-valued maps. The starting point was the following surprising "Lavrentiev phenomenon" discovered by Hardt and Lin [182]: there exists a smooth boundary condition $g : \mathbb{S}^2 \to \mathbb{S}^2$ such that $\deg g = 0$ and

$$\inf\left\{\int_{\mathbb{B}^3} |\nabla u|^2;\ u \in H^1_g(\mathbb{B}^3; \mathbb{S}^2)\right\} < \inf\left\{\int_{\mathbb{B}^3} |\nabla u|^2;\ u \in C^1_g(\overline{\mathbb{B}^3}; \mathbb{S}^2)\right\};$$

here, $\mathbb{B}^3$ is the unit ball in $\mathbb{R}^3$.
This example led Bethuel, Brezis, and Coron [30] to introduce the relaxed energy defined, for $u \in H^1_g(\mathbb{B}^3; \mathbb{S}^2)$, by

$$R_{2,g}(u) := \inf\left\{\liminf_{n\to\infty} \int_{\mathbb{B}^3} |\nabla u_n|^2;\ u_n \in C^1_g(\overline{\mathbb{B}}^3; \mathbb{S}^2),\ u_n \to u \text{ a.e.}\right\}.$$

They gave an explicit formula for $R_{2,g}(u)$, in the spirit of (4.49) . As we are going to see, there is a similar gap phenomenon for $\mathbb{S}^1$-valued maps and the $W^{1,p}$ energy with $1 \le p < 2$ (see Section 13.3). This leads naturally to the relaxed energy for $W^{1,1}(\Omega; \mathbb{S}^1)$ maps, given by

$$R_{1,g}(u) := \inf\left\{\liminf_{n\to\infty} \int_{\Omega} |\nabla u_n|;\ u_n \in C^1_g(\overline{\Omega}; \mathbb{S}^1),\ u_n \to u \text{ a.e.}\right\}.$$

Explicit formulas for $R_{1,g}(u)$ are discussed in Comment 9 above.

16. Theorem 1.11 with $N = 2$ appears in Brezis, Mironescu, and Ponce [92]. The argument there applies in any dimension.
17. With a different but equivalent formulation, Theorem 1.13 is due to Bethuel [26]; the formulation we present appears in Giaquinta and Mucci [161].
18. The first explicit formula for the least energy of maps having prescribed topological singularities, in the spirit of Section 1.8, was obtained by Brezis, Coron, and Lieb [79] (for maps in $H^1(\mathbb{B}^3; \mathbb{S}^2)$). We will come back to this in Chapter 4.

Chapter 2
The geometry of *Ju* and $\Sigma(u)$ in 2D; point singularities and minimal connections

2.0 Introduction

Throughout this chapter, $\Omega \subset \mathbb{R}^2$ is smooth, bounded, and simply connected. Most of this chapter is concerned with *nice* maps, i.e., maps $u \in W^{1,1}(\Omega; \mathbb{S}^1)$ that are continuous on Ω except at a finite number of points, say $a_1, \ldots, a_k$. With no loss of generality, we assume throughout this chapter that

$$a_\ell \neq a_m, \ \forall \, \ell \neq m. \tag{2.1}$$

We give geometrical interpretations of the main analytical objects associated with u in Chapter 1: the quantity $E(u)$, the distribution $Ju \in \mathscr{D}'(\Omega; \mathbb{R})$, and the quantity

$$\Sigma(u) = \min_{\varphi \in BV(\Omega;\mathbb{R})} \|u \wedge \nabla u - D\varphi\|_{\mathscr{M}}. \tag{2.2}$$

Recall that

$$\Sigma(u) = \inf\left\{ \int_\Omega |\nabla v|; \ v \in W^{1,1}(\Omega; \mathbb{S}^1), \ Jv = Ju \text{ in } \mathscr{D}'(\Omega) \right\}, \tag{2.3}$$

by Lemma 1.14.

Among other results, we show that Ju has a *very special* structure and that $\Sigma(u)$ can be computed explicitly. More precisely, the main result in Section 2.1 asserts that

$$Ju = \pi \sum_{j=1}^{k} d_j \delta_{a_j} \text{ in } \mathscr{D}'(\Omega), \tag{2.4}$$

where $d_j := \deg(u, a_j)$ is the degree (=winding number=Cauchy index) of u computed on a small circle around a_j. When $d_j \neq 0$, we say that a_j is a *topological*

H. Brezis and P. Mironescu, *Sobolev Maps to the Circle*, Progress in Nonlinear Differential Equations and Their Applications 96,
https://doi.org/10.1007/978-1-0716-1512-6_2

singularity of u. Formula (2.4) shows that Ju captures only the location and charge of the topological singularities.

In Section 2.2, we set aside for a while the original nice map u and concentrate on general pairs (a, d), where a denotes a collection $(a_1, \dots, a_k)$ of distinct points in Ω, and $d = (d_1, \dots, d_k)$ is a collection of integers in $\mathbb{Z}$. We introduce the quantity $L(a, d)$, which plays an important role throughout Chapter 2. We first define a pseudometric d on $\overline{\Omega}$; given $P, N \in \overline{\Omega}$, set

$$d(P, N) := \min\{|P - N|, \operatorname{dist}(P, \partial\Omega) + \operatorname{dist}(N, \partial\Omega)\}. \tag{2.5}$$

In particular, if $N \in \partial\Omega$ (resp. $P \in \partial\Omega$), then $d(P, N) = \operatorname{dist}(P, \partial\Omega)$ (resp. $d(P, N) = \operatorname{dist}(N, \partial\Omega)$), so that

$$d(P, N_1) = d(P, N_2), \ \forall P \in \overline{\Omega}, \ \forall N_1, N_2 \in \partial\Omega,$$

and similarly

$$d(P_1, N) = d(P_2, N), \ \forall P_1, P_2 \in \partial\Omega, \ \forall N \in \overline{\Omega}.$$

We then repeat the points a_i, $i = 1, \dots, k$, as $|d_i|$ "positive" points if $d_i > 0$, resp. $|d_i|$ "negative" points if $d_i < 0$. We omit from this list the points a_i with $d_i = 0$. If

$$\alpha := \sum_{d_i > 0} |d_i| - \sum_{d_i < 0} |d_i| = \sum d_i > 0,$$

we add to the list α negative points on $\partial\Omega$ (chosen arbitrarily). If $\alpha < 0$, we add to the list $|\alpha|$ positive points on $\partial\Omega$ (chosen arbitrarily). We rewrite these repeated points as a collection of m positive points (P_ℓ) in $\overline{\Omega}$ and m negative points (N_ℓ) in $\overline{\Omega}$, with $1 \le \ell \le m$ and $m := \max\left\{\sum_{d_i>0} |d_i|, \sum_{d_i<0} |d_i|\right\}$. Finally, we set

$$L(a, d) := \min_{\sigma \in S_m} \sum_{\ell=1}^{m} d(P_\ell, N_{\sigma(\ell)}), \tag{2.6}$$

where the minimum is taken over the set S_m of all the permutations of the integers $\{1, \dots, m\}$.

Here are a few examples.

Example 2.1. Let $a := (a_1)$ and $d := (+1)$. Then

$$L(a, d) = \operatorname{dist}(a_1, \partial\Omega).$$

Example 2.2. Let $a := (a_1, a_2)$ and $d := (+1, +1)$. Then

$$L(a, d) = \operatorname{dist}(a_1, \partial\Omega) + \operatorname{dist}(a_2, \partial\Omega).$$

Example 2.3. Let $a := (a_1, a_2)$ and $d := (+1, -1)$. Then

$$L(a, d) = \min\{|a_1 - a_2|, \operatorname{dist}(a_1, \partial\Omega) + \operatorname{dist}(a_2, \partial\Omega)\} = d(a_1, a_2).$$

Returning to the definition (2.6) of $L(a, d)$, we may write, up to a permutation, $L(a, d) = \sum_{\ell=1}^{m} d(P_\ell, N_\ell)$ for some collection $(P_\ell, N_\ell)_{\ell=1}^{m}$ of points in $\overline{\Omega}$. In view of (2.5), we have

$$d(P_\ell, N_\ell) = \min\{|P_\ell - N_\ell|, \operatorname{dist}(P_\ell, \partial\Omega) + \operatorname{dist}(N_\ell, \partial\Omega)\}.$$

We may thus always add points in $\partial\Omega$ to the collection $(P_\ell, N_\ell)_{\ell=1}^{m}$ and obtain a larger collection $(P_\ell, N_\ell)_{\ell=1}^{n}$ of points such that

$$L(a, d) = \sum_{\ell=1}^{n} |P_\ell - N_\ell|, \tag{2.7}$$

$$\sum_{\ell=1}^{n} \delta_{P_\ell} = \sum_{j=1}^{k} d_j^+ \, \delta_{a_j} \text{ in } \mathscr{D}'(\Omega), \tag{2.8}$$

and

$$\sum_{\ell=1}^{n} \delta_{N_\ell} = \sum_{j=1}^{k} d_j^- \, \delta_{a_j} \text{ in } \mathscr{D}'(\Omega). \tag{2.9}$$

In particular,

$$\sum_{\ell=1}^{n} \left(\delta_{P_\ell} - \delta_{N_\ell}\right) = \sum_{j=1}^{k} d_j \, \delta_{a_j} \text{ in } \mathscr{D}'(\Omega). \tag{2.10}$$

In addition, by construction we also have

$$\begin{aligned}&\text{for every } \ell = 1, \ldots, n,\ P_\ell \neq N_\ell, \text{ and the open interval } (P_\ell, N_\ell)\\ &\text{is contained in } \Omega.\end{aligned} \tag{2.11}$$

Note that the measures $\sum_{\ell=1}^{n} \delta_{P_\ell}$ and $\sum_{\ell=1}^{n} \delta_{N_\ell}$ do not "see" points in $\partial\Omega$, since (2.8) and (2.9) hold in the sense of $\mathscr{D}'(\Omega)$.

Any collection $(P_\ell, N_\ell)_{\ell=1}^{n}$ of line segments in $\overline{\Omega}$ satisfying (2.7), (2.8), (2.9), and (2.11) is called a *minimal configuration* associated with (a, d). In general, there are many distinct minimal configurations associated with (a, d). For a more complete discussion of minimal configurations, see Section 2.2.

The right-hand side of (2.6) has a flavor of OT. The main result in Section 2.2 (Theorem 2.3), which asserts that

$$L(a,d) = \max\left\{\sum_{j=1}^{k} d_j\,\zeta(a_j);\ \zeta \in W_0^{1,\infty}(\Omega),\ \|\nabla\zeta\|_{L^\infty} \le 1\right\}, \tag{2.12}$$

is closely related to the "Kantorovich duality" in OT theory, as explained in Section 2.12 (Comment 2).

In Section 2.3, we return to a nice u (as in Section 2.1) and prove (see Theorem 2.5) that

$$\Sigma(u) = 2\pi L(a,d). \tag{2.13}$$

This is an immediate consequence of three ingredients:

a) The fundamental formula giving $\Sigma(u)$ in terms of Ju when $N = 2$:

$$\Sigma(u) = 2\sup\left\{\langle Ju, \zeta\rangle;\ \zeta \in C_c^\infty(\Omega;\mathbb{R}),\ \|\nabla\zeta\|_{L^\infty} \le 1\right\}; \tag{2.14}$$

see Theorem 1.9 (whose proof relies on Hahn–Banach).

b) The fundamental formula

$$Ju = \pi\sum_{j=1}^{k} d_j\,\delta_{a_j} \text{ in } \mathscr{D}'(\Omega), \tag{2.15}$$

established in Section 2.1.

Combining (2.14) and (2.15) with a density argument (see Remark 1.30) we find that

$$\Sigma(u) = 2\pi\max\left\{\sum_{j=1}^{k} d_j\,\zeta(a_j);\ \zeta \in W_0^{1,\infty}(\Omega;\mathbb{R}),\ \|\nabla\zeta\|_{L^\infty} \le 1\right\}. \tag{2.16}$$

c) Finally, we invoke the OT formula (2.12) in conjunction with (2.16) and conclude that (2.13) holds.

In the next two sections, we present a totally different approach for the proof of Theorem 2.5, i.e., $\Sigma(u) = 2\pi L(a,d)$, which does *not* make use of the OT technology described in Section 2.2. This route is definitely longer (and more delicate) than the one presented above, but it involves two important geometric tools, the dipole construction and the coarea formula, which are useful in a variety of situations.

In Section 2.5, we establish the inequality $\Sigma(u) \le 2\pi L(a,d)$ via (2.3). More precisely, we exhibit, for any $\varepsilon > 0$, an *explicit* map $v \in W^{1,1}(\Omega;\mathbb{S}^1)$ such that $Jv = Ju$ and

$$\int_\Omega |\nabla v| \le 2\pi L(a,d) + \varepsilon.$$

This is the *dipole construction*.

In Section 2.4, we prove the reverse inequality $\Sigma(u) \geq 2\pi L(a, d)$. More precisely, we invoke the *coarea formula*

$$\int_\Omega |\nabla u| = \int_{\mathbb{S}^1} \mathscr{H}^1([u = \alpha])\, d\alpha$$

and establish that $\mathscr{H}^1([u = \alpha]) \geq L(a, d)$ for a.e. $\alpha \in \mathbb{S}^1$.

In Sections 2.6, 2.7, and 2.8, we return to *BV* liftings and minimal liftings of u and to the quantity

$$E(u) = \min\{|\varphi|_{BV};\ \varphi \in BV(\Omega; \mathbb{R}) \text{ is such that } u = e^{\imath\varphi} \text{ in } \Omega\},$$

extensively studied in Chapter 1.

In Chapter 1, we have established in full generality that

$$E(u) = \int_\Omega |\nabla u| + \Sigma(u)$$

(see Theorem 1.5). When u is nice, we know (see Theorem 2.3) that $\Sigma(u) = 2\pi L(a, d)$, and therefore

$$E(u) = \int_\Omega |\nabla u| + 2\pi L(a, d).$$

In Section 2.6, we present a *totally self-contained* and elementary proof of the inequality

$$E(u) \leq \int_\Omega |\nabla u| + 2\pi L(a, d), \tag{2.17}$$

when u is nice. For this purpose we construct a *natural* lifting $\varphi \in BV$ of u such that

$$|\varphi|_{BV} \leq \int_\Omega |\nabla u| + 2\pi L(a, d). \tag{2.18}$$

In the process we make use *only* of the formula

$$Ju = \pi \sum_j d_j \delta_{a_j} \text{ in } \mathscr{D}'(\Omega) \tag{2.19}$$

(see Theorem 2.1) and the concept of minimal configuration associated with (a, d).

More precisely, given a minimal configuration $(P_\ell, N_\ell)_{\ell=1}^n$ associated with (a, d), we exhibit a unique (mod 2π) minimal lifting $\varphi \in BV(\Omega; \mathbb{R})$ of u that is continuous on $\Omega \setminus \cup_{\ell=1}^n [P_\ell, N_\ell]$ and that jumps along $\bigcup_{\ell=1}^n [P_\ell, N_\ell]$, and satisfies (2.18) (see Theorem 2.6).

It turns out that *all* minimal liftings of u arise from such a construction. In other words, we have a full catalog of *all minimal liftings* of u once we know *all minimal configurations* associated with (a, d). The "cherry on the cake" is a *one-to-one correspondence* between analytical objects (minimal liftings) and geometric objects (minimal connections – a close "relative" of minimal configurations), which is made precise in Sections 2.7 and 2.8. The proof of this seemingly intuitive fact is non-trivial (see Sections 2.7, 2.8, and Appendix 15.8).

In Section 2.9, we extend some of the above results to the case of a *general* $u \in W^{1,1}(\Omega; \mathbb{S}^1)$. In particular, we show that, for every $u \in W^{1,1}(\Omega; \mathbb{S}^1)$, there exist sequences of points (P_j), $(N_j) \subset \overline{\Omega}$ such that

$$Ju = \pi \sum_{j=1}^{\infty} (\delta_{P_j} - \delta_{N_j}) \text{ in } \mathscr{D}'(\Omega) \tag{2.20}$$

and $\sum\limits_{j=1}^{\infty} |P_j - N_j| < \infty$; moreover

$$\Sigma(u) = 2\pi \inf \left\{ \sum_{j=1}^{\infty} |\widetilde{P}_j - \widetilde{N}_j|; \widetilde{P}_j, \widetilde{N}_j \in \overline{\Omega}, \sum_{j=1}^{\infty} |\widetilde{P}_j - \widetilde{N}_j| < \infty \right. \\ \left. \text{and } \sum_{j=1}^{\infty} (\delta_{\widetilde{P}_j} - \delta_{\widetilde{N}_j}) = \frac{1}{\pi} Ju \text{ in } \mathscr{D}'(\Omega) \right\}. \tag{2.21}$$

From (2.20) and (2.21), we deduce that $\Sigma(u) \le 2\pi \sum_{j=1}^{\infty} d(P_j, N_j)$. Moreover, for every $\varepsilon > 0$ it is possible to find (P_j), (N_j) such that (2.20) holds and

$$2\pi \sum_{j=1}^{\infty} d(P_j, N_j) \le \Sigma(u) + \varepsilon.$$

Unlike the case of nice u's, the infimum in (2.21) need not be achieved, i.e., it may be impossible to find $(\widetilde{P}_j, \widetilde{N}_j)$ as in (2.21) such that $\Sigma(u) = 2\pi \sum_{j=1}^{\infty} |\widetilde{P}_j - \widetilde{N}_j|$.

In Section 2.10, we present an integral formula for $\langle Ju, \zeta \rangle$. When u is nice and $\zeta \in C_c^\infty(\Omega; \mathbb{R})$, for a.e. $t \in \mathbb{R}$ the level set $[\zeta = t]$ consists of a finite union of smooth closed curves that do not contain the singularities of u. For such t, we may thus define $\deg(u, [\zeta = t])$. We prove that

$$\langle Ju, \zeta \rangle = \pi \int_{\mathbb{R}} \deg(u, [\zeta = t])\, dt \tag{2.22}$$

(see Theorem 2.15). We also extend (2.22) to general $u \in W^{1,1}(\Omega; \mathbb{S}^1)$ and $\zeta \in W_0^{1,\infty}(\Omega; \mathbb{R})$.

2.1 *Ju* as a sum of Dirac masses

Let $u \in W^{1,1}(\Omega; \mathbb{S}^1)$. Recall (see (1.35)) that, when $N = 2$, Ju is identified with the *scalar* distribution

$$Ju = \frac{1}{2}\left[\frac{\partial}{\partial x_1}\left(u_1\frac{\partial u_2}{\partial x_2} - u_2\frac{\partial u_1}{\partial x_2}\right) - \frac{\partial}{\partial x_2}\left(u_1\frac{\partial u_2}{\partial x_1} - u_2\frac{\partial u_1}{\partial x_1}\right)\right], \tag{2.23}$$

which belongs to $[W_0^{1,\infty}(\Omega; \mathbb{R})]^*$; moreover (see (1.152)),

$$\|Ju\|_W = \max\left\{\frac{1}{2}\int_\Omega (u \wedge \nabla u) \cdot \nabla^\perp \zeta,\ \zeta \in W_0^{1,\infty}(\Omega; \mathbb{R}),\ \|\nabla\zeta\|_{L^\infty} \le 1\right\}, \tag{2.24}$$

where $\nabla^\perp \zeta := (-\partial\zeta/\partial x_2, \partial\zeta/\partial x_1)$.

We start with an extension of the formula $Ju = \pi\delta_0$ for $u(x) = \dfrac{x}{|x|}$, discussed in Remark 1.9.

Theorem 2.1. *Let $u \in W^{1,1}(\Omega; \mathbb{S}^1) \cap C(\Omega \setminus \{a\})$, where $a \in \Omega$. Then*

$$Ju = \pi d \delta_a \text{ in } \mathscr{D}'(\Omega). \tag{2.25}$$

Here, $d :=$ deg(u, a) is the degree (=winding number=Cauchy index) of u computed on a small circle around a.

More generally, if $u \in W^{1,1}(\Omega; \mathbb{S}^1) \cap C(\Omega \setminus \{a_1, \dots, a_k\})$, where $a_1, \dots, a_k \in \Omega$, then

$$Ju = \pi \sum_j d_j \delta_{a_j} \text{ in } \mathscr{D}'(\Omega). \tag{2.26}$$

Here, $d_j :=$ deg(u, a_j).

Remark 2.1. The fact that deg(u, a_j) is well-defined, i.e., deg$(u, C(a_j, r))$ does not depend on small $r > 0$, follows from Lemma 1.2 applied to the family of maps $\mathbb{S}^1 \ni x \mapsto u_r(x) := u(a_j + rx)$, where $r > 0$ is small.

Remark 2.2. Equality (2.26) asserts that

$$-\int_\Omega (u \wedge \nabla u) \cdot \nabla^\perp \zeta = 2\pi \sum_j d_j \zeta(a_j), \ \forall \zeta \in C_c^\infty(\Omega; \mathbb{R}).$$

In fact, by Remark 1.30, a stronger statement holds, i.e.,

$$-\int_{\Omega}(u \wedge \nabla u) \cdot \nabla^{\perp}\zeta = 2\pi \sum_j d_j \zeta(a_j), \ \forall \zeta \in W_0^{1,\infty}(\Omega; \mathbb{R}). \tag{2.27}$$

First proof of Theorem 2.1. We start with the case where Ω is a disk centered at a_1, say $\Omega = \mathbb{D}$ and $a_1 = 0$. Let $\psi \in C_c^\infty(\mathbb{D})$. We need to show that

$$\langle Ju, \psi\rangle = \pi\, d\, \psi(0). \tag{2.28}$$

Since $u \in W^{1,1} \cap C(\mathbb{D} \setminus \{0\})$, we already know that

$$Ju = 0 \text{ in } \mathscr{D}'(\mathbb{D} \setminus \{0\}). \tag{2.29}$$

This is a consequence of Corollary 1.1 applied in $\mathbb{D} \setminus \{0\}$. [The fact that $\mathbb{D} \setminus \{0\}$ is not simply connected is irrelevant, since we have $Ju = 0$ in every disk $\omega \subset \mathbb{D} \setminus \{0\}$.]

Fix a smooth function $\gamma : [0, \infty) \to \mathbb{R}$ such that $\gamma(t) = \begin{cases} 1, & \text{for } t \le 1 \\ 0, & \text{for } t \ge 2 \end{cases}$ and set $\gamma^\varepsilon(x) := \gamma(|x|/\varepsilon)$, $x \in \mathbb{R}^2$. Using (2.29), we have

$$\langle Ju, \psi\rangle = \langle Ju, \psi\gamma^\varepsilon\rangle + \langle Ju, \psi(1-\gamma^\varepsilon)\rangle = \langle Ju, \psi\gamma^\varepsilon\rangle. \tag{2.30}$$

On the other hand, we have

$$\begin{aligned} 2\langle Ju, \psi\gamma^\varepsilon\rangle &= -\int_{\mathbb{D}}(u \wedge \nabla u) \cdot \nabla^{\perp}(\psi\gamma^\varepsilon) \\ &= -\int_{\mathbb{D}}(u \wedge \nabla u) \cdot (\psi\nabla^{\perp}\gamma^\varepsilon + \gamma^\varepsilon\nabla^{\perp}\psi). \end{aligned} \tag{2.31}$$

Next, we observe that

$$\left|\int_{\mathbb{D}}(u \wedge \nabla u) \cdot (\gamma^\varepsilon\nabla^{\perp}\psi)\right| \le C\int_{|x|<2\varepsilon}|\nabla u| \to 0 \text{ as } \varepsilon \to 0. \tag{2.32}$$

Similarly,

$$\int_{\mathbb{D}}(u \wedge \nabla u) \cdot (\psi\nabla^{\perp}\gamma^\varepsilon) = \psi(0)\int_{\mathbb{D}}(u \wedge \nabla u) \cdot \nabla^{\perp}\gamma^\varepsilon + o(1) \text{ as } \varepsilon \to 0. \tag{2.33}$$

Combining (2.30), (2.31), (2.32), and (2.33), we obtain

$$\langle Ju, \psi\rangle = -\frac{1}{2}\psi(0)\int_{\mathbb{D}}(u \wedge \nabla u) \cdot \nabla^{\perp}\gamma^\varepsilon + o(1) \text{ as } \varepsilon \to 0. \tag{2.34}$$

On the other hand, we have in polar coordinates

$$\langle Ju, \psi \rangle = -\frac{1}{2}\psi(0) \int_{r=\varepsilon}^{r=2\varepsilon} \int_{C(0,r)} (u \wedge \nabla u) \cdot \nabla^{\perp} \gamma^{\varepsilon} \, dr \, d\ell + o(1) \text{ as } \varepsilon \to 0. \tag{2.35}$$

But

$$(u \wedge \nabla u) \cdot \nabla^{\perp} \gamma^{\varepsilon}(x) = \frac{1}{\varepsilon} \left(u \wedge \frac{\partial u}{\partial \tau}(x) \right) \gamma'(|x|/\varepsilon), \tag{2.36}$$

where $\dfrac{\partial}{\partial \tau}$ stands for the tangential differentiation on the circle $C(0, r)$ with its standard (counterclockwise) orientation. Since $u \in W^{1,1}(C(0, r); \mathbb{S}^1)$ for a.e. r, we may apply Lemma 1.4 and assert that

$$\int_{C(0,r)} u \wedge \frac{\partial u}{\partial \tau} \, d\ell = 2\pi \deg(u, C(0, r)) \text{ for a.e. } r. \tag{2.37}$$

Moreover, the degree $\deg(u, C(0, r))$ is independent of r (see Remark 2.1). From (2.35), (2.36), and (2.37), we deduce that (2.28) holds.

In the general case, let $B_j := B(a_j, r_j)$, $j = 1, \dots, k$, be disjoint disks contained in Ω. Set $B_{k+1} := \Omega \setminus \cup_{j=1}^{k} \overline{B}(a_j, r_j/2)$, so that $\cup_{j=1}^{k+1} B_j = \Omega$.

Given $\varphi \in C_c^{\infty}(\Omega)$, the family $(B_j)_{1 \le j \le k+1}$ is a covering of $K := \operatorname{supp} \varphi$. Consider a partition of unity $(\zeta_j)_{0 \le j \le k+1}$ subordinate to this covering (see, for example, Brezis [75, Lemma 9.3]). We thus have

$$\begin{aligned} &\sup \zeta_j \subset B_j, \ j = 1, \dots, k+1, \\ &\operatorname{supp} \zeta_0 \subset \mathbb{R}^2 \setminus K, \\ &\sum_{j=0}^{k+1} \zeta_j = 1 \text{ on } \mathbb{R}^2, \end{aligned}$$

and, in particular,

$$\begin{aligned} &\zeta_j(a_j) = 1, \ \ j = 1, \dots, k, \\ &\zeta_0 \varphi = 0 \text{ on } \Omega. \end{aligned}$$

By the special case considered above, we have

$$\langle Ju, \zeta_j \varphi \rangle = \pi \ \deg(u, a_j) \, (\zeta_j \varphi)(a_j) = \pi \ \deg(u, a_j) \, \varphi(a_j), \ j = 1, \dots, k.$$

On the other hand, since u is continuous on B_{k+1},

$$\langle Ju, \zeta_{k+1} \varphi \rangle = 0, \text{ by Corollary 1.1.}$$

Thus,

$$\langle Ju, \varphi\rangle = \sum_{j=1}^{k}\langle Ju, \zeta_j\varphi\rangle = \pi\sum_{j=1}^{k}\deg(u, a_j)\,\varphi(a_j). \qquad \square \tag{2.38}$$

Second proof of Theorem 2.1. As above, it suffices to consider the special case where $k = 1$ and Ω is a disk centered at a_1, say $\Omega = \mathbb{D}$ and $a_1 = 0$. Set $v(z) := u(z)\left(\frac{z}{|z|}\right)^{-d}$; this v belongs to $W^{1,1}(\mathbb{D})$. Combining (1.68), (1.56), and (1.57), we see that (2.25) is equivalent to $Jv = 0$ in $\mathscr{D}'(\Omega)$. Since $v \in C(\mathbb{D} \setminus \{0\}; \mathbb{S}^1)$ and $\deg(v, 0) = 0$, we deduce from Lemma 1.6 that there exists some $\varphi \in C(\mathbb{D} \setminus \{0\}; \mathbb{R})$ such that $v = e^{\imath\varphi}$. Noting that, locally, we have $\varphi = -\imath \ln v$, where ln is a local smooth inverse of the exponential, we find that $\varphi \in W^{1,1}_{loc}(\mathbb{D} \setminus \{0\})$ and that $|\nabla\varphi| = |\nabla v|$. Using Lemma 1.10, we conclude that $\varphi \in W^{1,1}_{loc}(\mathbb{D})$. By (1.58), we obtain $Jv = 0$ in $\mathscr{D}'(\Omega)$. $\square$

Here is a more delicate result.

Theorem 2.2. *Let $u \in W^{1,1}(\Omega; \mathbb{S}^1) \cap VMO_{loc}(\Omega \setminus \{a_1, \ldots, a_k\})$, where $a_1, \ldots, a_k \in \Omega$. Then,*

$$Ju = \pi\sum_{j=1}^{k} d_j\delta_{a_j} \text{ in } \mathscr{D}'(\Omega). \tag{2.39}$$

[For the definition of *VMO*, see Section 5.2.] Here, d_j is again an appropriate integer, but this requires some explanation. Let $u \in W^{1,1}_{loc}(\Omega \setminus \{a_1, \ldots, a_k\})$. For fixed j and almost every small $\varepsilon > 0$, we have $u \in W^{1,1}(C(a_j, \varepsilon))$, and thus u has a degree on $C(a_j, \varepsilon)$. In general, this degree depends on ε (see Remark 2.3). However, under the additional assumption $u \in VMO_{loc}(\Omega \setminus \{a_1, \ldots, a_k\})$, the degree does not depend on ε (Corollary 2.2). This constant degree is denoted by $\deg(u, a_j)$ and appears in (2.39).

Lemma 2.1. *Let $u \in W^{1,1}_{loc}(\mathbb{D} \setminus \{0\}; \mathbb{S}^1)$ satisfy $Ju = 0$ in $\mathscr{D}'(\mathbb{D} \setminus \{0\})$. Then, $d := \deg(u, C(0, r))$ is well-defined for a.e. $r \in (0, 1)$ and is independent of r.*

If, in addition, $u \in W^{1,1}_{loc}(\mathbb{D}; \mathbb{S}^1)$, then $Ju = \pi\, d\, \delta_0$ in $\mathscr{D}'(\mathbb{D})$.

Proof. Let $v_r : \mathbb{S}^1 \to \mathbb{S}^1$, $v_r(z) := u(rz)$, $\forall\, r \in (0, 1)$, $\forall\, z \in \mathbb{S}^1$. By Lemma 1.4, the function $f : (0, 1) \to \mathbb{Z}$, defined by $f(r) := \deg v_r = \deg(u, C(0, r))$, is locally integrable. Let $\psi \in C^\infty_c(0, 1)$ and set $\zeta(x) := \psi(|x|) \in C^\infty_c(\mathbb{D} \setminus \{0\})$. If $x = rz$, with $r \in (0, 1)$ and $z \in \mathbb{S}^1$, then

$$(u \wedge \nabla u) \cdot \nabla^\perp\zeta(x) = \psi'(r)\, v_r \wedge \frac{\partial v_r}{\partial \tau}(z).$$

Combining this with Lemma 1.4, we obtain

$$0 = 2\langle Ju, \zeta\rangle = -\int_{\mathbb{D}} (u \wedge \nabla u) \cdot \nabla^{\perp}\zeta = -2\pi \int_0^1 \psi'(r) f(r)\, dr, \ \forall \psi \in C_c^{\infty}(0, 1).$$

Thus, $f' = 0$ in $\mathscr{D}'$, i.e., $\deg(u, C(0, r))$ is constant a.e.

For the second part of the lemma, we copy the first proof of Theorem 2.1. The only difference is that (2.37) is obtained using the first part of Lemma 2.1 (instead of Lemma 1.4). □

An immediate consequence is

Corollary 2.1. *Let $u \in W^{1,1}(\mathbb{D}; \mathbb{S}^1)$ satisfy $Ju = \pi\, d\, \delta_0$ for some $d \in \mathbb{R}$. Then, $d \in \mathbb{Z}$ and*

$\deg(u, C(0, r)) = d$ *for a.e.* $r \in (0, 1)$.

For a different proof of Corollary 2.1, see Brezis and Mironescu [87, Lemma 2].

Corollary 2.2. *Assume that $u \in W^{1,1}_{loc}(\mathbb{D} \setminus \{0\}; \mathbb{S}^1) \cap VMO_{loc}(\mathbb{D} \setminus \{0\}; \mathbb{S}^1)$. Then, $d := \deg(u, C(0, r))$ is well-defined for a.e. $r \in (0, 1)$ and is independent of r.*

If, in addition, $u \in W^{1,1}_{loc}(\mathbb{D}; \mathbb{S}^1)$, then $Ju = \pi\, d\, \delta_0$ in $\mathscr{D}'(\mathbb{D})$.

Proof. By Theorem 5.9, we know that $Ju = 0$ in $\mathscr{D}'(\mathbb{D} \setminus \{0\})$. □

Remark 2.3. Here is a "pathological" example illustrating the fact that, if $u \in W^{1,1}(\mathbb{D}; \mathbb{S}^1)$, then $\deg(u, C(0, r))$ (which is well-defined for a.e. $r \in (0, 1)$) can take distinct values in every neighborhood of $r = 0$. Consider two sequences of points $1 > N_1 > P_1 > N_2 > P_2 > \cdots > 0$ on the x_1-axis, and a sequence (ε_j) of positive numbers such that $\sum_{j=1}^{\infty} \varepsilon_j < \infty$. The dipole construction (see Section 2.5) yields a sequence of functions $(w_j) \subset W^{1,1}(\mathbb{D}; \mathbb{S}^1)$ such that:

1. $w_j \in C(\mathbb{R}^2 \setminus \{P_j, N_j\}; \mathbb{S}^1)$, $\deg(u, C(P_j, r)) = 1$ and $\deg(u, C(N_j, r)) = -1$, for r sufficiently small;
2. $\operatorname{supp}(w_j - 1)$ is contained in a lozenge of axis $[P_j, N_j]$;
3. $\int_{\mathbb{D}} |\nabla w_j| \leq 2\pi |P_j - N_j| + \varepsilon_j$.

The infinite product $u := \Pi_{j=1}^{\infty} w_j$ converges in $W^{1,1}(\mathbb{D}; \mathbb{S}^1)$ (since $\nabla \left(\Pi_{j=1}^{k} w_j\right) = \sum_{j=1}^{k} \nabla w_j$ and $\sum_{j=1}^{\infty} \int_{\mathbb{D}} |\nabla w_j| < \infty$).

Moreover, $u \in C(\mathbb{D} \setminus S)$, where $S := \cup_{j \geq 1} \{P_j, N_j\} \cup \{0\}$, and $\deg(u, C(0, r)) = 1$ if $r \in \cup_{j=1}^{\infty} (P_j, N_j)$, whereas $\deg(u, C(0, r)) = 0$ for the other values of $r \notin S$.

Proof of Theorem 2.2. Combine Corollary 2.2 with the argument leading to equality (2.38). □

2.2 Where optimal transport (=OT) enters: the quantity $L(a, d)$ and minimal configurations

In the previous section, we considered $u \in W^{1,1}(\Omega; \mathbb{S}^1) \cap C(\Omega \setminus \{a_1, \ldots, a_k\})$ and we established the formula $Ju = \pi \sum d_j \, \delta_{a_j}$.

In this section, we forget u for a while, and focus on the pair (a, d), where a denotes a collection $(a_1, \ldots, a_k)$ of distinct points in Ω and $d = (d_1, \ldots, d_k)$ is a collection of integers in $\mathbb{Z}$. [We will return to u in Section 2.3.]

We now introduce a quantity that plays a fundamental role in this chapter.

First, define on $\overline{\Omega}$ the following pseudometric. Given $P, N \in \overline{\Omega}$, set

$$d(P, N) := \inf\{\text{length}\,(\gamma \cap \Omega);\ \gamma \subset \overline{\Omega} \text{ is a rectifiable curve from } P \text{ to } N\}. \tag{2.40}$$

Clearly, we also have

$$d(P, N) = \inf\{\text{length}\,(\gamma \cap \Omega);\ \gamma \subset \mathbb{R}^2 \text{ is a rectifiable curve from } P \text{ to } N\}. \tag{2.41}$$

Note that the infimum in (2.40) is achieved either by $\gamma = [P, N]$, or by the union of $[P, N']$, $[P', N]$ and $\widetilde{\gamma}$, where $P', N' \in \partial\Omega$, $\widetilde{\gamma} \subset \partial\Omega$ connects N' and P', $\text{dist}(P, \partial\Omega) = |P - N'|$, $\text{dist}(N, \partial\Omega) = |N - P'|$. [This is possible since Ω is simply connected and thus $\partial\Omega$ is connected.] Clearly,

$$d(P, N) = \min\{|P - N|,\ \text{dist}(P, \partial\Omega) + \text{dist}(N, \partial\Omega)\}. \tag{2.42}$$

In particular, if $N \in \partial\Omega$ (resp. $P \in \partial\Omega$) then $d(P, N) = \text{dist}(P, \partial\Omega)$ (resp. $d(P, N) = \text{dist}(N, \partial\Omega)$). Moreover, if $P, N \in \overline{\Omega}$ then

$$d(P, N) = 0 \iff [\text{either } P = N \text{ or } P, N \in \partial\Omega]. \tag{2.43}$$

In view of (2.43), we may identify $\partial\Omega$ with a single point in $\overline{\Omega}$, denoted by Γ. Γ is added to the collection $(a_1, \ldots, a_k)$ as a_{k+1}, and is associated with the integer

$$d_{k+1} := -\sum_{j=1}^{k} d_j. \tag{2.44}$$

We then repeat the points a_i, $i = 1, \ldots, k + 1$, as $|d_i|$ "positive" points if $d_i > 0$ (resp. $|d_i|$ "negative" points if $d_i < 0$). We omit from this list the points a_i with

$d_i = 0$. By (2.44), we may rewrite these repeated points as a collection of m positive points (P_ℓ) in $\overline{\Omega}$ and m negative points (N_ℓ) in $\overline{\Omega}$, with $1 \le \ell \le m$. Finally, we set

$$L(a, d) := \min_{\sigma \in S_m} \sum_{\ell=1}^{m} d(P_\ell, N_{\sigma(\ell)}), \tag{2.45}$$

where the min is taken over the set S_m of all the permutations of the integers $\{1, \ldots, m\}$. [Strictly speaking, $L(a, d)$ depends also on Ω.] Note that the points P_ℓ and N_ℓ are defined up to a permutation; however, the quantity $L(a, d)$ is independent of the specific choice of (P_ℓ) and (N_ℓ).

Here are some examples.

Example 2.4. Consider the pair (a, d), with $a := (a_1)$ and $d := (+1)$. Then, $P_1 = a_1$ and $N_1 = \Gamma$. In this case, $L(a, d) = \text{dist}(a_1, \partial\Omega)$.

A straightforward generalization is the following

Example 2.5. If $a := (a_1, \ldots, a_k)$ and $d := (d_1, \ldots, d_k)$, with $d_j \ge 0$, $\forall j$, then $L(a, d) = \sum d_j \,\text{dist}(a_j, \partial\Omega)$.

Example 2.6. Consider the pair (a, d), with $a := (a_1, a_2)$, $a_1 \ne a_2$, and $d := (-1, +1)$. Then, $P_1 = a_2$ and $N_1 = a_1$. In this case,

$$L(a, d) = \min \{|a_1 - a_2|, \,\text{dist}(a_1, \partial\Omega) + \text{dist}(a_2, \partial\Omega)\}$$

(see Fig. 2.1).

Example 2.7. Consider the pair (a, d), with $a := (a_1, a_2)$, $a_1 \ne a_2$, and $d := (+3, -1)$. Then, $P_1 = P_2 = P_3 = a_1$, $N_1 = a_2$, and $N_2 = N_3 = \Gamma$. In this case,

$$L(a, d) = 2\,\text{dist}(a_1, \partial\Omega) + \min \{|a_1 - a_2|, \,\text{dist}(a_1, \partial\Omega) + \text{dist}(a_2, \partial\Omega)\}.$$

The main result in this section is the following

Theorem 2.3. *We have*

$$\begin{aligned} L(a, d) &= \left\| \sum_{j=1}^{k} d_j \delta_{a_j} \right\|_W \quad \textit{(where } \| \, \|_W \textit{ is the norm defined in } (1.152)\textit{)} \\ &= \max \left\{ \sum_{j=1}^{k} d_j \zeta(a_j); \; \zeta \in W_0^{1,\infty}(\Omega) \textit{ and } \|\nabla \zeta\|_{L^\infty} \le 1 \right\} \\ &= \sup \left\{ \sum_{j=1}^{k} d_j \zeta(a_j); \; \zeta \in C_c^\infty(\Omega) \textit{ and } \|\nabla \zeta\|_{L^\infty} \le 1 \right\}. \end{aligned} \tag{2.46}$$

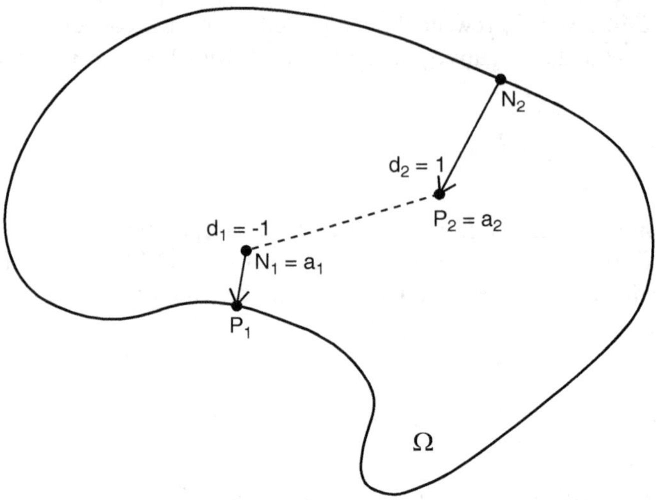

Fig. 2.1 How to determine $L(a, d)$ in Example 2.6: compare the total length of the solid lines with the length of the dashed line. Then, $L(a, d)$ is the least of these quantities.

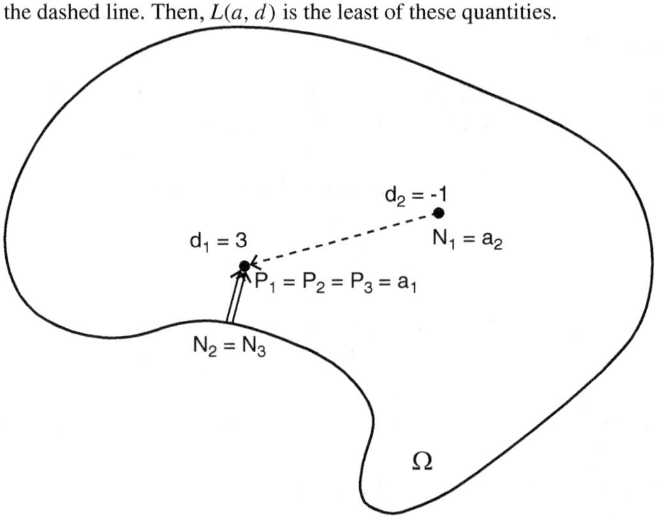

Fig. 2.2 In Example 2.7, $L(a, d) = 2\,|P_2 - N_2| + d(a_1, a_2)$.

The latter equality follows from Remark 1.30. The heart of the theorem consists of the first equality.

The main ingredient in the proof of Theorem 2.3 is the following basic result, closely related to OT (see Comment 2 in Section 2.12).

Theorem 2.4. *Let* (M, d) *be a pseudometric space. Let* $P_1, \ldots, P_m, N_1, \ldots, N_m$ *be* $2m$ *points in* M*. Then,*

$$\max\left\{\sum_{j=1}^{m}(\zeta(P_j)-\zeta(N_j));\ \zeta\in \mathrm{Lip}(M;\mathbb{R}),\ |\zeta|_{\mathrm{Lip}}\le 1\right\} = \min_{\sigma\in S_m}\sum_{j=1}^{m} d(P_j, N_{\sigma(j)}). \tag{2.47}$$

Proof. With no loss of generality, we may assume that

$$\sum_{j=1}^{m} d(P_j, N_j)\le \sum_{j=1}^{m} d(P_j, N_{\sigma(j)}),\ \forall\sigma\in S_m. \tag{2.48}$$

It is clear that

$$\sum_{j=1}^{m}(\zeta(P_j)-\zeta(N_j))\le \sum_{j=1}^{m} d(P_j, N_j)$$

for every function ζ with $|\zeta|_{\mathrm{Lip}}\le 1$. This yields "$\le$" in (2.47). The heart of the proof consists of proving the reverse inequality, i.e.,

$$\max\left\{\sum_{j=1}^{m}(\zeta(P_j)-\zeta(N_j));\ |\zeta|_{\mathrm{Lip}}\le 1\right\}\ge \sum_{j=1}^{m} d(P_j, N_j), \tag{2.49}$$

under the assumption (2.48).

Let

$$Y := (\cup_{j=1}^{m}\{P_j\})\cup(\cup_{j=1}^{m}\{N_j\}). \tag{2.50}$$

It suffices to construct a function $\zeta : Y\to\mathbb{R}$, 1-Lipschitz on Y and such that

$$\sum_{j=1}^{m}\big(\zeta(P_j)-\zeta(N_j)\big)=\sum_{j=1}^{m} d(P_j, N_j). \tag{2.51}$$

Indeed, any 1-Lipschitz function ζ on $A\subset M$ can be extended as a 1-Lipschitz function to the whole space M by setting

$$\widetilde{\zeta}(x) := \inf\{\zeta(a)+d(x,a);\ a\in A\},\ \forall x\in M. \tag{2.52}$$

Since, for any such ζ, we have $\zeta(P_j)-\zeta(N_j)\le d(P_j, N_j)$, this amounts to constructing a function ζ, 1-Lipschitz on Y, such that

$$\zeta(P_j)=\zeta(N_j)+d(P_j, N_j),\ 1\le j\le m. \tag{2.53}$$

Proof of (2.49). Set $d_j := d(P_j, N_j)$ and consider $\lambda_j := \zeta(N_j)$, $1 \le j \le m$, as being the unknowns to be determined. Combining (2.53) with the constraint $|\zeta|_{\text{Lip}} \le 1$, we are led to the following system of inequalities, which is equivalent to $|\zeta|_{\text{Lip}} \le 1$:

$$\begin{cases} |\lambda_i - \lambda_j| \le d(N_i, N_j) \\ |(\lambda_i + d_i) - (\lambda_j + d_j)| \le d(P_i, P_j), \ 1 \le i, j \le m. \\ |(\lambda_i + d_i) - \lambda_j| \le d(P_i, N_j) \end{cases} \tag{2.54}$$

This system is equivalent to

$$\lambda_i - \lambda_j \le d(P_i, N_j) - d_i, \ \ 1 \le i, j \le m; \tag{2.55}$$

all the other inequalities in (2.54) are consequences of (2.55) and of the triangle inequality. In other words, we have to find a solution $(\lambda_j)_{j=1}^m$ for the problem

$$\lambda_i - \lambda_j \le b_{ij}, \ \ 1 \le i, j \le m, \tag{2.56}$$

where $b_{ij} := d(P_i, N_j) - d_i$. Note that

$$b_{ii} = 0, \ \ 1 \le i \le m, \tag{2.57}$$

and that the hypothesis (2.48) reads

$$\sum_{j=1}^{m} b_{j\sigma(j)} \ge 0, \ \forall \sigma \in S_m. \tag{2.58}$$

We complete the proof of (2.49) (and thus of Theorem 2.4) via the next result. □

Lemma 2.2. *Assume* (2.57). *Then the system* (2.56) *has a solution if and only if* (2.58) *holds.*

Proof. Necessity of (2.58) follows from

$$0 = \sum_{j=1}^{m} (\lambda_j - \lambda_{\sigma(j)}) \le \sum_{j=1}^{m} b_{j\sigma(j)}.$$

Assume next that (2.58) holds. We first propose an ansatz for the λ_j's, and then prove that these λ_j's have all the required properties. This approach is also similar to the one leading to (2.52). A *chain* K connecting i to j is a finite sequence $K = (i_1, \dots i_k)$ such that $k \ge 2$, $i_l \in \{1, \dots, m\}$, $\forall\, l$, $i_1 = i$ and $i_k = j$. [We do not assume that $i_1, \dots, i_k$ are distinct.]

Given a chain K connecting $i = i_1$ to $j = i_k$, set

$$S_K := b_{i_1 i_2} + b_{i_2 i_3} + \cdots + b_{i_{k-1} i_k}. \tag{2.59}$$

Suppose now that a solution (λ_i) of (2.56) exists, and consider a chain K connecting i to j. We have

$$\begin{aligned}
&\lambda_{i_1} - \lambda_{i_2} \leq b_{i_1 i_2},\\
&\lambda_{i_2} - \lambda_{i_3} \leq b_{i_2 i_3},\\
&\dots\\
&\lambda_{i_{k-1}} - \lambda_{i_k} \leq b_{i_{k-1} i_k}.
\end{aligned}$$

Adding these inequalities yields

$$\lambda_i - \lambda_j \leq S_K, \tag{2.60}$$

and, in particular,

$$\lambda_i - \lambda_1 \leq \inf \{S_K;\ K \text{ is a chain connecting } i \text{ to } 1\}. \tag{2.61}$$

We now turn to the *existence* of a solution (λ_i) for (2.56). Since the λ_i's are defined modulo an additive constant it is tempting – in view of (2.61) – to set, for every $1 \leq i \leq m$,

$$\lambda_i := \inf \{S_K;\ K \text{ is a chain connecting } i \text{ to } 1\}. \tag{2.62}$$

A priori it may happen that $\lambda_i = -\infty$, but this will be excluded below.

Fix $1 \leq i, j \leq m$, and let $K = (i_1, \dots, i_k)$ be any chain connecting j to 1. Then $\widetilde{K} := (i, K)$ connects i to 1, and, therefore (by (2.62)),

$$\lambda_i \leq S_{\widetilde{K}} = b_{ij} + S_K. \tag{2.63}$$

Taking the inf over K in (2.63), we obtain

$$\lambda_i \leq b_{ij} + \lambda_j,\ \forall\, 1 \leq i, j \leq m, \tag{2.64}$$

which is the desired inequality (2.56). It remains to establish that $\lambda_j \neq -\infty$, $\forall j$; assumptions (2.57) and (2.58) enter here.

We will prove that

$$\lambda_1 = 0. \tag{2.65}$$

Combining (2.64) and (2.65), we may then deduce that

$$0 = \lambda_1 \leq b_{1j} + \lambda_j,\ \forall j,$$

and thus $\lambda_j \neq -\infty$, $\forall j$. We now turn to the proof of (2.65). First, we choose the chain $K = (1, 1)$ in (2.62), and obtain

$$\lambda_1 \le b_{11} = 0. \tag{2.66}$$

Next, we establish that $\lambda_1 \ge 0$. We start with some terminology. A chain K connecting i to $j = i$ is called a *loop* (or a cycle). A loop is *simple* if $i_1, \dots, i_{k-1}$ are distinct. We claim that for every loop K we have

$$S_K \ge 0. \tag{2.67}$$

Indeed, when K is a simple loop (2.67) follows from (2.58) (and ((2.57)) applied to the permutation $i_1 \to i_2 \cdots \to i_k$ (the other integers are invariant). By decomposing a general loop in simple loops we find that (2.67) holds for all loops.

Applying (2.67) to any chain connecting 1 to 1, we deduce from (2.62) that $\lambda_1 \ge 0$. □

The next result provides a description of functions that are 1-Lipschitz with respect to the pseudometric introduced in (2.42).

Lemma 2.3. *Let $\zeta : \overline{\Omega} \to \mathbb{R}$. Then, the following three properties are equivalent:*

$$\zeta - C \in W_0^{1,\infty}(\Omega; \mathbb{R}) \textit{ for some } C \in \mathbb{R} \textit{ and } \|\nabla\zeta\|_{L^\infty} \le 1. \tag{2.68}$$

$$\zeta \textit{ is constant on } \partial\Omega \textit{ and } |\zeta(x) - \zeta(y)| \le |x - y|,\ \forall x, y \in \overline{\Omega}. \tag{2.69}$$

$$|\zeta|_{\mathrm{Lip}(\overline{\Omega}, d))} \le 1,\ \textit{i.e.}, |\zeta(x) - \zeta(y)| \le d(x, y),\ \forall x, y \in \overline{\Omega}. \tag{2.70}$$

Proof. Let $x, y \in \overline{\Omega}$. Assume first (2.68). If $[x, y] \subset \overline{\Omega}$, then (2.69) follows from the mean value theorem. Otherwise, let $z, t \in [x, y] \cap \partial\Omega$ be such that $[x, z], [t, y] \subset \overline{\Omega}$. Then,

$$|\zeta(x) - \zeta(y)| \le |\zeta(x) - \zeta(z)| + |\zeta(t) - \zeta(y)| \le |x - z| + |t - y| \le |x - y|.$$

Assume next (2.69). If $d(x, y) = |x - y|$, then (2.70) holds. Otherwise, there are $z, t \in \partial\Omega$ such that $d(x, y) = |x - z| + |y - t|$, so that, again, (2.70) holds.

Finally, assume (2.70). Then, ζ has to be constant on $\partial\Omega$, and

$$|\zeta(x) - \zeta(y)| \le d(x, y) \le |x - y|,\ \forall x, y \in \overline{\Omega}. \tag{2.71}$$

We find that $\|\nabla\zeta\|_{L^\infty} \le 1$. □

Proof of Theorem 2.3. Without loss of generality, we may assume that the collection $(P_\ell, N_\ell)_{\ell=1}^m$ in (2.6) satisfies

$$\sum_{\ell=1}^m d(P_\ell, N_\ell) \le \sum_{\ell=1}^m d(P_\ell, N_{\sigma(\ell)}),\ \forall \sigma \in S_m. \tag{2.72}$$

From the definition of the points (P_ℓ) and (N_ℓ), we have

$$\sum_j d_j \zeta(a_j) = \sum_\ell (\zeta(P_\ell) - \zeta(N_\ell)), \ \forall \zeta \in W_0^{1,\infty}(\Omega). \tag{2.73}$$

On the other hand, by Theorem 2.4, we have

$$\begin{aligned} &\max\left\{\sum_\ell (\zeta(P_\ell) - \zeta(N_\ell));\ \zeta \in \mathrm{Lip}(\overline{\Omega}; \mathbb{R}) \text{ and } |\zeta|_{\mathrm{Lip}} \le 1\right\} \\ &= \min_{\sigma \in S_m} \sum_j d(P_j, N_{\sigma(j)}) = \sum_\ell d(P_\ell, N_\ell) \text{ (by (2.72))}. \end{aligned} \tag{2.74}$$

[Here, Lip refers to the pseudometric defined in (2.40).]
Applying Lemma 2.3, we see that

$$\begin{aligned} &\max\left\{\sum_\ell (\zeta(P_\ell) - \zeta(N_\ell));\ \zeta \in \mathrm{Lip}(\overline{\Omega}; \mathbb{R}),\ |\zeta|_{\mathrm{Lip}} \le 1\right\} \\ &= \max\left\{\sum_\ell (\zeta(P_\ell) - \zeta(N_\ell));\ \zeta \in W_0^{1,\infty}(\Omega) \text{ and } \|\nabla\zeta\|_{L^\infty} \le 1\right\}. \end{aligned} \tag{2.75}$$

Combining (2.73), (2.74), and (2.75) yields

$$\begin{aligned} &\max\left\{\sum_j d_j \zeta(a_j);\ \zeta \in W_0^{1,\infty}(\Omega) \text{ and } \|\nabla\zeta\|_{L^\infty} \le 1\right\} \\ &= \sum_\ell d(P_\ell, N_\ell) = L(a, d) \text{ (by (2.6))}. \end{aligned}$$

□

Theorems 2.3, 2.4, and Lemma 2.3 lead to the following

Corollary 2.3. *Let $(\overline{P}_\ell, \overline{N}_\ell)_{\ell=1}^{\overline{n}}$ be any collection of points in $\overline{\Omega}$ such that*

$$\sum_{\ell=1}^{\overline{n}} (\delta_{\overline{P}_\ell} - \delta_{\overline{N}_\ell}) = \sum_{j=1}^{k} d_j \delta_{a_j} \ \textit{in } \mathscr{D}'(\Omega).$$

Then

$$L(a, d) = \min_{\sigma \in S_{\overline{n}}} \sum_{\ell=1}^{\overline{n}} d(\overline{P}_\ell, \overline{N}_{\sigma(\ell)}). \tag{2.76}$$

Returning to the definition (2.6) of $L(a, d)$, we may write $L(a, d) = \sum_{\ell=1}^m d(P_\ell, N_\ell)$ for some collection $(P_\ell, N_\ell)_{\ell=1}^m$ in $\overline{\Omega}$ such that $\sum_j d_j \delta_{a_j} = \sum_{\ell=1}^m (\delta_{P_\ell} - \delta_{N_\ell})$ in $\mathscr{D}'(\Omega)$. Since, by (2.42),

$$d(P_\ell, N_\ell) = \min\{|P_\ell - N_\ell|,\ \mathrm{dist}(P_\ell, \partial\Omega) + \mathrm{dist}(N_\ell, \partial\Omega)\},$$

we may always add points in $\partial\Omega$ to the collection $(P_\ell, N_\ell)_{\ell=1}^m$, and obtain a larger collection, denoted by $(P_\ell, N_\ell)_{\ell=1}^n$, such that

$$L(a, d) = \sum_{\ell=1}^{n} |P_\ell - N_\ell| \tag{2.77}$$

and

$$\sum_{j=1}^{k} d_j \delta_{a_j} = \sum_{\ell=1}^{n} (\delta_{P_\ell} - \delta_{N_\ell}) \text{ in } \mathscr{D}'(\Omega). \tag{2.78}$$

Note that

$$n \le \sum_{j=1}^{k} |d_j|, \tag{2.79}$$

where n comes from (2.77). Indeed, set $A := \sum\limits_{d_j>0} |d_j|$ and $B := \sum\limits_{d_j<0} |d_j|$, so that (2.79) amounts to $n \le A + B$.

If $A = B$, we deduce from (2.42) and (2.6) that $n \le 2A = A + B$. Suppose, for example, that $A < B$. Following (2.44), we add to the collection $(a_1, \dots, a_k)$ a positive point $a_{k+1} := \Gamma \in \partial\Omega$ with multiplicity $-\sum_{j=1}^k d_j = B - A$. Write now $L(a, d) = \sum_{\ell=1}^m d(P_\ell, N_\ell)$ as in (2.6). Note that $m = B$.

We may assume that $P_\ell \neq \Gamma$ when $\ell \le A$ and $P_\ell = \Gamma$ when $\ell > A$. For $\ell > A$, we have $d(\Gamma, N_\ell) = \operatorname{dist}(N_\ell, \partial\Omega) = |N_\ell - P_\ell|$ for some $P_\ell \in \partial\Omega$. Therefore, we have again $n \le 2A + (B - A) = A + B$. Similarly when $B < A$.

For further use, let us summarize the above construction of points $(P_\ell, N_\ell)_{\ell=1}^n$.

Lemma 2.4. *Given (a,d), there exist points $P_\ell, N_\ell \in \{a_1, \dots, a_k\} \cup \partial\Omega$, $\ell=1, \dots, n$, such that:*

$$L(a, d) = \sum_{\ell=1}^{n} |P_\ell - N_\ell|, \tag{2.80}$$

$$\sum_{\ell=1}^{n} \delta_{P_\ell} = \sum_{j=1}^{k} d_j^+ \, \delta_{a_j} \textit{ in } \mathscr{D}'(\Omega), \tag{2.81}$$

$$\sum_{\ell=1}^{n} \delta_{N_\ell} = \sum_{j=1}^{k} d_j^- \, \delta_{a_j} \textit{ in } \mathscr{D}'(\Omega); \tag{2.82}$$

in particular,

$$\sum_{\ell=1}^{n} \left(\delta_{P_\ell} - \delta_{N_\ell}\right) = \sum_{j=1}^{k} d_j \, \delta_{a_j} \textit{ in } \mathscr{D}'(\Omega). \tag{2.83}$$

Moreover,

$$\text{for every } \ell = 1, \dots, n,\ P_\ell \neq N_\ell \text{ and the open interval } (P_\ell, N_\ell) \text{ is contained in } \Omega. \tag{2.84}$$

We will say that any such object (i.e., $(P_\ell, N_\ell)_{\ell=1}^n$) is a *minimal configuration* associated with (a, d). Here are some examples.

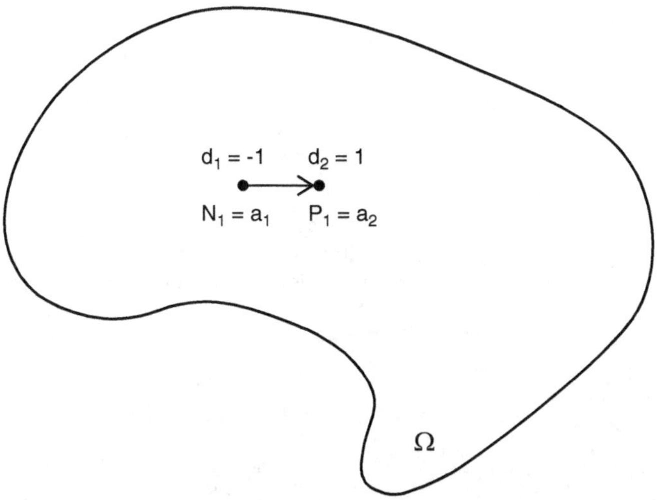

Fig. 2.3 Here, a_1, a_2 are well-inside Ω, and $d = (-1, +1)$. There is a unique minimal configuration: $N_1 = a_1, P_1 = a_2$.

Remark 2.4. In Fig. 2.7, the collection $((P_1, N_1), (P_2, N_2))$ with $P_1 = P_2 = a_1$, $N_1 = a_2$ and N_2 the projection of a_2 on $\partial\Omega$, is a minimal configuration, i.e., it satisfies the requirements (2.81) and (2.82), and $L(a, d) = 2|a_1 - a_2| + |a_2 - N_2|$.

On the other hand, consider the collection $((\overline{P}_1, \overline{N}_1), (\overline{P}_2, \overline{N}_2), (\overline{P}_3, \overline{N}_3))$, with $\overline{P}_1 = \overline{P}_2 = a_1$, $\overline{P}_3 = a_2$, $\overline{N}_1 = \overline{N}_2 = a_2$ and $\overline{N}_3 = N_2$. It satisfies $\overline{P}_\ell, \overline{N}_\ell \in \{a_1, a_2\} \cup \partial\Omega, \forall\, \ell$,

$$\sum_{\ell=1}^{3} |\overline{P}_\ell - \overline{N}_\ell| = 2|a_1 - a_2| + |a_2 - N_2| = L(a, d),$$

and also (2.83) and (2.84). However, it *does not* satisfy (2.81) and (2.82). Therefore, it is *not* a minimal configuration.

Thus, the definition of a minimal configuration through the requirements (2.80)–(2.82) and (2.84) is in general strictly more restrictive than a definition requiring (2.80), (2.83), and (2.84).

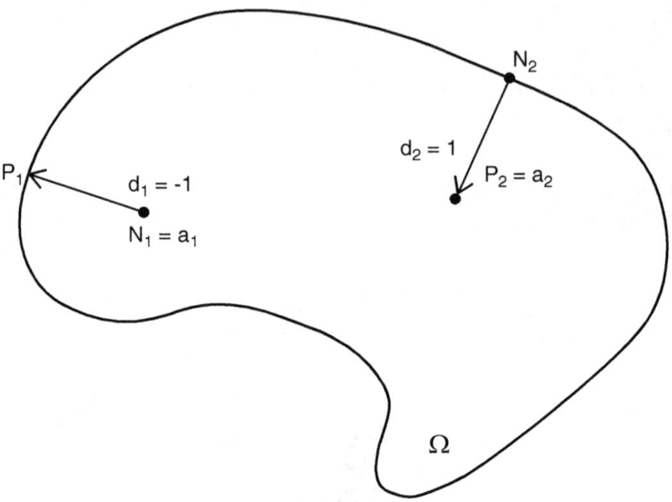

Fig. 2.4 Here, a_1, a_2 are sufficiently close to $\partial\Omega$, and $d = (-1, +1)$. There is a unique minimal configuration: $N_1 = a_1$, $P_2 = a_2$, and P_1 (resp. N_2) is the projection of N_1 (resp. P_2) on $\partial\Omega$.

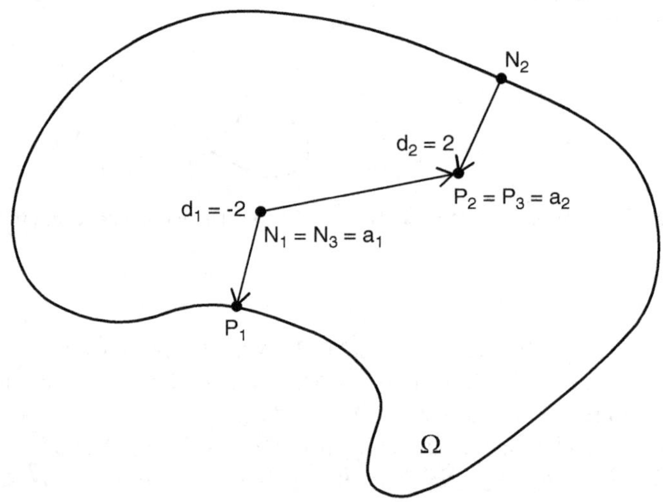

Fig. 2.5 Here, $|a_1 - a_2| = \text{dist}(a_1, \partial\Omega) + \text{dist}(a_2, \partial\Omega)$. When $d = (-2, +2)$, there are three minimal configurations; one of them is represented.

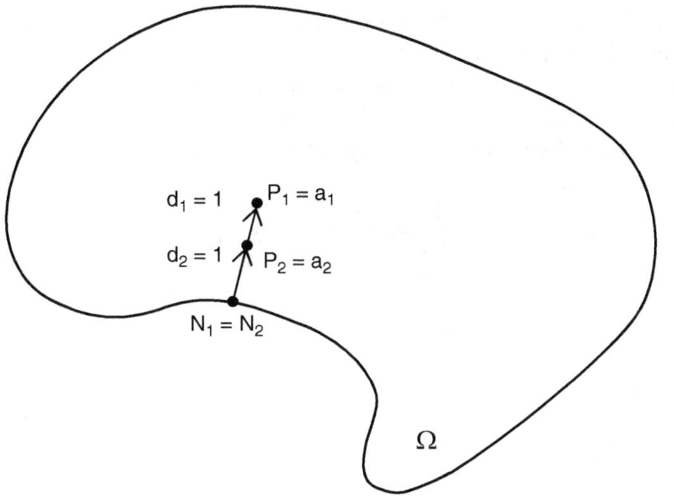

Fig. 2.6 Here, a_1, a_2 have the same projection on $\partial\Omega$, and $d = (+1, +1)$. There is one minimal configuration, and the segments $[P_1, N_1]$ and $[P_2, N_2]$ overlap.

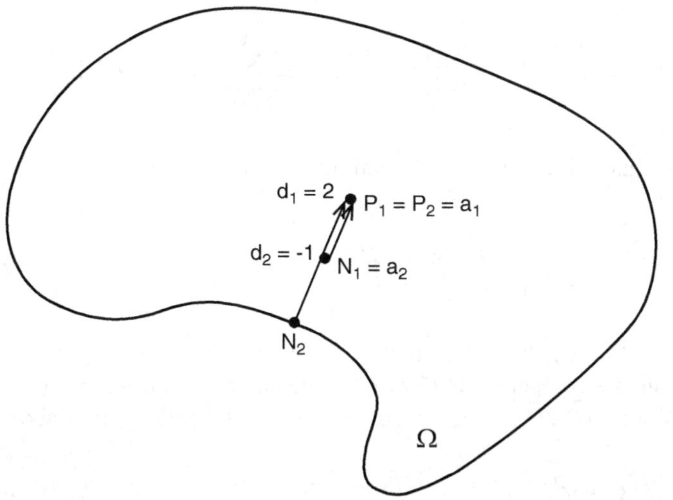

Fig. 2.7 Here, a_1, a_2 have the same projection on $\partial\Omega$, and $d = (+2, -1)$. There is one minimal configuration, and the segments $[P_1, N_1]$ and $[P_2, N_2]$ overlap.

Remark 2.5. Observe that there are usually many minimal configurations associated with a given (a, d). Note that, if $(P_\ell, N_\ell)_{\ell=1}^n$ is a minimal configuration, some of the line segments $[P_\ell, N_\ell]$ may intersect at their endpoints and/or overlap.

Remark 2.6. If we assume only (2.80)–(2.82), it might happen that $P_\ell = N_\ell$ for some ℓ. Indeed, fix a collection $(P_\ell, N_\ell)_{\ell=1}^n$ satisfying (2.80)–(2.82), take any point $b \in \partial\Omega$ and add $P_{n+1} = N_{n+1} = b$ to the above collection. The new collection $(P_\ell, N_\ell)_{\ell=1}^{n+1}$ still satisfies (2.80)–(2.82).

However, if we consider a collection $(P_\ell, N_\ell)_{\ell=1}^n$ such that (2.80)–(2.82) hold, and if we add the condition that $P_\ell \neq N_\ell$, $\forall\, \ell$, then
(i) For every ℓ, at most one of the points P_ℓ or N_ℓ belongs to $\partial\Omega$.
(ii) For every ℓ, the open interval (P_ℓ, N_ℓ) is contained in Ω.
Indeed, we have, by (2.80), Corollary 2.3, and the definition of d,

$$\sum_{\ell=1}^{n} |P_\ell - N_\ell| = L(a,d) \leq \sum_{\ell=1}^{n} d(P_\ell, N_\ell) \leq \sum_{\ell=1}^{n} |P_\ell - N_\ell|.$$

Therefore,

$$d(P_\ell, N_\ell) = |P_\ell - N_\ell|, \ \forall\, \ell = 1, \ldots, n. \tag{2.85}$$

Clearly, (2.85) implies (i) and (ii).

Remark 2.7. A final comment. Given (a, d), set

$$\widetilde{L} := \inf\Big\{ \sum_{\ell} \big|\widetilde{P}_\ell - \widetilde{N}_\ell\big| \,;\, \widetilde{P}_\ell, \widetilde{N}_\ell \in \overline{\Omega} \text{ and } \sum_{\ell} \big(\delta_{\widetilde{P}_\ell} - \delta_{\widetilde{N}_\ell}\big) = \sum_{j=1}^{k} d_j\, \delta_{a_j} \text{ in } \mathscr{D}'(\Omega)\Big\}, \tag{2.86}$$

where inf is taken over *all* finite collections $(\widetilde{P}_\ell, \widetilde{N}_\ell)$ satisfying

$$\sum_{\ell} \big(\delta_{\widetilde{P}_\ell} - \delta_{\widetilde{N}_\ell}\big) = \sum_{j=1}^{k} d_j\, \delta_{a_j} \text{ in } \mathscr{D}'(\Omega).$$

Then, $\widetilde{L} = L(a, d)$. It is clear that $\widetilde{L} \leq L(a, d)$, since we may use any minimal configuration as a competitor in (2.86). On the other hand, if we apply Corollary 2.3 with $(\overline{P}_\ell, \overline{N}_\ell) := (\widetilde{P}_\ell, \widetilde{N}_\ell)$, $\forall\, \ell$, (where $(\widetilde{P}_\ell, \widetilde{N}_\ell)$ is any collection as above), we find

$$L(a,d) \leq \sum_{\ell} d(\widetilde{P}_\ell, \widetilde{N}_\ell) \leq \sum_{\ell} |\widetilde{P}_\ell - \widetilde{N}_\ell|.$$

2.3 Returning to u: $\Sigma(u) = 2\pi L(a, d)$

In this short section, we return to the setting of Section 2.1, i.e., $u \in W^{1,1}(\Omega; \mathbb{S}^1) \cap C(\Omega \setminus \{a_1, \ldots, a_k\})$, and present an *explicit* formula for $\Sigma(u)$ involving the quantity $L(a, d)$ introduced in Section 2.2.

Theorem 2.5. *Let $u \in W^{1,1}(\Omega; \mathbb{S}^1) \cap C(\Omega \setminus \{a_1, \ldots, a_k\})$. Let d be the collection of integers $d_j := \deg(u, a_j)$, $j = 1, \ldots, k$. Then*

$$\Sigma(u) = 2\pi\, L(a, d). \tag{2.87}$$

Proof. From (1.144), we know that

$$\Sigma(u) = \max\left\{\int_\Omega (u \wedge \nabla u) \cdot \nabla^\perp \zeta;\ \zeta \in W_0^{1,\infty}(\Omega; \mathbb{R}),\ \|\nabla\zeta\|_{L^\infty} \le 1\right\} \tag{2.88}$$

(see also Comment 12 in Chapter 1 for a more direct proof). Next, we invoke (2.27) and (2.46) to conclude that

$$\begin{aligned}\Sigma(u) &= 2\pi \max\left\{\sum d_j \zeta(a_j);\ \zeta \in W_0^{1,\infty}(\Omega; \mathbb{R}),\ \|\nabla\zeta\|_{L^\infty} \le 1\right\} \\ &= 2\pi L(a, d).\end{aligned} \qquad \square \tag{2.89}$$

Remark 2.8. Let us pause to discuss the various facets of the quantity $\Sigma(u)$, and to explain the rationale behind the next two sections.

Recall (see (1.80)) that $\Sigma(u)$ is defined by

$$\Sigma(u) := \inf_{\eta \in W^{1,1}(\Omega; \mathbb{R})} \|u \wedge \nabla u - \nabla\eta\|_{L^1}, \tag{2.90}$$

which yields

$$\Sigma(u) = \min\{\|\mu\|_{\mathcal{M}};\ \mu \in \mathcal{M}(\Omega; \mathbb{R}^2),\ \operatorname{curl} \mu = 2Ju\}. \tag{2.91}$$

By duality, $\Sigma(u)$ becomes (see (1.133) and (2.14))

$$\Sigma(u) = 2 \sup\{\langle Ju, \zeta\rangle;\ \zeta \in C_c^\infty(\Omega; \mathbb{R}),\ \|\nabla\zeta\|_{L^\infty} \le 1\}. \tag{2.92}$$

Finally, enters $S_1(u)$, the least energy with prescribed Jacobian,

$$S_1(u) := \inf\left\{\int_\Omega |\nabla v|;\ v \in W^{1,1}(\Omega; \mathbb{S}^1),\ Jv = Ju\right\} \tag{2.93}$$

(see (1.185)), which is arguably one of the most interesting concepts in the book because of its physical interpretation. Indeed, when u, v are nice maps, the relation $Ju = Jv$ says that u and v have the same topological singularities (see Theorem 2.2). Thus, $S_1(u)$ can be viewed as a *least energy* required to generate a configuration having singularities of degrees d_j located at given points a_j, $j = 1, \ldots, k$. [More on the physical background in Comment 3 to Chapter 4.] We call the attention of the reader to the simplicity of the formula

$$S_1(u) = \Sigma(u) \tag{2.94}$$

(see Lemma 1.14); it asserts that the *physical* quantity $S_1(u)$ coincides, when $N = 2$, with the *geometrical* quantity $\Sigma(u) = 2\pi\, L(a, d)$, which represents the length of a minimal connection in 2D, or a least area in higher dimensions (see Chapter 3).

Incidentally, there is a natural variant of the quantity $S_1(u)$ when u is a nice map associated with the pair (a, d). Set

$$S_1^\sharp(u) := \inf\left\{\int_\Omega |\nabla v|;\ v \in W^{1,1}(\Omega; \mathbb{S}^1) \cap C(\Omega \setminus \{a_1, \dots, a_k\}),\right.$$
$$\left. \deg(v, a_j) = d_j,\ j = 1, \dots, k\right\}, \tag{2.95}$$

so that clearly $S_1^\sharp(u) \geq S_1(u)$. Actually, we have

$$S_1^\sharp(u) = S_1(u) \text{ for every nice map } u. \tag{2.96}$$

This is an immediate consequence of the density Lemma 2.5; alternatively, one could also use the dipole construction in Section 2.5 to establish that $S_1^\sharp(u) \leq 2\pi\, L(a, d) = S_1(u)$.

It turns out that the proof of (2.94) is remarkably simple, but relies heavily on *lifting* (see Lemma 1.14).

As we are going to see in Chapter 4, there are natural counterparts of (2.90)–(2.92) for maps $u \in W^{1,k}(\Omega; \mathbb{S}^k)$, $\Omega \subset \mathbb{R}^N$, $N \geq 2$, $1 \leq k \leq N - 1$, and they all coincide.

There is also a natural counterpart for (2.93), namely, given $u \in W^{1,k}(\Omega; \mathbb{S}^k)$, set

$$S_k(u) := \inf\left\{\int_\Omega |\nabla v|^k;\ v \in W^{1,k}(\Omega; \mathbb{S}^k),\ Ju = Jv\right\}$$

(for the definition of the distributional Jacobian Ju, see Section 4.0, more specifically (4.6) and (4.7)).

One may wonder what happens to (2.94), since lifting is out of the question when $k \geq 2$. On can show rather easily (using the (2.92) version of $\Sigma(u)$) that

$$S_k(u) \geq c_k\, \Sigma(u),\ \forall\, u \in W^{1,k}(\Omega; \mathbb{S}^k), \tag{2.97}$$

where $c_k := k^{k/2}$. Equality in (2.97) holds, of course, when $k = 1$; it also holds when $k = N - 1$ (any $N \geq 2$), but this requires another proof (see Chapter 4). However, strict inequality in (2.97) may occur in the intermediate range $1 < k < N - 1$ (see Chapter 4). When this happens, the "key player" is $S_k(u)$ and not $\Sigma(u)$; in fact, it is $S_k(u)$ that carries a geometrical flavor (as a least area), whereas $\Sigma(u)$ has no clear geometrical meaning.

In practice, it is of great interest to find methods for computing $S_k(u)$; the goal of Section 2.4 and 2.5 is to present efficient tools, which are of interest *even* when $k = 1$ (so that (2.94) holds). When $N = 2$, the combination of (2.94), (2.92), and OT provided us with the useful formula $S_1(u) = 2\pi\, L(a, d)$. By contrast, when $k = 1$

and $N \geq 3$ we still have (2.94) and (2.92), but we lack "standard" OT to conclude. [As we are going to see in Chapter 3, there is nevertheless some kind of "geometrical OT" that emerges; see also Brezis and Mironescu [88].] When $1 < k < N - 1$, we are definitely in need of new ideas for computing $S_k(u)$.

For simplicity, we take $k = 1$ and $N = 2$ in Sections 2.4–2.5, but we avoid (2.94) and make only unessential use of lifting.

In Section 2.4, we present a totally self-contained proof of the inequality $S_1(u) \geq 2\pi\, L(a,d)$. The main novelty consists of using the *coarea formula*. In higher dimensions, the details become more technical, but the basic geometric idea remains the same (see Section 3.3).

In Section 2.5, we present a totally self-contained proof of the inequality $S_1(u) \leq 2\pi\, L(a,d)$. The strategy involves a brute force construction of an *explicit* map $v \in W^{1,1}(\Omega;\mathbb{S}^1)$ with prescribed singularities and such that $\int_\Omega |\nabla v| \leq 2\pi L(a,d) + \varepsilon$ ($\varepsilon > 0$ arbitrarily small). This is done via a *dipole construction*. The map v is easily described when $\Omega \subset \mathbb{R}^2$ and the target is $\mathbb{S}^1$. In higher dimension the details become more technical, but the basic geometric idea remains the same (see Section 3.4).

2.4 $\Sigma(u) = S_1(u) \geq 2\pi\, L(a,d)$ via the coarea formula

Here, we present a technique originally introduced in Almgren, Browder, and Lieb [8] for the study of $\mathbb{S}^k$-valued maps. It relies heavily on the coarea formula, and we will apply it for deriving the inequality

$$S_1(u) \geq 2\pi L(a,d), \tag{2.98}$$

which complements (2.109).

Given $u \in W^{1,1}(\Omega;\mathbb{S}^1) \cap C(\Omega \setminus \{a_1,\ldots,a_k\})$, let

$$Z(u) := \{w \in W^{1,1}(\Omega;\mathbb{S}^1);\ Jw = Ju\}, \tag{2.99}$$

so that (2.98) amounts to

$$\int_\Omega |\nabla w| \geq 2\pi L(a,d),\ \ \forall\, w \in Z(u). \tag{2.100}$$

We next consider the following subclass of $Z(u)$:

$$Z^\infty(u) := \Big\{ w : \Omega \to \mathbb{S}^1;\ w \in C^\infty(\overline{\Omega} \setminus \{a_1,\ldots,a_k\}) \\ \text{and } w(z) = \left(\frac{z-a_j}{|z-a_j|}\right)^{d_j} \text{ near each } a_j \Big\}. \tag{2.101}$$

As we establish below, we have the following

Lemma 2.5. *$Z^\infty(u)$ is dense in $Z(u)$ for the $W^{1,1}$ topology.*

Granted this result, (2.98) amounts to the following variant of (2.100):

$$\int_\Omega |\nabla w| \geq 2\pi L(a,d), \ \forall\, w \in Z^\infty(u). \tag{2.102}$$

In turn, the proof of (2.102) relies on the following

Lemma 2.6. *Let $w \in Z^\infty(u)$. Then, for a.e. $\alpha \in \mathbb{S}^1$, the set $[w = \alpha]$ is a union of curves such that*

$$\mathscr{H}^1([w = \alpha]) \geq L(a,d). \tag{2.103}$$

Granted the two lemmas above, we proceed to the

Proof of (2.102). If $w \in Z^\infty(u)$, we claim that

$$\int_\Omega |\nabla w| = \int_{\mathbb{S}^1} \mathscr{H}^1([w = \alpha])\, d\alpha. \tag{2.104}$$

Indeed, when w is sufficiently smooth, say Lipschitz, equality (2.104) is the coarea formula (see, for example, Evans and Gariepy [146, Theorem 1, p. 112]). This shows the validity of (2.104) in $\Omega \setminus \cup_j \mathbb{D}(a_j, \varepsilon)$, with small $\varepsilon > 0$. On the other hand, in a small disk around a_j, the set $[w = \alpha]$ is a finite union of segments and the validity of (2.104) in this small disk is clear.

Combining Lemma 2.6 with (2.104), we obtain that (2.102) holds. □

Proof of Lemma 2.5. Fix any map $u_0 \in Z^\infty(u)$. If $w \in Z(u)$, then there exists some $\varphi \in W^{1,1}(\Omega; \mathbb{R})$ such that $w = e^{\imath\varphi} u_0$ (see (1.57) and Lemma 1.8). Consider a sequence $(\varphi_n) \subset C^\infty(\overline{\Omega}; \mathbb{R})$ such that $\varphi_n \to \varphi$ in $W^{1,1}$ and $\varphi_n = 0$ near each a_j; this is possible since a point has zero $W^{1,1}$-capacity [146, Section 4.7, Theorem 2]. Letting $w_n := e^{\imath\varphi_n} u_0$, we have $w_n \in Z^\infty(u)$ and $w_n \to w$ in $W^{1,1}$. □

Proof of Lemma 2.6. We may assume that every d_j is non-zero. This implies that, near each a_j and for every $\alpha \in \mathbb{S}^1$, the set $[w = \alpha]$ consists of $|d_j|$ segments and is, in particular, non-empty. Let α be a regular value of w in $\Omega \setminus \{a_1, \ldots, a_k\}$. By the above, the set $\mathscr{C}_\alpha := [w = \alpha]$ is non-empty, and consists of a locally finite union of curves.

Since α is a regular value of w, the implicit function theorem implies that there are four types of curves:

a) A curve connecting some a_i to some a_j
b) A curve connecting some a_i to $\partial\Omega$
c) A curve connecting some point on $\partial\Omega$ to some point on $\partial\Omega$
d) A closed loop in Ω.

We shall disregard the curves of type c) and d), since they only increase the value of $\mathscr{H}^1(\mathscr{C}_\alpha)$. We are thus left with curves of type a) and b). These curves have a natural orientation obtained as follows. Write, locally near some $z \in \mathscr{C}_\alpha$, $w = e^{\imath\varphi}$, with smooth φ. Then, $\nu := \nabla\varphi = w \wedge \nabla w$ is a non-zero vector (since α is a regular value of w), orthogonal to $\mathscr{C}_\alpha$ and independent of φ. We define on $\mathscr{C}_\alpha$ the orientation such that (ν, τ) is a direct basis of $\mathbb{R}^2$. With this convention, a curve of $\mathscr{C}_\alpha$ having a_j with $d_j > 0$ (resp. $d_j < 0$) as an endpoint goes towards a_j (resp. emanates from a_j). Consider the components C_ℓ of $\mathscr{C}_\alpha$ that have one of the a_j's as an endpoint. Using the special structure of w near a_j, we find that there are finitely many such (oriented) curves, say C_ℓ, $\ell = 1, \dots, n$, and that they satisfy:

1. C_ℓ connects Q_ℓ to R_ℓ for some $Q_\ell, R_\ell \in \{a_1, \dots, a_k\} \cup \partial\Omega$.
2. If $d_j > 0$, then there are exactly d_j values of ℓ such that $R_\ell = a_j$.
3. If $d_j < 0$, then there are exactly $|d_j|$ values of ℓ such that $Q_\ell = a_j$.

We obtain from the above properties that

$$\sum_{\ell=1}^{n} (\delta_{R_\ell} - \delta_{Q_\ell}) = \sum_{j=1}^{k} d_j\, \delta_{a_j} \text{ in } \mathscr{D}'(\Omega). \tag{2.105}$$

We find that

$$\mathscr{H}^1([w = \alpha]) \ge \sum_{\ell=1}^{n} \mathscr{H}^1(C_\ell) \ge \sum_{\ell=1}^{n} d(R_\ell, Q_\ell) \ge L(a, d),$$

the latter inequality following from Remark 2.7 (see Fig. 2.8). □

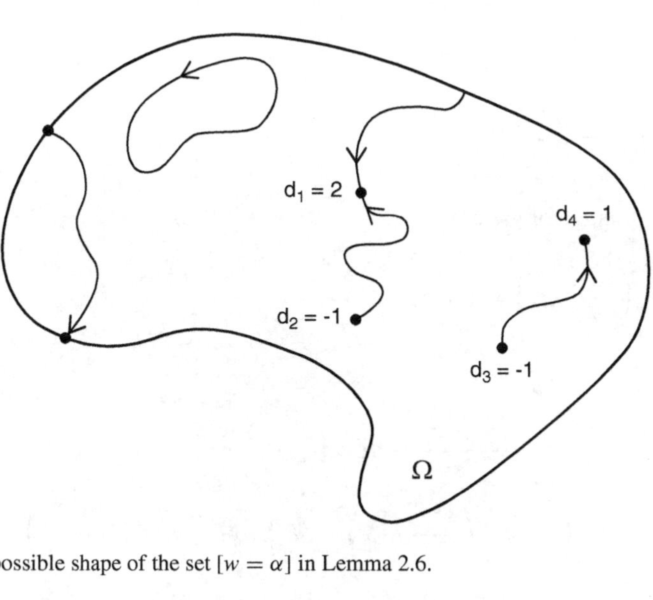

Fig. 2.8 A possible shape of the set $[w = \alpha]$ in Lemma 2.6.

2.5 $\Sigma(u) = S_1(u) \le 2\pi L(a, d)$ via the dipole construction

We first present the dipole construction in the simplest possible framework. In Chapter 1, we already mentioned a version of the dipole construction (see Proposition 1.1), but the setting was quite general and the simple geometric flavor was lost.

The dipole construction. Given $L > 0$ and $\varepsilon > 0$, consider the lozenge Q_ε in $\mathbb{R}^2$, defined by

$$Q_\varepsilon := \{(x_1, x_2);\ L|x_1| < \varepsilon x_2,\ 0 < x_2 < L\},$$

with height L and base of length 2ε.

Consider the function $v_\varepsilon : \{(x_1, x_2) \in \mathbb{R}^2;\ x_2 < L\} \to \mathbb{S}^1$, defined by

$$v_\varepsilon(x_1, x_2) := \begin{cases} -e^{-\imath\pi L x_1/(\varepsilon x_2)}, & \text{if } (x_1, x_2) \in Q_\varepsilon \\ 1, & \text{otherwise} \end{cases}.$$

Next, we consider the symmetry S with respect to the line (x_1, L), $x_1 \in \mathbb{R}$, and we define the map $v : \mathbb{R}^2 \to \mathbb{S}^1$, $v(x) := \begin{cases} v_\varepsilon(x), & \text{if } x_2 < L \\ v_\varepsilon(Sx), & \text{if } x_2 > L \end{cases}$.

Clearly, $(v - 1)$ has support in the region $\widetilde{Q}_\varepsilon := Q_\varepsilon \cup S(Q_\varepsilon)$. On the other hand, $v \in W^{1,\infty}_{loc}(\mathbb{R}^2 \setminus \{P, N\})$, where $P := (0, 0)$ and $N := (0, 2L)$. An easy computation shows that

$$\deg(v, P) = +1,\ \deg(v, N) = -1. \tag{2.106}$$

On the other hand, in Q_ε we have

$$|\nabla v_\varepsilon(x_1, x_2)| = \frac{\pi L}{\varepsilon} \frac{(x_1^2 + x_2^2)^{1/2}}{x_2^2}.$$

Using the change of variables

$$(t, x_2) \mapsto \left(\frac{\varepsilon x_2 t}{L}, x_2\right) = (x_1, x_2),$$

we obtain

$$\begin{aligned} \int_{Q_\varepsilon} |\nabla v_\varepsilon| &= \frac{\pi L}{\varepsilon} \int_{Q_\varepsilon} \frac{(x_1^2 + x_2^2)^{1/2}}{x_2^2} dx_1 dx_2 \\ &= \pi \int_0^L \left(\int_{-1}^1 \left(\frac{\varepsilon^2 t^2}{L^2} + 1\right)^{1/2} dt \right) dx_2 \le 2\pi L \left(1 + \frac{\varepsilon}{2L}\right). \end{aligned}$$

Consequently,

$$\int_{\mathbb{R}^2} |\nabla v| = 2 \int_{Q_\varepsilon} |\nabla v_\varepsilon| \le 2\pi |P - N| + 2\varepsilon\pi. \tag{2.107}$$

The same construction can be performed starting from arbitrary points $P, N \in \mathbb{R}^2$. In conclusion, the dipole construction yields the following

Proposition 2.1. *For every $P, N \in \mathbb{R}^2$ and $\varepsilon > 0$, there exists a map $w = w_{\varepsilon,P,N} \in W^{1,1}_{loc}(\mathbb{R}^2; \mathbb{S}^1) \cap C(\mathbb{R}^2 \setminus \{P, N\}; \mathbb{S}^1)$ such that $w - 1$ is supported in an ε-neighborhood of $[P, N]$ and, in addition*

$$Jw = \pi\,(\delta_P - \delta_N) \text{ and } \int_{\mathbb{R}^2} |\nabla w| \le 2\pi\,|P - N| + \varepsilon. \tag{2.108}$$

Proof of

$$S_1(u) \le 2\pi L(a, d). \tag{2.109}$$

Consider a minimal configuration $(P_\ell, N_\ell)_{\ell=1}^n$, associated with (a, d) in the sense of Lemma 2.4, and fix $\varepsilon > 0$. Using (1.56), (2.108), and the inequality

$$|u\, v|_{W^{1,1}} \le |u|_{W^{1,1}} + |v|_{W^{1,1}}, \ \forall\, u, v \in W^{1,1}_{loc}(\mathbb{R}^2; \mathbb{S}^1),$$

we obtain the existence of a map $w \in W^{1,1}_{loc}(\mathbb{R}^2; \mathbb{S}^1) \cap C(\mathbb{R}^2 \setminus \cup_\ell \{P_\ell, N_\ell\}; \mathbb{S}^1)$ such that

$$Jw = \pi \sum_{\ell=1}^{n} (\delta_{P_\ell} - \delta_{N_\ell}) = \pi \sum_{j=1}^{k} d_j \delta_{a_j} = Ju \tag{2.110}$$

and

$$\int_{\mathbb{R}^2} |\nabla w| \le 2\pi \sum_{\ell=1}^{n} |P_\ell - N_\ell| + n\varepsilon = 2\pi L(a, d) + n\varepsilon. \tag{2.111}$$

From (1.84), we know that

$$S_1(u) = \inf \left\{ \int_\Omega |\nabla w|;\ w \in W^{1,1}(\Omega; \mathbb{S}^1),\ Jw = Ju \right\}, \tag{2.112}$$

and therefore

$$S_1(u) \le 2\pi L(a, d) + n\varepsilon,$$

which implies (2.109), since $\varepsilon > 0$ is arbitrary. □

2.6 $E(u) \le \int_\Omega |\nabla u| + 2\pi L(a, d)$ via minimal configurations

In Chapter 1, we established in full generality that

$$E(u) = \int_\Omega |\nabla u| + \Sigma(u)$$

(see Theorem 1.5). When u is nice we know (see Theorem 2.3) that $\Sigma(u) = 2\pi L(a, d)$, and therefore

$$E(u) = \int_\Omega |\nabla u| + 2\pi L(a, d).$$

In this section, we present a *totally self-contained* and elementary proof of the inequality

$$E(u) \le \int_\Omega |\nabla u| + 2\pi L(a, d) \tag{2.113}$$

when u is nice. For this purpose we construct a *natural* lifting $\varphi \in BV$ of u such that

$$|\varphi|_{BV} \le \int_\Omega |\nabla u| + 2\pi L(a, d). \tag{2.114}$$

In the process, we make use *only* of the formula

$$Ju = \pi \sum_j d_j \delta_{a_j} \text{ in } \mathscr{D}'(\Omega) \tag{2.115}$$

(see Theorem 2.1) and of the concept of minimal configuration associated with (a, d) in the sense of Lemma 2.4.

We start with a simple lemma. Let $\gamma \subset \mathbb{R}^2$ be a smooth connected oriented curve going from N to P, with $N, P \in \overline{\Omega}$. Let τ be the unit tangent vector to γ and let $\nu := -\tau^\perp$, i.e., $\nu = (\tau_2, -\tau_1)$ is the oriented normal to γ.

Lemma 2.7. *Consider the $\mathbb{R}^2$-valued measure on Ω defined by*

$$\mu := \nu \mathscr{H}^1 \llcorner (\gamma \cap \Omega). \tag{2.116}$$

Then

$$\operatorname{curl} \mu = \delta_P - \delta_N \text{ in } \mathscr{D}'(\Omega). \tag{2.117}$$

Proof. If $\zeta \in C_c^\infty(\Omega)$, we have

$$\langle \operatorname{curl} \mu, \zeta \rangle = -\int_\gamma \nu \cdot \nabla^\perp \zeta = \int_\gamma \frac{\partial \zeta}{\partial \tau} = \zeta(P) - \zeta(N). \qquad \square$$

Given a minimal configuration $(P_\ell, N_\ell)_{\ell=1}^n$ associated with (a, d), set

$$\mu := \sum_{\ell=1}^n \nu_\ell \mathscr{H}^1 \llcorner [N_\ell, P_\ell], \tag{2.118}$$

where ν_ℓ is the normal associated with the oriented segment $[N_\ell, P_\ell]$. By Lemma 2.7 and (2.83), we have

$$\operatorname{curl} \mu = \sum_{j=1}^k d_j \delta_{a_j} \text{ in } \mathscr{D}'(\Omega). \tag{2.119}$$

The main result in this section is the following

Theorem 2.6. *Fix a nice $u \in W^{1,1}(\Omega; \mathbb{S}^1)$, and write $Ju = \pi \sum_j d_j \delta_{a_j}$. Given any minimal configuration associated with (a, d), there exists a unique (mod 2π) function $\varphi \in BV(\Omega; \mathbb{R})$ such that*

$$u = e^{\imath\varphi} \text{ in } \Omega \tag{2.120}$$

and

$$D\varphi = u \wedge \nabla u - 2\pi \mu \text{ in } \mathscr{D}'(\Omega), \tag{2.121}$$

where μ is defined in (2.118).
In particular,

$$|\varphi|_{BV} \leq \int_\Omega |\nabla u| + 2\pi L(a, d), \tag{2.122}$$

and consequently φ is a minimal lifting of u.

Proof. From (2.115), we know that

$$\operatorname{curl}(u \wedge \nabla u) = 2Ju = 2\pi \sum_{j=1}^{k} d_j \delta_{a_j} \text{ in } \mathscr{D}'(\Omega). \tag{2.123}$$

Combining (2.119) and (2.123), we see that

$$\operatorname{curl}(u \wedge \nabla u - 2\pi \mu) = 0 \text{ in } \mathscr{D}'(\Omega), \tag{2.124}$$

and therefore there exists $\varphi \in BV(\Omega; \mathbb{R})$ such that

$$D\varphi = u \wedge \nabla u - 2\pi \mu \text{ in } \mathscr{D}'(\Omega). \tag{2.125}$$

We claim that

$$e^{-\imath\varphi} u \text{ is constant in } \Omega; \tag{2.126}$$

this implies that $u = e^{\imath(\varphi + C)}$ for some appropriate constant $C \in \mathbb{R}$ and thus $\varphi + C$ satisfies (2.120) and (2.121).

Set $\Omega' := \Omega \setminus \bigcup_{\ell=1}^{n} [N_\ell, P_\ell]$. We deduce from (2.125) that

$$D\varphi = u \wedge \nabla u \text{ in } \mathscr{D}'(\Omega'), \tag{2.127}$$

and thus $\varphi \in W^{1,1}(\Omega'; \mathbb{R})$. If Ω' is connected, we conclude from (2.127) that

$$u = e^{\imath(\varphi + C)} \text{ in } \Omega',$$

for some constant C (see (1.53)) and therefore (2.126) holds.

When Ω' is not connected (and this can happen; see Fig. 2.5), the claim (2.126) is still valid, but the proof is slightly more complicated.

Let ω be an arbitrary disk such that

$$\overline{\omega} \subset \Omega \setminus \{a_1, \ldots, a_k\}. \tag{2.128}$$

Then the claim amounts to

$$e^{-\imath\varphi} u \text{ is constant in } \omega, \text{ for every } \omega \text{ as in (2.128)}. \tag{2.129}$$

Applying Corollary 1.1 in ω (recall that $u \in C(\overline{\omega})$), we may write

$$u = e^{\imath\eta} \text{ for some } \eta \in W^{1,1}(\omega; \mathbb{R}) \cap C(\overline{\omega}; \mathbb{R}) \tag{2.130}$$

so that, by (1.52),

$$\nabla\eta = u \wedge \nabla u \text{ in } \omega. \tag{2.131}$$

By (2.128) and Lemma 2.4, we have $N_\ell, P_\ell \notin \overline{\omega}$, $\ell = 1, \ldots, n$. Consequently, $[N_\ell, P_\ell] \cap \overline{\omega}$ is either empty, or a chord in $\overline{\omega}$.

Next, let

$$I := \{\ell \in \{1, \ldots, n\};\ [N_\ell, P_\ell] \cap \omega \neq \emptyset\}.$$

By the above, if $\ell \in I$, then we have $[N_\ell, P_\ell] \cap \overline{\omega} = [R_\ell, Q_\ell]$, with $R_\ell, Q_\ell \in \partial\omega$. The chord $[R_\ell, Q_\ell]$ cuts ω into two domains, $\omega_\ell^\pm$, such that ν_ℓ is an inward normal to ω_ℓ^+ (see Fig. 2.9 below). Let $\psi_\ell := \begin{cases} 0, & \text{in } \omega_\ell^- \\ 1, & \text{in } \omega_\ell^+ \end{cases}$, so that

$$D\psi_\ell = \nu_\ell \mathscr{H}^1 \llcorner ([R_\ell, Q_\ell]) \text{ in } \mathscr{D}'(\omega). \tag{2.132}$$

Fig. 2.9 .

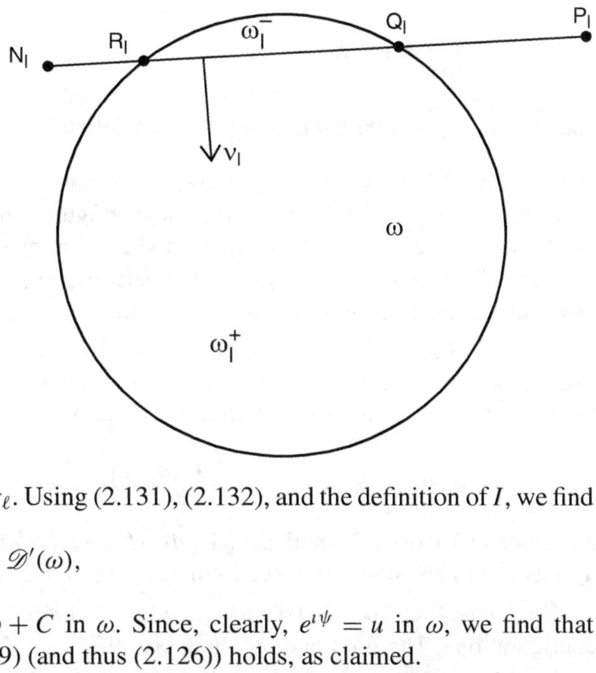

Set $\psi := \eta - 2\pi \sum_{\ell \in I} \psi_\ell$. Using (2.131), (2.132), and the definition of I, we find

$$D\psi = u \wedge \nabla u - 2\pi\mu \text{ in } \mathscr{D}'(\omega),$$

so that (by (2.125)) $\psi = \varphi + C$ in ω. Since, clearly, $e^{\imath\psi} = u$ in ω, we find that $e^{-\imath\varphi} u = e^{\imath C}$ in ω, i.e., (2.129) (and thus (2.126)) holds, as claimed.

Uniqueness of φ follows from (2.121) and the fact that Ω is connected.

Finally, we have

$$|\varphi|_{BV} = \|D\varphi\|_{\mathcal{M}} \leq \int_\Omega |u \wedge \nabla u| + 2\pi \|\mu\|_{\mathcal{M}} \leq \int_\Omega |\nabla u| + 2\pi \sum_{\ell=1}^{n} |P_\ell - N_\ell|$$
$$= \int_\Omega |\nabla u| + 2\pi L(a, d) \text{ (by (2.80))},$$

which implies (2.122).

From (2.122) and (2.87), we have

$$|\varphi|_{BV} \leq \int_\Omega |\nabla u| + \Sigma(u).$$

Applying the fundamental equality

$$E(u) = \int_\Omega |\nabla u| + \Sigma(u)$$

(see Theorem 1.5) we conclude that

$$|\varphi|_{BV} = \int_\Omega |\nabla u| + \Sigma(u), \tag{2.133}$$

and therefore φ is a *minimal* lifting of u, as defined in Section 1.4.2. □

Remark 2.9. We may now summarize the situation as follows. *Fix* a nice u and write $Ju = \pi \sum_j d_j \delta_{a_j}$. Given any minimal configuration $(P_\ell, N_\ell)_{\ell=1}^n$ associated with (a, d), we have constructed a unique (mod 2π) *minimal* lifting $\varphi \in BV(\Omega; \mathbb{R})$ of u such that $D\varphi = u \wedge \nabla u - 2\pi \mu$, or, equivalently, $D^j \varphi = -2\pi \mu$, i.e., φ jumps precisely along the given minimal configuration.

It turns out that *all* minimal liftings of u arise from such a construction. Surprisingly, this fact is non-trivial and requires an excursion via *minimal connections* (see Sections 2.7, 2.8, and Appendix 15.8). In other words, the mapping

$$\Phi : \text{Minimal configuration} \mapsto \text{Minimal lifting}$$

described in Theorem 2.6 and given by $\Phi\left((P_\ell, N_\ell)_{\ell=1}^n\right) := \varphi$, is surjective. However, Φ may fail to be one-to-one (see Remarks 2.10 and 2.12).

This "defect" can be fixed if we work with minimal connections instead of minimal configurations. The final outcome is a beautiful bijective correspondence between analytical objects (minimal liftings) and geometric objects (minimal connections); see Theorem 2.9.

2.7 Connections associated with (a, d). Minimal connections

Our next goal is to give a geometrical characterization of *all BV* liftings and minimal BV liftings corresponding to nice u's. By Theorem 2.2, for such u we have $Ju = \pi \sum d_j \delta_{a_j}$. If φ is a BV lifting of u, its jump set consists, roughly speaking, of a collection of curves "connecting" the points a_j. Our purpose in this section is to give a precise meaning to the concept of connection (resp. minimal connection) connecting the a_j's. For a while, we ignore u and focus on the pair (a, d) (as we have done in Section 2.2). Here, a is any collection of distinct points in Ω, and d is any collection of integers. We will return to u in Section 2.8.

Before presenting the definition of connections, let us give precise definitions of some types of arcs and curves in $\mathbb{R}^2$ we will work with *in this section.* A *C^1 arc* is $f((0,1))$, for some C^1 map $f : [0,1] \to \mathbb{R}^2$, one-to-one and such that $f'(t) \neq 0$, $\forall\, t \in [0,1]$. A C^1 *curve* is the image $f([0,1])$ of some C^1 map $f : [0,1] \to \mathbb{R}^2$ such that f is one-to-one on $[0,1)$, $f(0) = f(1)$, $f'(0) = f'(1)$, $f'(t) \neq 0$, $\forall\, t \in [0,1]$.

More general objects are the rectifiable Jordan arcs or curves. Following Young [347] (see also the more modern presentation in Ambrosio, Caselles, Masnou, and Morel [10, Section 8]), a *rectifiable Jordan arc* is $\gamma := f([0,1])$, with $f : [0,1] \to \mathbb{R}^2$ Lipschitz and one-to-one. A *rectifiable Jordan curve* is given by a Lipschitz function $f : [0,1] \to \mathbb{R}^2$ such that f is one-to-one on $[0,1)$, and $f(0) = f(1)$. A rectifiable Jordan arc (or curve) has finite length. Given a parametrization, we may define, for $\mathcal{H}^1$-a.e. $x = f(t) \in \gamma$, the tangent vector $\tau := f'(t)/|f'(t)|$ at x. This coincides with the classical tangent unit vector when $f \in C^1$.

Clearly, two equivalent parametrizations f_1, f_2 (i.e., two parametrizations f_1, f_2 such that $(f_2)^{-1} \circ f_1$ is increasing) yield, for $\mathcal{H}^1$-a.e. $x \in \gamma$, the same tangent vector τ. We then set $\nu := -\tau^\perp$, i.e.,

$$\nu(x) := \frac{(f_2'(t), -f_1'(t))}{|f'(t)|} \text{ for } \mathcal{H}^1 - \text{a.e. } x = f(t) \in \gamma. \tag{2.134}$$

We define $\overline{\mu} := \nu \mathcal{H}^1 \llcorner \gamma$. Up to a sign, $\overline{\mu}$ is independent of the parametrization f. The measure $\overline{\mu}$ has mass $\mathcal{H}^1(\gamma)$. It acts on vector fields $\xi \in C^\infty(\mathbb{R}^2; \mathbb{R}^2)$ as follows:

$$\begin{aligned}\langle \nu \mathcal{H}^1 \llcorner \gamma, \xi \rangle &= \int_\gamma \nu \cdot \xi \, d\mathcal{H}^1 = -\int_\gamma \tau^\perp \cdot \xi \, d\mathcal{H}^1 \\ &= \int_0^1 [\xi_1(f(t)) f_2'(t) - \xi_2(f(t)) f_1'(t)]\, dt.\end{aligned} \tag{2.135}$$

First, a definition.

Definition 2.1. A *connection* μ associated with (a, d) is an $\mathbb{R}^2$-valued measure μ on Ω such that:

$$\mu = \sum_i \nu_i \mathcal{H}^1 \llcorner (S_i \cap \Omega), \tag{2.136}$$

where (S_i) is a countable collection of Borel subsets S_i of C^1 oriented curves $\Gamma_i \subset \mathbb{R}^2$ with unit normals ν_i, such that $\sum \mathcal{H}^1(S_i \cap \Omega) < \infty$, and

$$\operatorname{curl} \mu = \sum_{j=1}^k d_j \delta_{a_j} \text{ in } \mathcal{D}'(\Omega). \tag{2.137}$$

We emphasize that the S_i's need not be disjoint and thus we allow multiplicities.

Here are some examples.

Example 2.8. Let $P, N \in \Omega$, and let $\Gamma \subset \mathbb{R}^2$ be a smooth connected curve oriented from N to P. Then, by Lemma 2.7,

$$\operatorname{curl} \left(\nu \mathcal{H}^1 \llcorner (\Gamma \cap \Omega)\right) = \delta_P - \delta_N \text{ in } \mathcal{D}'(\Omega). \tag{2.138}$$

In other words, $\mu := \nu \mathcal{H}^1 \llcorner (\Gamma \cap \Omega)$ is a connection associated with $a := (P, N)$ and the collection of integers $d := (+1, -1)$.

Example 2.9. If, in the above example, we let $P \in \Omega$ and $N \in \partial\Omega$, then

$$\operatorname{curl} \left(\nu \mathcal{H}^1 \llcorner (\Gamma \cap \Omega)\right) = \delta_P \text{ in } \mathcal{D}'(\Omega),$$

and thus $\nu \mathcal{H}^1 \llcorner (\Gamma \cap \Omega)$ is a connection associated with (P) and $(+1)$.

Example 2.10. Consider a finite collection of curves Γ_i from N_i to P_i, as above. Then, $\mu := \sum \nu_i \mathcal{H}^1 \llcorner \Gamma_i$ is a connection associated with (a, d) if and only if

$$\sum_i (\delta_{P_i} - \delta_{N_i}) = \sum_{j=1}^k d_j \delta_{a_j} \text{ in } \mathcal{D}'(\Omega). \tag{2.139}$$

This happens, for example, if for each j the number of curves ending at a_j minus the number of curves starting at a_j equals d_j.

The above examples are generic. Indeed, using deep results from geometric measure theory (see Fleming [151] and Federer [147, Section 4.2.25]; see also Ambrosio, Caselles, Masnou, and Morel [10]), one may prove the following

Theorem 2.7. *Every connection μ associated with (a, d) is a (possibly infinite) sum of the form*

$$\mu = \sum_i \nu_i \mathscr{H}^1 \llcorner \gamma_i,$$

where every γ_i *is an oriented rectifiable Jordan arc or curve contained in* $\overline{\Omega}$ *and*

$$\sum_i \mathscr{H}^1(\gamma_i) < \infty.$$

Moreover, for each i:

1. *Either* $\gamma_i \subset \Omega$ *and* γ_i *connects some* a_j *to some* a_m
2. *Or* $\gamma_i \subset \Omega \cup \{b\}$, *with* $b \in \partial\Omega$, *and* γ_i *connects some* a_j *to* b
3. *Or* $\gamma_i \subset \Omega \cup \{b, c\}$, *with* $b, c \in \partial\Omega$, $b \neq c$, *and* γ_i *connects* b *to* c
4. *Or* γ_i *is a closed curve in* $\overline{\Omega}$ *such that* $\gamma_i \cap \partial\Omega$ *consists of at most one point.*

The proof is presented in Appendix 15.8.

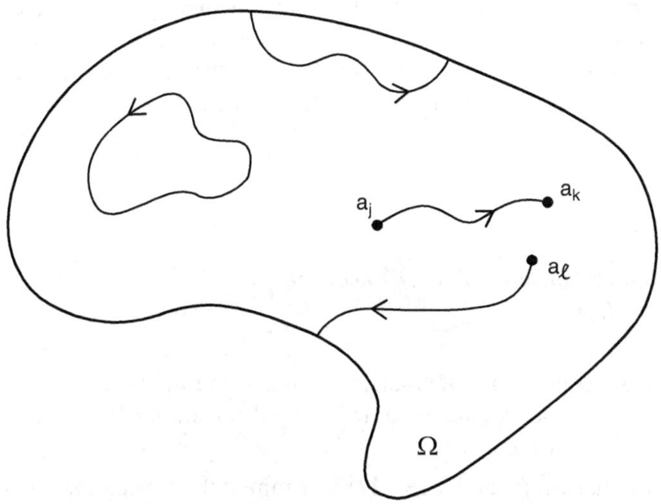

Fig. 2.10 A collection of curves associated with a connection.

Next, we present a useful property of connections.

Proposition 2.2. *We have*

$$\min\{\|\mu\|_{\mathcal{M}};\ \mu \text{ is a connection associated with } (a,d)\} = L(a,d). \tag{2.140}$$

Proof. Inequality "$\geq$" holds in greater generality: if μ is any $\mathbb{R}^2$-valued measure satisfying (2.137), then

$$\|\mu\|_{\mathcal{M}} \geq L(a,d). \tag{2.141}$$

Indeed, if $\zeta \in C_c^\infty(\Omega)$ and $\|\nabla\zeta\|_{L^\infty} \leq 1$, then we have

$$\sum_{j=1}^{k} d_j\zeta(a_j) = \langle \operatorname{curl}\mu, \zeta\rangle = \int \left(\frac{\partial\zeta}{\partial x_2}\,d\mu_1 - \frac{\partial\zeta}{\partial x_1}\,d\mu_2\right) \leq \|\mu\|_{\mathcal{M}}.$$

We conclude by taking sup over ζ in the above inequality and using Theorem 2.3.

For the reverse inequality, let $(P_\ell, N_\ell)_{\ell=1}^n$ be a minimal configuration associated with (a,d) in the sense of Lemma 2.4, and let γ_ℓ be the oriented line segments from N_ℓ to P_ℓ. Then the measure μ_0 defined in (2.118), i.e.,

$$\mu_0 = \sum_{\ell=1}^{n} \nu_\ell \mathcal{H}^1 \llcorner \gamma_\ell, \tag{2.142}$$

is a connection associated with (a,d) and $\|\mu_0\|_{\mathcal{M}} \leq \sum_{\ell=1}^n |P_\ell - N_\ell| = L(a,d)$, so that $\|\mu_0\|_{\mathcal{M}} = L(a,d)$. □

Fix (a,d). A special class of connections plays an important role. By definition, a *minimal connection* associated with (a,d) is a connection μ associated with (a,d) such that $\|\mu\|_{\mathcal{M}} = L(a,d)$.

Let us note that μ_0 given by (2.142) is a minimal connection. The converse is true: minimal connections are precisely of this form, but this is a deeper result.

Theorem 2.8. *Let μ be a minimal connection associated with (a,d). Then, $\mu = \sum_{\ell=1}^n \nu_\ell \mathcal{H}^1 \llcorner \gamma_\ell$, where γ_ℓ is the oriented line segment from N_ℓ to P_ℓ for some minimal configuration $(P_\ell, N_\ell)_{\ell=1}^n$ associated with (a,d).*

The proof of Theorem 2.8 is presented in Appendix 15.8.

Remark 2.10. Minimal configurations versus minimal connections. A word of warning. A minimal configuration is a collection of oriented line segments. A minimal connection is an $\mathbb{R}^2$-valued measure. In view of Theorem 2.8, the mapping

$$\Psi : \text{Minimal configuration} \mapsto \text{Minimal connection given by (2.118)},$$

i.e., $\Psi\left((P_\ell, N_\ell)_{\ell=1}^n\right) := \sum_{\ell=1}^{n} \nu_\ell \mathscr{H}^1 \llcorner [N_\ell, P_\ell]$ is surjective. However, Ψ is *not* one-to-one. In Fig. 2.11, we give an example of a pair (a, d) that admits two distinct minimal configurations generating the same minimal connection

$$\mu = \nu\,\mathscr{H}^1 \llcorner [a_1, a_2] + 2\,\nu\,\mathscr{H}^1 \llcorner [a_2, a_3] + \nu\,\mathscr{H}^1 \llcorner [a_3, a_4].$$

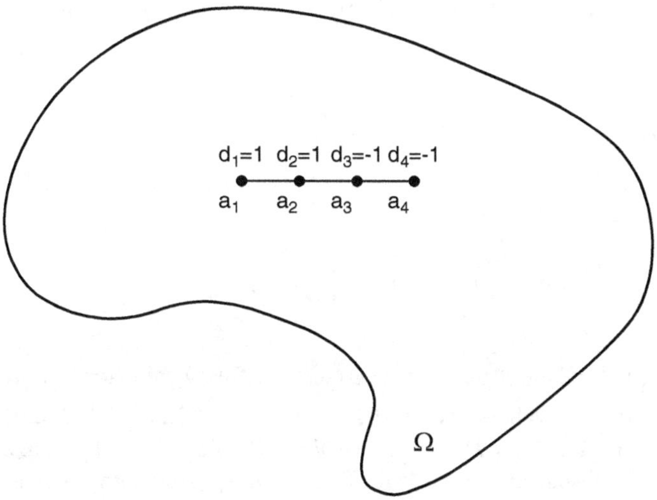

Fig. 2.11 Here, the points a_1, a_2, a_3, a_4 are well-inside Ω. There are two distinct minimal configurations: $[P_1, N_1] \cup [P_2, N_2]$, where $P_1 = a_1, P_2 = a_2, N_1 = a_3, N_2 = a_4$, resp. $[P_1, N_1'] \cup [P_2, N_2']$, where $N_1' = a_4, N_2' = a_3$.

2.8 Returning to u: a bijective correspondence between BV liftings and connections

Fix $u \in W^{1,1}(\Omega; \mathbb{S}^1) \cap C(\Omega \setminus \{a_1, \ldots, a_k\})$ and set $d := (d_1, \ldots, d_k)$, with $d_j := \deg(u, a_j)$.

The basic result of this section is the following

Theorem 2.9. *We have*

1. *Let* $\varphi \in BV(\Omega; \mathbb{R})$ *be a lifting of* u*. Then,* $\mu := -\dfrac{1}{2\pi} D^j\varphi$ *is a connection associated with* (a, d)*.*

2. *Conversely, given any connection* μ *associated with* (a, d), *there exists a unique (mod* 2π*) lifting* $\psi \in BV(\Omega; \mathbb{R})$ *of* u *such that* $D^j\psi = -2\pi\mu$, *or equivalently* $D\psi = u \wedge \nabla u - 2\pi\mu$.
3. *In this bijective correspondence, minimal liftings correspond to minimal connections.*

Remark 2.11. It is not difficult to see that the above theorem also holds if we merely assume that $u \in W^{1,1}(\Omega; \mathbb{S}^1)$ satisfies $Ju = \pi \sum d_j\delta_{a_j}$.

Proof. 1. Recall that (see (1.104)–(1.106)) $D\varphi = \nabla\varphi + D^c\varphi + D^j\varphi$, with $\nabla\varphi = u \wedge \nabla u$ and $D^c\varphi = 0$.

Thus,

$$D^j\varphi = D\varphi - u \wedge \nabla u,$$

so that

$$\operatorname{curl} \mu = \frac{1}{2\pi} \operatorname{curl}(u \wedge \nabla u) = \frac{1}{\pi} Ju = \sum_{j=1}^{k} d_j\delta_{a_j}. \tag{2.143}$$

Using (2.143) and the structure of $D^j\varphi$ (see (1.107)), we find that μ is a connection.

2. Let F be the $\mathbb{R}^2$-valued measure $F := u \wedge \nabla u - 2\pi\mu$. Then, $\operatorname{curl} F = 0$, by assumption, and thus $F = D\psi$ for some $\psi \in BV(\Omega; \mathbb{R})$. This ψ satisfies $\nabla\psi = u \wedge \nabla u$, $D^c\psi = 0$ and $D^j\psi = -2\pi\mu$. Taking the specific form of μ into account, we find that

$$\psi^+ - \psi^- \in 2\pi\mathbb{Z}\ \mathscr{H}^1 - \text{a.e. on } J_\psi. \tag{2.144}$$

This is intuitively clear, at least when there is only one S_i (see Definition 2.1); for a justification of (2.144) in general, see Lemma 15.31 in Appendix 15.7.

In order to conclude, it suffices to prove that $u = e^{\imath(\psi + C)}$ for appropriate C. Here is a proof. Volpert's chain rule (see (1.109) and Appendix 15.7) applied to $w := e^{i\psi}$ yields

$$D^a w = \imath\, w\nabla\psi = \imath\, w(u \wedge \nabla u),\ D^c w = 0,$$

and

$$D^j w = \sum_k \left(e^{\imath\psi^+} - e^{\imath\psi^-}\right) \nu_k \mathscr{H}^1 \llcorner (S_k \cap \Omega),$$

where ν_k and S_k come from (2.136). It follows from (2.144) that $D^j w = 0$. Hence, $w \in W^{1,1}(\Omega; \mathbb{S}^1)$, with $\nabla w = \imath\, w(u \wedge \nabla u)$ and

$$\nabla(u\overline{w}) = \overline{w}\nabla u + u\nabla\overline{w} = \overline{w}\nabla u - \imath u\overline{w}(u \wedge \nabla u) =$$
$$= -\imath u\overline{w}(\imath\overline{u}\nabla u + u \wedge \nabla u) = 0, \text{ by (1.46)}.$$

Therefore, $u\overline{w} = e^{\imath C}$ for some constant C, so that $u = e^{\imath(\psi + C)}$.

3. From the above we have (as in (1.97)),

$$\|D\psi\|_{\mathcal{M}} = \|u \wedge \nabla u\|_{L^1} + \|D^j\psi\|_{\mathcal{M}} = \|\nabla u\|_{L^1} + 2\pi\,\|\mu\|_{\mathcal{M}},$$

by (1.47). We conclude that minimal liftings are in correspondence with connections of minimal mass, that is, minimal connections. □

Remark 2.12. Minimal liftings, minimal configurations, and minimal connections. A final word. Recall that we have introduced in Remark 2.9 the map

$$\Phi : \text{Minimal configuration} \mapsto \text{Minimal lifting},$$

defined by $\Phi\left((P_\ell, N_\ell)_{\ell=1}^n\right) := \varphi$, where φ is the minimal lifting described in Theorem 2.6, so that φ satisfies $D^j\varphi = -2\pi \sum_{\ell=1}^{n} \nu_\ell \mathcal{H}^1 \llcorner [N_\ell, P_\ell]$.

In Remark 2.10, we have introduced the map

$$\Psi : \text{Minimal configuration} \mapsto \text{Minimal connection},$$

defined by $\Psi\left((P_\ell, N_\ell)_{\ell=1}^n\right) := \sum_{\ell=1}^{n} \nu_\ell \mathcal{H}^1 \llcorner [N_\ell, P_\ell]$.

Finally, in Theorem 2.9, we have constructed a map

$$\Theta : \text{Minimal lifting} \mapsto \text{Minimal connection},$$

defined by $\Theta(\varphi) := -(1/2\pi)D^j\varphi$. Clearly, $\Psi = \Theta \circ \Phi$. The map Θ is bijective; the maps Φ and Ψ are surjective, but they may fail to be one-to-one.

2.9 Describing Ju and $\Sigma(u)$ for a general $u \in W^{1,1}$

In this section, we investigate Ju, $\Sigma(u)$ and the connections (resp. minimal connections) associated with an arbitrary $u \in W^{1,1}(\Omega; \mathbb{S}^1)$, not necessarily nice. We first describe an object that plays a basic role and replaces the measure $\sum_{j=1}^{k} d_j\delta_{a_j}$.

2.9.1 The class $\mathscr{E}$

The class $\mathscr{E}$ is the subset of $\mathscr{D}'(\Omega)$ defined by

$$\mathscr{E} := \left\{ T = \sum_{j=1}^{\infty} (\delta_{P_j} - \delta_{N_j});\ P_j, N_j \in \overline{\Omega}, \sum_{j=1}^{\infty} |P_j - N_j| < \infty \right\}; \tag{2.145}$$

here, it may happen that $P_i = P_j$, or $N_i = N_j$, or $P_i = N_j$ when $i \neq j$.

Clearly, we have $\mathscr{E} \subset [W_0^{1,\infty}(\Omega)]^*$ and thus, for $T \in \mathscr{E}$ (see (1.152)),

$$\|T\|_W := \sup \left\{ \langle T, \zeta \rangle;\ \zeta \in W_0^{1,\infty}(\Omega), \|\nabla \zeta\|_{L^\infty} \leq 1 \right\} < \infty. \tag{2.146}$$

The next result is in the spirit of Section 2.2.

Proposition 2.3. *Let $T \in \mathscr{E}$. Then,*

$$\|T\|_W = \inf \left\{ \sum_{j=1}^{\infty} |\widetilde{P}_j - \widetilde{N}_j|;\ \widetilde{P}_j, \widetilde{N}_j \in \overline{\Omega}, \sum_{j=1}^{\infty} |\widetilde{P}_j - \widetilde{N}_j| < \infty \right.$$
$$\left. \text{and } \sum_{j=1}^{\infty} (\delta_{\widetilde{P}_j} - \delta_{\widetilde{N}_j}) = T \text{ in } \mathscr{D}'(\Omega) \right\}. \tag{2.147}$$

Proof. Let A be the right-hand side in (2.147). The inequality $\|T\|_W \leq A$ is obvious. To establish that $\|T\|_W \geq A$, let $T = \sum_{j=1}^{\infty} (\delta_{P_j} - \delta_{N_j})$ as in (2.145). We fix ε and let $k = k(\varepsilon)$ be such that

$$\sum_{j=k+1}^{\infty} |P_j - N_j| < \varepsilon. \tag{2.148}$$

Applying Lemma 2.4 to $T' := \sum_{j=1}^{k} (\delta_{P_j} - \delta_{N_j})$ we obtain:

a) Points P'_m, N'_m in $\overline{\Omega}$ with $m = 1, \ldots, k'$, $k' \geq k$, such that

$$\sum_{m=1}^{k'} (\delta_{P'_m} - \delta_{N'_m}) = T' \tag{2.149}$$

b) Some $\zeta_0 \in W_0^{1,\infty}(\Omega)$ such that $\|\nabla \zeta_0\|_{L^\infty} \leq 1$ and

$$\sum_{m=1}^{k'} (\zeta_0(P'_m) - \zeta_0(N'_m)) = \sum_{m=1}^{k'} |P'_m - N'_m|. \tag{2.150}$$

Denote by $(\widetilde{P}_\ell, \widetilde{N}_\ell)$, $1 \le \ell < \infty$, the collection consisting of the points P'_m, N'_m, $m = 1, \dots, k'$, and the points P_j, N_j for $j \ge k+1$.

Clearly,

$$\sum_{\ell=1}^{\infty} (\delta_{\widetilde{P}_\ell} - \delta_{\widetilde{N}_\ell}) = T$$

and

$$\begin{aligned}
\sum_{\ell=1}^{\infty} (\zeta_0(\widetilde{P}_\ell) - \zeta_0(\widetilde{N}_\ell)) &= \sum_{m=1}^{k'} (\zeta_0(P'_m) - \zeta_0(N'_m)) + \sum_{j=k+1}^{\infty} (\zeta_0(P_j) - \zeta_0(N_j)) \\
&\ge \sum_{m=1}^{k'} |P'_m - N'_m| - \varepsilon = \sum_{\ell=1}^{\infty} |\widetilde{P}_\ell - \widetilde{N}_\ell| - \sum_{j=k+1}^{\infty} |P_j - N_j| - \varepsilon \\
&\ge \sum_{\ell=1}^{\infty} |\widetilde{P}_\ell - \widetilde{N}_\ell| - 2\varepsilon,
\end{aligned}$$

by (2.148) and (2.150). Therefore,

$$\|T\|_W \ge \sum_{\ell=1}^{\infty} (\zeta_0(\widetilde{P}_\ell) - \zeta_0(\widetilde{N}_\ell)) \ge \sum_{\ell=1}^{\infty} |\widetilde{P}_\ell - \widetilde{N}_\ell| - 2\varepsilon \ge A - 2\varepsilon.$$

Since ε is arbitrary, we conclude that $\|T\|_W \ge A$. □

It is easy to see that the sup in (2.146) is achieved (for every $T \in \mathscr{E}$). We warn the reader that the inf in (2.147) need *not* be achieved.

Proposition 2.4. *There exists some $T \in \mathscr{E}$ such that for every $(\widetilde{P}_j, \widetilde{N}_j)$ satisfying*

$$\sum_{j=1}^{\infty} |\widetilde{P}_j - \widetilde{N}_j| < \infty \text{ and } \sum_{j=1}^{\infty} (\delta_{\widetilde{P}_j} - \delta_{\widetilde{N}_j}) = T,$$

we have $\sum_{j=1}^{\infty} |\widetilde{P}_j - \widetilde{N}_j| > \|T\|_W$.

An explicit example of such pathological T is presented in Appendix 15.9 (see Proposition 15.3). This example shows that a general $u \in W^{1,1}(\Omega; \mathbb{S}^1)$ may fail to admit minimal configurations (as in the case of nice u's). However, it *always* admits minimal connections.

We conclude this subsection with a useful property of $\mathscr{E}$.

Proposition 2.5. *The set $\mathscr{E}$ is closed in $[W_0^{1,\infty}(\Omega)]^*$.*

Proof. Let $(T_k) \subset \mathscr{E}$ be such that $T_k \to T$ in $[W_0^{1,\infty}(\Omega)]^*$. We may extract a subsequence (still denoted by (T_k)) such that

$$\|T_k - T_{k+1}\|_{[W_0^{1,\infty}(\Omega)]^*} \leq 2^{-k}, \ \forall k \geq 1.$$

On the one hand, we have

$$T_1 - T = \sum_{k=1}^{\infty}(T_k - T_{k+1}) \text{ in } [W_0^{1,\infty}(\Omega)]^*. \tag{2.151}$$

On the other hand, we may write, for each k,

$$T_k - T_{k+1} = \sum_{j=1}^{\infty}\left(\delta_{P_j^k} - \delta_{N_j^k}\right), \tag{2.152}$$

for some sequences $(P_j^k, N_j^k)_{j=1}^{\infty}$ such that

$$\sum_{j=1}^{\infty}\left|P_j^k - N_j^k\right| \leq \|T_k - T_{k+1}\|_{[W_0^{1,\infty}(\Omega)]^*} + 2^{-k} \leq 2^{-(k-1)}. \tag{2.153}$$

Thus,

$$\sum_{k=1}^{\infty}\sum_{j=1}^{\infty}\left|P_j^k - N_j^k\right| < \infty,$$

so that $\sum_{k=1}^{\infty}(T_k - T_{k+1}) \in \mathscr{E}$ (by (2.152) and (2.153)) and $T \in \mathscr{E}$ (by (2.151)). □

2.9.2 *Ju* is an infinite sum of Dirac masses. And conversely

We have seen in the previous sections that, if $u \in W^{1,1}(\Omega; \mathbb{S}^1) \cap C(\Omega \setminus \{a_1, \ldots, a_k\})$, then $Ju = \pi \sum d_j \delta_{a_j}$ for appropriate integers d_j. Recall that Ju is well-defined in $\mathscr{D}'(\Omega)$ for every $u \in W^{1,1}(\Omega; \mathbb{S}^1)$; more precisely $Ju \in [W_0^{1,\infty}(\Omega)]^*$. The next result gives the form of Ju for an arbitrary u.

Theorem 2.10. *Let $u \in W^{1,1}(\Omega; \mathbb{S}^1)$. Then there exist points $P_j, N_j \in \overline{\Omega}$ such that*

$$\sum |P_j - N_j| < \infty \tag{2.154}$$

and

$$Ju = \pi \sum (\delta_{P_j} - \delta_{N_j}) \text{ in } [W_0^{1,\infty}(\Omega)]^* \text{ (and thus in } \mathscr{D}'(\Omega)). \tag{2.155}$$

In addition, we have

$$\begin{aligned}\Sigma(u) &= 2\|Ju\|_W \\ &= 2\pi \inf \Big\{ \sum |\widetilde{P}_j - \widetilde{N}_j|;\ \widetilde{P}_j, \widetilde{N}_j \in \overline{\Omega} \text{ and } \sum(\delta_{\widetilde{P}_j} - \delta_{\widetilde{N}_j}) \\ &\qquad = \sum \left(\delta_{P_j} - \delta_{N_j}\right) \text{ in } \mathscr{D}'(\Omega)\Big\}.\end{aligned} \tag{2.156}$$

Proof. From the density result of Bethuel and Zheng [37] (see also Theorem 10.3 with $s = 1$ and $p = 1$), we know that nice maps are dense in $W^{1,1}(\Omega; \mathbb{S}^1)$. Let (u_k) be a sequence of nice maps such that $u_k \to u$ in $W^{1,1}$. By Remark 1.4, we have $Ju_k \to Ju$ in $[W_0^{1,\infty}(\Omega)]^*$, and thus $(1/\pi)\, Ju \in \mathscr{E}$, by Proposition 2.5. □

Theorem 2.11. *Conversely, given $P_j, N_j \in \overline{\Omega}$ such that $\sum |P_j - N_j| < \infty$, there exists some $u \in W^{1,1}(\Omega; \mathbb{S}^1)$ such that $Ju = \pi \sum(\delta_{P_j} - \delta_{N_j})$ in $\mathscr{D}'(\Omega)$.*

Proof. Let u_j satisfy

$$Ju_j = \pi(\delta_{P_j} - \delta_{N_j}),\ \int_\Omega |\nabla u_j| \leq 3\pi |P_j - N_j|, \text{ and } |A_j| \leq \frac{1}{2^j}, \tag{2.157}$$

where

$$A_j := [u_j \neq 1]. \tag{2.158}$$

[Existence of u_j follows from Proposition 2.1.] Assuming that

$$\prod_{j=1}^{\infty} u_j \text{ converges in } W^{1,1} \text{ to some } u \in W^{1,1}(\Omega; \mathbb{S}^1), \tag{2.159}$$

we have

$$Ju = \lim_{m \to \infty} J\left(\prod_{j=1}^{m} u_j\right) = \pi \sum_j (\delta_{P_j} - \delta_{N_j}).$$

It thus remains to prove (2.159). Set

$$F := \sum_{j=1}^{\infty} |\nabla u_j|,\ v_n := \prod_{j=1}^{n} u_j \text{ and, if } m > n,\ v_{n,m} := \prod_{j=n+1}^{m} u_j \text{ and } B_{n,m} := \bigcup_{j=n+1}^{m} A_j.$$

By (2.157) and the assumption $\sum_{j=1}^{\infty} |P_j - N_j| < \infty$, we have

$$F \in L^1(\Omega), \ |\nabla v_n| \leq F, \ \|\nabla v_{n,m}\|_{L^1} \leq 3\pi \sum_{j=n+1}^{m} |P_j - N_j|,$$

$|B_{n,m}| \leq 2^{-n}$ and $v_{n,m} = 1$ in $\Omega \setminus B_{n,m}$.

For $m > n$, we have

$$\begin{aligned} \|\nabla v_m - \nabla v_n\|_{L^1} &= \|\nabla[v_n(v_{n,m} - 1)]\|_{L^1} \leq 2\|\nabla v_n\|_{L^1(B_{n,m})} + \|\nabla v_{n,m}\|_{L^1} \\ &\leq 2\|F\|_{L^1(B_{n,m})} + 3\pi \sum_{j\geq n+1} |P_j - N_j| \to 0 \text{ as } n, m \to \infty. \end{aligned} \tag{2.160}$$

On the other hand,

$$\|v_m - v_n\|_{L^1} = \|v_n(v_{n,m} - 1)\|_{L^1} \leq 2|B_{n,m}| \to 0 \text{ as } n, m \to \infty. \tag{2.161}$$

We obtain (2.159) from (2.160) and (2.161). □

2.9.3 Connections and minimal connections revisited

In this section, we study the connections (resp. minimal connections) associated with an *arbitrary* $u \in W^{1,1}(\Omega; \mathbb{S}^1)$, not necessarily nice.

A *connection* μ associated with $T \in \mathscr{E}$ is an $\mathbb{R}^2$-valued measure μ on Ω satisfying:

$$\mu = \sum \nu_j \mathscr{H}^1 \llcorner (S_j \cap \Omega), \tag{2.162}$$

where (S_j) is a countable collection of Borel subsets S_j of C^1 oriented curves $\gamma_j \subset \mathbb{R}^2$ with normals ν_j, such that $\sum \mathscr{H}^1(S_j \cap \Omega) < \infty$, and

$$\operatorname{curl} \mu = T \text{ in } \mathscr{D}'(\Omega). \tag{2.163}$$

By the argument leading to (2.141), we have

$$\|\mu\|_{\mathscr{M}} \geq \|T\|_W, \ \forall \mu \in \mathscr{M}(\Omega; \mathbb{R}^2) \text{ such that } \operatorname{curl} \mu = T. \tag{2.164}$$

We claim that

$$\inf \{\|\mu\|_{\mathscr{M}}; \ \mu \text{ is a connection associated with } T\} = \|T\|_W. \tag{2.165}$$

Indeed, let $\varepsilon > 0$ and choose (P_j, N_j) such that $T = \sum(\delta_{P_j} - \delta_{N_j})$ and $\sum |P_j - N_j| \le \|T\|_W + \varepsilon$ (see Proposition 2.3). Let $\mu := \sum \nu_j \mathcal{H}^1 \llcorner [N_j, P_j]$. Then, $\|\mu\|_{\mathcal{M}} \le \|T\|_W + \varepsilon$ and $\operatorname{curl} \mu = T$.

By definition, a *minimal connection* is a connection μ_0 associated with T and such that $\|\mu_0\|_{\mathcal{M}} = \|T\|_W$. Surprisingly (in view of Proposition 2.4), minimal connections always exist, but this is not as straightforward as in the case where T is a finite sum of Dirac masses.

Proposition 2.6. *Let $T \in \mathcal{E}$. Then there exists a minimal connection associated with T.*

This is an immediate consequence of Theorem 2.11 and

Theorem 2.12. *Fix $u \in W^{1,1}(\Omega; \mathbb{S}^1)$ and set $T := (1/\pi)Ju \in \mathcal{E}$. Then,*

1. *For any lifting $\varphi \in BV(\Omega; \mathbb{R})$ of u, the measure $\mu := -\dfrac{1}{2\pi} D^j\varphi$ is a connection associated with T.*
2. *Conversely, given any connection μ associated with T, there exists a unique (mod 2π) lifting $\psi \in BV(\Omega; \mathbb{R})$ of u such that $D^j\psi = -2\pi\mu$, or, equivalently, $D\psi = u \wedge \nabla u - 2\pi\mu$.*
3. *In this bijective correspondence, minimal liftings correspond to minimal connections.*

The proof of Theorem 2.12 is identical to that of Theorem 2.9 and will be omitted.

We conclude this section with a useful compactness property of the set of connections. Given $T \in \mathcal{E}$ and $m \ge 0$, set

$$\mathcal{K}_m := \{\mu;\ \mu \text{ is a connection associated with } T \text{ and } \|\mu\|_{\mathcal{M}} \le m\}. \tag{2.166}$$

Theorem 2.13. *For every $m \ge 0$, $\mathcal{K}_m$ equipped with the weak* topology*

$$\sigma(\mathcal{M}(\Omega; \mathbb{R}^2), C_0(\Omega; \mathbb{R}^2))$$

is a compact metrizable space.

The conclusion of Theorem 2.13 applies in particular to

$$\mathcal{K}_{\|T\|_W} = \mathcal{K}_{min} := \{\mu;\ \mu \text{ is a minimal connection associated with } T\}. \tag{2.167}$$

Proof. $\mathcal{K}_m$ is a subset of the closed ball $\overline{B}(0, m)$ of $\mathcal{M}(\Omega; \mathbb{R}^2) = [C_0(\Omega; \mathbb{R}^2)]^*$, which is compact and metrizable for the weak* topology $\sigma(\mathcal{M}(\Omega; \mathbb{R}^2), C_0(\Omega; \mathbb{R}^2))$ (see, for example, Brezis [75, Chapter 3]). Thus, it suffices to prove that, if $(\mu_n) \subset \mathcal{K}_m$ and $\mu_n \rightharpoonup \mu$ weak*, then $\mu \in \mathcal{K}_m$. It is clear that $\|\mu\|_{\mathcal{M}} \le m$. Therefore, it remains to show that μ is a connection associated with T.

Fix any $u \in W^{1,1}(\Omega; \mathbb{S}^1)$ such that $Ju = \pi T$. Such a u exists, by Theorem 2.11. By Theorem 2.12 (item 2) applied to μ_n, there exists $\psi_n \in BV(\Omega; \mathbb{R})$ such that

$$u = e^{\imath \psi_n} \tag{2.168}$$

and

$$D\psi_n = u \wedge \nabla u - 2\pi \mu_n. \tag{2.169}$$

We may assume that $(1/|\Omega|) \int_\Omega \psi_n \in [0, 2\pi)$, and then (ψ_n) is bounded in BV. Up to a subsequence, there exists $\psi \in BV$ such that $\psi_n \to \psi$ in L^1 and a.e.; thus, $D\psi_n \rightharpoonup D\psi$ weak*. Passing to the limit in (2.168) and (2.169) yields

$$u = e^{\imath \psi} \tag{2.170}$$

and

$$D\psi = u \wedge \nabla u - 2\pi \mu. \tag{2.171}$$

From (2.170) and Lemma 1.16, we deduce that

$$D\psi = u \wedge \nabla u + D^j \psi = u \wedge \nabla u - 2\pi \mu, \text{ by (2.171)}.$$

Therefore,

$$\begin{aligned} 0 &= \operatorname{curl} D\psi = \operatorname{curl}(u \wedge \nabla u) - 2\pi \operatorname{curl} \mu = 2Ju - 2\pi \operatorname{curl} \mu \\ &= 2\pi T - 2\pi \operatorname{curl} \mu, \end{aligned}$$

so that $\mu = -(1/(2\pi))\, D^j \psi$ is a connection associated with T. □

There exists an explicit metric δ, defined on $\mathscr{K}$ (and thus independent of m), inducing the $\sigma(\mathscr{M}(\Omega; \mathbb{R}^2), C_0(\Omega; \mathbb{R}^2))$ weak* topology on $\mathscr{K}_m$ for every m. More specifically, if $\mu_1, \mu_2 \in \mathscr{K}$, then $\operatorname{curl}(\mu_2 - \mu_1) = 0$, and thus there exists some $\psi \in BV(\Omega; \mathbb{R})$ such that $\mu_2 = \mu_1 + D\psi$ in $\mathscr{D}'(\Omega)$. We then set, for every $\mu_1, \mu_2 \in \mathscr{K}$,

$$\delta(\mu_1, \mu_2) := \inf \left\{ \int_\Omega |\psi|;\ \psi \in BV(\Omega; \mathbb{R}) \text{ and } \mu_2 = \mu_1 + D\psi \right\}. \tag{2.172}$$

Clearly, δ is a distance on $\mathscr{K}$.

Lemma 2.8. *The metric δ induces on $\mathscr{K}_m$ the $\sigma(\mathscr{M}(\Omega; \mathbb{R}^2), C_0(\Omega; \mathbb{R}^2))$ weak* topology, $\forall\, m \geq 0$.*

The proof of Lemma 2.8 relies on the following simple

Lemma 2.9. *We have*

$$|\langle \mu_2, \chi \rangle - \langle \mu_1, \chi \rangle| \leq \delta(\mu_1, \mu_2)\, \| \operatorname{div} \chi \|_{L^\infty},\ \forall\, \mu_1, \mu_2 \in \mathscr{K},\ \forall\, \chi \in C^1_c(\Omega; \mathbb{R}^2).$$

Proof of Lemma 2.9. This follows from the fact that, if $\mu_2 = \mu_1 + D\psi$, then

$$|\langle \mu_2, \chi\rangle - \langle \mu_1, \chi\rangle| = |\langle D\psi, \chi\rangle| = \left|\int_\Omega \psi \operatorname{div} \chi\right| \le \|\psi\|_{L^1} \|\operatorname{div} \chi\|_{L^\infty}. \qquad \square$$

Proof of Lemma 2.8. We have to prove that, for $\mu_n, \mu \in \mathscr{K}_m$, we have

$$\mu_n \rightharpoonup \mu \text{ in } \mathscr{M}(\Omega; \mathbb{R}^2) \iff \delta(\mu_n, \mu) \to 0. \tag{2.173}$$

Implication " $\Longleftarrow$ " is clear, in view of Lemma 2.9. Conversely, write $\mu_n = \mu + D\psi_n$, with $\int_\Omega \psi_n = 0$. Then, (ψ_n) is bounded in $BV(\Omega)$. Let $\psi \in BV(\Omega)$ be such that $\int_\Omega \psi = 0$ and, possibly up to a subsequence, $\psi_n \to \psi$ in $L^1(\Omega)$. Then,

$$\int_\Omega \psi \operatorname{div} \chi = \lim_{n\to\infty} \int_\Omega \psi_n \operatorname{div} \chi = \lim_{n\to\infty} \langle \mu_n - \mu, \chi\rangle = 0, \ \forall\, \chi \in C_c^1(\Omega; \mathbb{R}^2). \tag{2.174}$$

It follows that $D\psi = 0$, and thus $\psi = 0$. Therefore, for the full original sequence (ψ_n), we have $\psi_n \to 0$ in $L^1(\Omega)$, so that $\delta(\mu_n, \mu) \to 0$. $\square$

2.9.4 *The same story told in the language of geometric measure theory (= GMT)*

Before interpreting the results from the previous section into the language of GMT, let us recall the notion of *ℓ-rectifiable current* in $\mathbb{R}^N$, with $0 \le \ell \le N-1$ an integer. Such a current $\mathscr{C}$ is given by a countable family (S_i), where:

1. Each S_i is a Borel subset of some C^1 oriented ℓ-submanifold Σ_i of $\mathbb{R}^N$.
2. The total mass $\sum_i \mathscr{H}^\ell(S_i)$ is finite.
3. The action of $\mathscr{C}$ on smooth compactly supported ℓ-forms $\omega \in C_c^\infty(\mathbb{R}^N; \Lambda^\ell)$ is given by

$$\langle \mathscr{C}, \omega\rangle := \sum_i \int_{S_i} \omega.$$

[Some authors use the terminology "ℓ-rectifiable currents with integer multiplicity" to denote the same objects.]

Rectifiable currents on $\Omega \subset \mathbb{R}^N$ are defined similarly.

In what follows, we will work with 1-*rectifiable currents* $\mathscr{C}$ in $\Omega \subset \mathbb{R}^2$. By the above definition, such a $\mathscr{C}$ is an element of $\mathscr{M}(\Omega; \Lambda^1)$, or, equivalently, an $\mathbb{R}^2$-valued measure that may be written as

$$\mathscr{C} = \sum_i \tau_i \mathscr{H}^1 \llcorner (S_i \cap \Omega), \tag{2.175}$$

where (S_i) is a countable collection of Borel subsets of oriented C^1 curves in $\mathbb{R}^2$, with unit tangent τ_i, and such that $\sum_i \mathscr{H}^1(S_i \cap \Omega) < \infty$.

We now return to the setting of Section 2.9.3. We start with a 0-current Γ in Ω such that $\Gamma = \partial\mathscr{C}$ for some 1-rectifiable current $\mathscr{C}$ in Ω. This amounts to

$$\Gamma = \sum_{j=1}^{\infty} (\delta_{P_j} - \delta_{N_j}) \tag{2.176}$$

for some sequences of points $P_j, N_j \in \overline{\Omega}$ such that $\sum_{j=1}^{\infty} |P_j - N_j| < \infty$ (see Remark 2.13 below).

The boundary $\partial\mathscr{C}$ is defined in the sense of GMT, i.e., $\partial\mathscr{C}$ is the 0-current satisfying

$$\begin{gathered}\langle \partial\mathscr{C}, \eta\rangle = \langle \mathscr{C}, d\eta\rangle = \int_{\mathscr{C}} d\eta = \sum_i \int_{S_i} \tau_i \cdot \nabla\eta, \\ \forall\, \eta \in C_c^\infty(\Omega; \Lambda^0) \simeq C_c^\infty(\Omega; \mathbb{R}),\end{gathered} \tag{2.177}$$

i.e.,

$$\langle \partial\mathscr{C}, \eta\rangle = -\langle \operatorname{div}\mathscr{C}, \eta\rangle, \ \forall\, \eta \in C_c^\infty(\Omega; \mathbb{R}).$$

The equality $\Gamma = \partial\mathscr{C}$ is understood in the sense of distributions. Therefore,

$$[\partial\mathscr{C} = \Gamma] \iff [\operatorname{div}\mathscr{C} = -\Gamma]. \tag{2.178}$$

Comparing (2.162) with (2.175), we see that $\mathscr{C}$ is a 1-rectifiable current satisfying $\partial\mathscr{C} = \Gamma$ if and only if $-\mathscr{C}^\perp$ is a connection associated with Γ (where $\perp$ denotes the rotation by $+\pi/2$ in $\mathbb{R}^2$).

Remark 2.13. Clearly, if Γ is of the form (2.176), then $\Gamma = \partial\mathscr{C}$, where $\mathscr{C}$ is the 1-rectifiable current defined by

$$\mathscr{C} := \sum_j \tau_j \mathscr{H}^1 \llcorner ([N_j, P_j] \cap \Omega);$$

here, $[N_j, P_j]$ denotes the oriented line segment from N_j to P_j.

Conversely, assume that $\Gamma = \partial\mathscr{C}$ for some 1-rectifiable current $\mathscr{C}$. Then, Γ can be represented as in (2.176) (see Proposition 15.4 in Appendix 15.9). Equivalently, we have (see (2.145))

$$\mathscr{E} = \{\partial\mathscr{C};\ \mathscr{C} \text{ is a 1-rectifiable current (in } \Omega)\}. \tag{2.179}$$

Associated with Γ are three quantities

$$A(\Gamma) := \inf \{|\mathscr{C}|;\ \mathscr{C} \text{ is a 1-rectifiable current and } \partial\mathscr{C} = \Gamma\}, \tag{2.180}$$

$$B(\Gamma) := \min \left\{\|\nu\|_{\mathscr{M}};\ \nu \in \mathscr{M}(\Omega; \Lambda^1) \text{ and } \partial\nu = \Gamma\right\}, \tag{2.181}$$

and

$$C(\Gamma) := \sup \left\{\langle \Gamma, \eta\rangle;\ \eta \in C_c^\infty(\Omega; \Lambda^0) \text{ and } \|d\eta\|_{L^\infty} \le 1\right\}. \tag{2.182}$$

Equivalently,

$$B(\Gamma) = \min \left\{\|\nu\|_{\mathscr{M}};\ \nu \in \mathscr{M}(\Omega; \mathbb{R}^2) \text{ and } \operatorname{div} \nu = \Gamma\right\} \tag{2.183}$$

and

$$C(\Gamma) = \sup \left\{\sum_j [\eta(P_j) - \eta(N_j)];\ \eta \in C_c^\infty(\Omega; \mathbb{R}) \text{ and } \|\nabla\eta\|_{L^\infty} \le 1\right\}. \tag{2.184}$$

The first result in this section is a reformulation of Corollary 1.8 when $N = 2$.

Corollary 2.4. *We have*

$$A(\Gamma) = B(\Gamma) = C(\Gamma). \tag{2.185}$$

In addition, the inf *in* (2.180) *is achieved. [The minimizers are called mass-minimizing (or area-minimizing)* 1*-rectifiable currents with boundary* Γ*.]*

Proof. From Theorem 2.11, we know that there exists some $u \in W^{1,1}(\Omega; \mathbb{S}^1)$ such that

$$Ju = \pi \sum_j (\delta_{P_j} - \delta_{N_j}) = \pi\, \Gamma. \tag{2.186}$$

Combining Remark 1.31, (2.147), Theorem 2.10, and (2.184) yields

$$\Sigma(u) = 2\|Ju\|_W = 2\pi\, \|\Gamma\|_W = 2\pi C(\Gamma). \tag{2.187}$$

From (1.85) and (2.186), we deduce that

$$\Sigma(u) = \min \left\{\|\mu\|_{\mathscr{M}};\ \mu \in \mathscr{M}(\Omega; \mathbb{R}^2) \text{ and } \operatorname{curl} \mu = 2\pi\, \Gamma\right\}. \tag{2.188}$$

Equalities (2.183), (2.188), and $\operatorname{curl} \mu^\perp = \operatorname{div} \mu$, yield

$$\Sigma(u) = 2\pi B(\Gamma). \tag{2.189}$$

Finally, we establish that

$$\Sigma(u) = 2\pi A(\Gamma). \tag{2.190}$$

From Theorem 2.12, we deduce that the following properties are equivalent:

$$\mathscr{C} \text{ is a 1-rectifiable current with } \partial\mathscr{C} = \Gamma \tag{2.191}$$

and

$$\mathscr{C} = -\frac{1}{2\pi}(D^j\varphi)^\perp \text{ for some lifting } \varphi \in BV(\Omega; \mathbb{R}) \text{ of } u. \tag{2.192}$$

Applying Corollary 1.5 and the above assertion, we are led to (2.190).

In this correspondence, mass-minimizing 1-rectifiable currents correspond precisely to minimal liftings. □

We now investigate the structure of minimizers of (2.180) and (2.181). Set

$$\mathscr{R}(\Gamma) := \{\mathscr{C};\ \mathscr{C} \text{ is a mass-minimizing 1-rectifiable current with } \partial\mathscr{C} = \Gamma\},$$
$$\mathscr{Q}(\Gamma) := \{\nu \in \mathscr{M}(\Omega; \Lambda^1);\ \partial\nu = \Gamma \text{ and } \|\nu\|_{\mathscr{M}} = B(\Gamma)\}.$$

Clearly, $\mathscr{R}(\Gamma) \subset \mathscr{Q}(\Gamma)$, $\mathscr{Q}(\Gamma)$ is convex and, in principle, could be much bigger than $\mathscr{R}(\Gamma)$.

Corollary 2.5. *We have*

$$\overline{\text{conv}(\mathscr{R}(\Gamma))}^{weak^*} = \mathscr{Q}(\Gamma). \tag{2.193}$$

Proof. Apply Corollary 1.9 and note that

$$\mathscr{R}(\Gamma) = \left\{\frac{1}{2\pi}\mu^\perp;\ \mu \in \mathscr{R}(u)\right\}$$

and

$$\mathscr{Q}(\Gamma) = \left\{\frac{1}{2\pi}\mu^\perp;\ \mu \in \mathscr{Q}(u)\right\}.$$

□

Remark 2.14. Corollary 2.5 follows from the inclusion $\text{Ext}(\mathscr{Q}(\Gamma)) \subset \mathscr{R}(\Gamma)$. We already know that this inclusion may be strict (see Remark 1.18). Here is another simple example of strict inclusion. Consider the situation in Fig. 2.5. Set $\Gamma := 2(\delta_{a_2} - \delta_{a_1})$ and

$$\nu_1 := 2\tau \mathscr{H}^1 \llcorner [N_3, P_3],$$
$$\nu_2 := \tau \mathscr{H}^1 \llcorner [N_1, P_1] + \tau \mathscr{H}^1 \llcorner [N_2, P_2] + \tau \mathscr{H}^1 \llcorner [N_3, P_3],$$
$$\nu_3 := 2\tau \mathscr{H}^1 \llcorner [N_1, P_1] + 2\tau \mathscr{H}^1 \llcorner [N_2, P_2].$$

Clearly, $\nu_j \in \mathscr{R}(\Gamma), j = 1, 2, 3$, but $\nu_2 \notin \mathrm{Ext}(\mathscr{Q}(\Gamma))$, since $\nu_2 = (\nu_1 + \nu_3)/2$.

2.9.5 *The case where Ju is a measure*

Given $u \in W^{1,1}(\Omega; \mathbb{S}^1)$, Ju, defined by (1.35), is a distribution. We address here the question whether Ju is better than just a distribution, e.g., a measure, i.e., we ask whether Ju satisfies the estimate

$$|\langle Ju, \zeta \rangle| \leq C \|\zeta\|_{L^\infty}, \ \forall \zeta \in C_c^\infty(\Omega). \tag{2.194}$$

Theorem 2.14. *Ju is a measure if and only if Ju may be written as a finite sum of the form* $\pi \sum_j (\delta_{P_j} - \delta_{N_j})$.

Proof. The if part is clear. Conversely, assume that Ju is a measure. By Theorem 2.10, we have $\dfrac{1}{\pi} Ju = \partial \mathscr{C}$, where $\mathscr{C}$ is a 1-rectifiable current. More specifically, we may take $\mathscr{C}$ of the form $\mathscr{C} = \sum \tau_l \mathscr{H}^1 \llcorner [M_l, R_l]$. The assumption that Ju is a measure is equivalent to the fact that $\partial \mathscr{C}$ is a measure. By the Federer–Fleming boundary rectifiability theorem [147, Theorem 4.2.16 (2), p. 413], $\partial \mathscr{C}$ is a finite sum of Dirac masses, and thus $\partial \mathscr{C}$ can be written as a finite sum of the form $\sum_j (\delta_{P_j} - \delta_{N_j})$. □

Corollary 2.6. *Given any* $u \in W^{1,1}(\Omega; \mathbb{S}^1)$ *such that Ju is a measure, there exists some* $v \in W^{1,p}(\Omega; \mathbb{S}^1)$, $\forall p < 2$, *such that* $Jv = Ju$ *and* $\|\nabla v\|_{L^p} \leq C(\Omega, p) \, \|Ju\|_{\mathscr{M}}$.

Proof. Apply Theorem 2.14 and write $Ju = \sum_{j=1}^k d_j \delta_{a_j}$, with $a_j \in \Omega$, $\forall j$, so that $\|Ju\|_{\mathscr{M}} = \sum_{j=1}^k |d_j|$. Then choose $v = \prod_{j=1}^k ((x - a_j)/|x - a_j|)^{d_j}$. □

Remark 2.15. The conclusion of Corollary 2.6 fails for $p = 2$. Take, for example, u such that $Ju = \pi \delta_a$ for some $a \in \Omega$ and recall that $Jv = 0$ if $v \in W^{1,2}(\Omega; \mathbb{S}^1)$.

2.10 An integral representation of the distribution *Ju*

Let $\zeta \in C_c^\infty(\Omega)$. Then, for a.e. $t \in \mathbb{R}$, the level set $\gamma_t := [\zeta = t]$ is a finite union of closed connected curves. In what follows, we adopt the following convention: if t is a regular value of ζ, then we define the tangent vector τ to γ_t such that

$$(\tau, \nabla\zeta) \text{ is a direct basis of } \mathbb{R}^2; \text{ equivalently, we let } \tau := -\frac{\nabla^\perp \zeta}{|\nabla^\perp \zeta|}. \tag{2.195}$$

If $u \in W^{1,1}(\Omega; \mathbb{S}^1)$, then (see below), for a.e. t, the restriction of u to γ_t belongs to $W^{1,1}$, and thus u has a degree on γ_t, given by (1.30). The next result gives a formula for $\langle Ju, \zeta\rangle$ in terms of this degree, even in the more general situation where $\zeta \in W_0^{1,\infty}(\Omega)$ and

$$\langle Ju, \zeta\rangle := -\frac{1}{2}\int_\Omega (u \wedge \nabla u)\cdot \nabla^\perp \zeta.$$

Theorem 2.15. *Let $\zeta \in W_0^{1,\infty}(\Omega)$ and $u \in W^{1,1}(\Omega; \mathbb{S}^1)$. Let $\gamma_t := [\zeta = t]$. Then:*

1. *For a.e. $t \in \mathbb{R}$, the restriction of u to γ_t belongs to $W^{1,1}(\gamma_t)$.*
2. *We have*

$$\langle Ju, \zeta\rangle = \pi \int_{\mathbb{R}} \deg(u, \gamma_t)\, dt. \tag{2.196}$$

Here, γ_t is oriented according to (2.195).

Proof. We start by presenting a proof in the special case where $\zeta \in C_c^\infty(\Omega)$. Using Sard's lemma and a local change of variables, item *1* amounts to the fact that, if $u \in W^{1,1}((0,1)^2)$, then $u(\cdot, y) \in W^{1,1}((0,1))$ for a.e. y.

By the coarea formula (see, for example, Evans and Gariepy [146, Theorem 2, p. 117]), we have

$$\begin{aligned}\langle Ju, \zeta\rangle &= -\frac{1}{2}\int_\Omega (u\wedge\nabla u)\cdot\nabla^\perp\zeta = -\frac{1}{2}\int_{\mathbb{R}}\left(\int_{\gamma_t}(u\wedge\nabla u)\cdot\frac{\nabla^\perp\zeta}{|\nabla\zeta|}\right)dt\\ &= \frac{1}{2}\int_{\mathbb{R}}\left(\int_{\gamma_t} u\wedge\frac{\partial u}{\partial\tau}\right)dt = \pi\int_{\mathbb{R}}\deg(u,\gamma_t)\,dt.\end{aligned} \tag{2.197}$$

For a "general" $\zeta \in W_0^{1,\infty}(\Omega)$, we invoke the following

Lemma 2.10. *Let $\zeta \in W_0^{1,\infty}(\Omega; \mathbb{R})$. Then, for a.e. t, the following facts hold true:*

a) γ_t is, up to an $\mathscr{H}^1$-null set, a countable union of rectifiable Jordan curves of finite total length.
b) $\nabla\zeta \neq 0$ $\mathscr{H}^1$-a.e. on γ_t.
c) τ defined by (2.195) is tangent to γ_t.
d) Let $\omega_t := [\zeta > t]$. Then, ω_t has finite perimeter.
e) The reduced boundary of ω_t coincides $\mathscr{H}^1$-a.e with γ_t.
f) The measure theoretic outward normal ν_t to ω_t equals $\nu_t := -\dfrac{\nabla\zeta}{|\nabla\zeta|}$ $\mathscr{H}^1$-a.e.

The proof of the lemma is postponed until the end of Appendix 15.8.1.

On the other hand, for any "good" t as above and a smooth u, we have:

$$\left|\frac{\partial u}{\partial \tau}\right| \leq |\nabla u|. \tag{2.198}$$

By (2.198) and the coarea formula, we have

$$\int_{\mathbb{R}} |u|_{W^{1,1}(\gamma_t)}\, dt \leq \int_{\mathbb{R}} \|\nabla u\|_{L^1(\gamma_t)}\, dt = \int_{\Omega} |\nabla u||\nabla \zeta|, \tag{2.199}$$

and thus

$$u \in W^{1,1}(\gamma_t) \text{ for a.e. } t. \tag{2.200}$$

Using (2.198), (2.199), and a standard approximation argument, we find that (2.200) holds for every $u \in W^{1,1}$. Therefore, for each $u \in W^{1,1}(\Omega; \mathbb{S}^1)$, $\deg(u, \gamma_t)$ makes sense for a.e. t. Finally, we note that (2.197) holds for a general u. □

As a consequence of (2.196), we have

$$\left|\int_{\mathbb{R}} \deg(u, \gamma_t)\, dt\right| \leq \frac{1}{\pi}\|Ju\|_W \, \|\nabla\zeta\|_{L^\infty}, \ \forall \zeta \in W_0^{1,\infty}(\Omega). \tag{2.201}$$

Estimate (2.201) can be strengthened as follows.

Theorem 2.16. *1. We have, for every $u \in W^{1,1}(\Omega; \mathbb{S}^1)$,*

$$\int_{\mathbb{R}} |\deg(u, \gamma_t)|\, dt \leq \frac{1}{\pi}\|Ju\|_W \|\nabla\zeta\|_{L^\infty}, \ \forall \zeta \in W_0^{1,\infty}(\Omega). \tag{2.202}$$

2. *If ζ is a maximizer in* (1.144), *then* $\deg(\zeta, \gamma_t) \geq 0$ *for a.e. t.*

Proof. We start with the case where u is smooth outside some finite set. In this case, by Theorem 2.1 and Lemma 2.4, $T := \dfrac{1}{\pi} Ju$ is a finite sum of the form $T = \sum_j (\delta_{P_j} - \delta_{N_j})$, with $P_j, N_j \in \overline{\Omega}$ and $\sum_j |P_j - N_j| = \|T\|_W$. Let t be as in the proof of the previous theorem. We further assume that $t \notin \cup_j \{\zeta(P_j), \zeta(N_j)\}$. Using the Gauss–Green formula in the set of finite perimeter

$$\omega := \omega_t \setminus (\cup_j \overline{\mathbb{D}}(P_j, \varepsilon) \cup \cup_j \overline{\mathbb{D}}(N_j, \varepsilon)),$$

formula (1.23), and the fact that $\operatorname{curl}(u \wedge \nabla u) = 0$ in ω, we see that, for a.e. small ε, we have

$$\begin{aligned}\int_{\gamma_t} u \wedge \frac{\partial u}{\partial \tau} =& \sum_{P_j \in \omega_t} \frac{1}{2\pi} \int_{C(P_j,\varepsilon)} u \wedge \frac{\partial u}{\partial \tau} \\ &- \sum_{N_j \in \omega_t} \frac{1}{2\pi} \int_{C(N_j,\varepsilon)} u \wedge \frac{\partial u}{\partial \tau} \\ =& \#\{j;\ P_j \in \omega_t\} - \#\{j;\ N_j \in \omega_t\}.\end{aligned} \tag{2.203}$$

We find that, for a.e. $t \notin \cup_j\{\zeta(P_j), \zeta(N_j)\}$, we have

$$\int_{\gamma_t} u \wedge \frac{\partial u}{\partial \tau} = \frac{1}{2} \sum_j \operatorname{sgn}(\zeta(P_j) - t) - \frac{1}{2} \sum_j \operatorname{sgn}(\zeta(N_j) - t). \tag{2.204}$$

Let Γ_j denote the (unoriented) segment $[N_j, P_j]$. Then,

$$\operatorname{sgn}(\zeta(P_j) - t) - \sum_j \operatorname{sgn}(\zeta(N_j) - t) \neq 0 \implies \Gamma_j \cap \gamma_t \neq \emptyset,$$

and this leads to

$$\frac{1}{2} \sum_j |\operatorname{sgn}(\zeta(P_j) - t) - \operatorname{sgn}(\zeta(N_j) - t)| \leq \sum_j \#(\Gamma_j \cap \gamma_t). \tag{2.205}$$

Combining (2.204), (2.205), and the area formula (see, for example, Evans and Gariepy [146, Theorem 1, p. 96]), we find that

$$\begin{aligned}\int_{\mathbb{R}} |\deg(u, \gamma_t)|\, dt &\leq \sum_j \int_{\mathbb{R}} \#(\Gamma_j \cap \gamma_t)\, dt = \sum_j \int_{\Gamma_j} \left| \frac{\partial \zeta}{\partial \tau} \right| \\ &\leq \sum |P_j - N_j| \|\nabla \zeta\|_{L^\infty} = \|T\|_W \, \|\nabla \zeta\|_{L^\infty}.\end{aligned} \tag{2.206}$$

Here, $\dfrac{\partial \zeta}{\partial \tau}$ is understood as the derivative of the restriction of ζ to Γ_j.

We next turn to the case of a general $u \in W^{1,1}(\Omega; \mathbb{S}^1)$. We consider a sequence $(u_n) \subset W^{1,1}(\Omega; \mathbb{S}^1)$ such that:

1. Each u_n is smooth outside a finite set.
2. $u_n \to u$ in $W^{1,1}$.

Such a sequence does exist, by Theorem 10.3.

Let $T_n := \dfrac{1}{\pi} Ju_n$ and $T := \dfrac{1}{\pi} Ju$. By Remark 1.4, we have

$$\|T_n\|_W \to \|T\|_W \text{ as } n \to \infty. \tag{2.207}$$

Using the formula

$$\int_{\mathbb{R}} |\deg(u, \gamma_t)|\, dt = \frac{1}{2\pi} \int_{\mathbb{R}} \left| \int_{\gamma_t} u \wedge \frac{\partial u}{\partial \tau} \right| dt,$$

the domination $\left| u \wedge \dfrac{\partial u}{\partial \tau} \right| \leq |\nabla u|$, the first part of the proof, and (2.207), we find that

$$\int_{\mathbb{R}} |\deg(u, \gamma_t)|\, dt = \lim_n \int_{\mathbb{R}} |\deg(u_n, \gamma_t)|\, dt \leq \lim_n \|T_n\|_W \, \|\nabla \zeta\|_{L^\infty} = \|T\|_W \, \|\nabla \zeta\|_{L^\infty}.$$

This completes the proof of item *1*.

We now turn to item *2*. If ζ is a maximizer in (1.144), then

$$\|T\|_W = \int_{\mathbb{R}} \deg\,(u, \gamma_t)\, dt \leq \int_{\mathbb{R}} |\deg\,(u, \gamma_t)|\, dt \leq \|T\|_W.$$

Therefore, $\deg(u, \gamma_t) = |\deg(u, \gamma_t)| \geq 0$ for a.e. t. □

2.11 Complements and open problems

2.11.1 Minimizing the $W^{1,1}$ energy with prescribed Jacobian

We return to the minimization problem (2.3), i.e.,

$$\inf \left\{ \int_{\Omega} |\nabla v|;\ v \in W^{1,1}(\Omega; \mathbb{S}^1),\ Jv = Ju \text{ in } \mathscr{D}'(\Omega) \right\}, \tag{2.208}$$

where $u \in W^{1,1}(\Omega; \mathbb{S}^1)$ is given. Recall that the value of the infimum in (2.208) equals $\Sigma(u) = 2\|Ju\|_W$ (see (1.151)).

In this subsection we first address, following Brezis and Mironescu [87], the question of the existence of a minimizer in (2.208). [We may always assume that $Ju \neq 0$, for otherwise the answer is trivial.] In Proposition 1.2, we established the fact that this question is equivalent to the existence of minimizers of various other problems.

There are two natural questions:

Question 1. Is the infimum in (2.208) achieved?

Question 2. Characterize the minimizers if they exist.

Questions 1 and 2 turn out to be much more delicate than expected. We do not know any characterization of the u's such that the infimum in (2.208) is achieved –

not even when u is nice! We have a full answer only when u admits *one singularity*. Here is our result in this case.

Without loss of generality we may assume that $a = 0 \in \Omega$ and $d \geq 1$, so that

$$Ju = \pi d\, \delta_0. \tag{2.209}$$

Suppose for simplicity that $\operatorname{dist}(0, \partial\Omega) = 1$ and set

$$X := \{x \in \partial\Omega;\ |x| = 1\} \subset \mathbb{S}^1.$$

Theorem 2.17. *The infimum in* (2.208)–(2.209) *is achieved if and only if* $|X| > 0$ *(where* $|X|$ *refers to the Lebesgue measure of* X *on* $\mathbb{S}^1$*).*

The proof of Theorem 2.17 is presented in [87], together with a complete characterization of all minimizers (when they exist).

Next is a partial result concerning non-existence.

Theorem 2.18. *Let* $u \in W^{1,1}(\Omega; \mathbb{S}^1)$ *and set* $T := (1/\pi)\, Ju$. *Assume that there are only finitely or countably many minimal connections associated with* T *(in the sense of Section 2.9.3). Then the infimum in* (2.208) *is not achieved.*

In the case where u is nice, minimal connections correspond to minimal configurations (by Theorem 2.8) and therefore it is easy to decide whether Theorem 2.18 applies. Returning, for example, to the framework of Theorem 2.17, we see that minimal configurations are of the form $(0, x)$ with $x \in X$. In this case, Theorem 2.18 is much weaker than Theorem 2.17.

The proof of Theorem 2.18 relies on two non-trivial ingredients.

First, a version of the coarea formula

$$\int_\Omega |\nabla u| = \int_{\mathbb{S}^1} \mathscr{H}^1([u = \alpha])\, d\alpha \tag{2.210}$$

valid for all maps $u \in W^{1,1}(\Omega; \mathbb{S}^1)$ provided we choose a convenient representative of u (see, for example, Hajłasz [173], Malý, Swanson, and Ziemer [234], and the references therein).

Second, one uses the fact that, for every $u \in W^{1,1}(\Omega; \mathbb{S}^1)$, and for a.e. $\alpha \in \mathbb{S}^1$, one has

$$\mathscr{H}^1([u = \alpha]) \geq \frac{1}{\pi}\, \|Ju\|_W. \tag{2.211}$$

Inequality (2.211) relies on a delicate generalization of Lemma 2.6, due to Alberti, Baldo, and Orlandi [6, Theorem 3.8] (with roots in Almgren, Browder, and Lieb [8, Appendices A.5-A.8] and Federer [147, Section 4.3]).

We refer the interested reader to [87] for a detailed proof of Theorem 2.18 (see also Lemma 4.3).

From Theorems 2.17 and 2.18, we deduce that the infimum in (2.208) is "rarely" achieved, and thus it is of interest to analyze the asymptotic behavior of minimizing sequences for (2.208). We have a complete answer for a general $u \in W^{1,1}(\Omega; \mathbb{S}^1)$.

Recall (see Section 2.9) that $T := (1/\pi)\, Ju \in \mathscr{E}$ and that

$$\mathscr{K}_{min} := \{\mu \in \mathscr{M}(\Omega; \mathbb{R}^2);\ \mu \text{ is a minimal connection associated with } T\}$$

is compact and metrizable for the weak* topology on $\mathscr{M}(\Omega; \mathbb{R}^2)$.

Theorem 2.19. *Let (v_n) be a minimizing sequence in* (2.208). *Then there exists a probability measure $\mathscr{P}$ on $\mathscr{K}_{min}$ such that, up to a subsequence,*

$$\int_\Omega (v_n \wedge \nabla v_n) \cdot \chi \to 2\pi \int_{\mu \in \mathscr{K}_{min}} \langle \mu, \chi \rangle \, d\mathscr{P}(\mu), \ \forall \chi \in C_0(\Omega; \mathbb{R}^2),$$

and

$$\int_\Omega |\nabla v_n| \zeta \to 2\pi \int_{\mu \in \mathscr{K}_{min}} \langle |\mu|, \zeta \rangle \, d\mathscr{P}(\mu), \ \forall \zeta \in C_0(\Omega; \mathbb{R}).$$

Moreover, any probability measure $\mathscr{P}$ on $\mathscr{K}_{min}$ can be reached via some minimizing sequence (v_n).

The proof of Theorem 2.19 is given in [87].

Remark 2.16. Clearly, the function $\mathscr{K}_{min} \ni \mu \mapsto \langle \mu, \chi \rangle$ (for fixed χ) is continuous on $\mathscr{K}_{min}$ equipped with the weak* topology. It is also true, but less obvious, that the function $\mathscr{K}_{min} \ni \mu \mapsto \langle |\mu|, \zeta \rangle$ is continuous on $\mathscr{K}_{min}$ equipped with the weak* topology (see [87, Lemma 10]).

Remark 2.17. In the special case where $\mathscr{K}_{min}$ is a finite set (or, equivalently, there are only finitely many minimal configurations), we have

$$(v_n \wedge \nabla v_n)\, dx \rightharpoonup 2\pi \sum_j \alpha_j \mu_j$$

and

$$|\nabla v_n|\, dx \rightharpoonup 2\pi \sum_j \alpha_j |\mu_j|,$$

with $\alpha_j \in [0, 1]$, $\forall j$, and $\sum_j \alpha_j = 1$. Moreover, any such convex combination can be reached via some minimizing sequence (v_n).

Remark 2.18. A similar question was originally investigated in Brezis, Coron, and Lieb [79] for the minimization problem

$$\inf\left\{\int_\Omega |\nabla v|^2;\ v \in H^1(\Omega;\mathbb{S}^2) \text{ has prescribed point singularities}\right\}, \tag{2.212}$$

with $\Omega \subset \mathbb{R}^3$. The main result in [79] concerning the behavior of minimizing sequences (v_n) for (2.212) asserts that $|\nabla v_n|^2$ concentrates along a *single* minimal configuration (and does *not* split over a convex combination of minimal configurations). This is in sharp contrast to the phenomenon described above.

2.11.2 Describing $J(W^{1,p}(\Omega;\mathbb{S}^1))$ (following Bousquet)

Theorems 2.10 and 2.11 establish the equality

$$\mathscr{A}_1 := \left\{\frac{1}{\pi}Ju;\ u \in W^{1,1}(\Omega;\mathbb{S}^1)\right\} = \mathscr{E}, \tag{2.213}$$

where

$$\mathscr{E} := \left\{T = \sum(\delta_{P_j} - \delta_{N_j});\ P_j,\ N_j \in \overline{\Omega},\ \sum|P_j - N_j| < \infty\right\}.$$

It is natural to investigate the analogue of (2.213) when u has a different Sobolev regularity. Consider, for example, the set

$$\mathscr{A}_p := \left\{\frac{1}{\pi}Ju;\ u \in W^{1,p}(\Omega;\mathbb{S}^1)\right\},\ 1 \le p < \infty.$$

By (1.61), we have $\mathscr{A}_p = \{0\}$ when $p \ge 2$. It remains to identify $\mathscr{A}_p$ when $1 < p < 2$.

We start with two simple remarks. On the one hand, we have $\mathscr{A}_p \subset \mathscr{A}_1 = \mathscr{E}$. On the other hand, it is clear from the definition (1.35) of Ju that, if $u \in W^{1,p}(\Omega;\mathbb{S}^1)$, then $\frac{1}{\pi}Ju \in [W_0^{1,q}(\Omega)]^*$ (in the sense that $\frac{1}{\pi}Ju$ extends continuously from $C_c^\infty(\Omega)$ to $W_0^{1,q}(\Omega)$). Here, $q := \frac{p}{p-1}$. Thus,

$$\mathscr{A}_p \subset \mathscr{E} \cap [W_0^{1,q}(\Omega)]^*. \tag{2.214}$$

Bousquet [61] proved that equality holds in (2.214).

Theorem 2.20. *Let $1 < p < 2$ and set $q := \frac{p}{p-1}$. Then,*

$$\mathscr{A}_p = \mathscr{E} \cap [W_0^{1,q}(\Omega)]^*.$$

Motivated by the fact that $\mathscr{A}_1 = \mathscr{E}$, we raised in Brezis, Mironescu, and Ponce [92, Open Problems 1 and 2] the question whether the equality

$$\mathscr{A}_p = \left\{ \sum (\delta_{P_j} - \delta_{N_j});\ P_j, N_j \in \overline{\Omega},\ \sum |P_j - N_j|^{2/p-1} < \infty \right\} \tag{2.215}$$

holds for $1 < p < 2$. The exponent $2/p - 1$ comes from a scaling argument. We prove in Appendix 15.10 that "$\supset$" holds in (2.215), but that "$\subset$" is wrong.

2.11.3 Micromagnetics and $\mathbb{S}^1$-valued maps

Various models used in the theory of thin ferromagnetic films involve a Ginzburg–Landau energy $E_\varepsilon(u)$ defined for maps $u : \Omega \subset \mathbb{R}^2 \to \mathbb{R}^2$, where $\varepsilon > 0$ is a small parameter proportional to the film thickness (see, for example, [11, 18, 134–136, 194, 195, 197, 202, 306] and the numerous references therein). These variational problems have in common that, in the limit as $\varepsilon \to 0$, configurations u_ε of bounded energy (i.e., such that $E_\varepsilon(u_\varepsilon) \leq C$) converge to a limit u satisfying

$$|u| = 1 \text{ a.e. on } \Omega, \tag{2.216}$$

$$\operatorname{div} u = 0 \text{ in } \mathscr{D}'(\Omega). \tag{2.217}$$

Maps in this class generally admit line and/or point (vortex) singularities corresponding in 3D micromagnetics to domain walls and Bloch lines. Here are two typical examples satisfying (2.216) and (2.217) in the unit disk $\mathbb{D}$.

Example 2.11. Let $U(x_1, x_2) := \begin{cases} (1, 0), & \text{if } x_2 > 0 \\ (-1, 0), & \text{if } x_2 < 0 \end{cases}.$

Example 2.12. Let $V(x_1, x_2) := \left(-\dfrac{x_2}{|x|}, \dfrac{x_1}{|x|} \right)$, where $|x| := (x_1^2 + x_2^2)^{1/2}$.

Clearly, $U \in BV(\mathbb{D}; \mathbb{S}^1)$ and $U \notin W^{1,1}(\mathbb{D}; \mathbb{S}^1)$, whereas $V \in W^{1,p}(\mathbb{D}; \mathbb{S}^1)$, $\forall p < 2$. Here is a striking result due to Ignat [195].

Theorem 2.21. *Let $u \in W^{1,1}(\Omega; \mathbb{S}^1)$ satisfy* (2.217). *Then, u is locally Lipschitz in Ω except at a locally finite family $\{P_i\}$ of points. Moreover, for every i,*

$$u(x) = \pm V(x - P_i),\ \text{in any convex neighborhood of } P_i \text{ in } \Omega. \tag{2.218}$$

Previously, Jabin, Otto, and Perthame [202] had established that (2.218) holds for any measurable u satisfying (2.216) and (2.217) and some appropriate "kinetic formulation." Ignat proves that the kinetic formulation holds ipso facto for any $u \in W^{1,1}(\Omega; \mathbb{S}^1)$ satisfying (2.217).

The conclusion of Theorem 2.21 still holds under the weaker assumption that $u \in W^{1/p,p}(\Omega; \mathbb{S}^1)$ for some $1 \le p \le 3$ [130, 194, 195].

It would be interesting to find another proof of this result that does not rely on the kinetic formulation. Incidentally, it is not known whether the conclusion of Theorem 2.21 holds when $u \in W^{1/p,p}(\Omega; \mathbb{S}^1)$, with $p > 3$.

2.11.4 Open problems

Open Problem 3. Let $u \in W^{1,1}(\Omega; \mathbb{S}^1)$ be a nice map. Find a proof of the lower bound for the relaxed energy

$$R_1(u) \ge \int_\Omega |\nabla u| + 2\pi L(a, d)$$

having a simple geometric flavor – involving, for example, the coarea formula. Our only proof (see Theorems 1.5 and 1.11, combined with Theorem 2.5) relies on lifting and the fine theory of BV functions.

2.12 Comments

1. Theorem 2.1 appears for the first time in Brezis, Coron, and Lieb [79]. This paper is mostly concerned with maps $u \in H^1(\Omega; \mathbb{S}^2)$, $\Omega \subset \mathbb{R}^3$; maps $u \in W^{1,N-1}(\Omega; \mathbb{S}^{N-1})$ with $\Omega \subset \mathbb{R}^N$, $N \ge 2$ are briefly discussed in [79, Appendix B].
2. The concept of minimal configuration and the quantity $L(a, d)$ were introduced by Brezis, Coron, and Lieb [79], in connection with their study of point defects occurring in liquid crystals. Theorem 2.4 is taken from [79]; the proof in [79] is very short but relies on three non-trivial tools:

   a) Birkhoff's theorem on doubly stochastic matrices.
   b) The Krein–Milman theorem on extreme points.
   c) Kantorovich's duality in OT, which asserts that $K = D$.

   The proof of $M = K$ involves a) and b).
   We call the attention of the reader to the straightforward proof of $K = D$ in Brezis and Mironescu [88], based on the analytic form of the Hahn–Banach theorem. By contrast, the proof presented here – taken from Brezis [69] (see also Brezis [77]) – is totally elementary and self-contained. The heart of the matter is the "brute force construction," described in Lemma 2.2, of a solution of a linear system of inequalities with the help of "chains" and "cycles." Such a strategy

was introduced by Afriat [5]. A similar device was rediscovered independently by Rockafellar [307] in his celebrated paper on cyclically monotone operators. This theme has become very popular in OT (see, for example, Villani [342, Part I, Section 5], [77] and the references therein). Presumably, Kantorovich's original (cryptic) self-contained proof of Theorem 2.4 [211] was in the same spirit as above.

The pseudometric $d(x, y)$ defined by (2.42) could be useful in OT. Suppose for example that Ω is a pristine island without roads. In order to transport a mass from $x \in \Omega$ to $y \in \Omega$ it might be advantageous to move it from x to the nearest point on the shore, then move it by boat along the shore to the point that is nearest to y.

3. Theorem 2.5 (i.e., $\Sigma(u) = 2\pi L(a, d)$) has its roots in Brezis, Coron, and Lieb [79], who proved a version of (2.87) for maps $u \in H^1(\Omega; \mathbb{S}^2)$, $\Omega \subset \mathbb{R}^3$. Their proof used a different strategy; we explain below what would be the analogue of their method in the $W^{1,1}(\Omega; \mathbb{S}^1)$ setting.

a) The lower bound $\Sigma(u) \geq 2\pi L(a, d)$ is established as follows. Clearly,

$$\int_\Omega |\nabla v| \geq \int_\Omega (v \wedge \nabla v) \cdot \nabla^\perp \zeta, \ \forall \zeta \in W_0^{1,\infty}(\Omega) \text{ such that } \|\nabla \zeta\|_{L^\infty} \leq 1. \tag{2.219}$$

Assuming in addition that $Jv = \pi \sum_j d_j \delta_{a_j} = Ju$, we deduce from (2.219) that

$$\int_\Omega |\nabla v| \geq 2\pi \sum_j d_j \zeta(a_j), \ \forall \zeta \in W_0^{1,\infty}(\Omega) \text{ such that } \|\nabla \zeta\|_{L^\infty} \leq 1. \tag{2.220}$$

Taking $\inf_v$ and $\sup_\zeta$ in (2.220) yields $\Sigma(u) \geq 2\pi L(a, d)$; here we have used only (2.3), Remark 2.2, and Theorem 2.3.

b) The upper bound $\Sigma(u) \leq 2\pi L(a, d)$ is established using an elementary dipole construction explained in Section 2.5. Recall that a fancier version of the dipole construction was already given in Proposition 1.1.

4. In Section 2.4, we present a proof of the lower bound $\Sigma(u) \geq 2\pi L(a, d)$ for nice u's, based on the coarea formula. The use of the coarea formula in the context of sphere-valued maps was originally introduced by Almgren, Browder, and Lieb [8]. Their paper deals in particular with maps from $\Omega \subset \mathbb{R}^3$ to $\mathbb{S}^1$ having a curve Γ of singularities, and they are interested in the least energy of maps that are singular on Γ. This is the higher dimensional analogue of the case of a finite number of singularities for maps from $\Omega \subset \mathbb{R}^2$ to $\mathbb{S}^1$ (see Chapter 3). The case where u is a general map (as opposed to nice maps) in $W^{1,1}(\Omega; \mathbb{S}^1)$, $\Omega \subset \mathbb{R}^2$, is more delicate. We already know (see Theorem 2.10) that $\Sigma(u) = 2\,\|Ju\|_W$, $\forall\, u \in W^{1,1}(\Omega; \mathbb{S}^1)$. The goal would be to deduce the lower bound

$$\Sigma(u) \geq 2\,\|Ju\|_W, \ \forall\, u \in W^{1,1}(\Omega; \mathbb{S}^1), \tag{2.221}$$

from a coarea-type argument. In view of (2.3), this amounts to showing that

$$\int_{\Omega} |\nabla v| \geq 2 \, \|Ju\|_{W}, \ \forall \, v \in W^{1,1}(\Omega; \mathbb{S}^1) \text{ such that } Jv = Ju \text{ in } \mathscr{D}'(\Omega). \tag{2.222}$$

In other words, one needs to prove that

$$\int_{\Omega} |\nabla v| \geq 2 \, \|Jv\|_{W}, \ \forall \, v \in W^{1,1}(\Omega; \mathbb{S}^1). \tag{2.223}$$

This can be done, but the price is high! One relies on the non-trivial ingredients (2.210) and (2.211). We will come back to this in Chapters 3 and 4.

5. The proof of the inequality $E(u) \leq \int_{\Omega} |\nabla u| + 2\pi L(a, d)$ for nice u's presented in Section 2.6 is new. However, proofs of related results, also using minimal configurations, were known before. For example, Demengel and Hadiji [133] established the inequality $R_1(u) \leq \int_{\Omega} |\nabla u| + 2\pi L(a, d)$ for nice u's, using the "dipole removing" technique due to Bethuel [25] – a strategy with roots in Bethuel, Brezis, and Coron [30]. Recall that the relaxed energy $R_1(u)$ defined in Section 1.7 coincides with $E(u)$.
6. Theorem 2.7 is essentially due to Fleming [151]. Its proof in Appendix 15.8 follows the modern presentation in Ambrosio, Caselles, Masnou, and Morel [10].
7. Theorem 2.8, which gives a precise description of minimal connections, was known in GMT (see, for example, Morgan [273, Chapter 10] when $\Omega = \mathbb{R}^2$).
8. Theorem 2.9 and its generalization Theorem 2.12, which establish the basic one-to-one correspondence between (minimal) liftings and (minimal) connections, are new.
9. Proposition 2.3 and Theorem 2.11 are taken from Bourgain, Brezis, and Mironescu [54, Lemma 12' and Lemma 16]. Theorem 2.10 comes from Brezis, Mironescu, and Ponce [92], with roots in [51, Theorem1]. Related results were known before (see, for example, Hang and Lin [176], and Brezis [73, Theorem 7']). These theorems can also be viewed as a special case of a more general result of Alberti, Baldo, and Orlandi [6, Theorem 3.8, Theorem 5.6] on the Jacobians of maps $u : \mathbb{R}^N \to \mathbb{S}^k$, with $N \geq k + 1$. We will come back to this in Chapter 4.
10. Theorem 2.13 is closely related to the Federer–Fleming compactness theorem (see Federer and Fleming [149], Federer [147, Theorem 4.2.17, p. 414], and Simon [324, Theorem 27.3]). Compactness is stated in the literature for rectifiable currents in terms of the flat norm topology, or, equivalently, in terms of the weak* topology. Such currents are related to connections through the following "dictionary" (see Section 2.9.4):

$$\begin{gathered} \mu \text{ is a connection associated with } T \iff \\ \mathscr{C} := \mu^{\perp} \text{ is a } 1-\text{rectifiable current such that } \partial \mathscr{C} = T. \end{gathered} \tag{2.224}$$

In addition, we have

$$\|\mathscr{C}\|_{\mathscr{M}} = \|\mu\|_{\mathscr{M}}. \tag{2.225}$$

Our presentation, which uses connections and the weak* topology, is a matter of convenience. Indeed, recall that, on $\mathscr{K}_m$, the $\sigma(\mathscr{M}(\Omega; \mathbb{R}^2); C_0(\Omega; \mathbb{R}^2))$ weak* topology is induced by the distance δ (Lemma 2.8). One may prove that the flat norm $\| \; \|$ satisfies, for some $c > 0$,

$$c\,\delta(\mu_1, \mu_2) \le \|\mu_2^\perp - \mu_1^\perp\| \le \delta(\mu_1, \mu_2), \ \forall\, \mu_1, \mu_2 \in \mathscr{K},$$

and thus results on currents can be restated as results on connections via (2.224). Lemmas 2.8 and 2.9 are from Brezis and Mironescu [87].

11. Proposition 2.4 is due to Ponce [298, Example 5].
12. Equality $A(\Gamma) = B(\Gamma)$ in Corollary 2.4 is originally due to Federer [148, Section 5.10] when $\Omega = \mathbb{R}^2$. Federer seemed to be unaware of its connections to the MK theory in OT. Equality $A(\Gamma) = B(\Gamma)$ corresponds to the equality $M = K$ in Theorem 1.10 (see (1.180) and (1.181) for the definitions of M and K). This result was extended by Federer to $\mathbb{R}^N$, $N \ge 3$, in two different directions, which are discussed in Chapter 3. The first one [148, Section 5.10] corresponds to the equality $A(T) = B(T)$ in (3.70) (for $N = 3$) and (3.108) for $N \ge 4$. The second one [148, Section 5.12] corresponds to equality $A^0(T) = B^0(T)$ in (4.69), and is basically similar to $M = K$ in $\mathbb{R}^N$.
    Equality $B(\Gamma) = C(\Gamma)$ in Corollary 2.4 is a straightforward consequence of convex duality à la Hahn–Banach (as in Section 1.5, see also Comment 12 in Chapter 1) and has nothing to do with GMT. It appears in the literature in various forms (see, for example, Federer [147, Section 4.1.12], Morgan [273, Section 4.5], Brezis, Coron, and Lieb [79, Theorem 5.1], and Giaquinta, Modica, and Souček [157, Section 4.2.5, Proposition 2, p. 414 and Section 4.2.6, Proposition 2, p. 421]).
    Corollary 2.5 is new.
13. Theorem 2.14 is originally due to Jerrard and Soner [204]. Their proof uses the fact that Ju has a Jacobian structure. Later, Giaquinta and Modica noted that this result is a special case of the Federer–Fleming boundary rectifiability theorem [147, Theorem 4.2.16 (2), p. 413] (or, more specifically, of its variant Lemma 15.39 in Appendix 15.11); a proof of Theorem 2.14 (in any dimension) using the argument attributed to Giaquinta and Modica is presented in Jerrard and Soner [205, Section 6].
    The boundary rectifiability theorem is related to the differentiable structure of $\mathbb{R}^N$. It is natural to ask whether the following generalization of Theorem 2.14:

$$S = \sum (\delta_{R_l} - \delta_{M_l}),\ \sum d(R_l, M_l) < \infty,\ |S(\zeta)| \le C\|\zeta\|_{L^\infty},\ \forall\, \zeta \in \mathrm{Lip}(X)$$
$$\Longrightarrow S = \sum_{\text{finite}} (\delta_{P_j} - \delta_{N_j})$$

holds when $\mathbb{R}^N$ is replaced by a general metric space (X, d). The first result in this direction is due to Smets [325], who proved that the implication holds if X is a locally compact space. The final answer is due to Ponce [295, Theorem 13], who obtained that the above implication holds in complete metric spaces. Easy examples prove that completeness is also a necessary condition.

14. Theorem 2.15 generalizes an earlier result from Bourgain, Brezis, and Mironescu [54] ([54] deals with $H^{1/2}(\Omega; \mathbb{S}^1)$ maps, but this is irrelevant). Theorem 2.16 appears in Brezis, Mironescu, and Ponce [92].

Chapter 3
The geometry of Ju and $\Sigma(u)$ in 3D (and higher); line singularities and minimal surfaces

3.0 Introduction

This chapter is the 3D (and higher) counterpart of the previous one. Here, geometry plays a crucial role. The (singular) points a_i are replaced by curves, connections are replaced by surfaces with prescribed boundaries or more general geometric objects (such as 2-rectifiable currents), minimal connections are replaced by area-minimizing surfaces spanned by curves (as in the classical Plateau problem).

For simplicity, we start with a smooth domain $\Omega \subset \mathbb{R}^3$. Nice maps from Chapter 2 (i.e., smooth except at a finite number of points) are replaced by maps $u \in W^{1,1}(\Omega; \mathbb{S}^1) \cap C(\Omega \setminus \Gamma)$, where $\Gamma \subset \Omega$ is a smooth closed connected oriented curve. We define $\deg(u, \Gamma)$ to be the degree of u restricted to any small circle linking Γ. The analogue of the 2D formula $Ju = \pi \sum_j d_j \delta_{a_j}$ becomes here $Ju = \pi \deg(u, \Gamma) \tau \mathscr{H}^1 \llcorner \Gamma$, where τ denotes the unit tangent vector to Γ (see Theorem 3.1).

In order to have a useful correspondence between maps $u \in W^{1,1}(\Omega; \mathbb{S}^1)$ and geometric objects, it is important to know (see Theorem 3.2) that *given any* Γ as above there exists some $u \in W^{1,1}(\Omega; \mathbb{S}^1) \cap C(\Omega \setminus \Gamma)$ such that $Ju = \pi\, \tau \mathscr{H}^1 \llcorner \Gamma$ (and thus $\deg(u, \Gamma) = 1$).

In Sections 3.2–3.4, we present an explicit formula for $\Sigma(u)$ that plays the role of the 2D formula $\Sigma(u) = 2\pi\, L(a, d)$. We prove that

$$\Sigma(u) = 2\pi\, A_0(\Gamma),\ \forall\, u \in W^{1,1}(\Omega; \mathbb{S}^1) \cap C(\Omega \setminus \Gamma) \text{ with } \deg(u, \Gamma) = 1,$$

where $A_0(\Gamma)$ denotes the least area spanned by Γ, i.e.,

$$\begin{aligned} A_0(\Gamma) := \inf\{|M|;\ & M \subset \Omega \setminus \Gamma \text{ is a smooth oriented} \\ & \text{2-submanifold of } \Omega \setminus \Gamma \text{ such that } \partial M = \Gamma\}. \end{aligned} \tag{3.1}$$

In Section 3.5, we discuss a 3D reformulation of Corollaries 1.8 and 2.4. Namely, we show that

H. Brezis and P. Mironescu, *Sobolev Maps to the Circle*, Progress in Nonlinear Differential Equations and Their Applications 96,
https://doi.org/10.1007/978-1-0716-1512-6_3

$$A_0(\Gamma) = B(\Gamma) = C(\Gamma), \tag{3.2}$$

where

$$B(\Gamma) := \min\left\{\|\mu\|_{\mathscr{M}};\ \mu \in \mathscr{M}(\Omega;\mathbb{R}^3),\ \mathrm{curl}\mu = \tau\mathscr{H}^1 \llcorner \Gamma \text{ in } \mathscr{D}'(\Omega)\right\} \tag{3.3}$$

and

$$C(\Gamma) := \sup\left\{\int_\Gamma \tau\cdot\zeta;\ \zeta \in C_c^\infty(\Omega;\mathbb{R}^3) \text{ and } \|\mathrm{curl}\zeta\|_{L^\infty} \le 1\right\}. \tag{3.4}$$

Note that the inequality $B(\Gamma) \le A_0(\Gamma)$ is obvious, since $\mu_0 := \nu\mathscr{H}^2 \llcorner M$, where M is as in (3.1) and ν is the unit normal vector to M orienting M, satisfies $\mathrm{curl}\mu_0 = \tau\mathscr{H}^1 \llcorner \Gamma$. Equality $A_0(\Gamma) = B(\Gamma)$ (or rather its variant $A(\Gamma) = B(\Gamma)$ – see (3.65) for the definition of $A(\Gamma)$), which is far from obvious, is known in GMT.

Formula (3.2) is reminiscent of the Monge–Kantorovich theorem (recalled in Theorem 1.10). Here is the correspondence to keep in mind.

Geometrical setting	Optimal transport setting
$A_0(\Gamma)$	Least cost of transport à la Monge
$B(\Gamma)$	Convexified least cost à la Kantorovich
$C(\Gamma)$	Dual problem à la Kantorovich

This correspondence between Geometry and OT "transits" via maps $u \in W^{1,1}(\Omega;\mathbb{S}^1)$ using the following dictionary (described in full details in Section 3.6).

$W^{1,1}(\Omega;\mathbb{S}^1)$ setting	Geometrical setting
"Singular set" of $u = (1/\pi)\,Ju$	Curve Γ
Jump set of a BV lifting φ of u	Surface spanned by Γ
Jump set of a minimal lifting φ of u	Minimal surface spanned by Γ
$\int \lvert D^j\varphi\rvert$ for a minimal lifting φ of u	$2\pi\times$ least area spanned by Γ

In Section 3.6 (for 3D) and Section 3.8.2 (for $N \ge 4$), we go beyond the situation described above, i.e., maps u with a curve of singularities. There, we can allow a *general* $u \in W^{1,1}(\Omega;\mathbb{S}^1)$. In this case, it is convenient to use the language of GMT and the correspondence takes the form:

$W^{1,1}(\Omega;\mathbb{S}^1)$ setting	Geometrical setting
"Singular set" of $u = (1/\pi)\,Ju$	$(N-2)$-currents T such that $T = \partial M$ for some $(N-1)$-rectifiable current M
Jump set of a BV lifting φ of u	$(N-1)$-rectifiable currents with prescribed boundary
Jump set of a minimal lifting φ of u	Area-minimizing $(N-1)$-rectifiable currents with prescribed boundary
$\int \lvert D^j\varphi\rvert$ for a minimal lifting φ of u	$2\pi\times$ least area of $(N-1)$-rectifiable currents with prescribed boundary

3.1 Examples. *Ju* as path integration

We start by generalizing to dimension 3 the quantity $\deg(u, a)$, where $\Omega \subset \mathbb{R}^2$, $a \in \Omega$, and $u \in C(\Omega \setminus \{a\}; \mathbb{S}^1)$ (see Section 2.1).

Assume that $\Omega \subset \mathbb{R}^3$ is smooth bounded and simply connected. Let $u \in C(\Omega \setminus \Gamma; \mathbb{S}^1)$, where $\Gamma \subset \Omega$ is a smooth compact connected oriented curve. Then we may define the integer $\deg(u, \Gamma)$ as follows. Let $x \in \Gamma$, and let N_x be the normal plane to Γ at x. On N_x, we have a natural orientation induced by the orientation of Γ (such that a direct basis of N_x, completed with the unit tangent vector to Γ, forms a direct basis of $\mathbb{R}^3$); see Fig. 3.1. Let, for small ε, $C(x, \varepsilon) := \{y \in N_x;\ |y - x| = \varepsilon\}$. This circle inherits an orientation from N_x, and does not intersect Γ. We let $\deg(u, \Gamma) := \deg(u, C(x, \varepsilon))$. By a homotopy argument (see Remark 2.1), this definition does not depend on x (here, we use the connectedness of Γ), or on small ε.

Another case of interest for us is when Γ is smooth oriented and connects two points of $\partial\Omega$ (when $\partial\Omega$ is connected, this is a limiting case of the previous one). In this case, we may still define $\deg(u, \Gamma)$ as above.

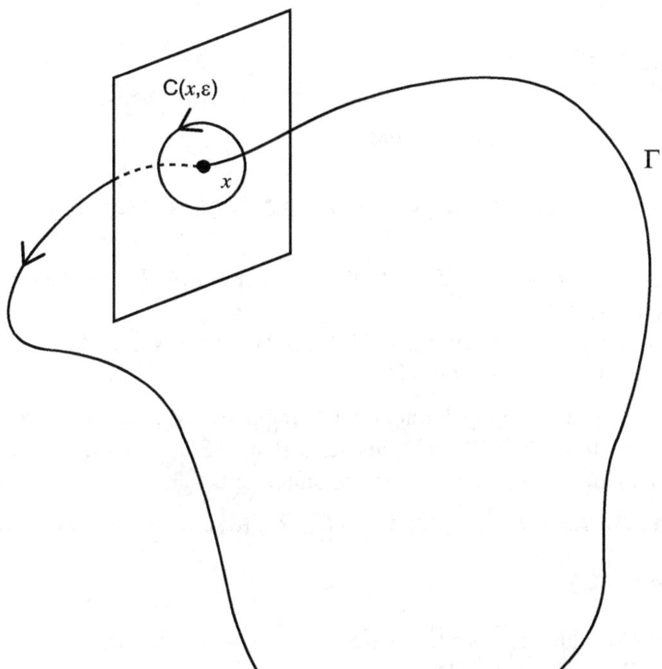

Fig. 3.1 The natural orientation on $C(x, \varepsilon)$.

To summarize, we may define $\deg(u, \Gamma)$ for $u \in C(\Omega \setminus \Gamma; \mathbb{S}^1)$, provided

$$\Gamma \subset \Omega \text{ is a smooth connected oriented curve without boundary (in } \Omega). \tag{3.5}$$

The first result is the 3D analogue of Theorem 2.1. Recall that Ju is a 2-form (or rather 2-current), and thus $*Ju$ is a 1-current, which we may naturally identify with a vector field in $\mathscr{D}'(\Omega;\mathbb{R}^3)$ (see (1.42)). With an abuse of notation (omitting the $*$), we write:

$$Ju \simeq ((Ju)_{23}, (Ju)_{31}, (Ju)_{12}), \tag{3.6}$$

where the coordinates $(Ju)_{ij}$ are given by (1.38).

Theorem 3.1. *Let $u \in W^{1,1}(\Omega;\mathbb{S}^1) \cap C(\Omega \setminus \Gamma)$, where $\Gamma \subset \Omega \subset \mathbb{R}^3$ satisfies* (3.5). *Then, with the identification* (3.6), *we have*

$$Ju = \pi \deg(u,\Gamma)\tau \mathscr{H}^1 \llcorner \Gamma, \tag{3.7}$$

where τ denotes the unit tangent vector to Γ.

More generally, let $u \in W^{1,1}(\Omega;\mathbb{S}^1) \cap C(\Omega \setminus \cup_{j=1}^m \Gamma_j)$, with $(\Gamma_j)_{1\le j\le m}$ disjoint curves in Ω satisfying (3.5). *Then,*

$$Ju = \pi \sum_{j=1}^m \deg(u,\Gamma_j)\tau_j \mathscr{H}^1 \llcorner \Gamma_j. \tag{3.8}$$

Remark 3.1. Equality (3.7) asserts that

$$\frac{1}{2}\int_\Omega (u \wedge \nabla u)\cdot \operatorname{curl}\zeta = \pi \deg(u,\Gamma)\int_\Gamma \tau\cdot\zeta,\ \forall \zeta \in C_c^\infty(\Omega;\mathbb{R}^3). \tag{3.9}$$

Hence, $\langle Ju, \zeta\rangle$ can be viewed, up to the factor $\pi \deg(u,\Gamma)$, as the circulation on Γ of the vector field ζ.

Using Corollary 1.12, we find that (3.9) holds for every $\zeta \in W^{1,\infty}(\Omega;\mathbb{R}^3)$ such that $\zeta_T = 0$ on $\partial\Omega$, i.e., $\zeta \parallel n$ on $\partial\Omega$.

Remark 3.2. The reader may wonder what happens to point singularities, which played such a central role in 2D. The answer is that, in 3D, they simply go undetected by the Jacobian Ju. Here is a more precise statement.

Proposition 3.1. *Let $u \in W^{1,1}(\Omega;\mathbb{S}^1) \cap C(\Omega \setminus \{a\})$, where $a \in \Omega \subset \mathbb{R}^3$. Then,*

$$Ju = 0 \text{ in } \mathscr{D}'(\Omega). \tag{3.10}$$

Proof. From Corollary 1.1, we know that $Ju = 0$ in $\mathscr{D}'(\Omega \setminus \{a\})$, and that u admits a lifting $\varphi \in W^{1,1}_{loc}(\Omega \setminus \{a\})$. [Here, we use the fact that $\Omega \setminus \{a\})$ is simply connected (since $\Omega \subset \mathbb{R}^3$ – this is not true in 2D!).] Since $|\nabla\varphi| = |\nabla u|$, we infer that $\int_{\Omega\setminus\{a\}} |\nabla\varphi| < \infty$. Applying Lemma 1.10 (see also Remark 1.7), we conclude that $\varphi \in W^{1,1}(\Omega)$, and therefore $Ju = 0$ in $\mathscr{D}'(\Omega)$ (by Theorem 1.2). □

Proof of Theorem 3.1. We consider only the case $m = 1$. It will be convenient to use the language of differential forms: $u \wedge \nabla u$ is identified with the 1-form

$$u \wedge du := \left(u \wedge \frac{\partial u}{\partial x_1}\right) dx_1 + \left(u \wedge \frac{\partial u}{\partial x_2}\right) dx_2 + \left(u \wedge \frac{\partial u}{\partial x_3}\right) dx_3,$$

ζ is identified with the 1-form $\zeta_1 dx_1 + \zeta_2 dx_2 + \zeta_3 dx_3$, and curl$\zeta$ is identified with the 2-form $d\zeta$.

In this formalism, we have

$$\langle Ju, \zeta\rangle = \frac{1}{2}\int_\Omega (u \wedge du) \wedge d\zeta,\ \forall \zeta \in C_c^\infty(\Omega; \Lambda^1), \tag{3.11}$$

and thus (3.8) amounts to

$$\int_\Omega (u \wedge du) \wedge d\zeta = 2\pi \ \deg(u, \Gamma) \int_\Gamma \zeta,\ \forall \zeta \in C_c^\infty(\Omega; \Lambda^1). \tag{3.12}$$

Using a partition of unity, we may "localize" (3.12), and thus it suffices to prove it in one of the two following model cases.

Case 1. Ω is a ball and $\Gamma = \emptyset$ (with the convention $\deg(u, \emptyset) = 0$).

Case 2. There exists an orientation-preserving bi-Lipschitz diffeomorphism $\Phi : (-1, 1)^3 \to \Omega$ such that $\Phi(\{(0, 0)\} \times (-1, 1)) = \Gamma$.

Case 1. In this case we have $u \in W^{1,1}(\Omega; \mathbb{S}^1) \cap C(\Omega)$, and thus, by Corollary 1.1, we may write $u = e^{\imath\varphi}$ for some $\varphi \in W^{1,1}(\Omega; \mathbb{R})$. By Theorem 1.2, we have $Ju = 0$, so that (by (3.11)) (3.12) holds.

Case 2. Using first the fact that Φ is orientation-preserving, and next the properties of the pullback of forms, we have

$$\begin{aligned}
\int_\Omega (u \wedge du) \wedge d\zeta &= \int_{(-1,1)^3} \Phi^\sharp[(u \wedge du) \wedge d\zeta] \\
&= \int_{(-1,1)^3} (\Phi^\sharp u \wedge \Phi^\sharp(du)) \wedge \Phi^\sharp(d\zeta) \\
&= \int_{(-1,1)^3} ((u \circ \Phi) \wedge d(u \circ \Phi)) \wedge d(\Phi^\sharp\zeta) \\
&= \int_{(-1,1)^3} (w \wedge dw) \wedge d\xi,
\end{aligned} \tag{3.13}$$

where $w := u \circ \Phi$ and $\xi := \Phi^\sharp\zeta$.

On the other hand, we have, with $\widetilde{\Gamma} := \{(0, 0)\} \times (-1, 1)$ (with the orientation induced by the natural orientation of $(-1, 1)$),

$$\int_\Gamma \zeta = \int_{\widetilde{\Gamma}} \xi. \tag{3.14}$$

Clearly, w belongs to $W^{1,1}((-1, 1)^3; \mathbb{S}^1) \cap C((-1, 1)^3 \setminus \widetilde{\Gamma})$ and satisfies $\deg(w, \widetilde{\Gamma}) = \deg(u, \Gamma)$, and ξ belongs to $C_c^\infty((-1, 1)^3; \Lambda^1)$. Combining this with (3.13) and

(3.14), we have thus reduced Case 2 to the special case where $\Omega = (-1, 1)^3$ and $\Gamma = \widetilde{\Gamma}$.

In order to complete the proof of the theorem, we first check the validity of (3.12) for a particular map U of given degree $d = \deg(U, \widetilde{\Gamma})$, and next we prove that, if (3.12) holds for *some* u (with given degree d), then it holds for *any* such u.

Consider, as a particular U, $U(r\cos\theta, r\sin\theta, x_3) := e^{\imath d\theta}$. Then, clearly,

$$U \in W^{1,1}((-1,1)^3; \mathbb{S}^1) \cap C((-1,1)^3 \setminus \widetilde{\Gamma})$$

and $\deg(U, \widetilde{\Gamma}) = d$. With $\zeta = \sum_{j=1}^3 \zeta_j dx_j$, we have

$$\begin{aligned}
\int_{(-1,1)^3} (U \wedge dU) \wedge d\zeta &= \int_{(-1,1)^3} \left(U \wedge \frac{\partial U}{\partial x_1}\frac{\partial \zeta_3}{\partial x_2} - U \wedge \frac{\partial U}{\partial x_2}\frac{\partial \zeta_3}{\partial x_1} \right) \\
&\quad + \int_{(-1,1)^3} \left(-U \wedge \frac{\partial U}{\partial x_1}\frac{\partial \zeta_2}{\partial x_3} + U \wedge \frac{\partial U}{\partial x_2}\frac{\partial \zeta_1}{\partial x_3} \right) \\
&= \int_{(-1,1)^3} \left(U \wedge \frac{\partial U}{\partial x_1}\frac{\partial \zeta_3}{\partial x_2} - U \wedge \frac{\partial U}{\partial x_2}\frac{\partial \zeta_3}{\partial x_1} \right) \\
&= 2\pi d \int_{-1}^{1} \zeta_3(0, x_3)\, dx_3 = 2\pi d \int_{\widetilde{\Gamma}} \zeta.
\end{aligned}$$

Here, we have used the fact that U does not depend on x_3 in order to obtain that the integral on the second line vanishes, resp. Theorem 2.1 (in the variables (x_1, x_2), for fixed x_3) and integration in x_3 in order to obtain the third equality. We find that (3.12) holds for U.

Consider now some u of degree d. Set $w := u/U$, so that w is of degree 0. By Lemma 3.1 below, there exists some $\varphi \in W^{1,1}((-1,1)^3; \mathbb{R})$ such that $w = e^{\imath\varphi}$. As in Case 1, we have $Jw = 0$. Since, on the other hand, we have $Ju = J(wU) = Jw + JU = JU$, we find that (3.12) holds for u.

The proof of Theorem 3.1 is complete. □

Lemma 3.1. *Let* $w \in W^{1,1}((-1,1)^3; \mathbb{S}^1) \cap C((-1,1)^3 \setminus \widetilde{\Gamma})$ *be such that* $\deg(w, \widetilde{\Gamma}) = 0$. *Then there exists some* $\varphi \in W^{1,1}((-1,1)^3; \mathbb{R})$ *such that* $w = e^{\imath\varphi}$.

Proof. We first construct $\varphi \in C((-1,1)^3 \setminus \widetilde{\Gamma})$ such that $w = e^{\imath\varphi}$. [Existence of φ is standard.]

By the second part of Lemma 1.1, we have $\varphi \in W^{1,1}_{loc}((-1,1)^3 \setminus \widetilde{\Gamma})$. Combining this with the equality $|\nabla\varphi| = |\nabla u|$ a.e. in $(-1,1)^3 \setminus \widetilde{\Gamma}$ and with Lemma 1.10, we find that $\varphi \in W^{1,1}((-1,1)^3)$. □

We next discuss the *existence* of $u \in W^{1,1}(\Omega; \mathbb{S}^1) \cap C(\Omega \setminus \Gamma)$ for a given Γ satisfying (3.5) and a given integer $d = \deg(u, \Gamma)$. [The analogue in 2D was trivial. Given (a, d), one can simply choose $u(x) = \prod_{j=1}^m ((x - a_j)/|x - a_j|)^{d_j}$.]

Theorem 3.2. *Assume that $\Gamma \subset \Omega$ is connected and of the form $\Gamma = \widetilde{\Gamma} \cap \Omega$, with $\widetilde{\Gamma} \subset \mathbb{R}^3$ a smooth closed oriented curve. Then there exists some $u \in C^\infty(\Omega \setminus \Gamma; \mathbb{S}^1)$ such that $u \in W^{1,p}_{loc}(\Omega; \mathbb{S}^1)$, $\forall\, 1 \le p < 2$, and* $\deg(u, \Gamma) = d$.

More generally, if $\Gamma_1, \dots, \Gamma_m$ are mutually disjoint connected curves of the form $\Gamma_j = \widetilde{\Gamma}_j \cap \Omega$, with $\widetilde{\Gamma}_j \subset \mathbb{R}^3$ smooth closed oriented curves, then there exists some $u \in C^\infty(\Omega \setminus \cup_{j=1}^m \Gamma_j; \mathbb{S}^1)$, such that $u \in W^{1,p}_{loc}(\Omega; \mathbb{S}^1)$, $\forall\, 1 \le p < 2$, and $\deg(u, \Gamma_j) = d_j$, $j = 1, \dots, m$.

Remark 3.3. We pause to discuss the assumption

$$\Gamma \text{ is connected and } \Gamma = \widetilde{\Gamma} \cap \Omega, \quad \text{with } \widetilde{\Gamma} \subset \mathbb{R}^3 \text{ a smooth closed oriented curve.} \tag{3.15}$$

Here, "closed" is in the sense of differential geometry, i.e., $\widetilde{\Gamma}$ is compact and without boundary; thus, by taking only the connected component of $\widetilde{\Gamma}$ containing Γ, we may assume that $\widetilde{\Gamma}$ is a knot, i.e., an embedding of $\mathbb{S}^1$ into $\mathbb{R}^3$.

The assumption (3.15) is clearly satisfied when Γ itself is a knot contained in Ω.

Let us consider the case where Γ connects two points of $\partial\Omega$. For simplicity, we assume that Γ is transversal to $\partial\Omega$. In general, such Γ *need not* be of the form $\Gamma = \widetilde{\Gamma} \cap \Omega$. Moreover, it may happen that there exists no u such that $\deg(u, \Gamma) \neq 0$.

Indeed, let $\Omega := B_{10}(0) \setminus \overline{B}_1(0)$ and let Γ be the oriented segment from $(0, 0, 1)$ to $(0, 0, 10)$. Then there exists no $u \in C(\Omega \setminus \Gamma; \mathbb{S}^1)$, such that $\deg(u, \Gamma) \neq 0$; in particular, Γ is not of the form $\Gamma = \widetilde{\Gamma} \cap \Omega$. Argue by contradiction, and consider some $u \in C(\Omega \setminus \Gamma; \mathbb{S}^1)$, such that $\deg(u, \Gamma) = d \neq 0$. We note that every circle $C(x, \varepsilon)$ as in the definition of $\deg(u, \Gamma)$ is homotopic, in $\Omega \setminus \Gamma$, to the circle $\begin{cases} (x_1)^2 + (x_2)^2 = 1 \\ x_3 = -2 \end{cases}$, which in turn is homotopic to a point. Via a homotopy argument, we find that $d = 0$ – a contradiction.

However, if we assume that Γ connects transversally two distinct points of $\partial\Omega$ that belong to the *same component* of $\partial\Omega$, then Γ is of the form $\Gamma = \widetilde{\Gamma} \cap \Omega$.

Proof of Theorem 3.2. It suffices to consider the case where $m = 1$. Indeed, if u_j corresponds to Γ_j as in Theorem 3.2, then $u := u_1 \cdots u_m$ has all the required properties (here, we use (1.56)).

When $m = 1$, by enlarging Ω, we have reduced the problem to the construction of u when $\widetilde{\Gamma}$ is connected and contained in Ω. Under this assumption, the existence of u is a consequence of the following result. □

Theorem 3.3. *Let $\Gamma \subset \mathbb{R}^3$ be a smooth closed connected oriented curve and let $d \in \mathbb{Z}$. Then there exists some $u \in C^\infty(\mathbb{R}^3 \setminus \Gamma; \mathbb{S}^1)$ such that $u \in W^{1,p}_{loc}(\mathbb{R}^3; \mathbb{S}^1)$, $\forall\, 1 \le p < 2$, and* $\deg(u, \Gamma) = d$.

Proof. We may assume that the degree equals 1. The plan is the following:

1. We construct some $F \in \mathscr{D}'(\mathbb{R}^3; \Lambda^1) \cap C^\infty(\mathbb{R}^3 \setminus \Gamma; \Lambda^1)$ such that its exterior differential dF satisfies $dF = 2\pi * \Gamma$ in $\mathscr{D}'(\mathbb{R}^3)$ and $F \in L^p_{loc}(\mathbb{R}^3), \forall\, 1 \le p < 2$. Throughout this chapter, Γ is often identified with the 1-current $\tau \mathscr{H}^1 \llcorner \Gamma$, so that

$$\langle \Gamma, \eta \rangle = \int_\Gamma \eta, \ \forall\, \eta \in C_c^\infty(\mathbb{R}^3; \Lambda^1). \tag{3.16}$$

Thus, $*\Gamma$ is the 2-current defined by

$$\langle * \Gamma, \zeta \rangle := \int_\Gamma *\zeta, \ \forall\, \zeta \in C_c^\infty(\mathbb{R}^3; \Lambda^2). \tag{3.17}$$

2. We prove that there exists some $u \in W^{1,p}_{loc}(\mathbb{R}^3; \mathbb{S}^1) \cap C^\infty(\mathbb{R}^3 \setminus \Gamma)$, $1 \le p < 2$, such that $u \wedge du = F$ in $\mathbb{R}^3 \setminus \Gamma$.
3. We prove that $\deg(u, \Gamma) = 1$.

Step 1. Construction of F
We first note that

$$*\Gamma \in W^{-1,p}(\mathbb{R}^3; \Lambda^2) = [W^{1,p'}(\mathbb{R}^3; \Lambda^2)]^*, \ \forall\, 1 < p < 2. \tag{3.18}$$

Actually, the fact that Γ has no boundary plays no role in (3.18), which holds for a compact curve, possibly with boundary. In order to obtain (3.18) for such a curve, it suffices to establish the *scalar* estimate

$$\int_\Gamma |f| \le C(\Gamma)\|f\|_{W^{1,q}}, \ \forall\, q > 2, \ \forall f \in C_c^\infty(\mathbb{R}^3; \mathbb{R}). \tag{3.19}$$

In turn, (3.19) is obtained as follows. Using a partition of unity and local changes of coordinates, it suffices to prove (3.19) when Γ is a (bounded) line segment. Let P be a plane containing Γ. By trace theory and the fact that $q > 2$, we have

$$\|f\|_{L^1(\Gamma)} \lesssim \|f\|_{L^q(\Gamma)} \lesssim \|f\|_{W^{1-2/q,q}(\Gamma)} \lesssim \|f\|_{W^{1-1/q,q}(P)} \lesssim \|f\|_{W^{1,q}(\mathbb{R}^3)},$$

whence (3.19).

Let G be the solution of $\Delta G = *\Gamma$ in $\mathscr{D}'(\mathbb{R}^3; \Lambda^2)$ given by the convolution of $*\Gamma$ with the fundamental solution $E(x) = -1/(4\pi|x|)$ of Δ. By (3.18) and elliptic regularity, we have $G \in W^{1,p}_{loc}(\mathbb{R}^3; \Lambda^2), \forall\, 1 \le p < 2$ and $G \in C^\infty(\mathbb{R}^3 \setminus \Gamma; \Lambda^2)$.

Set $F := 2\pi\, d^*G$, so that $F \in L^p_{loc}(\mathbb{R}^3; \Lambda^1), \forall\, 1 \le p < 2$, and $F \in C^\infty(\mathbb{R}^3 \setminus \Gamma; \Lambda^1)$. We claim that

$$dF = 2\pi * \Gamma \text{ in } \mathscr{D}'. \tag{3.20}$$

Indeed, denoting the convolution $*_{\mathbf{x}}$ (in order to avoid the confusion with the Hodge operator), we have:

$$dF = 2\pi\, dd^*\,(E *_{\mathbf{x}} (*\Gamma)) = 2\pi\, E *_{\mathbf{x}} (dd^*(*\Gamma)) = 2\pi\, E *_{\mathbf{x}} (\Delta(*\Gamma)) = 2\pi * \Gamma.$$

Here, we have used the fact that (from (1.141) and the assumption that Γ has no boundary) we have

$$d(*\Gamma) = (**)d(*\Gamma) = *(*d*)\Gamma = *(-\partial\Gamma) = 0,$$

and thus

$$\Delta(*\Gamma) = (dd^* + d^*d)(*\Gamma) = dd^*(*\Gamma).$$

Step 1 is complete.

Step 2. From F to a BV_{loc} "would-be-phase" φ

By the Frankl–Pontryagin theorem, since Γ is compact and has no boundary, there exists some smooth compact oriented ("Seifert") surface M with boundary Γ (see Frankl and Pontryagin [152], Seifert [319], and also Adams [2, Chapter 4] for an elementary presentation). M will often be identified with the 2-current given by the integration over M, i.e.,

$$\langle M, \omega\rangle = \int_M \omega, \ \forall \omega \in C_c^\infty(\mathbb{R}^3; \Lambda^2).$$

By Stokes' theorem, we have

$$\langle M, d\zeta\rangle = \int_\Gamma \zeta, \ \forall \zeta \in C_c^\infty(\mathbb{R}^3; \Lambda^1). \tag{3.21}$$

In the language of GMT, (3.21) asserts that $\partial M = \Gamma$. [Equivalently, we have

$$\operatorname{curl}(\nu\mathcal{H}^2 \llcorner M) = \tau\mathcal{H}^1 \llcorner \Gamma \text{ in } \mathcal{D}'(\Omega), \tag{3.22}$$

where ν denotes the unit normal orienting M and τ the oriented unit tangent to Γ.]

Using (1.141) and (3.21), we find that $d * M = *\Gamma$. Combining this with (3.20) yields

$$d(F - 2\pi * M) = 0 \text{ in } \mathcal{D}'(\mathbb{R}^3). \tag{3.23}$$

From (3.23), the fact that $F - 2\pi * M \in \mathcal{M}_{loc}(\mathbb{R}^3)$ and Poincaré's lemma, we obtain the existence of some $\varphi \in BV_{loc}(\mathbb{R}^3) \cap C^\infty(\mathbb{R}^3 \setminus (M \cup \Gamma))$ such that

$$d\varphi = F - 2\pi * M. \tag{3.24}$$

On the other hand, let ν be the unitary normal to M. Then (with the identification of 1-forms with vector fields)

$$\begin{aligned}\langle *M, \zeta\rangle &= \int_M *\zeta = \int_M \nu\cdot\zeta\, d\mathcal{H}^2 \\ &= \langle \nu\mathcal{H}^2 \llcorner M, \zeta\rangle, \ \forall \zeta \in C_c^\infty(\mathbb{R}^3; \Lambda^1).\end{aligned} \tag{3.25}$$

Combining (3.24) and (3.25), we obtain:

$$\nabla\varphi = F,\ D^c\varphi = 0,\ D^j\varphi = -2\pi\nu\,\mathscr{H}^2 \llcorner M\,. \tag{3.26}$$

In particular, we have, by (3.26),

$$\varphi^+(x) - \varphi^-(x) \in \mathbb{Z} \text{ for } \mathscr{H}^2 - \text{a.e. } x \in M\,. \tag{3.27}$$

Step 3. From φ to u
Set $u := e^{\imath\varphi} \in BV_{loc}(\mathbb{R}^3; \mathbb{S}^1)$. Combining (3.26) and (3.27) with Volpert's chain rule, we find that

$$du = \imath u F \text{ in } \mathscr{D}'(\mathbb{R}^3). \tag{3.28}$$

In particular, (3.28) implies that $u \in \mathrm{Lip}_{loc}(\mathbb{R}^3 \setminus \Gamma)$ and then (by bootstrap) that $u \in C^\infty(\mathbb{R}^3 \setminus \Gamma; \mathbb{S}^1)$.

Also from (3.28), we obtain that $u \in W^{1,p}_{loc}(\mathbb{R}^3; \mathbb{S}^1)$, $\forall\, 1 \le p < 2$, and

$$u \wedge du = F. \tag{3.29}$$

Combining (3.29) and (3.20), we obtain

$$Ju = \frac{1}{2}d(u \wedge du) = \frac{1}{2}dF = \pi * \Gamma \text{ in } \mathscr{D}'(\mathbb{R}^3). \tag{3.30}$$

By (3.30), the fact that $u \in W^{1,1}_{loc}(\mathbb{R}^3; \mathbb{S}^1) \cap C^\infty(\mathbb{R}^3 \setminus \Gamma)$ and Theorem 3.1, we find that $\deg(u, \Gamma) = 1$.

Thus, u has all the required properties. The proof of Theorem 3.3 is complete. □

3.2 $\Sigma(u)$ as a least area: $\Sigma(u) = 2\pi A_0(\Gamma)$

In the previous section, we have considered $u \in W^{1,1}(\Omega) \cap C(\Omega \setminus \cup_{j=1}^m \Gamma_j)$ and we have established the formula

$$Ju = \pi \sum_{j=1}^m \deg(u, \Gamma_j)\tau_j \mathscr{H}^1 \llcorner \Gamma_j = \pi \sum_{j=1}^m \deg(u, \Gamma_j)\, \Gamma_j.$$

Our next goal is to present an *explicit* formula for $\Sigma(u)$ that is the 3D counterpart of the 2D formula $\Sigma(u) = 2\pi L(a, d)$ in Section 2.3. It involves the quantity $A_0(\Gamma, d)$, which is the *least area spanned by* (Γ, d). For simplicity, we restrict our attention in this section to the special case of a single curve, i.e., $m = 1$, with associated degree $d = d_1 = 1$. [Results for the general case are presented in Section 3.8.1.]

We consider a smooth oriented curve $\Gamma \subset \Omega$, without boundary in Ω, of the form (3.15).

As in (3.16), we identify Γ with the 1-rectifiable current

$$\langle \Gamma, \zeta \rangle = \int_\Gamma \zeta, \ \forall \zeta \in C_c^\infty(\Omega; \Lambda^1).$$

Equivalently, with the identification $C_c^\infty(\Omega; \Lambda^1) \simeq C_c^\infty(\Omega; \mathbb{R}^3)$, we have

$$\langle \Gamma, \zeta \rangle = \int_\Gamma \tau \cdot \zeta, \ \forall \zeta \in C_c^\infty(\Omega; \mathbb{R}^3).$$

We now introduce the quantity $A_0(\Gamma)$, which plays a central role throughout this chapter. Set

$$\begin{aligned} A_0(\Gamma) := \inf\{|M|;\, M \subset \Omega \setminus \Gamma \text{ is a smooth oriented} \\ \text{2-submanifold of } \Omega \setminus \Gamma \text{ such that } \partial M = \Gamma\}. \end{aligned} \tag{3.31}$$

Here, $|M| = \mathscr{H}^2(M)$ is the area of M, and $M \cup \Gamma$ is a submanifold of Ω with boundary Γ (we emphasize that we consider only the part of the boundary of $M \cup \Gamma$ *inside* Ω). In order to keep notation simple, in the above situation, i.e., when the boundary of $M \cup \Gamma$ is Γ, we write $\partial M = \Gamma$ (in the geometrical sense). Strictly speaking, we should write $\partial(M \cup \Gamma) = \Gamma$.

By Stokes' theorem we have (as in (3.21) and (3.22))

$$\langle M, d\zeta \rangle = \int_M d\zeta = \int_\Gamma \zeta = \langle \Gamma, \zeta \rangle, \ \forall \zeta \in C_c^\infty(\mathbb{R}^3; \Lambda^1), \tag{3.32}$$

or, equivalently,

$$\operatorname{curl}(\nu \mathscr{H}^2 \llcorner M) = \tau \mathscr{H}^1 \llcorner \Gamma \text{ in } \mathscr{D}'(\Omega). \tag{3.33}$$

In the language of GMT, (3.32) asserts that $\partial M = \Gamma$, where this time ∂ stands for the boundary operator (not the geometric boundary). Let us recall the definition of ∂ (which coincides with d^*, the formal adjoint of d acting on forms). By definition, ∂M (i.e., ∂ acting on the 2-current M) is the 1-current satisfying

$$\begin{aligned} \langle \partial M, \zeta \rangle = \langle M, d\zeta \rangle = \int_M \nu \cdot \operatorname{curl}\zeta = \langle \operatorname{curl}(\nu \mathscr{H}^2 \llcorner M), \zeta \rangle, \\ \forall \zeta \in C_c^\infty(\Omega; \Lambda^1) \simeq C_c^\infty(\Omega; \mathbb{R}^3). \end{aligned} \tag{3.34}$$

Therefore,

$$\begin{aligned} &[\partial M = \Gamma \text{ in the sense of GMT}] \iff \\ &[\operatorname{curl}(\nu \mathscr{H}^2 \llcorner M) = \tau \mathscr{H}^1 \llcorner \Gamma \text{ in } \mathscr{D}'(\Omega)] \iff \text{(via Stokes)} \\ &[\partial M = \Gamma \text{ in the geometric sense}]. \end{aligned} \tag{3.35}$$

More generally, if $\Omega \subset \mathbb{R}^N$ and $T \in \mathscr{D}'(\Omega; \Lambda^k)$ for some $1 \leq k \leq N$, then $\partial T \in \mathscr{D}'(\Omega; \Lambda^{k-1})$ is defined by

$$\langle \partial T, \zeta \rangle = \langle T, d\zeta \rangle, \ \forall \zeta \in C_c^\infty(\Omega; \Lambda^{k-1}). \tag{3.36}$$

This applies in particular to the case where $M \subset \Omega$ is a k-dimensional submanifold of Ω with (geometric) boundary Γ. Then, M defines a k-current (still denoted by M) through the formula

$$\langle M, \xi \rangle = \int_M \xi, \ \forall \xi \in C_c^\infty(\Omega; \Lambda^k),$$

and

$$\langle \partial M, \zeta \rangle = \langle M, d\zeta \rangle = \int_M d\zeta = \int_\Gamma \zeta \text{ (by Stokes)} = \langle \Gamma, \zeta \rangle, \ \forall \zeta \in C_c^\infty(\Omega; \Lambda^{k-1}),$$

where Γ is viewed as a $(k-1)$-current.

Returning to (3.31), we note that the class of competitors is non-empty. Indeed, if $\widetilde{M}$ is a (smooth) "Seifert" surface in $\mathbb{R}^3$ with $\partial \widetilde{M} = \widetilde{\Gamma}$, then $M := \widetilde{M} \cap \Omega$ is admissible.

The main result in this section is the following

Theorem 3.4. *Let $u \in W^{1,1}(\Omega; \mathbb{S}^1) \cap C(\Omega \setminus \Gamma)$. We assume that Γ satisfies* (3.15) *and that* $\deg(u, \Gamma) = 1$. *Then,*

$$\Sigma(u) = 2\pi A_0(\Gamma). \tag{3.37}$$

The structure of the proof of Theorem 3.4 is the following.

Part 1. $\Sigma(u) \leq 2\pi A_0(\Gamma)$.

Part 2. $\Sigma(u) \geq 2\pi A_0(\Gamma)$.

Proof of Part 1. Starting from the definition of $\Sigma(u)$ and applying Hahn–Banach (or, more precisely, Theorem 1.9) yields

$$\Sigma(u) = 2 \sup\{\langle Ju, \zeta \rangle; \ \zeta \in C_c^\infty(\Omega; \mathbb{R}^3), \ \|\text{curl}\zeta\|_{L^\infty} \leq 1\}. \tag{3.38}$$

Inserting into (3.38) the value of Ju given by (3.8), we are led to

$$\Sigma(u) = 2\pi C(\Gamma), \tag{3.39}$$

where $C(\Gamma)$ is defined by

$$C(\Gamma) := \sup\left\{\langle \Gamma, \zeta \rangle = \int_\Gamma \tau \cdot \zeta; \ \zeta \in C_c^\infty(\Omega; \mathbb{R}^3) \text{ and } \|\text{curl}\zeta\|_{L^\infty} \leq 1\right\}. \tag{3.40}$$

Assuming that M is as in (3.31), we obtain (by (3.32)), $\forall \zeta \in C_c^\infty(\Omega; \mathbb{R}^3)$,

$$\langle \Gamma, \zeta \rangle = \int_M d\zeta = \int_M \nu \cdot \text{curl}\zeta \le |M| \, \|\text{curl}\zeta\|_{L^\infty}, \tag{3.41}$$

so that

$$C(\Gamma) \le |M|. \tag{3.42}$$

Taking the inf over M and using (3.31) gives

$$C(\Gamma) \le A_0(\Gamma). \tag{3.43}$$

From (3.39) and (3.43), we obtain

$$\Sigma(u) \le 2\pi A_0(\Gamma). \qquad \square \tag{3.44}$$

For Part 2, we do *not* follow the same strategy as in Section 2.3, since a priori we do not have at our disposal a 3D version of the OT Theorem 2.3 (or Theorem 2.4); see, however, the next paragraph. We use instead the analogue of Section 2.4, namely we establish the inequality $\Sigma(u) \ge 2\pi A_0(\Gamma)$ via the coarea formula (see Section 3.3).

Strictly speaking, there exists in the literature a "3D version" of the OT Theorem 2.3. Here is how it could be used. Let us return to the formula (3.39), i.e., $\Sigma(u) = 2\pi\, C(\Gamma)$. We might then invoke a result of Federer [148, 5.10] (see also Hardt and Pitts [187] and Almgren, Browder, and Lieb [8, Remark (3), p. 9–10]); it asserts that $A(\Gamma) = B(\Gamma)$ (see (3.65) and (3.66) for the definition of $A(\Gamma)$ and $B(\Gamma)$). Since $B(\Gamma) = C(\Gamma)$ via Hahn–Banach, we deduce that $A(\Gamma) = C(\Gamma)$, and thus $\Sigma(u) = 2\pi\, A(u)$. This can be viewed as a 3D version of Theorem 2.3. Clearly, $A(\Gamma)$ resembles $A_0(\Gamma)$ (defined in (3.31)) and $A(\Gamma) \le A_0(\Gamma)$. But we could only deduce that $\Sigma(u) \ge 2\pi\, A(\Gamma)$, which is *weaker* than the desired inequality $\Sigma(u) \ge 2\pi\, A_0(\Gamma)$. Therefore, we have decided to adopt a different strategy. It is indeed true that $A_0(\Gamma) = A(\Gamma)$ (see Corollary 3.1), but this requires some additional work.

In Section 3.5, we will show that Theorem 3.4 implies – a posteriori! – some form of MK "optimal transport" for the least area spanned by a curve.

3.3 $\Sigma(u) \ge 2\pi A_0(\Gamma)$ via the coarea formula

Here, we consider some Γ satisfying (3.15) and a map $u \in W^{1,1}(\Omega; \mathbb{S}^1)$ such that (3.7), which we rewrite in the condensed form $Ju = \pi\Gamma$, holds.

From Lemma 1.14, we know that

$$\Sigma(u) = \inf \left\{ \int_\Omega |\nabla w|;\ w \in W^{1,1}(\Omega; \mathbb{S}^1),\ Jw = Ju \text{ in } \mathscr{D}'(\Omega) \right\}.$$

Applying (3.8), we see that the desired inequality

$$\Sigma(u) \geq 2\pi A_0(\Gamma) \tag{3.45}$$

is an immediate consequence of the following

Lemma 3.2. *Assume* (3.15). *Let* $w \in W^{1,1}(\Omega; \mathbb{S}^1)$ *satisfy* $Jw = \pi\Gamma$ *in* $\mathscr{D}'(\Omega)$. *Then,*

$$\int_\Omega |\nabla w| \geq 2\pi A_0(\Gamma). \tag{3.46}$$

The proof of Lemma 3.2 relies on the following three-dimensional analogue of Lemma 2.5.

Lemma 3.3. *Let* $\widetilde{\Gamma} \subset \mathbb{R}^3$ *be a smooth closed connected oriented curve.*

1. *There exist some* $\varepsilon_1 > 0$ *and a diffeomorphism*

$$\Phi : \widetilde{\Gamma} \times \overline{\mathbb{D}}(0, \varepsilon_1) \to \{x \in \mathbb{R}^3;\ \mathrm{dist}(x, \widetilde{\Gamma}) \leq \varepsilon_1\}$$

such that:

a) $\Phi(x, z) \in N_x$ *(the normal plane at* x *to* $\widetilde{\Gamma}$*),* $\forall x \in \widetilde{\Gamma}$, $\forall z \in \overline{\mathbb{D}}(0, \varepsilon_1)$.
b) $|\Phi(x, z) - x| = |z|$, $\forall x \in \widetilde{\Gamma}$, $\forall z \in \overline{\mathbb{D}}(0, \varepsilon_1)$.

2. *Set* $\Gamma := \widetilde{\Gamma} \cap \Omega$,

$$Z := \left\{w \in W^{1,1}(\Omega; \mathbb{S}^1);\ Jw = \pi\Gamma \text{ in } \mathscr{D}'(\Omega)\right\}$$

and

$$\begin{aligned} Z^\infty := \Big\{ & w \in W^{1,1}(\Omega; \mathbb{S}^1) \cap C^\infty(\Omega \setminus \Gamma);\ \exists\, 0 < \varepsilon < \varepsilon_1,\ \exists f \in C^\infty(\overline{\Gamma}; \mathbb{S}^1) \\ & \text{such that } w(\Phi(x, re^{\imath\theta})) = f(x)\, e^{\imath\theta},\ \forall x \in \Gamma,\ \forall\, 0 < r < \varepsilon,\ \forall \theta \\ & \text{such that } x + \Phi(x, re^{\imath\theta}) \in \Omega \Big\}. \end{aligned}$$

Then, Z^∞ *is dense in* Z.

The proof of Lemma 3.3 is presented later in this section.

Proof of Lemma 3.2. By Lemma 3.3, we may assume that $w \in Z^\infty$. Let $\alpha = e^{\imath\xi} \in \mathbb{S}^1$ be a regular value of w in $\Omega \setminus \Gamma$ and set $M^\alpha := [w = \alpha]$, which is a 2-submanifold of $\Omega \setminus \Gamma$, oriented by $w \wedge \nabla w$. Since $w \in Z^\infty$, near each $x \in \Gamma$ we have

$$M^\alpha = \Phi\left(\{(y, re^{\imath\alpha(y)});\ y \in \Gamma,\ 0 < r < \varepsilon\}\right), \text{ with } \alpha(y) := \xi + \imath \ln f(y). \tag{3.47}$$

Here, $-\imath \ln f$ is a smooth local phase of f.

Clearly, (3.47) implies that $\partial M^\alpha = \Gamma$ (in the geometric sense), and thus M^α is a competitor in (3.31). Combining this with the *coarea formula*, we find that

$$\int_\Omega |\nabla w| = \int_{\mathbb{S}^1} |M^\alpha|\, d\alpha \geq \text{the right-hand side of (3.31)}. \tag{3.48}$$

Granted Lemma 3.3, the proof of Lemma 3.2 is complete. □

We now turn to Lemma 3.3. Its proof relies on the following simple global analogue of Lemma 3.1, whose proof is left to the reader.

Lemma 3.4. *Let $\widetilde{\Gamma}$ be a smooth manifold in $\mathbb{R}^N$ (any $N \geq 2$) and $\varepsilon > 0$. Let $w \in C^\infty(\widetilde{\Gamma} \times (\overline{\mathbb{D}}(0,\varepsilon) \setminus \{0\}); \mathbb{S}^1)$ be such that $\deg(w(x,\cdot), C(0,\varepsilon)) = 0$, $\forall\, x \in \widetilde{\Gamma}$.*

Set $f(x) := w(x,\varepsilon)$, $\forall\, x \in \widetilde{\Gamma}$. Then there exists some smooth function $\varphi : \widetilde{\Gamma} \times (\overline{\mathbb{D}}(0,\varepsilon) \setminus \{0\}) \to \mathbb{R}$ such that $w(x,z) = f(x)\, e^{\imath\varphi(x,z)}$, $\forall\, (x,z) \in \widetilde{\Gamma} \times (\overline{\mathbb{D}}(0,\varepsilon) \setminus \{0\})$.

Proof of Lemma 3.3. *Step 1.* Proof of item *1*
As in Step 2 in the proof of Theorem 3.3, we consider a smooth compact oriented surface M with (geometrical) boundary $\widetilde{\Gamma}$. Let, for $x \in \widetilde{\Gamma}$, $X(x)$ denote the outward unit normal to M at x. Let $\tau(x)$ be the unit tangent vector at $x \in \widetilde{\Gamma}$ (induced by a fixed orientation). Let $Y(x) \in \mathbb{R}^3$ be the unique unit vector such that $(X(x), Y(x), \tau(x))$ is a direct orthonormal basis of $\mathbb{R}^3$. Clearly, X and Y are smooth and, for each $x \in \widetilde{\Gamma}$, we have $X(x), Y(x) \in N_x$. By the inverse function theorem and the properties of the nearest point projection on $\widetilde{\Gamma}$, for sufficiently small $\varepsilon_1 > 0$ the map

$$\Phi(x,(y_1,y_2)) := x + y_1 X(x) + y_2 Y(x),\ x \in \widetilde{\Gamma},\ 0 \leq |(y_1,y_2)| \leq \varepsilon_1,$$

has all the required properties.
Step 2. Construction of a map in Z^∞
Let u_1 correspond to $\widetilde{\Gamma}$ and to $d = 1$ as in Theorem 3.3. Let Φ and $\varepsilon_1 > 0$ be as in item *1*. Define, for $x \in \widetilde{\Gamma}$, $0 < r \leq \varepsilon_1$ and $\theta \in \mathbb{R}$, $u_2(\Phi(x, re^{\imath\theta})) := e^{\imath\theta}$. We note that $\deg(u_2, \widetilde{\Gamma}) = 1$.

Set $\widetilde{\Omega} := \widetilde{\Gamma} \times (\overline{\mathbb{D}}(0,\varepsilon_1) \setminus \{0\})$ and $w := (u_1 \overline{u_2}) \circ \Phi : \widetilde{\Omega} \to \mathbb{S}^1$. Then, w satisfies the assumptions of Lemma 3.4. Therefore, there exist $f \in C^\infty(\widetilde{\Gamma}; \mathbb{S}^1)$ and $\varphi \in C^\infty(\widetilde{\Omega}; \mathbb{R})$ such that $w(x,z) = f(x)\, e^{\imath\varphi(x,z)}$, $\forall\,(x,z) \in \widetilde{\Omega}$. Consider some $\psi \in C^\infty(\widetilde{\Omega}; \mathbb{R})$ such that $\psi(x,z) = \varphi(x,z)$ if $|z| > \varepsilon_1/2$ and $\psi(x,z) = 0$ if $0 < |z| < \varepsilon_1/3$. Define $\widetilde{w}(x,z) := f(x)\, e^{\imath\psi(x,z)}$ and set, for $y \in \mathbb{R}^3 \setminus \widetilde{\Gamma}$,

$$u(y) := \begin{cases} \widetilde{w}(\Phi^{-1}(y))\, u_2(y), & \text{if } 0 < \operatorname{dist}(y,\widetilde{\Gamma}) < \varepsilon_1 \\ u_1(y), & \text{if } \operatorname{dist}(y,\widetilde{\Gamma}) \geq \varepsilon_1 \end{cases}.$$

Then, clearly, $u \in C^\infty$ and $u(\Phi(x, re^{\imath\theta})) = f(x)\, e^{\imath\theta}$ if $0 < r < \varepsilon_1/3$. We find that the restriction of u to Ω belongs to Z^∞.
Step 3. Proof of item *2*
Fix some $u_0 \in Z^\infty$. By Theorem 3.1, we have $u_0 \in Z$. Let $u \in Z$ and set $w := u\,\overline{u_0}$. Then, $Jw = 0$, and thus there exists some $\varphi \in W^{1,1}(\Omega; \mathbb{R})$ such that $w = e^{\imath\varphi}$. Consider a sequence $(\varphi_n)_{n\geq 1} \subset C^\infty(\overline{\Omega}; \mathbb{R})$ such that $\varphi_n \to \varphi$ in $W^{1,1}$ and, for each n, $\varphi_n \equiv 0$ in some neighborhood of $\overline{\Gamma}$. This is possible, since the $W^{1,1}$-capacity of $\overline{\Gamma}$ is zero. Set $u_n := e^{\imath\varphi_n} u_0$. Then, $(u_n) \subset Z^\infty$ and $u_n \to u$ in $W^{1,1}$. □

3.4 $\Sigma(u) \le 2\pi A_0(\Gamma)$ via the dipole construction

We consider Γ and u as in the previous section. An alternative proof of "$\Sigma(u) \le 2\pi A_0(\Gamma)$" in Theorem 3.4 relies on the dipole construction, in the spirit of Section 2.5. As explained in Remark 2.8, this much longer route is not of interest for $\mathbb{S}^1$-valued maps, but provides an effective approach in the case of $\mathbb{S}^k$-valued maps, with $k \ge 2$ (see Chapter 4).

Let M be a competitor in (3.31). For simplicity, we assume here that $M \cup \Gamma \subset \mathbb{R}^3$ is a smooth oriented surface with boundary Γ such that $M \cup \Gamma \Subset \Omega$; otherwise, the dipole construction becomes quite technical (when M meets $\partial\Omega$ tangentially).

Theorem 3.5. *Let $\Gamma \subset \mathbb{R}^3$ be a smooth closed connected oriented curve in $\mathbb{R}^3$. Let $M \subset \mathbb{R}^3 \setminus \Gamma$ be a smooth bounded oriented manifold such that $\partial M = \Gamma$. Let $\varepsilon > 0$. Then there exists some $w \in W^{1,1}_{loc}(\mathbb{R}^3; \mathbb{S}^1) \cap \mathrm{Lip}_{loc}(\mathbb{R}^3 \setminus \Gamma)$ such that*

1. $\mathrm{supp}\,(w-1) \subset \{x \in \mathbb{R}^3;\ \mathrm{dist}(x, M) < \varepsilon\}$.
2. $Jw = \pi\Gamma$ *in* $\mathscr{D}'(\mathbb{R}^3)$.
3. $\int_{\mathbb{R}^3} |\nabla w| \le 2\pi\,|M| + \varepsilon$.

Recall that the assumption $\partial M = \Gamma$ means that $M \cup \Gamma$ is a smooth compact manifold with geometric boundary Γ.

Proof of Theorem 3.5. Let $\nu(x)$ denote the orienting unit normal to M. Let $\delta(x) := \mathrm{dist}(x, \Gamma)$ and let $\varepsilon_1 > 0$ be such that δ is smooth on the set $\{x \in \mathbb{R}^3;\ 0 < \delta(x) \le \varepsilon_1\}$. Let $\varepsilon_2 > 0$ be such that the map $(x, t) \mapsto x + t\nu(x), x \in M, |t| < \varepsilon_2$ is a bi-Lipschitz diffeomorphism onto its image. [The existence of ε_2 follows from the fact that $M \cup \Gamma$ is a smooth compact manifold with boundary.]

For $0 < \varepsilon < \min\{\varepsilon_1, \varepsilon_2\}$, set

$$\delta_\varepsilon(x) := \min\{\delta(x), \varepsilon\},\ \forall x \in M,$$
$$U_\varepsilon := \{x + t\nu(x);\ x \in M,\ |t| \le \delta_\varepsilon(x)\},$$
$$\psi_\varepsilon : U_\varepsilon \to \mathbb{R},\ \psi_\varepsilon(x + t\nu(x)) := t/\delta_\varepsilon(x),\ \forall x \in M,\ \forall\,|t| \le \delta_\varepsilon(x),.$$
$$w_\varepsilon(y) := \begin{cases} -e^{-\imath\pi\psi_\varepsilon(y)}, & \text{if } y \in U_\varepsilon \\ 1, & \text{otherwise} \end{cases}.$$

Clearly, $w_\varepsilon \in \mathrm{Lip}_{loc}(\mathbb{R}^3 \setminus \Gamma)$. On the other hand, for each $x \in M$, the degree of w_ε on the line segment $[x - \varepsilon\nu(x), x + \varepsilon\nu(x)]$ is equal to one. Combining this with the fact that $w_\varepsilon = 1$ outside U_ε, we find that $\deg(w_\varepsilon, \Gamma_j) = 1, \forall j$.

It remains to estimate $\int_{\mathbb{R}^3} |\nabla w_\varepsilon|$. Let

$$V_\varepsilon := \{x + t\nu(x);\ x \in M,\ \delta(x) \ge \varepsilon,\ |t| \le \varepsilon\},\ W_\varepsilon := U_\varepsilon \setminus V_\varepsilon.$$

Set $f(x + t\nu(x)) := t, \forall x \in M, \forall\,|t| < \varepsilon_2$. Using the coarea formula, we find that

$$
\begin{aligned}
\int_{V_\varepsilon} |\nabla w_\varepsilon(y)|\, dy &= \frac{\pi}{\varepsilon}\int_{V_\varepsilon} |\nabla f(y)|\, dy = \frac{\pi}{\varepsilon}\int_{-\varepsilon}^{\varepsilon} \mathscr{H}^2([f=t])\, dt \\
&= \frac{\pi}{\varepsilon}\int_{-\varepsilon}^{\varepsilon} [|M| + O(\varepsilon)]\, dt = 2\pi|M| + O(\varepsilon).
\end{aligned}
\tag{3.49}
$$

On the other hand, when $0 < \delta(x) < \min\{\varepsilon_1, \varepsilon_2\}$ and $|t| \leq \delta(x)$, the function $x + t\nu(x) \mapsto g(x + t\nu(x)) := t/\delta(x)$ satisfies

$$
|(\nabla g)(x + t\nu(x))| \lesssim \frac{1}{\delta(x)} + \frac{|t||\nabla\delta(x)|}{\delta^2(x)} \leq \frac{2}{\delta(x)} \lesssim \frac{1}{\delta(x + t\nu(x))}.
$$

Since, in addition, for $x + t\nu(x) \in U_\varepsilon$ we have $\delta(x + t\nu(x)) \leq 2\varepsilon$, we find that

$$
\int_{W_\varepsilon} |\nabla w_\varepsilon| \lesssim \int_{\{y\in\mathbb{R}^3;\ \delta(y)\leq 2\varepsilon\}} \frac{1}{\delta(y)}\, dy = O(\varepsilon). \tag{3.50}
$$

Combining (3.49) with (3.50), we find that, for sufficiently small $c > 0$, $w := w_{c\varepsilon}$ has all the required properties. □

3.5 Least area spanned by a contour from the perspective of MK

In this section, we forget u for a while and focus on the curve Γ satisfying (3.15). [The map u will reappear later as a technical tool in the proof.] Recall the definitions of the quantities $A_0(\Gamma)$ and $C(\Gamma)$ introduced in Section 3.2:

$$
\begin{aligned}
A_0(\Gamma) = \inf\{|M|;\ M \subset \Omega \setminus \Gamma &\text{ is a smooth oriented} \\
&\text{2-submanifold of } \Omega\setminus\Gamma \text{ such that } \partial M = \Gamma\}
\end{aligned}
\tag{3.51}
$$

and

$$
C(\Gamma) = \sup\left\{\langle \Gamma, \zeta\rangle = \int_\Gamma \tau\cdot\zeta;\ \zeta \in C_c^\infty(\Omega;\mathbb{R}^3) \text{ and } \|\mathrm{curl}\zeta\|_{L^\infty} \leq 1\right\}.
$$

We now introduce a third quantity,

$$
B(\Gamma) := \min\left\{\|\mu\|_{\mathscr{M}};\ \mu \in \mathscr{M}(\Omega;\mathbb{R}^3),\ \mathrm{curl}\mu = \Gamma \text{ in } \mathscr{D}'(\Omega)\right\}. \tag{3.52}
$$

The main result in this section is a “cousin” of Corollaries 1.8 and 2.4.

Theorem 3.6. *Let Γ satisfy* (3.15). *We have*

$$
A_0(\Gamma) = B(\Gamma) = C(\Gamma). \tag{3.53}
$$

Proof. Assume that μ is as in (3.52). Then,

$$\langle \Gamma, \zeta \rangle = \int_\Omega \mu \cdot (\mathrm{curl}\zeta) \leq \|\mu\|_{\mathscr{M}} \|\mathrm{curl}\zeta\|_{L^\infty}, \ \forall \zeta \in C_c^\infty(\Omega; \mathbb{R}^3). \tag{3.54}$$

Taking the $\sup_\zeta$ and the $\inf_\mu$ in (3.54) yields $C(\Gamma) \leq B(\Gamma)$.

On the other hand, since $\nu \mathscr{H}^2 \llcorner M$ is an admissible measure in (3.52), we have $B(\Gamma) \leq A_0(\Gamma)$.

It remains to prove that $A_0(\Gamma) \leq C(\Gamma)$. Applying Theorems 3.1, 3.2, and 3.4, we obtain some $u_0 \in W^{1,1}(\Omega; \mathbb{S}^1)$ such that $Ju_0 = \pi \Gamma$ and $\Sigma(u_0) = 2\pi A_0(\Gamma)$. Since we also have $\Sigma(u_0) = 2\pi C(\Gamma)$ (by (3.39)), we conclude that $A_0(\Gamma) = C(\Gamma)$. □

Keeping in mind the MK theory as presented in Section 1.6, we can say that Theorem 3.6 provides a technique "à la Kantorovich" for computing the *least area* spanned by a curve, in the spirit of the *least cost* for a Monge transport problem.

3.6 The structure of *Ju* for a general $u \in W^{1,1}$. Where Federer encounters Kantorovich

This section is the 3D counterpart of Section 2.9. Here, we investigate Ju and $\Sigma(u)$ for an *arbitrary* $u \in W^{1,1}(\Omega; \mathbb{S}^1)$, as opposed to *special* u's considered in Sections 3.1 and 3.2, e.g., $u \in W^{1,1}(\Omega; \mathbb{S}^1) \cap C(\Omega \setminus \Gamma)$. The basic class $\mathscr{E}$ is now replaced by

$$\begin{aligned} \mathscr{F} := \{T;\ & T \text{ is a 1-current such that } T = \partial M \text{ for some} \\ & \text{2-rectifiable current } M \text{ on } \Omega\}, \end{aligned} \tag{3.55}$$

where the condition $T = \partial M$ is meant in the sense of *GMT*, i.e.,

$$\langle T, \zeta \rangle = \langle M, d\zeta \rangle = \int_M d\zeta, \ \forall \zeta \in C_c^\infty(\Omega; \Lambda^1). \tag{3.56}$$

[For the definition of 2-rectifiable currents, see Section 2.9.4.]

Note that, if $T \in \mathscr{F}$, then

$$|\langle T, \zeta \rangle| \leq |M| \, \|\mathrm{curl}\zeta\|_{L^\infty}, \ \forall \zeta \in C_c^\infty(\Omega; \mathbb{R}^3) \simeq C_c^\infty(\Omega; \Lambda^1).$$

Clearly, $\mathscr{F}$ is a subgroup of the space $\mathscr{D}'(\Omega; \Lambda^1)$, but $\mathscr{F}$ is *not* a subspace because, in general, $(1/2)T \notin \mathscr{F}$ when $T \in \mathscr{F}$.

By analogy with (2.146), we set, for every 1-current T,

$$\|T\|_W := \sup\{\langle T, \zeta \rangle;\ \zeta \in C_c^\infty(\Omega; \mathbb{R}^3) \text{ and } \|\mathrm{curl}\zeta\|_{L^\infty} \leq 1\} \tag{3.57}$$

(this quantity is somewhat similar to the flat norm in GMT). We have the following

Theorem 3.7. *Let $T \in \mathscr{F}$. Then,*

$$\|T\|_W = \inf\{|M|;\ M \text{ is a 2-rectifiable current such that } T = \partial M\}. \tag{3.58}$$

Moreover, the inf *in* (3.58) *is achieved.*

The minimizers of (3.58) are called *mass-minimizing* or *area-minimizing* 2-rectifiable currents with boundary Γ. Their existence is a priori not obvious. In our setting, this is an immediate consequence of the existence of optimal liftings (see item *3* in Theorem 3.9). In GMT, this follows from the Federer–Fleming compactness theorem (see Comment 10 in Chapter 2).

Proof of Theorem 3.7. By Theorem 3.8 below (converse part), there exists some $u \in W^{1,1}(\Omega; \mathbb{S}^1)$ such that $Ju = \pi T$. [Strictly speaking, as explained at the beginning of Section 3.1, we should write $*Ju = \pi T$.] By Theorem 3.9, the set on the right-hand side of (3.58) equals $\{(1/2\pi)\, \|D^j\varphi\|_{\mathcal{M}};\ \varphi \in BV(\Omega; \mathbb{R}) \text{ and } u = e^{\imath\varphi}\}$.

Equality (3.58) is thus a consequence of Corollary 1.8, and the existence of a minimizer in (3.58) follows from the existence of minimal liftings. □

The next result provides a description of Ju and $\Sigma(u)$ in the spirit of Theorems 2.10 and 2.11.

Theorem 3.8. *Assume that* $u \in W^{1,1}(\Omega; \mathbb{S}^1)$. *Then,*

$$\frac{1}{\pi} Ju \in \mathcal{F} \tag{3.59}$$

and

$$\Sigma(u) = 2\|Ju\|_W. \tag{3.60}$$

Conversely, given any $T \in \mathcal{F}$, *there exists some* $u \in W^{1,1}(\Omega; \mathbb{S}^1)$ *such that*

$$\frac{1}{\pi} Ju = T. \tag{3.61}$$

Proof. By the proof of Theorem 3.10 below, for every $u \in W^{1,1}(\Omega; \mathbb{S}^1)$ there exists a sequence $(\widetilde{M}_j)$ satisfying (3.74) and $Ju = \pi\, \partial M$, with M given by (3.76). Since clearly $\partial M \in \mathcal{F}$, we find that $(1/\pi)\, Ju \in \mathcal{F}$.

Equality (3.60) follows from Corollary 1.8.

It remains to prove that, if $T \in \mathcal{F}$, then we may write $T = (1/\pi)\, Ju$ (or rather $T = (1/\pi)\, *Ju$) for some $u \in W^{1,1}(\Omega; \mathbb{S}^1)$. The strategy consists of finding some $\psi \in BV(\Omega; \mathbb{R})$ such that $D^j\psi = -2\pi * M$ and $u = e^{\imath\psi} \in W^{1,1}(\Omega; \mathbb{S}^1)$. In view of Remark 1.15, for this to happen we must have $\psi \in SBV(\Omega; \mathbb{R})$. This suggests as key ingredient Lemma 3.5 below, whose proof (in any dimension) is presented in Appendix 15.7 (see Lemma 15.34 there).

Granted Lemma 3.5, we proceed as follows. Since M is a 2-rectifiable current, we have

$$*M = \sum_i \nu_i\, \mathcal{H}^2 \llcorner S_i, \tag{3.62}$$

with each S_i a Borel set contained in a C^1 surface Σ_i oriented by ν_i. Using (3.62), the equality $D^j\psi = -2\pi * M$ and Lemma 15.31 in Appendix 15.7, we find that the jumps of ψ are $2\pi\,\mathbb{Z}$-valued $\mathscr{H}^2$-a.e. on the jump set J_ψ of ψ. Combining this with the fact that $D^c\psi = 0$ and with Volpert's chain rule, we find that $u := e^{\imath\psi}$ satisfies $\nabla u = \imath u \nabla\psi$, $D^c u = 0$ and $D^j u = 0$. Therefore, u belongs to $W^{1,1}(\Omega; \mathbb{S}^1)$.

Finally, with the identification $\partial M \simeq \operatorname{curl}(*M)$, we have

$$T = \partial M = \operatorname{curl}(*M) = \operatorname{curl}(-1/(2\pi)\, D^j\psi) = \frac{1}{2\pi}\operatorname{curl}(-D\psi + \nabla\psi)$$
$$= \frac{1}{2\pi}\operatorname{curl}\nabla\psi = \frac{1}{2\pi}\operatorname{curl}(u \wedge \nabla u) \text{ (by (1.105))} = \frac{1}{\pi} Ju. \qquad \square$$

Lemma 3.5. *Let M be a 2-rectifiable current in $\Omega \subset \mathbb{R}^3$. Then there exists some $\psi \in BV(\Omega; \mathbb{R})$ such that $D^j\psi = -2\pi * M$ and $D^c\psi = 0$.*

[Equivalently, there exists some $\psi \in SBV(\Omega; \mathbb{R})$ such that the jump part of $D\psi$ is $-2\pi * M$.]

The 3D analogue of Theorem 2.12 is the following

Theorem 3.9. *Let $u \in W^{1,1}(\Omega; \mathbb{S}^1)$ and set $T := (1/\pi)\, Ju$. Then, the following hold.*

1. *For any lifting $\varphi \in BV(\Omega; \mathbb{R})$ of u, the measure $M := -(1/2\pi) * D^j\varphi$ is a 2-rectifiable current such that $\partial M = T$.*
2. *Given any 2-rectifiable current M such that $\partial M = T$, there exists a unique (mod 2π) lifting $\psi \in BV(\Omega; \mathbb{R})$ of u such that $*D^j\psi = -2\pi M$, or equivalently $D\psi = u \wedge \nabla u - 2\pi * M$.*
3. *In this bijective correspondence, minimal liftings correspond to minimizing M's in (3.58).*

The meaning of the equality $- * D^j\varphi = 2\pi M$ is

$$-\langle D^j\varphi, \zeta\rangle = 2\pi\,\langle M, *\zeta\rangle, \forall\, \zeta \in C_c^\infty(\Omega; \Lambda^1) \simeq C_c^\infty(\Omega; \mathbb{R}^3). \tag{3.63}$$

In view of the equality $\partial M = \Gamma$, equality (3.63) amounts to the familiar condition $\operatorname{curl}(-D^j\psi) = 2\pi\,\Gamma$. Thanks to this equivalence, the proof of Theorem 3.9 is identical to that of Theorem 2.9 and is omitted.

Remark 3.4. Let $u \in W^{1,1}(\Omega; \mathbb{S}^1)$ be such that $Ju = \pi\,\Gamma$, with Γ satisfying (3.15). Let M be a competitor for $A_0(\Gamma)$. Consider a solution $\psi_0 \in BV(\Omega; \mathbb{R})$ of $D\psi_0 = u \wedge \nabla u - 2\pi * M$. By Theorem 3.9 item 2 (or by direct verification using Volpert's chain rule), $\psi := \psi_0 + C$ (with C an appropriate constant) is a BV lifting of u. We find that

$$\|D\psi\|_{\mathscr{M}} = \int_\Omega |u \wedge \nabla u| + 2\pi\,|M| = \int_\Omega |\nabla u| + 2\pi\,|M|. \tag{3.64}$$

Taking the inf over M in (3.64), we obtain another proof of the inequality $E(u) \le \int_\Omega |\nabla u| + 2\pi\, A_0(\Gamma)$ (where $E(u)$ is as in (1.6)), already obtained in Section 3.2 in the equivalent form $\Sigma(u) \le 2\pi\, A_0(\Gamma)$.

Returning now to the elementary setting of Section 3.1, suppose that Γ is a smooth curve in Ω satisfying (3.15). Then Γ, viewed as a 1-current, belongs to $\mathscr{F}$, since $\Gamma = \partial M$, where $M := \widetilde{M} \cap \Omega$ and $\widetilde{M}$ is a smooth surface in $\mathbb{R}^3$ such that $\partial \widetilde{M} = \widetilde{\Gamma}$. By analogy with this simple case, we introduce the three quantities $A(T), B(T), C(T)$ for a *general* $T \in \mathscr{F}$:

$$\begin{aligned} A(T) &:= \inf\{|M|;\ M \subset \Omega \text{ is a 2-rectifiable current such that } \partial M = T\}, \\ &= \|T\|_W \text{ (by Theorem 3.7)}, \end{aligned} \tag{3.65}$$

$$B(T) := \inf\left\{\|\mu\|_{\mathscr{M}};\ \mu \in \mathscr{M}(\Omega;\ \Lambda^2) \text{ is a measure such that } \partial\mu = T\right\}, \tag{3.66}$$

and

$$\begin{aligned} C(T) &:= \sup\left\{\langle T, \zeta\rangle;\ \zeta \in C_c^\infty(\Omega;\ \Lambda^1),\ \|d\zeta\|_{L^\infty} \le 1\right\} \\ &= \|T\|_W \text{ (by definition (3.57))}. \end{aligned} \tag{3.67}$$

Some comments:

a) When $T = \Gamma$, we clearly have

$$A(\Gamma) \le A_0(\Gamma), \tag{3.68}$$

since the class of competitors in (3.65) is (much) bigger than in (3.31).

b) The condition $\partial\mu = T$ in (3.66) amounts to $\operatorname{curl}(*\mu) = T$, and thus

$$B(T) = \min\{\|\nu\|_{\mathscr{M}};\ \nu \in \mathscr{M}(\Omega;\ \Lambda^1) \text{ with } \operatorname{curl}\nu = T\}. \tag{3.69}$$

c) With the identification of 1-forms with vector fields, the condition $\|d\zeta\|_{L^\infty} \le 1$ in (3.67) corresponds to $\|\operatorname{curl}\zeta\|_{L^\infty} \le 1$.

d) Consequently, the definitions (3.52) and (3.66) are consistent. Similarly for (3.40) and (3.67).

Corollary 3.1. *For every $T \in \mathscr{F}$, we have*

$$A(T) = B(T) = C(T), \tag{3.70}$$

and the inf *in* (3.65) *is achieved. In fact, if we set*

$$\mathscr{Q}(T) := \{\mu \in \mathscr{M}(\Omega;\ \Lambda^2);\ \partial\mu = T \text{ and } \|\mu\|_{\mathscr{M}} = B(T)\},$$

then

any extreme point of $\mathscr{Q}(T)$ is an area-minimizing current in (3.65). (3.71)

Moreover, for every smooth curve Γ satisfying (3.15) *we have*

$$A_0(\Gamma) = A(\Gamma). \tag{3.72}$$

Proof. From Theorem 3.8, we know that $T = (1/\pi)\, Ju_0$ for some $u_0 \in W^{1,1}(\Omega; \mathbb{S}^1)$. Recall that in Section 1.6 we have introduced three quantities: $A(u_0)$, $B(u_0)$, and $C(u_0)$. In view of (1.161)–(1.163), (3.65)–(3.67), and Theorem 3.9, we see that

$$A(u_0) = 2\pi A(T),\; B(u_0) = 2\pi B(T) \text{ and } C(u_0) = 2\pi C(T). \tag{3.73}$$

We then deduce (3.70) from Corollary 1.8. Assertion (3.71) comes from Corollary 1.9.

On the other hand, when $T = \Gamma$ we have $\Sigma(u_0) = 2\pi A_0(\Gamma)$ (by Theorem 3.4) and $A(u_0) = \Sigma(u_0)$ (by Corollary 1.8). Combining this with (3.73), we obtain

$$A_0(\Gamma) = \frac{1}{2\pi}\Sigma(u_0) = \frac{1}{2\pi}A(u_0) = A(\Gamma).$$

□

We conclude this section with an approximation result for elements in $\mathscr{F}$.

Theorem 3.10. *Given any $T \in \mathscr{F}$, there exists a sequence of smooth compact oriented surfaces $\widetilde{M}_j \subset \mathbb{R}^3$ such that*

$$\sum_j |\widetilde{M}_j| < \infty \tag{3.74}$$

and

$$T = \partial M, \tag{3.75}$$

where M is the 2-rectifiable current defined by

$$M := \sum_j M_j, \text{ with } M_j := \widetilde{M}_j \cap \Omega. \tag{3.76}$$

Moreover, we have

$$A(T) = \inf\Big\{\sum |M_j|;\ \widetilde{M}_j \text{ are compact oriented surfaces such that (3.75) and (3.76) hold}\Big\}. \tag{3.77}$$

Remark 3.5. Clearly, if M is defined by (3.76) and (3.74) holds, then $\partial M \in \mathscr{F}$. Thus, Theorem 3.10 asserts that

$$\{\partial M;\ M \text{ is a 2-rectifiable current in } \Omega\}$$
$$= \Big\{\partial \sum_j (\widetilde{M}_j \cap \Omega);\ \widetilde{M}_j \text{ is a compact oriented surface with boundary in } \mathbb{R}^3 \text{ and } \sum |\widetilde{M}_j| < \infty\Big\}, \tag{3.78}$$

which is the three-dimensional analogue of (2.179). It is actually possible to obtain (3.78) by adapting the argument leading to (2.179) (i.e., via a three-dimensional analogue of Proposition 15.4 in Appendix 15.9), but we follow below a different route, using $W^{1,1}(\Omega; \mathbb{S}^1)$ maps.

Remark 3.6. In view of Proposition 2.4, presumably, for a general $T \in \mathscr{F}$, the inf in (3.77) is not achieved.

Proof of Theorem 3.10. We consider some $u \in W^{1,1}(\Omega; \mathbb{S}^1)$ such that $Ju = \pi T$ (see Theorem 3.8). The main step consists of writing an arbitrary $u \in W^{1,1}(\Omega; \mathbb{S}^1)$ as an infinite product of nice maps. Let R be some large number such that $\overline{\Omega} \subset B_R(0)$. We consider the following class:

$$\widetilde{Z} := \Big\{ w \in W^{1,1}_{loc}(\mathbb{R}^3; \mathbb{S}^1);\ w(x) = 1 \text{ when } |x| \geq R+1,\ \exists H_1, \ldots, H_m \text{ mutually disjoint smooth closed connected oriented curves with } w \in C^\infty(\mathbb{R}^3 \setminus \cup_{j=1}^m H_j),\ \deg(w, H_j) = 1,\ \forall j = 1, \ldots, m \Big\}.$$

Starting from $\widetilde{Z}$, we define $\widetilde{Z}^\infty$ in the spirit of Lemma 3.3, item 2. More specifically, let $w \in \widetilde{Z}$ and $H_1, \ldots, H_m$ as above. For each j, we consider ε_j and a diffeomorphism Φ_j,

$$\Phi_j : H_j \times \overline{\mathbb{D}}(0, \varepsilon_j) \to \{x \in \mathbb{R}^3;\ \text{dist}(x, H_j) \leq \varepsilon_j\},$$

as in Lemma 3.3 item *1*. We then set

$$\widetilde{Z}^\infty := \Big\{ w \in \widetilde{Z};\ \exists\, 0 < \varepsilon < \min_j \varepsilon_j,\ \exists f_j \in C^\infty(H_j; \mathbb{S}^1),\ \forall j, \text{ such that } w(\Phi_j(x, re^{\imath\theta})) = f(x)\, e^{\imath\theta},\ \forall j,\ \forall x \in H_j,\ \forall\, 0 < r < \varepsilon,\ \forall \theta \Big\}.$$

We have the following

Lemma 3.6. *For each $u \in W^{1,1}(\Omega; \mathbb{S}^1)$ and $\varepsilon > 0$, there exists a sequence $(\widetilde{u}_j)_{j \geq 1} \subset \widetilde{Z}^\infty$ such that the following hold.*

1. *With u_j the restriction of $\widetilde{u}_j$ to Ω, we have $u = \prod_{j \geq 1} u_j$ in $W^{1,1}(\Omega; \mathbb{S}^1)$.*
2. $\sum_{j \geq 1} |\widetilde{u}_j|_{W^{1,1}(\mathbb{R}^3)} < \infty$.
3. $\sum_{j \geq 1} |u_j|_{W^{1,1}(\Omega)} < |u|_{W^{1,1}(\Omega)} + \varepsilon$.

Proof of Theorem 3.10 *completed (granted Lemma* 3.6*).* Set $\widetilde{\Gamma}_j := H_1 \cup \ldots \cup H_m$, with $H_1, \ldots, H_m$ associated with $\widetilde{u}_j$ as in the definition of the class $\widetilde{Z}^\infty$. Thus, $\widetilde{\Gamma}_j$ is a smooth closed oriented curve (in general, not connected). By Theorem 3.1 and the definition of the class $\widetilde{Z}^\infty$, we have $\widetilde{\Gamma}_j = (1/\pi) J\widetilde{u}_j$. By the proof of (3.46) in Section 3.3, for each $j \geq 1$ there exists some $\alpha = \alpha_j \in \mathbb{S}^1$ such that $\widetilde{M}_j := [\widetilde{u}_j = \alpha]$ (with the orientation induced by $\widetilde{u}_j \wedge \nabla \widetilde{u}_j$) is a smooth surface such that $\widetilde{M}_j \cup \widetilde{\Gamma}_j$ is compact,

$$\partial\widetilde{M}_j = \widetilde{\Gamma}_j \text{ (in the geometrical sense) and } |\widetilde{M}_j| \le \frac{1}{2\pi}\int_{\mathbb{R}^3} |\nabla\widetilde{u}_j|. \tag{3.79}$$

From (3.79) and Lemma 3.6, item *2*, we find that $\sum |\widetilde{M}_j| < \infty$.

On the other hand, as in the proof of Theorem 2.11, we have

$$\pi T = Ju = \lim_{m\to\infty} J\left(\prod_{j=1}^{m} u_j\right) = \pi\sum_j \Gamma_j = \pi\sum_j \partial M_j = \pi\,\partial\sum_j M_j,$$

since clearly $Ju_j = \pi\Gamma_j$.

It remains to prove (3.77). Inequality "$\le$" is clear. For the reverse inequality, we proceed as follows. Let $u_0, u \in W^{1,1}(\Omega; \mathbb{S}^1)$ be such that $Ju_0 = Ju = \pi T$. We change the definition of $\widetilde{M}_j$ by choosing the level set $\widetilde{M}_j = [\widetilde{u}_j = \alpha]$ such that

$$|M_j| = |\widetilde{M}_j \cap \Omega| \le \frac{1}{2\pi}\int_{\Omega} |\nabla\widetilde{u}_j| = \frac{1}{2\pi}|u_j|_{W^{1,1}(\Omega)}. \tag{3.80}$$

With this choice we find (using Lemma 3.6, item *3*)

$$\sum_j |M_j| \le \frac{1}{2\pi}(|u|_{W^{1,1}(\Omega)} + \varepsilon). \tag{3.81}$$

Letting in (3.81) $\varepsilon \to 0$ and taking the inf over the u's such that $Ju = Ju_0 = \pi T$, we obtain (using (1.161) and (1.162), Corollaries 1.8 and 3.1, and Theorem 3.9)

$$\begin{aligned}\frac{1}{2\pi}\Sigma(u_0) &\ge \text{right-hand side of (3.77)} \ge \frac{1}{2\pi}B(u_0)\\ &= \frac{1}{2\pi}\Sigma(u_0) = \frac{1}{2\pi}A(u_0) = A(T),\end{aligned}$$

whence (3.77). □

Proof of Lemma 3.6. *Step 1.* Extension of u to $\mathbb{R}^3$

Let v be the trace of u on $\partial\Omega$. Let $\psi \in L^\infty(\partial\Omega; \mathbb{R})$ be a phase of v. We consider some $\varphi \in W^{1,1}(\mathbb{R}^3 \setminus \overline{\Omega}; \mathbb{R})$ such that $\varphi = \psi$ on $\partial\Omega$ and $\varphi(x) = 0$ when $|x| \ge R$. Let $\widetilde{u} := \begin{cases} u, & \text{in } \Omega \\ e^{\imath\varphi}, & \text{in } \mathbb{R}^3 \setminus \overline{\Omega}\end{cases}$. Then, clearly, $\widetilde{u} - 1 \in W^{1,1}(\mathbb{R}^3)$ and $\widetilde{u}(x) = 1$ if $|x| \ge R$.

In order to complete the proof of Lemma 3.6, it clearly suffices to prove that $\widetilde{u} = \prod_{j\ge 1} w_j$ in $W^{1,1}_{loc}(\mathbb{R}^3; \mathbb{S}^1)$ for a sequence $(w_j)_{j\ge 1} \subset \widetilde{Z}^\infty$ such that $\sum_{j\ge 1} |w_j|_{W^{1,1}(\mathbb{R}^3)} < \infty$ and $\sum_{j\ge 1} |u_j|_{W^{1,1}(\Omega)} < |u|_{W^{1,1}(\Omega)} + \varepsilon$ (with u_j the restriction of w_j to Ω).

Step 2. Approximation of $\widetilde{u}$ with maps in $\widetilde{Z}^\infty$

By the proof of Theorem 10.3 in Section 10.2, for each $\widetilde{u}$ as above and each $\delta > 0$, there exists some $w = w^\delta \in \widetilde{Z}$ such that $\|\widetilde{u} - w\|_{W^{1,1}(\mathbb{R}^3)} < \delta$ and $\operatorname{supp}(w-1) \subset \overline{B}_{R+\delta}(0)$. The fact that we may, moreover, pick such a w in the smaller class $\widetilde{Z}^\infty$ follows from the proof of Lemma 3.3; it suffices to repeat Steps 2 and 3 near each H_j.

Step 3. Construction of the sequence $(w_j)_{j\geq 1}$
Since $\widetilde{u}/w^\delta - 1 \to 0$ in $W^{1,1}(\mathbb{R}^3)$ as $\delta \to 0$ (by Lemma 15.41 in Appendix 15.12), there exists some $w_1 \in \widetilde{Z}^\infty$ such that

$$\begin{aligned}
&|\widetilde{u}/w_1|_{W^{1,1}(\mathbb{R}^3)} < \varepsilon/4,\\
&|\widetilde{u} - w_1|_{W^{1,1}(\mathbb{R}^3)} < \varepsilon/4,\\
&\operatorname{supp}(w_1 - 1) \subset \overline{B}_{R+1/2}(0).
\end{aligned}$$

We next argue as above starting from $\widetilde{u}_1/w_1$ instead of $\widetilde{u}$, and so on. We obtain inductively a sequence (w_j) satisfying

$$|\widetilde{u}/(w_1 \cdots w_j)|_{W^{1,1}(\mathbb{R}^3)} < \varepsilon/2^{j+1}, \tag{3.82}$$

$$|\widetilde{u} - w_1 \cdots w_j|_{W^{1,1}(\mathbb{R}^3)} < \varepsilon/2^{j+1}, \tag{3.83}$$

$$\operatorname{supp}(w_j - 1) \subset \overline{B}_{R+1-1/2^j}(0). \tag{3.84}$$

Property (3.84) allows the induction process to be continued.
Estimate (3.83) implies that $\widetilde{u} = \prod_{j\geq 1} w_j$ in $W^{1,1}_{loc}(\mathbb{R}^3; \mathbb{S}^1)$.
On the other hand, (3.82) yields, for each $j \geq 2$,

$$\begin{aligned}
|w_j|_{W^{1,1}(\mathbb{R}^3)} &= \left|\frac{\widetilde{u}/(w_1 \cdots w_{j-1})}{\widetilde{u}/(w_1 \cdots w_j)}\right|_{W^{1,1}(\mathbb{R}^3)}\\
&\leq |\widetilde{u}/(w_1 \cdots w_{j-1})|_{W^{1,1}(\mathbb{R}^3)} + |\widetilde{u}/(w_1 \cdots w_j)|_{W^{1,1}(\mathbb{R}^3)}\\
&< \varepsilon/2^j + \varepsilon/2^{j+1},
\end{aligned} \tag{3.85}$$

and thus $\sum_{j\geq 1} |w_j|_{W^{1,1}(\mathbb{R}^3)} < \infty$. This proves item *2*.

Finally, we establish item *3*. Let u_j be the restriction of w_j to Ω. By (3.83) with $j = 1$, we have $|u_1|_{W^{1,1}(\Omega)} < |u|_{W^{1,1}(\Omega)} + \varepsilon/4$. Combining this with (3.85), we obtain item *3*.

The proof of Lemma 3.6 is complete. □

3.7 Further properties when *Ju* is a measure

The first result here is the following extension of Theorem 2.14.

Theorem 3.11. *Let $u \in W^{1,1}(\Omega; \mathbb{S}^1)$. Then, Ju is a measure if and only if $(1/\pi)\,Ju$ is a 1-rectifiable current.*

Proof. We know that $(1/\pi)\,Ju = \partial M$ for some 2-rectifiable current M (Theorem 3.8). If, in addition, $(1/\pi)\,Ju$ is a measure, then $(1/\pi)\,Ju$ is 1-rectifiable, by the Federer–Fleming boundary rectifiability theorem [147, Theorem 4.2.16 (2), p. 413] (which we use in the version established in Appendix 15.11; see Lemma 15.39).

Conversely, any rectifiable current is a finite measure. □

Corollary 3.2. *Assume that $Ju \in L^1(\Omega; \mathbb{R}^3)$. Then, $Ju = 0$.*

Proof. By Theorem 3.11, Ju is rectifiable, and thus singular with respect to the Lebesgue measure. Since, on the other hand, Ju is absolutely continuous with respect to the Lebesgue measure, we find that $Ju = 0$. □

The next result is specific to dimension $N \geq 3$ (and fails for $N = 2$).

Theorem 3.12. *Let $N \geq 3$ and $\Omega \subset \mathbb{R}^N$. Let $u \in W^{1,1}(\Omega; \mathbb{S}^1)$ be such that $Ju \in \mathcal{M}$. Then there exists a constant $C(\Omega)$ such that*

$$\min\{\|u \wedge \nabla u - \nabla\varphi\|_{L^{N/(N-1)}};\ \varphi \in W^{1,1}\} \leq C(\Omega)\|Ju\|_{\mathcal{M}}. \tag{3.86}$$

Remark 3.7. A priori it is not clear that the class of competitors in (3.86) is non-empty, i.e., that there exists $\varphi \in W^{1,1}(\Omega; \mathbb{R})$ such that $u \wedge \nabla u - \nabla\varphi \in L^{N/(N-1)}$. This will be established in the course of the proof of Theorem 3.12, as a consequence of the assumption $Ju \in \mathcal{M}$. Once this is proved, then the inf in (3.86) is achieved. Indeed, if $(\varphi_n) \subset W^{1,1}$ is a minimizing sequence, then, up to some subsequence, $u \wedge \nabla u - \nabla\varphi_{n_k} \rightharpoonup h$ weakly in $L^{N/(N-1)}$ for some $h \in L^{N/(N-1)}$. On the other hand, we may always assume that $\int \varphi_{n_k} = 0$, and thus $\varphi_{n_k} \to \varphi$ in L^1 for some $\varphi \in BV$, with $\nabla\varphi_{n_k} \rightharpoonup \nabla\varphi$ weak* in $\mathcal{M}$. Since $u \wedge \nabla u - \nabla\varphi = h$, we conclude that $\varphi \in W^{1,1}$ is a desired minimizer.

Corollary 3.3. *Let $N \geq 3$ and $\Omega \subset \mathbb{R}^N$. Let $u \in W^{1,1}(\Omega; \mathbb{S}^1)$ be such that $Ju \in \mathcal{M}$. Then there exists $v \in W^{1,N/(N-1)}(\Omega; \mathbb{S}^1)$ such that $Jv = Ju$ and*

$$\|\nabla v\|_{L^{N/(N-1)}} \leq C(\Omega)\|Ju\|_{\mathcal{M}}.$$

Proof. Take $v := ue^{-\imath\varphi}$, with φ a minimizer in Theorem 3.12. Then, $Jv = Ju$ and

$$\|\nabla v\|_{L^{N/(N-1)}} = \|v \wedge \nabla v\|_{L^{N/(N-1)}} = \|u \wedge \nabla u - \nabla\varphi\|_{L^{N/(N-1)}} \leq C(\Omega)\|Ju\|_{\mathcal{M}}.$$ □

Remark 3.8. The conclusion of Corollary 3.3 (and thus Theorem 3.12) fails when $N = 2$ and $Ju \neq 0$. Indeed, if there exists $v \in W^{1,2}(\Omega; \mathbb{S}^1)$ such that $Jv = Ju$, then $Ju = 0$. [Note that the assumption $N \geq 3$ enters in the proof of Theorem 3.12 via the use of Theorem 5” in Bourgain and Brezis [49], which assumes $N \geq 3$.]

Proof of Theorem 3.12. The key point of the proof is the estimate

$$\langle Ju, *\zeta\rangle \leq C\,\|Ju\|_{\mathcal{M}}\,\|d\zeta\|_{L^N},\ \forall \zeta \in C_c^\infty(\Omega; \Lambda^{N-2}). \tag{3.87}$$

Granted (3.87), we complete the proof of the theorem as follows. Estimate (3.87) implies that there exists some $\psi \in L^{N/(N-1)}(\Omega; \Lambda^1) \simeq L^{(N/(N-1)}(\Omega; \mathbb{R}^N)$ such that $\|\psi\|_{L^{N/(N-1)}} \leq C\,\|Ju\|_{\mathcal{M}}$ and

$$\langle Ju, *\zeta\rangle = (-1)^{N+1}\,\langle\psi, *d\zeta\rangle,\ \forall\,\zeta\in C_c^\infty(\Omega;\Lambda^{N-2}).$$

This implies (via (1.141) and (1.142)) that $d(u\wedge du-\psi)=0$, and thus $u\wedge\nabla u-\psi=\nabla\varphi$ for some $\varphi\in W^{1,1}(\Omega;\mathbb{R})$ satisfying

$$\|u\wedge\nabla u-\nabla\varphi\|_{L^{N/(N-1)}}=\|\psi\|_{L^{N/(N-1)}}\le C\,\|Ju\|_{\mathcal{M}}.$$

We now turn to (3.87). Recalling that $Ju=(1/2)\,d(u\wedge du)$, (3.87) will follow from a more general estimate. When $N\ge 3$ and $\mu\in\mathcal{M}(\Omega;\Lambda^2)$ is of the form $\mu=dX$ for some $X\in L^1(\Omega;\Lambda^1)$, we have

$$\langle\mu, *\zeta\rangle\le C\,\|\mu\|_{\mathcal{M}}\,\|d\zeta\|_{L^N},\ \forall\,\zeta\in C_c^\infty(\Omega;\Lambda^{N-2}). \tag{3.88}$$

[The assumptions on μ can be weakened to $\mu\in\mathcal{M}(\Omega;\Lambda^2)$ and $d\mu=0$.]

Indeed, let μ and X be as above. We claim that

$$\langle\mu, *\zeta\rangle=0 \text{ if } \zeta\in W^{1,1}(\Omega;\Lambda^{N-2})\cap C_0 \text{ and } d\zeta=0 \text{ in } \mathscr{D}'(\Omega). \tag{3.89}$$

When $\zeta\in C_c^\infty$, (3.89) follows from

$$\langle\mu, *\zeta\rangle=\langle dX, *\zeta\rangle=\langle X, d^**\zeta\rangle=\pm\langle X, *d**\zeta\rangle=\pm\langle X, *d\zeta\rangle=0,$$

where we have used (1.141), (1.142), and the assumption $d\zeta=0$. Equality (3.89) still holds when $\zeta\in C_c$; this is obtained by smoothing, using the fact that μ is a measure and ζ is continuous and compactly supported.

The general case is obtained by approximation as follows. Let $\zeta\in W^{1,1}\cap C_0$. Set $\Omega_\varepsilon:=\{x\in\Omega;\ \mathrm{dist}(x,\partial\Omega)>\varepsilon\}$ and let, for small ε, $\Phi_\varepsilon:\overline{\Omega_\varepsilon}\to\overline{\Omega}$ be a C^1-diffeomorphism such that $\|\Phi_\varepsilon-\mathrm{Id}\|_{C^1}\le C\varepsilon$. Let $\zeta_\varepsilon:=(\Phi_\varepsilon)^\sharp\zeta$, so that $\zeta_\varepsilon\in W^{1,1}(\Omega_\varepsilon;\Lambda^{N-2})\cap C_0(\overline{\Omega_\varepsilon})$ and $d\zeta_\varepsilon=0$ in $\mathscr{D}'(\Omega_\varepsilon)$. We extend ζ_ε by the value 0 in $\Omega\setminus\overline{\Omega_\varepsilon}$; the extension is still denoted by ζ_ε. Then, $\zeta_\varepsilon\in W^{1,1}(\Omega;\Lambda^{N-2})\cap C_c$ and $d\zeta_\varepsilon=0$ in $\mathscr{D}'(\Omega)$ (here, we use the fact that $\zeta_\varepsilon\in W^{1,1}(\Omega_\varepsilon)$ and $\zeta_\varepsilon=0$ on $\partial\Omega_\varepsilon$). Letting $\varepsilon\to 0$ in (3.89) applied to ζ_ε and using the fact that μ is a measure and $\zeta_\varepsilon\to\zeta$ uniformly in $\overline{\Omega}$ as $\varepsilon\to 0$, we obtain (3.89) for every $\zeta\in W^{1,1}\cap C_0$.

We next invoke the following result (see Bourgain and Brezis [49, Theorem 5"]). There exists a constant C such that, for each $\zeta\in C_c^\infty(\Omega;\Lambda^{N-2})$, we may find some $\eta\in W^{1,N}(\Omega;\Lambda^{N-2})\cap C_0$ with $d\eta=d\zeta$ and $\|\eta\|_{L^\infty}\le C\|d\zeta\|_{L^N}$. [Strictly speaking, this result is stated when $\Omega=(0,1)^N$, but the proof there can be easily adapted for a smooth Ω.]

Using (3.88), we find that

$$\langle\mu, *\zeta\rangle=\langle\mu, *\eta\rangle\le\|\mu\|_{\mathcal{M}}\,\|\eta\|_{L^\infty}\le C\|\mu\|_{\mathcal{M}}\,\|d\zeta\|_{L^N},$$

whence (3.88). □

Remark 3.9. The exponent $N/(N-1)$ in Theorem 3.12 is optimal (see Lemma 15.40 in Appendix 15.11).

Remark 3.10. Let us note the following amusing consequence of Theorem 3.12 – valid only in 3D. Let T be a 1-rectifiable current compactly supported in $\mathbb{R}^3$ and such that $\partial T = 0$. The weak isoperimetric inequality asserts that $T = \partial M$ for some 2-rectifiable current such that $|M| \leq C\,|T|^2$ (see Federer [147, 4.2.10]). We will improve this to

$$\inf\{|M|;\ M \subset \mathbb{R}^3 \text{ is 2-rectifiable}, \partial M = T\} \leq C \operatorname{diam}(T)\,|T| \tag{3.90}$$

(presumably known to the experts). Note that, when T is a connected curve, we have $\operatorname{diam}(T) \leq (1/2)\,|T|$, and therefore (3.90) does indeed refine the isoperimetric inequality.

Estimate (3.90) is obtained as follows. By scaling, we may assume that $\operatorname{diam}(T) = 1$, and thus (up to a translation) T is compactly contained in the ball $B := B_2(0)$. If M is a minimizer in (3.90), then the support of M is contained in B (see Morgan [273, 5.6] for a sketch of the argument), and thus (3.90) amounts to $A(T) \leq C\,|T|$.

On the other hand, we know that $\pi\, T = Ju$ for some $u \in W^{1,1}(B; \mathbb{S}^1)$, and thus $A(T) = (1/2\pi)\,\Sigma(u)$ for this u. We have therefore reduced (3.90) to $\Sigma(u) \leq C\,|T|$, which follows from Theorem 3.12. □

3.8 Complements and open problems

3.8.1 More on the case $N = 3$

We generalize here the quantities $A_0(\Gamma), B(\Gamma), C(\Gamma)$ to the case of several curves, i.e., $\Gamma = (\Gamma_1, \ldots, \Gamma_m)$, and arbitrary degrees $d = (d_1, \ldots, d_m)$. Here, the Γ_j's are smooth disjoint oriented curves satisfying (3.15). By changing, if necessary, the orientation of Γ_j, *we may assume that* $d_j > 0$, $\forall j$. To start with, let us consider the case $m = 1$, $\Gamma_1 = \Gamma$ and $d_1 = 2$. If we define $A_0(\Gamma, 2)$ in the spirit of (3.51), the requirement $\partial M = 2\Gamma$ makes no sense as a geometric equality. There are two options to overcome this difficulty. The first one consists of considering 2-rectifiable currents instead of oriented surfaces, and to require $\partial M = 2\Gamma$ in the sense of GMT. This leads to the quantity $A(2\Gamma)$, with A as in (3.65).

A more geometric option is to require that M is (at least locally near Γ) the union of two oriented manifolds with boundary Γ. Here is an even more restrictive class of competitors. We introduce the following *ad hoc* definition. A *nice* surface with boundary $\sum_{j=1}^m d_j\,\Gamma_j$ is an M such that:

1. M is a smooth oriented 2-submanifold M of $\Omega \setminus \cup_{j=1}^m \Gamma_j$, without boundary (as submanifold of $\Omega \setminus \cup_{j=1}^m \Gamma_j$).
2. $|M| < \infty$ (and thus M defines a 2-rectifiable current in Ω).
3. For each j and near each $x \in \Gamma_j$, M consists of d_j mutually disjoint oriented surfaces M_ℓ, such that every $M_\ell \cup \Gamma_j$ has (geometric) boundary Γ_j.
4. The above M_ℓ's are not tangent on Γ_j, i.e., the outward normals of M_ℓ and M_n to Γ_j are distinct if $\ell \neq n$.

We set

$$A_0(\Gamma, d) := \inf \left\{ |M|;\ M \text{ is a nice surface with boundary } \sum_{j=1}^{m} d_j \Gamma_j \right\}. \tag{3.91}$$

Note that, when $m = 1$, $d_1 = 1$ and $\Gamma_1 = \Gamma$, nice surfaces with boundary Γ are precisely oriented 2-submanifolds with geometric boundary Γ, and thus definition (3.91) is consistent with (3.31). We note also that items 1–4 imply that $\partial M = \sum_{j=1}^{m} d_j \Gamma_j$ (in the sense of GMT), and thus $A_0(\Gamma, d) \geq A(\sum_{j=1}^{m} d_j \Gamma_j)$.

By analogy with Sections 3.5 and 3.6, we also define

$$\begin{aligned} A(\Gamma, d) := A\left(\sum_{j=1}^{m} d_j\, \Gamma_j\right) = \inf \Bigg\{ |M|;\ M \subset \Omega \text{ is a 2-rectifiable} \\ \text{current such that } \partial M = \sum_{j=1}^{m} d_j\, \Gamma_j \Bigg\}, \end{aligned} \tag{3.92}$$

$$B(\Gamma, d) := \min \left\{ \|\mu\|_{\mathscr{M}};\ \mu \in \mathscr{M}(\Omega; \mathbb{R}^3),\ \operatorname{curl}\mu = \sum_{j=1}^{m} d_j\, \Gamma_j \text{ in } \mathscr{D}'(\Omega) \right\}, \tag{3.93}$$

and

$$C(\Gamma, d) := \sup \left\{ \sum_{j=1}^{m} d_j \int_{\Gamma_j} \tau \cdot \zeta;\ \zeta \in C_c^\infty(\Omega; \mathbb{R}^3) \text{ and } \|\operatorname{curl}\zeta\|_{L^\infty} \leq 1 \right\}. \tag{3.94}$$

In what follows, for simplicity we strengthen the assumption that the Γ_j's are mutually disjoint to the following requirement, in the spirit of (3.15):

$$\begin{aligned} &\text{there exist } \widetilde{\Gamma}_1, \ldots, \widetilde{\Gamma}_m \subset \mathbb{R}^3 \text{ smooth closed oriented curves such that} \\ &\Gamma_j = \widetilde{\Gamma}_j \cap \Omega \text{ is connected, } \forall j, \text{ and } \operatorname{dist}(\Gamma_\ell, \widetilde{\Gamma}_j) > 0,\ \forall\, \ell, j \text{ with } \ell \neq j. \end{aligned} \tag{3.95}$$

The main result here is the following

Theorem 3.13. *Assume* (3.95). *Let* $\Gamma = (\Gamma_1, \ldots, \Gamma_m)$ *and* $d = (d_1, \ldots, d_m) \in \mathbb{N}^m$. *Then,*

$$A_0(\Gamma, d) = A(\Gamma, d) = B(\Gamma, d) = C(\Gamma, d). \tag{3.96}$$

Moreover, for each $u \in W^{1,1}(\Omega; \mathbb{S}^1)$ *such that* $Ju = \pi \sum_{j=1}^{m} d_j\, \Gamma_j$, *all the above quantities are equal to* $(1/(2\pi))\, \Sigma(u)$.

Proof. The equalities $A(\Gamma, d) = B(\Gamma, d) = C(\Gamma, d)$ follow from Corollary 3.1. The equality $\Sigma(u) = 2\pi B(\Gamma, d)$ is a consequence of Corollary 1.8. Since we clearly have $A_0(\Gamma, d) \geq A(\Gamma, d)$, it suffices to prove that $\Sigma(u) \geq 2\pi A_0(\Gamma, d)$. In the case of a single curve with degree one, this inequality has been established in Section 3.3 via the coarea formula and the density result provided by Lemma 3.3, item 2. The key ingredient is the fact that, whenever $u \in Z^\infty$, for a.e. $\alpha \in \mathbb{S}^1$, the level sets $[u = \alpha]$ are manifolds with geometric boundary Γ. In order to adapt this proof to our situation, we need a dense class consisting of maps u such that

$$\text{for a.e. } \alpha \in \mathbb{S}^1,\ [u = \alpha] \text{ is a nice manifold with boundary } \sum_{j=1}^{m} d_j\, \Gamma_j. \tag{3.97}$$

The existence of such a class $\overline{Z}^\infty$ is the content of the next lemma. It is straightforward that all maps in $\overline{Z}^\infty$ satisfy (3.97). □

Lemma 3.7. *Let Γ_j, $\widetilde{\Gamma}_j$, $j = 1, \ldots, m$, be as in* (3.95). *Let ε_j, Φ_j be associated with $\widetilde{\Gamma}_j$ as in Lemma 3.3, item 1. Set*

$$\overline{Z} := \left\{ w \in W^{1,1}(\Omega; \mathbb{S}^1);\ Jw = \sum_{j=1}^{m} d_j\, \Gamma_j \text{ in } \mathscr{D}'(\Omega) \right\},$$

$$\begin{aligned} \overline{Z}^\infty := \Big\{ & w \in W^{1,1}(\Omega; \mathbb{S}^1) \cap C^\infty(\Omega \setminus \cup_{j=1}^m \Gamma_j);\ \exists 0 < \varepsilon < \min\{\varepsilon_1, \ldots, \varepsilon_m\}, \\ & \exists f_j \in C^\infty(\overline{\Gamma}_j; \mathbb{S}^1),\ \forall j, \text{ such that } w(\Phi_j(x, re^{\imath\theta})) = f_j(x)\, e^{\imath d_j \theta}, \\ & \forall j,\ \forall x \in \Gamma_j,\ \forall 0 < r < \varepsilon,\ \forall \theta \text{ such that } x + \Phi_j(x, re^{\imath\theta}) \in \Omega \Big\}. \end{aligned}$$

Then, $\overline{Z}^\infty$ is dense in $\overline{Z}$.

Sketch of proof of Lemma 3.7. It suffices to construct a map u in $\overline{Z}^\infty$. [Then density is proved exactly as in Step 3 in the proof of Lemma 3.3.] Let $\varepsilon > 0$ be such that $3\varepsilon < \varepsilon_j$, $\forall j$, and $3\varepsilon < \operatorname{dist}(\Gamma_\ell, \widetilde{\Gamma}_j)$, $\forall \ell, j$ with $\ell \neq j$. Let u_j be associated with $\widetilde{\Gamma}_j$ as in Step 2 in the proof of Lemma 3.4. By mimicking the construction of the auxiliary function ψ in Step 2 in the proof of Lemma 3.3, we obtain the existence of a map $\widetilde{u}_j \in W^{1,1}_{loc}(\mathbb{R}^3; \mathbb{S}^1) \cap C^\infty(\mathbb{R}^3 \setminus \widetilde{\Gamma}_j)$ and of smooth functions $f_{j,\ell} : \overline{\Gamma_\ell} \to \mathbb{R}$, $\ell \neq j$, such that

$$\begin{aligned} &\widetilde{u}_j(y) = u_j(y),\ \forall y \in \Phi_j(\widetilde{\Gamma}_j \times (\overline{\mathbb{D}}(0, \varepsilon) \setminus \{0\})), \\ &\widetilde{u}_j(\Phi_\ell(x, z)) = f_{j,\ell}(x),\ \forall \ell \neq j,\ \forall x \in \overline{\Gamma_\ell},\ \forall |z| \leq \varepsilon/3. \end{aligned}$$

Then the restriction u to Ω of $\widetilde{u} := \widetilde{u}_1 \cdots \widetilde{u}_m$ has all the required properties. □

3.8.2 $N \geq 4$

We present here the counterparts of some of the results in Sections 3.1–3.8.1 in dimension $N \geq 4$. As a general rule, the three-dimensional results can be extended (with very similar proofs) to higher dimensions, using the following dictionary: 1-currents (resp. oriented curves) become $(N-2)$-currents (resp. oriented $(N-2)$-manifolds), and 2-rectifiable currents (resp. oriented surfaces) become $(N-1)$-rectifiable currents (resp. oriented $(N-1)$-manifolds).

Let $\Omega \subset \mathbb{R}^N$ be a smooth bounded simply connected domain. If Γ is a smooth connected oriented $(N-2)$-submanifold of Ω and $u \in C(\Omega \setminus \Gamma; \mathbb{S}^1)$, then we may define the integer $\deg(u, \Gamma)$ as in Section 3.1. Here, we use the fact that the (two-dimensional) normal plane N_x at some $x \in \Gamma$ has a natural orientation (such that a direct basis of N_x completed with a direct basis of $T_x(\Gamma)$ forms a direct basis of $\mathbb{R}^N$).

The analogue of the condition (3.5) is

$$\Gamma \subset \Omega \text{ is a smooth connected oriented } (N-2)\text{-submanifold without boundary (in } \Omega). \tag{3.98}$$

Remark 3.11. In 3D, Γ is a curve and its orientability is not an issue. We may have considered in (3.5) a smooth connected curve, then choose an orientation on it. However, when $N \geq 4$ we have to *assume* Γ orientable, since this "does not come with Γ." [Think of the Klein bottle.]

In order to generalize Theorem 3.1 to higher dimensions, it is more convenient to start from the definition $Ju = \frac{1}{2} d\,(u \wedge du)$ (and not to rely on (3.6)). Then the conclusion of Theorem 3.1 reads

$$\langle Ju, *\zeta \rangle = \pi \deg(u, \Gamma) \int_\Gamma \zeta, \ \forall \zeta \in C_c^\infty(\Omega; \Lambda^1). \tag{3.99}$$

The generalization of Theorem 3.1 to all dimensions $N \geq 3$ is the following

Theorem 3.14. *Let $\Gamma \subset \Omega \subset \mathbb{R}^N$ satisfy (3.98). Let $u \in W^{1,1}(\Omega; \mathbb{S}^1) \cap C(\Omega \setminus \Gamma)$. Then,*

$$\langle Ju, *\zeta \rangle = \pi \deg(u, \Gamma) \int_\Gamma \zeta, \ \forall \zeta \in C_c^\infty(\Omega; \Lambda^{N-2}).$$

More generally, let $\Gamma_1, \ldots, \Gamma_m$ be mutually disjoint $(N-2)$-manifolds satisfying (3.98). Let $u \in W^{1,1}(\Omega; \mathbb{S}^1) \cap C(\Omega \setminus \cup_{j=1}^m \Gamma_j)$. Then, $*Ju = \pi \sum_{j=1}^m \deg(u, \Gamma_j)\, \Gamma_j$, i.e.,

$$\langle Ju, *\zeta \rangle = \pi \sum_{j=1}^m \deg(u, \Gamma_j) \int_{\Gamma_j} \zeta, \ \forall \zeta \in C_c^\infty(\Omega; \Lambda^{N-2}). \tag{3.100}$$

Remark 3.12. In view of (1.143), equality (3.100) asserts that

$$\int_\Omega (u \wedge du) \cdot (*d\zeta) = (-1)^{N+1}\, 2\pi \sum_{j=1}^m \deg(u, \Gamma_j) \int_{\Gamma_j} \zeta, \ \forall \zeta \in C_c^\infty(\Omega; \Lambda^{N-2}).$$

This is equivalent to

$$\int_\Omega (u \wedge du) \wedge d\zeta = 2\pi \sum_{j=1}^m \deg(u, \Gamma_j) \int_{\Gamma_j} \zeta, \ \forall \zeta \in C_c^\infty(\Omega; \Lambda^{N-2}).$$

Using Corollary 1.12, we find that

$$\int_\Omega (u \wedge du) \wedge d\zeta = 2\pi \sum_{j=1}^m \deg(u, \Gamma_j) \int_{\Gamma_j} \zeta,$$
$$\forall \zeta \in W^{1,\infty}(\Omega; \Lambda^{N-2}) \text{ such that } \zeta_T = 0 \text{ on } \partial\Omega.$$

The proof of Theorem 3.14 is identical to that of Theorem 3.1.

Remark 3.13. The reader may wonder what happens to lower dimensional singularities, e.g., supported on a smooth submanifold $\Gamma \subset \Omega$ of dimension $\leq N-3$. They simply go undetected by Ju. More specifically, we have the following counterpart of Proposition 3.1.

Proposition 3.2. *Let $u \in W^{1,1}(\Omega; \mathbb{S}^1) \cap C(\Omega \setminus \Gamma)$, where $\Gamma \subset \Omega \subset \mathbb{R}^N$ is a smooth manifold of dimension $\leq N-3$. Then,*

$$Ju = 0 \text{ in } \mathscr{D}'(\Omega). \tag{3.101}$$

Proof. Fix a point $x_0 \in \Gamma$ and a small ball $B = B(x_0, r)$. It suffices to prove that $Ju = 0$ in $\mathscr{D}'(B)$. We claim that $B \setminus \Gamma$ is simply connected. This is clear when Γ is flat (since $\dim \Gamma \leq N-3$); the general case follows by straightening Γ. From here, one may proceed as in the proof of Proposition 3.1. □

We now turn to the generalization of Theorem 3.2. Its formulation relies on the higher dimensional analogue of (3.15), which is

$$\begin{aligned} &\Gamma \text{ is connected and } \Gamma = \widetilde{\Gamma} \cap \Omega, \\ &\qquad \text{with } \widetilde{\Gamma} \subset \mathbb{R}^N \text{ a smooth closed oriented } (N-2)\text{-manifold.} \end{aligned} \tag{3.102}$$

We have the following result.

Theorem 3.15. *Let $\Gamma_1, \ldots, \Gamma_m \subset \Omega$ be mutually disjoint and satisfy* (3.102). *Let $d_1, \ldots, d_m \in \mathbb{Z}$. Then there exists some $u \in C^\infty(\Omega \setminus \cup_{j=1}^m \Gamma_j; \mathbb{S}^1)$, such that $u \in W^{1,p}_{loc}(\Omega; \mathbb{S}^1)$, $\forall\, 1 \leq p < 2$, and* $\deg(u, \Gamma_j) = d_j$, $j = 1, \ldots, m$.

The proof of Theorem 3.15 is essentially the same as that of Theorem 3.2. The main additional difficulty consists of writing $\widetilde{\Gamma}_j$ as a (geometric) boundary. This is possible, since each closed oriented $(N-2)$-submanifold of $\mathbb{R}^N$ is the boundary of a "Seifert" smooth oriented hypersurface (see, for example, Kirby [213, Theorem 3, p. 50]).

The dipole construction (Theorem 3.5 in Section 3.4) can be extended, with essentially the same proof, to any dimension.

By analogy with Section 3.6, we introduce

$$\begin{aligned}\mathscr{F} := \{T;\, T \text{ is an } (N-2)\text{-current such that } T = \partial M \text{ for some}\\ (N-1)\text{-rectifiable current } M \text{ on } \Omega\},\end{aligned} \tag{3.103}$$

and then we have (compare with Theorem 3.8)

$$\mathscr{F} = \left\{\frac{1}{\pi} * Ju;\ u \in W^{1,1}(\Omega; \mathbb{S}^1)\right\}. \tag{3.104}$$

Given $T \in \mathscr{F}$, we define

$$\begin{aligned}A(T) := \inf\{|M|;\, M \subset \Omega \text{ is an } (N-1)\text{-rectifiable current}\\ \text{such that } \partial M = T\},\end{aligned} \tag{3.105}$$

$$\begin{aligned}B(T) := \inf\{\|\mu\|_{\mathscr{M}};\, \mu \in \mathscr{M}(\Omega; \Lambda^{N-1}) \text{ is a measure}\\ \text{such that } \partial\mu = T\},\end{aligned} \tag{3.106}$$

and

$$C(T) := \sup\left\{\langle T, \zeta\rangle;\ \zeta \in C_c^\infty(\Omega; \Lambda^{N-2}),\ \|d\zeta\|_{L^\infty} \le 1\right\} := \|T\|_W. \tag{3.107}$$

Then,

$$A(T) = B(T) = C(T),\ \forall\, T \in \mathscr{F}, \tag{3.108}$$

and the inf in (3.105) is achieved. In fact, if we set

$$\mathscr{Q}(T) := \{\mu \in \mathscr{M}(\Omega; \Lambda^{N-1});\ \partial\mu = T \text{ and } \|\mu\|_{\mathscr{M}} = B(T)\},$$

then

$$\begin{aligned}\text{any extreme point of } \mathscr{Q}(T) \text{ is an area-minimizing}\\ (N-1)\text{-rectifiable current in (3.105).}\end{aligned} \tag{3.109}$$

Moreover,

$$\begin{aligned}\frac{1}{2\pi}\Sigma(u) = A(T) = B(T) = C(T),\\ \forall\, u \in W^{1,1}(\Omega; \mathbb{S}^1) \text{ such that } * Ju = \pi T.\end{aligned} \tag{3.110}$$

We next discuss the generalizations of the results in Section 3.8.1. We assume that the following counterpart of (3.95) holds.

$$\begin{aligned}&\widetilde{\Gamma}_1, \dots, \widetilde{\Gamma}_m \subset \mathbb{R}^N \text{ are smooth closed oriented } (N-2)\text{-manifolds}\\&\text{such that, for every } j,\ \Gamma_j = \widetilde{\Gamma}_j \cap \Omega \text{ is connected}\\&\text{and } \operatorname{dist}(\Gamma_\ell, \widetilde{\Gamma}_j) > 0,\ \forall\, \ell, j \text{ with } \ell \neq j.\end{aligned} \tag{3.111}$$

When $m = 1$, $\Gamma_1 = \Gamma$ and $d_1 = 1$, we set

$$\begin{aligned}A_0(\Gamma) := \inf\{|M|;\ M \subset \Omega \setminus \Gamma \text{ is a smooth oriented}\\ (N-1)\text{-submanifold such that } \partial M = \Gamma\}.\end{aligned} \tag{3.112}$$

Here, $|M| = \mathcal{H}^{N-1}(M)$, and $\partial M = \Gamma$ is an equality in the geometrical sense, i.e., $M \cup \Gamma$ is a submanifold of Ω with boundary Γ.

In the general case, we set, with $d = (d_1, \dots, d_m) \in \mathbb{N}^m$,

$$A_0(\Gamma, d) := \inf\left\{|M|;\ M \subset \Omega \setminus \cup_{j=1}^m \Gamma_j \text{ is a nice } (N-1)-\text{manifold}\right.\\ \left.\text{with boundary } \sum_{j=1}^m d_j\, \Gamma_j\right\}. \tag{3.113}$$

The above definition is analogous to that adopted when $N = 3$ in Section 3.8.1. We also set

$$A(\Gamma, d) := \inf\left\{|M|;\ M \subset \Omega \text{ is an } (N-1)\text{-rectifiable}\right.\\ \left.\text{current such that } \partial M = \sum_{j=1}^m d_j\, \Gamma_j\right\}, \tag{3.114}$$

$$B(\Gamma, d) := \min\left\{\|\mu\|_{\mathcal{M}};\ \mu \in \mathcal{M}(\Omega; \Lambda^{N-1})\right.\\ \left.\text{and } \partial\mu = \sum_{j=1}^m d_j \Gamma_j \text{ in } \mathcal{D}'(\Omega)\right\}, \tag{3.115}$$

$$C(\Gamma, d) := \sup\left\{\sum_{j=1}^m d_j \int_{\Gamma_j} \zeta;\ \zeta \in C_c^\infty(\Omega; \Lambda^{N-2}),\ \|d\zeta\|_{L^\infty} \le 1\right\}. \tag{3.116}$$

We then have the following analogue of Theorem 3.13.

Theorem 3.16. *Assume* (3.111). *Let* $\Gamma = (\Gamma_1, \dots, \Gamma_m)$ *and* $d = (d_1, \dots, d_m) \in \mathbb{N}^m$. *Then,*

$$A_0(\Gamma, d) = A(\Gamma, d) = B(\Gamma, d) = C(\Gamma, d). \tag{3.117}$$

Moreover, for each $u \in W^{1,1}(\Omega; \mathbb{S}^1)$ *such that* $*Ju = \pi \sum_{j=1}^m d_j \Gamma_j$*, all the above quantities are equal to* $(1/(2\pi))\, \Sigma(u)$.

Finally, let us note the following *global* version of Theorems 3.13 and 3.16.

Theorem 3.17. *Assume* $N \geq 3$. *Let* $\Gamma_1, \dots, \Gamma_m \subset \mathbb{R}^N$ *be mutually disjoint smooth closed connected oriented* $(N-2)$*-manifolds. Let* $d_1, \dots, d_m \in \mathbb{N}^m$. *Then,*

$$A_0(\Gamma, d) = A(\Gamma, d) = B(\Gamma, d) = C(\Gamma, d), \tag{3.118}$$

where

$$A_0(\Gamma, d) := \inf \Big\{ |M|;\, M \subset \mathbb{R}^N \setminus \Gamma \text{ is a nice } (N-1) - \text{manifold with boundary } \sum_{j=1}^m d_j\, \Gamma_j \Big\},$$

$$A(\Gamma, d) := \inf \Big\{ |M|;\, M \subset \mathbb{R}^N \text{ is an } (N-1)\text{-rectifiable current such that } \partial M = \sum_{j=1}^m d_j\, \Gamma_j \Big\},$$

$$B(\Gamma, d) := \min \left\{ \|\mu\|_{\mathscr{M}};\ \mu \in \mathscr{M}(\mathbb{R}^N; \Lambda^{N-1}) \text{ and } \partial\mu = \sum_{j=1}^m d_j \Gamma_j \text{ in } \mathscr{D}'(\mathbb{R}^N) \right\},$$

$$C(\Gamma, d) := \sup \left\{ \sum_{j=1}^m d_j \int_{\Gamma_j} \zeta;\ \zeta \in C_c^\infty(\mathbb{R}^N; \Lambda^{N-2}),\ \|d\zeta\|_{L^\infty} \leq 1 \right\}.$$

Proof. For simplicity, we assume $N = 3$, but the argument is the same in any dimension.

Let us first note that $A_0(\Gamma, d) < \infty$. Indeed, by the dipole construction (Theorem 3.5) combined with the Frankl–Pontryagin theorem, we may construct some $u_j \in W^{1,1}_{loc}(\mathbb{R}^3; \mathbb{S}^1)$ such that $u_j - 1 \in W^{1,1}_c(\mathbb{R}^3; \mathbb{R}^2)$ and $Ju_j = \pi \Gamma_j$. The map $u := \prod_{j=1}^m (u_j)^{d_j}$ satisfies $u \in W^{1,1}_{loc}(\mathbb{R}^3; \mathbb{S}^1)$, $u - 1 \in W^{1,1}_c(\mathbb{R}^3; \mathbb{R}^2)$ and $Ju = \pi \sum_{j=1}^m d_j \Gamma_j$.

By the proof of Lemma 3.3 (see also Lemma 3.7), there exists some w with the same properties as u such that, in addition, $w \in \overline{Z}^\infty$. For any such w and almost each $\alpha \in \mathbb{S}^1$, the level set $[w = \alpha]$, oriented by $w \wedge \nabla w$, is a competitor for $A_0(\Gamma, d)$.

Since, clearly, $C(\Gamma, d) = B(\Gamma, d) \leq A(\Gamma, d) \leq A_0(\Gamma, d) < \infty$, it suffices to prove that $A_0(\Gamma, d) \leq C(\Gamma, d)$. Let $R_0 > 0$ be such that $\Gamma_j \subset B(0, R_0)$, $\forall j$. For $R > R_0$, let $A^R(\Gamma, d)$ be the quantity in (3.114) calculated with respect to $\Omega := B(0, R)$. We define similarly $B^R(\Gamma, d)$ and $C^R(\Gamma, d)$. Set also

$$A_0^R(\Gamma, d) := \inf \Big\{ |M|;\, M \subset \mathbb{R}^N \setminus \cup_{j=1}^m \Gamma_j \text{ is a nice } (N-1) - \text{manifold with boundary } \sum_{j=1}^m d_j\, \Gamma_j, \text{ supported in } B(0, R) \Big\}.$$

We have $C^R(\Gamma, d) \le C(\Gamma, d)$ and $A_0(\Gamma, d) \le A_0^R(\Gamma, d)$. It therefore suffices to prove that, for arbitrary $\varepsilon > 0$, there exists some $R > R_0$ such that

$$A_0^R(\Gamma, d) \le C^R(\Gamma, d) + C\varepsilon, \tag{3.119}$$

where $C > 0$ is a constant independent of ε.

We note that $C^R(\Gamma, d) \nearrow C(\Gamma, d) < \infty$. Consider some $R > 2R_0$ such that $C^R(\Gamma, d) - C^{R/2}(\Gamma, d) < \varepsilon$. Let $u \in W^{1,1}(B(0, R); \mathbb{S}^1)$ be such that $Ju = \pi \sum_{j=1}^m d_j \Gamma_j$ and $\operatorname{supp}(u - 1) \subset B(0, R/2)$ (the existence of u follows from the above discussion). By Theorem 3.13 and the definition (1.80) of $\Sigma(u)$, there exists some $\eta \in W^{1,1}(B(0, R); \mathbb{R})$ such that

$$\|u \wedge \nabla u - \nabla\eta\|_{L^1(B(0,R))} \le \Sigma^R(u) + \varepsilon = 2\pi C^R(\Gamma, d) + \varepsilon, \tag{3.120}$$

where $\Sigma^R(u)$ is as in (1.80), calculated with respect to the domain $B(0, R)$. [We define similarly $\Sigma^r(u)$, $\forall\, r > R_0$.]

Since, on the other hand,

$$\|u \wedge \nabla u - \nabla\eta\|_{L^1(B(0,R/2))} \ge \Sigma^{R/2}(u) = 2\pi C^{R/2}(\Gamma, d) \tag{3.121}$$

and

$$u \wedge \nabla u = 0 \text{ in } B(0, R) \setminus B(0, R/2), \tag{3.122}$$

we find (thanks to our choice of R) that

$$\int_{B(0,R)\setminus B(0,R/2)} |\nabla\eta| < 2\varepsilon, \tag{3.123}$$

and, with no loss of generality, we may assume that

$$\int_{B(0,R)\setminus B(0,R/2)} \eta = 0, \tag{3.124}$$

so that (from (3.123), (3.124), and Poincaré's inequality)

$$\int_{B(0,R)\setminus B(0,R/2)} |\eta| \le C\varepsilon. \tag{3.125}$$

[Here and in what follows, C denotes various constants independent of ε.]

Consider some $\zeta \in C_c^\infty(B(0, R); \mathbb{R})$ such that $\zeta = 1$ in $B(0, R/2)$ and $|\nabla\zeta| \le C/R$. If we set $\psi := \zeta\eta$, then, by (3.123) and (3.125),

$$\int_{B(0,R)\setminus B(0,R/2)} |\nabla\psi| \le C\varepsilon. \tag{3.126}$$

Set $v := e^{-\imath\psi} u \in W^{1,1}(B(0,R); \mathbb{S}^1)$. By (3.120), (3.126), and the definitions of ψ and v, we have $Jv = \pi \sum_{j=1}^{m} d_j \Gamma_j$, supp $(v-1) \subset B(0,R)$ and

$$\|v \wedge \nabla v\|_{L^1(B(0,R))} \leq 2\pi C^R(\Gamma, d) + C\varepsilon. \tag{3.127}$$

Arguing again as at the beginning of the proof of the theorem, we may, in addition, assume that $v \in \overline{Z}^{\infty}$. Then, for a.e. $\alpha \in \mathbb{S}^1$, $M^{\alpha} := [v = \alpha]$ is a competitor for $A_0^R(\Gamma, d)$ and, by the coarea formula (see (3.48))

$$\int_{B(0,R)} |\nabla v| = \int_{\mathbb{S}^1} |M^{\alpha}|\, d\alpha. \tag{3.128}$$

Combining (3.127) and (3.128), we find that (3.119) holds. □

3.8.3 *Describing* $J(W^{1,p}(\Omega; \mathbb{S}^1))$

As in 2D, we may ask which $(N-2)$-currents can be written as $*Ju$ with $u \in W^{1,p}(\Omega; \mathbb{S}^1)$, for some $1 < p < 2$. Theorem 2.20 can be extended to higher dimensions, as proved by Bousquet [61]. Recall (see (3.104)) that

$$\left\{\frac{1}{\pi} * Ju;\ u \in W^{1,1}(\Omega; \mathbb{S}^1)\right\} = \mathscr{F}.$$

Set

$$\mathscr{A}_p := \left\{\frac{1}{\pi} * Ju;\ u \in W^{1,p}(\Omega; \mathbb{S}^1)\right\},\ \forall\, 1 \leq p < \infty.$$

By (1.61), we have $\mathscr{A}_p = \{0\}$ when $p \geq 2$.

Theorem 3.18. *Let* $1 < p < 2$ *and set* $q := \dfrac{p}{p-1}$. *Then,*

$$\mathscr{A}_p = \mathscr{F} \cap [W_0^{1,q}(\Omega)]^*.$$

3.8.4 *Open problems*

Open Problem 4. Let $\Gamma \subset \mathbb{R}^3$ be a smooth closed oriented curve (not necessarily connected). We know, by Theorem 3.17, that

$$\begin{aligned}&\sup\left\{\int_{\Gamma} \tau \cdot \zeta;\ \zeta \in C_c^{\infty}(\mathbb{R}^3; \mathbb{R}^3) \text{ and } \|\mathrm{curl}\,\zeta\|_{L^\infty} \leq 1\right\}\\ &= \inf\{|M|;\ M \subset \mathbb{R}^3 \text{ is a smooth oriented surface with boundary } \Gamma\}.\end{aligned} \tag{3.129}$$

Find an elementary constructive proof of (3.129).

[Recall that the proof of Theorem 3.17 is indirect and requires an excursion via $W^{1,1}(\Omega; \mathbb{S}^1)$.] Note that the inequality "$\leq$" in (3.129) is an immediate consequence of Stokes' theorem. For the reverse inequality, "$\geq$," it would be instructive to construct a ζ such that $\int_\Gamma \tau \cdot \zeta \geq \inf \cdots - \varepsilon$, in the spirit of the proof of Theorem 2.4.

3.9 Comments

1. Theorem 3.1 first appeared in Brezis, Coron, and Lieb [79, equation (8.30)]. The proof we present is inspired by that in Bousquet [61].
2. Theorem 3.3 (for $N = 3$) and Theorem 3.15 (for $N \geq 4$) are essentially due to Bousquet [61]. A comparable earlier result, which appears in Alberti, Baldo, and Orlandi [6, Theorem 4.4], asserts only the existence of $u \in W^{1,p}_{loc}(\mathbb{R}^N; \mathbb{S}^1)$ with $p < N/(N-1)$ (see also Comment 11 below).
3. The type of question answered by Theorem 3.4 (and its extensions, Theorems 3.13 and 3.16) was originally introduced in Brezis, Coron, and Lieb [79], who obtained the conclusion of the theorem when Γ is a planar curve and $\Omega = \mathbb{R}^3$. Subsequently, the same assertion was established in Almgren, Browder, and Lieb [8] for a general closed curve $\Gamma \subset \mathbb{R}^3$. The main novelty in [8] was the use of the coarea formula, as in Part 2 of the proof of Theorem 3.4. All the subsequent generalizations of Theorem 3.4 (see, for example, Alberti, Baldo, and Orlandi [6], Molnar [271], and Bousquet and Mironescu [63]) rely heavily on the use of the coarea formula. As pointed out in [8], the use of the coarea formula leads to a new, simpler proof, of Federer's result [148, 5.10].
4. The auxiliary results in Section 3.3 are taken from Brezis and Mironescu [88].
5. The dipole construction in Section 3.4 is the higher dimensional analogue of that presented in Section 2.5. It was originally proposed in Brezis, Coron, and Lieb [79, Section VIII C] and carried out in Almgren, Browder, and Lieb [8] and Alberti, Baldo, and Orlandi [6].
6. Equality $A_0(\Gamma, d) = B(\Gamma, d)$ in Theorem 3.13 (for $N = 3$) and in Theorem 3.16 (for $N \geq 4$), and also in the global Theorem 3.17, is a "smooth version" of Federer's result [148, 5.10]. Note that $A_0(\Gamma, d)$ involves the minimization of the area over a class of *smooth* surfaces spanned by Γ. By contrast, under the technical assumption that Γ is compact (which need not be the case in Theorems 3.13 and 3.16), Federer's result asserts that $A(\Gamma, d) = B(\Gamma, d)$, where $A(\Gamma, d)$ involves the minimization of the area over the larger class of 2-rectifiable currents M such that $\partial M = \Gamma$. Equality $A_0(\Gamma, d) = B(\Gamma, d)$ could presumably be deduced from Federer's result by approximation. We have chosen a different route, which, at the end of the day, implies Federer's result.
   Equality $A_0(\Gamma, d) = C(\Gamma, d)$ in Theorem 3.17 is presumably known to experts of the Plateau problem, but we could not find an explicit reference in the literature.
7. Assertions (3.59) and (3.61) in 3D, and their extension (3.104) to $N \geq 4$, are due to Alberti, Baldo, and Orlandi [6, Theorem 3.8, Theorem 5.6].

8. As mentioned in Comment 6 (see also Comment 12 in Chapter 2), equality $A(T) = B(T)$ in Corollary 3.1 (and its extension to $N \geq 4$, Theorem 3.16) is a non-trivial fact originally due to Federer [148, 5.10].
   Interestingly, there are now four different proofs of this result, with different flavors:

   a) The original proof of Federer [148, 5.10].
   b) The proof of Almgren, Browder, and Lieb [8], via the coarea formula.
   c) The short and elegant proof of Hardt and Pitts [187], which shows, in addition, that the inf in the definition of A is achieved.
   d) The proof we have presented in Section 3.6 (see the proof of Corollary 3.1), which "transits" via some $u_0 \in W^{1,1}(\Omega; \mathbb{S}^1)$, relies on the equality (3.73) combined with the equality $A(u_0) = B(u_0)$, which admits a self-contained proof based on the Krein–Milman theorem and Poliakovsky's Theorem 1.7 (see the alternative proof of Corollary 1.8 in Section 1.6). The idea of using OT, extreme points, and the Krein-Milman theorem within the framework of the Plateau problem (as in Corollary 3.1 for $N = 3$ and in (3.109) for $N \geq 4$) comes from Brezis and Mironescu [88].

   Concerning the equality $B(T) = C(T)$ in Corollary 3.1 (and in Theorem 3.16), see Comment 12 in Chapter 2.
9. Theorem 3.9 is new.
10. As already mentioned in Comment 13 to Chapter 2, Theorem 3.11 is originally due to Jerrard and Soner [204] and can be viewed as a special case of the Federer–Fleming boundary rectifiability theorem [147, Theorem 4.2.16 (2), p. 413].
11. Corollary 3.3 is new and sharpens the result of Alberti, Baldo, and Orlandi [6, Theorem 4.4]; their v belongs only to $W^{1,p}(\Omega; \mathbb{S}^1)$, $\forall p < N/(N-1)$. Its proof relies heavily on results from Bourgain and Brezis [49].

Chapter 4
A digression: sphere-valued maps

4.0 Introduction

The purpose of this chapter is to explain rather informally how some of the tools and results presented in the previous chapters extend to maps $u : \Omega \to \mathbb{S}^k$, where $\Omega \subset \mathbb{R}^N$ is a smooth bounded open set, $N \geq 2$ and $1 \leq k \leq N-1$. To avoid any complication, we assume throughout this chapter that Ω is *diffeomorphic to a ball*.

Although the topic of lifting for Sobolev maps with values into manifolds has attracted some attention (see, for example, [21, 29, 32, 269], and the references therein), little has been done in the spirit of "minimal lifting" (with the noticeable exception of [196]). By contrast, the natural counterparts for the expressions Ju (introduced in Section 1.2), $\Sigma(u)$ (in Section 1.4.2), the least energy with prescribed Jacobian $S_p(u)$ (in (1.84) and in Section 1.8), and the relaxed energy $R_p(u)$ (in Section 1.7) are now well understood. As we are going to see, they exhibit similarities with the $\mathbb{S}^1$-valued case, but some new phenomena occur and several open problems remain.

We first discuss the concept of Jacobian Ju. For simplicity, we start with the "historical" case $N = 3$ and $k = 2$. For every $u \in H^1(\Omega; \mathbb{S}^2)$, set

$$Ju := \frac{1}{3}\text{div}D(u), \tag{4.1}$$

where $D(u) \in L^1(\Omega; \mathbb{R}^3)$ is the vector field

$$D(u) := \left(\det\left(u, \frac{\partial u}{\partial x_2}, \frac{\partial u}{\partial x_3}\right), \det\left(\frac{\partial u}{\partial x_1}, u, \frac{\partial u}{\partial x_3}\right), \det\left(\frac{\partial u}{\partial x_1}, \frac{\partial u}{\partial x_2}, u\right)\right). \tag{4.2}$$

Thus, $Ju \in \mathscr{D}'(\Omega; \mathbb{R})$ acts on $C_c^\infty(\Omega; \mathbb{R})$ via the formula

$$\langle Ju, \zeta\rangle = -\frac{1}{3}\int_\Omega [D(u)] \cdot \nabla\zeta, \ \forall \zeta \in C_c^\infty(\Omega; \mathbb{R}). \tag{4.3}$$

H. Brezis and P. Mironescu, *Sobolev Maps to the Circle*, Progress in Nonlinear Differential Equations and Their Applications 96,
https://doi.org/10.1007/978-1-0716-1512-6_4

Note that, if $u \in C^2(\Omega; \mathbb{R}^3)$, then

$$Ju = \det\left(\frac{\partial u}{\partial x_1}, \frac{\partial u}{\partial x_2}, \frac{\partial u}{\partial x_3}\right) \tag{4.4}$$

is the Jacobian of u. This is the analogue of (1.36). In particular, if $u \in C^2(\overline{\Omega}; \mathbb{S}^2)$, then

$$Ju = 0 \text{ in } \mathscr{D}'(\Omega). \tag{4.5}$$

The same holds under the weaker assumption $u \in W^{1,3}(\Omega; \mathbb{S}^2)$, or even $u \in H^1(\Omega; \mathbb{S}^2) \cap VMO\,(\Omega; \mathbb{S}^2)$.

When treating the general case, $N \geq 2$ and $1 \leq k \leq N-1$, it is more convenient to adopt the language of differential forms. Given $u \in W^{1,k}(\Omega; \mathbb{S}^k)$, set

$$D(u) := \sum_{j=1}^{k+1} (-1)^{j-1} u_j \, du_1 \wedge \ldots \wedge \widehat{du_j} \wedge \ldots \wedge du_{k+1} \in L^1(\Omega; \Lambda^k), \tag{4.6}$$

where du_j denotes, as usual, the 1-form

$$du_j := \sum_{\ell=1}^{N} \frac{\partial u_j}{\partial x_\ell} \, dx_\ell.$$

In other words,

$$D(u) = u^\sharp \, \omega_k,$$

where $u^\sharp \, \omega_k$ is the pullback by u of the standard volume form ω_k on $\mathbb{S}^k$, given by

$$\omega_k := \sum_{j=1}^{k+1} (-1)^{j-1} x_j \, dx_1 \wedge \ldots \wedge \widehat{dx_j} \wedge \ldots \wedge dx_{k+1}.$$

Then define

$$Ju := \frac{1}{k+1} dD(u) \in \mathscr{D}'(\Omega; \Lambda^{k+1}). \tag{4.7}$$

In the special case $N \geq 3$ and $k = 2$, (4.6) takes the form

$$D(u) = \sum_{1 \leq j < \ell \leq N} \det\left(u, \frac{\partial u}{\partial x_j}, \frac{\partial u}{\partial x_\ell}\right) dx_j \wedge dx_\ell. \tag{4.8}$$

In particular, when $N = 3$, (4.8) becomes

$$D(u) = \det\left(u, \frac{\partial u}{\partial x_1}, \frac{\partial u}{\partial x_2}\right) dx_1 \wedge dx_2 + \det\left(u, \frac{\partial u}{\partial x_1}, \frac{\partial u}{\partial x_3}\right) dx_1 \wedge dx_3$$
$$+ \det\left(u, \frac{\partial u}{\partial x_2}, \frac{\partial u}{\partial x_3}\right) dx_2 \wedge dx_3, \tag{4.9}$$

which is consistent with (4.2), modulo the Hodge $*$ operator, once vector fields have been identified with 1-forms.

For the purposes of comparison with the previous chapters, we also point out that, when $N \geq 2$ and $k = 1$, $D(u)$, given by (4.6), takes the form

$$D(u) = u_1\, du_2 - u_2\, du_1 = \sum_{\ell=1}^{N} \left(u_1 \frac{\partial u_2}{\partial x_\ell} - u_2 \frac{\partial u_1}{\partial x_\ell}\right) dx_\ell,$$

which is identified, as usual, with the vector field $u \wedge \nabla u$. For $k = 1$, formula (4.7) becomes

$$Ju = \frac{1}{2} d(u \wedge \nabla u),$$

which coincides with definition (1.40) of Ju, adopted throughout Chapters 1–3.

The other tool that played a central role in Chapters 1–3 was the quantity $\Sigma(u)$. Recall that $\Sigma(u)$ was defined in (1.80) as the "L^1 distance of $u \wedge \nabla u$ to gradients." Assuming now that $N \geq 3$ and $2 \leq k \leq N-1$, the counterpart of $u \wedge \nabla u$ is the k-form $D(u)$ defined by (4.6), and it is natural to introduce its distance to the exact k-forms. More precisely, given $u \in W^{1,k}(\Omega; \mathbb{S}^k)$, we set

$$\Sigma(u) := \inf\left\{\|D(u) - d\eta\|_{L^1};\ \eta \in W^{1,1}(\Omega; \Lambda^{k-1})\right\}. \tag{4.10}$$

Applying Theorem 1.15 to $F := D(u)$ we deduce, in full generality, that

$$\Sigma(u) = \min\{\|\mu\|_{\mathscr{M}};\ \mu \in \mathscr{M}(\Omega; \Lambda^k),$$
$$d\mu = (k+1)Ju \text{ in } \mathscr{D}'(\Omega; \Lambda^{k+1})\}, \tag{4.11}$$

$$= \sup\left\{\int_\Omega \langle D(u), *d\zeta\rangle;\ \zeta \in C_c^\infty(\Omega; \Lambda^{N-k-1}),\ \|d\zeta\|_{L^\infty} \leq 1\right\} \tag{4.12}$$

$$= (k+1) \sup\{\langle Ju, *\zeta\rangle;\ \zeta \in C_c^\infty(\Omega; \Lambda^{N-k-1}),\ \|d\zeta\|_{L^\infty} \leq 1\}, \tag{4.13}$$

which are the exact analogues of (1.85) and (1.135).

We next turn to the $\mathbb{S}^k$ counterpart of the least energy with prescribed Jacobian introduced in Section 1.8. By analogy, we set, for every $u \in W^{1,k}(\Omega; \mathbb{S}^k)$,

$$S_k(u) := \inf\left\{\int_\Omega |\nabla v|^k;\ v \in W^{1,k}(\Omega; \mathbb{S}^k),\ Jv = Ju \text{ in } \mathscr{D}'(\Omega; \Lambda^{k+1})\right\}. \tag{4.14}$$

Recall that, when $k = 1$, we have established that, for any $N \geq 2$,

$$S_1(u) = \Sigma(u), \ \forall u \in W^{1,1}(\Omega; \mathbb{S}^1). \tag{4.15}$$

The proof of (4.15) was quite elementary (see Lemma 1.14), but relied heavily on *lifting*. More precisely, it used the relations

$$[Jv = Ju] \Longleftrightarrow [v \wedge \nabla v = u \wedge \nabla u - \nabla \eta = D(u) - d\eta \text{ for some } \eta \in W^{1,1}(\Omega; \mathbb{R}) \simeq W^{1,1}(\Omega; \Lambda^0)]$$

and

$$|\nabla v| = |v \wedge \nabla v|.$$

One may wonder what happens to (4.15) when $k \geq 2$. It turns out that, when $N = k + 1$, one still has

$$S_k(u) = c_k \, \Sigma(u), \ \forall u \in W^{1,k}(\Omega; \mathbb{S}^k), \ \forall k \geq 2, \tag{4.16}$$

where

$$c_k := k^{k/2}, \tag{4.17}$$

but the proof is much more involved; it is presented in Section 4.1.2 for the case $k = 2$, $N = 3$ and sketched in Section 4.4.3 for the case $k \geq 3$, $N = k + 1$.

However, when $N > k + 1$, one can only assert that

$$S_k(u) \geq c_k \, \Sigma(u), \ \forall u \in W^{1,k}(\Omega; \mathbb{S}^k), \ \forall k \geq 2, \tag{4.18}$$

and it may indeed happen that $S_k(u) > c_k \, \Sigma(u)$. The correct value of $S_k(u)$ is described in Sections 4.3.3 and 4.4.3. This requires an excursion into the world of area-minimizing $(N - k)$-rectifiable currents whose boundaries coincide with $(1/|\mathbb{B}^{k+1}|)\,(*Ju)$ (see Section 4.2). When u is smooth on $\Omega \setminus \Gamma$, where Γ is an $(N - k - 1)$-submanifold of Ω, $S_k(u)$ corresponds to the least area spanned by Γ.

Finally, we investigate the $\mathbb{S}^k$ counterpart of the relaxed energy introduced in Section 1.7. By analogy we set, for every $u \in W^{1,k}(\Omega; \mathbb{S}^k)$,

$$R_k(u) := \inf \left\{ \liminf_{n\to\infty} \int_\Omega |\nabla u_n|^k; \ u_n \in C^\infty(\overline{\Omega}; \mathbb{S}^k), \ u_n \to u \text{ a.e.} \right\} \leq \infty. \tag{4.19}$$

This is a meaningful quantity, in view of the fact that $C^\infty(\overline{\Omega}; \mathbb{S}^k)$ is never dense in $W^{1,k}(\Omega; \mathbb{S}^k)$, since $J \not\equiv 0$ (see Section 4.4.1).

Recall (see Theorem 1.11) that, when $k = 1$, we established that, for any $N \geq 2$,

$$R_1(u) = \int_\Omega |\nabla u| + \Sigma(u), \ \forall u \in W^{1,1}(\Omega; \mathbb{S}^1). \tag{4.20}$$

Again, one may wonder what happens to (4.20) when $k \geq 2$, since the proof of Theorem 1.11 also relied heavily on lifting. We will prove in Section 4.1.3 that, when

$k = 2$ and $N = 3$,

$$R_2(u) = \int_\Omega |\nabla u|^2 + 2\,\Sigma(u), \ \forall\, u \in H^1(\Omega; \mathbb{S}^2). \tag{4.21}$$

When $k = 2$ and $N \geq 4$, one still has

$$R_2(u) = \int_\Omega |\nabla u|^2 + S_2(u), \ \forall\, u \in H^1(\Omega; \mathbb{S}^2), \tag{4.22}$$

but the proof is much more delicate (see Sections 4.3.4 and 4.4.4).

No formula is known for $R_k(u)$ when $k \geq 3$ and $N \geq k + 1$, even in the lowest possible situation: $N = 4$ and $k = 3$ (see Open Problem 6).

4.1 The "historical" case: $N = 3$ and $k = 2$, where everything fits into place!

Returning to the vector field notation, as at the beginning of the Introduction, we write, for every $u \in H^1(\Omega; \mathbb{S}^2)$,

$$D(u) = \left(\det\left(u, \frac{\partial u}{\partial x_2}, \frac{\partial u}{\partial x_3}\right), \det\left(\frac{\partial u}{\partial x_1}, u, \frac{\partial u}{\partial x_3}\right), \det\left(\frac{\partial u}{\partial x_1}, \frac{\partial u}{\partial x_2}, u\right)\right) \in L^1(\Omega; \mathbb{R}^3), \tag{4.23}$$

$$Ju = \frac{1}{3}\mathrm{div} D(u) \in \mathscr{D}'(\Omega), \tag{4.24}$$

$$\langle Ju, \zeta\rangle = -\frac{1}{3}\int_\Omega [D(u)] \cdot \nabla\zeta, \ \forall\, \zeta \in C_c^\infty(\Omega; \mathbb{R}), \tag{4.25}$$

$$\Sigma(u) = \inf\left\{\int_\Omega |D(u) - \mathrm{curl}\ \eta|;\ \eta \in W^{1,1}(\Omega; \mathbb{R}^3)\right\} \tag{4.26}$$

$$= \min\{\|\mu\|_{\mathscr{M}};\ \mu \in \mathscr{M}(\Omega; \mathbb{R}^3),\ \mathrm{div}\,\mu = 3\,Ju \text{ in } \mathscr{D}'(\Omega)\} \tag{4.27}$$

$$= \sup\left\{\int_\Omega [D(u)] \cdot \nabla\zeta;\ \zeta \in C_c^\infty(\Omega; \mathbb{R}),\ \|\nabla\zeta\|_{L^\infty} \leq 1\right\} \tag{4.28}$$

$$= 3\sup\{\langle Ju, \zeta\rangle;\ \zeta \in C_c^\infty(\Omega; \mathbb{R}),\ \|\nabla\zeta\|_{L^\infty} \leq 1\}. \tag{4.29}$$

As already mentioned in the Introduction, we know that

$$Ju = 0, \ \forall\, u \in W^{1,3}(\Omega; \mathbb{S}^2). \tag{4.30}$$

A typical example of an $\mathbb{S}^2$-valued map that does not belong to $W^{1,3}$ is $u(x) := x/|x|$, when $0 \in \Omega$; note that $u \in W^{1,p}(\Omega; \mathbb{S}^2)$, $\forall\, p < 3$. The analogue of (1.68) is

$$Ju = \frac{4\pi}{3}\delta_0 \text{ in } \mathscr{D}'(\Omega). \tag{4.31}$$

More generally, if $u \in H^1(\Omega; \mathbb{S}^2) \cap C(\Omega \setminus \{a_1, \dots, a_k\})$, then

$$Ju = \frac{4\pi}{3} \sum_{j=1}^{k} d_j \, \delta_{a_j} \text{ in } \mathscr{D}'(\Omega), \tag{4.32}$$

where d_j is the topological (Brouwer) degree of u on a small sphere around a_j. [For an introduction to the Brouwer degree, see, for example, Dinca and Mawhin [141, Chapter 1]. By homotopy invariance, $\deg(u, S(a_j, r))$ does not depend on small $r > 0$.] This is the analogue of (2.26).

Combining (4.32) and (4.29) yields

$$\Sigma(u) = 4\pi \sup \left\{ \sum_{j=1}^{k} d_j \, \zeta(a_j);\ \zeta \in C_c^\infty(\Omega; \mathbb{R}),\ \|\nabla \zeta\|_{L^\infty} \le 1 \right\}, \tag{4.33}$$

and thus we enter the familiar territory of Sections 2.2 and 2.3, so that

$$\Sigma(u) = 4\pi \, L(a, d) \tag{4.34}$$

(the only difference is that here $\Omega \subset \mathbb{R}^3$, whereas $\Omega \subset \mathbb{R}^2$ in Chapter 2, but the rest is unchanged).

By analogy with Section 2.9, we may even go one step further:

4.1.1 The structure of Ju for a general $u \in H^1(\Omega; \mathbb{S}^2)$ and one more formula for $\Sigma(u)$

The main results are the following.

Theorem 4.1. *Let $u \in H^1(\Omega; \mathbb{S}^2)$. Then there exist points $P_j, N_j \in \overline{\Omega}$ such that*

$$\sum_j |P_j - N_j| < \infty \tag{4.35}$$

and

$$Ju = \frac{4\pi}{3} \sum_j (\delta_{P_j} - \delta_{N_j}) \textit{ in } [W_0^{1,\infty}(\Omega)]^* \textit{ (and thus in } \mathscr{D}'(\Omega)). \tag{4.36}$$

In addition, we have

$$\Sigma(u) = 4\pi \inf \Big\{ \sum_j |\widetilde{P}_j - \widetilde{N}_j|;\ \widetilde{P}_j, \widetilde{N}_j \in \overline{\Omega} \textit{ and } \sum_j (\delta_{\widetilde{P}_j} - \delta_{\widetilde{N}_j}) = \sum_j \left(\delta_{P_j} - \delta_{N_j}\right) \textit{ in } \mathscr{D}'(\Omega) \Big\}. \tag{4.37}$$

The proof is similar to that of Theorem 2.10, and relies on the density of nice maps in $H^1(\Omega; \mathbb{S}^2)$ (see Bethuel and Zheng [37, Theorem 4]).

Theorem 4.2. *Conversely, given points $P_j, N_j \in \overline{\Omega}$ such that* (4.35) *holds, there exists some $u \in H^1(\Omega; \mathbb{S}^2)$ such that* (4.36) *holds.*

More precisely, for every $\varepsilon > 0$ there exists some $v \in H^1(\Omega; \mathbb{S}^2)$ such that

$$Jv = \frac{4\pi}{3} \sum_j (\delta_{P_j} - \delta_{N_j}) \tag{4.38}$$

and

$$\int_\Omega |\nabla v|^2 \leq 8\pi \sum_j |P_j - N_j| + \varepsilon. \tag{4.39}$$

Sketch of proof. When there are only finitely many points P_j, N_j, such a v can be obtained via a dipole construction, in the spirit of Section 2.5 [79, Section III]. Some care is needed when there are infinitely many points. This can be achieved via an iterative procedure (see Alberti, Baldo, and Orlandi [6, proof of Theorem 5.6] for a more general situation). □

4.1.2 Least energy with prescribed Jacobian: $S_2(u) = 2\Sigma(u)$

The main result of this section concerns the least energy, defined for any $u \in H^1(\Omega; \mathbb{S}^2)$ by

$$S_2(u) := \inf \left\{ \int_\Omega |\nabla v|^2;\ v \in H^1(\Omega; \mathbb{S}^2),\ Jv = Ju \text{ in } \mathscr{D}'(\Omega) \right\}. \tag{4.40}$$

Theorem 4.3. *For every $u \in H^1(\Omega; \mathbb{S}^2)$, we have*

$$S_2(u) = 2\,\Sigma(u). \tag{4.41}$$

Proof. Applying Theorem 4.1 we obtain, for every $\varepsilon > 0$, points $\widetilde{P}_j, \widetilde{N}_j \in \Omega$ such that

$$Ju = \frac{4\pi}{3} \sum_j (\delta_{\widetilde{P}_j} - \delta_{\widetilde{N}_j})$$

and

$$4\pi \sum_j |\widetilde{P}_j - \widetilde{N}_j| \leq \Sigma(u) + \varepsilon.$$

Theorem 4.2 then yields some $v \in H^1(\Omega; \mathbb{S}^2)$ such that

$$Jv = \frac{4\pi}{3}\sum_j (\delta_{\widetilde{P}_j} - \delta_{\widetilde{N}_j}) = Ju$$

and

$$\int_\Omega |\nabla v|^2 \le 8\pi \sum_j |\widetilde{P}_j - \widetilde{N}_j| + \varepsilon \le 2\,\Sigma(u) + 3\varepsilon.$$

Since $\varepsilon > 0$ is arbitrary, it follows that

$$S_2(u) \le 2\,\Sigma(u), \ \forall\, u \in H^1(\Omega; \mathbb{S}^2). \tag{4.42}$$

For the reverse inequality, we fix $u \in H^1(\Omega; \mathbb{S}^2)$ and take any $v \in H^1(\Omega; \mathbb{S}^2)$ such that

$$Jv = Ju. \tag{4.43}$$

We first claim that

$$|D(v)| \le \frac{1}{2}|\nabla v|^2 \text{ a.e. in } \Omega \tag{4.44}$$

(where $D(v)$ is given by (4.23)). Inequality (4.44) is a direct consequence of the following

Lemma 4.1. *Given $U, X, Y, Z \in \mathbb{R}^3$, set*

$$V := (\det(U, Y, Z),\ \det(X, U, Z),\ \det(X, Y, U)) \in \mathbb{R}^3. \tag{4.45}$$

Then,

$$|V| \le \frac{1}{2}|U|\,(|X|^2 + |Y|^2 + |Z|^2). \tag{4.46}$$

Proof of Lemma 4.1. Using the invariance with respect to isometries and the linearity of V with respect to U , we may assume that $U = (0, 0, 1)$. Then, V takes the form $V = A \wedge B$, where $A := (X_1, Y_1, Z_1)$ and $B := (X_2, Y_2, Z_2)$. Thus,

$$|V| \le |A|\,|B| \le \frac{1}{2}(|A|^2 + |B|^2) \le \frac{1}{2}(|X|^2 + |Y|^2 + |Z|^2).$$ □

Proof of Theorem 4.3 *continued.* Applying Lemma 4.1 with $U := v$, $X := \dfrac{\partial v}{\partial x_1}$, $Y := \dfrac{\partial v}{\partial x_2}$, and $Z := \dfrac{\partial v}{\partial x_3}$ yields (4.44). From (4.28), we then deduce that $\Sigma(v) \le (1/2)\int_\Omega |\nabla v|^2$. On the other hand, (4.43) and (4.29) imply that $\Sigma(u) = \Sigma(v)$. Therefore,

$$\Sigma(u) \le \frac{1}{2}\int_\Omega |\nabla v|^2, \ \forall\, v \in H^1(\Omega; \mathbb{S}^2) \text{ such that } Jv = Ju. \tag{4.47}$$

Taking the inf over v in (4.47), we obtain $2\,\Sigma(u) \le S_2(u)$. □

Remark 4.1. We do not know a "non-geometrical" proof of Theorem 4.3 (see Open Problem 5). Recall (as already mentioned in the Introduction) that, when $k = 1$, we established, for any $N \ge 2$, the formula $S_1(u) = \Sigma(u), \forall\, u \in W^{1,1}(\Omega; \mathbb{S}^1)$. The proof was quite elementary; it relied heavily on lifting and did not involve a geometrical detour via minimal connections.

4.1.3 The relaxed energy: $R_2(u) = \int_\Omega |\nabla u|^2 + 2\,\Sigma(u)$

The main result in this section concerns the relaxed energy defined, for any $u \in H^1(\Omega; \mathbb{S}^2)$, by

$$R_2(u) := \inf\left\{\liminf_{n\to\infty} \int_\Omega |\nabla u_n|^2;\ (u_n) \subset C^\infty(\overline{\Omega}; \mathbb{S}^2),\ u_n \to u \text{ a.e.}\right\} \le \infty. \tag{4.48}$$

We point out that it is easy to show that there exist sequences $(u_n) \subset C^\infty(\overline{\Omega}; \mathbb{S}^2)$ such that $u_n \to u$ a.e., even when $u : \Omega \to \mathbb{S}^2$ is merely measurable (see, for example, Escobedo [145]). However, it is not clear that there exists a sequence $(u_n) \subset C^\infty(\overline{\Omega}; \mathbb{S}^2)$ such that $u_n \to u$ a.e. *and* $\liminf_{n\to\infty} \int_\Omega |\nabla u_n|^2 < \infty$; this is part of the conclusion of our next result (see also (4.121) and Comment 6 in Chapter 10 for more general considerations).

Theorem 4.4. *For every $u \in H^1(\Omega; \mathbb{S}^2)$, we have*

$$R_2(u) = \int_\Omega |\nabla u|^2 + 2\,\Sigma(u). \tag{4.49}$$

Sketch of proof. We start by noting that the quantity $R_2(u)$ does not change if, in its definition (4.48), we replace "a.e. convergence" with "weak H^1 convergence."

Step 1. Proof of "$\ge$" in (4.49)

Let $(u_n) \subset C^\infty(\overline{\Omega}; \mathbb{S}^2)$ be such that $u_n \rightharpoonup u$ for some $u \in H^1(\Omega; \mathbb{S}^2)$, weakly in H^1. We need to prove that

$$\liminf_{n\to\infty} \int_\Omega |\nabla u_n|^2 \ge \int_\Omega |\nabla u|^2 + 2\,\Sigma(u). \tag{4.50}$$

In view of (4.28), it suffices to show that

$$\liminf_{n\to\infty} \int_\Omega |\nabla u_n|^2 \ge \int_\Omega |\nabla u|^2 + 2\int_\Omega [D(u)]\cdot\nabla\zeta, \quad \forall\,\zeta \in C_c^\infty(\Omega; \mathbb{R}),\ \|\nabla\zeta\|_{L^\infty} \le 1. \tag{4.51}$$

Set $v_n := u_n - u \rightharpoonup 0$ in H^1, so that

$$\int_\Omega |\nabla u_n|^2 = \int_\Omega |\nabla u|^2 + \int_\Omega |\nabla v_n|^2 + o(1). \tag{4.52}$$

Recall that $Ju_n = (1/3)\operatorname{div} D(u_n) = 0$ (by (4.5)), and thus

$$\int_\Omega [D(u_n)] \cdot \nabla\zeta = 0, \ \forall\, \zeta \in C_c^\infty(\Omega; \mathbb{R}). \tag{4.53}$$

Fix $\zeta \in C_c^\infty(\Omega; \mathbb{R})$ and note the identity

$$[D(u_n)] \cdot \nabla\zeta = A_n + B_n + C_n, \tag{4.54}$$

where

$$A_n := \sum_c \det\left(u_n, \frac{\partial u}{\partial x_2}, \frac{\partial u}{\partial x_3}\right) \frac{\partial \zeta}{\partial x_1}, \tag{4.55}$$

$$B_n := \sum_c \det\left(u_n, \frac{\partial v_n}{\partial x_2}, \frac{\partial v_n}{\partial x_3}\right) \frac{\partial \zeta}{\partial x_1}, \tag{4.56}$$

$$C_n := \sum_c \left[\det\left(u_n, \frac{\partial u}{\partial x_2}, \frac{\partial v_n}{\partial x_3}\right) + \det\left(u_n, \frac{\partial v_n}{\partial x_2}, \frac{\partial u}{\partial x_3}\right)\right] \frac{\partial \zeta}{\partial x_1}, \tag{4.57}$$

and Σ_c denotes a cyclic sum.

Clearly, we have

$$\int_\Omega A_n \to \int_\Omega [D(u)] \cdot \nabla\zeta \text{ as } n \to \infty \tag{4.58}$$

and

$$\int_\Omega C_n \to 0 \text{ as } n \to \infty; \tag{4.59}$$

for the last assertion, note, for example, that

$$\det\left(u_n, \frac{\partial u}{\partial x_2}, \frac{\partial v_n}{\partial x_3}\right) = \xi_n \cdot \frac{\partial v_n}{\partial x_3},$$

where (ξ_n) is strongly convergent in L^2, and $\partial v_n/\partial x_3 \rightharpoonup 0$ in L^2.

On the other hand, Lemma 4.1 yields the pointwise inequality

$$|B_n| \le \frac{1}{2}|\nabla v_n|^2 \, \|\nabla\zeta\|_{L^\infty} \le \frac{1}{2}|\nabla v_n|^2 \text{ if } \|\nabla\zeta\|_{L^\infty} \le 1. \tag{4.60}$$

Combining (4.53), (4.54), and (4.58)–(4.60), we obtain

$$\left|2\int_\Omega [D(u)] \cdot \nabla\zeta\right| \le \int_\Omega |\nabla v_n|^2 + o(1) \text{ as } n \to \infty, \tag{4.61}$$
$$\forall\, \zeta \in C_c^\infty(\Omega; \mathbb{R}) \text{ with } \|\nabla\zeta\|_{L^\infty} \le 1.$$

Inserting (4.61) into (4.52) yields (4.51), and thus (4.50).

Step 2. Proof of "$\le$" in (4.49)
Clearly, the map $u \mapsto D(u)$ is continuous from $H^1(\Omega; \mathbb{S}^2)$ into $L^1(\Omega; \mathbb{R}^3)$, and thus (by (4.28)) the function $u \mapsto \Sigma(u)$ is continuous on $H^1(\Omega; \mathbb{S}^2)$. On the other hand, the function $u \mapsto R_2(u)$ (defined by (4.48)) is clearly lower semicontinuous with respect to the strong H^1 convergence. Therefore, it suffices to establish "$\le$" when u belongs to a dense subset $\mathscr{R}$ of $H^1(\Omega; \mathbb{S}^2)$ (relative to the strong H^1 convergence). We will work with the class

$$\mathscr{R} := \{u \in H^1(\Omega; \mathbb{S}^2);\ u \text{ is smooth on } \overline{\Omega} \setminus \{a_1, \dots, a_k\}\},$$

which is dense in $H^1(\Omega; \mathbb{S}^2)$ (see Bethuel and Zheng [37, Theorem 4]). Applying the "dipole removing" construction of Bethuel [25, Theorem 2] we obtain, for any given $u \in \mathscr{R}$, a sequence $(u_n) \subset C^\infty(\overline{\Omega}; \mathbb{S}^2)$ such that $u_n \to u$ a.e. and

$$\limsup_{n\to\infty} \int_\Omega |\nabla u_n|^2 \le \int_\Omega |\nabla u|^2 + 8\pi\, L(a, d) = \int_\Omega |\nabla u|^2 + 2\,\Sigma(u). \qquad \square$$

Before proceeding with the analysis of the case $N \ge 4$ and $k = 2$, it is useful to discuss the counterparts of the classes $\mathscr{E}$ (introduced in Chapter 2; see (2.145)) and $\mathscr{F}$ (introduced in Chapter 3; see (3.55) for $N = 3$ and (3.103) for $N \ge 4$).

4.2 A distinguished class of currents. Definition of $\mathscr{F}^\ell$

Let $\Omega \subset \mathbb{R}^N$, $N \ge 2$. Given an integer ℓ, $0 \le \ell \le N - 2$, set

$$\begin{aligned} \mathscr{F}^\ell := \{T;\ & T \text{ is an } \ell\text{-current such that } T = \partial M \text{ for some} \\ & (\ell+1) - \text{rectifiable current } M \text{ on } \Omega\} \end{aligned} \tag{4.62}$$

(for the definition of ℓ-rectifiable currents, see Section 2.9.4).

Note that $\mathscr{E}$ in (2.145) corresponds to $N = 2$ and $\ell = 0$ in (4.62), and $\mathscr{F}$ defined in (3.55) (resp. (3.103)) corresponds to $N = 3$, $\ell = 1$ (resp. $N \ge 4$, $\ell = N - 2$) in (4.62). Note also that $T_1 + T_2 \in \mathscr{F}^\ell$, $\forall\, T_1, T_2 \in \mathscr{F}^\ell$, and $mT \in \mathscr{F}^\ell$, $\forall\, m \in \mathbb{Z}$, $\forall\, T \in \mathscr{F}^\ell$; but, in general, $\frac{1}{2}T \notin \mathscr{F}^\ell$ if $T \in \mathscr{F}^\ell$!

By analogy with (3.105)–(3.107), we define, $\forall\, T \in \mathscr{F}^\ell$,

$$\begin{aligned} A^\ell(T) := \inf\{|M|;\ & M \subset \Omega \text{ is an } (\ell+1)\text{-rectifiable current} \\ & \text{such that } \partial M = T\}, \end{aligned} \tag{4.63}$$

$$B^\ell(T) := \inf\{\|\mu\|_{\mathscr{M}};\ \mu \in \mathscr{M}(\Omega; \Lambda^{\ell+1}) \text{ is a measure such that } \partial\mu = T\}, \tag{4.64}$$

and

$$C^\ell(T) := \sup\{\langle T, \zeta\rangle;\ \zeta \in C_c^\infty(\Omega; \Lambda^\ell),\ \|d\zeta\|_{L^\infty} \le 1\}. \tag{4.65}$$

Keeping in mind that $\partial\mu = d^*\mu = \pm * d * \mu$ (see (1.141)), we deduce that

$$B^\ell(T) = \min\{\|\nu\|_{\mathscr{M}};\ \nu \in \mathscr{M}(\Omega; \Lambda^{N-\ell-1}),\ d\nu = *T\}. \tag{4.66}$$

Proposition 4.1. *We have*

$$A^\ell(T) \ge B^\ell(T) = C^\ell(T),\ \forall \ell \text{ with } 0 \le \ell \le N-2,\ \forall T \in \mathscr{F}^\ell. \tag{4.67}$$

Sketch of proof. It is clear from (4.63) and (4.64) that $A^\ell(T) \ge B^\ell(T)$, since $(\ell+1)$-rectifiable currents are particular elements of $\mathscr{M}(\Omega; \Lambda^{\ell+1})$. Equality $B^\ell(T) = C^\ell(T)$ is a consequence of Hahn–Banach, as in the proof of Theorem 1.9 (see also Comment 12 in Chapter 1). □

It is natural to raise the following question:

$$\text{Is } A^\ell(T) = B^\ell(T),\ \forall \ell \text{ with } 0 \le \ell \le N-2,\ \forall T \in \mathscr{F}^\ell? \tag{4.68}$$

Indeed, we know that the answer is positive in several cases:

1. $N = 2$, $\ell = 0$, by Corollary 2.4.
2. $N = 3$, $\ell = 1$, by Corollary 3.1.
3. $N \ge 4$, $\ell = N-2$, by (3.108).

More generally, we have

Proposition 4.2. *Assume $N \ge 2$, and $\ell = N-2$ or $\ell = 0$. Then,*

$$A^\ell(T) = B^\ell(T) = C^\ell(T),\ \forall T \in \mathscr{F}^\ell. \tag{4.69}$$

Proof. The case $\ell = N-2$ has been settled above, and thus it suffices to treat the case $\ell = 0$.

As explained in Section 2.9.4 (see also Proposition 15.4 in Appendix 15.9), elements $T \in \mathscr{F}^0$ are of the form $T = \sum_{j=1}^\infty (\delta_{P_j} - \delta_{N_j})$ for some sequences of points P_j, N_j in $\overline{\Omega}$ such that $\sum_{j=1}^\infty |P_j - N_j| < \infty$.

The proof of (4.69) in any dimension $N \ge 2$ is exactly the same as the proof of Corollary 2.4 (and the essential ingredient is the OT Theorem 2.3, which holds in any dimension $N \ge 2$). □

Remark 4.2. It turns out that (suprisingly) the answer to the above question (4.68) is negative for $N \ge 4$ and ℓ in the intermediate range, $1 \le \ell \le N-3$. See Young [348], White [343], and Morgan [272] for a counterexample concerning $N = 4$ and $\ell = 1$. The idea is to construct a smooth closed connected curve Γ in $\mathbb{R}^4$ such that

$$A^1(2\Gamma) < 2A^1(\Gamma), \tag{4.70}$$

where Γ is identified, as usual (see (3.16)) with a 1-current (which clearly belongs to $\mathscr{F}^1$). Inequality (4.70) implies that

$$A^1(\Gamma) > B^1(\Gamma). \tag{4.71}$$

Indeed, we have trivially

$$B^1(2\Gamma) = 2\,B^1(\Gamma), \tag{4.72}$$

whereas

$$B^1(2\Gamma) \le A^1(2\Gamma) \text{ by (4.67).} \tag{4.73}$$

Combining (4.70), (4.72), and (4.73) yields (4.71).

Presumably, a variant of their construction yields a counterexample for *any* $N \ge 4$ and *any* ℓ, $1 \le \ell \le N-3$.

In the special case where T is of the form

$$T = \Gamma = \partial M \text{ for some nice } (\ell+1)\text{-manifold } M \subset \Omega \\ \text{with finite mass and geometric boundary } \Gamma, \tag{4.74}$$

where $0 \le \ell \le N-2$ and M is identified, as usual, with an $(\ell+1)$-rectifiable current, we may also define, by analogy with (3.112),

$$A_0^\ell(\Gamma) := \inf\{|M|;\ M \text{ as in (4.74)}\}. \tag{4.75}$$

Proposition 4.3. *Let* $N \ge 2$ *and* $0 \le \ell \le N-2$. *If* Γ *is as in* (4.74), *then*

$$A_0^\ell(\Gamma) = A^\ell(\Gamma). \tag{4.76}$$

In other words, the least mass spanned by Γ *in the GMT sense and in the usual sense coincide.*

From Theorem 3.16, we know that (4.76) holds when $\ell = N-2$. As we are going to see below, it also holds when $\ell = 0$. The intermediate cases, $1 \le \ell \le N-3$ (with $N \ge 4$), were announced by Lin [228].

Proof of Proposition 4.3 *when* $\ell = 0$. In this case, Γ is a finite collection of points. As in Lemma 2.4, we may write $\Gamma = \sum_{j=1}^m (\delta_{P_j} - \delta_{N_j})$ and $A^0(\Gamma) = \sum_{j=1}^m |P_j - N_j|$ for some points P_j, N_j in $\overline{\Omega}$. In general, $M := \cup_{j=1}^m (N_j, P_j)$ is *not* a nice curve, since some of the segments may overlap. Therefore, we cannot use M as a competitor in (4.75). However, given an arbitrary $\varepsilon > 0$, it is not difficult to construct a nice curve C such that $|C| \le A^0(\Gamma) + \varepsilon$ and $\partial C = \sum_{j=1}^n (\delta_{P_j} - \delta_{N_j}) = \Gamma$. This implies that $A_0^0(\Gamma) \le A^0(\Gamma)$. □

4.3 The case $N = 4$ and $k = 2$; where complications appear

We already encounter at this level a new phenomenon, which will reappear later in all dimensions $N \geq 4$ and for all $2 \leq k \leq N-2$.

Throughout this section, we assume that $\Omega \subset \mathbb{R}^4$, and deal with maps $u \in H^1(\Omega; \mathbb{S}^2)$. Recall that we have introduced in (4.7) the Jacobian $Ju \in \mathscr{D}'(\Omega; \Lambda^3)$, and in (4.10) the quantity $\Sigma(u)$. They are both continuous under (strong) H^1 convergence.

It follows immediately from the definition of Ju that $Ju = 0$, $\forall\, u \in C^2(\overline{\Omega}; \mathbb{S}^2)$ and thus we also have $Ju = 0$, $\forall\, u \in H^1(\Omega; \mathbb{S}^2) \cap C(\Omega)$. Furthermore (using the density of $C^2(\overline{\Omega}; \mathbb{S}^2)$ in $W^{1,4}(\Omega; \mathbb{S}^2)$, which is a consequence of the embedding $W^{1,4} \hookrightarrow VMO$, see Section 15.5) we have

$$Ju = 0, \ \ \forall\, u \in W^{1,4}(\Omega; \mathbb{S}^2). \tag{4.77}$$

Actually, we have a better result:

$$Ju = 0, \forall\, u \in W^{1,3}(\Omega; \mathbb{S}^2). \tag{4.78}$$

This is the $\mathbb{S}^2$ counterpart of the fact that $Ju = 0$, $\forall\, u \in W^{1,2}(\Omega; \mathbb{S}^1)$, $\Omega \subset \mathbb{R}^N$ with $N \geq 2$ (see (1.61)), and the proof is similar. Namely, one uses the fact that, for any $u \in W^{1,3}(\Omega; \mathbb{S}^2)$, there exists a sequence $(u_n) \subset C^\infty(\overline{\Omega}; \mathbb{R}^3)$ (but *not necessarily in* $C^\infty(\overline{\Omega}; \mathbb{S}^2)$) such that $u_n \to u$ in $W^{1,3}$ and $\|u_n\|_{L^\infty} \leq 1$. One may then follow the same argument as in Lemma 1.7, noting that expressions like $\det\left(\frac{\partial u_n}{\partial x_j}, \frac{\partial u_n}{\partial x_k}, \frac{\partial u_n}{\partial x_l}\right)$ converge to 0 in L^1.

As a consequence, we also have

$$\Sigma(u) = 0, \ \ \forall\, u \in W^{1,3}(\Omega; \mathbb{S}^2). \tag{4.79}$$

4.3.1 A typical example

We follow the same approach as in Section 3.1, except that $\mathbb{S}^1$ is replaced by $\mathbb{S}^2$. First consider a map $u \in H^1(\Omega; \mathbb{S}^2)$, continuous except on a smooth closed connected oriented curve $\Gamma \subset \Omega$. Then we may define $\deg(u, \Gamma)$, the Brouwer degree of u restricted to a small sphere, centered at $x \in \Gamma$, in the 3D space normal to Γ at x and oriented by Γ. More generally, we may define $\deg(u, \Gamma)$ when $\Gamma \subset \Omega$ is smooth, connected, oriented, and has no boundary in Ω. By analogy with the results in Section 3.1, we have the following

Theorem 4.5. *1. Let $u \in H^1(\Omega; \mathbb{S}^2) \cap C(\Omega \setminus \Gamma)$. Then, with $m := \deg(u, \Gamma)$, we have*

$$*Ju = \frac{4\pi}{3} \, m\,\Gamma \text{ in } \mathscr{D}'(\Omega; \Lambda^1), \tag{4.80}$$

where Γ is identified with a 1*-current (as in* (3.16)*).*

2. *Let $\Gamma \subset \Omega$ satisfy* (3.15) *(in 4D). Then there exists some $u \in H^1(\Omega; \mathbb{S}^2)$ such that* (4.80) *holds.*

See Jerrard and Soner [205, Section 5]) for the first item, and Alberti, Baldo, and Orlandi [6, Theorem 5.6] for the second one. [The result in [6] applies since $\Gamma \in \mathscr{F}^1$ ($\mathscr{F}^1$ is defined in (4.62)). To see that $\Gamma \in \mathscr{F}^1$, we note that, since $\widetilde{\Gamma}$ is closed, there exists some 2-rectifiable current $\widetilde{M}$ such that $\partial \widetilde{M} = \widetilde{\Gamma}$ (see Federer [147, 4.2.10], where it is proved that we may choose $\widetilde{M}$ such that $|\widetilde{M}| \le C\,|\widetilde{\Gamma}|^2$), and then $M := \widetilde{M} \cap \Omega$ is a 2-rectifiable current satisfying $\partial M = \Gamma$.]

Remark 4.3. The reader may wonder what happens to point singularities, which played a central role when $N = 3$ (see Section 4.1, and in particular (4.31) and (4.32)). The answer is that in 4D point singularities go undetected by Ju. This echoes Proposition 3.1. More specifically, we have the following result, which includes Propositions 3.1 and 3.2 as special cases. It also implies that point singularities of $\mathbb{S}^2$-valued maps go undetected in 4D.

Proposition 4.4. *Let $u \in W^{1,k}(\Omega; \mathbb{S}^k) \cap C(\Omega \setminus \Gamma)$, where $\Gamma \subset \Omega \subset \mathbb{R}^N$ is a smooth manifold of dimension $\le N - k - 2$. Then,*

$$Ju = 0 \text{ in } \mathscr{D}'(\Omega). \tag{4.81}$$

Sketch of proof. For simplicity, we present the argument when $N = 4$, $k = 2$, $\Omega = (-1, 1)^4$ and $\Gamma = \{0\}$; the general case can be obtained according to similar lines.

Recall that, if $v \in W^{1,2}(\omega; \mathbb{S}^2) \cap C(\omega; \mathbb{S}^2)$, where $\omega \subset \mathbb{R}^3$ is an open set, then $Jv = 0$ in $\mathscr{D}'(\omega)$ (see, for example, the sentence following (4.5)). This follows by smoothing, the case where $v \in C^2$ being clear, by (4.4).

If $u \in W^{1,2}((-1, 1)^4; \mathbb{S}^2) \cap C((-1, 1)^4 \setminus \{0\})$, then, for a.e. $x_4 \in (-1, 1)$, the partial map $u_{4,x_4} := u(\cdot, x_4) : (-1, 1)^3 \to \mathbb{S}^2$ belongs to $W^{1,2}((-1, 1)^3; \mathbb{S}^2)$ and to $C((-1, 1)^3; \mathbb{S}^2)$ and thus, by the above, its Jacobian vanishes. Similarly for the partial maps u_{j,x_j}, $j = 1, 2, 3$.

We next invoke "slicing." The general idea is to apply Fubini, and to express the Jacobian of u in terms of the Jacobians of the u_{j,x_j}'s.

For this purpose, we note that every 3-form on $(-1, 1)^4$ can be written as

$$\zeta = \sum_{j=1}^{4} \zeta^j \widehat{dx_j}, \tag{4.82}$$

where $\widehat{dx_j} := dx_1 \wedge \ldots \wedge dx_{j-1} \wedge dx_{j+1} \wedge \ldots \wedge dx_4$. We claim that, for every

$$u \in W^{1,2}((-1, 1); \mathbb{R}^3) \cap L^\infty((-1, 1)^4; \mathbb{R}^3),$$

we have

$$\langle Ju, \zeta\rangle = \sum_{j=1}^{4} \int_{(-1,1)} \langle Ju_{j,x_j}, (\zeta^j)_{j,x_j}\rangle \, dx_j. \tag{4.83}$$

This is clear when $u \in C^\infty([0,1]^N; \mathbb{R}^{k+1})$. The general case follows by density and dominated convergence.

The conclusion follows from (4.83) and the first part of the proof. □

Remark 4.4. With more work, one may replace "Γ smooth" by "Γ rectifiable." This relies on the general version of (4.83) and the coarea formula. Same conclusion if we suppose, more generally, that $\mathscr{H}^{N-k-1}(\Gamma) = 0$.

Remark 4.5. In particular, Proposition 4.4 implies that, if $u(x) = H(x/|x|)$, where $x \in \Omega \subset \mathbb{R}^4$, $0 \in \Omega$, and $H \in C^1(\mathbb{S}^3; \mathbb{S}^2)$, then $Ju = 0$ in $\mathscr{D}'(\Omega)$. [Alternatively, we could note that $u \in W^{1,p}(\Omega; \mathbb{S}^2)$, $\forall p < 4$, and in particular that $u \in W^{1,3}(\Omega; \mathbb{S}^2)$, and infer that $Ju = 0$, by (4.78).] This implies in particular that Ju does not detect lower dimensional "topological singularities," since H may carry a non-trivial Hopf degree.

4.3.2 The structure of *Ju* for a general $u \in H^1(\Omega; \mathbb{S}^2)$

The main result in this direction is a far-reaching extension of Theorem 4.5, due to Alberti, Baldo, and Orlandi [6, Theorem 3.8, Theorem 5.6].

Theorem 4.6. *For any $u \in H^1(\Omega; \mathbb{S}^2)$, we have*

$$\frac{3}{4\pi} * Ju \in \mathscr{F}^1. \tag{4.84}$$

Conversely, given any $T \in \mathscr{F}^1$, there exists some $u \in H^1(\Omega; \mathbb{S}^2)$ such that

$$*Ju = \frac{4\pi}{3} T. \tag{4.85}$$

For the proof, we refer the reader to [6].

To every $u \in H^1(\Omega; \mathbb{S}^2)$, we may associate several quantities. First, we have $\Sigma(u)$ (defined in (4.10)) and its avatars (4.11)–(4.13), showing in particular that $\Sigma(u)$ depends only on Ju.

On the other hand, since $\dfrac{3}{4\pi} * Ju \in \mathscr{F}^1$, we may consider, as in (4.63)–(4.65), the three quantities

$$A^1\left(\frac{3}{4\pi} * Ju\right) \ge B^1\left(\frac{3}{4\pi} * Ju\right) = C^1\left(\frac{3}{4\pi} * Ju\right). \tag{4.86}$$

At the end of the day, we have only two quantities of interest attached to u (or in fact to Ju):

a) $\Sigma(u)$,

and

b) $$\Sigma^*(u) := 4\pi A^1\left(\frac{3}{4\pi} * Ju\right) \geq 4\pi C^1\left(\frac{3}{4\pi} * Ju\right) = \Sigma(u) \tag{4.87}$$

(by (4.86), (4.65), and (4.13)).

We know, from Remark 4.2, that *strict inequality* may occur in (4.86), and thus also in (4.87).

In the special case where $u \in H^1(\Omega; \mathbb{S}^2) \cap C(\Omega \setminus \Gamma)$, for some curve Γ satisfying (3.15) (in 4D), and $\deg(u, \Gamma) = 1$, the quantity $\Sigma^*(u)$ has a remarkable geometric interpretation:

$$\Sigma^*(u) = 4\pi \times \text{the least area spanned by } \Gamma. \tag{4.88}$$

Indeed, we know from Theorem 4.5 that $\frac{3}{4\pi} * Ju = \Gamma$. Therefore, $\Sigma^*(u) = 4\pi A^1(\Gamma)$, where $A^1(\Gamma)$ is the least area, in the sense of GMT, spanned by the curve Γ (see (4.63)).

As mentioned in Remark 4.2, it may happen that

$$\Sigma(u) < 4\pi \times \text{the least area spanned by } \Gamma.$$

This is in sharp contrast with the situation $N = 3$, $k = 1$, described in Section 3.2, where we always had

$$\Sigma(u) = 2\pi \times \text{the least area spanned by } \Gamma.$$

4.3.3 The least energy $S_2(u)$ as a least area: $S_2(u) = 2\,\Sigma^*(u)$

The next result provides the value of $S_2(u)$ defined in (4.14).

Theorem 4.7. *For any $u \in H^1(\Omega; \mathbb{S}^2)$, we have*

$$S_2(u) = 2\,\Sigma^*(u). \tag{4.89}$$

In particular, if $u \in H^1(\Omega; \mathbb{S}^2) \cap C(\Omega \setminus \Gamma)$, for some smooth, closed, connected, oriented curve $\Gamma \subset \Omega$, and $\deg(u, \Gamma) = 1$, *then*

$$S_2(u) = 8\pi \times \textit{the least area spanned by } \Gamma. \tag{4.90}$$

The proof of Theorem 4.7 is non-trivial; we refer the reader to Molnar [271], with roots in Almgren, Browder, and Lieb [8] and Alberti, Baldo, and Orlandi [6]. We sketch here the main ingredients.

Step 1. Proof of "$\geq$" in (4.89)
We cannot argue as in the proof of Theorem 4.3; this would lead to $S_2(u) \geq 2\Sigma(u)$, which is correct, but not optimal. The main new ingredient is the coarea formula. We explained this technique in Sections 2.4 and 3.3. The approach there extends with little change to the case where u is smooth except on a nice set [8]. The general case is more delicate (see [6]).

Step 2. Proof of "$\leq$" in (4.89)
This is done via a dipole construction in the spirit of Brezis, Coron, and Lieb [79], as described in Section 3.4. For more details, we refer the reader to [6] and [271].

4.3.4 *Relaxed energy: $R_2(u) = \int_\Omega |\nabla u|^2 + S_2(u)$*

The result of this section concerns the relaxed energy (see (4.19)). It is due to Giaquinta and Mucci [159], with roots in Pakzad [285] and Tarp-Ficenc [334].

Theorem 4.8. *For any $u \in H^1(\Omega; \mathbb{S}^2)$ we have*

$$R_2(u) = \int_\Omega |\nabla u|^2 + S_2(u). \tag{4.91}$$

Again, the proof is non-trivial; we refer the reader to [159].

4.4 The general case: $N \geq 2$ and $1 \leq k \leq N-1$

Throughout this section, we assume that $\Omega \subset \mathbb{R}^N$, $N \geq 2$, $1 \leq k \leq N-1$ and deal with maps $u \in W^{1,k}(\Omega; \mathbb{S}^k)$. This is a natural extension of the previous Sections 4.1 and 4.3, corresponding to $N = 3$, $k = 2$, resp. $N = 4$, $k = 2$.

Recall that we have defined $Ju \in \mathscr{D}'(\Omega; \Lambda^{k+1})$ in (4.7) and $\Sigma(u)$ in (4.10). We know (adapting the arguments in the introduction of Section 4.3) that

$$Ju = 0 \text{ in } \mathscr{D}', \ \forall\, u \in W^{1,k}(\Omega; \mathbb{S}^k) \cap C(\Omega), \text{ and } \forall\, u \in W^{1,k+1}(\Omega; \mathbb{S}^k), \tag{4.92}$$

and thus

$$\Sigma(u) = 0 \text{ for every } u \text{ as in (4.92)}. \tag{4.93}$$

We follow the same format as in Sections 4.1 and 4.3, emphasizing the similarities and mentioning an open direction (in Subsection 4.4.4).

4.4.1 A typical example

The objects of interest are maps u that are smooth except on a smooth connected oriented $(N-k-1)$-submanifold Γ without boundary (in Ω). If $u \in C(\Omega \setminus \Gamma; \mathbb{S}^k)$, we may define $\deg(u, \Gamma)$ to be the Brouwer degree of u restricted to a small sphere, centered at $x \in \Gamma$, in the $(k+1)$-dimensional space normal to Γ at x (with the orientation induced by that of Γ).

The counterpart of (4.32) and Theorem 4.5 is

Theorem 4.9. *1. Let Γ be a smooth connected oriented $(N-k-1)$-submanifold Γ without boundary (in Ω). Let $u \in W^{1,k}(\Omega; \mathbb{S}^k) \cap C(\Omega \setminus \Gamma)$. Then, with $m := \deg(u, \Gamma)$, we have*

$$*Ju = |\mathbb{B}^{k+1}|\, m\, \Gamma \text{ in } \mathscr{D}'(\Omega; \Lambda^{N-k-1}), \tag{4.94}$$

where Γ is identified, as usual, with an $(N-k-1)$-current.

2. Given any connected Γ of the form $\Gamma = \widetilde{\Gamma} \cap \Omega$, with $\widetilde{\Gamma}$ a smooth closed oriented $(N-2)$-manifold, and any $m \in \mathbb{Z}$, there exists $u \in W^{1,k}(\Omega; \mathbb{S}^k)$ such that (4.94) holds.

4.4.2 The structure of Ju for a general $u \in W^{1,k}(\Omega; \mathbb{S}^k)$

By analogy with Theorem 4.6, we have the following result, also due to Alberti, Baldo, and Orlandi [6].

Theorem 4.10. *For any $u \in W^{1,k}(\Omega; \mathbb{S}^k)$, we have*

$$\frac{1}{|\mathbb{B}^{k+1}|} * Ju \in \mathscr{F}^{N-k-1}. \tag{4.95}$$

Conversely, given any $T \in \mathscr{F}^{N-k-1}$, there exists some $u \in W^{1,k}(\Omega; \mathbb{S}^k)$ such that

$$*Ju = |\mathbb{B}^{k+1}|\, T. \tag{4.96}$$

To any $u \in W^{1,k}(\Omega; \mathbb{S}^k)$ we may thus associate (as in Section 4.2) three numbers:

$$A^\ell\left(\frac{1}{|\mathbb{B}^{k+1}|} * Ju\right) \geq B^\ell\left(\frac{1}{|\mathbb{B}^{k+1}|} * Ju\right) = C^\ell\left(\frac{1}{|\mathbb{B}^{k+1}|} * Ju\right), \tag{4.97}$$

where

$$\ell := N-k-1. \tag{4.98}$$

Comparing (4.13) and (4.65) yields

$$\Sigma(u) = (k+1)\,|\mathbb{B}^{k+1}|\,C^\ell\left(\frac{1}{|\mathbb{B}^{k+1}|} * Ju\right) = |\mathbb{S}^k|\,C^\ell\left(\frac{1}{|\mathbb{B}^{k+1}|} * Ju\right). \tag{4.99}$$

Summarizing, we have have only two distinct quantities attached to u (or in fact to Ju):

a) $\Sigma(u)$,

and

b) $$\Sigma^*(u) := |\mathbb{S}^k|\,A^\ell\left(\frac{1}{|\mathbb{B}^{k+1}|} * Ju\right) \geq \Sigma(u). \tag{4.100}$$

We know, from Proposition 4.2, that

$$\Sigma^*(u) = \Sigma(u) \text{ when } k = N-1\ (N \geq 2),\ \text{ or } k = 1\ (N \geq 2); \tag{4.101}$$

this corresponds, via (4.98), to $\ell = 0$, resp. $\ell = N-2$.

In the intermediate cases $2 \leq k \leq N-2$ $(N \geq 4)$, strict inequality in (4.100) may occur (see Remark 4.2).

When $u \in W^{1,k}(\Omega; \mathbb{S}^k) \cap C(\Omega \setminus \Gamma)$, where Γ is a smooth connected oriented $(N-k-1)$-submanifold Γ without boundary (in Ω), and $\deg(u, \Gamma) = 1$, then (4.94) holds and we have (from the definition (4.100))

$$\Sigma^*(u) = |\mathbb{S}^k| \times \text{the least area/mass spanned by } \Gamma, \tag{4.102}$$

in the sense of GMT. Similarly, for a finite collection $\Gamma = (\Gamma_1, \ldots, \Gamma_m)$ of disjoint such submanifolds Γ_j.

In particular, if $N \geq 2$ and $k = N-1$, then Γ is a finite collection of points and we have, by (4.101) and (4.102),

$$\Sigma(u) = \Sigma^*(u) = |\mathbb{S}^k| \times \text{the length of the minimal connection spanned by } \Gamma, \tag{4.103}$$

recovering (4.34).

While, if $N \geq 3$ and $k = 1$, Γ is connected, of the form $\Gamma = \widetilde{\Gamma} \cap \Omega$, with $\widetilde{\Gamma} \subset \mathbb{R}^N$ a smooth closed oriented $(N-2)$-dimensional manifold, then we have, by (4.101) and (4.102),

$$\Sigma(u) = \Sigma^*(u) = 2\pi \times \text{the least area/mass spanned by } \Gamma, \tag{4.104}$$

a result that coincides with Theorem 3.4 $(N = 3)$ and (3.110) $(N \geq 4)$.

4.4.3 The least energy $S_k(u)$ as a least area/mass

The main result provides the value of $S_k(u)$ (defined in (4.14)) and is a far-reaching extension of Theorems 4.3 and 4.7.

Theorem 4.11. *For any $u \in W^{1,k}(\Omega; \mathbb{S}^k)$, any $N \geq 2$, and any $1 \leq k \leq N-1$, we have*

$$S_k(u) = k^{k/2}\, \Sigma^*(u). \tag{4.105}$$

Sketch of proof of Theorem 4.11. The proof is non-trivial and follows the same strategy as that sketched for the proof of Theorem 4.7.
Step 1. Proof of "$\geq$" in (4.105)

Let $v \in W^{1,k}(\Omega; \mathbb{S}^k)$ satisfy $Jv = Ju$. The starting point is the following generalization of (4.44)

$$|D(v)| \leq \frac{1}{k^{k/2}} |\nabla v|^k, \ \forall N \geq 2, \ \forall\, 1 \leq k \leq N-1, \ \forall\, v \in W^{1,k}(\Omega; \mathbb{S}^k) \tag{4.106}$$

(see Lemma 4.3 below); here, $D(v)$ is the k-form defined by (4.6).

When $N = k+1$, inequality (4.106) leads, exactly as in the proof of Theorem 4.3, to

$$S_k(u) \geq k^{k/2}\, \Sigma(u).$$

We then use the fact that $\Sigma(u) = \Sigma^*(u)$.

When $N > k+1$, we have to rely on the coarea formula (see Almgren, Browder, and Lieb [8] for the special case where $(1/|\mathbb{B}^{k+1}|) * Jv$ is a manifold, and Alberti, Baldo, and Orlandi [6] for the general case). The key observation is that $|D(v)(x)|$ coincides pointwise with $|J_k v(x)|$, which is, by definition, the square root of the sum of the determinants of the $k \times k$ minors of the differential $dv(x)$ of v at $x \in \Omega$, viewed as a linear map from $\mathbb{R}^N$ to $T_{v(x)}\mathbb{S}^k$. Therefore, assuming that v has been replaced by its precise representative, the coarea formula yields (see [6, Appendix])

$$\int_\Omega |D(v)| = \int_\Omega |J_k v| = \int_{\mathbb{S}^k} \mathscr{H}^{N-k}([v = \alpha])\, ds_\alpha. \tag{4.107}$$

On the other hand, we rely on the following generalization of Lemma 2.6 (see [6, Theorem 3.8], with roots in [8, Appendices A.5-A.8] and Federer [147, Section 4.3]).

Lemma 4.2. *Assume that $v \in W^{1,k}(\Omega; \mathbb{S}^k)$ and that v equals its precise representative. Then, for a.e. $\alpha \in \mathbb{S}^k$, $M_\alpha := [v = \alpha]$, oriented by $D(v)$, is an $(N-k)$-rectifiable current such that $\partial M_\alpha = (1/|\mathbb{B}^{k+1}|) * Jv$.*

By Lemma 4.2 and (4.102), for a.e. $\alpha \in \mathbb{S}^k$ we have

$$\mathscr{H}^{N-k}([v = \alpha]) \geq \frac{1}{|\mathbb{S}^k|}\,\Sigma^*(v) = \frac{1}{|\mathbb{S}^k|}\,\Sigma^*(v). \tag{4.108}$$

Combining (4.106)–(4.108), we obtain "$\geq$" in (4.105).

Step 2. Proof of "$\leq$" in (4.105)
This is done via a dipole construction in the spirit of Brezis, Coron, and Lieb [79], as described in Section 3.4. For more details, we refer the reader to [6] and Molnar [271]. □

Lemma 4.3. *Let $\Omega \subset \mathbb{R}^N$, with $N \geq k+1$ and $v \in W^{1,k}(\Omega; \mathbb{S}^k)$. Then,*

$$|D(v)| \leq \frac{1}{k^{k/2}}\,|\nabla v|^k \text{ a.e. in } \Omega. \tag{4.109}$$

Proof. Consider some $x_0 \in \Omega$. By invariance with respect to isometries, we may assume that $v(x_0) = (0, 0, \dots, 0, 1)$, and then we have to prove that

$$|D(v)(x_0)| = |dv_1(x_0) \wedge \dots \wedge dv_k(x_0)| \leq \frac{1}{k^{k/2}}|\nabla v(x_0)|^k. \tag{4.110}$$

We actually claim that, when $\xi_1, \dots, \xi_k$ are 1-forms in $\mathbb{R}^N$, with $N \geq k$, we have the following Hadamard-type inequality (well-known in linear algebra):

$$|\xi_1 \wedge \dots \wedge \xi_k| \leq \prod_{j=1}^{k} |\xi_j|. \tag{4.111}$$

Assuming (4.111), we continue as follows:

$$|D(v)| \leq \prod_{j=1}^{k} |\nabla v_j| = \left(\prod_{j=1}^{k} |\nabla v_j|^2\right)^{1/2} \leq \left(\frac{\sum_{j=1}^{k} |\nabla v_j|^2}{k}\right)^{k/2} \leq \frac{1}{k^{k/2}}|\nabla v|^k.$$

In order to complete the proof of the lemma, let us establish (4.111). When $N = k$, identifying forms with vectors, this is the classical Hadamard inequality

$$|\det(\xi_1, \dots, \xi_k)| \leq \prod_{j=1}^{k} |\xi_j|. \tag{4.112}$$

When $N = k+1$, identifying again forms with vectors, $\xi_1 \wedge \dots \wedge \xi_k$ is the "vector product" defined by the property

$$\langle \xi_1 \wedge \dots \wedge \xi_k, \eta\rangle = \det(\xi_1, \dots, \xi_k, \eta), \ \forall\, \eta \in \mathbb{R}^{k+1}. \tag{4.113}$$

Applying Hadamard's inequality (4.112) to the right-hand side of (4.113) and taking sup over η such that $|\eta| \leq 1$, we obtain (4.111).

In the general case, we adapt the proof of Hadamard's inequality. By the Gram–Schmidt process, we may find an orthogonal family $(\eta_j)_{1 \leq j \leq k}$ such that:

$$\eta_1 = \xi_1,$$
$$|\eta_j| \leq |\xi_j|, \ \forall j \geq 2,$$
$$\xi_1 \wedge \ldots \wedge \xi_k = \eta_1 \wedge \ldots \eta_k,$$

and thus it suffices to prove (4.111) for the η's. We may therefore assume that we have an orthogonal family.

On the other hand, (4.111) is invariant by isometries; this can be seen, for example, from the Cauchy–Binet formula

$$|\xi_1 \wedge \ldots \wedge \xi_k|^2 = \det \big(\langle \xi_j, \xi_\ell \rangle \big)_{1 \leq j, \ell \leq k}$$

(see, for example, Evans and Gariepy [146, Theorem 4, p. 89]). Therefore, we may assume that $\xi_1 = (|\xi_1|, 0, \ldots, 0)$, $\xi_2 = (0, |\xi_2|, 0 \ldots, 0)$, etc., and then we see that, for an orthogonal family, (4.111) becomes an equality. □

4.4.4 The relaxed energy

We define, for every $k \geq 1$, $N \geq k+1$, and $u \in W^{1,k}(\Omega; \mathbb{S}^k)$,

$$R_k(u) := \inf \left\{ \liminf_{n \to \infty} \int_\Omega |\nabla u_n|^k; \ (u_n) \subset C^\infty(\overline{\Omega}; \mathbb{S}^k), \ u_n \to u \text{ a.e.} \right\} \leq \infty. \tag{4.114}$$

Its exact value is known only for $k = 1$ (Theorem 1.11) and $k = 2$ (Theorem 4.4 for $N = 3$ and Theorem 4.8 for $N = 4$, which still holds when $N \geq 5$; see Giaquinta and Mucci [159]). Summarizing, we have the following

Theorem 4.12. *Let $k = 1$ or $k = 2$, and let $N \geq k+1$. Then we have*

$$R_k(u) = \int_\Omega |\nabla u|^k + S_k(u), \ \forall u \in W^{1,k}(\Omega; \mathbb{S}^k). \tag{4.115}$$

The cases where $k \geq 3$ are open (see Open Problem 6).

4.5 Complements and open problems

4.5.1 Density/non-density of $C^\infty(\overline{\Omega}; \mathbb{S}^k)$ in $W^{1,p}(\Omega; \mathbb{S}^k)$

The question whether $C^\infty(\overline{\Omega}; \mathbb{S}^k)$ is dense (resp. weakly sequentially dense) in $W^{1,p}(\Omega; \mathbb{S}^k)$, $\Omega \subset \mathbb{R}^N$, $N \geq 1$, $k \geq 1$, and $p \geq 1$, is a very natural one. The answer turns out to be more subtle than expected, already when $k = 2$. This topic of research was initiated by Schoen and Uhlenbeck [318] in 1982, who proved that $C^\infty(\overline{\mathbb{B}^3}; \mathbb{S}^2)$ is not dense in $H^1(\mathbb{B}^3; \mathbb{S}^2)$, and Bethuel and Zheng [37], who obtained several results concerning the density of $C^\infty(\overline{\mathbb{B}^3}; \mathbb{S}^2)$ in $W^{1,p}(\mathbb{B}^3; \mathbb{S}^2)$ (see below). Even in the low dimensional case $N = 4$ and $k = 2$, the weak sequential density question was completely answered only very recently by Bethuel [29]!

The classes of interest are

$$\overline{C^\infty}^{W^{1,p}} := \overline{C^\infty(\overline{\Omega}; \mathbb{S}^k)}^{W^{1,p}},$$

$$\overline{C^\infty}^{w-W^{1,p}} := \{u \in W^{1,p}(\Omega; \mathbb{S}^k);\ u \text{ is the } L^p \text{ limit of a sequence } (u_n) \subset C^\infty(\overline{\Omega}; \mathbb{S}^k) \text{ such that } |u_n|_{W^{1,p}} \text{ is bounded}\}$$

(alternatively, $u_n \rightharpoonup u$ weakly in $W^{1,p}$ when $p > 1$).

The class

$$[J = 0] := \{u \in W^{1,p}(\Omega; \mathbb{S}^k);\ Ju = 0\},$$

defined for $p \geq k$, sometimes plays a role.

To simplify the presentation, we restrict ourselves to the case $k = 1$ (any $N \geq 2$, any $p \geq 1$) and $k = 2$ ($N \geq 4$, any $p \geq 1$). Further results are discussed in Sections 10.6.2 and 10.7.

4.5.1.1 The case $k = 1, N \geq 2, p \geq 1$

In this case the situation is quite simple and can easily be treated using lifting (see also Chapter 10, dealing with fractional Sobolev spaces).

The main result concerning $k = 1$ is the following

Proposition 4.5.

(i) When $p = 1$, we have

$$[J = 0] = \overline{C^\infty}^{W^{1,1}} \subsetneq \overline{C^\infty}^{w-W^{1,1}} = W^{1,1}(\Omega; \mathbb{S}^1). \tag{4.116}$$

(ii) When $1 < p < 2$, we have

$$[J = 0] = \overline{C^\infty}^{W^{1,p}} = \overline{C^\infty}^{w-W^{1,p}} \subsetneq W^{1,p}(\Omega; \mathbb{S}^1). \tag{4.117}$$

(iii) When $p \geq 2$, we have

$$\overline{C^\infty}^{W^{1,p}} = W^{1,p}(\Omega; \mathbb{S}^1). \tag{4.118}$$

4.5.1.2 The case $k = 2, N = 2, p \geq 1$

The main result is the following

Proposition 4.6. *When $k = 2$, $N = 2$, $p \geq 1$ we have*

$$\overline{C^\infty}^{W^{1,p}} = W^{1,p}(\Omega; \mathbb{S}^2). \tag{4.119}$$

Assertion (4.119) is due to Bethuel and Zheng [37] when $1 \leq p < 2$, and follows from the embedding $W^{1,2} \hookrightarrow VMO$ when $p \geq 2$.

4.5.1.3 The case $k = 2, N = 3, p \geq 1$

This case is already more complicated. The main result is the following

Theorem 4.13. *Assume $k = 2$ and $N = 3$.*

(i) When $1 \leq p < 2$, we have

$$\overline{C^\infty}^{W^{1,p}} = W^{1,p}(\Omega; \mathbb{S}^2). \tag{4.120}$$

(ii) When $p = 2$, we have

$$[J = 0] = \overline{C^\infty}^{W^{1,2}} \subsetneq \overline{C^\infty}^{w-W^{1,2}} = W^{1,2}(\Omega; \mathbb{S}^2). \tag{4.121}$$

(iii) When $2 < p < 3$, we have

$$[J = 0] = \overline{C^\infty}^{W^{1,p}} = \overline{C^\infty}^{w-W^{1,p}} \subsetneq W^{1,p}(\Omega; \mathbb{S}^2). \tag{4.122}$$

(iv) When $p \geq 3$, we have

$$\overline{C^\infty}^{W^{1,p}} = W^{1,p}(\Omega; \mathbb{S}^2). \tag{4.123}$$

Assertion (4.120) comes again from [37]. The two assertions in (4.121) are due to Bethuel [25]; note that the second equality in (4.121) could also be derived from the fact that $R_2(u) < \infty, \forall\, u \in W^{1,2}(\Omega; \mathbb{S}^2)$ (see Theorem 4.4). The first equality in (4.122) is due to Bethuel, Coron, Demengel, and Hélein [33] (see also Ponce and Van Schaftingen [301]), and the second equality in (4.122) is a special case of Bethuel [26, Theorem 3]. Assertion (4.123) comes again from the embedding $W^{1,3} \hookrightarrow VMO$.

4.5.1.4 The case $k = 2, N = 4, p \geq 1$

This case is already extremely delicate. The main result is the following

Theorem 4.14. *Assume $k = 2$ and $N = 4$.*

(i) When $1 \leq p < 2$, we have

$$\overline{C^\infty}^{W^{1,p}} = W^{1,p}(\Omega; \mathbb{S}^2). \tag{4.124}$$

(ii) When $p = 2$, we have

$$[J = 0] = \overline{C^\infty}^{W^{1,2}} \subsetneq \overline{C^\infty}^{w-W^{1,2}} = W^{1,2}(\Omega; \mathbb{S}^2). \tag{4.125}$$

(iii) When $2 < p < 3$, we have

$$[J = 0] = \overline{C^\infty}^{W^{1,p}} = \overline{C^\infty}^{w-W^{1,p}} \subsetneq W^{1,p}(\Omega; \mathbb{S}^2). \tag{4.126}$$

(iv) When $p = 3$, we have

$$\overline{C^\infty}^{W^{1,3}} \subsetneq \overline{C^\infty}^{w-W^{1,3}} \subsetneq W^{1,3}(\Omega; \mathbb{S}^2) = [J = 0]. \tag{4.127}$$

(v) When $3 < p < 4$, we have

$$\overline{C^\infty}^{W^{1,p}} = \overline{C^\infty}^{w-W^{1,p}} \subsetneq W^{1,p}(\Omega; \mathbb{S}^2) = [J = 0]. \tag{4.128}$$

(vi) When $p \geq 4$, we have

$$\overline{C^\infty}^{W^{1,p}} = W^{1,p}(\Omega; \mathbb{S}^2). \tag{4.129}$$

Sketch of proof. Assertion (4.124) comes again from [37].

The first equality in (4.125) is from [33] (see also [301]) and the second one is from [26, Theorem 5].

The first equality in (4.126) is from [33] (see also [301]) and the second one is from [26, Theorem 3]. For an example of a map $u \in W^{1,p}(\Omega; \mathbb{S}^2)$, $\forall p < 3$, such that $Ju \neq 0$, take, for example, $0 \in \Omega$ and $u(x) := x'/|x'|$, where $x = (x', x_4)$.

Concerning (4.127), we first need to find a map $u \in \overline{C^\infty}^{w-W^{1,3}}$ such that $u \notin \overline{C^\infty}^{W^{1,3}}$. A typical example is a map of the form $u(x) = H(x/|x|)$ (assuming $0 \in \Omega$)), where $H \in C^\infty(\mathbb{S}^3; \mathbb{S}^2)$ has a non-trivial Hopf degree. Clearly, such u belongs to $W^{1,p}(\Omega; \mathbb{S}^2)$, $\forall p < 4$. Moreover, $u \notin \overline{C^\infty}^{W^{1,3}}$; the proof is by contradiction and relies on the continuity of the Hopf degree under convergence in $W^{1,3}(\mathbb{S}^3)$. Next we claim that such u belongs to $\overline{C^\infty}^{w-W^{1,3}}$. This is established using a "bubbling" technique; see, for example, Bethuel [29, Proposition 1]. The final step in the proof of (4.127) consists of producing some $u \in W^{1,3}(\Omega; \mathbb{S}^2)$ with $u \notin \overline{C^\infty}^{w-W^{1,3}}$. Such a construction has been done by Bethuel in his remarkable recent paper [29].

Equality in (4.128) comes again from [26, Theorem 3]. For strict inclusion in (4.128), it suffices to take $u(x) = H(x/|x|)$ as above, which belongs to $W^{1,q}(\Omega; \mathbb{S}^2)$, $\forall\, q < 4$, but not to $\overline{C^\infty}^{W^{1,p}}$.

Finally, (4.129) is a consequence of the embedding $W^{1,4} \hookrightarrow VMO$. □

4.5.2 More on the least energy with prescribed Jacobian

This is the $\mathbb{S}^k$ counterpart of Section 1.8. We return here to the question of the least energy in the sense of (4.14), but involving $\int_\Omega |\nabla v|^p$, with a general p, instead of $\int_\Omega |\nabla v|^k$.

Throughout this subsection we assume that $0 \in \Omega \subset \mathbb{R}^N, N \geq 3, 2 \leq k \leq N-1$ and

$$k < p < k+1. \tag{4.130}$$

We fix some $u \in W^{1,p}(\Omega; \mathbb{S}^k)$ such that $Ju \neq 0$, for example $u(x) := x'/|x'|$, where $x = (x', x'')$, $x' \in \mathbb{R}^{k+1}$. The quantity of interest is

$$S_p(u) := \inf \left\{ \int_\Omega |\nabla v|^p;\ v \in W^{1,p}(\Omega; \mathbb{S}^k),\ Jv = Ju \in \mathscr{D}'(\Omega; \Lambda^{k+1}) \right\}. \tag{4.131}$$

The restriction (4.130) is natural, since we have already studied the case $p = k$ in the Subsections 4.1.2, 4.3.3, and 4.4.3; on the other hand, we know (see (4.92)) that $Jv = 0, \forall\, v \in W^{1,k+1}(\Omega; \mathbb{S}^k)$.

The main result is the following

Theorem 4.15. *Assume* (4.130). *Then the* inf *in* (4.131) *is achieved, i.e., there exists some* $v \in W^{1,p}(\Omega; \mathbb{S}^k)$ *such that* $Jv = Ju$ *and* $\int_\Omega |\nabla v|^p = S_p(u)$.

Remark 4.6. When $p = k$, it is expected that in general the inf in (4.131) is *not* achieved. The difficulty arises from the fact that J is *not* weakly sequentially continuous on $W^{1,k}(\Omega; \mathbb{S}^k)$ (for the same reason as in Remark 8.3).

Theorem 4.15 is an immediate consequence of the following basic

Lemma 4.4. *Assume* (4.130). *If* $(u_n) \subset W^{1,p}(\Omega; \mathbb{S}^k)$ *and* $u_n \rightharpoonup u$ *in* $W^{1,p}(\Omega; \mathbb{R}^{k+1})$, *then*

$$Ju_n \to Ju \text{ in } \mathscr{D}'(\Omega; \Lambda^{k+1}). \tag{4.132}$$

Sketch of proof. When $N = k+1$, this follows from the estimate (see Brezis and Nguyen [96])

$$\begin{aligned}&|\langle Ju - Jv, \varphi\rangle| \lesssim \|u - v\|_{L^q}(\|\nabla u\|_{L^p}^{N-1} + \|\nabla v\|_{L^p}^{N-1})\|\nabla\varphi\|_{L^\infty},\\ &\qquad \forall\, u, v \in W^{1,p}(\Omega; \mathbb{R}^{k+1}),\ \forall\, \varphi \in C_c^\infty(\Omega);\end{aligned} \tag{4.133}$$

here, q is defined by $\dfrac{N-1}{p} + \dfrac{1}{q} = 1$.

The general case $N \geq k+1$ is obtained by slicing (as in the proof of Proposition 4.4), via (4.133) and Hölder's inequality. We explain this when $N = 4$ and $k = 2$; the general case is similar. With the notation in the proof of Proposition 4.4, we have the identity

$$\langle Ju_n - Ju, \zeta\rangle = \sum_{j=1}^{4} \int_{(-1,1)} \langle Ju_{j,x_j} - Ju_{nj,x_j}, (\zeta^j)_{j,x_j}\rangle\, dx_j. \tag{4.134}$$

Applying, for fixed j and x_j, the estimate (4.133) to $Ju_{j,x_j} - Ju_{nj,x_j}$, we find that

$$\begin{aligned}|\langle Ju_n - Ju, \zeta\rangle| \lesssim \sum_{j=1}^{4} \int_{(-1,1)} &\|u_{j,x_j} - u_{nj,x_j}\|_{L^q}\, \|\nabla\varphi\|_{L^\infty}\\ &\times (\|\nabla u_{j,x_j}\|_{L^p}^3 + \|\nabla u_{nj,x_j}\|_{L^p}^3)\, dx_j.\end{aligned} \tag{4.135}$$

Applying Hölder's inequality with exponents q and $p/3$ in (4.135), we obtain the higher dimensional analogue of (4.133)

$$\begin{aligned}&|\langle Ju - Jv, \varphi\rangle| \lesssim \|u - v\|_{L^q}(\|\nabla u\|_{L^p}^{3} + \|\nabla v\|_{L^p}^{3})\|\nabla\varphi\|_{L^\infty},\\ &\qquad \forall\, u, v \in W^{1,p}(\Omega; \mathbb{R}^{3}),\ \forall\, \varphi \in C_c^\infty(\Omega; \Lambda^3).\end{aligned} \tag{4.136}$$

With the same proof, we obtain, for every k and every $N \geq k+1$,

$$\begin{aligned}&|\langle Ju - Jv, \varphi\rangle| \lesssim \|u - v\|_{L^q}(\|\nabla u\|_{L^p}^{k+1} + \|\nabla v\|_{L^p}^{k+1})\|\nabla\varphi\|_{L^\infty},\\ &\qquad \forall\, u, v \in W^{1,p}(\Omega; \mathbb{R}^{k+1}),\ \forall\, \varphi \in C_c^\infty(\Omega; \Lambda^{k+1}).\end{aligned} \tag{4.137}$$

This completes the proof of the lemma. □

4.5.3 *Open problems*

Open Problem 5. Assume that $k \geq 2$. Is there an "analytical" proof of Theorem 4.3, i.e., a proof that does not rely on the interpretation of the quantity $\Sigma(u)$ as the length of a minimal connection?

Open Problem 6. Let $k \geq 3$ and $N \geq k+1$. Let $\Omega \subset \mathbb{R}^N$ be smooth bounded diffeomorphic to a ball. Is it true that

$$R_k(u) = \int_\Omega |\nabla u|^k + S_k(u),\ \forall\, u \in W^{1,k}(\Omega; \mathbb{S}^k)?$$

It is plausible that a combination of the approaches in Pakzad [285] and Giaquinta and Mucci [159] could be effective in solving the above problem.

4.6 Comments

1. Formula (4.32), which asserts that

$$Ju = \frac{4\pi}{3} \sum_{j=1}^{k} \deg(u, a_j)\, \delta_{a_j} \text{ in } \mathscr{D}'(\Omega)$$

for every $u \in H^1(\Omega; \mathbb{S}^2) \cap C(\Omega \setminus \{a_1, \ldots, a_k\})$, is originally due to Brezis, Coron, and Lieb [79].
The fact that the distributional Jacobian of sphere-valued maps conveys information about the location and the "strength" of topological singularities (carried by points, lines, etc.) was used and amplified in numerous subsequent works.
2. Theorem 4.1 is due to Brezis and Mironescu and was announced in [73, Theorem 7']. It was inspired by an earlier result of Bourgain, Brezis, and Mironescu [51, Theorem 1], [54, Theorem 1] concerning $\mathbb{S}^1$-valued maps (similar to Theorem 2.10).
3. Theorem 4.3 is strongly anchored in the work of Brezis, Coron, and Lieb [79]. Their goal was to answer a question originally raised by Ericksen [143] in the context of liquid crystals. A nematic liquid crystal can be modeled by a vector field on a domain $\Omega \subset \mathbb{R}^3$ describing the *direction* $u(x)$ of the rod-like molecules at $x \in \Omega$, so that one can view $u(x)$ as a point in $\mathbb{S}^2$. Except for defects, which are points (or lines) in Ω, $u(x)$ varies continuously. A simple model energy can be taken to be (see, for example, Ericksen and Kinderlehrer [144])

$$\mathscr{E}(u) = \int_\Omega |\nabla u|^2.$$

A very natural quantity is the *least energy* required to generate a configuration having singularities of given degrees d_j located at given points $a_j, j = 1, \ldots, k$, starting from a configuration at rest (all molecules oriented in the same direction). Mathematically, this corresponds to the quantity

$$S_2^\sharp(a, d) := \inf \left\{ \int_\Omega |\nabla v|^2;\ v \in H^1(\Omega; \mathbb{S}^2) \cap C(\Omega \setminus \{a_1, \ldots, a_k\}), \right.$$
$$\left. \deg(v, a_j) = d_j,\ \forall j \right\}. \tag{4.138}$$

Brezis, Coron, and Lieb [79] discovered a simple, *explicit* formula for $S_2^\sharp(a, d)$ (which, surprisingly, had not been anticipated by physicists), namely

$$S_2^\sharp(a,d) = 8\pi\, L(a,d). \tag{4.139}$$

In the process of establishing (4.139), they designed two techniques, which have become popular tools in the analysis of sphere-valued maps with singularities:

(i) The *dipole construction*, which exhibits an *explicit* map $v \in H^1(\Omega;\mathbb{S}^2)\cap C(\Omega\setminus\{a_1,\ldots,a_k\})$ such that $\deg(v,a_j)=d_j$, $\forall j$, and $\int_\Omega |\nabla v|^2 \le 8\pi\, L(a,d)+\varepsilon$ (with $\varepsilon>0$ arbitrary), and yields the upper bound
$$S_2^\sharp(a,d) \le 8\pi\, L(a,d). \tag{4.140}$$

(ii) The *D-field plus duality approach*, which can be summarized as follows. Starting with (4.44), i.e.,
$$\int_\Omega |\nabla v|^2 \ge 2\int_\Omega |D(v)| \ge 2\int_\Omega [D(v)]\cdot\nabla\zeta,\ \forall\,\zeta\in C_c^\infty(\Omega;\mathbb{R}) \quad \text{with } \|\nabla\zeta\|_{L^\infty}\le 1,$$
and recalling that (by (4.1) and (4.32))
$$\operatorname{div} D(v) = 3\,Jv = 4\pi\sum_{j=1}^k d_j\,\delta_{a_j},\ \forall\, v\in H^1(\Omega;\mathbb{S}^2)\cap C(\Omega\setminus\{a_1,\ldots,a_k\}) \quad \text{with } \deg(v,a_j)=d_j,\ \forall j,$$
we are led to
$$\int_\Omega |\nabla v|^2 \ge 4\pi\sum_{j=1}^k d_j\,\zeta(a_j),\ \forall\,\zeta\in C_c^\infty(\Omega;\mathbb{R}) \text{ with } \|\nabla\zeta\|_{L^\infty}\le 1,$$
which yields (via OT) the lower bound
$$S_2^\sharp(a,d) \ge 8\pi\, L(a,d). \tag{4.141}$$

The passage from (4.138) and (4.139) to (4.40) and Theorem 4.3 involves two modifications:

(i) Enlarging the class of competitors, i.e., replacing the constraint
$$v\in H^1(\Omega;\mathbb{S}^2)\cap C(\Omega\setminus\{a_1,\ldots,a_k\}),\ \deg(v,a_j)=d_j,\ \forall j,$$
by the *weaker constraint* $Jv=(4\pi/3)\sum_{j=1}^k d_j\,\delta_{a_j}$, so that
$$S_2(u)\le S_2^\sharp(a,d),\ \forall\, u\in H^1(\Omega;\mathbb{S}^2)\cap C(\Omega\setminus\{a_1,\ldots,a_k\}) \quad \text{with } \deg(u,a_j)=d_j,\ \forall j. \tag{4.142}$$

(ii) Allowing in Theorem 4.3 possibly *infinitely many singularities*, which was not the case in [79].

Concerning (i), we know that the least energy remains the *same*, even though we have enlarged the class of testing functions.
Concerning (ii), we call the attention of the reader to a big advantage of the formulation (4.40), involving a *prescribed Jacobian*, as opposed to (4.138), which involves *prescribing finitely many singularities*. Note that the compact formula $Jv = Ju$ encapsulates the option of *prescribing infinitely many singularities*.
The proof of Theorem 4.3 we have presented follows closely the strategy of [79]. However, the possible presence of infinitely many singularities requires the help of two additional ingredients (which enter through the use of Theorems 4.1 and 4.2): the density of nice maps in $H^1(\Omega; \mathbb{S}^2)$, due to Bethuel and Zheng [37], and a refined adaptation of the dipole construction in the presence of infinitely many singularities, carefully explained and performed in Alberti, Baldo, and Orlandi [6].

4. Theorem 4.4 is due to Bethuel, Brezis, and Coron [30].
5. Concerning the results in Section 4.2, see Comment 12 to Chapter 2 and Comment 6 to Chapter 3.
6. At the very end of Brezis, Coron, and Lieb [79], the authors formulated a bold conjecture, which says, roughly speaking, that

   "least energy with prescribed topological singularities
   $= c \times$ least area spanned by the singular set."

   The conjecture turned out to be true, and its precise formulation is the main content of Theorem 4.7 ($N = 4$, $k = 2$) and Theorem 4.11 (any $N \geq 2$ and $1 \leq k \leq N - 1$).
   When the singular set is nice (e.g., as in Section 4.3.1), the result is due to Almgren, Browder, and Lieb [8]. In their remarkable paper, they explained that the upper bound $S_k(u) \leq k^{k/2}\, \Sigma^*(u)$ can be obtained by adapting the dipole construction.
   On the other hand, they pointed out that the *D-field plus duality* approach used in [79] and described in Section 4.1.2 and in Comment 3 above, would lead to the lower bound $S_k(u) \geq k^{k/2}\, \Sigma(u)$, which is correct, but (as explained in Section 4.2) *not optimal* in the intermediate range $1 < k < N - 1$.
   Their main contribution was to show that, by using the *coarea formula* instead, one is led to the optimal lower bound $S_k(u) \geq k^{k/2}\, \Sigma^*(u)$.
   In full generality, Theorem 4.11 is due to Molnar [271], but her proof relies heavily:

   a) On the program introduced by Almgren, Browder, and Lieb [8]
   b) On the refinements of the dipole construction and of the coarea formula developed by Alberti, Baldo, and Orlandi [6]. In fact, [6, Theorem 3.8 and Theorem 5.6] yield the relation $S_k(u) \simeq \Sigma^*(u)$.

7. Concerning the relaxed energy $R_2(u)$, Pakzad [285] noted that, when $\Omega \subset \mathbb{R}^N$ with $N \geq 4$, the approach in Bethuel, Brezis, and Coron [30] does not provide the optimal lower bound $R_2(u) \geq \int_\Omega |\nabla u|^2 + S_2(u)$ in (4.115). He also obtained

this lower bound via Cartesian currents techniques.
Tarp-Ficenc [334] obtained the upper bound $R_2(u) \le \int_\Omega |\nabla u|^2 + S_2(u)$ in (4.115) under a technical assumption on u (u is required to be "not too discontinuous"). This extra assumption on u was removed by Giaquinta and Mucci [159], who obtained Theorem 4.12 in full generality. The proof in [159] relies heavily on the theory of Cartesian currents developed by Giaquinta, Modica, and Souček [156, 157].

Chapter 5
Lifting in fractional Sobolev spaces and in *VMO*

5.0 Introduction

This chapter is a follow-up to Chapter 1. We consider again maps $u : \Omega \to \mathbb{S}^1$, with $\Omega \subset \mathbb{R}^N$, $N \geq 1$, smooth bounded simply connected, and our goal is to find a lifting φ "as smooth as u permits." The difference with Chapter 1 is that u belongs either to a fractional Sobolev space $W^{s,p}$ (whose definition is recalled in Section 5.1), or to the space *VMO* (functions of vanishing mean oscillation, whose definition is recalled in Section 5.2.1 below).

The first part of this chapter concerns maps $u \in W^{s,p}(\Omega; \mathbb{S}^1)$, $0 < s < \infty$, $1 \leq p < \infty$. As we have seen in Chapter 1, there are three natural questions.

Question 1. Characterize the values of s and p such that $W^{s,p}(\Omega; \mathbb{S}^1)$ has the *lifting property*, i.e., *every* $u \in W^{s,p}(\Omega; \mathbb{S}^1)$ can be written as $u = e^{\imath\varphi}$ for some $\varphi \in W^{s,p}(\Omega; \mathbb{R})$.

Assuming that $W^{s,p}(\Omega; \mathbb{S}^1)$ does *not* have the lifting property, we may then ask

Question 2. *Characterize* the u's in $W^{s,p}(\Omega; \mathbb{S}^1)$ that can be written as $u = e^{\imath\varphi}$ for some $\varphi \in W^{s,p}(\Omega; \mathbb{R})$.

Question 3. Assume that $u \in W^{s,p}(\Omega; \mathbb{S}^1)$ admits no lifting in $W^{s,p}$. Which are the "best possible" liftings of u (in a sense to be made precise)?

When $s > 1$, there is an additional phenomenon that pops out, namely that $\varphi \in W^{s,p}(\Omega; \mathbb{R})$ alone does *not* imply $e^{\imath\varphi} \in W^{s,p}(\Omega; \mathbb{S}^1)$ (see Appendix 15.3). A natural substitute is $[\varphi \in W^{s,p} \cap L^\infty] \implies e^{\imath\varphi} \in W^{s,p}$ or, more generally (via Gagliardo–Nirenberg) $[\varphi \in W^{s,p} \cap W^{1,sp}] \implies e^{\imath\varphi} \in W^{s,p}$. Therefore, when $s > 1$, it seems natural to raise the question of existence of a lifting $\varphi \in W^{s,p} \cap W^{1,sp}$ (instead of $\varphi \in W^{s,p}$).

We completely answer Question 1. This is the content of Theorems 5.1, 5.2, and 5.3.

H. Brezis and P. Mironescu, *Sobolev Maps to the Circle*, Progress in Nonlinear Differential Equations and Their Applications 96,
https://doi.org/10.1007/978-1-0716-1512-6_5

Next we turn to Question 2. Recall that, in Chapter 1, we established (see Theorem 1.2) that, in the "nasty" range $1 \le p < 2$ and $N \ge 2$, $u \in W^{1,p}(\Omega; \mathbb{S}^1)$ admits a lifting $\varphi \in W^{1,p}$ if and only if $Ju = 0$. In the $W^{s,p}$ setting, Ju is well-defined when $u \in W^{s,p}$ with $s \ge 1$ and $1 \le p < \infty$ (since $W^{s,p} \subset W^{1,1}$). Here, the "nasty" range is $1 \le sp < 2$ (see Theorem 5.3), and our main result (Theorem 5.6) asserts that $u \in W^{s,p}(\Omega; \mathbb{S}^1)$ admits a lifting $\varphi \in W^{s,p}$ if and only if $Ju = 0$ (and in this case the lifting belongs to $\varphi \in W^{s,p} \cap W^{1,sp}$).

When $s < 1$, the "nasty" range is $1 \le sp < N$. With much effort, we are still able to give a meaning to Ju when $u \in W^{s,p}(\Omega; \mathbb{S}^1)$, $0 < s < 1$, $sp \ge 1$. This is done in Chapter 8, where we also establish the fact that, for $u \in W^{s,p}(\Omega; \mathbb{S}^1)$, with $0 < s < 1$ and $sp \ge 1$, the existence of a lifting $\varphi \in W^{s,p}$ implies $Ju = 0$. Unfortunately, the converse is *not* true (see Remark 8.11).

We present some partial results concerning Question 3 in Section 8.2.

In the last part of this chapter, we also prove that maps in $VMO\,(\Omega; \mathbb{S}^1)$ admit a lifting in $VMO\,(\Omega; \mathbb{R})$. Furthermore, we prove that, for every s and every p, maps in $VMO \cap W^{s,p}$ admit a lifting in $VMO \cap W^{s,p}$.

5.1 Lifting of $\mathbb{S}^1$-valued maps in fractional Sobolev spaces

We first address Question 1.

Recall the definition of fractional Sobolev spaces (see Appendices 15.1–15.4 for some of their main properties). Let $\Omega \subset \mathbb{R}^N$ be a smooth bounded domain. For $0 < \sigma < 1$ and $1 \le p < \infty$, let

$$W^{\sigma,p}(\Omega) := \left\{ f \in L^p(\Omega);\ \int_\Omega \int_\Omega \frac{|f(x) - f(y)|^p}{|x-y|^{N+\sigma p}}\, dxdy < \infty \right\},$$

equipped with the standard seminorm

$$|f|^p_{W^{\sigma,p}} = \int_\Omega \int_\Omega \frac{|f(x) - f(y)|^p}{|x-y|^{N+\sigma p}}\, dxdy.$$

In order to obtain a norm, it suffices to add $\|f\|_{L^p}$ or $\left| \int_\Omega f \right|$ to $|f|_{W^{\sigma,p}}$.

If $0 < s < \infty$ is a real number and s is not an integer, write $s = m + \sigma$ where $m := [s]$ is the integer part of s and $\sigma := s - m \in (0, 1)$. Then set

$$W^{s,p}(\Omega) := \left\{ f \in W^{m,p}(\Omega);\ \partial^\alpha f \in W^{\sigma,p}(\Omega),\ \forall \alpha,\ |\alpha| = m \right\}.$$

When $s \ge 1$, a convenient seminorm in $W^{s,p}$ is $|f|_{W^{s,p}} := \|\nabla f\|_{W^{s-1,p}}$.

As in Section 1.2, we set

$$W^{s,p}(\Omega; \mathbb{S}^1) := \left\{ u \in W^{s,p}(\Omega; \mathbb{C});\ |u| = 1 \text{ in } \Omega \right\}.$$

Remark 5.1. We will often use the special case $s = 1/p$, with $1 < p < \infty$. Note that $W^{1/p,p}(\Omega; \mathbb{S}^1)$ *increases* with p, $1 < p < \infty$. This follows from

$$|u|^p_{W^{1/p,p}} \leq 2^{p-q}|u|^q_{W^{1/q,q}} \text{ if } 1 < q < p < \infty \text{ and } |u| = 1.$$

We present below the answer to Question 1. It is more convenient to divide it into three statements corresponding to three different cases.

Theorem 5.1. *Assume that $N = 1$, $0 < s < \infty$ and $1 \leq p < \infty$. Then, every $u \in W^{s,p}(\Omega; \mathbb{S}^1)$ may be written as $u = e^{\imath\varphi}$ for some $\varphi \in W^{s,p}(\Omega; \mathbb{R})$.*

Theorem 5.2. *Assume that $N \geq 2$, $0 < s < 1$ and $1 \leq p < \infty$. The answer to the lifting question is:*

a) Positive when $sp \geq N$
b) Negative when $1 \leq sp < N$
c) Positive when $sp < 1$.

Here, "positive" means that every $u \in W^{s,p}(\Omega; \mathbb{S}^1)$ may be written as $u = e^{\imath\varphi}$ for some $\varphi \in W^{s,p}(\Omega; \mathbb{R})$; "negative" means that for some u's in $W^{s,p}(\Omega; \mathbb{S}^1)$ there is no $\varphi \in W^{s,p}(\Omega; \mathbb{R})$ such that $u = e^{\imath\varphi}$.

Theorem 5.3. *Assume that $N \geq 2$, $1 \leq s < \infty$ and $1 \leq p < \infty$. The answer to the lifting question is:*

a) Positive when $sp \geq 2$
b) Negative when $1 \leq sp < 2$.

Furthermore, in case a), every $u \in W^{s,p}(\Omega; \mathbb{S}^1)$ admits a lifting $\varphi \in W^{s,p} \cap W^{1,sp}$.

In case $p = 2$ the above theorems for $H^s = W^{s,2}$ can be summarized as follows.

Corollary 5.1. *When $N = 1$, the answer to the lifting problem in H^s is always positive.*
When $N \geq 2$ the answer to the lifting problem in H^s is:

a) Positive when $0 < s < 1/2$
b) Negative when $1/2 \leq s < 1$
c) Positive when $s \geq 1$.

Remark 5.2. So far, this concerns the question of existence. Turning to the question of *uniqueness*, one may ask whether the difference $\varphi_1 - \varphi_2$ of two solutions is of the form $2\pi k$, for some $k \in \mathbb{Z}$. Clearly, $\dfrac{1}{2\pi}(\varphi_1 - \varphi_2)$ takes its values in $\mathbb{Z}$, but this does *not* imply that it is constant (since $\varphi_1 - \varphi_2$ need not be continuous). We will prove in Section 6.3 that uniqueness holds if $sp \geq 1$ (but fails if $sp < 1$).

We present here partial proofs of the above results and we refer the reader to Bourgain, Brezis, and Mironescu [50] for complete proofs. Again, for the convenience of the reader, we consider several cases.

We start with a case that is quite elementary, namely the case where $sp > N$, so that u is continuous, by the Sobolev embedding theorem.

Theorem 5.4. *Assume that* $N \geq 1$, $0 < s < \infty$, $1 \leq p < \infty$ *and* $sp > N$ *(or* $p = 1$ *and* $s = N$*). Then, any* $u \in W^{s,p}(\Omega; \mathbb{S}^1)$ *may be lifted as* $u = e^{\imath\varphi}$ *with* $\varphi \in W^{s,p}(\Omega; \mathbb{R})$.

Proof. Let φ be the lifting given by Lemma 1.1. Locally, we have $\varphi = -\imath \ln u$. We conclude via the next lemma about composition. □

Lemma 5.1. *Assume that* $N \geq 1$, $0 < s < \infty$ *and* $1 \leq p < \infty$*. Assume that* $v \in W^{s,p}(\Omega) \cap L^\infty(\Omega)$*, and let* $\Phi \in C^\infty$*. Then* $\Phi \circ v \in W^{s,p}(\Omega)$.

The above lemma is a special case of Theorem 15.7 in Appendix 15.3. The proof is very simple when $0 < s < 1$ (using the definition of $W^{s,p}$ and the fact that Φ is Lipschitz on the range of v). The result is also well-known when s is an integer, and in this case its proof relies on the Gagliardo–Nirenberg inequality. When $s > 1$ is not an integer the argument is more delicate, and the result is due to Meyer [248] (for different arguments, see Escobedo [145], Brezis and Mironescu [83], and Maz'ya and Shaposhnikova [242]).

We now turn to the cases that require more work.

5.1.1 Case 1: N = 1

We may assume that $\Omega = I := (0, 1)$.

Subcase $N = 1$, $sp \geq 1$

Proof of Theorem 5.1 *when* $sp \geq 1$. In view of Theorem 5.4, it suffices to consider the case where $p > 1$ and $s = 1/p$. We first extend u by a map, still denoted by u, in $W^{s,p}((-1, 2); \mathbb{S}^1)$; this is done by reflections across 0 and 1. By standard trace theory, there exists some $\widetilde{u} \in W^{s+1/p,p}(I^2; \mathbb{R}^2)$ such that $\widetilde{u}(x, 0) = u(x)$.

Since u takes its values in $\mathbb{S}^1$, one may expect that, near $I \times \{0\}$, $\widetilde{u}$ takes its values "close" to $\mathbb{S}^1$. This is not true for a *general* extension $\widetilde{u}$. However, since $W^{1/p,p}(I) \hookrightarrow VMO$ (see, for example, Brezis and Nirenberg [104, Section 1.2], and also Lemma 15.20 in Appendix 15.5), *special* extensions of u do have that property. For example, if $\rho \in C_c^\infty(\mathbb{R})$ is a normalized bump function (i.e., $\operatorname{supp} \rho \subset (-1, 1)$, $\rho \geq 0$ and $\int \rho = 1$), and if we set $\rho_\varepsilon(x) := (1/\varepsilon)\, \rho(x/\varepsilon)$, $\forall x \in \mathbb{R}$, $\forall \varepsilon > 0$, then

$$\widetilde{u}(x, \varepsilon) := u * \rho_\varepsilon(x), \ \ x \in I, \varepsilon \in I,$$

has the following properties (see Corollary 15.6 in Appendix 15.5):

a) $\widetilde{u} \in W^{s+1/p,p}(I^2; \mathbb{R}^2)$
b) $|\widetilde{u}|_{W^{s+1/p,p}(I^2)} \leq C|u|_{W^{s,p}(I)}$
c) $\widetilde{u}(\cdot, \varepsilon) \to u(\cdot)$ in $W^{s,p}(I)$ as $\varepsilon \to 0$

d) $|\widetilde{u}(x,\varepsilon)| \to 1$, *uniformly* in $x \in I$ as $\varepsilon \to 0$.

In particular, there exists some $\varepsilon_0 > 0$ such that $|\widetilde{u}(x,\varepsilon)| \geq 1/2$, $\forall x \in I$, $\forall \varepsilon \in (0, \varepsilon_0]$. If we consider $v := \widetilde{u}/|\widetilde{u}|$ in $\omega := I \times (0, \varepsilon_0)$, we have an extension v of u such that $v \in W^{s+1/p,p}(\omega; \mathbb{S}^1)$; here, we have again applied Lemma 5.1. [Alternatively, we could have used, instead of $\widetilde{u} = u * \rho_\varepsilon$, any harmonic extension of u in $(-2, 1)^2$ (see Brezis and Nirenberg [105, Appendix 3]).]

Let us now explain how to complete the proof of the theorem when $p = 2$, i.e., for $u \in H^{1/2}(I; \mathbb{S}^1)$. From the above discussion, u has an extension $v \in H^1(\omega; \mathbb{S}^1)$. Applying Theorem 1.1, we may write $v = e^{\imath\psi}$ for some $\psi \in H^1(\omega; \mathbb{R})$, and then $\varphi := \psi_{|I}$ has the required properties.

Remark 5.3. We call the attention of the reader to the fact that the above construction of φ does *not* yield an estimate for $|\varphi|_{H^{1/2}(I)}$ in terms of $|u|_{H^{1/2}(I)}$. Indeed, we have $|\widetilde{u}|_{H^1(I^2)} \leq C|u|_{H^{1/2}(I)}$ and $|v|_{H^1(\omega)} \leq C|\widetilde{u}|_{H^1(I^2)} \leq C'|u|_{H^{1/2}(I)}$. On the other hand,

$$|\varphi|_{H^{1/2}(I)} = |\psi_{|I}|_{H^{1/2}(I)} \leq C(\varepsilon_0)|\psi|_{H^1(\omega)} \leq C'C(\varepsilon_0)|u|_{H^{1/2}(I)},$$

where the constant $C(\varepsilon_0)$ (which comes from trace theory) blows up as $\varepsilon_0 \to 0$. Since we have no estimate for the rate of convergence in d) as $\varepsilon \to 0$, this implies that we have no control on $|\varphi|_{H^{1/2}}$ in terms of $|u|_{H^{1/2}}$. This is not just an artefact of the proof. In Proposition 9.2, we will present a surprising phenomenon: φ exists in $H^{1/2}$, but it is impossible to estimate $|\varphi|_{H^{1/2}}$ in terms of $|u|_{H^{1/2}}$.

We now return to the general case $p \neq 2$. Here, we shall use the following lemma about products in fractional Sobolev spaces. Its proof is given in Appendix 15.12 when $\Omega = \mathbb{R}^N$. The case of a smooth domain Ω follows by extending the functions to $\mathbb{R}^N$.

Lemma 5.2. *Assume that $s \geq 1$ and $1 \leq p < \infty$. Let $f, g \in W^{s,p}(\Omega; \mathbb{R}) \cap L^\infty(\Omega; \mathbb{R})$, where Ω is a smooth bounded domain in $\mathbb{R}^N$. Then, $f\nabla g \in W^{s-1,p}(\Omega)$. In addition, we have*

$$\|f\nabla g\|_{W^{s-1,p}} \leq C(\|f\|_{L^\infty}|g|_{W^{s,p}} + \|g\|_{L^\infty}|f|_{W^{s,p}}). \tag{5.1}$$

Proof of Theorem 5.1 *completed when $s = 1/p$.* Let v be as above, and extend it by reflection to $Q := I \times (-\varepsilon_0 \times \varepsilon_0)$; the extension, still denoted by v, belongs to $W^{s+1/p,p}(Q; \mathbb{S}^1)$ (see Lemma 15.27 in Appendix 15.6).

If $s + 1/p < 1$, applying the same construction once more, we obtain some extension $w \in W^{s+2/p,p}(U; \mathbb{S}^1)$ of v, where U is a neighborhood of $Q \times \{0\}$ in $\mathbb{R}^3$. This construction is possible since $(s + 1/p)p = 2$, so that we are again in a limiting case for the Sobolev embedding (more specifically, we have $W^{s+1/p,p}(Q) \hookrightarrow VMO$; see Lemma 15.20 in Appendix 15.5), and thus $v \in VMO$. Iterating this construction, if $s + (j-1)/p < 1$, we may find some extension $\zeta \in W^{s+j/p,p}(G; \mathbb{S}^1)$ of u, where G is a neighborhood of $I \times \{0\} \times \ldots \times \{0\}$ in $\mathbb{R}^{j+1}$.

Consider the first integer $k \geq 1$ such that $s + \dfrac{k}{p} \geq 1$. From the Gagliardo–Nirenberg embedding $W^{s+k/p,p} \cap L^\infty \hookrightarrow W^{1,k+1}$ (see Remark 15.5 in Appendix

15.3), we have $\zeta \in W^{1,k+1}(G; \mathbb{S}^1)$. By Theorem 1.1, we may write $\zeta = e^{\imath\psi}$ for some $\psi \in W^{1,k+1}(G; \mathbb{R})$. Moreover, by (1.52) and (1.46), we have $\nabla\psi = -\imath\overline{\zeta}\nabla\zeta$. By Lemma 5.2, we have $\nabla\psi \in W^{s+k/p-1,p}(G)$. Hence, $\psi \in W^{s+k/p,p}(G)$. Taking traces, we find that the trace φ of ψ on $I \times \{0\} \times \ldots \times \{0\} \simeq I$ satisfies $\varphi \in W^{s,p}(I)$ and $u = e^{\imath\varphi}$. □

Subcase $N = 1$, $sp < 1$

Sketch of the proof of Theorem 5.1 *when $sp < 1$.* We present the principal ingredients and refer the reader to [50] for detailed arguments. The main tool is formula (5.9), which is a characterization of $W^{s,p}$ functions in terms of their components in a Haar (or wavelet) basis. (5.9) is originally due to Bourdaud [43] (see also an earlier work of DeVore and Popov [137]). An alternative proof may be found in [50].

Recall that $\Omega = I = (0, 1)$. For each integer $j \geq 0$, consider the partition $\mathscr{P}_j$ of $[0, 1)$ into 2^j dyadic intervals $[a_k, b_k)$, each of length 2^{-j}. Denote by $\mathscr{E}_j$ the space of functions from I to $\mathbb{C}$ which are constant on each interval of $\mathscr{P}_j$. Given a function $f \in L^p(I)$, consider the function $f_j = E_j(f) \in \mathscr{E}_j$ defined as follows. Each $x \in I$ belongs exactly to one interval of the partition $\mathscr{P}_j$, say $Q_j(x)$. Set

$$f_j(x) := E_j(f)(x) = \fint_{Q_j(x)} f = 2^j \int_{Q_j(x)} f. \tag{5.2}$$

Clearly, we have

$$\|E_j(f)\|_{L^p(I)} \leq \|f\|_{L^p(I)}, \forall j, \tag{5.3}$$

and

$$E_j(f) \to f \text{ in } L^p \text{ and a.e. as } j \to \infty. \tag{5.4}$$

The main ingredients of the proof of Theorem 5.1 when $sp < 1$ are Lemmas 5.3 and 5.4 below.

Lemma 5.3. *For any $0 < s < 1$ and any $1 \leq p < \infty$, we have*

$$\sum_{j\geq 0} 2^{spj}\|f - f_j\|^p_{L^p(I)} \leq C|f|^p_{W^{s,p}(I)}, \tag{5.5}$$

where C depends only on s and p.

Lemma 5.4. *Assume that $sp < 1$, and let $(g_j)_{j\geq 0}$ be a sequence of functions on I such that*

$$g_j \in \mathscr{E}_j, \ \forall j = 0, 1, \ldots \tag{5.6}$$

and

$$\sum_{j\geq 1} 2^{spj} \|g_j - g_{j-1}\|^p_{L^p(I)} < \infty. \tag{5.7}$$

Then, (g_j) converges in $L^p(I)$ to some $g \in W^{s,p}(I)$ with

$$|g|^p_{W^{s,p}} \leq C \sum_{j\geq 1} 2^{spj} \|g_j - g_{j-1}\|^p_{L^p(I)}, \tag{5.8}$$

where the constant C depends only on s, p and blows up as $sp \nearrow 1$.

Remark 5.4. The proof of Lemma 5.3 is easy, but the proof of Lemma 5.4 is quite technical. This makes the complete proof of Theorem 5.1 when $sp < 1$ rather involved. In Section 8.2.2, we will present a totally different proof of Theorems 5.1 and 5.2 when $sp < 1$. It relies on the heavy machinery of factorization. It would be interesting to find a simpler proof, even in the case $p = 2$, i.e., in $H^s = W^{s,2}$, when $s < 1/2$ (see Open Problem 7 in Section 5.4.2).

Combining Lemmas 5.3 and 5.4 (applied with $g_j := E_j(f)$) we obtain, for $sp < 1$, the seminorm equivalence

$$|f|^p_{W^{s,p}(I)} \sim \sum_{j\geq 1} 2^{spj} \|E_j(f) - E_{j-1}(f)\|^p_{L^p(I)}. \tag{5.9}$$

Proof of Theorem 5.1 continued. Let $u \in W^{s,p}(I; \mathbb{S}^1)$. For each integer $j \geq 0$, define u_j as in (5.2) and

$$U_j(x) := \begin{cases} \dfrac{u_j(x)}{|u_j(x)|}, & \text{if } u_j(x) \neq 0 \\ 1, & \text{if } u_j(x) = 0 \end{cases}.$$

Clearly, $U_j \to u$ a.e. For each $j \geq 0$, we construct a function $\varphi_j : I \to \mathbb{R}$ in $\mathscr{E}_j$, such that

$$e^{\imath\varphi_j} = U_j \text{ in } I, \tag{5.10}$$
$$|\varphi_j - \varphi_{j-1}| \leq C|U_j - U_{j-1}|, \ j = 1, 2, \ldots \tag{5.11}$$

Note that (5.11) can be achieved by induction on j, for example with $C = \pi/2$. On the other hand, observe that for every $\xi, \eta, \zeta \in \mathbb{C}$ with $|\zeta| = 1$, we have

$$\left| \frac{\xi}{|\xi|} - \frac{\eta}{|\eta|} \right| \leq 4(|\zeta - \xi| + |\zeta - \eta|), \tag{5.12}$$

with the convention that $\dfrac{0}{0} = 1$. [In order to prove (5.12), consider separately the case where $|\xi|, |\eta| \geq 1/2$, and the case where either $|\xi| < 1/2$ or $|\eta| < 1/2$.]

Applying (5.12) to $\xi := E_j(u)(x)$, $\eta := E_{j-1}(u)(x)$ and $\zeta := u(x)$, we obtain

$$|U_j - U_{j-1}| \leq 4(|u - E_j(u)| + |u - E_{j-1}(u)|) \text{ a.e. in } \Omega. \tag{5.13}$$

Combining this with (5.11) yields

$$|\varphi_j - \varphi_{j-1}| \leq C(|u - E_j(u)| + |u - E_{j-1}(u)|), \tag{5.14}$$

and thus

$$\sum_{j\geq 1} 2^{spj} \|\varphi_j - \varphi_{j-1}\|^p_{L^p} \leq C \sum_{j\geq 0} 2^{spj} \|u - E_j(u)\|^p_{L^p}. \tag{5.15}$$

In view of Lemmas 5.3 and 5.4, we conclude that $\varphi_j \to \varphi$ in L^p with $\varphi \in W^{s,p}$, $e^{\imath\varphi} = u$ and

$$|\varphi|_{W^{s,p}(I)} \leq C|u|_{W^{s,p}(I)}. \qquad \square \tag{5.16}$$

We may always assume (by adding an integer multiple of 2π to φ) that the phase φ constructed above satisfies $\left|\int_I \varphi\right| \leq 2\pi$. Thus, to each $u \in W^{s,p}(\Omega; \mathbb{S}^1)$, where $sp < 1$, we have associated a phase φ in $W^{s,p}$ such that $e^{\imath\varphi} = u$ and

$$\|\varphi\|_{W^{s,p}(I)} \leq C\|u\|_{W^{s,p}(I)}. \tag{5.17}$$

Note that there is an interesting contrast between estimate (5.17) and the lack of estimate when $s = 1/p$ (see Proposition 9.2).

Remark 5.5. As we are going to see later in this chapter, it is possible to choose φ satisfying the extra regularity assumption $\varphi \in BMO$ (see Theorem 5.10).

Remark 5.6. A natural strategy to establish existence of lifting for $u \in W^{s,p}(\Omega; \mathbb{S}^1)$ could be the following: consider a sequence $(u_n) \subset C^\infty(\overline{\Omega}; \mathbb{S}^1)$ such that $u_n \to u$ in $W^{s,p}$. Let φ_n be a smooth lifting of u_n. Prove that (mod 2π) φ_n converges in $W^{s,p}$ to a lifting of u.

In the special case $sp < 1$, there are good and bad news: the good one is that such a sequence (u_n) does exist (see Theorems 10.1 and 10.2), the bad one is that the corresponding φ_n's need not converge in $W^{s,p}$; actually, it may happen that $|\varphi_n|_{W^{s,p}} \to \infty$ (see Proposition 9.1).

5.1.2 Case 2: $N \geq 2$, $0 < s < 1$

Subcase $N \geq 2, 0 < s < 1, sp \geq N$

Proof of Theorem 5.2 a). When $sp > N$, the conclusion follows from Theorem 5.4. When $sp = N$, the argument is exactly the same as in the proof of Theorem 5.1 (with $sp = 1$). $\square$

Subcase $N \geq 2, 0 < s < 1, sp < 1$

Proof of Theorem 5.2 c). We may assume that Ω is a cube. Indeed, by Lemma 5.5 below, if C is a cube containing Ω, then $v := \begin{cases} u, & \text{in } \Omega \\ 1, & \text{in } C \setminus \Omega \end{cases}$ belongs to $W^{s,p}(C; \mathbb{S}^1)$.

Lemma 5.5. *Let $N \geq 1$, $s \geq 0$ and $1 \leq p < \infty$ be such that $sp < 1$. Then the extension operator*

$$W^{s,p}(\Omega; \mathbb{R}) \ni u \mapsto v := \begin{cases} u, & \text{in } \Omega \\ 0, & \text{in } \mathbb{R}^N \setminus \Omega \end{cases}$$

is continuous from $W^{s,p}(\Omega; \mathbb{R})$ to $W^{s,p}(\mathbb{R}^N; \mathbb{R})$.

[This is a well-known result; see, for example, Mironescu [259, Lemma 2].]

When Ω is a cube, the existence of a lifting $\varphi \in W^{s,p}$ when $sp < 1$ is proved using the same argument as in the case $N = 1$. [The statements of Lemma 5.3 and 5.4 hold without any change when the interval I is replaced by a cube; see [50].] □

Remark 5.7. When $sp < 1$, estimate (5.17) holds in any dimension. The constant C in (5.17) depends on s, p and N and blows up as $sp \to 1$. Indeed, if the constant C were to remain bounded we would reach a contradiction with Proposition 9.2 when $N = 1$ and a contradiction with Theorem 5.2 *b*) when $N \geq 2$. The behavior of the best constant in (5.17) as $sp \to 1$ is discussed in Chapter 9.

Subcase $N \geq 2, 0 < s < 1, 1 \leq sp < N$

Proof of Theorem 5.2 b). Without loss of generality, we may assume that Ω is the unit ball. Let $\psi(x) := \dfrac{1}{|x|^\alpha}$, where $\dfrac{N - sp}{p} \leq \alpha < \dfrac{N - sp}{sp}$, and let $u := e^{\imath\psi}$.

Using Lemma 15.11 in Appendix 15.4, we see that for some $\varepsilon > 0$, we have

$$\psi \in W^{1,sp+\varepsilon} \text{ and } \psi \notin W^{s,p}. \tag{5.18}$$

Thus, $u \in W^{1,sp+\varepsilon}(\Omega; \mathbb{S}^1)$ and, by Gagliardo–Nirenberg (see Remark 15.5 in Appendix 15.2), we have $u \in W^{s,p}(\Omega; \mathbb{S}^1)$.

Next, we claim that there is no $\varphi \in W^{s,p}(\Omega; \mathbb{R})$ such that $u = e^{\imath\varphi}$. Assume by contradiction that such φ exists. Set $\eta := \dfrac{1}{2\pi}(\varphi - \psi)$. Clearly, η takes its values in $\mathbb{Z}$ and $\eta \in W^{1,sp} + W^{s,p}$. We deduce from Corollary 6.5 in Chapter 6 that η is a constant. Therefore, $\psi \in W^{s,p}$, which contradicts (5.18). □

Remark 5.8. The above example plays an important role and is useful in various situations. Let us emphasize its main features.

Assume $N \geq 2$. Given any $0 < s < 1$ and $1 < p < \infty$ such that $1 \leq sp < N$, there exists a function ψ satisfying:

$$\psi \in W^{1,sp}(\Omega;\mathbb{R}) \text{ (and even } \psi \in W^{1,sp+\varepsilon}(\Omega;\mathbb{R}) \text{ for some } \varepsilon > 0), \tag{5.19}$$

$$u := e^{\imath\psi} \text{ belongs to } W^{\sigma,q},\ \forall\, 0 < \sigma \le 1,\ \forall\, 1 \le q < \infty \text{ with } \sigma q \le sp, \text{ and, in particular, } u \in W^{s,p}, \tag{5.20}$$

$$\psi \notin W^{s,p}, \tag{5.21}$$

$$\text{there exists no } \varphi \in W^{s,p} \text{ such that } u = e^{\imath\varphi}. \tag{5.22}$$

We continue this section with another example of a map without lifting. Here, the non-existence of a lifting is induced by a topological obstruction.

In the special case $N = 2$, $\Omega =$ the unit disk, $s = 1/2$ and $p = 2$, we consider, as in Remark 1.6, the map $u(x) := \dfrac{x}{|x|}$. This u belongs to $H^{1/2}(\Omega;\mathbb{S}^1)$ (since $u \in W^{1,p}(\Omega)$, for every $p < 2$). We claim that there exists no $\varphi \in H^{1/2}(\Omega;\mathbb{R})$ such that $u = e^{\imath\varphi}$. Assume, by contradiction, that such φ exists. As already mentioned in Remark 5.2, the uniqueness of the lifting holds in $H^{1/2}$. Applying this fact in a disk away from the origin, we deduce that $\varphi \in C^\infty(\Omega \setminus \{0\})$. We conclude as in Remark 1.6.

When $N \ge 3$, the same construction as above, with $\Omega = B_1(0)$ and

$$u(x) := \frac{(x_1, x_2)}{(x_1^2 + x_2^2)^{1/2}},\quad x = (x_1, x_2, \dots, x_N),$$

provides an example of a function u that belongs to $W^{s,p}(\Omega;\mathbb{S}^1)$ for every $0 < s < 1$ and every $1 \le p < \infty$ with $sp < 2$, and that has no lifting in $W^{s,p}$ when $sp \ge 1$. However, this example does not cover the full "nasty" range $1 \le sp < N$ when $N \ge 3$.

Remark 5.9. The above examples show that a "loss of regularity" (in the $W^{\sigma,p}$ scale) may occur in the process of lifting. The exact value of the loss is presented in Chapter 8, Section 8.2.

5.1.3 *Case 3: $N \ge 2$, $s \ge 1$*

Subcase $N \ge 2$, $s \ge 1$, $sp \ge 2$

In this case, we present a slight improvement of Theorem 5.3 *a*).

Theorem 5.5. *Assume that $N \ge 2$, $s \ge 1$, $1 \le p < \infty$, and $sp \ge 2$. Then, any $u \in W^{s,p}(\Omega;\mathbb{S}^1)$ may be lifted as $u = e^{\imath\varphi}$ with $\varphi \in W^{s,p}(\Omega;\mathbb{R}) \cap W^{1,sp}(\Omega;\mathbb{R})$. And conversely.*

Proof. Observe that $W^{s,p} \cap L^\infty \subset W^{1,sp}$, by Remark 15.5 in Appendix 15.2. Since $sp \ge 2$, we may apply Theorem 1.1 and write $u = e^{\imath\varphi}$ for some $\varphi \in W^{1,sp}(\Omega;\mathbb{R})$. Using Lemma 5.2, we find that $\nabla\varphi = -\imath\bar{u}\,\nabla u \in W^{s-1,p}$, so that $\varphi \in W^{s,p}$.

Conversely, let $\varphi \in W^{s,p}(\Omega;\mathbb{R}) \cap W^{1,sp}(\Omega;\mathbb{R})$. Then, $u = e^{\imath\varphi}$ belongs to $W^{s,p}$. This is a special case of Theorem 15.8 in Appendix 15.3, asserting that $\Phi(u) \in W^{s,p}$ if $u \in W^{s,p} \cap W^{1,sp}$ and $\Phi \in C^\infty$ with $\partial^\alpha \Phi \in L^\infty$ for every α. □

Subcase $N \geq 2, s \geq 1, 1 \leq sp < 2$

Proof of Theorem 5.3 b). Consider the example of Remark 1.6. Assume $N = 2$ (the case $N \geq 3$ is handled as in Remark 1.6). By Lemma 15.12 in Appendix 15.4, the map $u(x) := \dfrac{x}{|x|}$ belongs to $W^{s,p}(\Omega;\mathbb{S}^1)$ for any $0 < s < \infty$ and any $1 \leq p < \infty$ with $sp < 2$. Since $s \geq 1$ we conclude, as in Remark 1.6, that u has no lifting in $W^{s,p}$. □

Now that we have found the answer to Question 1, we turn to Question 2, that is: assuming that the answer to Question 1 (existence of lifting) is negative, characterize the maps $u \in W^{s,p}(\Omega;\mathbb{S}^1)$ that admit a lifting in $W^{s,p}(\Omega;\mathbb{R})$. We distinguish two cases.

Case $N \geq 2, s \geq 1, 1 \leq sp < 2$

Theorem 5.6. *Assume that $N \geq 2$, $s \geq 1$, $1 \leq p < \infty$, and $1 \leq sp < 2$. Let $u \in W^{s,p}(\Omega;\mathbb{S}^1)$. Then, the following assertions are equivalent:*

a) u can be written as $u = e^{\imath\varphi}$ for some $\varphi \in W^{s,p}$
b) Ju (defined by (1.33)) satisfies $Ju = 0$.

Furthermore, in this case u admits a lifting $\varphi \in W^{s,p} \cap W^{1,sp}$.

The proof is exactly the same as the one of Theorem 1.2, using in addition Lemma 5.2.

Case $N \geq 2, 0 < s < 1, 1 \leq sp < N$

In this case, Question 2 is widely open. Note that a first difficulty is that the vector field $u \wedge \nabla u$ is not well-defined as a distribution when $s < 1$. However, in Chapter 8 we will give a meaning to $u \wedge \nabla u$ and to Ju as distributions, provided $u \in W^{s,p}(\Omega;\mathbb{S}^1)$, with $sp \geq 1$. If $u \in W^{s,p}(\Omega;\mathbb{S}^1)$ admits a lifting in $W^{s,p}$, then $Ju = 0$. As we are going to see in Chapter 8, the converse is false for each s and p in the range $0 < s < 1$, $1 \leq p < \infty$, and $1 \leq sp < N$. There is no known characterization of maps in $W^{s,p}(\Omega;\mathbb{S}^1)$ that admit a lifting in $W^{s,p}$.

5.2 Lifting of $\mathbb{S}^1$-valued maps in *VMO* and *BMO*

5.2.1 *BMO and VMO*

We start this section by recalling the definition of the space *BMO* (=bounded mean oscillation) of John and Nirenberg [206] on a smooth bounded domain $\Omega \subset \mathbb{R}^N$.

Definition 5.1. A function $f \in L^1_{loc}(\Omega)$ belongs to $BMO(\Omega)$ if

$$|f|_{BMO} = |f|_{BMO(\Omega)} := \sup_{\substack{B \text{ ball} \\ \overline{B}\subset\Omega}} \fint_B \fint_B |f(x) - f(y)| \, dxdy < \infty. \tag{5.23}$$

Here, we consider only Euclidean balls.

Following Sarason [316], we also introduce the space *VMO* (=vanishing mean oscillation).

Definition 5.2. A function $f \in BMO(\Omega)$ belongs to $VMO(\Omega)$ if

$$\lim_{|B|\to 0} \fint_B \fint_B |f(x) - f(y)| \, dxdy = 0, \tag{5.24}$$

where the limit is taken over all the balls B such that $\overline{B} \subset \Omega$.

We recall here a few properties of these spaces.

1. Clearly, $|\ |_{BMO}$ is a seminorm. An equivalent seminorm is obtained by considering $\fint_B \left| f - \fint_B f \right|$ instead of $\fint_B \fint_B |f(x) - f(y)| \, dxdy$.
   We have $BMO(\Omega) \hookrightarrow L^1(\Omega)$ and

$$\|f\|_{L^1(\Omega)} \le C_B \left(|f|_{BMO(\Omega)} + \left| \int_B f \right| \right), \text{ for every ball } \overline{B} \subset \Omega. \tag{5.25}$$

   The space $BMO(\Omega)$ is equipped with the norm

$$\|f\|_{BMO} = \|f\|_{BMO(\Omega)} := |f|_{BMO(\Omega)} + \|f\|_{L^1(\Omega)},$$

   or any equivalent norm

$$|f|_{BMO(\Omega)} + \left| \int_B f \right|,$$

   where B is any ball such that $\overline{B} \subset \Omega$.
   Moreover, the following Poincaré-type inequality holds:

$$\left\| f - \fint_\Omega f \right\|_{L^1(\Omega)} \leq C|f|_{BMO(\Omega)}. \tag{5.26}$$

For these properties, see Lemma 15.21 in Appendix 15.5.

2. For a function $f \in L^1(\Omega)$, for $1 \leq p < \infty$ and for $t > 0$, we set

$$M_{t,p}(f, \Omega) := \sup_{\substack{B \text{ ball of radius } \leq t \\ \overline{B} \subset \Omega}} \fint_B \fint_B |f(x) - f(y)|^p \, dxdy. \tag{5.27}$$

For simplicity, we denote

$$M_t(f) = M_t(f, \Omega) := M_{t,1}(f, \Omega). \tag{5.28}$$

Then, $f \in BMO(\Omega)$ is equivalent to $\sup_{t>0} M_t(f, \Omega) < \infty$. On the other hand, if $f \in BMO(\Omega)$, then $f \in VMO(\Omega)$ if and only if $\lim_{t \to 0} M_t(f, \Omega) = 0$.
An important inequality, which is an immediate consequence of the John–Nirenberg inequality [206] (see also Brezis and Nirenberg [104, Lemma B.4]) asserts that

$$M_{t,p}(f, \Omega) \leq C_p(M_t(f, \Omega))^p, \ \forall\, 1 < p < \infty, \ p \text{ an integer}, \ \forall\, t > 0, \tag{5.29}$$

where C_p does not depend on f, t or Ω.

3. Clearly, $L^\infty(\Omega) \hookrightarrow BMO(\Omega)$ and $C(\overline{\Omega}) \hookrightarrow VMO(\Omega)$. Moreover, $VMO(\Omega)$ is the closure of $C(\overline{\Omega})$ in $BMO(\Omega)$. This important result is due to Sarason [316].
4. As a consequence of the above result, $VMO(\Omega)$ is the closure of $C^\infty(\overline{\Omega})$ in $BMO(\Omega)$.
5. Let $f \in BMO(\Omega)$ and set $f_t(x) := \fint_{B_t(x)} f$ if $x \in \Omega_t := \{x \in \Omega;\ \text{dist}(x, \partial\Omega) > t\}$. Then

$$\|f - f_t\|_{BMO(\Omega_t)} \leq C\, M_t(f)$$

([316]; see also [104, Lemma 3']).

6. By the above, if $f \in VMO(\Omega)$, for every open subset $\omega \subset \Omega$ with $\text{dist}(\omega, \partial\Omega) > 0$, we have $f_t \to f$ in $BMO(\omega)$ as $t \to 0$ ([316]; see also [104, Corollary 1]).
7. If $s > 0$ and $1 \leq p < \infty$ are such that $sp = N$, then $W^{s,p}(\Omega) \hookrightarrow VMO(\Omega)$ (see, for example, [104, Section I.2] and Lemma 15.20 in Appendix 15.5).
8. Recall the following standard examples for a domain Ω containing the origin. The function $x \mapsto \ln|x|$ belongs to $BMO(\Omega)$, but not to $VMO(\Omega)$, and, for $0 < \alpha < 1$, $x \mapsto |\ln|x||^\alpha$ belongs to $VMO(\Omega)$.
9. The space *BMO* does not change if we consider, in the definition of the *BMO* seminorm, only balls of size $\leq \delta$ for some δ. More specifically, the new seminorm defined in this way is equivalent to that in Definition 5.2.

5.2.2 Lifting in BMO and VMO

In this section, we address the question of existence of a *VMO* lifting for maps in $VMO(\Omega;\mathbb{S}^1)$. Note that *BMO* maps do have *BMO* phases. Indeed, if $u \in BMO(\Omega;\mathbb{S}^1)$, then $\varphi := \arg u$ (where arg denotes the principal branch of the argument) is in L^∞ and thus in *BMO*. Therefore, in the *BMO* context, the issue will not be existence, but rather the control of φ in terms of u (see Theorem 5.8).

The main result in this section is

Theorem 5.7. *Let $u \in VMO(\Omega;\mathbb{S}^1)$. Then there exists some $\varphi \in VMO(\Omega;\mathbb{R})$ such that $u = e^{\imath\varphi}$.*

Moreover, there are constants $\delta, K > 0$ such that, if $u \in VMO(\Omega;\mathbb{S}^1)$ satisfies $|u|_{BMO} \le \delta$, then this lifting φ satisfies $|\varphi|_{BMO} \le K|u|_{BMO}$.

Here is the strategy of the proof. We may assume that u is defined in a smooth simply connected neighborhood $\widetilde{\Omega}$ of $\overline{\Omega}$. This follows from the fact that the extension by reflection across $\partial\Omega$ of a *VMO* map is still *VMO* (see Lemma 15.28 in Appendix 15.6). Next we set, for $x \in \overline{\Omega}$ and small ε, $u_\varepsilon(x) := \fint_{B_\varepsilon(x)} u$. As we are going to see in Step 1, we have $|u_\varepsilon| \to 1$ uniformly on $\overline{\Omega}$. Thus, for small $\varepsilon > 0$, we have a lifting φ_ε of $u_\varepsilon/|u_\varepsilon|$. In Step 2, we prove that $\varphi_\varepsilon \to \varphi$ a.e. for some lifting φ of u. The heart of the proof is in Step 3, where we prove that $\varphi \in VMO$.

The estimate of $|\varphi|_{BMO}$ will follow from the proof of Theorem 5.8.

Proof of Theorem 5.7. Step 1. We have $|u_\varepsilon| \to 1$ uniformly on $\overline{\Omega}$
This is the content of the next result.

Lemma 5.6. *Let $u \in VMO(\widetilde{\Omega};\mathbb{S}^1)$. Let $\varepsilon_0 := \operatorname{dist}(\partial\widetilde{\Omega}, \overline{\Omega})$. For $0 < \varepsilon < \varepsilon_0$ and $x \in \overline{\Omega}$, define $u_\varepsilon(x) := \fint_{B_\varepsilon(x)} u$. Then, $|u_\varepsilon| \to 1$ uniformly on $\overline{\Omega}$ as $\varepsilon \to 0$.*

Proof. For $x \in \overline{\Omega}$ we have

$$\fint_{B_\varepsilon(x)} |u(y) - u_\varepsilon(x)|\, dy = \fint_{B_\varepsilon(x)} \left| u(y) - \fint_{B_\varepsilon(x)} u(z)\, dz \right| dy$$
$$\le \fint_{B_\varepsilon(x)} \fint_{B_\varepsilon(x)} |u(y) - u(z)|\, dydz.$$

Clearly, for a.e. y,

$$|u(y) - u_\varepsilon(x)| \ge \operatorname{dist}(u_\varepsilon(x), \mathbb{S}^1) = ||u_\varepsilon(x)| - 1|,$$

and therefore

$$||u_\varepsilon(x)| - 1| \le \fint_{B_\varepsilon(x)} \fint_{B_\varepsilon(x)} |u(y) - u(z)|\, dydz.$$

We conclude using the definition of *VMO*. ◻

Remark 5.10. The same proof shows that, if $u \in VMO(\widetilde{\Omega}; F)$, where F is some arbitrary set, then $\operatorname{dist}(u_\varepsilon, F)$ tends to 0 uniformly on $\overline{\Omega}$ as $\varepsilon \to 0$.

Step 2. Construction and convergence of φ_ε
Let ε_1 be such that, for $\varepsilon < \varepsilon_1$, the map u_ε satisfies $1/2 \le |u_\varepsilon| \le 1$ in $\overline{\Omega}$. Let $\psi(x, \varepsilon)$ be a continuous lifting of $u_\varepsilon(x)/|u_\varepsilon(x)|$ in $\overline{\Omega} \times (0, \varepsilon_1)$, and set $\varphi_\varepsilon(x) := \psi(x, \varepsilon)$.

At each Lebesgue point of u, we have $\lim_{\varepsilon \to 0} u_\varepsilon(x) = u(x)$. This fact, combined with the continuity of $\varepsilon \mapsto \varphi_\varepsilon(x)$ and the next lemma (suggested by Guy David) implies that $\varphi(x) := \lim_{\varepsilon \to 0} \varphi_\varepsilon(x)$ is defined and finite a.e., measurable, and is a lifting of u.

Lemma 5.7. *Let* $\varepsilon_1 > 0$ *and let* $f \in C((0, \varepsilon_1); \mathbb{R})$. *If* $\lim_{\varepsilon \to 0} e^{\imath f(\varepsilon)}$ *exists, then* $\lim_{\varepsilon \to 0} f(\varepsilon)$ *exists and is finite.*

Proof. Assume, for example, that $u(\varepsilon) := e^{\imath f(\varepsilon)} \to 1$ as $\varepsilon \to 0$. Then, for small ε, we have $u = e^{\imath g}$, where $g := -\imath \ln u$, and ln stands for the principal branch of the logarithm. By assumption, g is continuous and $g(\varepsilon) \to 0$ as $\varepsilon \to 0$. We conclude using the fact that $f = g + C$ (by the continuity of f and g). □

Step 3. We have $\varphi \in VMO$
The starting point is the inequality

$$|\alpha - \beta| \le |e^{\imath\alpha} - e^{\imath\beta}| + \frac{1}{2}|\alpha - \beta|^2, \ \forall \alpha, \beta \in \mathbb{R}, \tag{5.30}$$

which follows from

$$|e^{\imath t} - 1 - \imath t| \le \frac{1}{2}t^2, \ \forall t \in \mathbb{R}.$$

Combined with the estimate

$$\left|\frac{\xi}{|\xi|} - \frac{\eta}{|\eta|}\right| \le C|\xi - \eta|, \ \forall \xi, \eta \in \mathbb{C} \text{ such that } |\xi| \ge \frac{1}{2}, |\eta| \ge \frac{1}{2},$$

and the fact that φ_ε is a lifting of $\dfrac{u_\varepsilon}{|u_\varepsilon|}$, (5.30) implies that

$$M_t(\varphi_\varepsilon, \Omega) \le C(M_t(u_\varepsilon, \Omega) + M_{t,2}(\varphi_\varepsilon, \Omega)). \tag{5.31}$$

In view of the inequality (5.29), (5.31) yields

$$M_t(\varphi_\varepsilon, \Omega) \le C(M_t(u_\varepsilon, \Omega) + (M_t(\varphi_\varepsilon, \Omega))^2). \tag{5.32}$$

The next step consists of estimating $M_t(u_\varepsilon, \Omega)$. This is done via the following

Lemma 5.8. *We have*

$$M_t(u_\varepsilon, \Omega) \le M_t(u, \Omega) + CM_\varepsilon(u, \Omega), \ \forall t > 0, \ \forall \varepsilon < \varepsilon_0. \tag{5.33}$$

Proof of Lemma 5.8. By Brezis and Nirenberg [104, Lemma A.5], we have

$$|u_\varepsilon - u|_{BMO(\Omega)} \le CM_\varepsilon(u, \Omega). \tag{5.34}$$

[This fact is not difficult, but not completely trivial.] This estimate implies (5.33) via the obvious inequalities

$$M_t(u_\varepsilon, \Omega) \le M_t(u, \Omega) + M_t(u_\varepsilon - u, \Omega) \le M_t(u, \Omega) + |u_\varepsilon - u|_{BMO(\Omega)}. \qquad \square$$

Step 3 completed. Lemma 5.8 combined with (5.32) implies that

$$M_t(\varphi_\varepsilon, \Omega) \le C(M_t(u, \Omega) + M_\varepsilon(u, \Omega) + (M_t(\varphi_\varepsilon, \Omega))^2), \ \forall t > 0, \ \forall \varepsilon < \varepsilon_1. \tag{5.35}$$

Consider the function $h(X) := X - CX^2$, $X \ge 0$, where C is the constant in (5.35). Clearly,

$$h(X) \le 1/(8C) \implies \left[X \le \frac{1}{2C} - \frac{1}{2\sqrt{2}C} \text{ or } X \ge \frac{1}{2C} + \frac{1}{2\sqrt{2}C} \right]. \tag{5.36}$$

Since $u \in VMO$, there exists some $\alpha < \varepsilon_1$ such that

$$M_t(u, \Omega) + M_\varepsilon(u, \Omega) \le \frac{1}{8C^2}, \ \forall t < \alpha, \ \forall \varepsilon < \alpha. \tag{5.37}$$

On the other hand, for every fixed ε, φ_ε is continuous in $\overline{\Omega}$ and thus

$$\lim_{t \to 0} M_t(\varphi_\varepsilon, \Omega) = 0.$$

In addition, the map $t \mapsto M_t(\varphi_\varepsilon, \Omega)$ is continuous, so that the set $\{M_t(\varphi_\varepsilon, \Omega);\ 0 < t < \alpha\}$ is connected. This fact, combined with (5.35)–(5.37), implies that

$$M_t(\varphi_\varepsilon, \Omega) \le \frac{1}{2C} - \frac{1}{2\sqrt{2}C} \le \frac{1}{2C}, \ \forall t < \alpha, \ \forall \varepsilon < \alpha. \tag{5.38}$$

Returning to (5.35) and using (5.38), we see that

$$\begin{aligned} M_t(\varphi_\varepsilon, \Omega) \le& C(M_t(u, \Omega) + M_\varepsilon(u, \Omega) + (M_t(\varphi_\varepsilon, \Omega))^2) \\ \le& C(M_t(u, \Omega) + M_\varepsilon(u, \Omega)) + \frac{1}{2} M_t(\varphi_\varepsilon, \Omega). \end{aligned}$$

Therefore,

$$M_t(\varphi_\varepsilon, \Omega) \le 2C(M_t(u, \Omega) + M_\varepsilon(u, \Omega)), \ \forall t < \alpha, \ \forall \varepsilon < \alpha. \tag{5.39}$$

Passing to the limit as $\varepsilon \to 0$ (with fixed t) we deduce by Fatou that

$$M_t(\varphi, \Omega) \le 2CM_t(u, \Omega), \ \forall t < \alpha. \tag{5.40}$$

By (5.40) and the fact that $u \in VMO$, we have $\varphi \in VMO$. □

Remark 5.11. We will see in Section 6.1 that this φ is unique (mod 2π).

We continue by discussing the question of a control of the *BMO* seminorm of some lifting φ of a map $u \in BMO(\Omega; \mathbb{S}^1)$. Since we always have $|u|_{BMO} \le 2$ and since the phase φ of u obtained using the principal branch of the logarithm satisfies $|\varphi|_{BMO} \le 2\pi$, we see that the only possibly interesting issue is the existence of a better control for $|\varphi|_{BMO}$ when $|u|_{BMO}$ is small.

Theorem 5.8. *There exists a constant $K > 0$ such that every $u \in BMO(\Omega; \mathbb{S}^1)$ has a lifting φ satisfying $|\varphi|_{BMO} \le K|u|_{BMO}$.*

Proof. The heart of the proof consists of finding φ when $|u|_{BMO} \le \gamma$, with γ some small constant to be fixed later. When $|u|_{BMO} > \gamma$, it suffices to consider the phase φ of u obtained using the principal branch of the logarithm, which satisfies

$$|\varphi|_{BMO} \le 2\pi \le \frac{2\pi}{\gamma}\,|u|_{BMO}.$$

The extension operator $u \mapsto \widetilde{u}$ by reflection across $\partial\Omega$ is a bounded linear operator in *BMO*, i.e., we have $|\widetilde{u}|_{BMO(\widetilde{\Omega})} \le C|u|_{BMO(\Omega)}$ (Lemma 15.28). Therefore, it suffices to work with *BMO* maps $\widetilde{u}$ defined in $\widetilde{\Omega}$ and satisfying $|\widetilde{u}|_{BMO(\widetilde{\Omega})} \le \widetilde{\gamma}$ for appropriate $\widetilde{\gamma}$. For simplicity, we write u, γ instead of $\widetilde{u}, \widetilde{\gamma}$.

The proof of Lemma 5.6 implies that

$$||u_\varepsilon(x)| - 1| \le M_\varepsilon(u, \widetilde{\Omega}), \ \forall \varepsilon < \varepsilon_0, \ \forall x \in \Omega.$$

Thus, if $|u|_{BMO(\widetilde{\Omega})} \le 1/2$, then we have $1/2 \le |u_\varepsilon| \le 1$ in Ω.

Assuming that $|u|_{BMO(\widetilde{\Omega})} \le 1/2$, we proceed as in the proof of Theorem 5.7: we construct φ_ε and φ. Repeating the argument leading from (5.35) to (5.39), we see that (5.39) (with $\alpha = \varepsilon_0$) still holds in our context provided $|u|_{BMO(\widetilde{\Omega})} \le \gamma$, with sufficiently small γ. Letting $\varepsilon \to 0$ in (5.39), we obtain

$$M_t(\varphi, \Omega) \le 4CM_t(u, \Omega), \ \forall t < \varepsilon_0,$$

and the theorem follows. □

We now return to lifting in *VMO* and combine it with lifting in $W^{s,p}$. Given $u \in VMO(\Omega; \mathbb{S}^1)$, we know that there exists a unique (mod 2π) $\varphi \in VMO$ such that $u = e^{\imath\varphi}$. If we assume in addition that $u \in W^{s,p}$, is it true that φ also belongs to $W^{s,p}$? The answer is positive.

Theorem 5.9. *Let $s > 0$ and $1 \le p < \infty$. Let $u \in W^{s,p} \cap VMO(\Omega; \mathbb{S}^1)$. Then there exists some $\varphi \in W^{s,p} \cap VMO(\Omega; \mathbb{R})$ such that $u = e^{\imath\varphi}$.*

If, in addition, $s \ge 1$, then we also have $\varphi \in W^{1,sp}$; and, if $N \ge 2$, then $Ju = 0$.

Proof. Step 1. Proof in the special case $s = 1, p = 1$
By Lemma 1.1, the conclusion is true when $u \in W^{1,1} \cap C$. For a general u, let u_ε, φ_ε be as in the proof of Theorem 5.7 (with u belonging to $W^{1,1} \cap VMO\,(\widetilde{\Omega}; \mathbb{S}^1)$). On the one hand, we know from the proof of Theorem 5.7 that $\varphi_\varepsilon \to \varphi$ a.e. as $\varepsilon \to 0$, where $\varphi \in VMO$ is a lifting of u. On the other hand, the identity

$$\nabla \varphi_\varepsilon = \frac{u_\varepsilon}{|u_\varepsilon|} \wedge \nabla \left(\frac{u_\varepsilon}{|u_\varepsilon|} \right)$$

(valid since φ_ε is a Lipschitz lifting of $\dfrac{u_\varepsilon}{|u_\varepsilon|}$) implies that $\nabla \varphi_\varepsilon \to u \wedge \nabla u$ in $L^1(\Omega)$ as $\varepsilon \to 0$.

Set $\psi_\varepsilon := \varphi_\varepsilon - \fint_\Omega \varphi_\varepsilon$, so that, by Poincaré's inequality, $\|\psi_\varepsilon\|_{L^1} \leq C \|\nabla \varphi_\varepsilon\|_{L^1} \leq C$. Thus, (ψ_ε) is relatively compact in L^1 and, along a subsequence $\varepsilon_n \to 0$, ψ_{ε_n} converges a.e. to some $\psi \in W^{1,1}(\Omega)$. Hence, $\fint_\Omega \varphi_{\varepsilon_n} \to C$ and $\psi = \varphi - C$.

Step 2. Proof in the case $s \geq 1$ and $p \geq 1$
By Step 1, we already know that u admits a lifting $\varphi \in W^{1,1}$. We may then argue as in the proof of Theorem 5.5 to conclude that $\varphi \in W^{s,p} \cap W^{1,sp}$.

Step 3. Proof in the case $0 < s < 1$ and $p \geq 1$
As in the proof of Theorem 5.7, we may assume that u is defined in a neighborhood $\widetilde{\Omega}$ of $\overline{\Omega}$; this relies on Lemmas 15.27 and 15.28 in Appendix 15.6. The strategy is somewhat similar to the one in Step 3 in the proof of Theorem 5.7, except that we replace the John–Nirenberg type inequality (5.29) (used in (5.32)) with a Gagliardo–Nirenberg type inequality involving *BMO*:

Lemma 5.9. *Let $0 < s < 1$ and $1 \leq p < \infty$. Then there exists some constant $C' = C'(s, p, N)$ such that, for every ball $B \subset \mathbb{R}^N$ and for every $f \in W^{s,p} \cap BMO\,(B; \mathbb{R})$, we have*

$$|f|^2_{W^{s/2,2p}(B)} \leq C' |f|_{BMO\,(B)} |f|_{W^{s,p}(B)}. \tag{5.41}$$

The proof of Lemma 5.9 is presented in Appendix 15.2 (see Lemma 15.7 there).

Granted Lemma 5.9, we complete Step 3 as follows. It suffices to prove that there exists some $\rho > 0$ such that, for every $x \in \overline{\Omega}$, we have $\varphi \in W^{s,p}(B_\rho(x))$. Let $B := B_r(x)$ (for some small r such that $B_r(x) \subset \widetilde{\Omega}$). The starting point is the following straightforward analogue of (5.31)

$$|\varphi_\varepsilon|^p_{W^{s,p}(B)} \leq C(|u_\varepsilon|^p_{W^{s,p}(B)} + |\varphi_\varepsilon|^{2p}_{W^{s/2,2p}(B)}), \ \forall x \in \overline{\Omega}, \ \forall \varepsilon < \varepsilon_2, \ \forall r < r_0. \tag{5.42}$$

By (5.41) and (5.42), we have, with x, ε, and r as above,

$$|\varphi_\varepsilon|^p_{W^{s,p}(B)} \leq C(|u_\varepsilon|^p_{W^{s,p}(B)} + (C')^p |\varphi_\varepsilon|^p_{BMO\,(B)} |\varphi_\varepsilon|^p_{W^{s,p}(B)}). \tag{5.43}$$

By repeating the argument leading to (5.39), for sufficiently small α and r_0 we have

$$M_t(\varphi_\varepsilon, B) \leq 2C''(M_t(u, B) + M_\varepsilon(u, B)), \ \forall x \in \overline{\Omega}, \ \forall r < r_0, \forall t < \alpha, \forall \varepsilon < \alpha,$$

and thus

$$|\varphi_\varepsilon|_{BMO\,(B)} = M_r(\varphi_\varepsilon, B) \leq 4C''M_r(u, B), \ \forall x \in \overline{\Omega}, \ \forall \varepsilon < r < r_0. \tag{5.44}$$

Combining (5.43) and (5.44), we obtain

$$|\varphi_\varepsilon|_{W^{s,p}(B)} \leq C'''|u_\varepsilon|_{W^{s,p}(B)}, \ \forall x \in \overline{\Omega}, \ \forall 0 < \varepsilon < r < r_1, \tag{5.45}$$

where r_1 is such that $C[4C'C''M_{r_1}(u, B)]^p < 1/2$. [This is possible since $u \in VMO$.]

Letting $\varepsilon \to 0$ in (5.45) and using Fatou, and the fact that $u_\varepsilon \to u$ in $W^{s,p}$ as $\varepsilon \to 0$, we find that $\varphi \in W^{s,p}(B)$ for every $r < r_1$. □

Remark 5.12. The above proof shows that there exist constants γ and K such that, if $u \in W^{s,p} \cap VMO\,(\Omega; \mathbb{S}^1)$ satisfies $|u|_{BMO} \leq \gamma$, then $|\varphi|_{W^{s,p}} \leq K|u|_{W^{s,p}}$.

5.3 Lifting in $W^{s,p}$, $sp < 1$, upgraded

The main result in this section is the following improvement of Theorems 5.1 and 5.2 when $sp < 1$.

Theorem 5.10. *Let $N \geq 1$, $0 < s < 1$ and $1 \leq p < \infty$ be such that $sp < 1$. Let $u \in W^{s,p}(\Omega; \mathbb{S}^1)$. Then, u has a lifting $\varphi \in W^{s,p}(\Omega; \mathbb{R}) \cap BMO$. In particular, this φ belongs to $W^{\theta s, p/\theta}$, $\forall 0 < \theta < 1$.*

Proof. When $p = 1$, the theorem follows from Lemma 5.10 below combined with the Gagliardo–Nirenberg embeddings $W^{s,1} \cap L^\infty \subset W^{\theta s, 1/\theta}$, $\forall 0 < \theta < 1$ (Lemma 15.6).

When $p > 1$, let $f \in W^{s,p}$ be a lifting of u. Write, as in Lemma 5.11 below, $f = g + h$, with $g \in W^{sp,1} \cap W^{s,p}$ and $h \in BMO \cap W^{s,p}$. Let $v := e^{\imath g} \in W^{sp,1}$. By Lemma 5.10, v has a lifting $\zeta \in W^{sp,1} \cap L^\infty$. Set $\varphi := h + \zeta$. In view of the Gagliardo–Nirenberg embedding $W^{sp,1} \cap L^\infty \subset W^{s,p}$ and of the fact that $L^\infty \hookrightarrow BMO$, we see that φ is a lifting of u satisfying $\varphi \in W^{s,p} \cap BMO$. The last assertion follows from Lemma 15.7. □

Remark 5.13. We do not know whether, in Theorem 5.10, it is possible to find a lifting $\varphi \in W^{s,p} \cap L^\infty$.

Lemma 5.10. *Let $0 < \sigma < 1$. Then, every $v \in W^{\sigma,1}(\Omega; \mathbb{S}^1)$ has a lifting $\zeta \in W^{\sigma,1} \cap L^\infty$ such that $\|\zeta\|_{L^\infty} \leq 2\pi$ and*

$$|\zeta|_{W^{\sigma,1}} \leq 2|v|_{W^{\sigma,1}}. \tag{5.46}$$

Proof of Lemma 5.10. We copy the argument used in the proof of Theorem 1.4. Let Θ_α be as in (1.75). By (1.76), we have

$$\int_0^{2\pi} |\Theta_\alpha(v)|_{W^{\sigma,1}}\, d\alpha \leq 4\pi |v|_{W^{\sigma,1}}$$

(this is an analogue of (1.77)). We find that, for some appropriate α, $\zeta := \Theta_\alpha(u)$ has all the required properties. □

Lemma 5.11. *Let $s > 0$ and $1 < p < \infty$. Then, every $f \in W^{s,p}(\Omega; \mathbb{R})$ can be written as $f = g + h$, with $g \in W^{sp,1} \cap W^{s,p}$ and $h \in BMO \cap W^{s,p}$.*

Lemma 5.11 is a special case of Mironescu [260, Theorem 1.7].

5.4 Complements and open problems

5.4.1 Lifting in covering spaces (following Bethuel and Chiron, Mironescu and Van Schaftingen)

One may address the lifting Questions 1 and 2 in the following more general context. Consider a smooth compact manifold $\mathscr{N}$, its universal covering $\mathscr{E}$, and the universal covering map $\pi : \mathscr{E} \to \mathscr{N}$. The standard case we have studied in Chapters 1 and 5 corresponds to $\mathscr{N} = \mathbb{S}^1$, $\mathscr{E} = \mathbb{R}$, and $\pi(t) = e^{\imath t}$, $\forall t \in \mathbb{R}$. We assume $\mathscr{N}$, resp. $\mathscr{E}$ smoothly embedded in the Euclidean spaces $\mathbb{R}^m$, resp. $\mathbb{R}^\ell$. For $\Omega \subset \mathbb{R}^N$ smooth bounded open set, $s \geq 0$, $1 \leq p \leq \infty$, set

$$W^{s,p}(\Omega; \mathscr{N}) := \{u : \Omega \to \mathbb{R}^m;\ u \in W^{s,p} \text{ and } u(x) \in \mathscr{N} \text{ a.e.}\},$$

resp.

$$W^{s,p}(\Omega; \mathscr{E}) := \{\varphi : \Omega \to \mathbb{R}^\ell;\ \varphi \in W^{s,p} \text{ and } \varphi(x) \in \mathscr{E} \text{ a.e.}\}.$$

[When $0 < s \leq 1$, one can prove that these spaces do not depend on the choice of the embeddings of $\mathscr{N}$ and $\mathscr{E}$ (see Bethuel and Chiron [32]).]

The analogue of our Question 1 becomes: let $0 < s < \infty$, $1 \leq p < \infty$ and let Ω be a smooth bounded simply connected domain in $\mathbb{R}^N$. Decide whether for each $u \in W^{s,p}(\Omega; \mathscr{N})$ there exists some $\varphi \in W^{s,p}(\Omega; \mathscr{E})$ such that $u = \pi \circ \varphi$. Question 2 is asked similarly.

We first present the main results of Bethuel and Chiron [32] concerning Question 1 in this general situation. A first result in [32] asserts the following: if $\mathscr{E}$ is *non-compact*, then the answer to Question 1 is *exactly the same* as in the standard case $\mathscr{N} = \mathbb{S}^1$, $\mathscr{E} = \mathbb{R}$, $\pi(t) = e^{\imath t}$.

The situation is more involved when $\mathscr{E}$ is compact. [A typical example is $\mathscr{N} = \mathbb{RP}^2$, $\mathscr{E} = \mathbb{S}^2$, $\pi(x) = \widehat{x} = \{x, -x\}$.] A second main result in [32] asserts that, when $\mathscr{E}$ is *compact*, the answer to Question 1 is *exactly the same* as in the standard case, possibly except when $0 < s < 1$, $1 \leq p < \infty$, $N \geq 3$, and $2 \leq sp < N$. In this case, left open in [32], Mironescu and Van Schaftingen [269] proved the existence of lifting. This contrasts with the standard case, where we have non-existence (Theorem 5.2 *b*)).

To summarize, the results obtained in [32, 269] lead to the following complete picture.

Theorem 5.11. *Assume that $\mathscr{E}$ is non-compact. Then, the answer to the lifting question is the following.*

A) $N = 1$: always positive
B) $N = 2$:

B1) Negative when $0 < s < \infty$ and $1 \leq sp < 2$
B2) Positive in all remaining cases

C) $N \geq 3$:

C1) Negative when $0 < s < 1$ and $1 \leq sp < N$
C2) Negative when $s \geq 1$ and $1 \leq sp < 2$
C3) Positive in all remaining cases.

Theorem 5.12. *Assume that $\mathscr{E}$ is compact. Then, the answer to the lifting question is the following.*

A) $N = 1$: always positive
B) $N \geq 2$:

B1) Negative when $0 < s < \infty$ and $1 \leq sp < 2$
B2) Positive in all remaining cases.

Note that the only difference between the two cases occurs when $N \geq 3, 0 < s < 1$ and $2 \leq sp < N$.

5.4.2 *Open problems*

Open Problem 7. Find a simple proof of Theorem 5.1 when $sp < 1$. In particular, find a simple proof of the existence of a lifting $\varphi \in H^s((0,1);\mathbb{R})$ for every $u \in H^s((0,1);\mathbb{S}^1)$ when $0 < s < 1/2$.

Open Problem 8. Let $N \geq 2$, $0 < s < 1$ and $1 < p < \infty$ be such that $1 \leq sp < N$. [Recall that in this range $W^{s,p}(\Omega;\mathbb{S}^1)$ does not have the lifting property.] Characterize those u's that admit a lifting within $W^{s,p}$.

5.5 Comments

1. Section 5.1 is entirely taken from Bourgain, Brezis, and Mironescu [50]. There, it was assumed that $p > 1$, but the proofs are still valid when $p = 1$.
2. The study of lifting in *BMO* goes back to Coifman and Meyer [119] and in *VMO* to Brezis and Nirenberg [104]. In the proofs of Theorems 5.7 and 5.8, we follow the strategy and the arguments in [104].
3. Theorem 5.9 is new.

4. Numerous authors have used Gagliardo–Nirenberg type inequalities where *BMO* replaces L^∞ in the spirit of Lemma 5.9 (see, for example, Adams and Frazier [3, Lemma 2.2], Meyer and Rivière [249], Strzelecki [333], Kozono and Wadade [216], Angelsberg and Pumberger [14], and the references therein). Under the special assumptions of Lemma 5.9, the result was communicated to us by Frazier [153] when B is replaced by $\mathbb{R}^N$.
5. Theorem 5.10 is new. Lemma 5.10 is from Mironescu and Molnar [263].

Chapter 6
Uniqueness of lifting and beyond

6.0 Introduction

So far, we have been concerned with the *existence* of a lifting $\varphi : \Omega \to \mathbb{R}$ for a given $u : \Omega \to \mathbb{S}^1$. A natural question is whether such φ is *unique* (mod 2π). More precisely, assume that we have two liftings, φ_1, φ_2. Then, $\varphi_1(x) - \varphi_2(x) = 2\pi k(x)$ for some $k : \Omega \to \mathbb{Z}$. Therefore, we are led to the question of finding minimal assumptions on a measurable function $k : \Omega \to \mathbb{Z}$ implying that k must be constant. Clearly, continuity is sufficient, but, as we are going to see, constancy holds for a surprisingly large class of functions.

In the previous chapters, we established the existence of a lifting in the $W^{s,p}$ or VMO setting. By the above, lifting in $W^{s,p}$ is unique if $sp > N$. It turns out that the condition $sp > N$ is much too strong for uniqueness, and can be weakened to $sp \geq 1$. Moreover, uniqueness holds in VMO (although VMO functions need not be continuous).

In the first three sections, we treat the cases $VMO(\Omega; \mathbb{Z})$, $W^{1,1}(\Omega; \mathbb{Z})$, and $W^{1/p,p}(\Omega; \mathbb{Z})$, with $1 < p < \infty$. For the enjoyment of the reader, we present three proofs of constancy in $W^{1/p,p}(\Omega; \mathbb{Z})$ with totally different flavors.

The first one is by dimensional reduction and uses the fact that, in 1D, $W^{1/p,p} \hookrightarrow VMO$.

The second proof relies on the fact that, by definition of $W^{1/p,p}$, we have

$$\int_\Omega \int_\Omega \frac{|\varphi(x) - \varphi(y)|^p}{|x - y|^{N+1}} \, dxdy < \infty,$$

and thus, since φ takes values in $\mathbb{Z}$,

$$\int_\Omega \int_\Omega \frac{|\varphi(x) - \varphi(y)|}{|x - y|^{N+1}} \, dxdy < \infty. \tag{6.1}$$

At this stage, one can discard the assumption that φ is integer-valued, and prove that any measurable function $\varphi : \Omega \to \mathbb{R}$ satisfying (6.1) must be constant. This

H. Brezis and P. Mironescu, *Sobolev Maps to the Circle*, Progress in Nonlinear Differential Equations and Their Applications 96,
https://doi.org/10.1007/978-1-0716-1512-6_6

assertion is established using the "BBM formula" (Theorem 6.2), which allows us to compute $\int_\Omega |\nabla\varphi|$ as a limit of finite differences of the type $\int_\Omega\int_\Omega \frac{|\varphi(x)-\varphi(y)|}{|x-y|}\rho_\varepsilon(x-y)\,dxdy$. Under the assumption (6.1), it yields $\nabla\varphi = 0$, so that φ is constant.

The third route is the shortest one. It involves a "magic trick" proving directly that any measurable function $\varphi : \Omega \to \mathbb{R}$ satisfying (6.1) must be constant, and which totally avoids the BBM technology.

As we are going to see later, it is also important to consider sums of functions in such spaces. More precisely, assume that $\varphi : \Omega \to \mathbb{Z}$ can be written as $\varphi = \varphi_1 + \varphi_2 + \varphi_3$, with $\varphi_1 \in VMO\,(\Omega;\mathbb{R})$, $\varphi_2 \in W^{1,1}(\Omega;\mathbb{R})$, and $\varphi_3 \in W^{1/p,p}(\Omega;\mathbb{R})$ for some $1 < p < \infty$. Can one conclude that φ is constant? It turns out that the answer is positive (Theorem 6.6), but the proof requires much more work and involves the introduction (in Section 6.5) of a new function space having the constancy property. Moreover, this space is sufficiently large to contain $VMO\,(\Omega;\mathbb{R})$, $W^{1,1}(\Omega;\mathbb{R})$, and $W^{1/p,p}(\Omega;\mathbb{R})$.

In Section 6.6, we give a proof of the BBM formula. We seize this opportunity to discuss in Section 6.7 variations on the theme of the BBM formula involving non-local energies, non-local perimeter, etc.

6.1 Constancy in $VMO\,(\Omega;\mathbb{Z})$

Proposition 6.1. *Any map φ in $VMO\,(\Omega;\mathbb{Z})$ is constant.*

Proof. We may assume that Ω is the unit ball, $\Omega = B_1(0)$. [It is well-known that, in a domain, a function that is locally constant is also globally constant.] Set, for $0 < \varepsilon < 1$, $\Omega_\varepsilon := B_{1-\varepsilon}(0)$, and $\varphi_\varepsilon(x) := \fint_{B_\varepsilon(x)} \varphi(y)\,dy$. Then, $\operatorname{dist}(\varphi_\varepsilon,\mathbb{Z}) \to 0$ uniformly on Ω_ε as $\varepsilon \to 0$ (see Remark 5.10). Since φ_ε is continuous and Ω_ε is connected, for small ε there exists some $k_\varepsilon \in \mathbb{Z}$ such that $\|\varphi_\varepsilon - k_\varepsilon\|_{L^\infty(\Omega_\varepsilon)} \to 0$ as $\varepsilon \to 0$. Using the fact that $\int_K \varphi_\varepsilon$ converges on each compact $K \subset \Omega$, we find that $k_\varepsilon = k$ is independent of ε for small ε. Thus, $\varphi = k$ a.e. in Ω. □

Corollary 6.1. *Let $0 < s < \infty$ and $1 \le p < \infty$ be such that $sp \ge N$. Assume that $\varphi \in W^{s,p}(\Omega;\mathbb{Z})$. Then, φ is constant.*

Proof. We may always assume that $sp = N$. [Indeed, if $sp > N$, then u is continuous, by the Sobolev embedding theorem, and the conclusion is obvious.] If $sp = N$, we rely on the embedding

$$W^{s,p} \hookrightarrow VMO \quad \text{if } sp = N \tag{6.2}$$

(see, for example, Brezis and Nirenberg [104, Example 2, p. 210] or Lemma 15.20 in Appendix 15.5). □

6.2 Constancy in $W^{1,1}(\Omega; \mathbb{Z})$

Proposition 6.2. *Assume that $\varphi \in W^{1,1}(\Omega; \mathbb{Z})$. Then, φ is constant.*

Proof. Let $k \in \mathbb{Z}$ be such that $|[\varphi = k]| > 0$. Consider $\Phi \in C^\infty(\mathbb{R})$ such that $\Phi(t) \equiv 1$ near $t = k$ and $\Phi \equiv 0$ outside a small neighborhood of k. Then, $\nabla(\Phi \circ \varphi) = \Phi'(\varphi)\nabla\varphi = 0$ a.e. (since $\Phi'(\ell) = 0,\ \forall \ell \in \mathbb{Z}$). Hence, $\Phi \circ \varphi$ is constant, so $\Phi \circ \varphi \equiv 1$. It follows that $\varphi = k$ a.e.

Alternatively, let $A_k := [\varphi = k]$, so that $\Omega = \bigcup_{k\in\mathbb{Z}} A_k$. By a well-known result of Stampacchia [329] (see also Gilbarg and Trudinger [164, Lemma 7.7]), $\nabla\varphi = 0$ a.e. in A_k and therefore $\nabla\varphi = 0$ a.e. in Ω. Hence, φ is a constant. □

6.3 Constancy in $W^{1/p,p}(\Omega; \mathbb{Z})$, $1 < p < \infty$

We will deduce constancy in $W^{1/p,p}(\Omega; \mathbb{Z})$, $1 < p < \infty$, from the following more general result.

Theorem 6.1. *Assume that $\varphi : \Omega \to \mathbb{Z}$ is measurable and satisfies*

$$\iint_{\varphi(x)\neq\varphi(y)} \frac{1}{|x-y|^{N+1}}\, dxdy < \infty. \tag{6.3}$$

Then, φ is constant.

The exponent $N + 1$ is optimal (see Remark 6.1).

Corollary 6.2. *Let $0 < s < \infty$ and $1 \leq p < \infty$ be such that $sp \geq 1$. Assume that $\varphi \in W^{s,p}(\Omega; \mathbb{Z})$. Then, φ is constant.*

Proof of Corollary 6.2 (assuming Proposition 6.2 and Theorem 6.1). The case where $s \geq 1$ follows from Proposition 6.2. The case where $0 < s < 1$ is an immediate consequence of Theorem 6.1 and of the fact that

$$|\varphi(x) - \varphi(y)| \geq 1 \text{ if } \varphi(x) \neq \varphi(y),$$

and therefore

$$\begin{aligned}\iint_{\varphi(x)\neq\varphi(y)} \frac{1}{|x-y|^{N+1}}\, dxdy &\leq \iint \frac{|\varphi(x)-\varphi(y)|^p}{|x-y|^{N+1}}\, dxdy \\ &\leq C \iint \frac{|\varphi(x)-\varphi(y)|^p}{|x-y|^{N+sp}}\, dxdy < \infty.\end{aligned}$$

□

Remark 6.1. It is surprising that the conclusion of Corollary 6.2 holds under the condition $sp \geq 1$, which is much weaker than the condition $sp \geq N$ needed for the injection of $W^{s,p}$ in *VMO*. Assumption $sp \geq 1$ is *optimal*. Indeed, the characteristic function, $\varphi = \chi_\omega$, of any smooth domain ω compactly contained in Ω belongs to $W^{s,p}(\Omega)$ for any s, p with $sp < 1$ (see Lemma 15.9 in Appendix 15.4).

Remark 6.2. One may wonder whether the conclusions of Propositions 6.1, 6.2, and Corollary 6.2 persist when $\mathbb{Z}$ is replaced by an arbitrary countable set $D \subset \mathbb{R}$. Clearly, any function $\varphi \in W^{1,1}(\Omega; D)$ is constant (the second proof of Proposition 6.2 remains unchanged). Surprisingly, the answer is negative in the two other cases (Bourgain [47]). Indeed, let C be the ternary Cantor set, and consider the Cantor staircase function φ, which satisfies $\varphi \in C^{0,\alpha}([0, 1]; \mathbb{R})$ for some $0 < \alpha < 1$ (see, for example, Leoni [220, Exercise 1.32]) and, in particular, $\varphi \in VMO$. Moreover, we have $\varphi(x) \in D$ for $x \in (0, 1) \setminus C$, for some countable set $D \subset \mathbb{R}$. We claim that φ belongs to $W^{1/p,p}((0, 1))$, $\forall p > 1$. Indeed, we clearly have $\varphi \in BV((0, 1)) \hookrightarrow W^{r,1}((0, 1))$, $\forall\, 0 < r < 1$. Choosing $r = 1 - (p - 1)\alpha$ (we may always assume that $\alpha < 1/(p - 1)$) and using Theorem 15.5, we find that $\varphi \in W^{r,1} \cap C^{0,\alpha} \hookrightarrow W^{1/p,p}$.

As mentioned in the Introduction, we present three proofs of Theorem 6.1, with totally different flavors.

First proof of Theorem 6.1. We start by establishing Corollary 6.2 from scratch.

Direct proof of Corollary 6.2. It suffices to prove that φ is locally constant and thus we may assume that $\Omega = (0, 1)^N$. Recall that, if $\varphi \in W^{s,p}(\Omega; \mathbb{R})$ (any s and any p), then, for each $1 \leq i \leq N$ and almost every $x' := (x_1, x_2, ..., x_{i-1}, x_{i+1}, ..., x_N) \in (0, 1)^{N-1}$, we have

$$(0, 1) \ni a \mapsto v(a) = \varphi(x_1, x_2, ..., x_{i-1}, a, x_{i+1}, ..., x_N) \in W^{s,p}((0, 1))$$

(see, for example, Adams [4, Lemma 7.44] and Corollary 15.1 in Appendix 15.1). Combining this with the embedding (6.2), we deduce, from Proposition 6.1, that for a.e. $x' \in (0, 1)^{N-1}$ the function v is constant a.e. To complete the proof of Corollary 6.2, we rely on the following purely measure-theoretical result. □

Lemma 6.1. *Let φ be a real-valued measurable function on $(0, 1)^N$ such that, for each $1 \leq i \leq N$ and for a.e. $(x_1, x_2, ...x_{i-1}, x_{i+1}, ..., x_N) \in (0, 1)^{N-1}$, the function*

$$a \mapsto v(a) = \varphi(x_1, x_2, ...x_{i-1}, a, x_{i+1}, ..., x_N)$$

is constant a.e. on $(0, 1)$. Then, φ is constant a.e. on $(0, 1)^N$.

Proof. We may always assume that φ is also bounded (and thus integrable), since otherwise we may replace φ by $\arctan \varphi$. By the triangle inequality, we have, with $\lambda = (\lambda_1, ..., \lambda_N)$ and $\mu = (\mu_1, ..., \mu_N)$:

$$\begin{aligned}|\varphi(\lambda) - \varphi(\mu)| \leq &|\varphi(\lambda_1, \lambda_2, ..., \lambda_{N-1}, \lambda_N) - \varphi(\lambda_1, \lambda_2, ..., \lambda_{N-1}, \mu_N)|\\ &+ |\varphi(\lambda_1, \lambda_2, ..., \lambda_{N-1}, \mu_N) - \varphi(\lambda_1, \lambda_2, ..., \mu_{N-1}, \mu_N)|\\ &+ ... + |\varphi(\lambda_1, \mu_2, ..., \mu_{N-1}, \mu_N) - \varphi(\mu_1, \mu_2, ..., \mu_{N-1}, \mu_N)|.\end{aligned}$$

It follows from the assumption that

$$\int_{(0,1)^N} \int_{(0,1)^N} |\varphi(\lambda) - \varphi(\mu)| \, d\lambda d\mu = 0.$$

Consequently, $\varphi(\lambda) - \varphi(\mu) = 0$ a.e. on $(0, 1)^N \times (0, 1)^N$, which implies that φ is constant a.e. on $(0, 1)^N$. □

First proof of Theorem 6.1 completed. Applying Corollary 6.2 with $s := 1/2, p := 2$ to χ_A, the characteristic function of a measurable set, we obtain the following

Corollary 6.3. *Let A be a measurable subset of Ω. If*

$$\int_A \int_{^cA} \frac{1}{|x-y|^{N+1}} \, dxdy < \infty, \tag{6.4}$$

then either A or ^cA is a null set.

We may now complete the first proof of Theorem 6.1. Let φ satisfy (6.3). Let $k \in \mathbb{Z}$ be such that the set $A := [\varphi = k]$ has positive measure. Applying Corollary 6.3 to this A, we find that $\varphi \equiv k$. □

The second proof of Theorem 6.1 relies on the following

Theorem 6.2. (the BBM formula). *Let $u \in L^1_{loc}(\Omega; \mathbb{R})$ with $|u|_{BV} = \|Du\|_{\mathcal{M}} < \infty$. Then,*

$$\lim_{\varepsilon \to 0} \int_\Omega \int_\Omega \frac{|u(x) - u(y)|}{|x-y|} \rho_\varepsilon(x-y) \, dxdy = K(1, N)|u|_{BV}. \tag{6.5}$$

Here, (ρ_ε) is a family of radial mollifiers in the sense of (6.8)–(6.10) *below, and $K(1, N)$ is a constant depending only on N, given by*

$$K(1, N) := \fint_{\mathbb{S}^{N-1}} |\omega \cdot e| \, d\sigma(\omega) \ (\text{any } e \in \mathbb{S}^{N-1}) = 2\frac{|\mathbb{B}^{N-1}|}{|\mathbb{S}^{N-1}|}, \tag{6.6}$$

where $|\mathbb{S}^0| = 2$ and $|B^0| = 1$, so that $K(1, 1) = 1$.

Conversely, if $u \in L^1_{loc}(\Omega)$ is such that

$$\liminf_{\varepsilon \to 0} \int_\Omega \int_\Omega \frac{|u(x) - u(y)|}{|x-y|} \rho_\varepsilon(x-y) \, dxdy < \infty, \tag{6.7}$$

then $|u|_{BV} < \infty$ and (6.5) *applies.*

In the above, (ρ_ε) is a family of radial functions on $\mathbb{R}^N$ such that

$$\rho_\varepsilon \geq 0 \text{ in } \mathbb{R}^N, \ \forall \varepsilon > 0, \tag{6.8}$$

$$\int_{\mathbb{R}^N} \rho_\varepsilon = 1, \ \forall \varepsilon > 0, \tag{6.9}$$

$$\lim_{\varepsilon \to 0} \int_{|x|>\delta} \rho_\varepsilon = 0 \text{ in } \mathbb{R}^N, \ \forall \delta > 0. \tag{6.10}$$

The value of $K(1, N)$ with $N \geq 2$ is obtained as follows. We may assume that $e = (0, \dots, 0, 1)$, and then

$$\begin{aligned} |\mathbb{S}^{N-1}|\, K(1, N) &= \int_{\mathbb{S}^{N-1}} |\omega_N|\, d\sigma = 2 \int_0^{\pi/2} \cos\theta\, (\sin\theta)^{N-2}\, |\mathbb{S}^{N-2}|\, d\theta \\ &= \frac{2}{N-1}\, |\mathbb{S}^{N-2}|. \end{aligned} \tag{6.11}$$

Noting that $(N-1)\,|B^{N-1}| = |\mathbb{S}^{N-2}|$, we obtain (6.6) from (6.11).
The proof of Theorem 6.2 is presented in Section 6.6.

Second proof of Theorem 6.1 (via the BBM formula). This relies on an alternative argument leading to Corollary 6.3. Let A be as in Corollary 6.3 and let $\rho_\varepsilon := C_N \varepsilon^{-N} \chi_{B_\varepsilon(0)}$. Then,

$$\begin{aligned} \iint \frac{|\chi_A(x) - \chi_A(y)|}{|x-y|} \rho_\varepsilon(x-y)\, dxdy =& \frac{2C_N}{\varepsilon^N} \iint_{\substack{x\in A,\, y\in {}^cA \\ |x-y|<\varepsilon}} \frac{1}{|x-y|}\, dxdy \\ \leq& 2C_N \iint_{\substack{x\in A,\, y\in {}^cA \\ |x-y|<\varepsilon}} \frac{1}{|x-y|^{N+1}}\, dxdy, \end{aligned}$$

and the latter quantity tends to 0 as $\varepsilon \to 0$ (by (6.4)). From Theorem 6.2 applied with $u := \chi_A$ and with $\rho_\varepsilon := C_N \varepsilon^{-N} \chi_{B_\varepsilon(0)}$, we obtain $D\chi_A = 0$. This implies Corollary 6.3 and thus, as above, Theorem 6.1. □

Third proof of Theorem 6.1. We rely on the following

Theorem 6.3. *Let $\varphi : \Omega \to \mathbb{R}$ be a measurable function. Assume that*

$$\int_\Omega \int_\Omega \frac{|\varphi(x) - \varphi(y)|}{|x-y|^{N+1}}\, dxdy < \infty. \tag{6.12}$$

Then, φ is constant.

We emphasize that here φ takes its values in $\mathbb{R}$ (and not $\mathbb{Z}$).
Theorem 6.3 applied with $\varphi := \chi_A$ implies Corollary 6.3, and thus Theorem 6.1. □

Proof of Theorem 6.3 (from scratch). As in the second proof of Theorem 6.1, Theorem 6.3 can be obtained by applying Theorem 6.2 with $\rho_\varepsilon := C_N \varepsilon^{-N} \chi_{B_\varepsilon(0)}$. Here is an alternative proof, which is more direct and does not rely on Theorem 6.2. We may assume that Ω is the unit ball. Set, for $0 < r \leq 2$, $U_r := \{(x, y) \in \Omega^2;\ |x - y| < r\}$ and

$$I_r := \iint_{U_r} \frac{|\varphi(x) - \varphi(y)|}{|x - y|^{N+1}}\, dxdy.$$

Then,

$$I_2 = \int_\Omega \int_\Omega \frac{|\varphi(x) - \varphi(y)|}{|x - y|^{N+1}}\, dxdy < \infty. \tag{6.13}$$

We will prove that

$$I_r \leq I_{r/2},\ \forall\, r \in (0, 2]. \tag{6.14}$$

Granted (6.14), we have, by (6.13), $I_2 = 0$, and thus φ is constant.

In turn, assertion (6.14) is obtained as follows. Set, for $(x, y) \in \Omega^2$, $z := (x + y)/2$. Let us note that, if $(x, y) \in U_r$, then $(x, z), (y, z) \in U_{r/2}$. We find that

$$\begin{aligned} I_r &= \iint_{U_r} \frac{|\varphi(x) - \varphi(y)|}{|x - y|^{N+1}}\, dxdy \leq \iint_{U_r} \frac{|\varphi(x) - \varphi(z)| + |\varphi(y) - \varphi(z)|}{|x - y|^{N+1}}\, dxdy \\ &= \frac{1}{2^{N+1}} \iint_{U_r} \frac{|\varphi(x) - \varphi(z)|}{|x - z|^{N+1}}\, dxdy + \frac{1}{2^{N+1}} \iint_{U_r} \frac{|\varphi(y) - \varphi(z)|}{|y - z|^{N+1}}\, dxdy \\ &\leq \frac{1}{2} \iint_{U_{r/2}} \frac{|\varphi(x) - \varphi(z)|}{|x - z|^{N+1}}\, dxdz + \frac{1}{2} \iint_{U_{r/2}} \frac{|\varphi(y) - \varphi(z)|}{|y - z|^{N+1}}\, dydz = I_{r/2}. \quad \square \end{aligned}$$

The above argument leads (as in Ranjbar-Motlagh [303]) to the following extension of Theorem 6.3.

Theorem 6.4. *Let $\Phi : [0, \infty) \to [0, \infty)$ be a convex increasing function such that $\Phi(0) = 0$. Let $\varphi : \Omega \to \mathbb{R}$ be a measurable function. Assume that*

$$\int_\Omega \int_\Omega \Phi\left(\frac{|\varphi(x) - \varphi(y)|}{|x - y|}\right) \frac{1}{|x - y|^N}\, dxdy < \infty. \tag{6.15}$$

Then, φ is constant.

Of particular interest is the case $\Phi(t) = t^p$ with $1 \leq p < \infty$.

Corollary 6.4. *Let $u : \Omega \to \mathbb{R}$ be a measurable function. Assume that*

$$\int_\Omega \int_\Omega \frac{|u(x) - u(y)|^p}{|x - y|^{N+p}}\, dxdy < \infty \text{ for some } 1 \leq p < \infty.$$

Then, u is constant.

Remark 6.3. There is a stronger variant of Corollary 6.4 (see Brezis, Van Schaftingen, and Yung [106, Theorem 2] and the reference therein to a preprint of Poliakovsky).

Proposition 6.3. *Let $u : \Omega \to \mathbb{R}$ be a measurable function. Assume that*

$$\lim_{\lambda\to\infty} \lambda \text{ meas} \left\{(x, y) \in \Omega \times \Omega;\ \frac{|u(x) - u(y)|^p}{|x - y|^{N+p}} \geq \lambda\right\} = 0$$

for some $1 \leq p < \infty$.

Then, u is constant.

We do not know whether the conclusion of Proposition 6.3 is still valid when "lim" is replaced by "lim inf" (see [106, Open Problem 1]).

6.4 Connectedness of the essential range

In a different but related direction, we mention the following fact. Recall that, if $\varphi : \Omega \to \mathbb{R}^d$ is measurable, then the essential range $\text{essR}(\varphi)$ of φ is defined as the smallest closed set F in $\mathbb{R}^d$ such that $\varphi(x) \in F$ a.e.

Theorem 6.5. *Let $0 < s < \infty$, $p \geq 1$ be such that $sp \geq 1$. Assume that $\varphi \in W^{s,p}(\Omega; \mathbb{R}^d) \cap L^\infty(\Omega; \mathbb{R}^d)$. Then, $\text{essR}(\varphi)$ is connected.*

The assumption $\varphi \in L^\infty$ can be removed in the following cases:

a) $sp > 1$
b) $d = 1$
c) $s = 1, p = 1$.

Proof. We may assume that $\Omega = (0, 1)^N$. We have to prove that, if F_1, F_2 are two disjoint closed sets such that $\varphi(x) \in F_1 \cup F_2$ a.e., then either $\varphi(x) \in F_1$ a.e. or $\varphi(x) \in F_2$ a.e.

If φ is bounded, then we may assume that F_1, F_2 are compact. Let $\Phi \in C^\infty(\mathbb{R}^d; \mathbb{R})$ have bounded derivatives and equal 1 in F_1, and 0 in F_2. Since the map $\Phi \circ \varphi$ belongs to $W^{s,p}(\Omega; \mathbb{Z})$, it is constant, by Corollary 6.2. The conclusion of the theorem is now clear.

If φ is not assumed bounded anymore, then there still exists some $\Phi \in C^\infty(\mathbb{R}^d; \mathbb{R})$ such that $\Phi = 1$ in F_1 and $\Phi = 0$ in F_2, but Φ need not be Lipschitz anymore. In cases *a)–c)*, we argue as follows:

a) For a.e. $x' \in (0, 1)^{N-1}$, the map

$$a \mapsto \zeta(a) := \varphi(x_1, x_2, ...x_{i-1}, a, x_{i+1}, ..., x_N)$$

belongs to $W^{s,p}$ and thus (after modification on a null set) is bounded and continuous (since $sp > 1$). For any such x', the map $\Phi \circ \zeta$ is continuous with values in $\{0, 1\}$, thus constant. We conclude via Lemma 6.1. The same argument works in case *c)*.

b) If $d = 1$, then we may assume that F_1, F_2 are disjoint closed half lines, and in this case we may take a Φ with bounded derivatives. □

Remark 6.4. The restriction $u \in L^\infty(\Omega; \mathbb{R}^d)$ is important when $1 < p < \infty$, $s = \dfrac{1}{p}$ and $d \geq 2$. For example, on $\Omega := (-1, 1)$, the function

$$t \mapsto \left(t, \left|\ln|t|\right|^\alpha\right), \ 0 < \alpha < 1/2,$$

belongs to $H^{1/2}(\Omega; \mathbb{R}^2)$ and its essential range is disconnected.

6.5 A new function space. Applications to sums

In this section, we present a "common roof" to the function spaces considered in Sections 6.1–6.3. The main result (Theorem 6.7) yields in particular the following

Theorem 6.6. *Assume that $\varphi : \Omega \to \mathbb{Z}$ is measurable and can be written as $\varphi = \varphi_1 + \varphi_2 + \varphi_3$, with $\varphi_1 \in VMO(\Omega; \mathbb{R})$, $\varphi_2 \in W^{1,1}(\Omega; \mathbb{R})$, and $\varphi_3 \in W^{1/p,p}(\Omega; \mathbb{R})$ for some $1 < p < \infty$. Then, φ is constant.*

This theorem is not only for "academic" purposes. It will be used in situations where the natural function space for the phase is not a single Sobolev space, but a sum of Sobolev spaces (see, for example, Theorems 8.5 and 8.8 in Section 8.2).

The proof of Theorem 6.6 relies heavily on the introduction of a new space of functions.

We may clearly assume that $\Omega = Q := (0, 1)^N$. For $0 < \varepsilon < 1$, let $Q_\varepsilon(a)$ be the ε-cube centered at a, with sides parallel to the axes.

Given $\varphi \in L^1(Q; \mathbb{R})$ and an ε-cube $Q_\varepsilon \subset Q$, we set

$$M(\varphi, Q_\varepsilon) := \fint_{Q_\varepsilon} |\varphi - \varphi_{Q_\varepsilon}|, \text{ where } \varphi_{Q_\varepsilon} := \fint_{Q_\varepsilon} \varphi, \tag{6.16}$$

and

$$M^*(\varphi, Q_\varepsilon) := \fint_{Q_\varepsilon}\fint_{Q_\varepsilon} |\varphi(y) - \varphi(z)|\, dydz. \tag{6.17}$$

Clearly, we have

$$M(\varphi, Q_\varepsilon) \leq M^*(\varphi, Q_\varepsilon) \leq 2M(\varphi, Q_\varepsilon). \tag{6.18}$$

Note that if $\varphi := \chi_A$, with $A \subset Q$ measurable, then

$$M(\varphi, Q_\varepsilon) = M^*(\varphi, Q_\varepsilon) = \frac{2\,|A \cap Q_\varepsilon|\,(|Q_\varepsilon| - |A \cap Q_\varepsilon|)}{|Q_\varepsilon|^2} \leq \frac{1}{2}. \tag{6.19}$$

The following quantity plays an important role:

$$[\varphi]_\varepsilon := \sup_{\mathcal{F}} \left\{ \varepsilon^{N-1} \sum_{j \in J} M(\varphi, Q_\varepsilon(a_j)) \right\}. \tag{6.20}$$

Here, $\mathcal{F}$ denotes a collection of mutually disjoint ε-cubes, $\mathcal{F} = (Q_\varepsilon(a_j))_{j\in J}$, such that $\#J$ (the cardinality of J) is $\le 1/\varepsilon^{N-1}$, and the sup in (6.20) is taken over all such collections. [Instead of $\#J$, we sometimes write $\#\mathcal{F}$.]

We then introduce the space

$$B := \left\{ \varphi \in L^1(Q; \mathbb{R});\ \sup_{0<\varepsilon<1} [\varphi]_\varepsilon < \infty \right\},$$

and the corresponding norm (modulo constants)

$$\|\varphi\|_B := \sup_{0<\varepsilon<1} [\varphi]_\varepsilon. \tag{6.21}$$

The definition of B is inspired by the celebrated *BMO* space of John–Nirenberg [206], equipped with the norm (modulo constants)

$$\|\varphi\|_{BMO} := \sup_{0<\varepsilon<1} \sup_{a\in Q} \{M(\varphi, Q_\varepsilon(a));\ Q_\varepsilon(a) \subset Q\}. \tag{6.22}$$

Here are several examples of functions in B.

Example 1. $BMO \subset B$, with continuous injection.
Indeed, using (6.22) we find that $\|\varphi\|_B \le \|\varphi\|_{BMO}$.

When $N = 1$, we clearly have $B = BMO$; however, when $N \ge 2$, B is strictly bigger than BMO (see, for example, Example 2 below).

Example 2. $BV \subset B$, with continuous injection.
Indeed, by Poincaré's inequality,

$$\fint_{Q_\varepsilon} |\varphi - \varphi_{Q_\varepsilon}| \le \frac{c_N}{\varepsilon^{N-1}} \int_{Q_\varepsilon} |D\varphi|,$$

so that

$$\sum_{j\in J} M(\varphi, Q_\varepsilon(a_j)) \le \frac{c_N}{\varepsilon^{N-1}} \int_{\cup_{j\in J} Q_\varepsilon(a_j)} |D\varphi|, \tag{6.23}$$

and thus

$$[\varphi]_\varepsilon \le c_N \|D\varphi\|_{\mathcal{M}}. \tag{6.24}$$

Example 3. $W^{1/p,p} \subset B$, $1 < p < \infty$, with continuous injection.
Indeed, for every fixed $\alpha > 0$ we have

$$\int_{Q_\varepsilon}\int_{Q_\varepsilon} |\varphi(y) - \varphi(z)|\, dydz \leq N^{\alpha/2}\varepsilon^{\alpha} \int_{Q_\varepsilon}\int_{Q_\varepsilon} \frac{|\varphi(y) - \varphi(z)|}{|y-z|^{\alpha}}\, dydz.$$

Choosing $\alpha := (N+1)/p$ and applying Hölder's inequality gives

$$M^*(\varphi, Q_\varepsilon) \leq \frac{c_N}{\varepsilon^{(N-1)/p}} \left[\int_{Q_\varepsilon}\int_{Q_\varepsilon} \frac{|\varphi(y) - \varphi(z)|^p}{|y-z|^{N+1}}\, dydz\right]^{1/p}, \text{ with } c_N := N^{(N+1)/2}.$$

Since $\#J \leq 1/\varepsilon^{N-1}$, we obtain, with $Q_\varepsilon := Q_\varepsilon(a_j)$,

$$\varepsilon^{N-1} \sum_{j \in J} M^*(\varphi, Q_\varepsilon) \leq c_N \left[\sum_{j\in J} \int_{Q_\varepsilon}\int_{Q_\varepsilon} \frac{|\varphi(y) - \varphi(z)|^p}{|y-z|^{N+1}}\, dydz\right]^{1/p}. \tag{6.25}$$

Therefore,

$$[\varphi]_\varepsilon \leq c_N |\varphi|_{W^{1/p,p}}.$$

An important quantity, associated with every $\varphi \in B$, is defined by

$$[\varphi] := \limsup_{\varepsilon \to 0}\, [\varphi]_\varepsilon. \tag{6.26}$$

The subspace

$$B_0 := \{\varphi \in B;\ [\varphi] = 0\} \tag{6.27}$$

plays a key role in this section.

Example 1'. $VMO \subset B_0$.
This is clear, since VMO functions (see Sarason [316]) are characterized by

$$\lim_{\varepsilon \to 0} \sup_{a \in Q} \{M(f, Q_\varepsilon(a));\ Q_\varepsilon(a) \subset Q\} = 0.$$

Moreover, $VMO = B_0$ when $N = 1$.

Example 2'. $W^{1,1} \subset B_0$.
This is clear from (6.23) and the fact that $\left|\cup_{j\in J} Q_\varepsilon(a_j)\right| \leq \varepsilon$.

Example 3'. $W^{1/p,p} \subset B_0$, $1 < p < \infty$.
This is an immediate consequence of (6.25) and the fact that

$$\left|\cup_{j\in J} Q_\varepsilon(a_j) \times Q_\varepsilon(a_j)\right| \leq \varepsilon^{N+1}.$$

In particular, we see that

$$VMO + W^{1,1} + W^{1/p,p} \subset B_0. \tag{6.28}$$

In view of (6.28), Theorem 6.6 is a special case of the following

Theorem 6.7. *Let φ be an integer-valued function on Q such that $\varphi \in B_0$. Then, φ is constant.*

When $N = 1$, we have $B_0 = VMO$ and we may then invoke the fact that functions in $VMO\,(Q; \mathbb{Z})$ are constant, by Proposition 6.1. Therefore, it suffices to prove Theorem 6.7 when $N \geq 2$. Next, we observe that it suffices to prove Theorem 6.7 when $\varphi := \chi_A$ for some $A \subset Q$. Indeed, let $k \in \mathbb{Z}$ be such that the set $A := [\varphi = k]$ satisfies $|A| > 0$. Consider $\Phi \in C^\infty(\mathbb{R})$ such that $\Phi(k) = 1$ and $\Phi \equiv 0$ outside a small neighborhood of k. Set $\psi := \Phi \circ \varphi$, so that $\psi \equiv \chi_A$. Clearly,

$$M\,(\psi, Q_\varepsilon) \leq M^*(\psi, Q_\varepsilon) \leq 2|\Phi|_{\text{Lip}}\, M\,(\varphi, Q_\varepsilon).$$

Since $\varphi \in B_0$, we deduce that $\psi \in B_0$. If the theorem holds for $\psi = \chi_A$, then $\psi \equiv 1$ and thus $\varphi \equiv k$.

Hence, it remains to prove Theorem 6.7 when $N \geq 2$ and $f := \chi_A$. In this case we have the following quantitative improvement of Theorem 6.7.

Theorem 6.8. *Let $N \geq 2$. There exists a constant C_N such that, if $\varphi = \chi_A$, with $A \subset Q$ measurable, then*

$$\left\| \varphi - \fint_Q \varphi \right\|_{L^{N/(N-1)}(Q)} \leq C_N[\varphi]. \tag{6.29}$$

Theorem 6.8 relies on the isoperimetric inequality (or equivalently, the embedding $W^{1,1} \hookrightarrow L^{N/(N-1)}$) and we refer the reader to Bourgain, Brezis, and Mironescu [57] for its proof.

Remark 6.5. A much more precise result (see Ambrosio, Bourgain, Brezis, and Figalli [9]) asserts that there exist two constants $0 < \underline{c}_N \leq \overline{c}_N < \infty$ such that, if $\varphi = \chi_A$, then

$$\underline{c}_N \min\{1, \|D\varphi\|_{\mathcal{M}}\} \leq [\varphi] \leq \overline{c}_N \min\{1, \|D\varphi\|_{\mathcal{M}}\}, \tag{6.30}$$

with the convention that $\|D\varphi\|_{\mathcal{M}} = \infty$ if $\varphi \notin BV(Q)$.

Note that

$$\left\| \varphi - \fint_Q \varphi \right\|_{L^{N/(N-1)}(Q)} \leq C\, \|D\varphi\|_{\mathcal{M}}, \tag{6.31}$$

by the Sobolev embedding, and that clearly

$$\left\| \varphi - \fint_Q \varphi \right\|_{L^{N/(N-1)}(Q)} \leq 2 \text{ when } \varphi = \chi_A. \tag{6.32}$$

Therefore,

$$\left\| \varphi - \fint_Q \varphi \right\|_{L^{N/(N-1)}(Q)} \le C \min\{1, \|D\varphi\|_{\mathscr{M}}\} \le C'[\varphi] \text{ by (6.30).}$$

In fact, using a variant of the definition (6.20) involving ε-cubes of general orientation, one obtains a quantity $[\varphi]_\varepsilon^*$ satisfying

$$[\varphi]_\varepsilon \le [\varphi]_\varepsilon^* \le C_1[\varphi]_{C_2\varepsilon}$$

for some constants $C_1 > 1$, $C_2 > 1$ depending only on N (see [9]). The main result in [9] asserts that, if $\varphi = \chi_A$, then

$$\lim_{\varepsilon\to 0} [\varphi]_\varepsilon^* = \frac{1}{2}\min\{1, \|D\varphi\|_{\mathscr{M}}\} = \frac{1}{2}\min\{1, \mathrm{Per}(A, Q)\}, \tag{6.33}$$

where $\mathrm{Per}(A, Q)$ denotes the perimeter of A relative to Q (i.e., "inside" Q).

The ingredients of the proof of (6.33) are much more sophisticated than those needed in the proof of Theorem 6.8.

Here is an immediate consequence of Theorem 6.6.

Corollary 6.5. *Let $s_1, \ldots, s_n > 0$, $1 \le p_1, \ldots, p_n < \infty$ be such that $s_j p_j \ge 1$, $j = 1, \ldots, n$. Assume that $\varphi_j \in W^{s_j,p_j}(\Omega; \mathbb{R})$, $j = 1, \ldots, n$, and that $\varphi := \varphi_1 + \ldots + \varphi_n$ takes its values in $\mathbb{Z}$. Then, φ is constant.*

Note that, although Corollary 6.5 follows from Theorem 6.6, the full strength of this theorem is not required in the proof of this corollary. Indeed, it suffices to argue by dimensional reduction as in the direct proof of Corollary 6.2. [Warning: this would not be possible if the sum also involves a *VMO* function, since dimensional reduction does not work in *VMO*.]

We next present another result, based on Proposition 6.1 combined with dimensional reduction. It will be used in the proof of Theorem 7.1.

Lemma 6.2. *Let $s_1, \ldots, s_n > 0$ and $1 \le p_1, \ldots, p_n < \infty$ be such that $s_j p_j \ge 1$, $j = 1, \ldots, n$. Let $\varphi \in W^{s_1,p_1} + \cdots + W^{s_n,p_n}(\Omega; \mathbb{R})$ be such that $e^{\imath\varphi} \in W^{s,p}$ for some $s \ge 1$ and $1 \le p < \infty$. Then, $\varphi \in W^{s,p}$.*

Proof. We start by proving that $\varphi \in W^{1,1}$. It suffices to consider the case where $\Omega = (0, 1)^N$. Let $u := e^{\imath\varphi} \in W^{s,p} \subset W^{1,1}$. Consider, as above, the functions

$$a \mapsto v(a) = u(x_1, x_2, \ldots, x_{i-1}, a, x_{i+1}, \ldots, x_N),$$

resp.

$$a \mapsto f(a) = \varphi(x_1, x_2, \ldots, x_{i-1}, a, x_{i+1}, \ldots, x_N).$$

For a.e. $x' = (x_1, x_2, \ldots, x_{i-1}, x_{i+1}, \ldots, x_N) \in (0, 1)^{N-1}$, these functions belong to

$$W^{1,1}((0,1);\mathbb{S}^1)\subset C([0,1]),$$

resp. to

$$W^{s_1,p_1}+\cdots+W^{s_n,p_n}((0,1);\mathbb{R})\subset VMO\,((0,1)).$$

For any such x', v has a $W^{1,1}$ lifting ψ, whose derivative is $v\wedge v'$. By Proposition 6.1, we have $f=\psi+2k\pi$ a.e. on $(0,1)$, for some constant $k\in\mathbb{Z}$.

So far, we have established that, for each $i\in\{1,\ldots,N\}$ and almost every $x'\in(0,1)^{N-1}$, f is absolutely continuous and we have

$$\int_{(0,1)^{N-1}}\int_0^1 |f'(a)|\,da\,dx'<\infty. \tag{6.34}$$

Property (6.34) implies that $\varphi\in W^{1,1}$ (Ziemer [349, Theorem 2.1.4]).

We next prove that $\varphi\in W^{s,p}$. Indeed, we know that $\varphi\in W^{1,1}$. Therefore, we have $\nabla\varphi=u\wedge\nabla u$. We conclude via Lemma 5.2. □

Remark 6.6. When $N\geq 2$, the assumption $s\geq 1$ is essential in Lemma 6.2. More precisely, for any $0<s<1$ and any $1<p<\infty$ with $1\leq sp<N$, there exists some $\psi\in W^{1,sp}\setminus W^{s,p}$ such that $u=e^{\imath\psi}\in W^{s,p}$ (see Remark 5.8). Thus, the conclusion of Lemma 6.2 does not hold for $s_1=1$, $p_1=sp$.

However, the lemma holds when $0<s<1$, $1<p<\infty$ and $sp\geq N$. [In particular, when $N=1$, we may drop the assumption $s\geq 1$ in Lemma 6.2.] Indeed, in this case it suffices to combine existence of a lifting in $W^{s,p}$ (Theorems 5.1 and 5.2) with Corollary 6.5.

6.6 Proof of Theorem 6.2 (the BBM formula)

We treat only the case where $\Omega=\mathbb{R}^N$. The presence of the boundary creates additional (mild) technical difficulties, which will not be discussed here (see Dávila [128]). When $\Omega=\mathbb{R}^N$, the relevant space is not $BV(\mathbb{R}^N)$, but the space $X:=\{u\in L^1_{loc}(\mathbb{R}^N);\ Du\in\mathscr{M}(\mathbb{R}^N)\}$, endowed with the natural seminorm $|u|_{BV}=\|Du\|_{\mathscr{M}}$.

For $u\in L^1_{loc}(\mathbb{R}^N;\mathbb{R})$, set

$$I_\varepsilon(u):=\int_{\mathbb{R}^N}\int_{\mathbb{R}^N}\frac{|u(x)-u(y)|}{|x-y|}\rho_\varepsilon(x-y)\,dxdy, \tag{6.35}$$

where (ρ_ε) is a family of radial mollifiers in the sense of (6.8)–(6.10). Since ρ_ε is radially symmetric, we will sometimes write (with abuse of notation) $\rho_\varepsilon(|x-y|)$ instead of $\rho_\varepsilon(x-y)$.

A basic ingredient of the proof of Theorem 6.2 is the following

Lemma 6.3. *Let $u \in L^1_{loc}(\mathbb{R}^N)$. Let ρ be a non-negative radial function of integral 1. Then,*

$$I(u) := \int_{\mathbb{R}^N}\int_{\mathbb{R}^N} \frac{|u(x)-u(y)|}{|x-y|}\rho(x-y)\,dxdy \leq K(1,N)|u|_{BV}. \tag{6.36}$$

Here and elsewhere in the remaining sections, we use the convention $|u|_{BV} = \infty$ if $Du \notin \mathscr{M}$.

Proof. We start with the case where $u \in C^\infty$. Then,

$$\begin{aligned} I(u) &= \int_{\mathbb{R}^N}\int_{\mathbb{R}^N} \frac{1}{|h|}\left|\int_0^1 (\nabla u(x+th))\cdot h\,dt\right| \rho(h)\,dxdh \\ &\leq \int_{\mathbb{R}^N}\int_{\mathbb{R}^N}\int_0^1 \frac{|(\nabla u(x+th))\cdot h|}{|h|}\rho(h)\,dxdh \\ &= \int_{\mathbb{R}^N}\int_{\mathbb{R}^N} \frac{|(\nabla u(y))\cdot h|}{|h|}\rho(h)\,dydh = K(1,N)\|\nabla u\|_{L^1}. \end{aligned} \tag{6.37}$$

Let now $u \in L^1_{loc}(\mathbb{R}^N)$. Let (ζ_δ) be a family of standard mollifiers and set $u^\delta := u * \zeta_\delta$. Consider a sequence $\delta_n \to 0$ such that $u^{\delta_n} \to u$ a.e. If we apply (6.37) to u^{δ_n} and then let $n \to \infty$, we obtain (6.36) (via Fatou's lemma). □

Sketch of proof of Theorem 6.2 *when* $\Omega = \mathbb{R}^N$. We assume that $\operatorname{supp}\rho_\varepsilon \subset B_1(0)$, $\forall\, \varepsilon > 0$, although this is not relevant for the conclusion. In view of Lemma 6.3, it suffices to prove that, for any $u \in L^1_{loc}(\mathbb{R}^N)$, we have

$$\liminf_{\varepsilon\to 0} I_\varepsilon(u) \geq K(1,N)|u|_{BV}. \tag{6.38}$$

Let u^δ be as in the proof of Lemma 6.3. Since

$$\|\tau_h u^\delta - u^\delta\|_{L^1(\mathbb{R}^N)} \leq \|\tau_h u - u\|_{L^1(\mathbb{R}^N)},$$

we find that

$$I_\varepsilon(u^\delta) \leq I_\varepsilon(u). \tag{6.39}$$

Next, we use the following lemma, applied to $v := u^\delta$.

Lemma 6.4. *Let $v \in C^\infty(\mathbb{R}^N)$. Then,*

$$\liminf_{\varepsilon\to 0} I_\varepsilon(v) \geq K(1,N)|v|_{W^{1,1}}. \tag{6.40}$$

We obtain (6.38) by combining Lemma 6.4 with (6.39) and the fact that $\lim_{\delta\to 0}|u^\delta|_{W^{1,1}} = |u|_{BV}$. □

Proof of Lemma 6.4. It suffices to prove that

$$\liminf_{\varepsilon\to 0} I_\varepsilon(v) \geq K(1,N)\|\nabla v\|_{L^1(B_R(0))}, \tag{6.41}$$

and then let $R\to\infty$. Inequality (6.41) is obtained by plugging the Taylor expansion

$$v(y) = v(x) + (y-x)\cdot\nabla v(x) + O(|y-x|^2)$$

into the formula of $I_\varepsilon(v)$. We find that

$$\lim_{\varepsilon\to 0}\int_{B_R(0)}\int_{\mathbb{R}^N}\frac{|v(x)-v(y)|}{|x-y|}\rho_\varepsilon(x-y)\,dxdy = K(1,N)\|\nabla v\|_{L^1(B_R(0))}. \quad \square \tag{6.42}$$

We end this section by presenting two immediate consequences of Theorem 6.2.

Example 1. With the choice $\rho_\varepsilon(x) := \begin{cases}\dfrac{\varepsilon}{|\mathbb{S}^{N-1}||x|^{N-\varepsilon}}, & \text{if } |x|<1\\ 0, & \text{if } |x|>1\end{cases}$,

we find that, $\forall\, u\in L^1_{loc}(\Omega)$,

$$\lim_{\varepsilon\to 0}\varepsilon\iint_{|x-y|<1}\frac{|u(x)-u(y)|}{|x-y|^{N+1-\varepsilon}}\,dxdy = |\mathbb{S}^{N-1}|\,K(1,N)\,|u|_{BV} = 2|B^{N-1}|\,|u|_{BV}.$$

In particular,

$$\text{if } \int_\Omega\int_\Omega\frac{|u(x)-u(y)|}{|x-y|^{N+1-\varepsilon}}\,dxdy = o(1/\varepsilon) \text{ as } \varepsilon\to 0,\ \text{ then } u \text{ is constant.} \tag{6.43}$$

This is an improvement of Theorem 6.3.

Example 2. The choice $\rho_\varepsilon(x) := \begin{cases}\dfrac{(N+1)|x|}{|\mathbb{S}^{N-1}|\,\varepsilon^{N+1}}, & \text{if } |x|<\varepsilon\\ 0, & \text{if } |x|>\varepsilon\end{cases}$ leads to

$$\lim_{\varepsilon\to 0}\frac{1}{\varepsilon^{N+1}}\iint_{|x-y|<\varepsilon}|u(x)-u(y)|\,dxdy = |\mathbb{S}^{N-1}|\frac{K(1,N)}{N+1}|u|_{BV} = \frac{2\,|\mathbb{B}^{N-1}|}{N+1}|u|_{BV}.$$

6.7 Complements and open problems

6.7.1 Non-local energies and perimeter

Let $E\subset\mathbb{R}^N$ be measurable and $0<\varepsilon<1$. Set

$$P_\varepsilon(E) := \int_E\int_{\mathbb{R}^N\setminus E}\frac{dxdy}{|x-y|^{N+1-\varepsilon}} \in [0,\infty] \tag{6.44}$$

and

$$\operatorname{Per} E := |\chi_E|_{BV} = \int_{\mathbb{R}^N} |D(\chi_E)| = \|D\chi_E\|_{\mathcal{M}}.$$

Corollary 6.6. *Assume that $N \geq 2$. Then,*

$$\lim_{\varepsilon \to 0} \varepsilon P_\varepsilon(E) = \begin{cases} |\mathbb{B}^{N-1}| \operatorname{Per} E, & \text{if } \operatorname{Per} E < \infty \\ +\infty, & \text{otherwise} \end{cases}. \tag{6.45}$$

Remark 6.7. The conclusion of Corollary 6.6 fails when $N = 1$. [Take, for example, $E = (0, \infty)$, so that $P_\varepsilon(E) = \infty, \forall \varepsilon > 0$.] However, under the additional assumption that $|E| < \infty$ (or $|\mathbb{R} \setminus E| < \infty$) one has $\lim_{\varepsilon \to 0} \varepsilon P_\varepsilon(E) = \operatorname{Per} E$. This follows from the proof of Corollary 6.6.

Proof of Corollary 6.6. Write

$$P_\varepsilon(E) = \iint\limits_{\substack{E \times (\mathbb{R}^N \setminus E) \\ |x-y|<1}} \cdots + \iint\limits_{\substack{E \times (\mathbb{R}^N \setminus E) \\ |x-y| \geq 1}} \cdots := A_\varepsilon + B_\varepsilon. \tag{6.46}$$

Choosing ρ_ε as in Example 1 above, we see that

$$\int_{\mathbb{R}^N} \int_{\mathbb{R}^N} \frac{|\chi_E(x) - \chi_E(y)|}{|x-y|} \rho_\varepsilon(x-y)\, dxdy = \frac{2\varepsilon}{|\mathbb{S}^{N-1}|} A_\varepsilon.$$

Applying Theorem 6.2, we obtain that

$$\lim_{\varepsilon \to 0} \varepsilon A_\varepsilon = \frac{|\mathbb{S}^{N-1}|}{2} K(1, N) \operatorname{Per} E \ \text{ if } \operatorname{Per} E < \infty. \tag{6.47}$$

From (6.6), we know that

$$\frac{|\mathbb{S}^{N-1}|}{2} K(1, N) = |\mathbb{B}^{N-1}|.$$

On the other hand, we claim that, if $\operatorname{Per} E < \infty$, then either $|E| < \infty$ or $|\mathbb{R}^N \setminus E| < \infty$. Indeed, by Sobolev's inequality applied to $f_n(x) := \chi_E(x/n)$ on $B := \mathbb{B}^N$, we have

$$\left\| f_n - \fint_B f_n \right\|_{L^{N/(N-1)}(B)} \leq C_N \int_B |Df_n|,$$

i.e.,

$$\left(\int_{|y|<n} |\chi_E(y) - c_n|^{N/(N-1)}\, dy \right)^{(N-1)/N} \leq c_N \int_{|y|<n} |D\chi_E|,$$

where $c_n := \fint_B f_n \in [0, 1]$. Passing to the limit along a subsequence (n_k) such that $c_{n_k} \to c$, we obtain $\int_{\mathbb{R}^N} |\chi_E - c|^{N/(N-1)}\, dy < \infty$, and the conclusion follows.

We now return to (6.46). Assuming, for example, that $|E| < \infty$, we see that

$$B_\varepsilon \le \iint\limits_{\substack{E\times\mathbb{R}^N\\ |x-y|\ge 1}} \frac{dxdy}{|x-y|^{N+1-\varepsilon}} \le C_N\, |E|\, \frac{1}{1-\varepsilon},$$

and therefore $\lim_{\varepsilon\to 0} \varepsilon\, B_\varepsilon = 0$. Combining this with (6.47), we obtain (6.45) in the case $\operatorname{Per} E < \infty$.

When $\operatorname{Per} E = \infty$, we note that $P_\varepsilon(E) \ge A_\varepsilon$, and that, by (6.47), we have $\lim_{\varepsilon\to 0} \varepsilon\, A_\varepsilon = \infty$. □

6.7.2 *Γ-convergence of I_ε*

Let us recall the definition of Γ-convergence in $L^p(\Omega)$, $p \ge 1$. We say that a sequence (I_n) of functionals in $L^p(\Omega)$ Γ-converges to a limiting functional I_0 if the two following properties hold:

(Γ_1) $\forall\, u$, $\forall\, u_n \in L^p(\Omega)$ such that $u_n \to u$ in $L^p(\Omega)$, one has

$$\liminf_{n\to\infty} I_n(u_n) \ge I_0(u),$$

(Γ_2) $\forall\, u \in L^p(\Omega)$, $\exists\, (\widetilde{u_n}) \subset L^p(\Omega)$ such that $\widetilde{u_n} \to u$ in $L^p(\Omega)$ and

$$\limsup_{n\to\infty} I_n(\widetilde{u_n}) \le I_0(u).$$

The definition goes over to a family $(I_\varepsilon)_{\varepsilon>0}$.

The main result in this section is

Theorem 6.9. *Let I_ε be as in* (6.35). *Then, $I_\varepsilon \to I_0$ in the sense of Γ-convergence in $L^1(\mathbb{R}^N)$ as $\varepsilon \to 0$, where*

$$I_0(u) := K(1, N)|u|_{BV}.$$

Theorem 6.9 is taken from Ponce [297]; the proof presented below is somewhat simpler than the original one.

Proof. Property (Γ_2) follows from Theorem 6.2 with the choice $\widetilde{u_\varepsilon} :\equiv u$. We now turn to the proof of (Γ_1) and, for simplicity, we treat only the case $\Omega = \mathbb{R}^N$.

Let (ζ_δ) be a family of standard mollifiers. By (6.39), we have

$$I_\varepsilon(\zeta_\delta * u_\varepsilon) \le I_\varepsilon(u_\varepsilon),\ \forall\, \delta > 0. \tag{6.48}$$

The triangle inequality combined with Lemma 6.3 gives

$$\begin{aligned}|I_\varepsilon(\zeta_\delta * u_\varepsilon) - I_\varepsilon(\zeta_\delta * u)| &\le I_\varepsilon(\zeta_\delta * (u_\varepsilon - u)) \\ &\le K(1,N)\|\nabla(\zeta_\delta * (u_\varepsilon - u))\|_{L^1} \\ &\le K(1,N)\|\nabla\zeta_\delta\|_{L^1}\|u_\varepsilon - u\|_{L^1}.\end{aligned} \tag{6.49}$$

On the other hand, if we apply Theorem 6.2 to $\zeta_\delta * u$ we see that, for *fixed* δ,

$$\lim_{\varepsilon\to 0} I_\varepsilon(\zeta_\delta * u) = K(1,N)\|\nabla(\zeta_\delta * u)\|_{L^1}. \tag{6.50}$$

Combining (6.49), (6.50), and the assumption that $u_\varepsilon \to u$ in L^1, we conclude that, for fixed $\delta > 0$,

$$\lim_{\varepsilon\to 0} I_\varepsilon(\zeta_\delta * u_\varepsilon) = K(1,N)\|\nabla(\zeta_\delta * u)\|_{L^1}. \tag{6.51}$$

Putting (6.48) and (6.51) together, we find

$$\liminf_{\varepsilon\to 0} I_\varepsilon(u_\varepsilon) \ge K(1,N)\|\nabla(\zeta_\delta * u)\|_{L^1}, \ \forall\,\delta > 0. \tag{6.52}$$

Finally, we let $\delta \to 0$ and use the fact that

$$\lim_{\delta\to 0} \|\nabla(\zeta_\delta * u)\|_{L^1} = |u|_{BV}. \tag{6.53}$$

From (6.52) and (6.53), we conclude that

$$\liminf_{\varepsilon\to 0} I_\varepsilon(u_\varepsilon) \ge I_0(u).$$

□

6.7.3 Another convex non-local approximation of the BV norm

Inspired by Gilboa and Osher [165, 166], Leoni and Spector [221] introduced a variant of I_ε : for any given $q \ge 1$ and $u \in L^1(\Omega)$, set

$$\Psi_{\varepsilon,q}(u) := \int_\Omega dx \left[\int_\Omega \frac{|u(x)-u(y)|^q}{|x-y|^q}\rho_\varepsilon(x-y)dy\right]^{1/q}, \tag{6.54}$$

$$\Psi_{0,q}(u) := \begin{cases} [K(q,N)]^{1/q}|u|_{BV}, & \text{if } u \in BV(\Omega) \\ \infty, & \text{otherwise}\end{cases}, \tag{6.55}$$

where (ρ_ε) is a family of radial mollifiers and

$$K(q,N) := \fint_{\mathbb{S}^{N-1}} |\omega\cdot e|^q\, d\sigma(\omega) \ (\text{any } e \in \mathbb{S}^{N-1}). \tag{6.56}$$

Following the computation in Section 6.6, we see that when $\Omega = \mathbb{R}^N$ and $q \geq 1$,

$$\lim_{\varepsilon \to 0} \Psi_{\varepsilon,q}(u) = [K(q,N)]^{1/q} \int_{\mathbb{R}^N} |\nabla u|, \ \forall u \in C_c^\infty(\mathbb{R}^N). \tag{6.57}$$

However, as noted in Leoni and Spector [222] and Brezis and Nguyen [100], the analogue of (6.36) *may fail* when $q > 1$. In fact, the properties of the functional $\Psi_{\varepsilon,q}$ are *very sensitive* to the choice of ρ_ε. For every $q > 1$, there exist (see [100]) *special* families of radial mollifiers (ρ_ε) such that for every w, and every ε,

$$\Psi_{\varepsilon,q}(w) \leq C \int_\Omega |\nabla w(x)| dx, \tag{6.58}$$

for some constant C. In this case we have

$$\lim_{\varepsilon \to 0} \Psi_{\varepsilon,q}(u) = \Psi_{0,q}(u), \ \forall u \in L^1(\Omega). \tag{6.59}$$

On the other hand, for every $q > 1$, it is possible to construct families (ρ_ε) of radial mollifiers and a function $u \in W^{1,1}(\Omega)$ such that

$$\lim_{\varepsilon \to 0} \Psi_{\varepsilon,q}(u) = \infty \tag{6.60}$$

(see [100]). In other words, *pointwise convergence may fail* when $q > 1$: it is *not always* true that

$$\lim_{\varepsilon \to 0} \Psi_{\varepsilon,q}(u) = \Psi_{0,q}(u), \ \forall u \in BV(\Omega).$$

Interestingly, within the framework of Γ-convergence, the "natural" expected conclusion is "restored" for all families of radial mollifiers:

Theorem 6.10. *For every $q \geq 1$, we have*

$$\Psi_{\varepsilon,q} \to \Psi_{0,q} \text{ in the sense of } \Gamma\text{-convergence in } L^1 \text{ as } \varepsilon \to 0. \tag{6.61}$$

We refer the reader to the proofs in [222] and [100]. Note that Theorem 6.10 implies, in particular, in view of property (Γ_1) (see the definition of Γ-convergence in Section 6.7.2), that

$$\liminf_{\varepsilon \to 0} \Psi_{\varepsilon,q}(u) \geq \Psi_{0,q}(u), \ \forall u \in BV, \tag{6.62}$$

and we know from (6.60) that this inequality can be strict for some functions $u \in BV$.

6.7.4 A non-convex non-local approximation of the BV norm

We now present a different kind of approximation of the BV norm via non-convex functionals introduced by Brezis and Nguyen [76, 101, 102].

Throughout this section we asume that $\varphi : [0, \infty) \to [0, \infty)$ satisfies the following properties

$$\varphi \text{ is continuous and non-decreasing,} \tag{6.63}$$

$$\varphi(t) \le a t^2, \ \forall t \in [0, 1], \ \text{for some positive constant } a, \tag{6.64}$$

$$\varphi(t) \le b, \ \forall t \ge 0, \ \text{for some positive constant } b, \tag{6.65}$$

and the normalization condition

$$|\mathbb{S}^{N-1}| \int_0^\infty \frac{\varphi(t)}{t^2} \, dt = 1. \tag{6.66}$$

A typical example (which occurs in image processing) is

$$\varphi(t) := c \left(1 - e^{-t^2}\right), \ \forall t \ge 0, \tag{6.67}$$

where the constant $c = c_N$ is adjusted such that (6.66) holds.

Given a measurable function u on Ω and a small parameter $\delta > 0$, set

$$\Lambda(u) := \int_\Omega \int_\Omega \frac{\varphi(|u(x) - u(y)|)}{|x - y|^{N+1}} dxdy \text{ and } \Lambda_\delta(u) := \delta \Lambda(u/\delta). \tag{6.68}$$

Note that $\Lambda_\delta(u) < \infty$, e.g., if $u \in C_c^1(\Omega)$. It turns out that $\Lambda_\delta(u)$ converges as $\delta \to 0$ to a multiple of $\int_\Omega |\nabla u|$. But the mode of convergence is extremely delicate.

We start with the simple case where $u \in C_c^1(\Omega)$.

Theorem 6.11. *We have*

$$\lim_{\delta \to 0} \Lambda_\delta(u) = K(1, N) \int_\Omega |\nabla u|, \ \forall u \in C_c^1(\Omega), \tag{6.69}$$

where $K(1, N)$ *is defined in* (6.6).

In view of Theorem 6.11, one might have expected that

$$\lim_{\delta \to 0} \Lambda_\delta(u) = K(1, N) \, |u|_{BV}, \ \forall u \in L^1(\Omega). \tag{6.70}$$

Assertion (6.70) is definitely wrong. Actually, the study of the asymptotic behavior of Λ_δ as $\delta \to 0$ is *quite intricate*. Two basic properties satisfied by I_ε are *not* fulfilled by Λ_δ:

1. Despite the fact that (6.69) holds, there is *no* constant C such that

$$\Lambda_\delta(u) \le C \int_\Omega |\nabla u|, \ \forall \delta > 0, \ \forall u \in C_c^\infty(\Omega).$$

2. Λ_δ is *not* a convex functional.

The following theorem summarizes what is known about the pointwise convergence of Λ_δ.

Theorem 6.12. *One has*

$$\lim_{\delta \to 0} \Lambda_\delta(u) = K(1, N) \int_\Omega |\nabla u|, \text{ if } u \in W^{1,p}(\Omega) \text{ for some } p > 1, \tag{6.71}$$

$$\liminf_{\delta \to 0} \Lambda_\delta(u) \ge K(1, N) \int_\Omega |\nabla u|, \ \forall u \in W^{1,1}(\Omega), \tag{6.72}$$

and

$$\limsup_{\delta \to 0} \Lambda_\delta(u) \ge K(1, N)\, |u|_{BV}, \ \forall u \in L^1(\Omega). \tag{6.73}$$

The proofs of Theorems 6.11 and 6.12 are presented in Brezis and Nguyen [102].

Remark 6.8. Many pathologies may occur (see [102]).

1. One can construct a function $u \in W^{1,1}(\Omega)$ with compact support such that

$$\lim_{\delta \to 0} \Lambda_\delta(u) = \infty.$$

2. One can construct functions φ and $u \in BV(\Omega) \cap C(\overline{\Omega})$ (with $u \notin W^{1,1}(\Omega)$) such that

$$\liminf_{\delta \to 0} \Lambda_\delta(u) < K(1, N) \int_\Omega |\nabla u|.$$

It turns out that Γ-convergence is perfectly suited to analyze the asymptotic behavior of Λ_δ as $\delta \to 0$. The principal result in [102] is the following

Theorem 6.13. *There exists a constant $k \in (0, 1]$ such that, as $\delta \to 0$,*

$$\Lambda_\delta \to \Lambda_0 \text{ in the sense of } \Gamma\text{-convergence in } L^1(\Omega),$$

where

$$\Lambda_0(u) := \begin{cases} k\, K(1, N)\, |u|_{BV}, & \text{if } u \in BV(\Omega), \\ \infty, & \text{otherwise} \end{cases}.$$

Remark 6.9. This result provides an interesting situation where the pointwise limit and the Γ-limit are different. The proof of Theorem 6.13 is quite complex and relies on earlier results of Nguyen [278, 281]. The appearance of the constant k, with possibly $k < 1$, remains mysterious. It is not known whether $k < 1$ when φ is given by (6.67). On the other hand, Antonucci, Gobbino, and Picenni [15] have constructed an interesting example of φ satisfying (6.63)–(6.66), and such that $k = 1$.

6.7.5 L^p versions

Theorem 6.2 and all the results in Sections 6.7.2–6.7.4 admit L^p versions with $1 < p < \infty$. For example, here is an L^p version of Theorem 6.2, established in Bourgain, Brezis, and Mironescu [52].

For $u \in L^1_{loc}(\Omega)$ and $1 < p < \infty$, set

$$I_{\varepsilon,p}(u) := \int_\Omega \int_\Omega \frac{|u(x) - u(y)|^p}{|x - y|^p} \rho_\varepsilon(x - y)\, dxdy,$$

where (ρ_ε) is a family of radial mollifiers.

Theorem 6.14. *Let $u \in W^{1,p}(\Omega)$. Then,*

$$\lim_{\varepsilon \to 0} I_{\varepsilon,p}(u) = K(p, N) \int_\Omega |\nabla u|^p, \tag{6.74}$$

where $K(p, N)$ is given by (6.56).

Conversely, if $u \in L^p(\Omega)$ is such that $\liminf_{\varepsilon\to 0} I_{\varepsilon,p}(u) < \infty$, *then $u \in W^{1,p}(\Omega)$.*

Remark 6.10. An L^p version of Theorem 6.10 is presented in Leoni and Spector [221, 222] for the functional

$$\Psi_{\varepsilon,p,q}(u) := \int_\Omega dx \left[\int_\Omega \frac{|u(x) - u(y)|^{pq}}{|x - y|^{pq}} \rho_\varepsilon(x - y)\, dxdy \right]^{1/q},$$

with $1 \le p, q < \infty$. An L^p version of Theorem 6.11 is presented in Brezis and Nguyen [103] (with roots in works by Bourgain and Nguyen [60, 277]). The main result asserts that the functionals

$$\Lambda_{\delta,p}(u) := \delta^p \int_\Omega \int_\Omega \varphi\left(\frac{|u(x) - u(y)|}{\delta}\right) \frac{1}{|x - y|^{N+p}}\, dxdy,$$

with $1 < p < \infty$, Γ-converge, as $\delta \to 0$, to a multiple of $\int_\Omega |\nabla u(x)|^p\, dx$.

6.7.6 Open problems

Open Problem 9. Let $\varphi \in B_0$ be such that $u := e^{\imath\varphi} \in W^{1,1}$. Is it true that $\varphi \in W^{1,1}$?

This question is motivated by Lemma 6.2. Recall that the "typical" examples of functions in B_0 are those in *VMO*, $W^{1,1}$ and $W^{1/p,p}$. Note that, by Lemma 6.2, the answer is positive if $\varphi \in W^{1/p,p}$. It is also positive if $\varphi \in VMO$, since in that case we have $u \in VMO$ and (by Theorem 5.9), u has some phase in $W^{1,1} \cap VMO$. It then suffices to invoke Proposition 6.1.

The answer is also positive if $N = 2$ and u is smooth except at one point a. Indeed, in this case u has a local smooth phase away from a, and then Proposition 6.1 implies that φ is smooth in $\Omega \setminus \{a\}$. By differentiating the equality $u = e^{\imath\varphi}$ we find that $|\nabla\varphi| = |\nabla u|$ in $\Omega \setminus \{a\}$, and thus $\varphi \in W^{1,1}$ (by Lemma 1.10).

6.8 Comments

1. Proposition 6.1 is taken from Brezis and Nirenberg [104, Proof of Theorem 3, Step 5]. A *VMO* version of Theorem 6.5 appears in [104, Section I.5.3]. A result somewhat related to Theorem 6.5 can be found in Giaquinta, Modica, and Souček [155, Theorem 1].
2. As stated, Theorem 6.1 is new, but trivially equivalent to Corollary 6.2, which was known before. Corollary 6.2 with $N = 1$ was initially proved by Garsia and Rodemich [154, Corollary 7.1]. For $s = 1/2$ and $p = 2$, it appears in Hardt, Kinderlehrer, and Lin [181, Lemma 1.1]. They indicate a very sketchy argument when $N = 1$ and then proceed by dimensional reduction. The dimensional reduction was carefully justified in Brezis, Li, Mironescu, and Nirenberg [81], and our proof of Lemma 6.1 follows that paper. Bethuel and Demengel [34, Lemma A.1] have stated Corollary 6.2 when $sp > 1$. Their argument is also based on a dimensional reduction; the one-dimensional case is trivial. In full generality, Corollary 6.2 appears for the first time in Bourgain, Brezis, and Mironescu [50, Appendix B]. Corollary 6.2 is used, for example, in the study of turbulence (see Kuksin and Shirikyan [217, Lemma 5.2.16, Appendix 7.15]).
3. A weaker form of Theorem 6.2, where "=" in (6.5) is replaced by "∼," is originally due to Bourgain, Brezis, and Mironescu [52]. Equality in (6.5) was obtained by Dávila [128]. Our presentation incorporates a suggestion of Stein [332], as in Brezis [72], and ideas from Van Schaftingen and Willem [341]. The L^p version, Theorem 6.14, is from [52]. The special case $p = 2$, $\rho_\varepsilon(x) = C_N\,\varepsilon/|x|^{N-\varepsilon}\,\chi_{B_1(0)}(x)$, has been originally discovered by Maz'ya and Nagel [241]. Independently of [52], Gobbino and Mora [168] obtained Theorems 6.2 and 6.14 in the very special case where $\rho_\varepsilon(x) = \varepsilon^{-N} f(|x|/\varepsilon)$, with $f : [0,\infty) \to [0,\infty)$ continuous satisfying $\int_0^\infty r^{a-p} f(r)\,dr < \infty$, $\forall\, a \geq 0$.
   One of the consequences of Theorems 6.2 and 6.14 is the following: we have $\lim_{s \nearrow 1} (1-s)^{1/p} |u|_{W^{s,p}} = C(p,N)|u|_{W^{1,p}}$. Thus the factor $(1-s)^{1/p}$ in front of the $W^{s,p}$ seminorm restores the continuity of the mapping $s \mapsto |u|_{W^{s,p}}$ as $s \nearrow s_0 = 1$. The analogue of this result for $s \searrow s_0 = 0$ was obtained by Maz'ya and Shaposhnikova [243, 245]. The case of an arbitrary integer s_0 (in the scale of Besov spaces) was obtained by Milman [250], and by Karadzhov, Milman, and Xiao [212].

Extensions of Theorems 6.2 and 6.14 to higher order derivatives are presented in Borghol [41], Bojarski [38], Cui, Lam, and Lu [122], and Tian and Du [335]. Theorems 6.2 and 6.14 have been extended to non-smooth domains by Leoni and Spector [221]. We also mention an interesting variant of the BBM formula for vector fields, obtained by Mengesha [246].

4. Theorem 6.3 and its generalizations, Corollary 6.4 and formula (6.43), are due to Bourgain, Brezis, and Mironescu [52] (see also Brezis [72]). Actually, the BBM formula (Theorems 6.2 and 6.14) was discovered while searching for a proof of assertion (6.43). Corollary 6.4 when $p = N$ was obtained independently of [52] by Bourdon and Pajot [46].
The direct proof of Theorem 6.3 ("from scratch") we present here was discovered in 2013 by an anonymous author (see https://math.stackexchange.com/q/545051) and later rediscovered by Ranjbar-Motlagh [303]. Another direct proof of Theorem 6.3 was obtained by De Marco, Mariconda, and Solimini [131]. Theorem 6.4 is a special case of a result of Ignat [192], who studied in great generality the validity of Theorem 6.4 for continuous $\Phi : [0, \infty) \to [0, \infty)$ such that $\Phi(0) = 0$ and $\Phi(t) > 0$ when $t > 0$ (see also Garsia and Rodemich [154, Section 7] for earlier results).
Analogues of Corollary 6.4 on measured metric spaces have been investigated by Bourdon and Pajot [46], Pietruska-Pałuba [290], and Ranjbar-Motlagh [302]. The case of self-similar sets (such as the Sierpiński carpet) was studied in greater detail by Pietruska-Pałuba [289], Grigor'yan, Hu, and Lau [170], and a BBM-type formula in the spirit of Theorem 6.14 was obtained by Gu and Lau [171].
A characterization of polynomials in the spirit of Theorem 6.3 was obtained by Borghol [41] (see also Bojarski, Ihnatsyeva, and Kinnunen [39]).
Far-reaching extensions of Theorem 6.3 are presented in Bourgain, Brezis, and Mironescu [53] (see also Maz'ya and Shaposhnikova [243, 245]. A typical estimate in the spirit of [53] asserts that

$$\left\| \varphi - \fint_{\Omega} \varphi \right\|_{L^1} \leq C(\Omega)\varepsilon \int_{\Omega}\int_{\Omega} \frac{|\varphi(x) - \varphi(y)|}{|x - y|^{N+1-\varepsilon}}\, dxdy, \ \forall\, \varepsilon \in (0, 1/2].$$

Other topics related to the BBM formula appear in [99, 100, 291, 296, 300, 326–328, 335].

5. Most of Section 6.5 is taken from Bourgain, Brezis, and Mironescu [57]. Corollary 6.5 is from Bourgain, Brezis, and Mironescu [50]. Lemma 6.2 is new.
6. Some of the non-local functionals introduced in Sections 6.6 and 6.7 are related to various "denoising" tools used in image processing. The original filter of Rudin, Osher, and Fatemi [310] made extensive use of the *BV* norm. Subsequently, experts realized that it was more efficient to work with non-local functionals "resembling" the *BV* norm, see Buades, Coll, and Morel [108]. The functional I_ε defined by (6.35) occurs in Aubert and Kornprobst [16]. The functional $\Psi_{\varepsilon,q}$ defined in (6.54) with $q = 2$, is reminiscent of filters used in Gilboa and Osher [165, 166]. The functional Λ_δ defined by (6.68) is reminiscent of the "neigh-

borhood filters" presented in [107] and [108] (see also Brezis [76], Brezis and Nguyen [102], and the references therein).

7. A number of authors have extended the concept of Sobolev spaces to maps defined on a measured metric space X with values in a metric space Y. We mention in particular Ambrosio, Cheeger, Franchi, Gromov, Hajłasz, Heinonen, Jost, Kinnunen, N. Korevaar, Koskela, Reshetnyak, Schoen, and Shanmugalingam; see, for example, the monograph Heinonen [189] and the references therein.
   The results in Section 6.7 suggest a new definition of BV and Sobolev maps from X to Y. This direction of research has been investigated by Chiron [118], Bojarski [38], Barbieri [22], Cui, Lam, and Lu [121], Munnier [276], and Di Marino and Squassina [138].
8. In recent years, there has been much activity concerning the study of minimal surfaces associated with non-local energies of the type $P_\varepsilon(E)$ defined in (6.44). In view of Corollary 6.6, these minimal surfaces converge as $\varepsilon \to 0$ to classical minimal surfaces. We refer the reader to the papers [12, 109, 113, 114, 150, 238] and the references therein. Some of these authors have rediscovered Corollary 6.6 under additional restrictive assumptions on the regularity of E.
9. In a different but related direction, Peletier, Planqué, and Röger [288] proved that $W^{1,2}$ functions are characterized by the property

$$\int (\varphi * \rho_\varepsilon)^2 = \int \varphi^2 + O(\varepsilon^2), \tag{6.75}$$

   where $\rho_\varepsilon(x) = \varepsilon^{-N}\,\rho(|x|/\varepsilon)$. More precisely, if $\varphi \in W^{1,2}$, (6.75) becomes

$$\int (\varphi * \rho_\varepsilon)^2 = \int \varphi^2 - \varepsilon^2\, C(\rho) \int |\nabla\varphi|^2 + o(\varepsilon^2).$$

Chapter 7
Factorization

7.0 Introduction

In Chapter 1, we saw that a map in $W^{1,1}(\Omega; \mathbb{S}^1)$, where $N \geq 2$, need not have a phase in $W^{1,1}$. Recall that a typical obstruction arises when $N = 2$ and $0 \in \Omega$, with $v_0(x) := x/|x|$. We know that

$$v_0 \in W^{1,q}, \ \forall q < 2, \tag{7.1}$$

and that there exists no $\psi_0 \in W^{1,1}(\Omega; \mathbb{R})$ such that $v_0 = e^{\imath\psi_0}$. Fix any $\varphi_0 \in W^{1,1}$ with $\varphi_0 \notin W^{1,q}$ when $q > 1$. Set

$$u_0 := e^{\imath\varphi_0} v_0. \tag{7.2}$$

Note that u_0 is in $W^{1,1}$, but not in $W^{1,q}$ when $q > 1$. Clearly, there exists no $\psi \in W^{1,1}$ such that $u_0 = e^{\imath\psi}$. Nevertheless, by (7.2), u_0 is a product, where the first factor admits a lifting in the natural space $W^{1,1}$, whereas the second factor admits no lifting in $W^{1,1}$, but has better regularity than u_0 (by (7.1)).

The main result in this chapter shows that this is a very general phenomenon: given any s and p, given $u \in W^{s,p}(\Omega; \mathbb{S}^1)$, we may write $u = e^{\imath\varphi} v$, with $\varphi \in W^{s,p}$ and v "smoother" than just $W^{s,p}$.

Theorem 7.1. *Assume that $N \geq 1$, $s > 0$, and $1 \leq p < \infty$. Let $\Omega \subset \mathbb{R}^N$ be a smooth bounded domain (not necessarily simply connected). Then:*

1. *Every $u \in W^{s,p}(\Omega; \mathbb{S}^1)$ can be written as $u = e^{\imath\varphi}v$, where $\varphi \in W^{s,p}(\Omega; \mathbb{R})$ and $v \in W^{sp,1}(\Omega; \mathbb{S}^1)$ satisfy the estimates*

$$\|\varphi\|_{W^{s,p}} \leq C|u|_{W^{s,p}} \tag{7.3}$$

and

$$\left\| v - \fint_\Omega u \right\|_{W^{sp,1}} \leq C|u|^p_{W^{s,p}} + C|u|_{W^{s,p}}. \tag{7.4}$$

H. Brezis and P. Mironescu, *Sobolev Maps to the Circle*, Progress in Nonlinear Differential Equations and Their Applications 96,
https://doi.org/10.1007/978-1-0716-1512-6_7

2. *If, in addition, $sp \geq 1$, then in particular we have $v \in W^{1,sp}(\Omega;\mathbb{S}^1)$. In this case, v satisfies the additional estimate*

$$|v|_{W^{1,sp}} \leq C|u|_{W^{s,p}}^{1/s}. \tag{7.5}$$

3. *When $sp < 1$, estimate (7.4) can be refined to*

$$\left\| v - \fint_{\Omega} u \right\|_{L^1} \leq C|u|_{W^{s,p}} \tag{7.6}$$

and

$$|v|_{W^{sp,1}} \leq C|u|_{W^{s,p}}^{p}. \tag{7.7}$$

4. *Moreover, if $u \in VMO_{loc}(\Omega;\mathbb{S}^1)$, then we may choose $v \in C^\infty(\Omega;\mathbb{S}^1)$, and φ "as smooth as u permits": $\varphi \in C^\infty(\Omega;\mathbb{R})$ (resp. $\varphi \in C(\Omega;\mathbb{R})$, resp. $\varphi \in VMO_{loc}(\Omega;\mathbb{R})$) if $u \in C^\infty(\Omega;\mathbb{S}^1)$ (resp. $u \in C(\Omega;\mathbb{S}^1)$, resp. $u \in VMO_{loc}(\Omega;\mathbb{S}^1)$).*

In the above, C denotes various constants depending only on s, p, and Ω.

Remark 7.1. Theorem 7.1 is, in some sense, a first step towards an $\mathbb{S}^1$ version of the celebrated Weierstrass factorization theorem for holomorphic functions. This will be explained in Section 8.7.1.

Remark 7.2. When $N \geq 2$ and $1 \leq sp < 2$, the exponent sp, which appears in the regularity of v, is optimal, in the sense that in general we cannot choose $v \in W^{1,sp+\varepsilon}$, and thus, by Gagliardo–Nirenberg, we cannot choose v in $W^{sp+\varepsilon,1}$ (see Proposition 8.3).

However, *in general*, there is no optimal regularity for v. Consider, for example, the space $W^{1,p}(\Omega;\mathbb{S}^1)$, with $p \geq 2$, and assume Ω simply connected. If $u \in W^{1,p}(\Omega;\mathbb{S}^1)$, write $u = e^{\imath\psi}$, with $\psi \in W^{1,p}$, let $C := \fint_\Omega \psi$, and set $\varphi := \psi - C$, $v := e^{\imath C}$. Clearly, (7.3) holds.

The validity of (7.4) follows from the inequality

$$\left| F\left(\fint_\Omega \psi\right) - \fint_\Omega F(\psi) \right| \leq |F|_{\text{Lip}} \fint_\Omega \left| \psi - \fint_\Omega \psi \right|,$$

which leads to

$$\left\| v - \fint_\Omega u \right\|_{W^{p,1}} \sim \left| e^{\imath \fint_\Omega \psi} - \fint_\Omega e^{\imath\psi} \right| \leq \fint_\Omega \left| \psi - \fint_\Omega \psi \right| \lesssim \|\nabla\psi\|_{L^p} = |u|_{W^{1,p}}.$$

Remark 7.3. In Chapter 5, we already fully answered the question of existence of a lifting in $W^{s,p}$. In particular, when $N = 1$, we established the existence of a lifting for any s and p (see Theorem 5.1). We call the attention of the reader to the fact that Theorem 7.1 is of another nature. Let us explain the difference. Assume, for example, that $s = 1/p$, with $1 < p < \infty$. Then, every $u \in W^{1/p,p}((0,1);\mathbb{S}^1)$ has a unique (mod 2π) phase $\psi \in W^{1/p,p}((0,1);\mathbb{R})$. However, there is no bound for $|\psi|_{W^{1/p,p}}$ in terms of $|u|_{W^{1/p,p}}$ (see Proposition 9.2). By contrast, the factorization in Theorem 7.1 comes with the bounds (7.3)–(7.5). Similarly, when $N \geq 2, 0 < s < 1$, and $sp \geq N$.

Remark 7.4. The proof of Theorem 7.1 yields a stronger result; namely, the φ we construct satisfies $\varphi \in \bigcap_{0<\theta\leq 1} W^{\theta s,p/\theta}$ and

$$\|\varphi\|_{W^{\theta s,p/\theta}} \leq C_{s,p,\theta}|u|^{\theta}_{W^{s,p}}. \tag{7.8}$$

Indeed, when $s = 1$ and $p = 1$ we may take $\varphi = 0$. For the other values of s and p, it suffices to note that the construction of φ in the proof of Theorem 7.1 does not depend on s, p, and u. If $u \in W^{s,p}(\Omega; \mathbb{S}^1)$ and $0 < \theta < 1$, then $u \in W^{\theta s,p/\theta}(\Omega; \mathbb{S}^1)$ (Lemma 15.6), and thus φ satisfies $\varphi \in W^{\theta s,p/\theta}(\Omega; \mathbb{R})$ and the corresponding estimates.

However, we do not know whether *this* φ, given by formula (7.19) below, belongs to *BMO* (see Open Problem 11).

As we just pointed out, the case $s = 1, p = 1$ in Theorem 7.1 is trivial. However, in this case one can establish a factorization where v enjoys an additional property, namely $v \in W^{1/p,p}$, $\forall p \geq 1$ (see Theorem 7.2 and Remarks 7.7 and 7.8).

Theorem 7.2. *Assume that $N \geq 1$ and let $u \in W^{1,1}(\Omega; \mathbb{S}^1)$. Then, given any $1 < p < \infty$, u can be written as $u = e^{\imath\varphi}v$, where $\varphi \in W^{1,1}(\Omega; \mathbb{R})$ and $v \in W^{1/p,p}(\Omega; \mathbb{S}^1)$ satisfy the estimates*

$$|\varphi|_{W^{1,1}} \leq C|u|_{W^{1,1}} \tag{7.9}$$

and

$$|v|^p_{W^{1/p,p}} \leq C_p|u|_{W^{1,1}}. \tag{7.10}$$

7.1 Proof of Theorem 7.2

We start by introducing a class of maps that will play a crucial role in the next chapters; a similar class was first considered by Bethuel [26] in his seminal work on density questions. We let U be an "appropriate" relatively compact open neighborhood of Ω; "appropriate" is meant in the sense of Lemma 15.27, i.e., extension by reflection across $\partial\Omega$ preserves $W^{s,p}$ if $s < 1 + 1/p$. Set

$$\begin{aligned}\mathscr{R} := \{u_{|\Omega};\ & u \in C^\infty(\overline{U} \setminus S; \mathbb{S}^1) \text{ for some } (N-2)-\text{submanifold } S \text{ of } U,\\ & \text{and } |D^\ell u(x)| \leq C(\ell)\,[\operatorname{dist}(x, S)]^{-\ell},\ \forall \ell \geq 1,\ \forall x \in \overline{U} \setminus S\}.\end{aligned} \tag{7.11}$$

We emphasize the fact that the singular set S and the constants $C(\ell)$ depend on u. The main interest of $\mathscr{R}$ is that it is a dense class in $W^{s,p}(\Omega; \mathbb{S}^1)$, provided $1 \leq sp < 2$ (Theorem 10.3 in Chapter 10).

Remark 7.5. A typical example of a map in $\mathscr{R}$ when $N = 2$ is $u_{|\Omega}$, where

$$u(x) := \prod_{i=1}^{m} g_i((x - a_i)/|x - a_i|), \ x \in \mathbb{R}^2,$$

$(a_i)_{1 \leq i \leq m}$ are points in Ω and $(g_i)_{1 \leq i \leq m}$ belong to $C^\infty(\mathbb{S}^1; \mathbb{S}^1)$.

For the needs of the proof of Theorem 7.2, we define a class slightly larger than $\mathscr{R}$ as follows:

$$\begin{aligned}\widetilde{\mathscr{R}} := \{u_{|\Omega};\ & u \in C^\infty(\overline{U} \setminus S; \mathbb{S}^1) \text{ for some finite union } S \text{ of } (N-2)-\\ & \text{submanifolds of } U, \text{ and } |D^\ell u(x)| \leq C(\ell)[\operatorname{dist}(x, S)]^{-\ell}, \\ & \forall \ell \geq 1, \ \forall x \in \overline{U} \setminus S\}.\end{aligned} \tag{7.12}$$

We have $\widetilde{\mathscr{R}} \subset W^{1,1}(\Omega; \mathbb{S}^1)$, and more generally $\widetilde{\mathscr{R}} \subset W^{s,p}(\Omega; \mathbb{S}^1)$ when $sp < 2$ (see Remark 15.11).

Remark 7.6. A finite union of *disjoint* $(N-2)$-submanifolds of U is still an $(N-2)$-submanifold of U. However, a finite union of submanifolds that intersect is *not*, in general, a submanifold of U. In the definition of $\widetilde{\mathscr{R}}$, the submanifolds may intersect. Note that, if $u, v \in \mathscr{R}$, then the product uv need not belong to $\mathscr{R}$, but $uv \in \widetilde{\mathscr{R}}$.

The proof of Theorem 7.2 relies on the following

Lemma 7.1. *Let $u \in \widetilde{\mathscr{R}}$ and let $\delta > 0$. Then, for every $1 < p < \infty$, there exist some constants C_1 and C_p and some $v : \Omega \to \mathbb{S}^1$ (depending on u, δ and p) such that*

a) $v \in W^{1,1} \cap W^{1/p,p}(\Omega; \mathbb{S}^1)$
b) $|v|^p_{W^{1/p,p}} \leq C_p |u|_{W^{1,1}}$ *and* $|v|_{W^{1,1}} \leq C_1 |u|_{W^{1,1}}$
c) $|[v \neq 1]| \leq \delta$
d) $Jv = Ju$.

We defer the proof of Lemma 7.1 to Appendix 15.4 (see Lemma 15.15 there) and proceed to the

Proof of Theorem 7.2. Let $u \in W^{1,1}(\Omega; \mathbb{S}^1)$. By the density of $\mathscr{R}$ in $W^{1,1}(\Omega; \mathbb{S}^1)$ (Theorem 10.3), there exists a sequence $(u_n) \subset \mathscr{R}$ such that $u_n \to u$ in $W^{1,1}$. Since $W^{1,1} \cap L^\infty$ is an algebra (Lemma 15.41), we find that $\lim_{m,n \to \infty} u_m \overline{u_n} = 1$ (the limit is taken in $W^{1,1}$). Possibly after extraction and relabelling, we may assume that $|u_0|_{W^{1,1}} \leq 2|u|_{W^{1,1}}$ and $|u_n \overline{u_{n-1}}|_{W^{1,1}} \leq 2^{1-n}\, |u|_{W^{1,1}}$, $\forall n \geq 1$. We set $v_0 := u_0$ and, for $n \geq 1$, $v_n := u_n \overline{u_{n-1}}$. Then, $v_n \in \widetilde{\mathscr{R}}$, $u_k = \prod_{n=0}^{k} v_n \to u$ in $W^{1,1}$ as $k \to \infty$, and $|v_n|_{W^{1,1}} \leq 2^{1-n}\, |u|_{W^{1,1}}, \forall n \geq 0$.

By Lemma 7.1, for each n there exists some $w_n \in W^{1,1} \cap W^{1/p,p}(\Omega; \mathbb{S}^1)$ such that

$|[w_n \neq 1]| \leq 2^{-n}$, $Jw_n = Jv_n$, $|w_n|^p_{W^{1/p,p}} \leq 2^{1-n} C_p |u|_{W^{1,1}}$,
and $|w_n|_{W^{1,1}} \leq 2^{1-n} C_1 |u|_{W^{1,1}}$.

We claim that the infinite product $\prod_{n\geq 0} w_n$ converges in $W^{1/p,p}(\Omega; \mathbb{S}^1)$ and in $W^{1,1}(\Omega; \mathbb{S}^1)$. Indeed, set $F(x, y) := \sum_{n\geq 0} |w_n(x) - w_n(y)|$, so that

$$\left(\iint_{\Omega^2} \frac{[F(x, y)]^p}{|x - y|^{N+1}} dxdy\right)^{1/p} \leq \sum_{n\geq 0} |w_n|_{W^{1/p,p}} < \infty. \tag{7.13}$$

Set $f_k := w_0 \cdots w_k$. Clearly, we have

$$|f_k(x) - f_k(y)| \leq |w_0(x) - w_0(y)| + \cdots + |w_k(x) - w_k(y)| \leq F(x, y). \tag{7.14}$$

For $\ell > k$, set $g_{k,\ell} := w_{k+1} \cdots w_\ell$. Then,

$$f_\ell - f_k = f_k g_{k,\ell} - f_k \text{ and the set } A_{k,\ell} := [g_{k,\ell} \neq 1] \text{ satisfies } |A_{k,\ell}| \leq 2^{1-k}. \tag{7.15}$$

We next note the following straightforward inequalities, valid for $f, g, h, w : \Omega \to \mathbb{S}^1$, $0 < s < 1$ and $1 \leq q < \infty$:

$$|fg - f|_{W^{s,q}} \leq 2\left(\iint_{[g(x)\neq 1]} \frac{|f(x) - f(y)|^q}{|x - y|^{N+sq}} dxdy\right)^{1/q} + |g|_{W^{s,q}} \tag{7.16}$$

and

$$|hw|_{W^{s,q}} \leq |h|_{W^{s,q}} + |w|_{W^{s,q}}. \tag{7.17}$$

Using (7.13)–(7.17), we obtain

$$|f_\ell - f_k|_{W^{1/p,p}} \leq 2\left(\iint_{[g_{k,\ell}(x)\neq 1]} \frac{[F(x, y)]^q}{|x - y|^{N+sq}} dxdy\right)^{1/q} + 2^{1-k} \to 0$$

uniformly in ℓ as $k \to \infty$, as claimed.

Convergence of the product in $W^{1,1}$ is established similarly, using the appropriate substitutes of (7.16) and (7.17).

Since

$$J\left(\prod_{n=0}^{k} w_n\right) = \sum_{n=0}^{k} Jw_n = \sum_{n=0}^{k} Jv_n = J\left(\prod_{n=0}^{k} v_n\right) = Ju_k,$$

we find that $J\left(\prod_{n\geq 0} w_n\right) = \lim_{k\to\infty} Ju_k = Ju$. Thus, if we set $v := \prod_{n\geq 0} w_n$, then $v \in W^{1/p,p}(\Omega; \mathbb{S}^1) \cap W^{1,1}(\Omega; \mathbb{S}^1)$ and $Jv = Ju$. Since $Jv = Ju$, we may write $u = e^{\imath\varphi} v$

for some $\varphi \in W^{1,1}(\Omega; \mathbb{R})$. The decomposition $u = e^{\imath\varphi} v$ has all the required properties. □

Remark 7.7. With more work, it is possible to improve the conclusion of Theorem 7.2 to $v \in \bigcap_{q \geq 1} W^{1/q,q}(\Omega; \mathbb{S}^1)$ (see Bousquet and Mironescu [63, proof of Theorem 1.3], and also Remark 7.8).

7.2 Outline of the proof of Theorem 7.1

The case $0 < s < 1$ is at the heart of the proof. The case $s \geq 1$ reduces to the case $0 < s < 1$; this rather straightforward reduction relies essentially on the Gagliardo–Nirenberg inequalities (see Appendix 15.2).

Let $0 < s < 1$ and consider some $u \in W^{s,p}(\Omega; \mathbb{S}^1)$. We start by extending u to $\mathbb{R}^N$ such that the extension (still denoted by u) $u : \mathbb{R}^N \to \mathbb{R}^2$ is constant at infinity. We next extend u to $\mathbb{R}^N \times (0, \infty)$ by the formula $w(x, \varepsilon) := u * \rho_\varepsilon(x), \forall x \in \mathbb{R}^N, \forall \varepsilon > 0$; here, (ρ_ε) is a family of standard mollifiers. We continue by (almost) projecting w onto $\mathbb{S}^1$. For this purpose, we consider an "approximate projection" Π onto $\mathbb{S}^1$, i.e., a smooth map

$$\Pi \in C^\infty(\mathbb{R}^2; \mathbb{R}^2) \text{ such that } \Pi(z) = \frac{z}{|z|} \text{ when } |z| \geq \frac{1}{2}. \tag{7.18}$$

One may view the new map $V := \Pi \circ w$ as an approximate projection of w onto $\mathbb{S}^1$.

We then define φ by the formula

$$\varphi(x) := -\int_0^\infty V(x, \varepsilon) \wedge \partial_\varepsilon V(x, \varepsilon)\, d\varepsilon, \ \text{ for } x \in \Omega. \tag{7.19}$$

Notation. If $V = V(x, \varepsilon)$, then $\partial_\varepsilon V$ (resp. $\partial_{x_j} V$) stands for the partial derivative of V in the variable ε (resp. x_j). A "generic" partial derivative is denoted by $\partial^\alpha V$, with $\alpha \in \mathbb{N}^{N+1}$.

At this stage, it is not even clear that the integral in (7.19) converges for a.e. x; this will be part of the proof.

Moreover, we will prove that, if we set $v := e^{-\imath\varphi} u$, then φ and v satisfy respectively(7.3) and (7.4). In addition to some "geometric" ingredients such as Lemma 7.8, the proof of (7.3) and (7.4) relies heavily on the theory of weighted Sobolev spaces (see Uspenskiĭ [340]) in a version, adapted to the proof of Theorem 7.1, that we briefly recall in the next section.

7.3 A glimpse of the theory of weighted Sobolev spaces

We present here only the results relevant for our needs; much more general results, for which we refer the reader to Mironescu and Russ [265], hold true. For the "global" results (in the whole of $\mathbb{R}^N$) we follow Uspenskiĭ [340] and [265]. The local versions (on bounded domains), which can be obtained by adapting the global ones, are established in Mironescu, Russ, and Sire [266].

Let $\rho \in C_c^\infty(B_1(0))$ satisfy $\int \rho = 1$. Given $f \in L^1_{loc}(\mathbb{R}^N)$, we let

$$F(x, \varepsilon) := f * \rho_\varepsilon(x), \ \forall x \in \mathbb{R}^N, \ \forall \varepsilon > 0,$$

where

$$\rho_\varepsilon(x) := \frac{1}{\varepsilon^N}\rho(x/\varepsilon). \tag{7.20}$$

The "inverse" theory of weighted Sobolev spaces asserts the following.

Lemma 7.2. *Let $0 < s < 1$, $1 \le p < \infty$ and let ℓ be an integer ≥ 1. Then, for every $f \in W^{s,p}(\mathbb{R}^N)$ and for every $\alpha \in \mathbb{N}^{N+1}$ with $|\alpha| = \ell$, we have*

$$\int_{\mathbb{R}^N}\int_0^\infty \varepsilon^{p(\ell-s)-1}|\partial^\alpha F(x,\varepsilon)|^p \, d\varepsilon dx \le C_\ell |f|^p_{W^{s,p}(\mathbb{R}^N)}. \tag{7.21}$$

More generally, if $s > 0$ is not an integer and if $\ell > s$ is an integer, then

$$\int_{\mathbb{R}^N}\int_0^\infty \varepsilon^{p(\ell-s)-1}|\partial^\alpha F(x,\varepsilon)|^p \, d\varepsilon dx \le C_\ell |f|^p_{W^{s,p}(\mathbb{R}^N)}. \tag{7.22}$$

The "direct" theory asserts the following.

Lemma 7.3. *Let $s > 0$. Assume that $1 \le p < \infty$ if s is not an integer, resp. $1 \le p \le 2$ if s is an integer.*

Let ℓ be an integer $> s$. Consider some $G \in C^\infty(\mathbb{R}^N \times (0,\infty))$ such that the limit $f(x) := \lim_{\varepsilon\to 0} G(x,\varepsilon)$ exists for a.e. $x \in \mathbb{R}^N$, and belongs to $L^p(\mathbb{R}^N)$.

Assume, in addition, that

$$\int_{\mathbb{R}^N}\int_0^\infty \varepsilon^{p(\ell-s)-1}|\partial^\alpha G(x,\varepsilon)|^p \, d\varepsilon dx < \infty, \ \forall \alpha \in \mathbb{N}^{N+1} \text{ with } |\alpha| = \ell. \tag{7.23}$$

Then, $f \in W^{s,p}(\mathbb{R}^N)$, and we have

$$\begin{aligned}\|f\|^p_{W^{s,p}(\mathbb{R}^N)} \le & C_\ell \sum_{|\alpha|=\ell}\int_{\mathbb{R}^N}\int_0^\infty \varepsilon^{p(\ell-s)-1}|\partial^\alpha G(x,\varepsilon)|^p \, d\varepsilon dx \\ & + C_\ell \|f\|^p_{L^p(\mathbb{R}^N)}.\end{aligned} \tag{7.24}$$

We will also need the following local version of Lemma 7.3.

Lemma 7.4. *Let $\Omega \subset \mathbb{R}^N$ be a smooth bounded domain. Let s, p, and ℓ be as in Lemma 7.3. Consider some $G \in C^\infty(\Omega \times (0, \infty))$ such that the limit $f(x) := \lim_{\varepsilon \to 0} G(x, \varepsilon)$ exists for a.e. $x \in \Omega$, and belongs to $L^p(\Omega)$.*

Assume, in addition, that

$$\int_\Omega \int_0^\infty \varepsilon^{p(\ell-s)-1} |\partial^\alpha G(x,\varepsilon)|^p \, d\varepsilon dx < \infty, \ \forall \alpha \in \mathbb{N}^{N+1} \text{ with } |\alpha| = \ell. \tag{7.25}$$

Then, $f \in W^{s,p}(\Omega)$, and we have

$$\|f\|^p_{W^{s,p}(\Omega)} \le C_\ell \sum_{|\alpha|=\ell} \int_\Omega \int_0^\infty \varepsilon^{p(\ell-s)-1} |\partial^\alpha G(x,\varepsilon)|^p \, d\varepsilon dx + C_\ell \|f\|^p_{L^p(\Omega)}. \tag{7.26}$$

If, in addition, $0 < s < 1$, then we also have

$$|f|^p_{W^{s,p}(\Omega)} \le C \sum_{|\alpha|=1} \int_\Omega \int_0^\infty \varepsilon^{p(1-s)-1} |\partial^\alpha G(x,\varepsilon)|^p \, d\varepsilon dx. \tag{7.27}$$

We may now proceed to the

7.4 Proof of Theorem 7.1

Let us note that the theorem is trivial when $p = 1$, since in that case we may take $\varphi = 0$ and $v = u$. Therefore, we may always assume that $1 < p < \infty$; however, most of the estimates we establish are also valid for $p = 1$.

In Steps 1 and 2, we construct φ and v, and define the main functions used in the proof. The construction does not depend on s or p. In Steps 3 to 9, we assume that $0 < s < 1$. The reduction of the general case $s > 0$ to the case $0 < s < 1$ is performed in Steps 10 to 13; a specific argument is needed in the case where s is an integer. In Step 14, we establish estimate (7.5) (which does not follow from (7.4)). The case of a "smoother" u is investigated in Steps 15 and 16. Part of the estimates we establish do not use the fact that u is $\mathbb{S}^1$-valued; u could simply be bounded (see, for example, Step 3). The fact that u is $\mathbb{S}^1$-valued is used in a crucial way in Step 5 and some of the subsequent steps. This is especially relevant in Steps 5, 8, and 12, which are at the heart of the proof.

Step 1. Extension of u

Let $u \in W^{s,p}(\Omega; \mathbb{S}^1)$. We first explain how to extend u to $\mathbb{R}^N$. In this paragraph, the extension is denoted by u^*; later, we will omit the * sign. Let X be a smooth open neighborhood of $\overline{\Omega}$ and let $P : W^{s,p}(\Omega) \to W^{s,p}(\mathbb{R}^N)$ be a linear continuous extension operator such that supp $Pv \subset X$ and $\|Pv\|_{L^\infty} \le C\|v\|_{L^\infty}$, $\forall v$. Let a be the average of u in Ω (so that $|a| \le 1$). Then the extension u^* of u, defined as $u^* := a + P(u - a)$, satisfies

$$u^* - a \in W^{s,p}(\mathbb{R}^N) \text{ and } \|u^* - a\|_{W^{s,p}(\mathbb{R}^N)} \le C|u|_{W^{s,p}}, \tag{7.28}$$

$$u^* = u \text{ in } \Omega; \text{ in particular, } u^* \text{ is } \mathbb{S}^1\text{-valued in } \Omega, \tag{7.29}$$

$$u^* \text{ is constant outside } X, \tag{7.30}$$

$$|u^*| \le C, \text{ with } C \text{ an absolute constant.} \tag{7.31}$$

In (7.28) and in what follows, we let

$$|u|_{W^{s,p}} := |u|_{W^{s,p}(\Omega)}. \tag{7.32}$$

Let us note that P (and thus the subsequent constructions in Steps 1 and 2) can be chosen independent of s and p (see, for example, Stein [330, Section VI.3, Theorem 5, p. 181]).

We next extend u^* (from now on denoted by u) to $\mathbb{R}^N \times (0, \infty)$ as follows. With $\rho \in C_c^\infty(\mathbb{R}^N)$ and ρ_ε as in (7.20), we let

$$w(x, \varepsilon) := u * \rho_\varepsilon(x), \ \forall x \in \mathbb{R}^N, \ \forall \varepsilon > 0. \tag{7.33}$$

We let, with Π as in (7.18),

$$V(x, \varepsilon) := \Pi \circ w(x, \varepsilon), \ \forall x \in \mathbb{R}^N, \ \forall \varepsilon > 0. \tag{7.34}$$

Note that, since $u \in L^1_{loc}$, Lebesgue's differentiation theorem implies that

$$\lim_{\varepsilon \to 0} w(x, \varepsilon) = u(x) \text{ for a.e. } x \in \mathbb{R}^N, \tag{7.35}$$

and therefore

$$\lim_{\varepsilon \to 0} V(x, \varepsilon) = u(x) \text{ for a.e. } x \in \Omega. \tag{7.36}$$

Step 2. Factorization of u and extension of φ and v
We let φ be as in (7.19), with $x \in \mathbb{R}^N$. [Recall that, at this stage, we do not know that φ is well-defined for a.e. $x \in \mathbb{R}^N$; we will prove, in Step 4, that this is the case.] Set $v := e^{-\imath\varphi} u$. Note that, although v is defined a.e. on $\mathbb{R}^N$, $u = e^{\imath\varphi} v$ is a factorization of u in the sense of Theorem 7.1 only on Ω (where u and v are $\mathbb{S}^1$-valued).

We will estimate φ and v using the theory of weighted Sobolev spaces; this requires extending φ and v to $\mathbb{R}^N \times (0, \infty)$. Set

$$\psi(x, t) := -\int_t^\infty V(x, \varepsilon) \wedge \partial_\varepsilon V(x, \varepsilon)\, d\varepsilon, \ \forall x \in \mathbb{R}^N, \ \forall t > 0, \tag{7.37}$$

so that formally (by (7.19))

$$\lim_{t \to 0} \psi(x, t) = \varphi(x). \tag{7.38}$$

We will prove in Step 4 that ψ is well-defined everywhere and that (7.38) holds.

We also set

$$U(x,t) := e^{-\imath\psi(x,t)}\, V(x,t),\ \forall x \in \mathbb{R}^N,\ \forall t > 0. \tag{7.39}$$

Reminder. In Steps 3 to 9, we assume that $0 < s < 1$.

Step 3. Some basic estimates satisfied by w and V

By (7.28) and Lemma 7.2, for every $\alpha \in \mathbb{N}^{n+1}$ such that $\ell := |\alpha| \geq 1$, we have

$$\int_{\mathbb{R}^N}\int_0^\infty \varepsilon^{p(\ell-s)-1}|\partial^\alpha w(x,\varepsilon)|^p\, dxd\varepsilon \leq C_\ell |u|^p_{W^{s,p}}. \tag{7.40}$$

We next prove that (7.40) still holds when w is replaced by V.

Lemma 7.5. *For every $\alpha \in \mathbb{N}^{N+1}$ such that $\ell := |\alpha| \geq 1$, we have*

$$\int_{\mathbb{R}^N}\int_0^\infty \varepsilon^{p(\ell-s)-1}|\partial^\alpha V(x,\varepsilon)|^p\, dxd\varepsilon \leq C_\ell |u|^p_{W^{s,p}}. \tag{7.41}$$

Proof. This is clear when $\ell = 1$, since $V = \Pi \circ w$ and Π is Lipschitz. Assume next that $\ell = 2$ and let $\partial^\alpha = \partial_{x_j}\partial_{x_k}$. Then,

$$|\partial_{x_j}\partial_{x_k} V| \leq C|\partial_{x_j}\partial_{x_k} w| + C|\partial_{x_j} w||\partial_{x_k} w|,$$

and it suffices to prove that

$$\int_{\mathbb{R}^N}\int_0^\infty \varepsilon^{p(2-s)-1}|\partial_{x_j}\partial_{x_k} w(x,\varepsilon)|^p\, dxd\varepsilon \leq C|u|^p_{W^{s,p}} \tag{7.42}$$

and

$$\int_{\mathbb{R}^N}\int_0^\infty \varepsilon^{p(2-s)-1}|\partial_{x_j} w(x,\varepsilon)|^p\, |\partial_{x_k} w(x,\varepsilon)|^p\, dxd\varepsilon \leq C|u|^p_{W^{s,p}}. \tag{7.43}$$

The first estimate is clear (by (7.40)). For the second one, we note that, by Cauchy–Schwarz and Lemma 7.2,

$$\begin{aligned}&\int_{\mathbb{R}^N}\int_0^\infty \varepsilon^{p(2-s)-1}|\partial_{x_j} w(x,\varepsilon)|^p\, |\partial_{x_k} w(x,\varepsilon)|^p\, dxd\varepsilon\\ &\leq \left(\int_{\mathbb{R}^N}\int_0^\infty \varepsilon^{p(2-s)-1}|\nabla w(x,\varepsilon)|^{2p}\, dxd\varepsilon\right)^2 \leq C\,|u|^{2p}_{W^{s/2,2p}(\mathbb{R}^N)}.\end{aligned} \tag{7.44}$$

Combining (7.44) with the obvious Gagliardo–Nirenberg inequality

$$|u|_{W^{s/2,2p}(\mathbb{R}^N)} \leq C|u|^{1/2}_{W^{s,p}(\mathbb{R}^N)}\|u\|^{1/2}_{L^\infty(\mathbb{R}^N)}$$

and with (7.28), we obtain (7.43).

The case of a general partial derivative $\partial^\alpha V$ with $\ell = |\alpha| \geq 2$ is similar. As above, it relies on the family of estimates

$$\int_{\mathbb{R}^N}\int_0^\infty \varepsilon^{p(\ell-s)-1}|\partial^\beta w(x,\varepsilon)|^p\,|\partial^\gamma w(x,\varepsilon)|^p\cdots\,dxd\varepsilon \le C|u|^p_{W^{s,p}}, \tag{7.45}$$

valid for $\beta, \gamma, \ldots \in \mathbb{N}^{N+1}\setminus\{0\}$ satisfying $|\beta|+|\gamma|+\cdots=\ell$. Estimates (7.45) follow from the Gagliardo–Nirenberg inequalities (in $\mathbb{R}^N$)

$$|u|_{W^{(|\beta|/\ell)s,(\ell/|\beta|)p}} \le C|u|^{|\beta|/\ell}_{W^{s,p}}\,\|u\|^{1-|\beta|/\ell}_{L^\infty},\quad |u|_{W^{(|\gamma|/\ell)s,(\ell/|\gamma|)p}} \le C|u|^{|\gamma|/\ell}_{W^{s,p}}\,\|u\|^{1-|\gamma|/\ell}_{L^\infty},\ldots,$$

combined with Hölder's inequality, Lemma 7.2, and (7.28). □

We also note the following consequences of (7.28)–(7.31):

$$|\partial^\alpha w(x,\varepsilon)| \le C\varepsilon^{-|\alpha|},\ \forall\,\alpha\in\mathbb{N}^{N+1}; \tag{7.46}$$

$$|\partial^\alpha w(x,\varepsilon)| \le C\varepsilon^{-|\alpha|-N},\ \forall\,\alpha\in\mathbb{N}^{N+1}\setminus\{0\}. \tag{7.47}$$

[The first estimate follows from (7.31) and (7.33). For the second one, we rely on the fact that $u-a$ is supported in X and bounded by an absolute constant and that, when $\alpha\ne 0$, we have $\partial^\alpha w=(u-a)*(\partial^\alpha\rho_\varepsilon)$.]

Combining (7.46) and (7.47) with the chain rule and with (7.34), we obtain, by a straightforward induction on $|\alpha|$,

$$|\partial^\alpha V(x,\varepsilon)| \le C\varepsilon^{-|\alpha|},\ \forall\,\alpha\in\mathbb{N}^{N+1}; \tag{7.48}$$

$$|\partial^\alpha V(x,\varepsilon)| \le C\varepsilon^{-|\alpha|-N}+C\varepsilon^{-|\alpha|(N+1)},\ \forall\,\alpha\in\mathbb{N}^{N+1}\setminus\{0\}. \tag{7.49}$$

Step 4. Some basic estimates satisfied by ψ and φ

By (7.49) and the fact that V is bounded, the integral defining ψ in (7.37) is absolutely convergent, and the partial derivatives of ψ are obtained via the usual rules of differentiation of integrals depending on a parameter. In particular,

$$\psi \text{ and } U \text{ are smooth in } \mathbb{R}^N\times(0,\infty). \tag{7.50}$$

Lemma 7.6. *For a.e. $x\in\mathbb{R}^N$, the integral in* (7.19) *defining $\varphi(x)$ is absolutely convergent.*

Consequently, (7.38) *holds for a.e. x, and (by* (7.38) *and* (7.36)*),*

$$\textit{for a.e. } x\in\mathbb{R}^N \textit{ we have } \lim_{t\to 0} U(x,t)=e^{-\imath\varphi(x)}\,u(x):=v(x). \tag{7.51}$$

Proof. Since the integral defining $\psi(x,1)$ is absolutely convergent, it suffices to note the trivial estimate

$$\int_{\mathbb{R}^N}\left(\int_0^1 |\partial^\alpha V(x,\varepsilon)|\,d\varepsilon\right)^p dx \le C\int_{\mathbb{R}^N}\int_0^1 \varepsilon^{p-sp-1}|\partial^\alpha V(x,\varepsilon)|^p\,d\varepsilon dx, \tag{7.52}$$

valid for every $\alpha\in\mathbb{N}^{N+1}$, and to apply Lemma 7.5 with $|\alpha|=1$. □

One of our final goals is to prove that $\varphi\in W^{s,p}(\Omega)$. We start by establishing a weaker property.

Lemma 7.7. *For every compact $K \subset \mathbb{R}^N$, we have $\varphi \in L^p(K)$, and*

$$\|\varphi\|_{L^p(K)} \leq C_K |u|_{W^{s,p}}. \tag{7.53}$$

Proof. In view of (7.52) and Lemma 7.5, it suffices to prove the estimate

$$|\psi(x,1)| \leq C|u|_{W^{s,p}}, \forall x \in \mathbb{R}^N. \tag{7.54}$$

In turn, (7.54) is obtained by taking a closer look at the estimate (7.49) with $|\alpha| = 1$. Recalling (7.20), we first note that

$$\partial^\alpha \rho_\varepsilon = \frac{1}{\varepsilon}\eta_\varepsilon, \text{ for some } \eta = \eta(\rho, \alpha) \in C_c^\infty(\mathbb{R}^N) \text{ such that } \int_{\mathbb{R}^N} \eta = 0. \tag{7.55}$$

Since $u - a$ is supported in X, it then follows (using (7.28) and (7.55)) that

$$\begin{aligned} |\partial^\alpha w(x,\varepsilon)| =& \frac{1}{\varepsilon}|u * \eta_\varepsilon(x)| = \frac{1}{\varepsilon}|[u-a] * \eta_\varepsilon(x)| \\ \leq& \frac{C}{\varepsilon^{N+1}}\|u-a\|_{L^1(X)} \leq \frac{C}{\varepsilon^{N+1}}\|u-a\|_{L^p} \leq \frac{C}{\varepsilon^{N+1}}|u|_{W^{s,p}}. \end{aligned} \tag{7.56}$$

Since Π is Lipschitz and $|\alpha| = 1$, we also have (via (7.34))

$$|\partial^\alpha V(x,\varepsilon)| \leq \frac{C}{\varepsilon^{N+1}}|u|_{W^{s,p}}. \tag{7.57}$$

We conclude via (7.37) and (7.57). □

Step 5. The crucial estimate
Up to now, we did not take into account the fact that u is $\mathbb{S}^1$-valued in Ω. The estimates in the previous steps are valid for $u \in W^{s,p} \cap L^\infty$. In this step, we establish a fundamental estimate, (7.64) below, valid for $\mathbb{S}^1$-valued maps.

We define the open set

$$\mathscr{W} := \{(x,\varepsilon);\ x \in \mathbb{R}^N,\ \varepsilon > 0 \text{ and } |w(x,\varepsilon)| > 1/2\} \tag{7.58}$$

and the quantity

$$d(x) := \inf\{\varepsilon > 0;\ |w(x,\varepsilon)| \leq 1/2\},\ \forall x \in \mathbb{R}^N \tag{7.59}$$

(so that $0 \leq d(x) \leq \infty$).

Let us note that, by (7.18) and (7.34),

$$|V(x,\varepsilon)| \equiv 1,\ \forall (x,\varepsilon) \in \mathscr{W}. \tag{7.60}$$

This implies that, for every $j, k \in \{1,\ldots,N\}$, we have

$$\partial_{x_j} V \wedge \partial_{x_k} V \equiv 0 \text{ and } \partial_{x_j} V \wedge \partial_\varepsilon V \equiv 0 \text{ in } \mathscr{W}, \tag{7.61}$$

and in particular

$$\partial_{x_j} V \wedge \partial_{x_k} V(x, \varepsilon) = \partial_{x_j} V \wedge \partial_\varepsilon V(x, \varepsilon) = 0, \ \forall x \in \mathbb{R}^N, \ \forall 0 < \varepsilon < d(x). \quad (7.62)$$

The above hold in $\mathbb{R}^N$. If we consider, more specifically, points $x \in \Omega$ then, by (7.35) and (7.29), we have

$$d(x) > 0 \text{ for a.e. } x \in \Omega. \quad (7.63)$$

Lemma 7.8. *With $d(x)$ as in (7.59), we have*

$$\int_\Omega \frac{1}{d(x)^{sp}}\, dx \le C|u|^p_{W^{s,p}}. \quad (7.64)$$

Proof. We consider only the points $x \in \Omega$ such that $d(x) > 0$ and $\lim_{\varepsilon\to 0} w(x, \varepsilon) = u(x)$. In view of (7.35) and (7.63), a.e. $x \in \Omega$ satisfies these conditions. For such an x, we have either $d(x) = \infty$, or $d(x) < \infty$ and then $|w(x, d(x))| = 1/2$. Assume that $d(x) < \infty$. Using Hölder's inequality when $p > 1$ (and a trivial argument when $p = 1$) we find that

$$\begin{aligned}(1/2)^p &= ||w(x, d(x))| - |u(x)||^p \le |w(x, d(x)) - u(x)|^p \\ &\le \left(\int_0^{d(x)} |\partial_\varepsilon w(x, \varepsilon)|\, d\varepsilon\right)^p \\ &\le C_{s,p} d(x)^{sp} \int_0^{d(x)} \varepsilon^{p(1-s)-1} |\partial_\varepsilon w(x, \varepsilon)|^p\, d\varepsilon,\end{aligned} \quad (7.65)$$

and thus

$$\frac{1}{d(x)^{sp}} \le C_{s,p} \int_0^\infty \varepsilon^{p(1-s)-1} |\partial_\varepsilon w(x, \varepsilon)|^p\, d\varepsilon. \quad (7.66)$$

We obtain (7.64) by combining (7.66) with (7.40) (applied with $\ell = 1$). □

Step 6. Proof of (7.3) for $\varphi_{|\Omega}$
Using, on the one hand, Lemmas 7.6 and 7.7 (with $K := \overline{\Omega}$) and, on the other hand, Lemma 7.4 (with $G(x, t) := \psi(x, t)$ defined in (7.37)), estimate (7.3) will be a consequence of

$$\int_\Omega \int_0^\infty t^{p(1-s)-1} |\partial^\alpha \psi(x, t)|^p\, dt dx \le C|u|^p_{W^{s,p}}, \ \forall \alpha \in \mathbb{N}^{N+1} \text{ with } |\alpha| = 1, \quad (7.67)$$

which we now prove.

For this purpose, we start by calculating (using (7.37))

$$\partial_t \psi(x, t) = V(x, t) \wedge \partial_t V(x, t). \quad (7.68)$$

Thus, when $\partial^\alpha = \partial_t$, estimate (7.67) follows from (7.41). [Note that, for this term, we could have considered, in (7.67), the integral over $\mathbb{R}^N$ instead of Ω.]

Let us now estimate $\partial_{x_j}\psi$, again using (7.37). We have the following identities:

$$\begin{aligned}\partial_{x_j}\psi(x,t) =& -\int_t^\infty \partial_{x_j}V(x,\varepsilon)\wedge\partial_\varepsilon V(x,\varepsilon)\,d\varepsilon\\ &-\int_t^\infty V(x,\varepsilon)\wedge\partial_{x_j}\partial_\varepsilon V(x,\varepsilon)\,d\varepsilon\\ =&V(x,t)\wedge\partial_{x_j}V(x,t)-2\int_t^\infty \partial_{x_j}V(x,\varepsilon)\wedge\partial_\varepsilon V(x,\varepsilon)\,d\varepsilon\end{aligned}\tag{7.69}$$

(the integration by parts being justified via (7.49)).

The term $V(x,t)\wedge\partial_{x_j}V(x,t)$ in (7.69) is estimated as we did for $\partial_t\psi$. It remains to prove that

$$I := \int_\Omega\int_0^\infty t^{p(1-s)-1}\,F(x,t)^p\,dtdx \le C|u|^p_{W^{s,p}};\tag{7.70}$$

here,

$$F(x,t) := \int_t^\infty |\partial_{x_j}V(x,\varepsilon)\wedge\partial_\varepsilon V(x,\varepsilon)|\,d\varepsilon.$$

Let us note that, by (7.62) and (7.48), we have

$$F(x,t)\le\begin{cases}C/t, & \text{if } t\ge d(x)\\ C/d(x), & \text{if } t<d(x)\end{cases}.\tag{7.71}$$

By (7.71), the left-hand side I of (7.70) satisfies

$$I\le C\int_\Omega\int_0^{d(x)}\frac{t^{p(1-s)-1}}{d(x)^p}\,dtdx + C\int_\Omega\int_{d(x)}^\infty t^{-sp-1}\,dtdx \le C\int_\Omega\frac{1}{d(x)^{sp}}\,dx.\tag{7.72}$$

We obtain (7.70) using (7.64) and (7.72). [Note that, in the estimate of I, we cannot replace the integral over Ω by the integral over $\mathbb{R}^N$.]

Step 7. An L^1 estimate satisfied by v

We claim that

$$\left\|v-\fint_\Omega u\right\|_{L^1(\Omega)} \le C\,|u|_{W^{s,p}}.\tag{7.73}$$

Indeed, since $v = e^{-\imath\varphi}\,u$, we have

$$\begin{aligned}\left|v - \fint u\right| &\le |v-u| + \left|u - \fint u\right| = |e^{-\imath\varphi} - 1| + \left|u - \fint u\right| \\ &\le |\varphi| + \left|u - \fint u\right|,\end{aligned} \tag{7.74}$$

and thus (using (7.53) with $K := \overline{\Omega}$, (7.74) and Poincaré's inequality),

$$\left\|v - \fint u\right\|_{L^1(\Omega)} \le C \left\|v - \fint u\right\|_{L^p(\Omega)} \le C\,|u|_{W^{s,p}}. \tag{7.75}$$

Step 8. Proof of (7.4)
In view of (7.73), of Lemma 7.4 (with $G(x,t) := U(x,t)$ defined in (7.39)) and of (7.51), estimate (7.4) will follow provided we establish, for some integer $\ell > sp$, the inequalities

$$\int_\Omega \int_0^\infty t^{\ell-sp-1} |\partial^\alpha U(x,t)|\, dtdx \le C_\ell\, |u|^p_{W^{s,p}},\ \forall \alpha \in \mathbb{N}^{N+1} \text{ with } |\alpha| = \ell. \tag{7.76}$$

As a preliminary to the proof of (7.76), we establish some identities and estimates. Let us first note the following straightforward estimate, obtained, via (7.37) and (7.48), by induction on $|\alpha| \ge 1$:

$$|\partial^\alpha \psi(x,t)| \le \frac{C_\alpha}{t^{|\alpha|}},\ \forall x \in \mathbb{R}^N,\ \forall t > 0,\ \forall \alpha \in \mathbb{N}^{N+1} \setminus \{0\}. \tag{7.77}$$

Using (7.39), (7.48), and (7.77), we find that

$$|\partial^\alpha U(x,t)| \le \frac{C_\alpha}{t^{|\alpha|}},\ \forall x \in \mathbb{R}^N,\ \forall t > 0,\ \forall \alpha \in \mathbb{N}^{N+1} \setminus \{0\}. \tag{7.78}$$

Combining (7.78) and (7.64), we obtain that

$$\int_\Omega \int_{d(x)}^\infty t^{\ell-sp-1} |\partial^\alpha U(x,t)|\, dtdx \le C_\ell\, |u|^p_{W^{s,p}},\ \forall \alpha \in \mathbb{N}^{N+1} \text{ with } |\alpha| = \ell. \tag{7.79}$$

In view of (7.76) and (7.79), it thus remains to establish the estimate

$$\int_\Omega \int_0^{d(x)} t^{\ell-sp-1} |\partial^\alpha U(x,t)|\, dtdx \le C_\ell |u|^p_{W^{s,p}},\ \forall \alpha \in \mathbb{N}^{N+1} \text{ with } |\alpha| = \ell. \tag{7.80}$$

In turn, (7.80) clearly follows from (7.64) combined with the next result.

Lemma 7.9. *We have*

$$|\partial^\alpha U(x,t)| \le \frac{C_\alpha}{d(x)^{|\alpha|}},\ \forall x \in \mathbb{R}^N,\ \forall t \in (0, d(x)),\ \forall \alpha \in N^{N+1} \setminus \{0\}. \tag{7.81}$$

Proof. Let us first calculate $\partial_t U$. By (7.39) and (7.68), we have

$$\partial_t U = e^{-\imath\psi}\,[\partial_t V - \imath\,(V \wedge \partial_t V)\,V]. \tag{7.82}$$

Since $|V|^2 \equiv 1$ in $\mathscr{W}$, we have $V \perp \partial_t V = 0$ in $\mathscr{W}$, and then the reader may check that

$$\partial_t V = \imath\,(V \wedge \partial_t V)\,V \text{ in } \mathscr{W}. \tag{7.83}$$

We find that $\partial_t U = 0$ in $\mathscr{W}$, and thus all the partial derivatives $\partial^\alpha U$ involving at least one derivative in t vanish in $\mathscr{W}$. In particular, we have $\partial^\alpha U(x,t) = 0$ if $\alpha_{N+1} \neq 0$ and $0 < t < d(x)$.

We next estimate $\partial_{x_j} U$, given (in view of (7.39) and (7.69)) by the formula

$$\begin{aligned}\partial_{x_j} U(x,t) =& e^{-\imath\psi(x,t)}\,[\partial_{x_j} V(x,t) - \imath\,(V \wedge \partial_{x_j} V)(x,t)\,V(x,t)]\\ &+ 2\imath\,e^{-\imath\psi(x,t)}\,V(x,t)\int_t^\infty H(x,\varepsilon)\,d\varepsilon;\end{aligned} \tag{7.84}$$

here,

$$H(x,\varepsilon) := \partial_{x_j} V(x,\varepsilon) \wedge \partial_\varepsilon V(x,\varepsilon). \tag{7.85}$$

Arguing as for $\partial_t U$ and using (7.39), we see that, in $\mathscr{W}$, (7.84) simplifies to

$$\partial_{x_j} U(x,t) = 2\imath\,U(x,t)\int_t^\infty H(x,\varepsilon)\,d\varepsilon,\ \forall\,(x,t) \in \mathscr{W}. \tag{7.86}$$

Combining (7.86) with (7.48) and the fundamental observation (7.61), we find that, $\forall\,x \in \mathbb{R}^N$, $\forall\,0 < t < d(x)$,

$$|\partial_{x_j} U(x,t)| = \left|2\imath\,U(x,t)\int_{d(x)}^\infty H(x,\varepsilon)\,d\varepsilon\right| \le C\int_{d(x)}^\infty \varepsilon^{-2}\,d\varepsilon \le \frac{C}{d(x)}. \tag{7.87}$$

For a general partial derivative ∂^α not involving ∂_t, estimate (7.81) is obtained by induction on $|\alpha|$, by differentiating $\partial_{x_j} U$ in $\mathscr{W}$ (starting from the identity (7.86)), using (7.61), (7.48), and (7.77). We detail this when $\partial^\alpha = \partial_{x_k}\partial_{x_j}$, the general case being similar.

We have (by (7.86))

$$\begin{aligned}\partial_{x_k}\partial_{x_j} U(x,t) =& 2\imath\,\partial_{x_k} U(x,t)\int_t^\infty \partial_{x_j} V(x,\varepsilon) \wedge \partial_\varepsilon V(x,\varepsilon)\,d\varepsilon\\ &+ 2\imath\,U(x,t)\int_t^\infty \partial_{x_k}(\partial_{x_j} V \wedge \partial_\varepsilon V)(x,\varepsilon)\,d\varepsilon,\ \forall\,(x,t) \in \mathscr{W}.\end{aligned} \tag{7.88}$$

From (7.88), (7.81) (for $\partial_{x_k} U$), (7.61), and (7.48), we obtain

$$|\partial_{x_k}\partial_{x_j} U(x,t)| \le \frac{C}{d(x)}\int_{d(x)}^\infty \varepsilon^{-2}\,d\varepsilon + C\int_{d(x)}^\infty \varepsilon^{-3}\,d\varepsilon,\ \forall\,x \in \Omega,\ \forall\,0 < t < d(x),$$

which implies (7.81) for $\partial_{x_k}\partial_{x_j}U$. □

Step 9. Proof of (7.6) and (7.7)
Estimate (7.6) was established in Step 7 (see (7.73)). Combining (7.76) (with $\ell = 1$) with (7.27) (applied to $G(x,\varepsilon) := U(x,\varepsilon)$, $f(x) := v(x)$, $\ell := 1$, s replaced by sp and p replaced by 1) and (7.51), we obtain (7.7).

Step 10. Factorization when $s \geq 1$
We may assume that $(s,p) \neq (1,1)$ (see Remark 7.4), and thus $sp > 1$. Let $0 < \sigma < 1$ and $1 \leq q < \infty$ satisfy $\sigma q = sp > 1$. By the Gagliardo–Nirenberg embeddings (Appendix 15.2), we have $u \in W^{\sigma,q}$. By the previous steps, we have $\varphi \in W^{\sigma,q}$ and $e^{-\imath\varphi} u \in W^{sp,1}$. Again by the Gagliardo–Nirenberg embeddings, we have $e^{-\imath\varphi} u \in W^{s,p}$, so that $e^{\imath\varphi} = u\,\overline{e^{-\imath\varphi} u} \in W^{s,p}$ (here, we use the fact that $W^{s,p} \cap L^\infty$ is an algebra; see Lemma 15.41). Thus, $\varphi \in W^{\sigma,q}$ and $e^{\imath\varphi} \in W^{s,p}$. By Lemma 6.2, this implies that $\varphi \in W^{s,p}$. In conclusion, the "qualitative" part of Theorem 7.1 holds: we may write $u = e^{\imath\varphi} v$ with $\varphi \in W^{s,p}$ and $v \in W^{sp,1}$.

Step 11. Proof of (7.4) when $s \geq 1$
With σ and q as above, we have proved in Step 8 that

$$\left\| v - \fint_\Omega u \right\|_{W^{sp,1}} \leq C|u|^q_{W^{\sigma,q}} + C|u|_{W^{\sigma,q}}. \tag{7.89}$$

In (7.89), the term $C|u|_{W^{\sigma,q}}$ comes from (7.75) (with s replaced by σ and p replaced by q). A closer look at the proof of (7.75) shows that we may replace $C|u|_{W^{\sigma,q}}$ by $C_\tau|u|_{W^{\tau,p}}$, with arbitrary $0 < \tau < 1$. Therefore, the following version of (7.89) holds:

$$\left\| v - \fint_\Omega u \right\|_{W^{sp,1}} \leq C|u|^q_{W^{\sigma,q}} + C|u|_{W^{\tau,p}},\ \forall 0 < \tau < 1. \tag{7.90}$$

By (7.90) and Lemmas 15.1 and 15.6, we find that v satisfies (7.4).

Step 12. Proof of (7.3) when $s > 1$ is not an integer
Let $0 < \tau < 1$. Then we have (see Lemma 7.7)

$$\|\varphi\|_{L^p(\Omega)} \leq C|u|_{W^{\tau,p}} \leq C|u|_{W^{s,p}}, \tag{7.91}$$

the latter inequality following from Lemma 15.1.

In order to estimate the full norm $\|\varphi\|_{W^{s,p}}$, it remains to establish the estimate

$$I_\ell := \sum_{|\alpha|=\ell} \int_\Omega \int_0^\infty \varepsilon^{p(\ell-s)-1} |\partial^\alpha \psi(x,\varepsilon)|^p \, dx d\varepsilon \leq C|u|^p_{W^{s,p}}; \tag{7.92}$$

here, ℓ is any integer $> s$. [Here, we use Lemma 7.4 and the assumption that s is not an integer.]

To start with, let us note that Lemma 7.8 still holds when $s \geq 1$ and $(s,p) \neq (1,1)$. This is a consequence of the Gagliardo–Nirenberg inequalities (Lemma 15.6).

On the other hand, we have the family of inequalities

$$\int_{\Omega}\int_{0}^{\infty}\varepsilon^{p(\ell-s)-1}|\partial^{\beta}V(x,\varepsilon)|^{p}\,|\partial^{\gamma}V(x,\varepsilon)|^{p}\cdots\,dxd\varepsilon\le C_{\ell}|u|^{p}_{W^{s,p}},\tag{7.93}$$

when $\beta,\gamma,\ldots\in\mathbb{N}^{N+1}\setminus\{0\}$ and $|\beta|+|\gamma|+\cdots=\ell$. [See the proof of Lemma 7.5.]

Therefore, in order to prove (7.92), it suffices to establish the estimate

$$\begin{aligned}I_{\ell}\le&C\int_{\Omega}\frac{1}{d(x)^{sp}}\,dx\\&+C\sum\int_{\Omega}\int_{0}^{\infty}\varepsilon^{p(\ell-s)-1}|\partial^{\beta}V(x,\varepsilon)|^{p}\,|\partial^{\gamma}V(x,\varepsilon)|^{p}\cdots\,dxd\varepsilon,\end{aligned}\tag{7.94}$$

where the sum is over all $\beta,\gamma,\ldots$ as above.

For this purpose, we combine (7.61), (7.48), (7.68), and (7.69) and obtain (by a straightforward induction on the integer $\ell=|\alpha|\ge 1$) that, for every $x\in\mathbb{R}^N$, we have

$$\begin{aligned}|\partial^{\alpha}\psi(x,\varepsilon)|\le&C\sum|\partial^{\beta}V(x,\varepsilon)|\,|\partial^{\gamma}V(x,\varepsilon)|\cdots\\&+C\begin{cases}1/d(x)^{|\alpha|},&\text{if }\varepsilon<d(x)\\1/\varepsilon^{|\alpha|},&\text{if }\varepsilon\ge d(x)\end{cases}.\end{aligned}\tag{7.95}$$

[This is very similar to (7.71).]

Estimate (7.94) is an immediate consequence of (7.95).

Step 13. Proof of (7.3) when s is an integer

We may assume that $(s,p)\ne(1,1)$. Clearly, (7.91) still holds. Therefore, it remains to establish the estimate

$$\|\partial^{\alpha}\varphi\|_{L^p(\Omega)}\le C|u|_{W^{s,p}},\ \forall\alpha\in\mathbb{N}^N\text{ with }|\alpha|=s.\tag{7.96}$$

[Warning. Here, we take derivatives only in the x variable.]

We establish (7.96), for example, when $s=2$. [The case $s=1$ is simpler, and we will sketch below the proof for $s\ge 3$.] Let us first note that

$$\int_{\Omega}\frac{1}{d(x)^{2p}}\,dx\le C|u|^{p}_{W^{2,p}}.\tag{7.97}$$

Indeed, let $0<\sigma<1$ and $1<q<\infty$ be such that $\sigma q=2p$. By Lemma 7.8 and Gagliardo–Nirenberg (see Lemma 15.6 in Appendix 15.2), we have

$$\int_{\Omega}\frac{1}{d(x)^{2p}}\,dx=\int_{\Omega}\frac{1}{d(x)^{\sigma q}}\,dx\le C'|u|^{q}_{W^{\sigma,q}}\le C|u|^{p}_{W^{2,p}}.$$

Let $\partial^{\alpha}=\partial_{x_k}\partial_{x_j}$. We claim that the following identity holds for a.e. $x\in\Omega$:

$$\begin{aligned}\partial^\alpha \varphi(x) =& u \wedge \partial_{x_k}\partial_{x_j}u(x) + 2\int_0^\infty \partial_{x_k}(\partial_\varepsilon V \wedge \partial_{x_j}V)(x,\varepsilon)\,d\varepsilon \\ =& u \wedge \partial_{x_k}\partial_{x_j}u(x) + 2\int_{d(x)}^\infty \partial_{x_k}(\partial_\varepsilon V \wedge \partial_{x_j}V)(x,\varepsilon)\,d\varepsilon\end{aligned} \tag{7.98}$$

(where the second line in (7.98) uses (7.62)).

In order to obtain (7.98), we start from (7.69), which implies (using (7.49))

$$\begin{aligned}\partial_{x_k}\partial_{x_j}\psi(x,t) =& \partial_{x_k}V \wedge \partial_{x_j}V(x,t) + V \wedge \partial_{x_k}\partial_{x_j}V(x,t) \\ &+ 2\int_t^\infty \partial_{x_k}(\partial_\varepsilon V \wedge \partial_{x_j}V)(x,\varepsilon)\,d\varepsilon,\ \forall x \in \mathbb{R}^N,\ \forall t > 0.\end{aligned} \tag{7.99}$$

Let us note that

$$\psi(x,t) \to \varphi(x) \text{ in } L^p(\Omega) \text{ as } t \to 0; \tag{7.100}$$

this is a consequence of the proof of Lemma 7.6, and, more specifically, of (7.52).

We claim that, when $t \to 0$, the right-hand side of (7.99) converges, as a function of x and in $L^p(\Omega)$, to the right-hand side of (7.98). This is obtained as follows.

The (dominated) convergence

$$\int_t^\infty \partial_{x_k}(\partial_\varepsilon V \wedge \partial_{x_j}V)(x,\varepsilon)\,d\varepsilon \to \int_{d(x)}^\infty \partial_{x_k}(\partial_\varepsilon V \wedge \partial_{x_j}V)(x,\varepsilon)\,d\varepsilon \tag{7.101}$$

in $L^p(\Omega)$ as $t \to 0$ is a consequence of (7.62), (7.48), and (7.97).

We next prove that

$$\partial_{x_k}V \wedge \partial_{x_j}V(x,t) \to 0 \tag{7.102}$$

in $L^p(\Omega)$ as $t \to 0$. Pointwise convergence for a. e. $x \in \Omega$ is a consequence of (7.62) and of the fact that (by (7.97)) we have $d(x) > 0$ for a.e. $x \in \Omega$. On the other hand, we have the following domination:

$$|\partial_{x_k}V \wedge \partial_{x_j}V(x,t)|^p \lesssim |\nabla_x w(x,t)|^{2p} = |(\nabla u) * \rho_t(x)|^{2p} \lesssim \mathscr{M}|\nabla u|^{2p}; \tag{7.103}$$

here, $\mathscr{M}$ stands for the maximal function operator, and we have used the standard estimate $|f * \rho_t(x)| \le C_\rho \mathscr{M}f(x)$, $\forall x$, $\forall \varepsilon$, $\forall f$ (see, for example, Stein [331, I.6.1, Proposition, p. 24]).

Combining (7.103) with the maximal function theorem and the Gagliardo–Nirenberg embedding $W^{2,p}(\Omega;\mathbb{S}^1) \hookrightarrow W^{1,2p}(\Omega;\mathbb{S}^1)$ (see Remark 15.5 in Appendix 15.2), we obtain (7.102).

The convergence

$$V \wedge \partial_{x_k}\partial_{x_j}V(x,t) \to u \wedge \partial_{x_k}\partial_{x_j}u(x) \tag{7.104}$$

in $L^p(\Omega)$ as $t \to 0$ relies on the fact that, by the chain rule, we have

$$V \wedge \partial_{x_k}\partial_{x_j} V(x,t) = V \wedge [(D\Pi) \circ V](x,t)\partial_{x_k}\partial_{x_j} w(x,t) + O(|\nabla_x V|^2)(x,t). \tag{7.105}$$

Combining (7.105) with the fact that $\partial_{x_k}\partial_{x_j} u \in L^p(\Omega)$, (7.36), the identities

$$(D\Pi)(\xi) = \mathrm{Id},\ \forall \xi \in \mathbb{S}^1,$$

$$\partial_{x_k}\partial_{x_j} w(x,t) = (\partial_{x_k}\partial_{x_j} u) * \rho_t(x),\ \forall x \in \mathbb{R}^N,\ \forall t > 0,$$

(7.103), and using, as above, the maximal function theorem in conjunction with Gagliardo–Nirenberg, we obtain (7.104).

The claim follows from (7.101), (7.102), and (7.104). Combined with (7.100), the claim yields the rigorous justification of (7.98).

From (7.48) and (7.98), we obtain that

$$|\partial^\alpha \varphi(x)| \le |\partial^\alpha u(x)| + \frac{C}{d(x)^2},\ \text{for a.e. } x \in \Omega. \tag{7.106}$$

Combining (7.106) and (7.97), we obtain

$$\int_\Omega |\partial^\alpha \varphi(x)|^p\,dx \le C\int_\Omega |\partial^\alpha u(x)|^p\,dx + C\int_\Omega \frac{1}{d(x)^{2p}}\,dx \le C|u|^p_{W^{2,p}},$$

i.e., (7.96) holds.

For a general integer s, the argument is similar. For example, when $s = 3$, we start, as above, from (7.69), which leads to

$$\begin{aligned}\partial_{x_\ell}\partial_{x_k}\partial_{x_j}\psi(x,t) =& \partial_{x_\ell}\partial_{x_k}(V \wedge \partial_{x_j} V)(x,t)\\ &+2\int_t^\infty \partial_{x_\ell}\partial_{x_k}(\partial_\varepsilon V \wedge \partial_{x_j} V)(x,\varepsilon)\,d\varepsilon,\\ &\qquad\qquad\qquad\qquad \forall x \in \mathbb{R}^N,\ \forall t > 0.\end{aligned} \tag{7.107}$$

Using arguments similar to the ones above, in particular the maximal function theorem and Gagliardo–Nirenberg, we can pass to the limit $t \to 0$ in (7.107) and obtain

$$\begin{aligned}\partial_{x_\ell}\partial_{x_k}\partial_{x_j}\varphi(x) =& \partial_{x_\ell}\partial_{x_k}(u \wedge \partial_{x_j} u)(x)\\ &+2\int_{d(x)}^\infty \partial_{x_\ell}\partial_{x_k}(\partial_\varepsilon V \wedge \partial_{x_j} V)(x,\varepsilon)\,d\varepsilon,\ \text{a.e. } x \in \Omega.\end{aligned} \tag{7.108}$$

Using the analogue of (7.108) for a general partial derivative ∂^α, we find that

$$|\partial^\alpha \varphi(x)| \le C\sum |\partial^\beta u(x)|\,|\partial^\gamma u(x)| \cdots + \frac{C}{d(x)^{|\alpha|}},\ \text{a.e. } x \in \Omega, \tag{7.109}$$

where the sum is over $\beta, \gamma, \ldots \in \mathbb{N}^{N+1} \setminus \{0\}$ such that $|\beta| + |\gamma| + \cdots = |\alpha| = s$.

We then obtain (7.96) by combining (7.109) with (7.64), Hölder's inequality, and the Gagliardo–Nirenberg inequalities

$$\|\partial^\beta u\|_{L^{sp/|\beta|}} \leq C\,|u|_{W^{s,p}}^{|\beta|/s}\,\|u\|_{L^\infty}^{1-|\beta|/s},\quad \|\partial^\gamma u\|_{L^{sp/|\gamma|}} \leq C\,|u|_{W^{s,p}}^{|\gamma|/s}\,\|u\|_{L^\infty}^{1-|\gamma|/s},\ldots$$

[Strictly speaking, we have established (7.64) only for $0 < s < 1$. However, by Gagliardo–Nirenberg, it also holds when $s \geq 1$, provided $(s, p) \neq (1, 1)$; see the proof of (7.97).]

Step 14. Proof of (7.5)
We combine (7.78) and (7.81) (with $\partial^\alpha = \partial_{x_j}$), and find that

$$|\partial_{x_j} U(x,t)| \leq \frac{C}{d(x)},\ \forall x \in \Omega,\ \forall t > 0. \tag{7.110}$$

From (7.110), (7.51), and the fundamental estimate (7.64), we find that $v \in W^{1,sp}$ (when $sp > 1$), resp. $v \in BV$ (when $sp = 1$), and that $|v|_{W^{1,sp}} \leq C|u|_{W^{s,p}}^{1/s}$ (when $sp > 1$), resp. $|v|_{BV} \leq C|u|_{W^{s,p}}^{1/s}$ (when $sp = 1$). Since, when $sp \geq 1$, we already know that $v \in W^{1,1}$ (by (7.4)), we find that (7.5) holds in all cases.

Step 15. If $u \in VMO_{loc}(\Omega)$, then $v \in C^\infty(\Omega; \mathbb{S}^1)$
If $u \in VMO_{loc}(\Omega; \mathbb{S}^1)$, then for each ball $\overline{B} \subset \Omega$ there exists some $C_{\overline{B}} > 0$ such that $d(x) \geq C_{\overline{B}} > 0$ for $x \in \overline{B}$ (see Corollary 15.6). By Lemma 7.9, we have

$$|\partial^\alpha U(x,t)| \leq C_{\alpha,\overline{B}},\ \forall x \in \overline{B},\ \forall t \in (0, C_{\overline{B}}),\ \forall \alpha \in \mathbb{N}^{N+1} \setminus \{0\}. \tag{7.111}$$

Using (7.111) and the definition (7.51) of v, we easily find that $v \in C^\infty(\Omega; \mathbb{S}^1)$, and also that the map $\widetilde{U}(x,t) := \begin{cases} U(x,t), & \text{if } x \in \overline{B} \text{ and } 0 < t < C_{\overline{B}} \\ v(x), & \text{if } x \in \overline{B} \text{ and } t = 0 \end{cases}$ is smooth in $\overline{B} \times [0, C_{\overline{B}})$.

Step 16. If $u \in VMO_{loc}(\Omega)$, then φ is "as smooth as u permits"
When $sp \geq 1$ and $u \in VMO_{loc}(\Omega; \mathbb{S}^1)$, we may argue as follows. Since $v \in C^\infty(\Omega; \mathbb{S}^1)$ (Step 15), we find that $u\overline{v} \in VMO_{loc}(\Omega; \mathbb{S}^1)$ (Corollary 15.4 in Appendix 15.5). On each ball B such that $\overline{B} \subset \Omega$, we may then write $u\overline{v} = e^{\imath\psi}$, with $\psi \in VMO(B)$ (Theorem 5.7). Since $e^{\imath\varphi} = e^{\imath\psi}$ in B, we find that $\varphi = \psi + k$ in B, for some $k \in \mathbb{Z}$ (Theorem 6.6), and thus $\varphi \in VMO_{loc}$.

When $sp \geq 1$ and $u \in C(\Omega; \mathbb{S}^1)$ or $u \in C^\infty(\Omega; \mathbb{S}^1)$, we argue as above.

Assume next that $sp < 1$ and $u \in VMO_{loc}(\Omega; \mathbb{S}^1)$. Write, in $\overline{B} \times (0, C_{\overline{B}})$, $V = e^{\imath\eta}$, with η smooth. Similarly, write, in $\overline{B} \times [0, C_{\overline{B}})$, $\widetilde{U} = e^{\imath\zeta}$, with ζ smooth. Then, up to a constant multiple of 2π, we have $\psi = \eta - \zeta$ in $\overline{B} \times (0, C_{\overline{B}})$. Assuming the limit as $\varepsilon \to 0$ exists, we have

$$\varphi(x) = \lim_{\varepsilon \to 0} \eta(x, \varepsilon) - \zeta(x, 0)$$

(with $\zeta(\cdot, 0)$ smooth). We now claim that $\lim_{\varepsilon\to 0} \eta(x,\varepsilon) \in VMO(B)$. The claim follows from the definition of V, that of η, and the proof of Theorem 5.7. The phase η is constructed in the same way as the phase ψ in Step 2 in the proof of Theorem 5.7 (whence our claim, which follows from the proof of Theorem 5.7). The only difference with the proof of Theorem 5.7 occurs in the definition of V. In our case, we let $V := \Pi \circ w$, with $w(x, \varepsilon) := u * \rho_\varepsilon(x)$ and (ρ_ε) is a family of standard mollifiers. In the proof of Theorem 5.7, the role of V is played by $u_\varepsilon/|u_\varepsilon|$, where $u_\varepsilon(x) := \fint_{B_\varepsilon(x)} u$. However, the proof of Theorem 5.7 applies, with minor changes, to our V.

We assume now that $sp < 1$ and $u \in C(\Omega)$. In particular, we have $u \in VMO_{loc}(\Omega)$ and thus, by the above, $\varphi \in VMO_{loc}(\Omega)$. Write, in B, $u = e^{\imath\xi}$, with $\xi \in C(B; \mathbb{R})$. Since $e^{\imath\varphi} = e^{\imath\xi}$ in B, Proposition 6.1 implies that $\varphi - \xi$ is constant in B, and thus φ is (globally) continuous.

The case where $sp < 1$ and $u \in C^\infty(\Omega)$ is similar.

The proof of Theorem 7.1 is complete. □

7.5 Complements and open problems

7.5.1 *An improvement of Theorem 7.1 when sp is an integer*

Let us take a closer look at the proof of Theorem 7.1 when $1 < p < \infty$ and $s = 1/p$. In that case, we established the estimate (see (7.76))

$$\sum_{|\alpha|=2} \int_\Omega \int_0^\infty |\partial^\alpha U(x, t)|\, dtdx < \infty. \tag{7.112}$$

This implies that U belongs to $W^{2,1}(\Omega \times (0, 1))$, and thus the trace v of U on Ω belongs to $W^{1,1}$. However, the trace of a $W^{2,1}$ map is better than "just" $W^{1,1}$: it belongs to the Besov space $B^1_{1,1}$, which is a strict subspace of $W^{1,1}$, continuously embedded in $W^{1,1}$ (see, for example, Mironescu and Russ [265, Theorem 1.1]).

More generally, we have the following improvement of Theorem 7.1 (see Mironescu [262]).

Theorem 7.3. *Assume that $N \geq 1$, $s > 0$, $1 \leq p < \infty$, and that sp is an integer. Then, every $u \in W^{s,p}(\Omega; \mathbb{S}^1)$ can be written as $u = e^{\imath\varphi} v$, where $\varphi \in W^{s,p}(\Omega; \mathbb{R})$ and $v \in B^{sp}_{1,1}(\Omega; \mathbb{S}^1)$ satisfy the estimates*

$$\|\varphi\|_{W^{s,p}} \leq C|u|_{W^{s,p}} \tag{7.113}$$

and

$$\left\| v - \fint u \right\|_{B^{sp}_{1,1}} \leq C|u|^p_{W^{s,p}} + |u|_{W^{s,p}}. \tag{7.114}$$

Note that, unlike Theorem 7.1, Theorem 7.3 is non-trivial even when $s = 1$ and $p = 1$.

Remark 7.8. Combining Theorem 7.3 with the Gagliardo–Nirenberg type embeddings $B^1_{1,1} \cap L^\infty \hookrightarrow W^{1/p,p}$, $1 < p < \infty$ (plus the corresponding estimates), we recover and improve the property discussed in Remark 7.7: if $u \in W^{1,1}(\Omega; \mathbb{S}^1)$, then we may write $u = e^{\imath\varphi} v$, with $\varphi \in W^{1,1}$ and $v \in \bigcap_{p\geq 1} W^{1/p,p}$ satisfying the estimates

$$\|\varphi\|_{W^{1,1}} \leq C|u|_{W^{1,1}}$$

and

$$\left\| v - \fint u \right\|^p_{W^{1/p,p}} \leq C_p |u|_{W^{1,1}}.$$

7.5.2 Approach by duality

An immediate consequence of Theorem 7.1, applied with $s = 1/2$ and $p = 2$, is the following. Let $\varphi \in C^\infty(\overline{\Omega}; \mathbb{R})$ and set $u := e^{\imath\varphi}$. Then,

$$|\varphi|_{H^{1/2}+W^{1,1}} \leq C\, |u|_{H^{1/2}}\, (1 + |u|_{H^{1/2}}). \tag{7.115}$$

We propose here an alternative strategy for proving (7.115), without relying on factorization. It has not yet come to fruition, but raises an interesting open problem.

For simplicity, we consider functions that are periodic in each coordinate, that we identify with maps on the (flat) torus $\mathbb{T}^N$. Let $\varphi \in C^\infty(\mathbb{T}^N; \mathbb{R})$, with $N \geq 2$, and set $u := e^{\imath\varphi}$. The goal is to establish the estimate (7.115) using the following idea. Recall that $\nabla\varphi = u \wedge \nabla u$ and thus, $\forall\, \zeta \in C^\infty(\mathbb{T}^N; \mathbb{R}^N)$,

$$\begin{aligned}\left|\int_\Omega (\nabla\varphi)\,\zeta\right| = \left|\int_\Omega (u \wedge \nabla u)\,\zeta\right| &\leq C\, |u|_{H^{1/2}}\, |u\zeta|_{H^{1/2}} \\ &\leq C\, |u|_{H^{1/2}}\, (|u|_{H^{1/2}}\, \|\zeta\|_{L^\infty} + |\zeta|_{H^{1/2}});\end{aligned} \tag{7.116}$$

here, we use the fact that $H^{-1/2}$ and $H^{1/2}$ are in duality, the inequality $|\nabla u|_{H^{-1/2}} \leq C\, |u|_{H^{1/2}}$ and the fact that $H^{1/2} \cap L^\infty$ is an algebra (see Lemma 15.41). Hence,

$$\nabla\varphi \in (H^{1/2} \cap L^\infty)^* \text{``}=\text{''} H^{-1/2} + L^1,$$

and $\nabla\varphi$ admits a decomposition

$$\nabla\varphi = F_1 + F_2, \tag{7.117}$$

with

$$|F_1|_{H^{-1/2}} \leq C\, |u|_{H^{1/2}} \text{ and } \|F_2\|_{L^1} \leq C\, |u|^2_{H^{1/2}}. \tag{7.118}$$

One may be tempted to continue and write $F_1 = \nabla\varphi_1$, $F_2 = \nabla\varphi_2$, with

$$|\varphi_1|_{H^{1/2}} \le C\,|F_1|_{H^{-1/2}} \le C\,|u|_{H^{1/2}} \tag{7.119}$$

and

$$|\varphi_2|_{W^{1,1}} \le C\,\|F_2\|_{L^1} \le C\,|u|^2_{H^{1/2}}. \tag{7.120}$$

Then, φ would satisfy (7.115). Of course, the argument is flawed because F_1, F_2 given by (7.117) need not be (separately) gradients. [But this *is* the case if $N = 1$.] However, one might still hope to conclude that (7.115) holds provided the following problem has a positive answer.

Let $N \ge 2$. Let $\psi \in C^\infty(\mathbb{T}^N; \mathbb{R})$. Is it true that

$$\left\| \psi - \fint_{\mathbb{T}^N} \psi \right\|_{H^{1/2}+W^{1,1}} \le C\,\|\nabla\psi\|_{H^{-1/2}+L^1}? \tag{7.121}$$

This question was raised in Bourgain and Brezis [48, equation (1.16)] and is still open, even when $N = 2$.

By duality, (7.121) is equivalent to the solvability of the equation

$$\operatorname{div} Y = f \in H^{-1/2} \cap W^{-1,\infty} \tag{7.122}$$

with $Y \in H^{1/2} \cap L^\infty$, and the corresponding estimates.

An inequality that bears some resemblance to (7.121) is known to be true.

Theorem 7.4. *([48, Theorem 1], [49, Corollary 6]) Let $N \ge 2$. We have*

$$\left\| \psi - \fint_{\mathbb{T}^N} \psi \right\|_{L^{N/(N-1)}} \le C\,\|\nabla\psi\|_{W^{-1,N/(N-1)}+L^1}. \tag{7.123}$$

By duality, (7.123) is equivalent to the solvability of the equation

$$\operatorname{div} Y = f \in L^N \tag{7.124}$$

with $Y \in W^{1,N} \cap L^\infty$, and the corresponding estimates.

When $N = 2$, there is a direct proof of (7.123) via Fourier analysis [48] (see also Da Lio and Rivière [125], Da Lio, Rivière, and Wettstein [126], Maz'ya [239], and Mironescu [255]). When $N \ge 3$, the only known proof of Theorem 7.4 relies on a "brute force" construction of a solution Y of (7.124), involving a non-linear technique.

Note that (7.124) can be used to establish a consequence of (7.115), namely

$$\left\| \varphi - \fint_{\mathbb{T}^N} \varphi \right\|_{L^{N/(N-1)}} \le C\,|u|_{H^{1/2}}\,(1 + |u|_{H^{1/2}}) \tag{7.125}$$

(recall that $W^{1,1} \hookrightarrow L^{N/(N-1)}$ and $H^{1/2} \hookrightarrow L^{N/(N-1)}$). Indeed, returning to (7.116), we have

$$\left| \int_{\mathbb{T}^N} \varphi \operatorname{div} X \right| \leq C\, |u|_{H^{1/2}} \left(|u|_{H^{1/2}} \|X\|_{L^\infty} + |X|_{H^{1/2}} \right), \quad \forall X \in H^{1/2} \cap L^\infty(\mathbb{T}^N; \mathbb{R}^N). \tag{7.126}$$

We then insert in (7.126) $X = Y$, where $Y \in W^{1,N} \cap L^\infty$ is a solution of (7.124) satisfying the corresponding estimates, and use the obvious fact that $|X|_{H^{1/2}} \leq C\, |X|_{W^{1,N}}$.

See also Corollary 9.1 applied with $s = 1/2, p = 2$, for a more general version of (7.125).

7.5.3 *Open problems*

We recall the open problem from the previous section.

Open Problem 10. Let $N \geq 2$. Let $\psi \in C^\infty(\mathbb{T}^N; \mathbb{R})$. Is it true that

$$\left\| \psi - \fint_{\mathbb{T}^N} \psi \right\|_{H^{1/2}+W^{1,1}} \leq C\, \|\nabla \psi\|_{H^{-1/2}+L^1}?$$

Open Problem 11. Let $u \in W^{s,p}(\Omega; \mathbb{S}^1)$, and let φ be defined by (7.19). Is it true that $\varphi \in BMO$?

7.6 Comments

1. The ancestor of factorization is a result due to Bourgain and Brezis [48, Theorem 4] (see also [51, Theorem 3] and [54, Theorem 3]). It was formulated as follows: given $u \in C^\infty(\overline{\Omega}; \mathbb{S}^1)$ (with Ω simply connected), there exist $\varphi_1, \varphi_2 \in C^\infty(\overline{\Omega}; \mathbb{R})$ such that $u = e^{\imath(\varphi_1+\varphi_2)}$ and

$$|\varphi_1|_{H^{1/2}} \leq C|u|_{H^{1/2}}, \tag{7.127}$$

$$|\varphi_2|_{W^{1,1}} \leq C|u|_{H^{1/2}}^2. \tag{7.128}$$

   Note that this is a special case of Theorem 7.1, corresponding to $s = 1/2, p = 2$, $\varphi = \varphi_1$, and $v = e^{\imath\varphi_2}$. The original proof in [48] is based on paraproducts. The more general formulation as stated in Theorem 7.1 – starting with an arbitrary $u \in H^{1/2}(\Omega; \mathbb{S}^1)$ – appears in [54, Theorem 5].
2. The proof in [48] extends to $W^{1/p,p}$ when $1 < p \leq 2$: if $u \in C^\infty(\overline{\Omega}; \mathbb{S}^1)$, then we may write $u = e^{\imath(\varphi_1+\varphi_2)}$, where $\varphi_1, \varphi_2 \in C^\infty(\overline{\Omega}; \mathbb{R})$ satisfy

$$|\varphi_1|_{W^{1/p,p}} \leq C|u|_{W^{1/p,p}}, \tag{7.129}$$

$$|\varphi_2|_{W^{1,1}} \leq C|u|_{W^{1/p,p}}^p. \tag{7.130}$$

However, when $p > 2$, it is not clear how to adapt the proof in [48] in order to obtain (7.129) and (7.130), and this was raised as an open problem in Bourgain, Brezis, and Mironescu [56, Open Problem 1]. Nguyen [280] proved the validity of (7.129) and (7.130) when $p > 2$ and $N = 1$. His approach consists of establishing the following generalization of (7.116):

$$\left|\int (u \wedge \nabla u)\zeta\right| \leq C(|u|^p_{W^{1/p,p}}\|\zeta\|_{L^\infty} + |u|_{W^{1/p,p}}|\zeta|_{W^{1/p',p'}}). \tag{7.131}$$

Once (7.131) is established, we find that $u \wedge \nabla u$ splits as $u \wedge \nabla u = F_1 + F_2$, where $|F_1|_{W^{1/p-1,p}} \leq C|u|_{W^{1/p,p}}$ and $\|F_2\|_{L^1} \leq C|u|^p_{W^{1/p,p}}$. Since $N = 1$, we may take φ_j to be a primitive of F_j, $j = 1, 2$; these phases satisfy (modulo a constant) $u = e^{\imath(\varphi_1+\varphi_2)}$ and the estimates (7.129) and (7.130). Estimate (7.131) still holds when $N \geq 2$ but, as noted in Section 7.5.2, it does not imply Theorem 7.1 when $N \geq 2$, $1 < p < \infty$, and $s = 1/p$.

3. Theorem 7.1 is due to Mironescu [253, 254, 256, 262] when $sp \geq 1$ and to Mironescu and Molnar [263] when $sp < 1$.
The original proof of Theorem 7.1 in the range $0 < s < 1$ and $sp \geq 2$, presented in [253, 254, 256], involved paraproducts. The proof in [263] of Theorem 7.1 when $sp < 1$ (see also Mironescu, Russ, and Sire [266]) suggested that the theory of weighted Sobolev spaces (see Section 7.3) could be used in the proof of Theorem 7.1. This led to the proof we present here (see also [262]). This approach yields Theorem 7.3, which is an improvement of Theorem 7.1 even in the "historical" case $s = 1/2$, $p = 2$, $u \in C^\infty(\overline{\Omega}; \mathbb{S}^1)$, considered in [48].
4. The definition of the quantity $d(x)$ and the crucial estimate (7.64) (when $sp \geq 1$ and w is a suitable harmonic extension of u) are due to Bourgain, Brezis, and Mironescu [56]. The proof presented here, also valid when $sp < 1$ and w is any extension by smoothing, is new.
5. The second main theorem in this chapter, Theorem 7.2, is an easy consequence of Lemma 7.1. This auxiliary result was established by Bousquet and Mironescu [63, Lemma 6.15].

Chapter 8
Applications of the factorization

8.0 Introduction

In this chapter, we present various applications of the factorization.

1. So far, we have defined, in Chapter 1, the distribution Ju only for $u \in W^{1,1}(\Omega; \mathbb{S}^1)$, through the formula $Ju = \dfrac{1}{2} d(u \wedge \nabla u)$, where $u \wedge \nabla u \in L^1(\Omega; \mathbb{R}^N)$. When $u \in W^{s,p}(\Omega; \mathbb{S}^1)$ and $0 < s < 1$, the meaning of $u \wedge \nabla u$ is not clear even in the sense of distributions (see, however, Remark 8.5). We propose below "robust" definitions of the Jacobian Ju and of the quantity $\Sigma(u)$ for maps $u \in W^{1/p,p}(\Omega; \mathbb{S}^1)$, with $1 < p < \infty$. As a consequence, Ju and $\Sigma(u)$ make sense for maps $u \in W^{s,p}(\Omega; \mathbb{S}^1)$ with $s > 0$, $1 \leq p < \infty$, and $sp \geq 1$. It turns out that $Ju = 0$ and $\Sigma(u) = 0$ when $sp \geq 2$, and thus the range of interest is often assumed to be $1 \leq sp < 2$.
2. Factorization leads to a substantial refinement of the lifting theory already discussed in Chapters 1 and 5. Recall that, if $N \geq 2$ and $u \in W^{s,p}(\Omega; \mathbb{S}^1)$, then u does not necessarily admit a lifting $\varphi \in W^{s,p}(\Omega; \mathbb{R})$ in the following cases

    a) $0 < s < 1$, $1 \leq sp < N$
    b) $s \geq 1$, $1 \leq sp < 2$.

    Our goal is to establish the existence of a lifting φ in a space larger than $W^{s,p}$, and in some sense optimal. This will play an essential role in Chapter 10 (density) and Chapter 11 (traces).
3. In Sections 8.3–8.5, we adapt to $W^{s,p}$ various results from Chapter 1, concerning, for example, the least energy with prescribed Jacobian, the relaxed energy, and the square root problem.

H. Brezis and P. Mironescu, *Sobolev Maps to the Circle*, Progress in Nonlinear Differential Equations and Their Applications 96,
https://doi.org/10.1007/978-1-0716-1512-6_8

8.1 Existence of Ju for u in $W^{1/p,p}$

We start by discussing a way of defining Ju (when $N \geq 2$) and $u \wedge \nabla u$ (when $N \geq 1$), where $u \in W^{s,p}(\Omega; \mathbb{S}^1)$ with $0 < s < 1$ and $sp \geq 1$. Of course, the most interesting case is $s = 1/p$, since $W^{s,p} \hookrightarrow W^{1/p,p}$ when $sp \geq 1$. As we are going to see, the definitions of Ju and $u \wedge \nabla u$ are consistent with those in Chapter 1: if $u \in W^{1,1} \cap W^{s,p}$ for some $0 < s < 1$ and $1 \leq p < \infty$ such that $sp \geq 1$, then the new Ju and $u \wedge \nabla u$ coincide with the old ones.

We start with the definitions of Ju and $\Sigma(u)$ (when $N \geq 2$) for $u \in W^{s,p}(\Omega; \mathbb{S}^1)$, with $0 < s < 1$ and $sp \geq 1$. Using the factorization Theorem 7.1, we may write

$$u = e^{\imath\varphi} v, \text{ with } \varphi \in W^{s,p}(\Omega; \mathbb{R}) \text{ and } v \in W^{1,sp}. \tag{8.1}$$

Set

$$Ju := Jv = \frac{1}{2} d(v \wedge \nabla v) \tag{8.2}$$

and

$$\Sigma(u) := \Sigma(v). \tag{8.3}$$

These definitions do not depend on the choice of φ and v. Indeed, write

$$u = e^{\imath\varphi_1} v_1 = e^{\imath\varphi_2} v_2,$$

with $v_1, v_2, \varphi_1, \varphi_2$ as above. Let $v_3 := v_1 \overline{v_2}$ and $\varphi_3 := \varphi_2 - \varphi_1$. Then, $v_3 = e^{\imath\varphi_3} \in W^{1,1}$ and $\varphi_3 \in W^{s,p}$. Lemma 6.2 implies that $\varphi_3 \in W^{1,1}$. Using (1.56)–(1.58), we find that $Jv_3 = 0$, and thus $Jv_1 = Jv_2$. Consequently, by (1.11), $\Sigma(v_1) = \Sigma(v_2)$.

We also note that the definition of Ju does not depend on s and p: if $u \in W^{s_1,p_1} \cap W^{s_2,p_2}(\Omega; \mathbb{S}^1)$, with $s_1 p_1 \geq 1$ and $s_2 p_2 \geq 1$, then the two possible definitions of Ju coincide. Here is a proof of this fact. Let $u \in W^{s_1,p_1} \cap W^{s_2,p_2}$, and write $u = v_j e^{\imath\varphi_j}$, $j = 1, 2$, with $v_j \in W^{1,1}$ and $\varphi_j \in W^{s_j,p_j}$. Then,

$$v_1\overline{v_2} = e^{\imath(\varphi_2 - \varphi_1)}, \text{ with } v_1\overline{v_2} \in W^{1,1} \text{ and } \varphi_2 - \varphi_1 \in W^{s_1,p_1} + W^{s_2,p_2}. \tag{8.4}$$

By (8.4) and Lemma 6.2, we have $\varphi_2 - \varphi_1 \in W^{1,1}$. Using this fact, and also (8.4), (1.57), and (1.58), we find that $J(v_1\, \overline{v_2}) = 0$, so that $Jv_1 = Jv_2$, and thus the distributional Jacobian Ju of u does not depend on s and p.

Similarly, if $u \in W^{s,p} \cap W^{1,1}$, then the new Ju coincides with that defined in Chapter 1.

From the definitions of Ju and $\Sigma(u)$, and (1.86), we see that, given $u \in W^{s,p}(\Omega; \mathbb{S}^1)$ with $sp \geq 1$, we have

$$\Sigma(u) = 0 \iff Ju = 0. \tag{8.5}$$

However, the full analogue of (1.86) fails. Obviously,

$$\text{if } \varphi \in W^{s,p} \text{ with } sp \geq 1 \text{ and } u := e^{\imath\varphi}, \text{ then } Ju = 0. \tag{8.6}$$

[Just use the factorization $u = e^{\imath\varphi}\, 1$.] We warn the reader that, by contrast with the $W^{1,1}$-case, the converse is not true (see Remark 8.11).

In the same vein, we are able to define $u \wedge \nabla u$ when $N \geq 1$ and $u \in W^{s,p}(\Omega; \mathbb{S}^1)$ with $0 < s < 1$ and $sp \geq 1$. More precisely, one writes $u = e^{\imath\varphi}\, v$ as above, and sets

$$u \wedge \nabla u := v \wedge \nabla v + \nabla\varphi \in \mathscr{D}'(\Omega; \mathbb{R}^N). \tag{8.7}$$

As in the case of Ju, the definition of $u \wedge \nabla u$ is independent of the decomposition $u = e^{\imath\varphi}\, v$. [Here, we use (1.49)–(1.52).] Note that, by contrast with the $W^{1,1}$ case, $u \wedge \nabla u$ is a distribution, and not an L^1 vector field. Also note that, if $u \in W^{s,p} \cap W^{1,1}$, then the new $u \wedge \nabla u$ coincides with the L^1 vector field $u \wedge \nabla u$.

Remark 8.1. We recall here some properties of Ju, initially established for $W^{1,p}$ maps, and explain why they are (trivially) still true for $W^{s,p}$ maps with $sp \geq 1$.

We start by noting the following obvious consequences of the definitions of Ju and $u \wedge \nabla u$:

$$Ju = \frac{1}{2} d(u \wedge \nabla u), \ \forall\, u \in W^{1/p,p}(\Omega; \mathbb{S}^1), \tag{8.8}$$

$$(uv) \wedge \nabla(uv) = u \wedge \nabla u + v \wedge \nabla v, \ \forall\, u,\ v \in W^{1/p,p}(\Omega; \mathbb{S}^1), \tag{8.9}$$

$$J(uv) = Ju + Jv,\ J(u\overline{v}) = Ju - Jv, \ \forall\, u,\ v \in W^{1/p,p}(\Omega; \mathbb{S}^1), \tag{8.10}$$

and

$$Jf(u) = (\deg f)\, Ju, \ \forall\, u \in W^{1/p,p}(\Omega; \mathbb{S}^1), \ \forall f \in \mathrm{Lip}(\mathbb{S}^1; \mathbb{S}^1). \tag{8.11}$$

[For the proof of (8.11), it suffices to repeat the proof of Theorem 1.3 and use (8.10).]

Another property:

$$Ju = 0 \text{ if } u \in W^{s,p}(\Omega; \mathbb{S}^1) \text{ with } sp \geq 2. \tag{8.12}$$

Indeed, if u is as in (8.12), then $Ju = Jv$ for some $v \in W^{sp,1}(\Omega; \mathbb{S}^1) \subset W^{1,sp}(\Omega; \mathbb{S}^1)$. By (1.61), we have $Jv = 0$.

Yet another clear property: if ω is an open subset of Ω, and $u, \widetilde{u} \in W^{1/p,p}(\Omega; \mathbb{S}^1)$, then

$$u = \widetilde{u} \text{ in } \omega \implies Ju = J\widetilde{u} \text{ in } \mathscr{D}'(\omega), \tag{8.13}$$

i.e., Ju is a "local" object. The validity of (8.13) is obtained by repeating the proof of the fact that the definition of Ju does not depend on the factorization of u (see the beginning of this section).

In the same vein, we have the following: if $u \in W^{1/p,p}(\Omega;\mathbb{S}^1)$ and if we set $w := u_{|\omega}$, then

$$Jw \text{ is the restriction to } \omega \text{ of } Ju. \tag{8.14}$$

Our main result in this section is the following

Theorem 8.1. *Let $N \geq 1$. Assume that $1 < p < \infty$. Then the map*

$$u \mapsto u \wedge \nabla u \text{ from } W^{1/p,p}(\Omega;\mathbb{S}^1) \text{ to } \mathscr{D}'(\Omega;\mathbb{R}^N)$$

is continuous. More precisely, if $|u_n - u|_{W^{1/p,p}} \to 0$ then $u_n \wedge \nabla u_n \to u \wedge \nabla u$ in $\mathscr{D}'(\Omega)$.

In view of (8.8), we obtain the following

Corollary 8.1. *Let $N \geq 2$. Assume that $1 < p < \infty$. Then the map*

$$u \mapsto Ju \text{ from } W^{1/p,p}(\Omega;\mathbb{S}^1) \text{ to } \mathscr{D}'(\Omega;\Lambda^2)$$

is continuous. More precisely, if $|u_n - u|_{W^{1/p,p}} \to 0$ then $Ju_n \to Ju$ in $\mathscr{D}'(\Omega)$.

Remark 8.2. The conclusion of Theorem 8.1 is far from obvious, since φ and v are not unique. The above theorem relies on the fact that, in the factorization theorem, we may choose φ and v depending continuously on u.

The theorem also shows that we have given a proper meaning to $u \wedge \nabla u$ and to Ju when $u \in W^{1/p,p}$. Indeed, these definitions coincide with the usual ones when $u \in W^{1/p,p} \cap W^{1,1}$. On the other hand, $u \in W^{1/p,p}(\Omega;\mathbb{S}^1) \cap W^{1,1}$ is dense in $W^{1/p,p}(\Omega;\mathbb{S}^1)$. [This is a consequence of Theorem 10.3 in Chapter 10.]

Proof of Theorem 8.1. Set $j(u) := u \wedge \nabla u$. In Step 1, we establish continuity of j under the stronger assumption that $\|u_n - u\|_{W^{1/p,p}} \to 0$. In Step 2, we prove continuity assuming only $|u_n - u|_{W^{1/p,p}} \to 0$.

Step 1. Continuity of j for strong convergence
By Theorem 7.1,

$$j \text{ is continuous at } u \equiv 1. \tag{8.15}$$

Let now $u_n, u \in W^{1/p,p}(\Omega;\mathbb{S}^1)$ be such that $u_n \to u$ in $W^{1/p,p}$. By (8.9), we have

$$j(u_n) = j(u_n\overline{u}) + j(u). \tag{8.16}$$

Since $u_n\overline{u} \to 1$ in $W^{1/p,p}$ (Lemma 15.41), (8.15) implies that

$$j(u_n\overline{u}) \to j(1) = 0. \tag{8.17}$$

We obtain the continuity of j at u from (8.16) and (8.17).

Step 2. Suppose that $u_n \to u$ in the sense that

$$|u_n - u|_{W^{1/p,p}} \to 0. \tag{8.18}$$

We claim that, up to a subsequence,

- Either $\|u_n - u\|_{W^{1/p,p}} \to 0$
- Or there exist constants $C_1, C_2 \in \mathbb{S}^1$ such that $u = C_1$ and $\|u_n - C_2\|_{W^{1/p,p}} \to 0$.

In both cases, we conclude from the previous analysis that $j(u_n) \to j(u)$ in $\mathscr{D}'(\Omega)$, and this completes the proof of Theorem 8.1.

Proof of the claim. Since $\|u_n\|_{W^{1/p,p}} \le C$, we can extract a subsequence (still denoted by (u_n)) such that $u_n \to \widetilde{u}$ in L^p and a.e. for some $\widetilde{u} \in W^{1/p,p}(\Omega; \mathbb{S}^1)$. From (8.18) and Fatou's lemma, we see that $|\widetilde{u} - u|_{W^{1/p,p}} = 0$, and thus $\widetilde{u} - u = \alpha$ for some constant α. If $\alpha = 0$, we conclude that $\widetilde{u} = u$, so that $\|u_n - u\|_{W^{1/p,p}} \to 0$.

If $\alpha \ne 0$, we note that $\widetilde{u}$ takes its values in the set $\mathbb{S}^1 \cap (\mathbb{S}^1 + \alpha)$, which consists of at most two points. From Theorem 6.5, we deduce that $\widetilde{u} = C_2$ for some constant C_2, and thus $u = \widetilde{u} - \alpha$ is also constant. Finally, we have $\|u_n - C_2\|_{W^{1/p,p}} \to 0$. $\square$

Corollary 8.2. *Let $N \ge 2$. Assume that $1 < p < \infty$. Then the map*

$$u \mapsto \Sigma(u) \text{ from } W^{1/p,p}(\Omega; \mathbb{S}^1) \text{ to } \mathbb{R}$$

is continuous.

Proof. Let $u_n, u \in W^{1/p,p}(\Omega; \mathbb{S}^1)$ be such that $u_n \to u$ in $W^{1/p,p}$. Take $v \in W^{1,1}$ such that $Ju = Jv$. Let $w_n := u_n \overline{u}$, so that $w_n \to 1$ in $W^{1/p,p}$. By the proof of the Theorem 8.1, we may choose $v_n \in W^{1,1}$ such that $Jw_n = Jv_n$ and $v_n \to 1$ in $W^{1,1}$. Then, $Ju_n = Jv + Jv_n$ (by (8.10)), and thus

$$|\Sigma(u_n) - \Sigma(u)| \le \Sigma(u_n \overline{u}) = \Sigma(v_n) \le |v_n|_{W^{1,1}} \to 0 \text{ as } n \to \infty. \qquad \square$$

Remark 8.3. The map $u \mapsto u \wedge \nabla u$ is *not* continuous for the weak convergence in $W^{1/p,p}$. Indeed, let u_ε be the dipole constructed in Proposition 2.1. By a scaling argument, it is easy to see that u_ε is bounded in $W^{1/p,p}$, $\forall p \ge 1$. On the other hand, we have $Ju_\varepsilon = \pi(\delta_P - \delta_N)$ and $u_\varepsilon \to 1$ a.e. We find that $u_\varepsilon \rightharpoonup u \equiv 1$ in $W^{1/p,p}$, $\forall p > 1$, but $Ju_\varepsilon \not\to Ju$.

Remark 8.4. It is possible to give a definition of Ju for $u \in H^{1/2}(\Omega; \mathbb{S}^1)$, with $\Omega \subset \mathbb{R}^2$, which does not rely on factorization (see Bourgain, Brezis, and Mironescu [54]). The definition below extends easily to higher dimensions (see Bousquet and Mironescu [63]), but for simplicity we discuss here only the case where $N = 2$. We let $\widetilde{\Omega} := \Omega \times (0, \infty)$. Given $u \in H^{1/2}(\Omega; \mathbb{S}^1)$ and $\zeta \in C_c^\infty(\Omega)$, we consider an extension $U \in H^1(\widetilde{\Omega}; \mathbb{R}^2)$ of u and an extension $\Phi \in C_c^\infty(\Omega \times [0, \infty); \mathbb{R})$ of ζ. We consider the vector field

$$H := -(\partial_y U \wedge \partial_z U, \partial_z U \wedge \partial_x U, \partial_x U \wedge \partial_y U), \tag{8.19}$$

and then set

$$\langle Ju, \zeta \rangle := \int_{\widetilde{\Omega}} H \cdot \nabla \Phi. \tag{8.20}$$

This definition does not depend on the choice of U or Φ [54, Lemma 3]. The definitions (8.2) and (8.20) yield the same distribution Ju. Indeed, the two objects coincide when $u \in H^{1/2}(\Omega; \mathbb{S}^1) \cap W^{1,1}$ (Remark 8.1 and [54, Lemma 1]) and they depend continuously on the $H^{1/2}$ norm (Theorem 8.1 and [54, Lemma 9]). We conclude using the density of the class $H^{1/2}(\Omega; \mathbb{S}^1) \cap W^{1,1}$ in $H^{1/2}(\Omega; \mathbb{S}^1)$ (Theorem 10.3).

A similar procedure allows us to define Ju when $u \in W^{1/p,p}(\Omega; \mathbb{S}^1)$, with $1 < p < \infty$; however, when $p > 2$ the construction becomes more involved. We recall that, by standard trace theory (see, for example, Lemma 15.47 in Appendix 15.14), each $u \in W^{1/p,p}(\Omega)$ has an extension $U = U(u) \in W^{2/p,p} \cap C^\infty(\widetilde{\Omega})$ such that

$$u \mapsto U \text{ is continuous from } W^{1/p,p}(\Omega) \text{ to } W^{2/p,p}(\widetilde{\Omega}). \tag{8.21}$$

In addition, we may choose U such that

$$\text{if } u \in W^{1/p,p} \cap H^{1/2}, \text{ then } U \in H^1. \tag{8.22}$$

The key (and non-trivial) observation is that, when $u \in W^{1/p,p}(\Omega; \mathbb{S}^1)$, it is always possible to choose U such that, in addition to (8.21) and (8.22), we have (see Bourgain, Brezis, and Mironescu [56, proof of Lemma 2.1 and formula (2.12)])

$$\int_{\widetilde{\Omega}} |H| < \infty \tag{8.23}$$

(even though, in general, U does not belong to H^1), and

$$\text{the map } u \mapsto H \text{ is continuous from } W^{1/p,p}(\Omega; \mathbb{S}^1) \text{ to } L^1(\widetilde{\Omega}; \mathbb{R}^3); \tag{8.24}$$

here, H is given by (8.19).

In view of (8.23), we may still define Ju as in (8.20). By (8.22) and our discussion on $H^{1/2}$ maps, this Ju coincides with the usual one when $u \in H^{1/2}(\Omega; \mathbb{S}^1) \cap W^{1,1}$. By (8.24), the map $u \mapsto Ju$ is continuous, and thus (as explained for $H^{1/2}$ maps) coincides, for every $u \in W^{1/p,p}(\Omega; \mathbb{S}^1)$, with the Jacobian Ju defined in this chapter.

Remark 8.5. When $u = u_1 + \imath u_2 \in W^{1/p,p}(\Omega; \mathbb{S}^1)$ with $1 < p \le 2$, it is possible to give a meaning to $u \wedge \nabla u$ (and thus to Ju) without invoking factorization (or extensions as in Remark 8.4). Indeed, assume first that $p = 2$. If $\zeta \in C_c^\infty(\Omega; \mathbb{R}^N)$, then $u_j \zeta \in H_c^{1/2}(\Omega; \mathbb{R}^N)$, and $\nabla u_j \in [H_c^{1/2}(\Omega; \mathbb{R}^N)]^*, j = 1, 2$. We may thus define

$$(u \wedge \nabla u)(\zeta) = \langle \nabla u_2, u_1 \zeta \rangle - \langle \nabla u_1, u_2 \zeta \rangle, \ \forall \zeta \in C_c^\infty(\Omega; \mathbb{R}^N), \tag{8.25}$$

where $\langle \, , \, \rangle$ denotes the duality bracket between $[H_c^{1/2}(\Omega; \mathbb{R}^N)]^*$ and $H_c^{1/2}(\Omega; \mathbb{R}^N)$. Using Lemma 15.41 in Appendix 15.12, it is easy to see that $u \wedge \nabla u$ defined as in (8.25) is a distribution (or rather a current) depending continuously on u. Since clearly $u \wedge \nabla u$ coincides with the "usual" $u \wedge \nabla u$ when $u \in H^{1/2}(\Omega; \mathbb{S}^1) \cap W^{1,1}$, we obtain that (8.25) yields the same $u \wedge \nabla u$ as the one defined by (8.7), and thus the two approaches lead to the same Ju.

When $1 < p < 2$, we have $W^{1/p,p}(\Omega; \mathbb{S}^1) \subset H^{1/2}(\Omega; \mathbb{S}^1)$, and thus for $u \in W^{1/p,p}(\Omega; \mathbb{S}^1)$ we may still define $u \wedge \nabla u$ via (8.25).

Remark 8.6. When $sp < 1$, we have $W^{1,1}(\Omega; \mathbb{S}^1) \subset W^{s,p}(\Omega; \mathbb{S}^1)$, and thus one may ask whether it is possible to extend continuously $u \mapsto Ju$ from $W^{1,1}(\Omega; \mathbb{S}^1)$ to $W^{s,p}(\Omega; \mathbb{S}^1)$. The answer is negative (and thus there is no "natural" Ju in $W^{s,p}(\Omega; \mathbb{S}^1)$ with $sp < 1$).

A first reason is that $C^\infty(\overline{\Omega}; \mathbb{S}^1)$ is dense in $W^{s,p}(\Omega; \mathbb{S}^1)$ (Theorem 10.2) and $Ju = 0$ if $u \in C^\infty(\overline{\Omega}; \mathbb{S}^1)$. Thus, a continuous extension has to vanish for every u. However, if $u \in W^{1,1}$ we need not have $Ju = 0$.

A second reason is that, if we consider a dipole $u = u_\varepsilon$ as in Remark 8.3 and let $\varepsilon \to 0$, then $Ju_\varepsilon = \pi(\delta_P - \delta_N) \neq 0$ does not depend on ε, whereas the reader may check that $u_\varepsilon \to 1$ as $\varepsilon \to 0$ strongly in $W^{s,p}$ (since $sp < 1$). Thus, $u \mapsto Ju$, defined on $W^{1,1}$, is not continuous for the $W^{s,p}$ norm.

8.2 Lifting revisited

8.2.1 What we already know

Let us recall what we have established so far about lifting in Chapters 1 and 5. We examine separately the Cases 1, 2, and 3 considered in Section 5.1. We assume throughout that Ω is a smooth bounded simply connected domain in $\mathbb{R}^N$.

Case 1: $N = 1, 0 < s < \infty, 1 \le p < \infty$

Theorem 8.2. *Assume that $N = 1$, $0 < s < \infty$, and $1 \le p < \infty$. Then, every $u \in W^{s,p}(\Omega; \mathbb{S}^1)$ may be written as $u = e^{\imath\varphi}$ for some $\varphi \in W^{s,p}(\Omega; \mathbb{R})$.*

Remark 8.7. The converse is also true, i.e, if $\varphi \in W^{s,p}$ then $u \in W^{s,p}$. This follows from the fact that the composition with $\Phi(t) = e^{\imath t}$ transforms $W^{s,p}$ maps into $W^{s,p}$ maps provided either $s \le 1$ or $sp \ge N$ (see Corollary 15.3 in Appendix 15.3).

We now turn to

Case 2: $N \ge 2, 0 < s < 1, 1 \le p < \infty$

Theorem 8.3. *Assume that $N \ge 2$, $0 < s < 1$, and $1 \le p < \infty$. Then, the answer to the lifting question is:*

a) Positive when $sp \ge N$
b) Negative when $1 \le sp < N$
c) Positive when $sp < 1$.

Case 3: $N \geq 2$, $1 \leq s < \infty$, $1 \leq p < \infty$

Recall that Theorem 5.3 combined with Theorem 5.5 reads

Theorem 8.4. *Assume that $N \geq 2$, $1 \leq s < \infty$, and $1 \leq p < \infty$. Then, the answer to the lifting question is:*

a) Positive when $sp \geq 2$. Moreover, the (unique) lifting φ belongs to $W^{s,p} \cap W^{1,sp}$
b) Negative when $1 \leq sp < 2$.

Remark 8.8. The "converses" to Theorem 8.3 *a), c)* and to Theorem 8.4 *a)* are true. For example, when $s \geq 1$ and $sp \geq 2$, if $\varphi \in W^{s,p} \cap W^{1,sp}$, then $e^{\imath\varphi} \in W^{s,p}$. This follows from the results on composition presented in Appendix 15.3.

8.2.2 Old result, new proof: lifting in $W^{s,p}$ when $sp < 1$

We give here a new proof, relying on factorization, of the fact that, if $sp < 1$, then any $u \in W^{s,p}(\Omega; \mathbb{S}^1)$ has a phase $\varphi \in W^{s,p}$ such that $|\varphi|_{W^{s,p}} \leq C|u|_{W^{s,p}}$; see Theorems 5.1, 5.2 *c)*, and (5.16). [Recall that the proof in this case was the most delicate of all.] Set $\sigma := sp$ and let $u \in W^{s,p}(\Omega; \mathbb{S}^1)$. Using Theorem 7.1 *3*, write $u = e^{\imath\psi}\, v$, with $\|\psi\|_{W^{s,p}} \lesssim |u|_{W^{s,p}}$ and $|v|_{W^{\sigma,1}} \lesssim |u|^p_{W^{s,p}}$. By Lemma 5.10, v has a lifting ζ such that $|\zeta| \leq 2\pi$ and $|\zeta|_{W^{\sigma,1}} \leq 2|v|_{W^{\sigma,1}}$. We find that

$$|\zeta|_{W^{s,p}} \leq (2\|\zeta\|_{L^\infty})^{1-1/p}\, |\zeta|^{1/p}_{W^{\sigma,1}} \lesssim |u|_{W^{s,p}},$$

and thus $u = e^{\imath\varphi}$, with $\varphi := \psi + \zeta$ satisfying $|\varphi|_{W^{s,p}} \lesssim |u|_{W^{s,p}}$.

8.2.3 The new results

In this section, we return to all the cases where the answer to the lifting question is negative (i.e., $W^{s,p}$ maps need not have $W^{s,p}$ phases). Our goal is to find a lifting of $u \in W^{s,p}(\Omega; \mathbb{S}^1)$ in a "minimal" space containing $W^{s,p}$. We present here several results in this direction.

It is convenient to split the "nasty" zone into three regions.

Region A: $N \geq 3$, $0 < s < 1$, $1 \leq p < \infty$, $2 \leq sp < N$

A statement covering Region A is the following.

Theorem 8.5. *Assume that $N \geq 3$, $0 < s < 1$, $1 \leq p < \infty$, $2 \leq sp < N$. Then, every $u \in W^{s,p}(\Omega; \mathbb{S}^1)$ can be written as $u = e^{\imath\varphi}$, where $\varphi = \psi + \zeta$, with $\psi \in W^{s,p}(\Omega; \mathbb{R})$ and $\zeta \in W^{sp,1} \cap W^{1,sp}(\Omega; \mathbb{R})$. Conversely, if $\varphi = \psi + \zeta$, with $\psi \in W^{s,p}(\Omega; \mathbb{R})$ and $\zeta \in W^{1,sp}(\Omega; \mathbb{R})$, then $u = e^{\imath\varphi} \in W^{s,p}(\Omega; \mathbb{S}^1)$.*

Proof. By Theorem 7.1, we may write $u = e^{\imath\psi} v$, with $\psi \in W^{s,p}$ and $v \in W^{sp,1}$. By Theorem 8.4 *a)*, there exists some $\zeta \in W^{sp,1} \cap W^{1,sp}$ such that $v = e^{\imath\zeta}$. Thus, $u = e^{\imath\varphi}$, where $\varphi := \psi + \zeta$ has all the required properties.

The converse is straightforward: if $\varphi = \psi + \zeta$, with $\psi \in W^{s,p}$ and $\zeta \in W^{1,sp}$, then $e^{\imath\psi} \in W^{s,p}$ and $e^{\imath\zeta} \in W^{1,sp} \cap L^\infty \subset W^{s,p}$ (Appendix 15.2). We conclude using the fact that $W^{s,p} \cap L^\infty$ is an algebra (Lemma 15.41 in Appendix 15.12). □

Remark 8.9. As an amusing byproduct of the factorization Theorem 7.1, we obtain the following (additive!) decomposition result concerning real-valued fractional Sobolev functions.

Under the assumptions of Theorem 8.5, any function $\zeta \in W^{1,sp}(\Omega; \mathbb{R})$ can be decomposed as $\zeta = \zeta_1 + \zeta_2$, with $\zeta_1 \in W^{s,p} \cap W^{1,sp}(\Omega; \mathbb{R})$ and $\zeta_2 \in W^{sp,1} \cap W^{1,sp}(\Omega; \mathbb{R})$.

Indeed, consider the map $u := e^{\imath\zeta} \in W^{1,sp}(\Omega; \mathbb{S}^1) \subset W^{s,p}(\Omega; \mathbb{S}^1)$ (see above). By Theorem 8.5, we may write $u = e^{\imath(\zeta_1+\zeta_2)}$, with $\zeta_1 \in W^{s,p}$ and $\zeta_2 \in W^{sp,1} \cap W^{1,sp}$. Thus, up to a constant, $\zeta = \zeta_1 + \zeta_2$ (Corollary 6.5), and ζ_1, ζ_2 have all the required properties.

The above property is a special case of the following more general result (applied with $\sigma := sp$).

Theorem 8.6. *Let $0 < s < 1$ and $1 < \sigma < \infty$. Then, every $f \in W^{1,\sigma}(\Omega; \mathbb{R})$ can be decomposed as $f = \psi + \zeta$, where $\psi \in W^{s,\sigma/s} \cap W^{1,\sigma}$ and $\zeta \in W^{\sigma,1} \cap W^{1,\sigma}$ satisfy*

$$\|\psi\|_{W^{1,\sigma}} + \|\zeta\|_{W^{1,\sigma}} \lesssim \|f\|_{W^{1,\sigma}}, \ \|\psi\|_{W^{s,\sigma/s}} \lesssim \|f\|^{s}_{W^{1,\sigma}}, \ \|\zeta\|_{W^{\sigma,1}} \lesssim \|f\|^{\sigma}_{W^{1,\sigma}}. \tag{8.26}$$

Theorem 8.6 is a special case of Mironescu [260, Theorem 1.4].

Region B: $N \geq 2$, $0 < s < 1$, $1 \leq p < \infty$, $1 \leq sp < 2$

This region is very much a *terra incognita*, even when $N = 2$ and $u \in H^s$, where $\frac{1}{2} \leq s < 1$. The analogue of the previous theorem fails: given $u \in W^{s,p}(\Omega; \mathbb{S}^1)$, there need not exist a $\varphi \in W^{s,p} + W^{1,sp}$ such that $u = e^{\imath\varphi}$. For example, the canonical map $u(x) := \dfrac{x}{|x|}$ belongs to $W^{s,p}(\Omega; \mathbb{S}^1)$ when $sp < 2$, but has no lifting in $W^{s,p} + W^{1,sp}$, and not even in $W^{s,p} + W^{1,1}$. Indeed, such a lifting, if it exists, is unique (mod 2π), by Corollary 6.5. Therefore, it coincides with θ, which is impossible (see Remark 1.6). All we can say is the following.

Theorem 8.7. *Assume that $N \geq 2$, $0 < s < 1$, and $1 \leq sp < 2$. Then, each map $u \in W^{s,p}(\Omega; \mathbb{S}^1)$ can be written as $u = e^{\imath\varphi}$ for some $\varphi \in W^{s,p} + BV$.*

Unfortunately, the converse is not true. Clearly, if $\zeta \in BV$, then $e^{\imath\zeta}$ belongs to $W^{s,p}$ when $sp < 1$, but need not belong to $W^{s,p}$ if $sp \geq 1$. The example $u(x) = \dfrac{x}{|x|}$ shows that we cannot get rid of the BV part in the above.

Proof of Theorem 8.7. Argue as in the previous proof, but this time write (using Theorem 1.4) $v = e^{\imath\zeta}$, with $\zeta \in BV$. □

Region C: $N \geq 2$, $s \geq 1$, $1 \leq p < \infty$, $1 \leq sp < 2$

For the same reason as in Region B, in this case we cannot hope for better than the following assertion: if $u \in W^{s,p}(\Omega; \mathbb{S}^1)$, then u has a BV phase; here, we use the fact that $u \in W^{1,1}$, combined with Theorem 1.4.

8.2.4 Where $Ju = 0$ enters

As we already noticed in Chapter 1, the condition $Ju = 0$ leads to the existence of a "good" lifting of u. The next result is an extension of Theorem 1.2 to the $W^{s,p}$ setting. Recalling (8.12) and the fact that J is defined only when $sp \geq 1$, we obtain that the relevant range is $1 \leq sp < 2$.

Theorem 8.8. *Assume that $N \geq 2$, $s > 0$, $1 \leq p < \infty$, and $1 \leq sp < 2$. Let $u \in W^{s,p}(\Omega; \mathbb{S}^1)$. Then, the following assertions are equivalent:*

a) $Ju = 0$
b) $u = e^{\imath\varphi}$, *where φ belongs to*

b1) $W^{s,p} + W^{1,sp}$ *if* $0 < s < 1$
b2) $W^{s,p} \cap W^{1,sp}$ *if* $s \geq 1$.

Remark 8.10. We will return to Theorem 8.8 in Chapter 10, and present another characterization of maps satisfying $Ju = 0$ (see Theorem 10.4).

Remark 8.11. An important *warning*. When $0 < s < 1$, $N \geq 2$ and $1 \leq sp < 2$, it may happen that $Ju = 0$, although u is not of the form $u = e^{\imath\varphi}$ for some $\varphi \in W^{s,p}$. Indeed, let ψ be as in Remark 5.8. Take $u := e^{\imath\psi}$. Then, $u \in W^{1,sp} \cap W^{s,p}$ and, moreover, there is no $\varphi \in W^{s,p}$ such that $u = e^{\imath\varphi}$. However, we have $Ju = 0$, by Theorem 1.2.

Proof of Theorem 8.8. When $s \geq 1$, the equivalence between *a)* and *b2)* was established in Theorem 5.6.

Assume that $0 < s < 1$. Let $u \in W^{s,p}(\Omega; \mathbb{S}^1)$ be such that $Ju = 0$. Using successively Theorem 7.1, (8.2), and Theorem 5.6, we obtain that $u = e^{\imath\psi} v = e^{\imath\psi} e^{\imath\zeta} = e^{\imath\varphi}$, with $\psi \in W^{s,p}$, $\zeta \in W^{1,sp}$, and $\varphi := \psi + \zeta \in W^{s,p} + W^{1,sp}$.

Conversely, assume that $u = e^{\imath(\psi+\zeta)}$, with $\psi \in W^{s,p}$ and $\zeta \in W^{1,sp}$. Set $v := e^{\imath\zeta} \in W^{1,sp}$, so that $Ju = Jv$. By Theorem 1.2, we have $Jv = 0$. □

8.3 Least energy with prescribed Jacobian

We continue here the study of the least energy, performed in the $W^{1,p}$ context in Section 1.8. Throughout this section, we assume that $N \geq 2$. We define, for $0 < s < \infty$ and $1 \leq p < \infty$ such that $1 \leq sp < 2$, and $u \in W^{s,p}(\Omega; \mathbb{S}^1)$,

$$S_{s,p}(u) := \inf\{|w|^p_{W^{s,p}};\ Jw = Ju\}. \tag{8.27}$$

When $1 \le sp < 2$ (which is the relevant range for the existence of a non-trivial Jacobian), we have the two following results, complementing Lemma 1.14, resp. Proposition 1.3.

Theorem 8.9. *Assume that $N \ge 2$ and $1 < p < \infty$. Then there exist $C_1, C_2 > 0$ such that*

$$C_1 \Sigma(u) \le S_{1/p,p}(u) \le C_2 \Sigma(u), \quad \forall\, u \in W^{1/p,p}(\Omega; \mathbb{S}^1). \tag{8.28}$$

Proof. *Step 1.* Proof of "$C_1 \Sigma(u) \le S_{1/p,p}(u)$"
We have to prove that $|w|^p_{W^{1/p,p}} \ge C_1 \Sigma(w)$ for every $w \in W^{1/p,p}(\Omega; \mathbb{S}^1)$. Write $w = e^{\imath\varphi} v$, with φ as in v as in Theorem 7.1 2. Then, $\Sigma(w) = \Sigma(v) \le |v|_{W^{1,1}} \lesssim |w|^p_{W^{1/p,p}}$.
Step 2. Proof of "$S_{1/p,p}(u) \le C_2 \Sigma(u)$"
The above inequality amounts to: for every $v \in W^{1,1}(\Omega; \mathbb{S}^1)$, there exists some $w \in W^{1/p,p}(\Omega; \mathbb{S}^1)$ such that $Jw = Jv$ and $|w|^p_{W^{1/p,p}} \lesssim \Sigma(v)$. This follows from Theorem 7.2, in particular estimate (7.10). [Here, we are also using Lemma 1.14.] □

Proposition 8.1. *Assume that $1 < sp < 2$. Then, for every $u \in W^{s,p}(\Omega; \mathbb{S}^1)$, there exists some $v \in W^{s,p}(\Omega; \mathbb{S}^1)$ such that*

$$Jv = Ju, \ |v|^p_{W^{s,p}} = S_{s,p}(u). \tag{8.29}$$

Proof. By Corollary 8.1 and the compactness of the embedding $W^{s,p} \hookrightarrow W^{1/p,p}$, the map $u \mapsto Ju$ is weakly sequentially continuous on $W^{s,p}(\Omega; \mathbb{S}^1)$, and thus the infimum in (8.27) is achieved. □

8.4 Relaxed energy

We continue here the study of the relaxed energy initiated in Section 1.7. Let $u \in W^{s,p}(\Omega; \mathbb{S}^1)$. We set

$$R_{s,p}(u) := \inf\left\{\liminf_{n\to\infty} |u_n|^p_{W^{s,p}};\ u_n \in C^\infty(\overline{\Omega}; \mathbb{S}^1),\ u_n \to u \text{ a.e.}\right\} \le \infty.$$

We start by observing that the cases $N = 1$, or $sp < 1$, or $N \ge 2$ and $sp \ge 2$ are not of interest for the study of the relaxed energy. Indeed, we will see that in all these cases $C^\infty(\overline{\Omega}; \mathbb{S}^1)$ is dense in $W^{s,p}(\Omega; \mathbb{S}^1)$ (Theorems 10.1 and 10.2), so that $R_{s,p}(u) = |u|^p_{W^{s,p}}, \forall\, u \in W^{s,p}(\Omega; \mathbb{S}^1)$.

From now on, we consider the interesting case $N \ge 2$, $1 \le sp < 2$. Our purpose is to generalize Theorems 1.11 and 1.13. The corresponding results are the following.

Theorem 8.10. *Let $N \ge 2$ and $1 < p < \infty$. Then there exist $C_1, C_2 > 0$ such that*

$$|u|^p_{W^{1/p,p}} + C_1\Sigma(u) \le R_{1/p,p}(u) \le |u|^p_{W^{1/p,p}} + C_2\Sigma(u),\ \forall u \in W^{1/p,p}(\Omega;\mathbb{S}^1).$$

Theorem 8.11. *Assume that* $N \ge 2, s > 0, 1 \le p < \infty$, *and* $1 < sp < 2$. *Let* $u \in W^{s,p}(\Omega;\mathbb{S}^1)$. *Then,*

$$R_{s,p}(u) = \begin{cases} |u|^p_{W^{s,p}}, & \text{if } Ju = 0 \\ \infty, & \text{otherwise} \end{cases}.$$

Proof of Theorem 8.10. Let us first note that the relaxed energy does not change if we replace, in its definition, the condition $u_n \in C^\infty(\overline{\Omega};\mathbb{S}^1)$ with $u_n \in \overline{C^\infty(\overline{\Omega};\mathbb{S}^1)}^{W^{1/p,p}}$. In turn, this latter condition is equivalent to $Ju_n = 0$ (Theorem 10.4).

Step 1. Proof of

$$R_{1/p,p}(u) \le |u|^p_{W^{1/p,p}} + C_2\Sigma(u). \tag{8.30}$$

It is clear that $R_{1/p,p}$ is lower semicontinuous on $W^{1/p,p}(\Omega;\mathbb{S}^1)$, whereas, by Corollary 8.2, the map $u \mapsto \Sigma(u)$ is continuous on $W^{1/p,p}(\Omega;\mathbb{S}^1)$. The class $\mathscr{R}$ defined in (7.11) being dense in $W^{1/p,p}(\Omega;\mathbb{S}^1)$ (Theorem 10.3), it thus suffices to prove (8.30) when $u \in \mathscr{R}$. For such u, Proposition 15.2 in Appendix 15.4 yields a sequence $(v_n) \subset W^{1/p,p}(\Omega;\mathbb{S}^1)$ such that $Jv_n = Ju$ and $v_n \to 1$ a.e. and $|v_n|^p_{W^{1/p,p}} \lesssim \Sigma(u)$. Set $u_n := u\overline{v_n}$, so that $Ju_n = 0$ (by (8.10)).

Combining the identity

$$|u_n(x) - u_n(y)| = |u(x) - u(y) + u(y)\overline{v_n}(y)\,[v_n(y) - v_n(x)]|$$

with the fact that $v_n \to 1$ a.e. and the Brezis–Lieb lemma [82], we find that

$$\begin{aligned} |u_n|^p_{W^{1/p,p}} &= |u|^p_{W^{1/p,p}} + \int_\Omega\int_\Omega \frac{|u(y)\overline{v_n}(y)\,[v_n(y) - v_n(x)]|^p}{|x-y|^{N+1}}\,dxdy + o(1) \\ &= |u|^p_{W^{1/p,p}} + |v_n|^p_{W^{1/p,p}} + o(1) \text{ as } n \to \infty. \end{aligned} \tag{8.31}$$

Therefore, for large n we have $|u_n|^p_{W^{1/p,p}} \le |u|^p_{W^{1/p,p}} + C_2\Sigma(u)$ provided $\Sigma(u) \ne 0$. If $\Sigma(u) = 0$, the same holds if we take $v_n = 1$.

Step 2. Proof of

$$|u|^p_{W^{1/p,p}} + C_1\Sigma(u) \le R_{1/p,p}(u). \tag{8.32}$$

Let $(u_n) \subset C^\infty(\overline{\Omega};\mathbb{S}^1)$ be such that $u_n \to u \in W^{1/p,p}(\Omega;\mathbb{S}^1)$ a.e. Then, $J(u_n\overline{u}) = -Ju$ (by (8.10)) and thus (as above, with $v_n := u_n\,\overline{u}$)

$$|u_n|^p_{W^{1/p,p}} = |u|^p_{W^{1/p,p}} + |u_n\overline{u}|^p_{W^{1/p,p}} + o(1) \ge |u|^p_{W^{1/p,p}} + C_1\Sigma(u) + o(1),$$

the latter inequality following from the first inequality in (8.28). □

Let us note the following unusual consequence of Theorem 8.11.

Corollary 8.3. *Let* $N \ge 2$ *and* $1 \le p < \infty$. *Then,* $R_{1/p,p}$ *is continuous on* $W^{1/p,p}(\Omega;\mathbb{S}^1)$.

Note that, in general, a relaxed energy is only lower semicontinuous. Theorem 8.11 shows that $R_{s,p}$ is not continuous when $1 < sp < 2$.

Proof. The case $p = 1$ follows from Theorem 1.5 and Corollary 1.7.

Assume that $1 < p < \infty$. The map $R_{1/p,p}$ being lower semicontinuous, it suffices to prove that it is upper semicontinuous. We start by noting that

$$|v|_{W^{1/p,p}} \leq |u|_{W^{1/p,p}} + |v\,\overline{u}|_{W^{1/p,p}}, \ \forall\, u, v \in W^{1/p,p}(\Omega; \mathbb{S}^1),$$

and thus

$$[R_{1/p,p}(v)]^{1/p} \leq [R_{1/p,p}(u)]^{1/p} + [R_{1/p,p}(v\,\overline{u})]^{1/p}, \ \forall\, u, v \in W^{1/p,p}(\Omega; \mathbb{S}^1). \quad (8.33)$$

Let us next note that (8.30) and the first inequality in (8.28) imply that

$$R_{1/p,p}(v\,\overline{u}) \lesssim |v\,\overline{u}|^p_{W^{1/p,p}}, \ \forall\, u,\ v \in W^{1/p,p}(\Omega; \mathbb{S}^1). \quad (8.34)$$

Upper semicontinuity is a consequence of (8.33), (8.34), and of the fact that $v\,\overline{u} \to 1$ if $v \to u$ in $W^{1/p,p}(\Omega; \mathbb{S}^1)$. □

Proof of Theorem 8.11. As in the proof of Proposition 8.1, if $(u_n) \subset W^{s,p}(\Omega; \mathbb{S}^1)$ is such that $u_n \to u$ a.e. and $|u_n|_{W^{s,p}} \leq C$, then $Ju_n \to Ju$. In particular, if $R_{s,p}(u) < \infty$, then we may choose such u_n in $C^\infty(\overline{\Omega}; \mathbb{S}^1)$, and thus $Ju_n = 0$, so that $Ju = 0$.

Conversely, assume that $Ju = 0$. Then, $u \in \overline{C^\infty(\overline{\Omega}; \mathbb{S}^1)}^{W^{s,p}}$ (Theorem 10.4) and then clearly $R_{s,p}(u) = |u|^p_{W^{s,p}}$. □

8.5 Square root

We extend here Corollary 1.3 to the $W^{s,p}$ setting.

Theorem 8.12. *Let $s > 0$ and $1 \leq p < \infty$.*

1. *When $N = 1$ or $sp \geq 2$, every map $u \in W^{s,p}(\Omega; \mathbb{S}^1)$ can be written as $u = v^2$ with $v \in W^{s,p}(\Omega; \mathbb{S}^1)$.*
2. *Assume next that $N \geq 2$ and $1 < sp < 2$. Let $u \in W^{s,p}(\Omega; \mathbb{S}^1)$. Then, the following conditions are equivalent:*

   a) u can be written as $u = v^2$ for some $v \in W^{s,p}(\Omega; \mathbb{S}^1)$
   b) There exists some $\widetilde{v} \in W^{s,p}(\Omega; \mathbb{S}^1)$ such that $Ju = 2J\widetilde{v}$
   c) There exists some $w \in W^{1,1}(\Omega; \mathbb{S}^1)$ such that $Ju = 2Jw$.

The above result does not cover the spaces $W^{1/p,p}(\Omega; \mathbb{S}^1)$ with $N \geq 2$ and $1 < p < \infty$. In that case, it follows from (8.10) that $a) \implies b)$. On the other hand, by Theorems 7.1 and 7.2, we have $b) \iff c)$. However, the implication $c) \implies a)$ is not true (see Proposition 8.2 below).

Proof. In case *1*, we write (using one of the Theorems 8.2, 8.3, 8.4, or 8.5) $u = e^{\imath\varphi}$ for some appropriate φ. Let $v := e^{\imath\varphi/2}$. Then, $v^2 = u$ and (by applying Remark 8.8 and the converse part in Theorem 8.5), $v \in W^{s,p}$.

In case 2, "*a* $\implies$ *b)* $\implies$ *c)*" is an obvious consequence of Theorem 7.1, (8.2), and (8.10).

Assume that *c)* holds for some $u \in W^{s,p}(\Omega; \mathbb{S}^1)$ with $1 < sp < 2$. We recall the results in Section 3.8.3, and also refer to Section 8.7.1 below. More specifically, Theorem 8.14 clearly implies the following. Let $s > 0$ and $1 \le p < \infty$ be such that $1 < sp < 2$. Set $q := (sp)/(sp - 1) > 2$. If $w \in W^{1,1}(\Omega; \mathbb{S}^1)$ is such that $Jw \in [W_0^{1,q}(\Omega)]^*$, then there exists some $y \in W^{s,p}(\Omega; \mathbb{S}^1)$ such that $Jy = Jw$. Theorem 8.14 also implies that, if $u \in W^{s,p}(\Omega; \mathbb{S}^1)$ (with $1 < sp < 2$), then $Ju \in [W_0^{1,q}(\Omega)]^*$. Combining these two facts, we find that there exists some $y \in W^{s,p}(\Omega; \mathbb{S}^1)$ such that $Ju = 2Jy$. As in case $s = 1$ and $p = 1$, we find (using (8.10) and Theorems 8.8 and 15.8) that $u\overline{y}^2 = e^{\imath\varphi}$ for some φ in an appropriate space, and then $v := e^{\imath\varphi/2}y$ satisfies $v \in W^{s,p}(\Omega; \mathbb{S}^1)$ and $u = v^2$. □

Proposition 8.2. *Let $N \ge 2$ and $1 < p < \infty$. Then there exists some $u \in W^{1/p,p}(\Omega; \mathbb{S}^1)$ such that:*

a) $u \in C^\infty(\Omega; \mathbb{S}^1)$ *(and thus $Ju = 0$)*
b) *u cannot be written as $u = v^2$ for some $v \in W^{1/p,p}(\Omega; \mathbb{S}^1)$.*

For a proof of Proposition 8.2, we refer the reader to Mironescu and Van Schaftingen [269, proof of Theorem 4].

8.6 Minimizing the BV part of the phase

We present here another typical example of how we can transpose to $W^{1/p,p}$ results that were originally established for $W^{1,1}$.

Theorem 8.13. *Assume that $1 < p < \infty$. Let $u \in W^{1/p,p}(\Omega; \mathbb{S}^1)$. Then we have*

$$\inf\{\|D\psi\|_{\mathscr{M}};\ u = e^{\imath(\varphi+\psi)} \text{ with } \varphi \in W^{1/p,p} \text{ and } \psi \in BV\} = 2\Sigma(u). \tag{8.35}$$

This is analogous to Theorem 1.8. Before proceeding to the proof of the theorem, we note that it is always possible to write u as above (see Theorem 8.7).

Proof. The key ingredient is the following result.

Lemma 8.1. *Let $f \in W^{1/p,p}(\Omega; \mathbb{R})$ and $g \in BV(\Omega; \mathbb{R})$ be such that $g(x) - f(x) \in 2\pi\mathbb{Z}$ a.e. Then,*

1. *$f \in BV$ (and thus $g - f \in BV$).*
2. $D^j f = 0$.
3. *$D(g - f) = D^j(g - f)$.*

Granted Lemma 8.1, whose proof is presented at the end of Appendix 15.7, we continue as follows.

Step 1. If $u = e^{\imath(\varphi+\psi)}$, with $\varphi \in W^{1/p,p}(\Omega;\mathbb{R})$ and $\psi \in BV(\Omega;\mathbb{R})$, then curl $(-D^j\psi) = 2Ju$

Indeed, write (see Theorems 7.1 and 1.4) $u = e^{\imath\varphi_1}v = e^{\imath(\varphi_1+\psi_1)}$, with $\varphi_1 \in W^{1/p,p}(\Omega;\mathbb{R})$, $v \in W^{1,1}(\Omega;\mathbb{S}^1)$, and $\psi_1 \in BV(\Omega;\mathbb{R})$.

Let $f := \varphi_1 - \varphi$ and $g := \psi - \psi_1$. By Lemma 8.1 items *2* and *3*, we have

$$D^j\psi - D^j\psi_1 = D^j g = D^j(g-f) = D(g-f),$$

so that (by (1.108) applied to $v := e^{\imath\psi_1}$)

$$\operatorname{curl}(-D^j\psi) = \operatorname{curl}(-D^j\psi_1 - D(g-f)) = \operatorname{curl}(-D^j\psi_1) = 2Jv = 2Ju.$$

Step 2. Proof of "$\geq$" in (8.35)

Write, as above, $u = e^{\imath(\varphi+\psi)}$. By Step *1* and (1.85), we have $\|D^j\psi\|_{\mathcal{M}} \geq \Sigma(u)$. On the other hand, we have

$$\operatorname{curl}(D\psi - D^j\psi) = \operatorname{curl}(-D^j\psi) = 2Ju,$$

and thus (again by (1.85)), $\|D\psi - D^j\psi\|_{\mathcal{M}} \geq \Sigma(u)$. We obtain

$$\|D\psi\|_{\mathcal{M}} = \|D\psi - D^j\psi\|_{\mathcal{M}} + \|D^j\psi\|_{\mathcal{M}} \geq 2\Sigma(u).$$

Step 3. Proof "$\leq$" in (8.35)

We write, as in Step *1*, $u = e^{\imath\varphi_1}v$. By the definition (1.80), for each $\varepsilon > 0$ there exists some $\eta = \eta_\varepsilon \in C^\infty(\overline{\Omega};\mathbb{R})$ such that

$$\int_\Omega |v \wedge \nabla v - \nabla\eta| < \Sigma(v) + \varepsilon = \Sigma(u) + \varepsilon. \tag{8.36}$$

Set $w := e^{-\imath\eta}v \in W^{1,1}(\Omega;\mathbb{S}^1)$ and $\varphi := \varphi_1 + \eta \in W^{1/p,p}(\Omega;\mathbb{R})$. Then, $u = e^{\imath\varphi}w$, so that $\Sigma(w) = \Sigma(v) = \Sigma(u)$.

By (8.36), we have

$$\int_\Omega |\nabla w| = \int_\Omega |w \wedge \nabla w| = \int_\Omega |v \wedge \nabla v - \nabla\eta| < \Sigma(u) + \varepsilon. \tag{8.37}$$

Let ψ be an optimal BV lifting of w. Then, $u = e^{\imath(\varphi+\psi)}$ and, by Theorem 1.5 and (8.37),

$$|\psi|_{BV} = E(w) = \int_\Omega |\nabla w| + \Sigma(w) < 2\Sigma(u) + \varepsilon. \qquad \square$$

8.7 Complements and open problems

8.7.1 Describing $J(W^{s,p}(\Omega;\mathbb{S}^1))$

Recall what we already know about the image of the map $u \mapsto Ju$, when $u \in W^{1,p}$ and $N \geq 2$. As in Section 3.8.3, define

$$\mathscr{A}_p := \left\{\frac{1}{\pi} * Ju;\ u \in W^{1,p}(\Omega;\mathbb{S}^1)\right\},\quad 1 \leq p < \infty. \tag{8.38}$$

1. We have (see Section 2.9.4 when $N = 2$, Theorem 3.8 when $N = 3$, (3.104) when $N \geq 4$)

$$\mathscr{A}_1 = \Big\{\sum \Gamma_j;\ \Gamma_j = \partial M_j \cap \Omega,\ \text{with } M_j \subset \mathbb{R}^N \text{ closed}$$
$$\text{compact oriented hypersurface and } \sum |M_j| < \infty\Big\}. \tag{8.39}$$

2. When $1 < p < 2$, we have, with $q := p/(p-1)$ (Theorem 3.18), $\mathscr{A}_p = \mathscr{F} \cap [W_0^{1,q}(\Omega)]^*$.
3. When $p \geq 2$, we have (see (1.61)) $\mathscr{A}_p = \{0\}$.

The next result is a straightforward consequence of the factorization theory and of the definition of Ju.

Theorem 8.14. *Let $s > 0$ and $1 \leq p < \infty$ be such that $sp \geq 1$. Then,*

$$\left\{\frac{1}{\pi} * Ju;\ u \in W^{s,p}(\Omega;\mathbb{S}^1)\right\} = \mathscr{A}_{sp}.$$

Proof. By Theorem 7.1 and the Gagliardo–Nirenberg embeddings, whenever $u \in W^{s,p}(\Omega;\mathbb{S}^1)$, with s, p as above, we have $u = e^{\imath\varphi} v$ (and thus $Ju = Jv$) for some $v \in W^{1,sp}$ and some $\varphi \in W^{s,p}$. Conversely, if $v \in W^{1,sp}(\Omega;\mathbb{S}^1)$, then there exists some $u \in W^{s,p}(\Omega;\mathbb{S}^1)$ such that $Ju = Jv$. Indeed, when $sp > 1$, this follows again from Theorem 7.1 and the Gagliardo–Nirenberg embeddings. In the more delicate case where $sp = 1$ and $s < 1$, we rely on Theorem 7.2. □

Using Theorem 8.14, we are in a position to justify Remark 7.2. This is the content of the next result.

Proposition 8.3. *Let $N \geq 2$, $s > 0$, and $1 \leq p < \infty$ be such that $1 \leq sp < 2$. Let $\varepsilon > 0$. Then there exists some $u \in W^{s,p}(\Omega;\mathbb{S}^1)$ that cannot be written as $u = e^{\imath\varphi} v$ with $\varphi \in W^{s,p}$ and $v \in W^{1,sp+\varepsilon}$.*

Proof. By Theorem 8.14 and the definition (8.2) of Ju, it suffices to prove the following: if $1 \leq q < r < 2$, then $\mathscr{A}_r$ is strictly included in $\mathscr{A}_q$. By the monotonicity of

$\mathscr{A}_q$ in q, we may further assume that $1 < q < r < 2$. When $N = 2$, the proof of this fact is presented in Appendix 15.10 (see Lemma 15.38), but the argument applies to any N (see Remark 15.18). □

We next give an informal description of the "Weierstrass factorization type" result describing $W^{s,p}(\Omega; \mathbb{S}^1)$ when $1 \le sp < 2$ (alluded to in Remark 7.1). Let us recall that :

1. Given an entire holomorphic function $f : \mathbb{C} \to \mathbb{C}, f \not\equiv 0$, its zero set $\mathscr{Z}(f) := \{z \in \mathbb{C}; f(z) = 0\}$ can be written (counted with multiplicities) as $\mathscr{Z}(f) = \{z_j\}$, where either the sequence (z_j) is finite, or $|z_j| \to \infty$.
2. Given a sequence (z_j) as above, there always exists an entire holomorphic function $g(z) =$"$\prod_j (z - z_j)$" such that its zero set is $\mathscr{Z}(g) = \{z_j\}$. Here, "" is a warning that, if the sequence is infinite, then the product need not converge, and it has to be corrected with appropriate Weierstrass factors in order to converge.
3. Finally, if f is as in item 1 and g is as in item 2, then we may write $f = e^h g$ for some entire h, and thus

$$f = e^h \text{``}\prod_j (z - z_j).\text{''} \tag{8.40}$$

Equality (8.40) is the Weierstrass factorization theorem for entire functions. Moreover, this generalizes to open subsets $\Omega \subset \mathbb{C}$, and often the construction of factors is canonical, leading to the Hadamard factorization theorem (see Rudin [311, Chapter 15] for a detailed discussion).

Factorization combined with Theorem 8.14 and with the proofs of Theorems 2.20 and 3.18 lead to a similar description of maps in $W^{s,p}(\Omega; \mathbb{S}^1)$ with $1 \le sp < 2$. We explain this when $N = 2$, but a similar description holds for any $N \ge 2$ (one has to replace points by $(N-2)$-manifolds). The analogue of zeros of f are the topological singularities of u, and the analogue of $\mathscr{Z}(f)$ is $(1/\pi)\, Ju$. More specifically, let $r := sp \in [1, 2)$ and $u \in W^{s,p}(\Omega; \mathbb{S}^1)$. Then:

1′. By Theorems 2.10 and 7.1 and the definition (8.2) of Ju, we have $(1/\pi)\, Ju = \sum_j (\delta_{P_j} - \delta_{N_j}) \in \mathscr{A}_r$, with $\mathscr{A}_r$ as in (8.38).

2′. Conversely, consider points P_j, N_j as in item 1′.
When $1 < r < 2$, there exists a map $W^{r,1}(\Omega; \mathbb{S}^1)$, of the form

$$w(z) = \text{``}\prod_j \left(\frac{z - P_j}{|z - P_j|}\right) \left(\frac{\overline{z - N_j}}{|z - N_j|}\right)\text{''} \tag{8.41}$$

satisfying $Jw = Ju$.
Moreover, we may perform a "canonical" construction of w (leading to a Hadamard-type factorization), in the spirit of the proof of Theorem 3.3. See Bousquet [61], where it is proved that $w \in W^{1,r}(\Omega; \mathbb{S}^1)$, and also Mironescu [256], where it is established that the same construction leads to the stronger regularity $w \in W^{r,1}(\Omega; \mathbb{S}^1)$.

When $r = 1$ (and thus $s = 1/p$), a map w of the form (8.41) can be constructed as an infinite product of dipoles (adapted to the sequences (P_j), (N_j)), as in the inclusion "$\supset$" in (8.39) (see the proof of Theorem 2.11). Moreover, we may construct w such that, in addition, $w \in W^{1/p,p}(\Omega;\mathbb{S}^1)$ (see the proof of Theorem 7.2). This is the analogue of the basic Weierstrass factorization.

3′. Finally, if u is as in item 1′ and w as in item 2′, then $u\overline{w} \in W^{s,p}(\Omega;\mathbb{S}^1)$ and $J(u\overline{w}) = 0$ and therefore we may write $u = e^{\imath\varphi} w$, with φ as in Theorem 8.8. Thus, f admits the "Weierstrass-type factorization"

$$f = e^{\imath\varphi} \text{“}\prod_j \left(\frac{z-P_j}{|z-P_j|}\right)\left(\frac{\overline{z-N_j}}{|z-N_j|}\right).\text{”} \tag{8.42}$$

8.7.2 *Sphere-valued maps*

This is a micro introduction to the world of fractional Sobolev $\mathbb{S}^k$-valued maps. It is a natural continuation of both Chapter 4 (dealing with $s = 1$ and $k \ge 2$) and Section 8.1 (dealing with $0 < s < 1$ and $k = 1$). We explain briefly how the Jacobian Ju can be defined for maps $u \in W^{s,p}(\Omega;\mathbb{S}^k)$, $\Omega \subset \mathbb{R}^N$, within a range of parameters N, k, s, and p to be determined. As typical examples of applications, we present two results. The first one concerns the range of J. The second one concerns the least energy with prescribed Jacobian. We have inserted this section here, even though it is not related to factorization, simply because the main topic we discuss (i.e., giving a meaning to Ju when $u \in W^{s,p}$ with $0 < s < 1$) is a natural continuation of Section 8.1.

By analogy with the framework of Chapter 4, we assume here that

$$N \ge k+1,\ 0 < s < \infty,\ 1 \le p < \infty, \text{ and } sp \ge k. \tag{8.43}$$

Within this range, it is possible to give a robust definition of the distributional Jacobian Ju. This is clear when $s \ge 1$, since we then have $W^{s,p}(\Omega;\mathbb{S}^k) \hookrightarrow W^{1,k}(\Omega;\mathbb{S}^k)$ (see Theorem 15.5 in Appendix 15.2), and Ju can be defined as in (4.6) and (4.7).

The main result concerning the case $0 < s < 1$ is the following.

Theorem 8.15. *Assume that* $k \ge 2$, $N \ge k+1$, $0 < s < 1$, $1 \le p < \infty$, *and* $sp = k$. *There exists a unique continuous map* $J : W^{s,p}(\Omega;\mathbb{S}^k) \to \mathscr{D}'(\Omega;\Lambda^{k+1})$ *such that* Ju *is given by* (4.6) *and* (4.7) *when* $u \in W^{1,k}(\Omega;\mathbb{S}^k) \subset W^{s,p}(\Omega;\mathbb{S}^k)$.

The uniqueness is a consequence of the fact that, when s, p, and k are as above, $W^{1,k}(\Omega;\mathbb{S}^k)$ is a dense subset of $W^{s,p}(\Omega;\mathbb{S}^k)$. More specifically, the embedding $W^{1,k}(\Omega;\mathbb{S}^k) \hookrightarrow W^{s,p}(\Omega;\mathbb{S}^k)$ is continuous, by Remark 15.5. On the other hand, Theorem 6 in Brezis and Mironescu [85] implies that each map in $W^{s,p}(\Omega;\mathbb{S}^k)$ can be approximated (strongly in $W^{s,p}$) by maps in $W^{1,k}(\Omega;\mathbb{S}^k)$ (and even in $W^{1,q}(\Omega;\mathbb{S}^k)$, $\forall\, q < k+1$).

In the spirit of Remark 8.4, it is still possible to define the Jacobian when $s \ge 1 - 1/(k+1)$ (but not below $s = 1 - 1/(k+1)$), via trace theory and a definition similar to (8.20) (for such s, see also Hang and Lin [176] for an approach based on the coarea formula). For the existence of J within the full range $0 < s < 1$, we refer the reader to Bourgain, Brezis, and Mironescu [56, Theorem 0.8] when $N = k + 1$, and Bousquet and Mironescu [63, Theorem 1.1] when $N > k + 1$.

In this setting, we have the following analogue of Theorem 4.10 (see [56, Theorem 0.8], [63, Theorem 1.2, Theorem 1.3]).

Theorem 8.16. *Under the same assumptions as in Theorem 8.15, for any $u \in W^{s,p}(\Omega; \mathbb{S}^k)$, we have*

$$\frac{1}{|\mathbb{B}^{k+1}|} * Ju \in \mathscr{F}^{N-k-1}. \tag{8.44}$$

Conversely, given any $T \in \mathscr{F}^{N-k-1}$, there exists some $u \in W^{s,p}(\Omega; \mathbb{S}^k)$ such that

$$*Ju = |\mathbb{B}^{k+1}|\, T. \tag{8.45}$$

In the above, $\mathscr{F}^{N-k-1}$ is defined as in (4.62).

Next we introduce (under the same assumptions as in Theorem 8.15) the least energy with prescribed Jacobian,

$$S_{s,p}(u) := \inf\{|v|^p_{W^{s,p}};\ Jv = Ju \text{ in } \mathscr{D}'\},\ \forall\, u \in W^{s,p}(\Omega; \mathbb{S}^k). \tag{8.46}$$

and we consider (with A^{N-k-1} as in (4.63)) the quantity

$$\Sigma^*(u) := |\mathbb{S}^k|\, A^{N-k-1}\left(\frac{1}{|\mathbb{B}^{k+1}|} * Ju\right). \tag{8.47}$$

The following result is inspired by Theorems 4.11 and 8.9.

Theorem 8.17. *([63]) Under the same assumptions as in Theorem 8.15, for any $u \in W^{s,p}(\Omega; \mathbb{S}^k)$, we have*

$$S_{s,p}(u) \sim \Sigma^*(u). \tag{8.48}$$

8.8 Comments

1. The possibility of defining a distributional Jacobian for maps in $W^{1-1/(k+1),k+1}(\Omega; \mathbb{S}^k)$, with $k \geq 1$ and $N \geq k + 1$, was originally pointed out by Hang and Lin [176] (who were motivated by an earlier work by Jerrard and Soner [205]). They also proved that if this Jacobian has finite total mass, then it is an $(N - k - 1)$-rectifiable current. The special case $N = 2$, $k = 1$ was subsequently investigated in great detail in Bourgain, Brezis, and Mironescu [51, 54]. In full generality, the existence of the Jacobian for a map $u \in W^{k/p,p}(\Omega; \mathbb{S}^k)$, with $1 \leq k < p < \infty$ and

$N = k + 1$ was proved by Bourgain, Brezis, and Mironescu [56]. An extension of this result to arbitrary $k \geq 1$, arbitrary $N \geq k + 1$, and arbitrary $p \geq 1$ is due to Bousquet and Mironescu [63]. This is the content of Theorem 8.15.
In a *totally different* direction, distributional Jacobians have been studied for maps $u \in W^{s,p}(\Omega; \mathbb{R}^N)$ (resp. $u \in W^{s,p}(\Omega; \mathbb{S}^N)$), under suitable assumptions on s and p (see Brezis and Nguyen [96, 98] and the references therein). In this case the Jacobian does not have a special geometric structure as above.

2. The definition of Ju and $u \wedge \nabla u$ via factorization, as well as the proofs of the main properties of these objects, presented in Section 8.1, are new.
3. Lifting in $W^{s,p}(\Omega; \mathbb{S}^1)$ with $sp < 1$ via the factorization method (Section 8.2.2) is from Mironescu and Molnar [263].
4. Theorems 8.5, 8.7, and 8.8 are new.
5. Theorem 8.9 was obtained by Bourgain, Brezis, and Mironescu [56, Theorem 2.4] when $N = k + 1$ and $p > k$. It was extended to $W^{k/p,p}(\Omega; \mathbb{S}^k)$ with arbitrary $k \geq 1$, arbitrary $N \geq k + 1$, and arbitrary $p \geq 1$ in Bousquet and Mironescu [63].
6. Theorems 8.12 and 8.14 were announced in Mironescu [256] with a sketch of proof.
7. Proposition 8.2 is from Mironescu and Van Schaftingen [269].
8. Theorem 8.13 with $N = 2$ and $p = 2$ is from Bourgain, Brezis, and Mironescu [54].
9. Proposition 8.3 is new.

Chapter 9
Estimates of phases: positive and negative results

9.0 Introduction

In the previous chapters, we addressed the question whether, given $u \in X(\Omega; \mathbb{S}^1)$ (with $X = W^{1,p}, W^{s,p}, VMO$, etc.), there *exists* some $\varphi \in X(\Omega; \mathbb{R})$ such that $u = e^{\imath\varphi}$. In this chapter we discuss the question of *estimates*. Very often, in PDEs the two topics are closely related:

- Existence comes with estimates
- Formal a priori estimates imply existence.

As we are going to see, this philosophy is shattered in our setting. Here are two simple examples.

Existence without estimates. We know that every $u \in H^{1/2}((0,1); \mathbb{S}^1)$ admits a unique lifting $\varphi \in H^{1/2}((0,1); \mathbb{R})$ (Theorem 5.1). However, it is impossible to bound $|\varphi|_{H^{1/2}}$ in terms of $|e^{\imath\varphi}|_{H^{1/2}}$ (Proposition 9.2). Similarly in $W^{1/p,p}((0,1); \mathbb{S}^1)$, $1 < p < \infty$.

Estimates without existence. For any $1 \le p < \infty$ and $\varphi \in W^{1,p}$, we have $|\varphi|_{W^{1,p}} = |e^{\imath\varphi}|_{W^{1,p}}$. However, if $N \ge 2$ and $1 \le p < 2$, there exists some $u \in W^{1,p}(\Omega; \mathbb{S}^1)$ that does not admit a lifting φ within $W^{1,p}$ (Remark 1.6).

If we want to dissociate matters of existence and estimates, then the question we address is: given $\varphi \in X(\Omega; \mathbb{R})$, is it true that

$$\inf\left\{\left\|\psi - \fint \psi\right\|_X ;\ \psi \in X \text{ and } e^{\imath\psi} = e^{\imath\varphi}\right\} \le F\left(\|e^{\imath\varphi}\|_X\right)$$

for some function F? We formulate the question in this form because non-uniqueness may happen.

We may ask similar questions when $\left\|\psi - \fint \psi\right\|_X$ is replaced by $\left\|\psi - \fint \psi\right\|_Y$, where $\|\ \|_Y$ is weaker than $\|\ \|_X$.

Here are three striking results established in this chapter.

H. Brezis and P. Mironescu, *Sobolev Maps to the Circle*, Progress in Nonlinear Differential Equations and Their Applications 96,
https://doi.org/10.1007/978-1-0716-1512-6_9

1. When $N = 1$, $s = 1/p$, and $1 < p < \infty$, which is the "existence without estimates" situation mentioned above, we *do have the weaker control* $|\varphi|_{BMO} \le F(|u|_{W^{1/p,p}})$ (Theorem 9.3).
2. When $N \ge 2$, $0 < s < 1$, and $sp = N$, which is formally the higher dimensional analogue of the above case, we *do have the natural control* $|\varphi|_{W^{s,p}} \le F(|u|_{W^{s,p}})$ (Theorem 9.6).
3. When $N \ge 2$, $0 < s < 1$, and $1 \le sp < N$ (the "nasty" range where lifting in the natural class fails – see Theorem 5.2 *b)*), we do not have the natural control $|\varphi|_{W^{s,p}} \le F(|e^{\imath\varphi}|_{W^{s,p}})$, but we *do have a weaker control* $|\varphi|_{W^{s,p}+W^{1,sp}} \le F(|e^{\imath\varphi}|_{W^{s,p}})$ (Proposition 9.3 and Theorem 9.7).

In what follows, C_p denotes a constant depending on p, and possibly on N, but not on s or u, and $C_{s,p}$ denotes a constant depending on s, p, and possibly on N, but not on u.

9.1 $N \ge 1$ and $s \ge 1$

This case is elementary. We know (see Theorems 5.1, 5.5, and 5.6) that $u \in W^{s,p}(\Omega; \mathbb{S}^1)$ admits a lifting $\varphi \in W^{s,p}(\Omega; \mathbb{R})$ if and only if

1. Either $N = 1$
2. Or $N \ge 2$ and $sp \ge 2$
3. Or $N \ge 2$, $1 \le sp < 2$, and $Ju = 0$.

Theorem 9.1. *With N, s, p, and u as above, the phase $\varphi \in W^{s,p}$ of u satisfies $|\varphi|_{W^{s,p}} \le C_{s,p}|u|_{W^{s,p}}$.*

Moreover, we have $\varphi \in W^{1,sp}$ and $|\varphi|_{W^{1,sp}} \le C_{s,p}|u|_{W^{s,p}}^{1/s}$.

Proof. Examine the proof of Theorem 5.5. There, $N \ge 2$ and $sp \ge 2$, but the same arguments and estimates apply in all three of the above situations. □

Remark 9.1. Using explicit examples of φ and u, one may prove that the above estimates are optimal as $|u|_{W^{s,p}} \to 0$ or $|u|_{W^{s,p}} \to \infty$ (see Mironescu and Molnar [263, Section 5]).

9.2 $N \ge 1$ and $sp < 1$

As explained in the Introduction, and as Proposition 9.1 below shows, we can at best hope for an estimate for *some* phase φ. In this case, we have the following

Theorem 9.2. *Let $N \ge 1$, $s > 0$, and $1 \le p < \infty$ be such that $sp < 1$. Given any $u \in W^{s,p}(\Omega; \mathbb{S}^1)$, there exists some lifting $\varphi \in W^{s,p}(\Omega; \mathbb{R})$ satisfying*

$$|\varphi|_{W^{s,p}} \le \frac{C_p}{s(1-sp)}|u|_{W^{s,p}}. \tag{9.1}$$

Proof. "Monitor" the constants in the proof of Theorem 5.1. There, $N = 1$, but the same arguments and estimates are valid for any N. □

Remark 9.2. The constant in front of $|u|_{W^{s,p}}$ in (9.1) is not optimal. In the case $p = 2$, a more sophisticated argument (see Bourgain, Brezis, and Mironescu) [50, Theorem 4]) provides a better bound

$$|\varphi|_{H^s} \leq \frac{C_2}{s(1-2s)^{1/2}}|u|_{H^s}, \ \forall\, 0 < s < 1/2, \tag{9.2}$$

which is optimal (see Bourgain, Brezis, and Mironescu [53, Theorem 2]).

More generally, let $1 < p < \infty$ and $0 < s < 1/p$. Then we have the improved bound

$$|\varphi|_{W^{s,p}} \leq \frac{C_p}{s(1-sp)^{1/p}}|u|_{W^{s,p}}, \tag{9.3}$$

which is optimal (see Mironescu and Molnar [263, Theorem 1.3]).

However, when $p = 1$ and $0 < s < 1$, the above estimate can be dramatically improved to the optimal estimate

$$|\varphi|_{W^{s,1}} \leq 2|u|_{W^{s,1}}$$

(see Lemma 5.10).

It is also possible to recover (9.3) via the factorization technology. Here is a sketch of proof of this fact when s is close to $1/p$. We use the notation in the proof of Theorem 7.1. First, it is straighforward to see that, as $sp \nearrow 1$, the seminorm equivalence

$$K_1 \int_{\mathbb{R}^N}\int_0^\infty \varepsilon^{p(1-s)-1}|\nabla w(x,\varepsilon)|^p\, d\varepsilon dx \leq |u|^p_{W^{s,p}}$$
$$\leq K_2 \int_{\mathbb{R}^N}\int_0^\infty \varepsilon^{p(1-s)-1}|\nabla V(x,\varepsilon)|^p\, d\varepsilon dx$$

holds with K_1 and K_2 independent of s. [Here, $w(x,\varepsilon) := u * \rho_\varepsilon(x)$, and V could be any extension of u.] Combining this with Steps 3–6 in the proof of Theorem 7.1, we find that $|\varphi|_{W^{s,p}} \leq K_3|u|_{W^{s,p}}$. On the other hand, an inspection of Step 8 in the proof of Theorem 7.1, and more specifically, of the calculation leading to (7.80), shows that

$$\int_\omega\int_0^\infty \varepsilon^{-sp}|\nabla U(x,\varepsilon)|\, d\varepsilon dx \leq \frac{K_4}{1-sp}|u|^p_{W^{s,p}}.$$

In view of the above, we have $|v|_{W^{sp,1}} \leq \dfrac{K_5}{1-sp}|u|^p_{W^{s,p}}$. By Lemma 5.10,

$$v \text{ has a } W^{sp,1} \text{ phase } \psi \text{ with } |\psi|_{W^{sp,1}} \leq \frac{K_6}{1-sp}|u|^p_{W^{s,p}} \text{ and } |\psi| \leq 2\pi. \tag{9.4}$$

We conclude using (9.4) and the Gagliardo–Nirenberg inequalities.

Always with $sp < 1$, we now address a related question. Suppose that $u \in C^\infty(\overline{\Omega}; \mathbb{S}^1)$. Then, u admits a unique (mod 2π) lifting $\varphi = \varphi_C \in C^\infty(\overline{\Omega}; \mathbb{R})$ and we may ask whether one can estimate $|\varphi_C|_{W^{s,p}}$ in terms of $|u|_{W^{s,p}}$. The answer is "very" negative.

Proposition 9.1. *Let $s > 0$ and $1 \le p < \infty$ be such that $sp < 1$. Then there is no estimate of the form*

$$|\varphi_C|_{W^{s,p}} \le F(|u|_{W^{s,p}}),\ \forall\, u \in C^\infty(\overline{\Omega}; \mathbb{S}^1). \tag{9.5}$$

In fact, there is even no estimate of the form

$$\left\| \varphi_C - \fint_\Omega \varphi_C \right\|_{L^1} \le F(|u|_{W^{s,p}}),\ \forall\, u \in C^\infty(\overline{\Omega}; \mathbb{S}^1).$$

Proof. We let $N = 1$ and $\Omega = (0, 1)$, but the argument can be adapted to arbitrary Ω and N. Given numbers $M > 0$ and $\varepsilon > 0$, we claim that there exists some $\psi \in C_c^\infty(\Omega)$ such that $|\psi|_{W^{s,p}} > M$ and $|e^{\imath\psi}|_{W^{s,p}} < \varepsilon$. Indeed, let $I \subset \Omega$ be a compact interval and let $\zeta := 2\pi k \chi_I \in W^{s,p}$, with k an integer. Let (ρ_ε) be a family of standard mollifiers and set $\psi := \zeta * \rho_\varepsilon$. Then, as $\varepsilon \to 0$, $\psi \to \zeta$ in $W^{s,p}$, and $e^{\imath\psi} \to e^{\imath\zeta} = 1$ in $W^{s,p}$. We find that, for large k and small ε, ψ has the required properties. Similarly in L^1. □

9.3 $N = 1, s = 1/p$, and $p > 1$

In this case, we consider a map $u \in W^{1/p,p}(\Omega; \mathbb{S}^1)$ and let $\varphi \in W^{1/p,p}(\Omega; \mathbb{R})$ be its phase (see Theorem 5.1 and Corollary 6.2).

We start with a striking non-estimate result.

Proposition 9.2. *Let $N = 1$ and $1 < p < \infty$. Then there is no bound for $|\varphi|_{W^{1/p,p}}$ in terms of $|u|_{W^{1/p,p}}$. More precisely, there exists a sequence $(\varphi_n) \subset C^\infty(\overline{\Omega}; \mathbb{R})$ such that $|\varphi_n|_{W^{1/p,p}} \to \infty$ and $|e^{\imath\varphi_n}|_{W^{1/p,p}} \le C$.*

Proof. We construct such a sequence $(\varphi_n) \subset \mathrm{Lip}(\overline{\Omega})$. By smoothing, we can then produce a sequence in $C^\infty(\overline{\Omega})$ with the same properties.

Assume that $\Omega = (0, 1)$. We define φ_n as follows:

$$\varphi_n(x) := \begin{cases} 0, & \text{for } x \le 1/2 \\ 2\pi n(x - 1/2), & \text{for } 1/2 < x \le 1/2 + 1/n \\ 2\pi, & \text{for } x > 1/2 + 1/n \end{cases},\ \forall\, n \ge 2.$$

Clearly, $|\varphi_n|_{W^{1/p,p}} \to \infty$ (since $\varphi_n \to \varphi := 2\pi\, \chi_{(1/2,1)}$ in L^p and φ does not belong to $W^{1/p,p}$, by Corollary 6.2). On the other hand, if we extend $u_n := e^{\imath\varphi_n}$ by the value 1 outside Ω, then (by scaling)

$$|u_n|_{W^{1/p,p}((0,1))} \leq |u_n|_{W^{1/p,p}(\mathbb{R})} = |u_1|_{W^{1/p,p}(\mathbb{R})} < \infty. \qquad \square$$

Next, we present two results complementing Proposition 9.2 in positive directions.

Theorem 9.3. *Let $N = 1$ and $1 < p < \infty$. Then there exists a constant C_p such that*

$$|\varphi|_{BMO} \leq C_p(|u|_{W^{1/p,p}} + |u|^p_{W^{1/p,p}}), \ \forall u \in W^{1/p,p}(\Omega; \mathbb{S}^1). \tag{9.6}$$

Proof. Write, using factorization, lifting when $N = 1$ and uniqueness (Theorems 7.1 and 9.1 and Corollary 6.5), $\varphi = \psi + \zeta$, with $|\psi|_{W^{1/p,p}} \leq C|u|_{W^{1/p,p}}$ and $|\zeta|_{W^{1,1}} \leq C|u|^p_{W^{1/p,p}}$. Then use the embeddings $W^{1/p,p} \hookrightarrow VMO$ (Lemma 15.20) and $W^{1,1} \hookrightarrow VMO$. □

Remark 9.3. The reader might have been tempted to prove (9.6) using the fact that $W^{1/p,p} \hookrightarrow BMO$ and thus $|u|_{BMO} \leq C|u|_{W^{1/p,p}}$. Then try to bound $|\varphi|_{BMO}$ in terms of $|u|_{BMO}$. However, this is impossible. Indeed, let $\varphi \in C(\overline{\Omega})$ and set $u = e^{\imath\varphi}$. Then, φ is *the VMO* phase of u (Proposition 6.1). Note that $|u|_{BMO} \leq 2$, whereas $|\varphi|_{BMO}$ can be arbitrarily large.

Theorem 9.4. *Let $N = 1$ and $1 < p < \infty$. Then there are two constants γ_p and K_p such that, if $|u|_{W^{1/p,p}} \leq \gamma_p$, then $|\varphi|_{W^{1/p,p}} \leq K_p|u|_{W^{1/p,p}}$.*

Proof. It suffices to combine Remark 5.12 with the embedding $W^{1/p,p} \hookrightarrow VMO$ (see above). □

9.4 $N = 1, 0 < s < 1$, and $sp > 1$

Recall that every $u \in W^{s,p}(\Omega; \mathbb{S}^1)$ admits a unique (mod 2π) lifting $\varphi \in W^{s,p}(\Omega; \mathbb{R})$ (Theorem 5.1 and Corollary 6.2).

Theorem 9.5. *Let $N = 1$, $0 < s < 1$, and $1 < p < \infty$ be such that $sp > 1$. Then we have*

$$|\varphi|_{W^{s,p}} \leq C_{s,p}(|u|_{W^{s,p}} + |u|^{1/s}_{W^{s,p}}), \ \forall u \in W^{s,p}(\Omega; \mathbb{S}^1). \tag{9.7}$$

Proof. Write, using Theorem 7.1, $u = e^{\imath\psi} v$, with

$$|\psi|_{W^{s,p}} \leq C|u|_{W^{s,p}} \tag{9.8}$$

and

$$|v|_{W^{1,sp}} \leq C|u|^{1/s}_{W^{s,p}}.$$

Write, using Lemma 1.1, $v = e^{\imath\zeta}$, with

$$|\zeta|_{W^{1,sp}} = |v|_{W^{1,sp}} \le C|u|_{W^{s,p}}^{1/s}. \tag{9.9}$$

Then, $\varphi = \psi + \zeta \pmod{2\pi}$. By (9.8), (9.9), and the Sobolev embedding (Theorem 15.4) $W^{1,sp} \hookrightarrow W^{s,p}$ (which holds, since $sp > 1$), we find that φ satisfies (9.7). □

9.5 $N \ge 2$, $0 < s < 1$, and $sp \ge N$

We are again in an existence and uniqueness situation. Each $u \in W^{s,p}(\Omega; \mathbb{S}^1)$ admits a unique (mod 2π) lifting $\varphi \in W^{s,p}(\Omega; \mathbb{R})$ (Theorem 5.2 and Corollary 6.2).

Theorem 9.6. *Let $N \ge 2$, $0 < s < 1$, and $1 < p < \infty$ be such that $sp \ge N$. Then we have*

$$|\varphi|_{W^{s,p}} \le C_{s,p}(|u|_{W^{s,p}} + |u|_{W^{s,p}}^{1/s}). \tag{9.10}$$

Proof. Same proof as above, but use Theorem 1.1 instead of Lemma 1.1. Note that the Sobolev embedding $W^{1,sp} \hookrightarrow W^{s,p}$ holds, since $sp \ge N > 1$ (see Theorem 15.4). □

Remark 9.4. Estimate (9.10) when $N \ge 2$ and $sp = N$ is quite unexpected, and we do not know an elementary proof. There is a striking contrast with the case $N = 1$ and the non-estimate in Proposition 9.2. If we try to apply the above proof when $N = 1$, $s = 1/p$, and $1 < p < \infty$, we are stuck because of the non-embedding $W^{1,1} \not\hookrightarrow W^{1/p,p}$. When $N \ge 2$, the corresponding embedding $W^{1,N} \hookrightarrow W^{N/p,p}$, $N < p < \infty$, is true and allows us to derive Theorem 9.6.

9.6 $N \ge 2$, $0 < s < 1$, and $1 \le sp < N$

Recall that in this range a map $u \in W^{s,p}(\Omega; \mathbb{S}^1)$ need not have a lifting within $W^{s,p}$ (Theorem 5.2 *b*)).

We start with a non-estimate.

Proposition 9.3. *Let $N \ge 2$, $0 < s < 1$, and $1 < p < \infty$ be such that $1 \le sp < N$. Then there exists a sequence $(\varphi_n) \subset C^\infty(\overline{\Omega}; \mathbb{R})$ such that $|\varphi_n|_{W^{s,p}} \to \infty$ and $|e^{\imath\varphi_n}|_{W^{s,p}} \le C$.*

Proof. We argue as in the proof of Theorem 5.2 *b*). Assuming $0 \in \Omega$, let $\varphi_n(x) := \dfrac{1}{(|x|^2 + 1/n)^{\alpha/2}}$, with

$$\frac{N - sp}{p} \le \alpha < \frac{N - sp}{sp}. \tag{9.11}$$

Since $1/|x|^\alpha \notin W^{s,p}$ (Lemma 15.11), we have $|\varphi_n|_{W^{s,p}} \to \infty$. On the other hand, for sufficiently small $\varepsilon > 0$ we have (by (9.11)) $(\alpha + 1)(sp + \varepsilon) < N$. For any such

ε, we have $|e^{\imath\varphi_n}|_{W^{1,sp+\varepsilon}} = |\varphi_n|_{W^{1,sp+\varepsilon}} \leq C_\varepsilon < \infty$, and thus $|e^{\imath\varphi_n}|_{W^{s,p}} \leq C'_\varepsilon < \infty$, by Gagliardo–Nirenberg. □

In view of Proposition 9.3, the reader may wonder whether one can estimate $|\varphi|_Y$ in terms of $|e^{\imath\varphi}|_{W^{s,p}}$ for some seminorm $|\ |_Y$ weaker than $|\ |_{W^{s,p}}$. Here is a typical result.

Let $Y := W^{s,p} + W^{1,sp}$ equipped with its usual seminorm

$$|f|_Y := \inf\{|g|_{W^{s,p}} + |h|_{W^{1,sp}};\ f = g + h\}.$$

Theorem 9.7. *Let $N \geq 1$, $0 < s < 1$, and $1 < p < \infty$ be such that $sp \geq 1$. Let $\varphi \in W^{s,p}(\Omega; \mathbb{R})$ and set $u := e^{\imath\varphi}$. Then,*

$$|\varphi|_Y \leq C(|u|_{W^{s,p}} + |u|_{W^{s,p}}^{1/s}).$$

More precisely, we may find some $\psi \in W^{s,p}$ and some $\zeta \in W^{1,sp}$ such that $\varphi = \zeta + \psi$ and

$$|\psi|_{W^{s,p}} \leq C_{s,p}|u|_{W^{s,p}},\quad |\zeta|_{W^{1,sp}} \leq C_{s,p}|u|_{W^{s,p}}^{1/s}. \tag{9.12}$$

Note that the above result is in particular valid in the range $N \geq 2$, $0 < s < 1$, $1 \leq sp < N$ we discuss in this subsection.

Proof. Using factorization, we have $u = e^{\imath\psi}\, v$, with $\psi \in W^{s,p}$ and $v \in W^{1,sp}$. Since $Ju = 0$, we also have $Jv = 0$. Thus, we may write $v = e^{\imath\zeta}$, with $\zeta \in W^{1,sp}$. By uniqueness, $\varphi = \psi + \zeta \pmod{2\pi}$. Estimate (9.12) follows from Theorem 7.1. □

Corollary 9.1. *Let $N \geq 2$, $0 < s < 1$, and $1 < p < \infty$ be such that $1 \leq sp < N$. Let $\varphi \in W^{s,p}(\Omega; \mathbb{R})$ and set $u = e^{\imath\varphi}$. Then,*

$$\left\| \varphi - \fint \varphi \right\|_{L^{Nsp/(N-sp)}} \leq C(|u|_{W^{s,p}} + |u|_{W^{s,p}}^{1/s}). \tag{9.13}$$

Proof. Let $q := Nsp/(N - sp)$ and $q_1 := Np/(N - sp)$. We combine (9.12) with the Sobolev embeddings $W^{s,p} \hookrightarrow L^{q_1} \hookrightarrow L^q$ and $W^{1,sp} \hookrightarrow L^q$ (Theorem 15.4). □

Remark 9.5. The exponent $Nsp/(N - sp)$ is optimal. Indeed, let $r > Nsp/(N - sp)$. If we take, in the spirit of the proof of Proposition 9.3, $\varphi(x) := 1/|x|^\alpha$, where $N/r \leq \alpha < (N - sp)/(sp)$, then this φ satisfies $e^{\imath\varphi} \in W^{s,p}$ and $\varphi \notin L^r$.

Remark 9.6. In the special case $s = 1/2$, $p = 2$, estimate (9.13) becomes, for any $N \geq 2$,

$$\left\| \varphi - \fint \varphi \right\|_{L^{N/(N-1)}} \leq C(|u|_{H^{1/2}} + |u|_{H^{1/2}}^2). \tag{9.14}$$

We call the attention of the reader to the fact that, despite its simple appearance, estimate (9.14) has (so far) no elementary proof. We are aware of three distinct routes.

Derive it, as above, from Theorem 9.7, which can itself be established either via factorization as done here, or via paraproducts as in [48] (see also [54]). Alternatively, derive it from Theorem 7.4, as explained in Section 7.5.2. However, none of these routes is short.

9.7 Complements and open problems

9.7.1 Further *BMO*-type estimates for the phase

Recall that, for $N \geq 1$, $s = N/p$, and $p > N$, we have established that

$|\varphi|_{BMO} \leq C(|u|_{W^{s,p}} + |u|^{1/s}_{W^{s,p}})$ (Theorems 9.3 and 9.6).

By scaling, this is equivalent to

$$\iint_{Q\,Q} |\varphi(x) - \varphi(y)|\, dxdy \leq C\left(|u|_{W^{s,p}(Q)} + |u|^{1/s}_{W^{s,p}(Q)}\right), \tag{9.15}$$

with Q the unit cube. On the other hand, we have

$$|u|^p_{W^{s,p}} = \iint_{Q\,Q} \frac{|u(x) - u(y)|^p}{|x - y|^{2N}}\, dxdy = \iint_{|u(x)-u(y)|<\delta} \cdots + \iint_{|u(x)-u(y)|\geq\delta} \cdots$$
$$\geq \delta^p \iint_{\substack{Q\,Q \\ |u(x)-u(y)|\geq\delta}} \frac{1}{|x-y|^{2N}}\, dxdy := I.$$

Interestingly, a control on I alone yields a *BMO*-type estimate for φ. Here is a typical ingredient when $N = 1$.

Theorem 9.8. *Let $Q = (0, 1)$ and $0 < \delta < \sqrt{3}$. Let $\varphi \in VMO\,(Q; \mathbb{R})$ and set $u := e^{\imath\varphi}$. Then,*

$$\iint_{Q\,Q} |\varphi(x) - \varphi(y)|\, dxdy \leq C_\delta \left(\iint_{\substack{Q\,Q \\ |u(x)-u(y)|\geq\delta}} \frac{1}{|x-y|^2}\, dxdy + 1 \right). \tag{9.16}$$

The bound $\sqrt{3}$ is optimal, in the sense that, for any $\delta > \sqrt{3}$, the conclusion of the theorem fails (see Brezis and Nguyen [98, Appendix A]). The limiting case $\delta = \sqrt{3}$ is open.

The assumption that φ is *VMO* cannot be replaced by φ measurable: take $\varphi = 2k\pi\,\chi_A$, with A a non-trivial set, and let $k \to \infty$.

Theorem 9.8 was announced in Brezis and Nguyen [97], with a sketch of proof. A full (quite elaborate!) proof of Theorem 9.8 appears in Brezis and Nguyen [98].

A weaker version of Theorem 9.8 (with δ sufficiently small) was originally due to Bourgain, Brezis, and Mironescu [55, Theorem 3]. In a different but related direction, control of the degree in terms of I is presented in Section 12.5 (see Bourgain and Nguyen [58] for more details). The role of $\sqrt{3}$ was discovered by Nguyen [279].

9.7.2 *Profile decomposition in $W^{1/p,p}(\mathbb{S}^1;\mathbb{S}^1)$*

Recall the example in Proposition 9.2: for $1 < p < \infty$, there exists a sequence $\varphi_n : (0, 1) \to \mathbb{R}$ of smooth functions such that $|e^{\imath\varphi_n}|_{W^{1/p,p}} \lesssim 1$, whereas $|\varphi_n|_{W^{1/p,p}} \to \infty$. Such a φ_n is a "kink": a function moving quickly from the value 0 to the value 2π. We explain here why this loss of control mechanism is the only possible one. In order to have simpler statements, we work instead with maps $f : \mathbb{S}^1 \to \mathbb{S}^1$. This allows us to describe kinks using Moebius transforms. A Moebius transform M_a is given by $M_a(z) := \dfrac{a-z}{1-\overline{a}\,z}$, $z \in \overline{\mathbb{D}}$ and $a \in \mathbb{D}$. M_a is a self map of $\overline{\mathbb{D}}$ such that its restriction to $\mathbb{S}^1$, still denoted by M_a, is a smooth map $M_a : \mathbb{S}^1 \to \mathbb{S}^1$ of degree 1.

Let $f^a(z) := \overline{z}\, M_a(z)$, $z \in \mathbb{S}^1$, so that $f^a : \mathbb{S}^1 \to \mathbb{S}^1$ is smooth and $\deg f^a = 0$. We have $|M_a|_{W^{1/p,p}} = |\mathrm{Id}|_{W^{1/p,p}}$ (see the basic example in Section 12.8.1), and thus f^a is bounded in $W^{1/p,p}$. However, if $a \in \mathbb{D}$ approaches a point $\alpha = e^{\imath\xi} \in \mathbb{S}^1$, then the smooth phase φ^a of f^a converges a.e. to $\varphi(e^{\imath\theta}) := \begin{cases} \xi - \theta, & \text{if } \xi - \pi < \theta < \xi \\ 2\pi + \xi - \theta, & \text{if } \xi < \theta < \xi + \pi \end{cases}$, which does not belong to $W^{1/p,p}$. Thus, φ^a is not bounded in $W^{1/p,p}$ as $a \to \alpha \in \mathbb{S}^1$. On the other hand, the graph of φ^a shows that φ^a has a "kink shape," and thus we have here the analogue of the example on (0, 1). The following results, from Mironescu [258], show that every unimodular map "consists of" kinks and a phase that is under control.

Theorem 9.9. *Let $1 < p < \infty$. Then there exists a constant c_p and some increasing function $F : (0,\infty) \to (0,\infty)$ such that: every map $f \in W^{1/p,p}(\mathbb{S}^1;\mathbb{S}^1)$ can be written as $f = e^{\imath\varphi}\prod_{j=1}^{K}(M_{a_j})^{\pm 1}$, with $\varphi \in W^{1/p,p}(\mathbb{S}^1;\mathbb{R})$,*

$$K \leq c_p\, |f|^p_{W^{1/p,p}} \tag{9.17}$$

and

$$|\varphi|_{W^{1/p,p}} \leq F\left(|f|_{W^{1/p,p}}\right). \tag{9.18}$$

When $p = 2$, we may take $c_2 = 1/(4\pi^2)$, and this constant is optimal.

Corollary 9.2. *Let* $1 < p < \infty$ *and let* $f_n, f \in W^{1/p,p}(\mathbb{S}^1; \mathbb{S}^1)$ *be such that* $f_n \rightharpoonup f$ *in* $W^{1/p,p}$. *Then, up to a subsequence, there exist* $K \in \mathbb{N}$, $\varepsilon_j \in \{-1, 1\}$, $a_{j_n} \in \mathbb{D}$, $\alpha_j \in \mathbb{S}^1$, $j = 1, \ldots, K$, $\varphi_n \in W^{1/p,p}(\mathbb{S}^1; \mathbb{R})$, *and a constant C, such that:*
i) $f_n = e^{\imath\varphi_n} \prod_{j=1}^{K} (M_{a_{j_n}})^{\varepsilon_j} f$
ii) $a_{j_n} \to \alpha_j$ *as* $n \to \infty$
iii) $\varphi_n \rightharpoonup C$ *in* $W^{1/p,p}$ *as* $n \to \infty$.

Remark 9.7. Let us comment on the connection between Theorem 7.1 and the above theorem. A Moebius map satisfies $|M_a|_{W^{1,1}} = 2\pi$, and thus

$$\left| \prod_{j=1}^{K} (M_{a_j})^{\pm 1} \right|_{W^{1,1}} \le 2\pi\, K \le 2\pi\, c_p\, M\, |f|^p_{W^{1/p,p}}. \tag{9.19}$$

Estimate (9.19) is a refinement of the estimate of v in Theorem 7.1. On the other hand, (9.18) is weaker than the estimate of φ in Theorem 7.1, since F need not have a linear growth (and actually we do not know of any control on F). This suggests Open Problem 12.

9.7.3 Open problems

Open Problem 12. Does Theorem 9.9 hold with $F(M) \le d_p M$ for every $M > 0$ and some constant $d_p > 0$ depending only on p?

9.8 Comments

1. Theorems 9.1 and 9.2 are from Bourgain, Brezis, and Mironescu [50].
2. As explained in Remark 9.2, the conclusion of Theorem 9.2 is not optimal when $p > 1$. The improved estimate (9.2) was obtained in [50] using an L^2-Fourier analysis argument. The optimality of (9.2) when $N \ge 2$ was established in [50]. Its optimality when $N = 1$ required a different argument; a proof based on the asymptotic behavior of the best Sobolev constants was devised by Bourgain, Brezis, and Mironescu [53]. When $p \ne 2$, the counterpart of (9.2) is (9.3), which appears in Mironescu and Molnar [263]. Optimality of (9.3) is established in [263] using rearrangement arguments.
3. Proposition 9.1 is new.
4. Proposition 9.2 is from [50].
5. Theorem 9.3 was obtained by Bourgain, Brezis, and Mironescu [56] using a different argument. We do not know an elementary proof of this result, even when $p = 2$.
6. Strictly speaking, Theorem 9.4 is new. However, it can also be proved using the elementary arguments developed at the beginning of Section 5.1.1.

7. Theorem 9.5 is due to Merlet [247], who used elementary arguments.
8. Theorem 9.6 is from Mironescu and Molnar [263]. If $N \geq 2$, $0 < s < 1$, and $sp > N$, the method in [247] would give an estimate similar to (9.10), but possibly with an exponent larger than $1/s$. However, the method of [247] could not reach the borderline case $sp = N$, since it relies heavily on Morrey's embedding $W^{s,p} \hookrightarrow C^{0,\alpha}$.
9. Another attempt to prove Theorem 9.6 when $sp = N$ could consist of following carefully the existence proof described in Sections 5.1.1 and 5.1.2. However, this approach is doomed to fail, since it would apply to all $N \geq 1$, and in particular to $N = 1$. However, we know that (9.10) fails when $N = 1$, $1 < p < \infty$, and $sp = 1$ (Proposition 9.2).
10. In the special case $s = 1/2$ and $p = 2$, Theorem 9.7 is due to Bourgain and Brezis [48, Theorem 4] (see also [51, 54]). The argument there extends to $s = 1/p$ when $1 < p \leq 2$. However, the case $s = 1/p$, with $p > 2$, was raised as an open problem, even for $N = 1$ (see [56, Open Problem 1] and Comment 2 in Section 7.6). When $N = 1$, $s = 1/p$, and $1 < p < \infty$, with $p > 2$, Theorem 9.7 was established by Nguyen [280]. In full generality, Theorem 9.7 is new.

Chapter 10
Density

10.0 Introduction

Recall that, for every measurable function $u : \Omega \to \mathbb{S}^1$, there exists a sequence of smooth maps $u_j \in C^\infty(\overline{\Omega}; \mathbb{S}^1)$ converging to u a.e. (and thus in every L^p with $p < \infty$; see Lemma 1.19). Therefore, we will be concerned with stronger notions of convergence.

We first present the optimal conditions on s and p so that $C^\infty(\overline{\Omega}; \mathbb{S}^1)$ is dense in $W^{s,p}(\Omega; \mathbb{S}^1)$. When $N = 1$, this happens for every $s > 0$ and every p, $1 \le p < \infty$ (see Theorem 10.1). When $N \ge 2$, $s > 0$, and $1 \le p < \infty$, this happens if either $sp < 1$ or $sp \ge 2$ (see Theorem 10.2).

Next, we turn to the "nasty" range: $N \ge 2$, $s > 0$ and $1 \le sp < 2$. We may ask three natural questions.

Question 1. Find an appropriate substitute $\mathscr{R}$ for $C^\infty(\overline{\Omega}; \mathbb{S}^1)$, which is dense in $W^{s,p}(\Omega; \mathbb{S}^1)$ (for the strong topology).

As we are going to see in Section 10.2, maps in $W^{s,p}(\Omega; \mathbb{S}^1)$ that belong to $C^\infty(\overline{\Omega} \setminus \Sigma; \mathbb{S}^1)$, where Σ is a smooth $(N-2)$-submanifold of Ω, play an important role. More specifically, we will prove that such maps are dense in $W^{s,p}(\Omega; \mathbb{S}^1)$ (see Theorem 10.3, which contains more precise information about Σ). [Here, the location of the manifold Σ is not fixed.] When $N = 2$, we may let Σ be a finite subset of Ω, and thus maps in $W^{s,p}(\Omega; \mathbb{S}^1)$ can be approximated with maps that are smooth except at a finite number of points.

Question 2. *Characterize* the closure of $C^\infty(\overline{\Omega}; \mathbb{S}^1)$ in $W^{s,p}(\Omega; \mathbb{S}^1)$.

In Section 10.3, we present a simple answer, namely,

$$\overline{C^\infty(\overline{\Omega}; \mathbb{S}^1)}^{W^{s,p}} = \{u \in W^{s,p}(\Omega; \mathbb{S}^1);\ Ju = 0\}$$

(see Theorem 10.4). Recall that Ju has been defined (as a distribution) in Section 1.2 when $s = 1$ (and thus for $s \ge 1$), and in Section 8.1 in the remaining case $1/p \le s < 1$.

H. Brezis and P. Mironescu, *Sobolev Maps to the Circle*, Progress in Nonlinear Differential Equations and Their Applications 96,
https://doi.org/10.1007/978-1-0716-1512-6_10

Question 3. Is it true that every $u \in W^{s,p}(\Omega; \mathbb{S}^1)$ is a limit of a sequence $(u_j) \subset C^\infty(\overline{\Omega}; \mathbb{S}^1)$ for the weak $W^{s,p}$ topology?

We provide a full answer to Question 3 in Section 10.4.

In cases where $C^\infty(\overline{\Omega}; \mathbb{S}^1)$ is not dense in $W^{s,p}(\Omega; \mathbb{S}^1)$ (for the strong topology), we investigate the $W^{s,p}$ distance of any given $u \in W^{s,p}(\Omega; \mathbb{S}^1)$ to $C^\infty(\overline{\Omega}; \mathbb{S}^1)$ (see Section 10.5). When $s = 1$ and $p = 1$, we establish that this distance is comparable with $\Sigma(u)$ (Theorem 10.5).

Many of these results have their roots in the pioneering works of Bethuel [25–27, 37].

10.1 When are smooth maps dense?

Theorem 10.1. *When $N = 1$, $C^\infty(\overline{\Omega}; \mathbb{S}^1)$ is dense in $W^{s,p}(\Omega; \mathbb{S}^1)$, for every $s > 0$ and every p, $1 \le p < \infty$.*

Proof. By Theorem 5.1, every $u \in W^{s,p}(\Omega; \mathbb{S}^1)$ may be lifted as $u = e^{\imath\varphi}$, with $\varphi \in W^{s,p}$. Let $u_j = e^{\imath\varphi_j}$, with $\varphi_j \in C^\infty(\overline{\Omega}; \mathbb{R})$ and $\varphi_j \to \varphi$ in $W^{s,p}$. Then, $u_j \in C^\infty(\overline{\Omega}; \mathbb{S}^1)$ and $u_j \to u$ in $W^{s,p}$ (by Corollary 15.3). □

Theorem 10.2. *Assume that $N \ge 2$, $s > 0$, and $1 \le p < \infty$. Then, $C^\infty(\overline{\Omega}; \mathbb{S}^1)$ is dense in $W^{s,p}(\Omega; \mathbb{S}^1)$ if and only if $sp < 1$ or $sp \ge 2$.*

Proof. **Case 1:** $sp < 1$
In this case, we may write $u = e^{\imath\varphi}$ with $\varphi \in W^{s,p}$ (Theorem 5.2 *c)*). We continue as in the proof of Theorem 10.1.

Case 2: $1 \le sp < 2$
Assume that $0 \in \Omega$. Consider, as in Section 1.2, the topological counterexample $u(x) = \dfrac{(x_1, x_2)}{|(x_1, x_2)|}$, $x = (x_1, \dots, x_N)$, so that $u \in W^{s,p} \cap W^{1,1}$ (Lemma 15.12). By Remark 1.6 and Lemma 1.8, we know that $Ju \ne 0$. On the other hand, we have $Jv = 0$ if $v \in \overline{C^\infty(\overline{\Omega}; \mathbb{S}^1)}^{W^{s,p}}$ (by (1.61) and Corollary 8.1). Thus, $u \in W^{s,p}(\Omega; \mathbb{S}^1) \setminus \overline{C^\infty(\overline{\Omega}; \mathbb{S}^1)}^{W^{s,p}}$. We will return to this kind of argument involving Ju in Theorem 10.4.

Case 3: $s \ge 1$ and $sp \ge 2$
Let $u \in W^{s,p}(\Omega; \mathbb{S}^1)$. By Theorem 5.5, we may lift $u = e^{\imath\varphi}$, with $\varphi \in W^{s,p} \cap W^{1,sp}$. Let $\varphi_j \in C^\infty(\overline{\Omega}; \mathbb{R})$ be such that $\varphi_j \to \varphi$ in $W^{s,p} \cap W^{1,sp}$, and set $u_j = e^{\imath\varphi_j}$. By Theorem 15.8, we have $u_j \to u$ in $W^{s,p}$.

Case 4: $0 < s < 1$ and $sp \ge 2$
When $sp \ge N$, we may lift any $u \in W^{s,p}(\Omega; \mathbb{S}^1)$ as $u = e^{\imath\varphi}$, with $\varphi \in W^{s,p}(\Omega; \mathbb{R})$ (by Theorem 5.2 *a)*). Let $(\varphi_j) \subset C^\infty(\overline{\Omega}; \mathbb{R})$ be such that $\varphi_j \to \varphi$ in $W^{s,p}$ (see, for example, Corollary 15.10), and set $u_j = e^{\imath\varphi_j}$. By Theorem 15.6, we have $u_j \to u$ in $W^{s,p}$.

When $2 \le sp < N$, this is an interesting and unusual situation where lifting (within $W^{s,p}$) fails (see Theorem 5.2 *b)*), but density holds. This case is treated via factorization. Let $u \in W^{s,p}(\Omega; \mathbb{S}^1)$. By Theorem 8.5, we may write $u = e^{\imath\zeta}$, with $\zeta = \varphi + \psi$, $\varphi \in W^{s,p}$ and $\psi \in W^{1,sp}$. If we smooth φ and ψ, then $e^{\imath(\varphi_j+\psi_j)} \to u$ in $W^{s,p}$; this follows from Corollary 15.3, Remark 15.5, and Lemma 15.41. For an alternative treatment of Case 4 (without factorization), see Brezis and Mironescu [85]. $\square$

10.2 Density of $\mathscr{R}$. Answer to Question 1

In this section we assume that $N \ge 2$, $s > 0$, $1 \le p < \infty$, and $1 \le sp < 2$.

In view of the results of Bethuel [26] concerning $W^{1,p}$, a natural candidate for a dense class in $W^{s,p}(\Omega; \mathbb{S}^1)$ is

$$\mathscr{R}_{s,p} = \{u \in W^{s,p}(\Omega; \mathbb{S}^1);\ u \text{ is smooth in } \Omega \text{ outside a finite union of } (N-2) - \text{submanifolds of } \Omega\}.$$

[Here, the submanifolds may intersect.]

We will actually prove a stronger density result. For this purpose, we recall the definition of the class $\mathscr{R}$ already introduced in (7.11). Let U be an "appropriate," relatively compact, open neighborhood of $\overline{\Omega}$ ("appropriate" is meant in the sense of Lemma 15.27, which applies, since $sp < 2 \le p + 1$). Set

$$\mathscr{R} := \{u_{|\Omega};\ u \in C^\infty(\overline{U} \setminus S; \mathbb{S}^1) \text{ for some } (N-2) - \text{submanifold } S \text{ of } U, \text{ and } |D^\ell u(x)| \le C(\ell)\, [\text{dist}(x, S)]^{-\ell},\ \forall \ell \ge 1,\ \forall x \in \overline{U} \setminus S\}. \tag{10.1}$$

We emphasize the fact that the singular set S and the constants $C(\ell)$ depend on u. Note that $\mathscr{R}$ does not depend on s or p, and that $\mathscr{R} \subset \mathscr{R}_{s,p}$ (by Lemma 15.13 applied with $d = N - 2$).

Theorem 10.3. *The class $\mathscr{R}$ is dense in $W^{s,p}(\Omega; \mathbb{S}^1)$. In particular, $\mathscr{R}_{s,p}$ is dense in $W^{s,p}(\Omega; \mathbb{S}^1)$.*

When $N = 2$, $S \cap \overline{\Omega}$ is a finite set. Moreover, by considering a slightly smaller U, we may assume that S is finite.

Proof. Let $u \in W^{s,p}(\Omega; \mathbb{S}^1)$. With no loss of generality, we may assume that u is defined in a neighborhood U of $\overline{\Omega}$, independent of u (see Appendix 15.6).

Given a point $a \in \mathbb{R}^2$ with $|a| < 1/10$, let $\pi_a : \mathbb{R}^2 \setminus \{a\} \to \mathbb{S}^1$ be the radial projection onto $\mathbb{S}^1$ with vertex at a, i.e.,

$$\pi_a(\xi) := a + \lambda(\xi - a),\ \forall \xi \in \mathbb{R}^2 \setminus \{a\},$$

where $\lambda \in \mathbb{R}$ is the unique positive solution of $|a + \lambda(\xi - a)| = 1$. Clearly, we have $\pi_a(\xi) = \xi$ if $|\xi| = 1$.

It is convenient to note that

$$\pi_a(\xi) = j_a^{-1}\left(\frac{\xi - a}{|\xi - a|}\right) \text{ for } \xi \neq a,$$

where $j_a : \mathbb{S}^1 \to \mathbb{S}^1$, $j_a(z) := \dfrac{z-a}{|z-a|}$, is a smooth diffeomorphism with C^k bounds independent of a (since $|a| < 1/10$).

In particular,

$$|D^k \pi_a(\xi)| \leq \frac{C_k}{|\xi - a|^k}, \ \forall k \in \mathbb{N}, \ \forall \xi \in \mathbb{R}^2 \setminus \{a\}, \tag{10.2}$$

and π_a is Lipschitz on $\{|\xi| \geq 1/2\}$, with a uniform Lipschitz constant independent of a.

Let $u_\varepsilon := u * \rho_\varepsilon$, with (ρ_ε) a family of standard mollifiers, so that $u_\varepsilon \to u$ in $W^{s,p}(\Omega)$ (see Lemma 15.46). For small ε, u_ε is smooth in a fixed neighborhood of $\overline{\Omega}$, neighborhood still denoted by U. Set

$$V_{a,\varepsilon} := \pi_a \circ u_\varepsilon : U \setminus S_{a,\varepsilon} \to \mathbb{S}^1 \text{ and } v_{a,\varepsilon} := V_{a,\varepsilon|\Omega},$$

where $S_{a,\varepsilon} := \{x \in U;\ u_\varepsilon(x) = a\}$.

Note that, for every regular value of u_ε, $S_{a,\varepsilon}$ is an $(N-2)$-submanifold of U; in particular, when $N = 2$, $S_{a,\varepsilon}$ consists of a locally finite set of points in U, and thus $S_{a,\varepsilon} \cap \overline{\Omega}$ is finite. In addition, $v_{a,\varepsilon}$ belongs to $\mathscr{R}$. Here, we use (10.2). Thus (by Sard's lemma), for every ε, the map $v_{a,\varepsilon}$ belongs to $\mathscr{R}$ for a outside a null set (depending on ε).

We claim that there exists some a with $|a| < 1/10$ and some sequence $\varepsilon_n \to 0$ such that:

1. a is a regular value of u_{ε_n} for every n.
2. $v_{a,\varepsilon_n} \to u$ in $W^{s,p}$ as $n \to \infty$.

For this purpose, we introduce a smooth function $\psi : [0,\infty) \to [0,1]$ such that

$$\psi(t) = \begin{cases} 0, & \text{if } t \leq 1/4 \\ 1, & \text{if } t \geq 1/2 - 1/10 \end{cases}, \tag{10.3}$$

and we write

$$v_{a,\varepsilon} = \psi(|u_\varepsilon|)\pi_a(u_\varepsilon) + (1 - \psi(|u_\varepsilon|))\pi_a(u_\varepsilon) := w_{a,\varepsilon} + y_{a,\varepsilon}. \tag{10.4}$$

Clearly, the function

$$\Phi_a : \mathbb{R}^2 \to \mathbb{C},\ z \mapsto \psi(|z|)\pi_a(z),$$

satisfies

$$|D^\ell \Phi_a| \le C_\ell, \ \forall \ell, \ \forall a \text{ such that } |a| < 1/10. \tag{10.5}$$

Combining (10.5) with the fact that $u_\varepsilon \to u$ in $W^{s,p}$ as $\varepsilon \to 0$ and with Theorems 15.6 and 15.7 in Appendix 15.3, we obtain

$$w_{a,\varepsilon} = \Phi_a(u_\varepsilon) \to \Phi_a(u) = u \text{ in } W^{s,p} \text{ uniformly in a as } \varepsilon \to 0. \tag{10.6}$$

Therefore, it suffices to prove that, for some a and some subsequence $\varepsilon_n \to 0$, we have

$$|y_{a,\varepsilon_n}|_{W^{s,p}} \to 0 \text{ in } W^{s,p} \text{ as } n \to \infty \tag{10.7}$$

and

$$a \text{ is a regular value of } u_{\varepsilon_n}, \ \text{ for every } n. \tag{10.8}$$

More specifically, we prove that

$$\int_{[|a|<1/10]} |y_{a,\varepsilon}|^p_{W^{s,p}}\, da \to 0 \text{ as } \varepsilon \to 0, \tag{10.9}$$

which clearly implies (10.7) and (10.8).

It remains to prove that (10.9) holds. It turns out that the proof of (10.9) relies on similar but distinct estimates according to the cases $s < 1$, $s = 1$, and $s > 1$. For the convenience of the reader, we further split the case where $0 < s < 1$ into two subcases: $p < 2$ and $p \ge 2$; the first subcase is slightly easier than the second one. We will rely on several Gagliardo–Nirenberg estimates, for which we refer the reader to Appendix 15.2.

Case 1.1: Proof of (10.9) **when** $0 < s < 1$ **and** $1 \le p < 2$
Let $A_\varepsilon := [|u_\varepsilon| < 1/2]$. Note that, by (10.4) and (10.3), we have $\operatorname{supp} y_{a,\varepsilon} \subset A_\varepsilon$. In view of (10.2) and the properties of ψ, for every regular value a of u_ε and for every $x \in \Omega \setminus [u_\varepsilon = a]$, we have the estimate

$$|\nabla y_{a,\varepsilon}(x)|^p \lesssim \frac{|\nabla u_\varepsilon(x)|^p}{|u_\varepsilon(x) - a|^p} \chi_{A_\varepsilon}(x). \tag{10.10}$$

Integrating (10.10), we find that

$$\begin{aligned} \int_{[|a|<1/10]} \int_\Omega |\nabla y_{a,\varepsilon}(x)|^p\, dxda &\lesssim \int_{[|a|<1/10]} dx \int_{A_\varepsilon} \frac{|\nabla u_\varepsilon(x)|^p}{|u_\varepsilon(x) - a|^p}\, da \\ &\lesssim \int_{A_\varepsilon} |\nabla u_\varepsilon|^p\, dx; \end{aligned} \tag{10.11}$$

here, we use the fact that $p < 2$ and thus the integral with respect to a is bounded uniformly in x.

By (10.11) and Lemma 15.49 in Appendix 15.14, we find that

$$\int_{[|a|<1/10]} \|\nabla y_{a,\varepsilon}\|_{L^p}^p \, da \lesssim \int_{A_\varepsilon} |\nabla u_\varepsilon|^p \, dx = o(\varepsilon^{(s-1)p}) \text{ as } \varepsilon \to 0, \tag{10.12}$$

and clearly the above estimate still holds if we replace $\|\nabla y_{a,\varepsilon}\|_{L^p}$ by $\|y_{a,\varepsilon}\|_{W^{1,p}}$.

On the other hand, we have the Gagliardo–Nirenberg type estimate (Appendix 15.2)

$$|y_{a,\varepsilon}|_{W^{s,p}} \lesssim \|y_{a,\varepsilon}\|_{L^p}^{1-s} \|y_{a,\varepsilon}\|_{W^{1,p}}^s \lesssim |A_\varepsilon|^{(1-s)/p} \|y_{a,\varepsilon}\|_{W^{1,p}}^s. \tag{10.13}$$

In view of (10.12) and (10.13), in order to prove (10.9), it suffices to have

$$|A_\varepsilon| = O(\varepsilon^{sp}) \text{ as } \varepsilon \to 0. \tag{10.14}$$

In turn, estimate (10.14) follows by combining

$$|A_\varepsilon| \lesssim \int_{A_\varepsilon} |u_\varepsilon - u|^p$$

with Lemma 15.50 in Appendix 15.14.

Case 1.2: Proof of (10.9) **when** $0 < s < 1$ **and** $2 \le p < \infty$
Let q, r be specified later, such that $1 \le sp < q < 2 \le p < r < \infty$. As in the proof of (10.11), we have

$$\int_{[|a|<1/10]} |\nabla y_{a,\varepsilon}|^q \, da \lesssim \int_{A_\varepsilon} |\nabla u_\varepsilon|^q. \tag{10.15}$$

By (10.12), (10.14), (10.15), and Hölder's inequality, we find that

$$\int_{[|a|<1/10]} \|\nabla y_{a,\varepsilon}\|_{L^q}^q \, da = o(\varepsilon^{sp-q}) \text{ as } \varepsilon \to 0; \tag{10.16}$$

this estimate still holds if we replace $\|\nabla y_{a,\varepsilon}\|_{L^q}$ by the full norm $\|y_{a,\varepsilon}\|_{W^{1,q}}$.

Assume next that q and r satisfy

$$\frac{1}{p} = \frac{1-s}{r} + \frac{s}{q}. \tag{10.17}$$

For such q and r, we have the Gagliardo–Nirenberg inequality

$$\|y_{a,\varepsilon}\|_{W^{s,p}} \lesssim \|y_{a,\varepsilon}\|_{L^r}^{1-s} \|y_{a,\varepsilon}\|_{W^{1,q}}^s \lesssim |A_\varepsilon|^{(1-s)/r} \|y_{a,\varepsilon}\|_{W^{1,q}}^s. \tag{10.18}$$

We complete this case by combining (10.14), (10.16), and (10.18). We just have to make sure that we may choose q and r satisfying (10.17). If we choose q such that $sp < q < 2$, and we define r through (10.17), then q and r have all the required properties.

Case 2: Proof of (10.9) **when** $s = 1$ **(and** $1 \le p < 2$)
The argument leading to (10.11) implies that

$$\int_{[|a|<1/10]} |\nabla y_{a,\varepsilon}|^p \, da \lesssim \int_{A_\varepsilon} |\nabla u_\varepsilon|^p. \tag{10.19}$$

Since, clearly, $|A_\varepsilon| \to 0$ as $\varepsilon \to 0$ and $u_\varepsilon \to u$ in $W^{1,p}$, estimate (10.19) implies (10.9).

Case 3: Proof of (10.9) **when** $1 < s < 2/p$
We rely on the pointwise estimates (10.20) and (10.21), which are rather straightforward (we refer the reader to Bousquet [61, Section 5] for the detailed proofs).

$$\begin{aligned}|\nabla y_{a,\varepsilon}(y) - \nabla y_{a,\varepsilon}(x)| \lesssim & |\nabla u_\varepsilon(x)| \frac{|u_\varepsilon(y) - u_\varepsilon(x)|}{|u_\varepsilon(y) - a| \, |u_\varepsilon(x) - a|} \\ & + \frac{|\nabla u_\varepsilon(y) - \nabla u_\varepsilon(x)|}{|u_\varepsilon(y) - a|},\end{aligned} \tag{10.20}$$

and, for $|X| \le 1$, $|Y| \le 1$, we have

$$\int_{[|a|<1/10]} \frac{1}{|X - a|^p \, |Y - a|^p} \, da \lesssim \begin{cases} 1/|X - Y|^{2p-2}, & \text{if } p > 1 \\ 1 + |\ln |X - Y||, & \text{if } p = 1 \end{cases}. \tag{10.21}$$

We complete Case 3 when $p > 1$; the case $p = 1$ is similar. By (10.10), (10.20), and (10.21), we have (with $\sigma := s - 1 \in (0, 1)$ and using the fact that $p < 2$)

$$\begin{aligned}\int_{[|a|<1/10]} |y_{a,\varepsilon}|^p_{W^{s,p}} \, da &= \int_{[|a|<1/10]} |\nabla y_{a,\varepsilon}|^p_{W^{\sigma,p}} \, da \\ &\lesssim \underbrace{\int_\Omega \int_{A_\varepsilon} |\nabla u_\varepsilon(x)|^p \frac{|u_\varepsilon(y) - u_\varepsilon(x)|^{2-p}}{|y - x|^{N+\sigma p}} \, dxdy}_{I_\varepsilon} \\ &\quad + \int_\Omega \int_{A_\varepsilon} \frac{|\nabla u_\varepsilon(y) - \nabla u_\varepsilon(x)|^p}{|y - x|^{N+\sigma p}} \, dxdy.\end{aligned} \tag{10.22}$$

Using the fact that $|A_\varepsilon| \to 0$ and $\nabla u_\varepsilon \to \nabla u$ in $W^{\sigma,p}$, we find that the last integral in (10.22) tends to 0 with ε. In order to estimate I_ε, we first extend u_ε and u to $\mathbb{R}^N$ such that the extensions satisfy $\|u_\varepsilon\|_{L^\infty} \le 1$, $\|u\|_{L^\infty} \le 1$, and $u_\varepsilon \to u$ in $W^{s,p}$. We then rely on Lemma 15.45 (applied with $\beta := 2 - p$, $\gamma := \sigma p$, $f := u_\varepsilon$ – here, we use the assumption $sp < 2$), which implies that

$$I_\varepsilon \lesssim \int_{A_\varepsilon} |\nabla u_\varepsilon(x)|^p \, [\mathscr{M}|\nabla u_\varepsilon|(x)]^{\sigma p} \, dx \le \int_{A_\varepsilon} [\mathscr{M}|\nabla u_\varepsilon|(x)]^{sp} \, dx,$$

where $\mathscr{M}f$ denotes the maximal function of f.

Recall that, by Gagliardo–Nirenberg, we have $u \in W^{1,sp}$, since $u \in W^{s,p} \cap L^\infty$ and $s > 1$. We complete the proof using the fact that $|A_\varepsilon| \to 0$ and $\mathscr{M}|\nabla u_\varepsilon| \to \mathscr{M}|\nabla u|$ in L^{sp}. □

Let us note the following consequence of the proof of Theorem 10.3 (but not of the theorem itself).

Corollary 10.1. *Let s_1, s_2, p_1, p_2 be such that $1 \le s_j p_j < 2$, $j = 1, 2$. Let $u \in W^{s_1,p_1}(\Omega; \mathbb{S}^1) \cap W^{s_2,p_2}$. Then there exists a sequence $(u_k) \subset \mathscr{R}$ such that $u_k \to u$ in W^{s_j,p_j}, $j = 1, 2$.*

10.3 Characterization of $\overline{C^\infty(\overline{\Omega}; \mathbb{S}^1)}^{W^{s,p}}$. Answer to Question 2

Theorem 10.4. *Assume that $N \ge 2$, $s > 0$, $1 \le p < \infty$, and $1 \le sp < 2$. Let $u \in W^{s,p}(\Omega; \mathbb{S}^1)$. Then, the following assertions are equivalent:*
a) $Ju = 0$
b) $\Sigma(u) = 0$
c) $u \in \overline{C^\infty(\overline{\Omega}; \mathbb{S}^1)}^{W^{s,p}}$.

Proof. The equivalence "*a*) $\iff$ *b*)" comes from (1.86) when $s = 1$ and $p = 1$, and from (8.5) when $s = 1/p$, $1 < p < \infty$. The implication "*c*) $\implies$ *a*)" is clear, since $Ju = 0$ when $u \in C^\infty(\overline{\Omega}; \mathbb{S}^1)$, and the map $u \mapsto Ju$ is continuous, by Corollary 8.1. We now turn to "*a*) $\implies$ *c*)."

Case 1: $s \ge 1$ **and** $1 \le sp < 2$

By Theorem 8.8, we may write $u = e^{\imath\varphi}$, with $\varphi \in W^{s,p} \cap W^{1,sp}$. Consider a sequence $(\varphi_n) \subset C^\infty(\overline{\Omega}; \mathbb{R})$ such that $\varphi_n \to \varphi$ in $W^{s,p} \cap W^{1,sp}$. Set $u_n := e^{\imath\varphi_n}$. Then, $u_n \to u$ in $W^{s,p}$, by Theorem 15.8. We find that $u \in \overline{C^\infty(\overline{\Omega}; \mathbb{S}^1)}^{W^{s,p}}$.

Case 2: $0 < s < 1$ **and** $1 < sp < 2$

This is an interesting and unusual situation (somewhat similar to Case 4 in the proof of Theorem 10.2), where, for $u \in W^{s,p}$, the condition $Ju = 0$ does not imply existence of a lifting within $W^{s,p}$ (see Remark 8.11). We rely instead on Theorem 8.8, and we write $u = e^{\imath(\varphi+\psi)}$, with $\varphi \in W^{s,p}$ and $\psi \in W^{1,sp}$. Consider sequences (φ_n), $(\psi_n) \subset C^\infty(\overline{\Omega}; \mathbb{R})$ such that $\varphi_n \to \varphi$ in $W^{s,p}$ and $\psi_n \to \psi \in W^{1,sp}$. Set $u_n := e^{\imath(\varphi_n+\psi_n)}$. As in the proof of Theorem 10.2 (Case 4), we find that $u_n \to u$ in $W^{s,p}$, and thus $u \in \overline{C^\infty(\overline{\Omega}; \mathbb{S}^1)}^{W^{s,p}}$.

Case 3: $1 < p < \infty$ **and** $s = 1/p$

This is again an interesting and unusual case. Using Theorem 8.8, we may write $u = e^{\imath(\varphi+\psi)}$, where $\varphi \in W^{1/p,p}$ and $\psi \in W^{1,1}$. If we argue as in Case 2, we can say that $e^{\imath\psi_n} \to e^{\imath\psi}$ in $W^{1,1}$, but this does not imply that $e^{\imath\psi_n} \to e^{\imath\psi}$ in $W^{1/p,p}$ (see Remark 15.5). We therefore use a different argument.

Let $u \in W^{1/p,p}(\Omega; \mathbb{S}^1)$ satisfy $Ju = 0$. We need to show that $u \in \overline{C^\infty(\overline{\Omega}; \mathbb{S}^1)}^{W^{1/p,p}}$. Assume temporarily that $u \in W^{1,1}$. Consider a sequence $(u_k) \subset \mathscr{R}$ such that $u_k \to u$

in $W^{1/p,p}$ and in $W^{1,1}$ (see Corollary 10.1). Using (1.153) and the fact that $\Sigma(u)=0$, we find that $\Sigma(u_k)\to 0$. Consider some $\varphi_k\in C^\infty(\overline{\Omega};\mathbb{R})$ such that

$$\|u_k\wedge\nabla u_k-\nabla\varphi_k\|_{L^1}\le\Sigma(u_k)+2^{-k}.$$

Set $v_k:=u_k e^{-\imath\varphi_k}$. Since φ_k has a smooth extension to $\mathbb{R}^N$, we deduce that $v_k\in\mathscr{R}$. We have thus obtained a sequence $(v_k)\subset\mathscr{R}$ such that $Jv_k=Ju_k$ and $|v_k|_{W^{1,1}}\to 0$ as $k\to\infty$.

By Lemma 15.17, for each $\lambda>0$ there exists some $w_{k,\lambda}\in\mathscr{R}$ such that $Jw_{k,\lambda}=Jv_k$, $|w_{k,\lambda}|^p_{W^{1/p,p}}\le C_p|v_k|_{W^{1,1}}$ and $|[w_{k,\lambda}\ne 1]|\le\lambda$. We claim that

$$|u_k-u_k\overline{w_{k,\lambda}}|^p_{W^{1/p,p}}\le|w_{k,\lambda}|^p_{W^{1/p,p}}+r_{k,\lambda}\le C_p|v_k|_{W^{1,1}}+r_{k,\lambda},\ \forall\,k,\ \forall\,\lambda>0, \tag{10.23}$$

where $r_{k,\lambda}\to 0$ for each fixed k, as $\lambda\to 0$.

Indeed, write

$$\begin{aligned}&u_k(x)-u_k(x)\overline{w_{k,\lambda}}(x)-u_k(y)+u_k(y)\overline{w_{k,\lambda}}(y)\\&=u_k(x)\,[-\overline{w_{k,\lambda}}(x)+\overline{w_{k,\lambda}}(y)]+[u_k(x)-u_k(y)]\,[-\overline{w_{k,\lambda}}(y)+1],\end{aligned}$$

so that

$$|u_k-u_k\overline{w_{k,\lambda}}|_{W^{1/p,p}}\le|w_{k,\lambda}|_{W^{1/p,p}}+s_{k,\lambda},$$

where $s_{k,\lambda}\to 0$ for each fixed k, as $\lambda\to 0$.

On the other hand, we have $J(u_k\overline{w_{k,\lambda}})=0$ and $u_k\overline{w_{k,\lambda}}\in\widetilde{\mathscr{R}}$ (where $\widetilde{\mathscr{R}}$ is as in (7.12)). Let $1/p<\sigma<\min\{1,2/p\}$. By Remark 15.11, we have $u_k\overline{w_{k,\lambda}}\in W^{\sigma,p}$. By Case 2 above, we also have $u_k\overline{w_{k,\lambda}}\in\overline{C^\infty(\overline{\Omega};\mathbb{S}^1)}^{W^{\sigma,p}}$, and thus $u_k\overline{w_{k,\lambda}}\in\overline{C^\infty(\overline{\Omega};\mathbb{S}^1)}^{W^{1/p,p}}$. Combining this with (10.23), we find that

$$\operatorname{dist}_{W^{1/p,p}}\left(u_k,\overline{C^\infty(\overline{\Omega};\mathbb{S}^1)}^{W^{1/p,p}}\right)\le C_p|v_k|_{W^{1,1}}. \tag{10.24}$$

We obtain the desired conclusion, i.e., $u\in\overline{C^\infty(\overline{\Omega};\mathbb{S}^1)}^{W^{1/p,p}}$, by letting $k\to\infty$ in (10.24).

Finally, we remove the assumption $u\in W^{1,1}$. For this purpose, we factorize $u=v\,e^{\imath\varphi}$, with $v\in W^{1,1}$ and $\varphi\in W^{1/p,p}$. Then, $v\in W^{1/p,p}$, and $Jv=0$. Thus, we may apply the first part of the proof and find $(v_k)\subset C^\infty(\overline{\Omega};\mathbb{S}^1)$ such that $v_k\to v$ in $W^{1/p,p}$. If $(\varphi_k)\subset C^\infty(\overline{\Omega};\mathbb{R})$ is such that $\varphi_k\to\varphi$ in $W^{1/p,p}$, then clearly $v_k e^{\imath\varphi_k}\to u$ in $W^{1/p,p}$, and thus *c)* holds. □

Remark 10.1. If $u\in C\cap W^{s,p}(\Omega;\mathbb{S}^1)$ (or, more generally, if $u\in VMO\cap W^{s,p}(\Omega;\mathbb{S}^1)$), then $u\in\overline{C^\infty(\overline{\Omega};\mathbb{S}^1)}^{W^{s,p}}$, $\forall\,s>0$, $\forall\,1\le p<\infty$. Indeed, the relevant range for this conclusion is $N \geq 2$, $1 \leq sp < 2$. In view of Theorem 10.4, it suffices to prove that $Ju = 0$. When $s \geq 1$, this is a consequence of Theorem 5.9. When $s < 1$, we follow the procedure in Step 3 in the proof of Theorem 5.9. We first extend u to a neighborhood of $\overline{\Omega}$. We next note that $u_\varepsilon \to u$ in $W^{s,p}(\Omega)$, and that $|u_\varepsilon| \to 1$ uniformly on $\overline{\Omega}$. Thus, if we set $v_\varepsilon := u_\varepsilon/|u_\varepsilon|$, then $v_\varepsilon \to u$ in $W^{s,p}(\Omega)$ (Theorem 15.6). We conclude as follows: v_ε is Lipschitz, and thus $Jv_\varepsilon = 0$ (Corollary 1.1). Therefore, $Ju = 0$ (Theorem 8.1).

10.4 Weak sequential density. Answer to Question 3

Here, we assume that we do not have strong density, i.e., $N \geq 2$, $s > 0$, $1 \leq p < \infty$ and $1 \leq sp < 2$. We ask instead whether a weaker property holds. Given $u \in W^{s,p}(\Omega; \mathbb{S}^1)$, consider the following property.

$$(P_u) \quad \begin{cases} \text{There exists a sequence } (u_k) \subset C^\infty(\overline{\Omega}; \mathbb{S}^1), \\ \text{bounded in } W^{s,p} \text{ and such that } u_k \to u \text{ a.e.} \end{cases}$$

When $p > 1$, (P_u) is equivalent to the existence of a sequence $(u_k) \subset C^\infty(\overline{\Omega}; \mathbb{S}^1)$ such that $u_k \rightharpoonup u$ weakly in $W^{s,p}(\Omega; \mathbb{S}^1)$. Combining Theorems 1.11, 8.10, and 8.11, we obtain the following

Corollary 10.2. *1. When $sp = 1$, $1 \leq p < \infty$, (P_u) holds for every $u \in W^{s,p}(\Omega; \mathbb{S}^1)$.*
2. When $sp > 1$, (P_u) holds if and only if $Ju = 0$.

In view of Theorem 10.4, we see that, if $sp > 1$, (P_u) holds if and only if there exists a sequence $(u_k) \subset C^\infty(\overline{\Omega}; \mathbb{S}^1)$ such that $u_k \to u$ strongly in $W^{s,p}$. In particular, when $s = 1$ and $1 < p < 2$, we find that the strong and the weak sequential closure of $C^\infty(\overline{\Omega}; \mathbb{S}^1)$ in $W^{1,p}(\Omega; \mathbb{S}^1)$ coincide. A similar result holds for more general target manifolds (see Bethuel [26]).

10.5 Distance to smooth maps

Assume that $N \geq 2$. Set $X := C^\infty(\overline{\Omega}; \mathbb{S}^1)$. Recall that, if $1 \leq p < \infty$ and $u \in W^{1/p,p}(\Omega; \mathbb{S}^1)$, then $\Sigma(u) = 0$ if and only if the distance of u to X in $W^{1/p,p}$ is zero (Theorem 10.4). It is natural to ask whether $\Sigma(u)$ "measures" the distance of u to X in $W^{1/p,p}$, defined as

$$\text{dist}_{W^{1/p,p}}(u, X) = \inf\{|u - v|_{W^{1/p,p}};\ v \in X\}, \text{ with } 1 \leq p < \infty.$$

Here are some answers.

Theorem 10.5. *Let $u \in W^{1,1}(\Omega;\mathbb{S}^1)$. Then,*

$$\frac{2}{\pi}\Sigma(u) \le \mathrm{dist}_{W^{1,1}}(u,X) \le \Sigma(u). \tag{10.25}$$

Moreover, the constants $2/\pi$ and 1 are optimal in any domain.

Proposition 10.1. *If $1 < p < \infty$, then there exists a constant C_p such that*

$$\mathrm{dist}^p_{W^{1/p,p}}(u,X) \le C_p\,\Sigma(u),\ \forall\, u \in W^{1/p,p}(\Omega;\mathbb{S}^1). \tag{10.26}$$

Remark 10.2. By contrast, the "reverse" to (10.26) is wrong. We sketch the argument. By the proof of Proposition 12.13 (see Brezis, Mironescu, and Shafrir [93]), there exist sequences (f_n), $(g_n) \subset \mathrm{Lip}(\mathbb{S}^1;\mathbb{S}^1)$ such that

$$\deg f_n = 1,\ \deg g_n = 0, \tag{10.27}$$

and

$$|f_n - g_n|^p_{W^{1/p,p}} \to 0. \tag{10.28}$$

Let $\Omega := \mathbb{D}$ and set

$$u_n(x) := f_n(x/|x|),\ v_n(x) := g_n(x/|x|),\ \forall\, x \in \mathbb{D}\setminus\{0\}.$$

Then we have [95]

$$|u_n - v_n|_{W^{1/p,p}(\mathbb{D})} \lesssim |f_n - g_n|_{W^{1/p,p}(\mathbb{S}^1)}. \tag{10.29}$$

Clearly (by (10.27)), $Ju_n = \pi\delta_0$, $Jv_n = 0$. By Theorem 10.4, we find that

$$\mathrm{dist}_{W^{1/p,p}}(v_n,X) = 0, \tag{10.30}$$

and thus (using (10.28) and (10.29))

$$\mathrm{dist}^p_{W^{1/p,p}}(u_n,X) \le |u_n - v_n|^p_{W^{1/p,p}} \to 0,$$

whereas $\Sigma(u_n) = 2\pi$, by Theorem 2.5.

Proof of Theorem 10.5. We first establish the upper bound in (10.25) via a "dipole construction." Pick u_n such that $u_n \to 1$ a.e., $Ju_n = Ju$, and $\int |\nabla u_n| \to \Sigma(u)$ (see Proposition 1.1). Since $J(u\,\overline{u}_n) = 0$, we have, by Theorem 10.4,

$$\mathrm{dist}_{W^{1,1}}(u,X) \le \liminf_{n\to\infty}\int_\Omega |\nabla u - \nabla(u\overline{u_n})| = \lim_{n\to\infty}\int_\Omega |\nabla u_n| = \Sigma(u).$$

Concerning the lower bound in (10.25), we first observe that a weaker lower bound, $(1/2)\Sigma(u) \le \mathrm{dist}_{W^{1,1}}(u,X)$, is a trivial consequence of Corollary 1.7 and Theorem 10.4. The main ingredient in the proof of the sharp lower bound $(2/\pi)\Sigma(u) \le \mathrm{dist}_{W^{1,1}}(u,X)$ is the following

Lemma 10.1. *For any $w \in W^{1,1}(\Omega; \mathbb{S}^1)$ we have*

$$\int_\Omega |\nabla(|w-1|)| \geq \frac{2}{\pi}\Sigma(w). \tag{10.31}$$

Proof. We define $T : \mathbb{S}^1 \to \mathbb{S}^1$ by

$$T(e^{\iota\varphi}) := e^{\iota\theta} \text{ with } \theta = \theta(\varphi) = \pi \sin(\varphi/2),\ \forall \varphi \in (-\pi, \pi], \tag{10.32}$$

so that

$$|e^{\iota\varphi} - 1| = 2|\sin(\varphi/2)| = \frac{2}{\pi}|\theta|. \tag{10.33}$$

Clearly T is of class C^1 and its degree equals one. We claim that

$$|\nabla(|w-1|)| = \frac{2}{\pi}|\nabla(T \circ w)| \text{ a.e.} \tag{10.34}$$

This is a consequence of the standard fact that, if $F \in \mathrm{Lip}(\mathbb{S}^1; \mathbb{R}^2) \cap C^1(\mathbb{S}^1 \setminus \{1\})$ and $w \in W^{1,1}(\Omega; \mathbb{S}^1)$, then $F \circ w \in W^{1,1}(\Omega; \mathbb{S}^1)$ and, moreover,

$$\nabla(F \circ w) = \begin{cases} F'(w)\nabla w, & \text{a.e. in } [w \neq 1] \\ 0, & \text{a.e. in } [w = 1] \end{cases}$$

(see the discussion after the statement of Theorem 1.3).

Integration of (10.34) leads to

$$\int_\Omega |\nabla|w-1|| = \frac{2}{\pi}\int_\Omega |\nabla(T \circ w)|. \tag{10.35}$$

By Theorem 1.3, we have $J(T \circ w) = Jw$ and thus $\Sigma(T \circ w) = \Sigma(w)$. Therefore, (10.31) follows from (10.35).

Proof of Theorem 10.5 continued. Let $u \in W^{1,1}(\Omega; \mathbb{S}^1)$ and $v \in X$. Then, $w := u\bar{v}$ satisfies $Jw = Ju - Jv = Ju$, and thus $\Sigma(w) = \Sigma(u)$. Applying (10.31) yields

$$\int_\Omega |\nabla(u-v)| \geq \int_\Omega |\nabla(|u-v|)| = \int_\Omega |\nabla(|w-1|)| \geq \frac{2}{\pi}\Sigma(w) = \frac{2}{\pi}\Sigma(u).$$

It remains to prove the optimality of constants $2/\pi$ and 1. We present the argument for $2/\pi$. [The optimality of 1 is much more involved, and we refer the reader to Brezis, Mironescu, and Shafrir [93] for its proof.] We claim that it suffices to prove the following:

$$\forall u \in W^{1,1}(\Omega; \mathbb{S}^1),\ \exists v \in X \text{ such that } \int_\Omega |\nabla(u-v)| \leq \frac{2}{\pi}\int_\Omega |\nabla u|. \tag{10.36}$$

Indeed, assume that (10.36) holds. Let $u \in W^{1,1}(\Omega;\mathbb{S}^1)$ be such that $Ju \neq 0$. By (1.84), there exists a sequence $(u_n) \subset W^{1,1}(\Omega;\mathbb{S}^1)$ such that $Ju_n = Ju$ and $\int_\Omega |\nabla u_n| \to \Sigma(u)$. Let v_n correspond to u_n as in (10.36). Then,

$$\limsup_{n\to\infty} \frac{\mathrm{dist}_{W^{1,1}}(u_n, X)}{\Sigma(u_n)} \le \limsup_{n\to\infty} \frac{\int_\Omega |\nabla(u_n - v_n)|}{\Sigma(u_n)} \le \frac{2}{\pi},$$

whence the optimality of $2/\pi$.

It remains to prove (10.36). We start by introducing some notation. If $\zeta \in \mathbb{S}^1$, let I_ζ denote the half-circle having ζ as initial point (in the direct sense on $\mathbb{S}^1$) and not containing $-\zeta$. We note that the complement of I_ζ in $\mathbb{S}^1$ is $I_{-\zeta}$.

We define the map $P_\zeta : \mathbb{S}^1 \to I_\zeta \cup \{-\zeta\}$ by

$$P_\zeta(z) := \begin{cases} z, & \text{if } z \in I_\zeta \\ \zeta^2 \bar{z}, & \text{if } z \in I_{-\zeta} \end{cases}, \tag{10.37}$$

so that, for $z \in I_{-\zeta}$, $P_\zeta(z)$ is the reflection of z with respect to the line $\zeta\,\mathbb{R}$.

For each $u \in W^{1,1}(\Omega;\mathbb{S}^1)$ and $\zeta \in \mathbb{S}^1$, set

$$v_\zeta := P_\zeta \circ u. \tag{10.38}$$

Let us note that P_ζ is Lipschitz and $\deg P_\zeta = 0$. By Theorem 1.3, we have $J(v_\zeta) = 0$, and therefore $v_\zeta \in X$ (by Theorem 10.4). Assertion (10.36) is a consequence of the following identity.

$$\int_{\mathbb{S}^1} \left(\int_\Omega |\nabla(u - v_\zeta)|\, dx \right) d\zeta = 4 \int_\Omega |\nabla u|\, dx, \ \forall\, u \in W^{1,1}(\Omega;\mathbb{S}^1). \tag{10.39}$$

Indeed, assuming (10.39), by a mean value argument we may find some $\zeta \in \mathbb{S}^1$ such that $\int_\Omega |\nabla(u - v_\zeta)| \le (4/|\mathbb{S}^1|) \int_\Omega |\nabla u| \le (2/\pi) \int_\Omega |\nabla u|$.

It thus suffices to prove (10.39).

We note that

$$z - P_\zeta(z) = \begin{cases} 0, & \text{if } z \in I_\zeta \\ z - \zeta^2 \bar{z}, & \text{if } z \in I_{-\zeta} \end{cases}. \tag{10.40}$$

Set $w_\zeta := u - v_\zeta$. Using (10.40), we find that for every $\zeta = e^{\imath\varphi}$ and a.e. $x \in \Omega$ we have

$$\nabla w_\zeta(x) = \begin{cases} 0, & \text{if } u(x) \in I_\zeta \\ \nabla u(x) - \zeta^2\, \nabla \bar{u}(x), & \text{if } u(x) \in I_{-\zeta} \end{cases}. \tag{10.41}$$

Therefore, for a.e. $x \in \Omega$ we have

$$|\nabla w_\zeta(x)| = \begin{cases} 0, & \text{if } u(x) \in I_\zeta \\ 2|\cos(\theta(x) - \varphi)||\nabla u(x)|, & \text{if } u(x) = e^{\imath\theta(x)} \in I_{-\zeta} \end{cases}. \tag{10.42}$$

Indeed, we justify (10.42), for example, when $\zeta = 1$. In view of (10.41), we have to prove that

$$|\nabla \operatorname{Im} u(x)| = |\operatorname{Re} u(x)|\, |\nabla u(x)| \text{ for a.e. } x. \tag{10.43}$$

If we differentiate the identity $|u|^2 \equiv 1$, we obtain

$$\operatorname{Re} u\, \nabla(\operatorname{Re} u) + \operatorname{Im} u\, \nabla(\operatorname{Im} u) = 0 \text{ a.e.};$$

this easily implies (10.43).

Using (10.42) we find that, with $u(x) = e^{\imath\theta(x)}$ and

$$A(x) = \{\varphi \in (-\pi, \pi];\ u(x) \notin I_{e^{\imath\varphi}}\},$$

we have

$$\begin{aligned} \int_{\mathbb{S}^1} \int_\Omega |\nabla w_\zeta(x)|\, dx\, d\zeta &= \int_{-\pi}^{\pi} \int_\Omega \chi_{A(x)}(\varphi)\, 2|\cos(\theta(x) - \varphi)||\nabla u(x)|\, dx\, d\varphi \\ &= \int_\Omega |\nabla u(x)| \left(\int_{\theta(x)}^{\theta(x)+\pi} 2|\cos(\theta(x) - \varphi)|\, d\varphi \right) dx \\ &= 4 \int_\Omega |\nabla u(x)|\, dx, \end{aligned}$$

which is (10.39).

Here, we have used

$$\int_{\theta(x)}^{\theta(x)+\pi} 2|\cos(\theta(x) - \varphi)|\, d\varphi = \int_0^{\pi} 2|\cos t|\, dt = 4, \ \forall \theta \in \mathbb{R}. \qquad \square$$

Proof of Proposition 10.1. We sketch the argument. By density, it suffices to consider $u \in \mathscr{R}$. For such u, we argue as in the first part of the proof of Theorem 10.5, but we use Proposition 15.2 instead of Proposition 1.1. □

10.6 Complements and open problems

10.6.1 Density in $W_g^{1,p}(\Omega; \mathbb{S}^1)$

We address here the following loose question. Let $g : \partial\Omega \to \mathbb{S}^1$. Assuming that $u \in W^{s,p}(\Omega; \mathbb{S}^1)$ is such that $u = g$ on $\partial\Omega$, is it possible to approximate u with a "smoother" sequence $(u_j) \subset W^{s,p}(\Omega; \mathbb{S}^1)$ such that $u_j = g$ on $\partial\Omega$, $\forall j$? For the sake of simplicity, we consider only the case where $\Omega \subset \mathbb{R}^N$, $N \geq 2$, is a ball, $s = 1$ and g is smooth. For $1 \leq p < \infty$ and $g \in C^\infty(\partial\Omega; \mathbb{S}^1)$, set

$$W^{1,p}_g(\Omega; \mathbb{S}^1) := \{u \in W^{1,p}(\Omega; \mathbb{S}^1);\ u_{|\partial\Omega} = g\},$$

and

$$C^\infty_g(\overline{\Omega}; \mathbb{S}^1) = \{u \in C^\infty(\overline{\Omega}; \mathbb{S}^1);\ u_{|\partial\Omega} = g\}.$$

When $N \geq 3$, $C^\infty_g(\overline{\Omega}; \mathbb{S}^1)$ is non-empty. [This is a consequence of the fact that $\partial\Omega$ is simply connected and thus g admits a smooth lifting; see Section 1.1.] When $N = 2$, we have $W^{1,p}_g(\Omega; \mathbb{S}^1) \neq \emptyset$ if and only if $[1 \leq p < 2]$ or $[p \geq 2$ and $\deg g = 0]$ (see Proposition 12.2). Moreover, $C^\infty_g(\overline{\Omega}; \mathbb{S}^1) \neq \emptyset$ if and only if $\deg g = 0$.

Theorem 10.6. *Let $N \geq 2$ and $1 \leq p < \infty$. Let $g \in C^\infty(\partial\Omega; \mathbb{S}^1)$ and $u \in W^{1,p}_g(\Omega; \mathbb{S}^1)$.*

1. *If $p \geq 2$, then there exists a sequence $(u_j) \subset C^\infty(\overline{\Omega}; \mathbb{S}^1)$ such that $u_j \to u$ in $W^{1,p}$ and $u_j = g$ on $\partial\Omega$, $\forall j$.*
2. *If $1 \leq p < 2$, then there exists a sequence $(u_j) \subset \mathscr{R}$ such that $u_j \to u$ in $W^{1,p}$, u_j is smooth near $\partial\Omega$, and $u_j = g$ on $\partial\Omega$, $\forall j$.*

Proof. We consider first the case $N \geq 3$.

Proof of 1 when $N \geq 3$. We write $g = e^{\imath\psi_0}$ for some $\psi_0 \in C^\infty(\partial\Omega; \mathbb{R})$ and $u = e^{\imath\varphi}$ with $\varphi \in W^{1,p}(\Omega; \mathbb{R})$ (see Theorem 1.1). Thus, $e^{\imath\psi_0} = e^{\imath\varphi_{|\partial\Omega}}$ and therefore $(1/(2\pi))\,(\varphi_{|\partial\Omega} - \psi_0) \in W^{s,p}(\partial\Omega; \mathbb{Z})$, with $s = 1 - 1/p$. Since $sp = p - 1 \geq 1$, we deduce from Corollary 6.2 that $\varphi_{|\partial\Omega} - \psi_0 = 2\pi k$ on $\partial\Omega$, for some $k \in \mathbb{Z}$. Adding a constant to ψ_0, we may assume that $\varphi_{|\partial\Omega} - \psi_0 = 0$ on $\partial\Omega$. Let ψ be any smooth extension of ψ_0 to $\overline{\Omega}$. Then, $\varphi - \psi \in W^{1,p}_0(\Omega; \mathbb{R})$. Hence, there exists a sequence $(\zeta_j) \subset C^\infty_c(\Omega; \mathbb{R})$ such that $\zeta_j \to \varphi - \psi$ in $W^{1,p}$. Clearly, the sequence $u_j = e^{\imath(\zeta_j + \psi)}$ has all the required properties.

Proof of 2 when $N \geq 3$. Set $h := e^{\imath\psi}$ (where ψ denotes here any smooth extension of ψ_0 to a fixed neighborhood U of $\overline{\Omega}$) and $v := u\overline{h} - 1$, so that $v \in W^{1,p}_0(\Omega; \mathbb{R}^2)$. Hence, there exists a sequence $(v_j) \subset C^\infty_c(\Omega; \mathbb{R}^2)$ such that $v_j \to v$ in $W^{1,p}(\Omega)$. Let $w_j := (v_j + 1)h \in W^{1,p}(U; \mathbb{R}^2)$. Then, $w_{j|\Omega} \to u$ in $W^{1,p}(\Omega)$ and $w_j = h$ near $\partial\Omega$.

From here, we proceed as in the proof Theorem 10.3, using $V_{a,j} := \pi_a \circ w_j$ in place of $V_{a,\varepsilon} = \pi_a \circ u_\varepsilon$. Note that $V_{a,j} = h$ in U near $\partial\Omega$ (since $\pi_a(\xi) = \xi$ when $|\xi| = 1$); in particular, $V_{a,j|\partial\Omega} = g$ and $V_{a,j}$ is smooth near $\partial\Omega$.

We now turn to the case $N = 2$.

Proof of 1 when $N = 2$. Since $W^{1,g}_p(\Omega; \mathbb{S}^1) \neq \emptyset$ and $p \geq 2$, we have $\deg g = 0$ and we proceed as for item *1* with $N \geq 3$.

Proof of 2 when $N = 2$. Set $\alpha(x) := g(x/|x|)$, $x \in \mathbb{R}^2$, so that $\alpha_{|\Omega} \in \mathscr{R}$, and then $\widetilde{u} := u\overline{\alpha} \in W^{1,p}_1(\Omega; \mathbb{S}^1)$. We may now proceed as for item 2 with $N \geq 3$ and obtain a sequence $(\widetilde{u}_j) \subset \mathscr{R}$ such that $\widetilde{u}_j \to \widetilde{u}$ in $W^{1,p}$, with $\widetilde{u}_j$ smooth near $\partial\Omega$ and $\widetilde{u}_j = 1$ on $\partial\Omega$, $\forall j$. Clearly, the sequence $u_j := \widetilde{u}_j\alpha$ has all the required properties. ☐

Remark 10.3. The following "natural" assertion is *wrong*.

$$\text{Assume that } 1 \le p < 2, \text{ and } u \in W^{1,p}_g(\Omega; \mathbb{S}^1) \text{ satisfies } Ju = 0 \text{ and}$$
$$g \in C^\infty(\partial\Omega; \mathbb{S}^1). \text{ Then there exists a sequence } (u_j) \subset C^\infty_g(\overline{\Omega}; \mathbb{S}^1) \tag{10.44}$$
$$\text{such that } u_j \to u \text{ in } W^{1,p}.$$

Indeed, let $\omega \subset \partial\Omega$ be a smooth domain such that $\omega \neq \partial\Omega$ and $\omega \neq \emptyset$. Set $\zeta := 2\pi\chi_\omega$, so that $\zeta \in W^{s,p}(\partial\Omega)$, $\forall s < 1/p$ (by Lemma 15.8). In particular, $\zeta \in W^{1-1/p,p}(\partial\Omega)$, since $p < 2$. Let $\gamma \in W^{1,p}(\Omega)$ be such that $\gamma_{|\partial\Omega} = \zeta$. Set $u := e^{\imath\gamma}$, which belongs to $W^{1,p}_1(\Omega; \mathbb{S}^1)$, since $u_{\partial\Omega} = e^{\imath\zeta} = 1$. By Theorem 1.2, we have $Ju = 0$. Therefore, u satisfies the assumptions in (10.44) with $g = 1$.

We claim that there exists no sequence $(u_j) \subset C^\infty_1(\overline{\Omega}; \mathbb{S}^1)$ such that $u_j \to u$ in $W^{1,p}$. Assume by contradiction that such a sequence exists. We may write $u_j = e^{\imath\varphi_j}$, with $\varphi_j \in C^\infty(\overline{\Omega}; \mathbb{R})$ and $\varphi_j = 0$ on $\partial\Omega$. Then, $\nabla\varphi_j = u_j \wedge \nabla u_j \to u \wedge \nabla u = \nabla\gamma$ in $L^p(\Omega)$, and thus $\varphi_j \to \gamma + C$ for some constant C. Since $\varphi_j = 0$ on $\partial\Omega$, we deduce that $\gamma + C = 0$ on $\partial\Omega$, i.e., $\zeta + C = 0$ on $\partial\Omega$. Impossible.

10.6.2 Open problems

It is natural to try to formulate variants of statements in this chapter where $\mathbb{S}^1$ is replaced by $\mathbb{S}^k$, $k \ge 2$. Many results are known, but several questions remain open. Here are two typical examples. The first one is related to Theorem 10.2.

Open Problem 13. Let $0 < s < \infty$, $1 \le p < \infty$ be such that $k + 1 \le sp < N$. Let Ω be a ball in $\mathbb{R}^N$. Assume that $\pi_{[sp]}(\mathbb{S}^k) = 0$. Is $C^\infty(\overline{\Omega}; \mathbb{S}^k)$ dense in $W^{s,p}(\Omega; \mathbb{S}^k)$?

[$\pi_j(\mathbb{S}^k)$ denotes the jth homotopy group of $\mathbb{S}^k$.]

The answer is known and positive in the following special cases: $s = 1$ (Bethuel [26]), $s = 2, 3, \ldots$ (Bousquet, Ponce, and Van Schaftingen [66]), and $0 < s < 1$ (Brezis and Mironescu [85]).

The next one is related to Theorem 10.4.

Open Problem 14. Assume that $N \ge k + 1$. Let $0 < s < \infty$ and $1 \le p < \infty$ be such that $k \le sp < k + 1$. Let $\Omega \subset \mathbb{R}^N$. For $u \in W^{s,p}(\Omega; \mathbb{S}^k)$, is it true that

$$Ju = 0 \iff u \in \overline{C^\infty(\overline{\Omega}; \mathbb{S}^k)}^{W^{s,p}} \ ?$$

The answer is known to be positive when $s = 1$ (see Bethuel [25] and Bethuel, Coron, Demengel, and Hélein [33]; see also Ponce and Van Schaftingen [301]). It is very plausible that Bethuel's "dipole removing" technique [25] can be adapted to the case where $sp = k$, but the case where $2 \le k < sp < k + 1$ is widely open.

In view of Theorem 10.5, it is natural to ask whether equality

$$\operatorname{dist}_{W^{1,1}}(u, X) = \frac{2}{\pi} \Sigma(u) \tag{10.45}$$

holds for some domain Ω and some $u \in W^{1,1}(\Omega; \mathbb{S}^1)$ such that $\Sigma(u) \neq 0$. The answer is positive. Take, for example, $\Omega := \mathbb{D}$ and $u(x) := x/|x|$. In this case, we have $\Sigma(u) = 2\pi$ and it is easy to check that all the maps v_ζ as in (10.38) satisfy $\int_{\mathbb{D}} |\nabla(u - v_\zeta)| = 4 = (2/\pi)\Sigma(u)$.

By contrast, we do not know any non-trivial example of equality in

$$\operatorname{dist}_{W^{1,1}}(u, X) = \Sigma(u). \tag{10.46}$$

Open Problem 15. Are there some Ω and $u \in W^{1,1}(\Omega; \mathbb{S}^1)$ with $\Sigma(u) \neq 0$ such that

$$\operatorname{dist}_{W^{1,1}}(u, X) = \Sigma(u)? \tag{10.47}$$

10.7 Comments

1. Schoen and Uhlenbeck [318, Section 4] were the first to point out that the question of the density of smooth maps in Sobolev spaces with values in a compact manifold is much more delicate than for real-valued maps. They made two important observations, which paved the way for numerous subsequent investigations:

   a) When $\Omega \subset \mathbb{R}^2$, smooth maps are dense in $H^1(\Omega; \mathscr{N})$ (for every target $\mathscr{N}$).
   b) When $\Omega \subset \mathbb{R}^3$, smooth maps are *not* dense in $H^1(\Omega; \mathbb{S}^2)$.

2. In full generality, the question of density was originally investigated by Bethuel for maps in $W^{1,p}(\Omega; \mathscr{N})$, where $\Omega \subset \mathbb{R}^N, N \geq 2$, and $\mathscr{N}$ is a compact manifold without boundary. A beautiful result of Bethuel [26] is the following

Theorem 10.7. *Let $1 \leq p < \infty$. Then*

a) The class $\mathscr{R}$ is always dense in $W^{1,p}(\Omega; \mathscr{N})$
b) When $p \geq N$, $C^\infty(\overline{\Omega}; \mathscr{N})$ is dense in $W^{1,p}(\Omega; \mathscr{N})$
c) When $1 \leq p < N$ and, for example, Ω is a ball, we have that $C^\infty(\overline{\Omega}; \mathscr{N})$ is dense in $W^{1,p}(\Omega; \mathscr{N})$ if and only if $\pi_{[p]}(\mathscr{N}) = 0$.

Here, $\mathscr{R}$ is an appropriate analogue of the class $\mathscr{R}$ given by (10.1), and $\pi_j(\mathscr{N})$ is the jth homotopy group of $\mathscr{N}$.
In our situation, $\mathscr{N} = \mathbb{S}^1$ and $\pi_{[p]}(\mathbb{S}^1) = 0$ if and only if $p \geq 2$. Thereby, Theorem 10.7 corresponds to Theorems 10.3 and 10.2 when $s = 1$. [However, the classes $\mathscr{R}$ in Theorems 10.7 *a)* and 10.3 are not the same.]
For $s = 1$, an important subsequent work is due to Hang and Lin [177] (for general target manifolds $\mathscr{N}$ and general Ω).
For special $\mathscr{N}$'s, the arguments in [26] were simplified by Hajłasz [172].

Bousquet, Ponce, and Van Schaftingen [66] extended the results in [26] and some of the results in [177] to $s = 2, 3, \ldots$ In some situations, their long and difficult proof can be simplified [64]. They also considered the case of non-integer $s > 1$, but only for some special $\mathscr{N}$'s [65].
Brezis and Mironescu [85] extended the results in [26] to $0 < s < 1$. Special cases of the results in [85] were obtained by Mucci [275].

3. Comments on Theorem 10.2. When $sp < 1$, this is due to Escobedo [145]. When $s = 1$, this is due to Bethuel and Zheng [37]. The case $s \geq 1$ and $sp \geq 2$ is due to Brezis and Mironescu [84]. Bethuel [27] investigated the case $s = 1 - \frac{1}{p}$. The proof of non-density is basically that of Schoen and Uhlenbeck [318]. The case $0 < s < 1, 2 \leq sp < N$ is new; it was open for a while.
4. Comments on Theorem 10.3. The case $s = 1$, $1 \leq p < 2$ is due to Bethuel and Zheng [37] for $N = 2$ and to Bethuel [26] for $N \geq 3$. The case $s = 1 - 1/p$, $N < p < N + 1, N \geq 2$, is from Bethuel [27]. The case $s = 1/2, p = 2, N = 2$ is from Rivière [305], with a refinement in Bourgain, Brezis, and Mironescu [54]. The case $0 < s < 1$, $1 \leq p < \infty$, $1 \leq sp < 2$, $N = 2$ is from Bourgain, Brezis, and Mironescu [56]. Our proof follows essentially [54, 56]. The case $s > 1$ was obtained by Bousquet [61].
The "projection-averaging" technique, which plays a central role in the proof of Theorem 10.3, was first applied by Hardt and Lin [183, Section 6.2] and by Hardt, Kinderlehrer, and Lin [180, Appendix] for a different purpose (see Comment 1 in Chapter 11), and by Bethuel and Zheng [37] in the study of density problems. This method has roots in Federer and Fleming [149] (see also Federer [147, Sections 4.2.7–4.2.9]) and White [344, 345].
5. A first result in the spirit of Theorem 10.4 is due to Bethuel [25], in the context of liquid crystals; there, $s = 1$, $p = 2$, and the target is $\mathbb{S}^2$. When $s = p = 1$, Theorem 10.4 was obtained by Demengel [132]. In full generality, the theorem is new.
6. When $\mathscr{N} = \mathbb{S}^1$, Corollary 10.2 was proved by Rivière [305] for $N = 2, s = 1/2$, and $p = 2$, and by Giaquinta and Mucci [161] for $s = 1$ and $p = 1$, and for $s = 1/2$ and $p = 2$. The case where $sp > 1$ follows essentially the proof of Bethuel [26]. The remaining cases are new.
More generally, let $\mathscr{N}$ be a compact manifold without boundary. If there is weak, but not strong density of smooth maps in $W^{1,p}(\Omega; \mathscr{N})$, then p has to be an integer [26]. Assume now that p is an integer. When $p = 1$, Hang [175] proved that there is "density" in the sense of Corollary 10.2 (for a related result, see Pakzad [286]). Pakzad and Rivière [287] examined the case $p = 2$. Surprisingly, Bethuel [29] proved that weak density does not hold when $p = 3$, $\Omega = B^4$, and $\mathscr{N} = \mathbb{S}^2$.
7. Distance to smooth maps. Theorem 10.5 is due to Brezis, Mironescu, and Shafrir [94]. The upper bound (10.25) and Proposition 10.1 are reminiscent of a result of Bethuel [25] concerning $H^1(B^3; \mathbb{S}^2)$. For "non-estimates" similar to those in Remark 10.2 (in the context of $\mathbb{S}^N$-valued maps), see Levi and Shafrir [223].

8. Approximation with fixed boundary conditions (as in Section 10.6.1) has been investigated by Bethuel and Zheng [37], Bethuel, Coron, Demengel, and Hélein [33], and Bethuel [26].

Chapter 11
Traces

11.0 Introduction

Set $Q := (-1, 1)^N$ and $\widetilde{Q} := (-1, 1)^N \times (0, 1)$. Let $s > 0$ and $1 \le p < \infty$. We recall two classical facts about traces [338].

Fact 1. Every $v \in W^{s+1/p,p}(\widetilde{Q}; \mathbb{R})$ has a trace $u = \operatorname{tr} v \in W^{s,p}(Q; \mathbb{R})$ (here, we identify $Q \times \{0\}$ with Q).

Fact 2. Conversely, if s is not an integer, then every $u \in W^{s,p}(Q; \mathbb{R})$ is the trace of some $v \in W^{s+1/p,p}(\widetilde{Q}; \mathbb{R})$.

In addition, we may choose a bounded linear extension operator $u \mapsto v$ independent of s and p.

Throughout this chapter we assume that s is not an integer, and we investigate similar facts when the target $\mathbb{R}$ is replaced by $\mathbb{S}^1$. Clearly, Fact 1 above still holds. Our purpose is to investigate the analogue of Fact 2, more precisely

$$(E_{s,p,N}) \quad \text{Given an arbitrary } u \in W^{s,p}(Q; \mathbb{S}^1), \text{ is there some } v \in W^{s+1/p,p}(\widetilde{Q}; \mathbb{S}^1) \text{ such that } \operatorname{tr} v = u?$$

Fact 2 is usually established by constructing v as an average of u. Such a procedure does not yield an $\mathbb{S}^1$-valued map v, even if u is $\mathbb{S}^1$-valued. As we are going to see, this is not just a technical issue. The answer to question $(E_{s,p,N})$ is sometimes negative. Our goal is to characterize precisely the values of s and p such that the answer to $(E_{s,p,N})$ is positive. The main result in this direction is the following

Theorem 11.1. *The answer to $(E_{s,p,N})$ is positive if and only if $sp < 1$ or $sp \ge N$.*

When $1 \le sp < N$, we characterize the range of the mapping $v \mapsto \operatorname{tr} v$, with $v \in W^{s+1/p,p}(\widetilde{Q}; \mathbb{S}^1)$ (see Theorem 11.2).

All the proofs in this chapter rely on results and techniques from the previous chapters. In particular, lifting plays an important role in the proofs. Note, for example, that, if $0 < s < 1$, the answers to lifting and to $(E_{s,p,N})$ are similar (compare Theorem

H. Brezis and P. Mironescu, *Sobolev Maps to the Circle*, Progress in Nonlinear Differential Equations and Their Applications 96,
https://doi.org/10.1007/978-1-0716-1512-6_11

5.2 with Theorem 11.1). However, when $N \geq 3$, $s \geq 1$ and $2 \leq sp < N$, the answer to lifting is positive, whereas the answer to $(E_{s,p,N})$ is negative (compare Theorem 5.3 with Theorem 11.1).

11.1 Proof of Theorem 11.1

Proof. The positive cases are not surprising. When $sp < 1$, we rely on lifting. When $sp > N$, maps are continuous, and averaging leads to almost $\mathbb{S}^1$-valued extensions. It then suffices to project the extensions on $\mathbb{S}^1$. The same holds when $sp = N$, via *VMO*.

Step 1. The answer to $(E_{s,p,N})$ is positive when $sp < 1$ and $s \leq 1 - 1/p$
Indeed, since $sp < 1$, we may write $u = e^{\imath\varphi}$, with $\varphi \in W^{s,p}(Q; \mathbb{R})$ (see Theorems 5.1 and 5.2). Let $\psi \in W^{s+1/p,p}(\widetilde{Q}; \mathbb{R})$ be an extension of φ to $\widetilde{Q}$. Since $s + 1/p \leq 1$, we have $v := \exp(\imath\psi) \in W^{s+1/p,p}$. Clearly, v is the desired extension.

Step 2. The answer to $(E_{s,p,N})$ is positive when $sp < 1$ and $s > 1 - 1/p$
The condition $s > 1 - 1/p$ implies that $\sigma := 1 - \dfrac{1}{sp+1} < s$. By Theorem 5.10 with $\theta = \left(1 - \dfrac{1}{sp+1}\right)/s \in (0, 1)$, we may write $u = \exp(\imath\varphi)$, with $\varphi \in W^{s,p} \cap W^{1-1/(sp+1),sp+1}$. Pick an extension $\psi \in W^{s+1/p,p}(\widetilde{Q}) \cap W^{1,sp+1}(\widetilde{Q})$ of φ. This regularity of ψ ensures that $v := \exp(\imath\psi)$ belongs to $W^{s+1/p,p}$ (Theorem 15.8 in Appendix 15.3). Clearly, $\operatorname{tr} v = u$.

Step 3. The answer to $(E_{s,p,N})$ is positive when $sp \geq N$
Indeed, we may write $u = \exp(\imath\varphi)$, with $\varphi \in W^{s,p}$ (see Theorems 5.1, 5.2, and 5.3). Let $\psi \in W^{s+1/p,p}(\widetilde{Q})$ be an extension of φ. By Corollary 15.3 in Appendix 15.3, we have $v \in W^{s+1/p,p}$. Clearly, we have $\operatorname{tr} v = u$.

Step 4. The answer to $(E_{s,p,N})$ is negative when $N \geq 2$, $1 \leq sp < N$ and $s \leq 1 - 1/p$
Indeed, set

$$\varphi(x) := |x|^{-\alpha}, \text{ with } \alpha = \frac{N - sp}{sp + 1}. \tag{11.1}$$

Applying Lemma 15.11 and the assumption $1 \leq sp < N$, we see that, for some $\varepsilon > 0$, we have

$$\varphi \in W^{1,sp+\varepsilon}(Q), \tag{11.2}$$

$$\varphi \notin W^{1-1/(sp+1),sp+1}(Q), \tag{11.3}$$

$$\varphi \notin L^q(Q), \text{ with } q := \frac{N(sp+1)}{N - sp}. \tag{11.4}$$

Set $u := e^{\imath\varphi}$, so that, by (11.2),

$$u \in W^{1,sp+\varepsilon}(Q; \mathbb{S}^1). \tag{11.5}$$

From (11.5), the assumption $s < 1$, and the (suboptimal) Gagliardo–Nirenberg embeddings (see Remark 15.2), we deduce that

$$u \in W^{s,p}(Q; \mathbb{S}^1). \tag{11.6}$$

We claim that u has no extension. Argue by contradiction, and let

$$v \in W^{s+1/p,p}(\widetilde{Q}; \mathbb{S}^1) \tag{11.7}$$

be an extension of u. Applying Theorem 8.5, we obtain a lifting ψ for v, with $\psi \in W^{s+1/p,p}(\widetilde{Q}) + W^{1,sp+1}(\widetilde{Q})$. [Here we use the fact that $s + 1/p \le 1$ and $(s + 1/p)p \ge 2$.]

Set $\zeta := \operatorname{tr} \psi \in W^{s,p}(Q) + W^{1-1/(sp+1),sp+1}(Q)$. Clearly, $(\zeta - \varphi)/(2\pi)$ is $\mathbb{Z}$-valued and belongs to $W^{s,p} + W^{1-1/(sp+1),sp+1} + W^{1,sp}$. We deduce from Corollary 6.5 that $\zeta - \varphi$ is a constant C, and thus

$$\varphi = \zeta - C \in W^{s,p}(Q) + W^{1-1/(sp+1),sp+1}(Q). \tag{11.8}$$

On the other hand, by the Sobolev embedding, we have (with q as in (11.4))

$$\zeta \in W^{s,p}(Q) + W^{1-1/(sp+1),sp+1}(Q) \subset L^q(Q). \tag{11.9}$$

[Here, we use $s \le 1 - 1/p$.]

Properties (11.8) and (11.9) contradict (11.4).

Step 5. The answer to $(E_{s,p,N})$ is negative when $N \ge 2$, $1 \le sp < N$ and $s > 1 - 1/p$ We consider φ as in Step 4, and note that (11.1)–(11.4) still hold. From $s > 1 - 1/p$ and Lemma 15.11, we deduce that

$$\varphi \in W^{s,p}(Q). \tag{11.10}$$

We assert that $u \in W^{s,p}(Q; \mathbb{S}^1)$. This is clear from (11.10) when $s \le 1$; if $s > 1$ we rely on (11.2), (11.10), and Theorem 15.8.

We claim that u has no extension. Argue again by contradiction, and let $v \in W^{s+1/p,p}(\widetilde{Q}; \mathbb{S}^1)$ be an extension of u. Applying Theorem 5.5, we obtain a lifting ψ for v, with $\psi \in W^{s+1/p,p}(\widetilde{Q}) \cap W^{1,sp+1}(\widetilde{Q})$. Set $\zeta := \operatorname{tr} \psi \in W^{1-1/(sp+1),sp+1}(Q)$. As above, $\zeta - \varphi$ is constant. This contradicts (11.3).

The proof of Theorem 11.1 is complete. □

11.2 A sharp form of the extension problem

Theorem 11.1 asserts that tr $W^{s+1/p,p}(\widetilde{Q};\mathbb{S}^1) = W^{s,p}(Q;\mathbb{S}^1)$ if and only if $sp < 1$ or $sp \geq N$, but does not yield the precise expression of tr $W^{s+1/p,p}(\widetilde{Q};\mathbb{S}^1)$ when $N \geq 2$ and $1 \leq sp < N$. We have the following

Theorem 11.2. *Assume that $N \geq 2$, $s > 0$, $1 \leq p < \infty$, and $1 \leq sp < N$. Then,*

$$tr\, W^{s+1/p,p}(\widetilde{Q};\mathbb{S}^1) = \begin{cases} \{e^{\imath\varphi};\ \varphi \in W^{s,p} \cap W^{1-1/(sp+1),sp+1}\}, & \text{if } s \geq 1 - 1/p \\ \{e^{\imath\varphi};\ \varphi \in W^{s,p} + W^{1-1/(sp+1),sp+1}\}, & \text{if } s < 1 - 1/p \end{cases}.$$

Proof. Step 1. Proof of the theorem when $s \geq 1 - 1/p$
Let $\varphi \in W^{s,p} \cap W^{1-1/(sp+1),sp+1}$. Then, φ has an extension $\psi \in W^{s+1/p,p} \cap W^{1,sp+1}$. If $v := e^{\imath\psi}$, then $v \in W^{s+1/p,p}$ (Theorem 15.8). Clearly, tr $v = e^{\imath\varphi}$. This proves "$\supseteq$."

For the opposite inclusion, consider some $v \in W^{s+1/p,p}(\widetilde{Q};\mathbb{S}^1)$. Since $s + 1/p \geq 1$ and $\left(s + \dfrac{1}{p}\right)p \geq 2$, we may write $v = e^{\imath\psi}$ for some $\psi \in W^{s+1/p,p} \cap W^{1,sp+1}$ (Theorem 5.5). By taking traces, we find that tr $v = e^{\imath\varphi}$, where $\varphi \in W^{s,p} \cap W^{1-1/(sp+1),sp+1}$. This proves "$\subseteq$."

Step 2. Proof of the theorem when $s < 1 - 1/p$
Let $\varphi \in W^{s,p} + W^{1-1/(sp+1),sp+1}$. Consider an extension $\psi \in W^{s+1/p,p} + W^{1,sp+1}$ of φ. If $v := e^{\imath\psi}$, then $v \in W^{s+1/p,p}$ (see the "converse part" in Theorem 8.5). Clearly, tr $v = e^{\imath\varphi}$. This proves "$\supseteq$."

For the opposite inclusion, let $v \in W^{s+1/p,p}(\widetilde{Q};\mathbb{S}^1)$. Since $s + 1/p < 1$ and $\left(s + \dfrac{1}{p}\right)p \geq 2$, Theorem 8.5 implies that we may write $v = e^{\imath\psi}$ for some $\psi \in W^{s+1/p,p} + W^{1,sp+1}$. By taking traces, we find that tr $v = e^{\imath\varphi}$, where $\varphi \in W^{s,p} + W^{1-1/(sp+1),sp+1}$. This proves "$\subseteq$." □

11.3 Complements and open problems

11.3.1 An intermission

Let us pause briefly to compare the various properties discussed in the previous chapters that can be associated with a given $u \in W^{s,p}(Q;\mathbb{S}^1)$. Assume, for example, that $N \geq 2$, $0 < s < 1$, $1 \leq sp < 2$. Set

$$A := \{u \in W^{s,p}(Q;\mathbb{S}^1);\ u = e^{\imath\varphi} \text{ for some } \varphi \in W^{s,p}(Q;\mathbb{R})\},$$

$$B := \{u \in W^{s,p}(Q;\mathbb{S}^1);\ u = \text{tr}\, v \text{ for some } v \in W^{s+1/p,p}(\widetilde{Q};\mathbb{S}^1)\},$$

$$C := \{u \in W^{s,p}(Q;\mathbb{S}^1);\ Ju = 0\}$$
$$= \{u \in W^{s,p}(Q;\mathbb{S}^1);\ \exists\,(u_n) \subset C^\infty(\overline{\Omega};\mathbb{S}^1) \text{ such that } u_n \to u \text{ in } W^{s,p}\}.$$

The next result describes the inclusion relations between these sets.

Theorem 11.3. *Let $N \geq 2$, $0 < s < 1$, $1 \leq sp < 2$. Then,*

1. *If $s > 1 - 1/p$, we have*

$$B \subsetneqq A \subsetneqq C \subsetneqq W^{s,p}(\Omega;\mathbb{S}^1).$$

2. *If $s < 1 - 1/p$, we have*

$$A \subsetneqq B \subsetneqq C \subsetneqq W^{s,p}(\Omega;\mathbb{S}^1).$$

3. *If $s = 1 - 1/p$, we have*

$$A = B \subsetneqq C \subsetneqq W^{s,p}(\Omega;\mathbb{S}^1).$$

Proof. Step 1. $C \neq W^{s,p}(\Omega;\mathbb{S}^1)$
If $0 \in \Omega$ and we take $u(x) := (x_1, x_2)/|(x_1, x_2)|$, then $u \in W^{s,p}(\Omega;\mathbb{S}^1) \setminus C$ (see Case 2 in the proof of Theorem 10.2).

Step 2. If $s \geq 1 - 1/p$, then $B \subset A$
Indeed, let $u \in B$ and consider $v \in W^{s+1/p,p}(\widetilde{Q};\mathbb{S}^1)$ such that $\operatorname{tr} v = u$. Since $s + 1/p \geq 1$ and $(s + 1/p)p \geq 2$, we may write $v = e^{\imath\zeta}$, with $\zeta \in W^{s+1/p,p}$ (Theorem 5.3). Then, $u = e^{\imath\varphi}$, with $\varphi := \operatorname{tr}\zeta \in W^{s,p}$.

Step 3. If $s > 1 - 1/p$, then $B \neq A$
We go back to Step 5 in the proof of Theorem 11.1. Let φ and $u = e^{\imath\varphi}$ be as there. As proved in Step 5, we have $\varphi \in W^{s,p}$ and thus $u \in A$, but $u \notin B$.

Step 4. If $s \leq 1 - 1/p$, then $A \subset B$
Indeed, if $u = e^{\imath\varphi}$ with $\varphi \in W^{s,p}$, we let $\zeta \in W^{s+1/p,p}$ be an extension of φ, then $v := e^{\imath\zeta} \in W^{s+1/p,p}$ and $\operatorname{tr} v = u$.

Step 5. If $s < 1 - 1/p$, then $A \neq B$
Let $\psi : Q \to \mathbb{R}$ be such that $\psi \in W^{1-1/(sp+1),sp+1}$ but $\psi \notin W^{s,p}$. [Take for example $\psi(x) := 1/|x|^\alpha$ with $\alpha := N/p - s$.] Set $u := e^{\imath\psi}$. Let $\eta \in W^{1,sp+1}(\widetilde{Q};\mathbb{R})$ be an extension of ψ. Then, $u = \operatorname{tr} v$, where $v := e^{\imath\eta} \in W^{1,sp+1}(\widetilde{Q};\mathbb{S}^1)$, and thus $v \in W^{s+1/p,p}$, by Gagliardo–Nirenberg. Consequently, $u \in B$. By Corollary 6.5 and the fact that $\psi \notin W^{s,p}$, we see that $u \notin A$.

Step 6. We have $A \subset C$ and $B \subset C$
The first inclusion is trivial. For the second inclusion, let $u \in B$ and let v be an extension of u. Then there exists a sequence $t_n \to 0$ such that $U_n := v(\cdot, t_n) \in W^{s+1/p,p}(Q;\mathbb{S}^1)$ and $U_n \to u$ in $W^{s,p}$. By Theorem 10.2, we have $U_n \in C$, and thus $u \in C$.

Step 7. We have $A \neq C$
Indeed, let u be as in Remark 8.11. As noted there, we have $u \in C$ but $u \notin A$.

Step 8. We have $B \neq C$

Let φ and $u = e^{\imath\varphi}$ be as in Steps 4 and 5 in the proof of Theorem 11.1. We know that $u \in W^{s,p}$ and $u \notin B$, and moreover $Ju = 0$ (since $\varphi \in W^{1,sp}$), so that $u \in C$. □

11.3.2 Sphere-valued maps

It is natural to formulate a version of $(E_{s,p,N})$ where $\mathbb{S}^1$ is replaced by $\mathbb{S}^k$, $k \geq 1$, namely

$(E_{s,p,N,k})$ Given an arbitrary $u \in W^{s,p}(Q; \mathbb{S}^k)$, is there some $v \in W^{s+1/p,p}(\widetilde{Q}; \mathbb{S}^k)$ such that $\operatorname{tr} v = u$?

Problem $(E_{s,p,N,k})$ has been completely settled when $s = 1 - 1/p$ and $1 < p < \infty$.

Theorem 11.4. *Let $N \geq 1$, $k \geq 1$, $1 < p < \infty$, and $s = 1 - 1/p$.*

1. *If $k \geq N$, the answer to $(E_{s,p,N,k})$ is positive for every $1 < p < \infty$.*
2. *If $k < N$, the answer to $(E_{s,p,N,k})$ is positive if and only if $p < k + 1$ or $p \geq N + 1$.*

When $k = 1$ and $s = 1 - 1/p$, Theorem 11.4 corresponds exactly to Theorem 11.1 (since $sp = p - 1$). For the proof of Theorem 11.4, we refer the reader to Bethuel [28], Hardt, Kinderlehrer, and Lin [180], and Hardt and Lin [183]. [The existence of an extension follows from [180] and [183], whereas the obstruction is established in [28].] The proofs in [28, 180, 183] are quite different from the proofs presented in this chapter, since we cannot use lifting when $k > 1$.

In view of Theorems 11.1 and 11.4, it is natural to raise the following

Open Problem 16. Let $N \geq 1$, $k \geq 2$, $1 \leq p < \infty$, and $s > 0$.

a) If $k \geq N$, is the answer to $(E_{s,p,N,k})$ positive for every $s > 0$ and $1 \leq p < \infty$?
b) If $k < N$, is the answer to $(E_{s,p,N,k})$ positive if and only if $sp < k$ or $sp \geq N$?

11.4 Comments

1. The extension problem was originally tackled by Hardt and Lin [183, Section 6.2] and Hardt, Kinderlehrer, and Lin [180, Appendix], when $s = 1 - 1/p$ and the target $\mathbb{S}^1$ is replaced by $\mathbb{S}^k$ or a more general compact manifold without boundary. The existence of an extension in [183] and [180, Appendix] is obtained via the "projection-averaging" technique used in the proof of Theorem 10.3. Obstructions to the extension were originally described in [183, Section 6.3] (with $s = 1/2$, $p = 2$, $N = 2$, and $k = 1$) and in Bethuel and Demengel [34]. Decisive progress concerning the obstruction was made in Bethuel [28] (see also the review by Mironescu and Van Schaftingen [270]). Existence/non-existence of an extension for the target $\mathbb{S}^1$ was considered in Bourgain, Brezis, and Mironescu [54] ($s = 1/2$,

$p = 2, N = 2$) and in Brezis and Mironescu [84, Appendix] (where Theorem 11.1 is established for the case $sp < 1$). In full generality, Theorems 11.1 and 11.2 are new.

2. The theorems considered in this chapter concern the *local* extension problem. The conclusions hold if we replace Q, for example, by a closed hypersurface Σ and $\widetilde{Q}$ is replaced by a small tubular neighborhood of Σ. The situation becomes quite different if we deal with *global* extensions, because topological obstructions may occur. [Note that, by contrast, for real-valued maps, local extensions yield global extensions via partitions of unity and cutoff functions.] Here are two examples. Let Ω be the unit ball in $\mathbb{R}^{N+1}$ and $\Sigma = \partial\Omega = \mathbb{S}^N$. In this case, the global extension problem consists of extending $u \in W^{s,p}(\partial\Omega; \mathbb{S}^1)$ by some $v \in W^{s+1/p,p}(\Omega; \mathbb{S}^1)$. If $N \geq 2$, the conclusion of Theorem 11.1 is unchanged. Similarly, if $N = 1$ and $sp < 1$.
   However, when $N = 1$ and $sp \geq 1$, Theorem 11.1 needs to be modified as follows: a map $u \in W^{s,p}(\mathbb{S}^1; \mathbb{S}^1)$ has an extension $v \in W^{s+1/p,p}(\Omega; \mathbb{S}^1)$ if and only if $\deg u = 0$. When $sp > 1$, $\deg u$ is the degree of the continuous map u, as in (1.19). For the definition of $\deg u$ when $sp = 1$, we refer the reader to Chapter 12. For more complete results concerning the local trace problem vs the global trace problem in $W^{1-1/p,p}$, see Bethuel [28] and Isobe [200].
3. The case $s = 1/2, p = 2, N = 2$ in Theorem 11.3 is due to Bourgain, Brezis, and Mironescu [54].

Chapter 12
Degree

12.0 Introduction

The topological degree (also called winding number or index) is a classical concept: with every map $f \in C(\mathbb{S}^1; \mathbb{S}^1)$, one associates a number, $\deg f \in \mathbb{Z}$, which counts "how many times $f(\mathbb{S}^1)$ covers $\mathbb{S}^1$, taking into account algebraic multiplicity." We have recalled some of its basic properties in Section 1.1. As explained in Section 12.1, the concept of degree has been extended by Brezis and Nirenberg to maps $f \in VMO(\mathbb{S}^1; \mathbb{S}^1)$. As a consequence of the embedding $W^{1/p,p}(\mathbb{S}^1) \hookrightarrow VMO(\mathbb{S}^1)$, $1 < p < \infty$, we deduce that $\deg f$ is well-defined for $f \in W^{1/p,p}(\mathbb{S}^1; \mathbb{S}^1)$ and it is stable under convergence in $W^{1/p,p}$.

The degree can be expressed as an integral in two different ways (see, for example, Dinca and Mawhin [141, Sections 1.1.2 and 1.1.4]):

1. The Cauchy formula asserts that

$$\deg f = \frac{1}{2\imath\pi}\int_{\mathbb{S}^1}\frac{\dot{f}}{f} = \frac{1}{2\imath\pi}\int_{\mathbb{S}^1}\overline{f}\dot{f} = \frac{1}{2\pi}\int_{\mathbb{S}^1} f\wedge\dot{f},\ \forall f \in C^1(\mathbb{S}^1; \mathbb{S}^1). \tag{12.1}$$

2. The Kronecker formula asserts that

$$\deg f = \frac{1}{\pi}\int_{\mathbb{D}} \operatorname{Jac} u,\ \forall f \in C^1(\mathbb{S}^1; \mathbb{S}^1),\ \forall u \in C^1(\overline{\mathbb{D}}; \mathbb{R}^2) \text{ with } u_{|\partial\mathbb{D}} = f. \tag{12.2}$$

Note that there is a discrepancy: on the one hand, $\deg f$ is well-defined when $f \in VMO(\mathbb{S}^1; \mathbb{S}^1)$, and on the other hand, the Cauchy and Kronecker formulas require some additional regularity on f. We address this question in Section 12.3. We first justify the above formulas when $f \in H^{1/2}(\mathbb{S}^1; \mathbb{S}^1)$, and then give a meaning to (12.2) when f is merely in $VMO(\mathbb{S}^1; \mathbb{S}^1)$.

H. Brezis and P. Mironescu, *Sobolev Maps to the Circle*, Progress in Nonlinear Differential Equations and Their Applications 96,
https://doi.org/10.1007/978-1-0716-1512-6_12

Next, we investigate bounds of $|\deg f|$ in terms of some natural norms. Clearly, it is impossible to bound $|\deg f|$ in terms of $|f|_{BMO}$ (and even $\|f\|_C$). As we are going to see in Theorem 12.3, we have

$$|\deg f| \leq C_p\, |f|^p_{W^{1/p,p}},\ \forall f \in W^{1/p,p}(\mathbb{S}^1;\mathbb{S}^1),\ \forall\, 1 < p < \infty. \tag{12.3}$$

The proof relies heavily on estimates for the lifting of f (see Theorem 9.3).

A striking improvement of (12.3) (Theorem 12.5), due to Nguyen [279], asserts that

$$|\deg f| \leq C \iint\limits_{\substack{\mathbb{S}^1\times\mathbb{S}^1 \\ |f(x)-f(y)|\geq\sqrt{3}}} \frac{1}{|x-y|^2}\, dxdy. \tag{12.4}$$

The proof of (12.4) relies on an appropriate choice of u in the Kronecker formula and on a refinement of techniques developed in Chapter 7 (see Lemma 7.8).

In Section 12.6, we prove that the connected components of $W^{s,p}(\mathbb{S}^1;\mathbb{S}^1)$ with $sp \geq 1$ are classified by their degree.

In Section 12.7, we present a remarkable formula connecting the degree of f and the Fourier coefficients a_n of f:

$$\deg f = \sum_{n\in\mathbb{Z}} n\, |a_n|^2, \tag{12.5}$$

which can be easily derived from the Cauchy formula when $f \in H^{1/2}(\mathbb{S}^1;\mathbb{S}^1)$.

Again, there is a discrepancy between the fact that $\deg f$ is well-defined when $f \in C(\mathbb{S}^1;\mathbb{S}^1)$, or even $f \in VMO(\mathbb{S}^1;\mathbb{S}^1)$, and the fact that the right-hand side of (12.5) has no obvious meaning when $f \notin H^{1/2}$. This leads to delicate questions concerning the validity of (12.5) when $f \in C$ or $f \in W^{1/p,p}$ with $p > 2$.

In Section 12.8.1, we return to (12.3), and investigate the least $W^{1/p,p}$ energy required to "produce" a map of given degree d.

12.1 Degree and *VMO*

In Section 1.1, we recalled the definition of the degree (using lifting) and some straightforward properties:

$$\text{If } f_n, f \in C(\mathbb{S}^1;\mathbb{S}^1) \text{ and } \|f_n - f\|_{L^\infty} \to 0, \text{ then } \deg f_n \to \deg f; \tag{12.6}$$

$$\deg(f\,g) = \deg f + \deg g,\ \forall f, g \in C(\mathbb{S}^1;\mathbb{S}^1); \tag{12.7}$$

$$\deg \overline{f} = -\deg f,\ \forall f \in C(\mathbb{S}^1;\mathbb{S}^1); \tag{12.8}$$

$$[f \in C(\mathbb{S}^1;\mathbb{S}^1) \text{ and } \deg f = 0] \Leftrightarrow [f = e^{\imath\varphi} \text{ for some } \varphi \in C(\mathbb{S}^1;\mathbb{R})]. \tag{12.9}$$

The notion of degree extends to the larger class of maps $f \in VMO(\mathbb{S}^1;\mathbb{S}^1)$; the space $VMO(\Omega;\mathbb{R})$ was defined in Section 5.2.1 when $\Omega \subset \mathbb{R}^N$ is a smooth bounded domain in $\mathbb{R}^N$, but the definition can be adapted easily to maps, for example, from $\mathbb{S}^N$ to $\mathbb{R}^k$ (see Brezis and Nirenberg [104] for a detailed exposition). The main result in this section is the following

Theorem 12.1. *There exists a well-defined function $VMO(\mathbb{S}^1;\mathbb{S}^1) \to \mathbb{Z}$, still called degree and denoted by* deg, *satisfying*

$$\text{If } f \in C(\mathbb{S}^1;\mathbb{S}^1),\ \text{the new degree coincides with the classical degree;} \tag{12.10}$$

$$\text{If } f_n, f \in VMO(\mathbb{S}^1;\mathbb{S}^1) \text{ and } |f_n - f|_{BMO} \to 0 \text{ as } n \to \infty, \\ \text{then } \deg f_n \to \deg f \text{ as } n \to \infty; \tag{12.11}$$

$$\deg(f\,g) = \deg f + \deg g,\ \forall f, g \in VMO(\mathbb{S}^1;\mathbb{S}^1); \tag{12.12}$$

$$\deg \overline{f} = -\deg f,\ \forall f \in VMO(\mathbb{S}^1;\mathbb{S}^1); \tag{12.13}$$

$$[f \in VMO(\mathbb{S}^1;\mathbb{S}^1) \text{ and } \deg f = 0] \Leftrightarrow [f = e^{\imath\varphi} \text{ with } \varphi \in VMO(\mathbb{S}^1;\mathbb{R})]. \tag{12.14}$$

Remark 12.1. We point out that property (12.11) is of interest even in the setting of the classical degree. Indeed, if $f_n, f \in C(\mathbb{S}^1;\mathbb{S}^1)$ satisfy $|f_n - f|_{BMO} \to 0$, then $\deg f_n \to \deg f$. [Note that the assumption $|f_n - f|_{BMO} \to 0$ is much weaker than $\|f_n - f\|_{L^\infty} \to 0$.]

Remark 12.2. As a consequence of the embedding $W^{s,p}(\mathbb{S}^1) \hookrightarrow VMO(\mathbb{S}^1)$ when $sp \geq 1$ (see Lemma 15.20), we are able to define $\deg f$ for every $f \in W^{s,p}(\mathbb{S}^1;\mathbb{S}^1)$ when $sp \geq 1$. Moreover, this degree is stable under *strong* $W^{s,p}$ convergence. However, in the limiting case $W^{1/p,p}(\mathbb{S}^1;\mathbb{S}^1)$, $1 < p < \infty$, we warn the reader that the degree is *not stable under weak* $W^{1/p,p}$ convergence. Consider, for example, the sequence

$$f_n(e^{\imath\theta}) := \begin{cases} e^{\imath n\theta}, & \text{if } 0 \leq \theta \leq \dfrac{2\pi}{n} \\ 1, & \text{if } \dfrac{2\pi}{n} < \theta < 2\pi \end{cases}.$$

One easily checks that $\|f_n\|_{W^{1/p,p}} \leq C_p$, $\forall\, n \geq 1$, $\forall\, 1 < p < \infty$, and that (f_n) converges, weakly in $W^{1/p,p}$, to $f \equiv 1$. However, $\deg f_n \equiv 1$ and $\deg f = 0$.

See also the basic example in Section 12.8.2.

By contrast, there is no "reasonable" notion of degree in $W^{s,p}(\mathbb{S}^1;\mathbb{S}^1)$ when $sp < 1$. More precisely, one can construct a sequence (f_n) in $C^\infty(\mathbb{S}^1;\mathbb{S}^1)$ such that $\|f_n - 1\|_{W^{s,p}} \to 0$ for every $s > 0$ and every $p \geq 1$ with $sp < 1$ and $\deg f_n \equiv 1$ (see again the basic example in Section 12.8.2).

Proof of Theorem 12.1. As in Section 5.2.2, we introduce, for $\varepsilon < 1$,

$$f_\varepsilon(x) := \fint_{I_\varepsilon(x)} f(t)\,dt,\ \forall\, x \in \mathbb{S}^1, \tag{12.15}$$

where $I_\varepsilon(x)$ is the arc on $\mathbb{S}^1$ of length 2ε centered at x. By the proof of Lemma 5.6, if $f \in VMO(\mathbb{S}^1; \mathbb{S}^1)$ then $|f_\varepsilon(x)| \to 1$ *uniformly* on $\mathbb{S}^1$ as $\varepsilon \to 0$. An easy inspection of that proof shows that, if $f_n, f \in VMO(\mathbb{S}^1; \mathbb{S}^1)$ satisfy $|f_n - f|_{BMO} \to 0$ as $n \to \infty$, then $f_{n,\varepsilon}(x) := \fint_{I_\varepsilon(x)} f_n(t)\,dt$ satisfies

$$|f_{n,\varepsilon}(x)| \to 1 \text{ as } \varepsilon \to 0, \text{ uniformly in } x \in \mathbb{S}^1 \text{ and uniformly in } n. \tag{12.16}$$

We may now define $\deg f$ for $f \in VMO(\mathbb{S}^1; \mathbb{S}^1)$. Fix $\varepsilon_0 > 0$ such that $|f_\varepsilon(x)| \geq 1/2$, $\forall x \in \mathbb{S}^1$, $\forall \varepsilon < \varepsilon_0$. Clearly, the map $(x, \varepsilon) \mapsto f_\varepsilon(x)$ is continuous on $\mathbb{S}^1 \times (0, \varepsilon_0)$; and similarly for the map $(x, \varepsilon) \mapsto \widehat{f_\varepsilon}(x) := f_\varepsilon(x)/|f_\varepsilon(x)|$. Therefore, $\deg \widehat{f_\varepsilon}$ is well-defined (since $\widehat{f_\varepsilon} \in C(\mathbb{S}^1; \mathbb{S}^1)$), and $\deg \widehat{f_\varepsilon}$ is *independent* of $\varepsilon \in (0, \varepsilon_0)$ (since the classical degree is continuous under uniform convergence). We set

$$\deg f := \deg \widehat{f_\varepsilon},\ 0 < \varepsilon < \varepsilon_0. \tag{12.17}$$

Note that, if $f \in C(\mathbb{S}^1; \mathbb{S}^1)$, then $\widehat{f_\varepsilon} \to f$ uniformly as $\varepsilon \to 0$, and thus $\deg \widehat{f_\varepsilon}$ converges to the classical degree as $\varepsilon \to 0$. From (12.17), we deduce (12.10).

We now turn to the proof of (12.11). We will use the following ingredient similar to Step 2 in the proof of Theorem 8.1.

Lemma 12.1. *Assume that* $f_n, f \in VMO(\mathbb{S}^1; \mathbb{S}^1)$ *satisfy*

$$|f_n - f|_{BMO} \to 0 \text{ as } n \to \infty. \tag{12.18}$$

Then,

$$\text{either } \|f_n - f\|_{BMO} \to 0 \text{ as } n \to \infty, \tag{12.19}$$

$$\begin{aligned} &\text{or there exist two constants } C_1, C_2 \in \mathbb{S}^1 \text{ and a subsequence } n_k \to \infty \\ &\text{such that } f \equiv C_1 \text{ and } \|f_{n_k} - C_2\|_{BMO} \to 0 \text{ as } k \to \infty. \end{aligned} \tag{12.20}$$

Here, $\|\ \|_{BMO}$ denotes the full norm, i.e., $\|h\|_{BMO} = |h|_{BMO} + \|h\|_{L^1}$. Recall (see Lemma 15.21 in Appendix 15.5) that

$$\|h\|_{L^1} \leq C\left(|h|_{BMO} + \left|\int_{\mathbb{S}^1} h\right|\right),\ \forall h \in BMO. \tag{12.21}$$

Proof of Lemma 12.1. Assume that $\|f_n - f\|_{BMO}$ does not tend to zero. Then, for some subsequence (f_{n_k}), we have

$$\|f_{n_k} - f\|_{BMO} \geq \alpha > 0,\ \forall k, \tag{12.22}$$

and $\fint_{\mathbb{S}^1} f_{n_k} \to \ell$.

From (12.21), we deduce that (f_{n_k}) is Cauchy in L^1, and thus $f_{n_k} \to \widetilde{f}$ in $L^1(\mathbb{S}^1)$. In particular, $|\widetilde{f}| = 1$ a.e. On the other hand, passing to the limit in (12.18) using Fatou, we see that $|\widetilde{f} - f|_{BMO} = 0$. Thus, $\widetilde{f} - f = C$. Moreover, $C \neq 0$ by (12.22). Hence, f takes its values in $\mathbb{S}^1 \cap (\mathbb{S}^1 - C)$, which is reduced to at most two points. We deduce from Proposition 6.1 that $f \equiv C_1$ and $\widetilde{f} \equiv C_2$, i.e., (12.20) holds. □

Proof of Theorem 12.1 *continued.* In the case that (12.19) holds, we see that, for every fixed $\varepsilon < \varepsilon_0, \widehat{f}_{n,\varepsilon}(x) \to \widehat{f}_\varepsilon(x)$ as $n \to \infty$ uniformly in x. From the properties of the classical degree, we deduce that $\deg \widehat{f}_{n,\varepsilon} \to \deg \widehat{f}_\varepsilon$ as $n \to \infty$, so that $\deg f_n \to \deg f$ (by (12.17)).

In the case that (12.20) holds, we see that $\deg f = 0$ and (as above) $\deg f_{n_k} \to 0$. By uniqueness of the limit, we conclude in all cases that (12.11) holds.

We now turn to (12.12). By Corollaries 15.5 and 15.4 in Appendix 15.5, we have $\widehat{f}_\varepsilon \widehat{g}_\varepsilon \to f\, g$ in *BMO*. From the properties of the classical degree, for small ε we have

$$\deg \left(\widehat{f}_\varepsilon \widehat{g}_\varepsilon\right) = \deg \widehat{f}_\varepsilon + \deg \widehat{g}_\varepsilon = \deg f + \deg g.$$

Passing to the limits using (12.11), we deduce (12.12).

Property (12.13) is obvious; we turn to (12.14). For this purpose, we will present an alternative approach to the *VMO*-degree via lifting, modeled on what we did for continuous functions in Chapter 1.

Assume that $f \in VMO(\mathbb{S}^1; \mathbb{S}^1)$. Consider the function $\mathbb{R} \ni t \mapsto h(t) := f(e^{\imath t})$. It is easy to check that $h \in VMO((a, b); \mathbb{S}^1)$ for every bounded interval (a, b). By Theorem 5.7, we may write $h(t) = e^{\imath \psi(t)}$, $t \in \mathbb{R}$, for some function ψ such that $\psi \in VMO((a, b); \mathbb{R})$ for every bounded interval (a, b). Since $h(t + 2\pi) = h(t)$ for a.e. $t \in \mathbb{R}$, we deduce that $(\psi(t + 2\pi) - \psi(t))/(2\pi) \in \mathbb{Z}$ for a.e. $t \in \mathbb{R}$. From Proposition 6.1, we infer that there exists some $k \in \mathbb{Z}$ such that

$$\psi(t + 2\pi) - \psi(t) = 2\pi\, k \text{ for a.e. } t \in \mathbb{R}. \tag{12.23}$$

We claim that

$$k = \deg f \text{ (defined in (12.17)).} \tag{12.24}$$

Indeed, set

$$\psi_\varepsilon(t) := \frac{1}{2\varepsilon} \int_{t-\varepsilon}^{t+\varepsilon} \psi(s)\, ds$$

and $h_\varepsilon := e^{\imath \psi_\varepsilon}$. From (12.23), we deduce that $\psi_\varepsilon(t + 2\pi) - \psi_\varepsilon(t) = 2\pi\, k, \forall\, t$, so that $h_\varepsilon(t + 2\pi) = h_\varepsilon(t), \forall\, t$. Hence, we may write $h_\varepsilon(t) = \widetilde{f}_\varepsilon(e^{\imath t})$ for some $\widetilde{f}_\varepsilon \in C(\mathbb{S}^1; \mathbb{S}^1)$ such that $\deg \widetilde{f}_\varepsilon = k$.

As $\varepsilon \to 0$, we have $\psi_\varepsilon \to \psi$ in *BMO* (item 6 in Section 5.2.1) and $h_\varepsilon \to h$ (resp. $\widetilde{f}_\varepsilon \to f$) in *BMO* (Theorem 15.9). From (12.11), we deduce that $\deg \widetilde{f}_\varepsilon \to \deg f$ and thus, (12.24) holds.

Formula (12.24) implies "$\Longleftarrow$" in (12.14). Conversely, if $\deg f = 0$, then the above ψ is 2π-periodic, and thus there exists some $\varphi \in VMO(\mathbb{S}^1; \mathbb{R})$ such that $\varphi(e^{\imath t}) = \psi(t), \forall\, t \in \mathbb{R}$. We thus have $f = e^{\imath \varphi}$, with $\varphi \in VMO$. □

Corollary 12.1. *Let $k \in \mathbb{Z}$. Then,*

$$\{f \in VMO(\mathbb{S}^1; \mathbb{S}^1);\ \deg f = k\} = \{z^k\, e^{\imath \varphi(z)};\ \varphi \in VMO(\mathbb{S}^1; \mathbb{R})\}.$$

Proof. It suffices to note that $g(z) := z^{-k} f(z)$ satisfies $\deg g = 0$ (by (12.12)) and to apply (12.14) to g. □

Remark 12.3. Warning. The continuous dependence of $\deg f$ in the *VMO* setting is somewhat delicate. Let us compare properties (12.6) and (12.11). It is well-known that if $f, g \in C(\mathbb{S}^1; \mathbb{S}^1)$ satisfy $\|f - g\|_{L^\infty} < 2$, then $\deg f = \deg g$; this is obtained by considering the homotopy

$$h_t = \frac{(1-t)f + tg}{|(1-t)f + tg|}, \quad t \in [0, 1].$$

By contrast, a similar property *fails* for the *BMO* norm. All we can say (by (12.11)) is that, for every $f \in VMO(\mathbb{S}^1; \mathbb{S}^1)$, there exists $\delta > 0$ *depending* on f such that, if $g \in VMO(\mathbb{S}^1; \mathbb{S}^1)$ and $|g - f|_{BMO} < \delta$, then $\deg g = \deg f$.

Moreover, one can construct sequences (f_n) and (g_n) in $C^\infty(\mathbb{S}^1; \mathbb{S}^1)$ such that

$$\|g_n - f_n\|_{BMO} \to 0 \text{ as } n \to \infty, \ \deg f_n = 1, \ \forall n, \text{ and } \deg g_n = 0, \ \forall n \tag{12.25}$$

(see Brezis and Nirenberg [104, Lemma 6]).

In fact, a stronger claim holds: one can even find (f_n) and (g_n) as above such that $\|f_n - g_n\|_{W^{1/p,p}} \to 0$ for every $1 < p < \infty$, as $n \to \infty$ (see Proposition 12.13 and Brezis, Mironescu, and Shafrir [93, Theorem 2]).

However, we mention that the degree is uniformly continuous on bounded sets of $W^{1/p,p}(\mathbb{S}^1; \mathbb{S}^1)$ (see Mironescu and Shafrir [267]). More specifically, let $1 < p < \infty$ and $M > 0$. Then there exists some $\delta = \delta(p, M) > 0$ such that

$$[u, v \in W^{1/p,p}(\mathbb{S}^1; \mathbb{S}^1), \ |u|_{W^{1/p,p}} \le M, \ |u - v|_{W^{1/p,p}} \le \delta] \implies \deg u = \deg v.$$

12.2 Degree, lifting, and traces in $W^{s,p}(\mathbb{S}^1; \mathbb{S}^1)$

In this section, we revisit some questions concerning lifting and trace in $W^{s,p}(\mathbb{S}^1; \mathbb{S}^1)$. As we are going to see, the degree plays a role, since the domain of the maps is now $\mathbb{S}^1$ instead of $(0, 1)$.

We start with lifting.

Proposition 12.1. *Let $s > 0$ and $1 \le p < \infty$. Then,*

1. *If $sp < 1$, every $f \in W^{s,p}(\mathbb{S}^1; \mathbb{S}^1)$ admits a lifting in $W^{s,p}$, i.e., $f = e^{\imath\varphi}$ with $\varphi \in W^{s,p}(\mathbb{S}^1; \mathbb{R})$.*
2. *If $sp \ge 1$, f admits a lifting in $W^{s,p}$ if and only if $\deg f = 0$. In addition, for every $k \in \mathbb{Z}$,*

$$\{f \in W^{s,p}(\mathbb{S}^1; \mathbb{S}^1); \ \deg f = k\} = \{z^k e^{\imath\varphi(z)}; \ \varphi \in W^{s,p}(\mathbb{S}^1; \mathbb{R})\}. \tag{12.26}$$

Proof. Step 1. Proof of item 1

Let $\mathbb{S}^1_+$ and $\mathbb{S}^1_-$ denote the upper and lower half unit circle. On $\mathbb{S}^1_\pm$, we may write

$f = e^{\imath\varphi_\pm}$, with $\varphi_\pm \in W^{s,p}(\mathbb{S}^1_\pm)$ (Theorem 5.1). We next glue together φ_+ and φ_-, by setting $\varphi := \begin{cases} \varphi_+, & \text{on } \mathbb{S}^1_+ \\ \varphi_-, & \text{on } \mathbb{S}^1_- \end{cases}$. Then, $\varphi \in W^{s,p}(\mathbb{S}^1)$ (Lemma 15.10).

Step 2. Proof of item 2
Assume first that $f \in W^{s,p}(\mathbb{S}^1;\mathbb{S}^1)$ admits a lifting $\varphi \in W^{s,p}$. Then, $\varphi \in VMO$, and thus $\deg f = 0$ (Corollary 12.1). The reverse is a consequence of the argument below.

We now turn to (12.26). The inclusion "$\supset$" follows from Corollary 12.1. For the opposite inclusion, if $f \in W^{s,p}(\mathbb{S}^1;\mathbb{S}^1)$ and $\deg f = k$, then (by Corollary 12.1) we may write $f(z) = z^k e^{\imath\varphi(z)}$, with $\varphi \in VMO$. By Theorem 5.6 and Proposition 6.1, we have $\varphi \in W^{s,p}$. □

Next, we address the question of traces. In Chapter 11, we discussed only the *local* aspect of trace theory for $\mathbb{S}^1$-valued maps (see Comment 2 in Section 11.4). We investigate here the *global* aspect in a very elementary setting; we will return to this topic in Section 12.8.5 for more complex geometries. Let $0 < s < \infty$ be a non-integer, and $1 \le p < \infty$. Given $f \in W^{s,p}(\mathbb{S}^1;\mathbb{S}^1)$, we ask whether $f = \operatorname{tr} u$ for some $u \in W^{s+1/p,p}(\mathbb{D};\mathbb{S}^1)$, where $\mathbb{D}$ is the unit disk of $\mathbb{R}^2$.

Proposition 12.2. *We have*

1. *If $sp < 1$, every $f \in W^{s,p}(\mathbb{S}^1;\mathbb{S}^1)$ admits an extension in $W^{s+1/p,p}$, i.e., $f = \operatorname{tr} u$ for some $u \in W^{s+1/p,p}(\mathbb{D};\mathbb{S}^1)$.*
2. *If $sp \ge 1$, a map $f \in W^{s,p}(\mathbb{S}^1;\mathbb{S}^1)$ can be written as $f = \operatorname{tr} u$ for some $u \in W^{s+1/p,p}(\mathbb{D};\mathbb{S}^1)$ if and only if $\deg f = 0$.*

Step 1. Proof of item 1
Apply Proposition 12.1, item *1*, and follow Steps *1* and 2 from the proof of Theorem 11.1.

Step 2. Proof of item 2
"$\Longrightarrow$" Since $(s+1/p)p \ge 2$, we may approximate u in $W^{s+1/p,p}$ by a sequence $(u_n) \subset C^\infty(\overline{\mathbb{D}};\mathbb{S}^1)$ (Theorem 10.2). Then, $f_n := \operatorname{tr} u_n \to \operatorname{tr} u = f$ in $W^{s,p}(\mathbb{S}^1;\mathbb{S}^1)$ and therefore $0 = \deg f_n \to \deg f$, by Remark 12.2.
"$\Longleftarrow$" Applying Proposition 12.1, item 2, we may write $f = e^{\imath\varphi}$ for some $\varphi \in W^{s,p}(\mathbb{S}^1;\mathbb{R})$. Let $\psi \in W^{s+1/p,p}(\mathbb{D};\mathbb{R})$ be an extension of φ. Then, $u = e^{\imath\psi}$ has the desired properties (by Corollary 15.3) since $(s+1/p)p \ge 2 = N$. □

12.3 Integral representations for the degree

Let us recall that, when $f \in W^{1,1}(\mathbb{S}^1;\mathbb{S}^1)$, we have the Cauchy formula

$$\deg f = \frac{1}{2\pi}\int_{\mathbb{S}^1} f \wedge \dot{f} = \frac{1}{2\imath\pi}\int_{\mathbb{S}^1} \overline{f}\dot{f}; \tag{12.27}$$

see (1.23).

In this section, we investigate several extensions and variants of (12.27).

There is a striking "discrepancy" between the fact that $\deg f$ is well-defined for every $f \in C(\mathbb{S}^1; \mathbb{S}^1)$ – even $f \in VMO\,(\mathbb{S}^1; \mathbb{S}^1)$ – and the Cauchy formula, which requires more regularity, i.e., $f \in W^{1,1}(\mathbb{S}^1; \mathbb{S}^1)$.

Our goal in this section is twofold:

a) We first give a meaning to the Cauchy formula for $f \in H^{1/2}(\mathbb{S}^1; \mathbb{S}^1)$ (recall however that $H^{1/2}$ is not larger than $W^{1,1}$; see Lemma 15.4). This requires a little detour via the theory of distributions.
b) We next present an appropriate version of the Kronecker formula (12.2), valid for *special* extensions u of f. As we are going to see, this approach provides an integral representation of the degree when f is merely *VMO*.

We start with the Cauchy formula for $f \in H^{1/2}(\mathbb{S}^1; \mathbb{S}^1)$. Note that $\dot{f} \in H^{-1/2}(\mathbb{S}^1; \mathbb{R}^2)$, which is the dual space of $H^{1/2}(\mathbb{S}^1; \mathbb{R}^2)$. Therefore, the right-hand side of (12.27) can be interpreted as the duality bracket $\langle \cdot, \cdot \rangle$ between $[H^{1/2}]^*$ and $H^{1/2}$. We then have the following variant of (12.27).

Proposition 12.3. *For every $f = (f_1, f_2) \in H^{1/2}(\mathbb{S}^1; \mathbb{S}^1)$, we have*

$$\deg f = \frac{1}{2\pi}[\langle \dot{f}_2, f_1 \rangle - \langle \dot{f}_1, f_2 \rangle]. \tag{12.28}$$

For simplicity, we will write (12.28) in the form (12.27), even when $f \in H^{1/2}$.

Remark 12.4. As we are going to see in Section 12.7, one can use Fourier series instead of the theory of distributions to give a meaning to the Cauchy formula when $f \in H^{1/2}(\mathbb{S}^1; \mathbb{S}^1)$.

Remark 12.5. From the trivial embedding $W^{1/p,p} \cap L^\infty \hookrightarrow H^{1/2}$ when $1 < p < 2$, we deduce that (12.28) also holds for $f \in W^{1/p,p}(\mathbb{S}^1; \mathbb{S}^1)$ with $1 < p < 2$.

However, this approach does not make sense when $f \in W^{1/p,p}(\mathbb{S}^1; \mathbb{S}^1)$ with $2 < p < \infty$.

Proof of Proposition 12.3. Formula (12.28) holds when f is smooth, since in that case it is the same as (12.27).

Combining the continuity of both sides of (12.28) with respect to strong $H^{1/2}$ convergence and the density of $C^\infty(\mathbb{S}^1; \mathbb{S}^1)$ in $H^{1/2}(\mathbb{S}^1; \mathbb{S}^1)$ (Corollary 15.8 in Appendix 15.5), we obtain that (12.28) holds when f is merely $H^{1/2}$. □

We next recall another classical formula for the degree. If $f \in C^\infty(\mathbb{S}^1; \mathbb{S}^1)$ and if $U \in C^\infty(\overline{\mathbb{D}}; \mathbb{R}^2)$ is an extension of f, then we have the *Kronecker formula*

$$\deg f = \frac{1}{\pi} \int_{\mathbb{D}} \operatorname{Jac} U. \tag{12.29}$$

[Warning: the extension U is not supposed to be $\mathbb{S}^1$-valued.] Formula (12.29) follows from (12.27) thanks to the identity

$$\int_{\mathbb{D}} \operatorname{Jac} U = \frac{1}{2} \int_{\mathbb{S}^1} U \wedge \dot{U}, \ \forall\, U \in C^\infty(\overline{\mathbb{D}}; \mathbb{R}^2), \tag{12.30}$$

that we now prove. Let, more generally, $\Omega \subset \mathbb{R}^2$ be smooth bounded, and let $U \in C^\infty(\overline{\Omega}; \mathbb{R}^2)$. Then,

$$\begin{aligned}
\int_\Omega \operatorname{Jac} U &= \int_\Omega \frac{1}{2} \left[(U \wedge U_y)_x + (U_x \wedge U)_y \right] \\
&= \frac{1}{2} \int_{\partial\Omega} [(U \wedge U_y)\, n_x + (U_x \wedge U)\, n_y] \\
&= \frac{1}{2} \int_{\partial\Omega} U \wedge (-U_x\, n_y + U_y\, n_x) \\
&= \frac{1}{2} \int_{\partial\Omega} U \wedge (U_x\, \tau_x + U_y\, \tau_y) = \frac{1}{2} \int_{\partial\Omega} U \wedge U_\tau .
\end{aligned} \tag{12.31}$$

Here, $n = (n_x, n_y)$ denotes the outward normal unit vector, τ denotes the tangent unit vector oriented in such a way that (n, τ) is direct, and we have used the fact that $n_x = \tau_y$ and $n_y = -\tau_x$. When $\Omega = \mathbb{D}$, (12.31) becomes (12.30).

In the same spirit, we note the identity

$$\int_{\mathbb{D}} \operatorname{Jac} U = -\int_{\mathbb{S}^1} \dot{U}_1\, U_2, \ \forall\, U \in C^\infty(\overline{\mathbb{D}}; \mathbb{R}^2), \tag{12.32}$$

obtained from the following calculation in the special case where $\Omega = \mathbb{D}$:

$$\begin{aligned}
\int_\Omega \operatorname{Jac} U &= \int_\Omega \left[((U_1)_x\, U_2)_y - ((U_1)_y\, U_2)_x \right] \\
&= \int_{\partial\Omega} [(U_1)_x\, U_2\, n_y - (U_1)_y\, U_2\, n_x] \\
&= -\int_{\partial\Omega} U_2\, [(U_1)_x\, \tau_x + (U_1)_y\, \tau_y] = -\int_{\partial\Omega} (U_1)_\tau\, U_2 .
\end{aligned} \tag{12.33}$$

For further use, let us note that, in (12.29) and (12.30), the regularity assumption on U can be weakened to $U \in \operatorname{Lip}(\mathbb{D}; \mathbb{R}^2)$.

As in the case of the Cauchy formula (12.27), it is possible to extend the Kronecker formula (12.29) to $f \in H^{1/2}(\mathbb{S}^1; \mathbb{S}^1)$.

Proposition 12.4. *Let $f \in H^{1/2}(\mathbb{S}^1; \mathbb{S}^1)$ and let $U : \mathbb{D} \to \mathbb{R}^2$ be any H^1 map such that* $\operatorname{tr} U = f$. *Then,* (12.29) *holds.*

Proof. Step 1. The right-hand side of (12.29) depends only on f

Let $U = (U_1, U_2)$, $V = (V_1, V_2) \in H^1(\mathbb{D}; \mathbb{R}^2)$ be such that $\operatorname{tr} U = \operatorname{tr} V$. Let $W = (W_1, W_2) := V - U \in H^1_0(\mathbb{D}; \mathbb{R}^2)$. Then, $\operatorname{Jac} V - \operatorname{Jac} U = \operatorname{Jac}(W_1, U_2) - \operatorname{Jac}(W_2, V_1)$, and thus it suffices to prove that

$$\int_{\mathbb{D}} \operatorname{Jac}(\xi, \eta) = 0, \ \forall \xi \in H_0^1(\mathbb{D}; \mathbb{R}), \ \forall \eta \in H^1(\mathbb{D}; \mathbb{R}). \tag{12.34}$$

In turn, it suffices to prove (12.34) when $\xi \in C_c^\infty(\mathbb{D}; \mathbb{R})$ and $\eta \in C^\infty(\overline{\mathbb{D}}; \mathbb{R})$, and in that case we rely on (12.32).

Step 2. Conclusion
Identity (12.29) holds when f and U are smooth. In particular, it holds when f is smooth and U is the harmonic extension $\widetilde{f}$ of f:

$$\deg f = \frac{1}{\pi} \int_{\mathbb{D}} \operatorname{Jac} \widetilde{f}, \ \forall f \in C^\infty(\mathbb{S}^1; \mathbb{S}^1). \tag{12.35}$$

By the density of $C^\infty(\mathbb{S}^1; \mathbb{S}^1)$ in $H^{1/2}(\mathbb{S}^1; \mathbb{S}^1)$ (Corollary 15.8 in Appendix 15.5), the continuity of the degree with respect to strong $H^{1/2}$ convergence, and the continuity of the map $H^{1/2}(\mathbb{S}^1) \ni f \mapsto \widetilde{f} \in H^1(\mathbb{D})$, we find that (12.35) holds for any $f \in H^{1/2}(\mathbb{S}^1; \mathbb{S}^1)$. We complete the proof of the proposition using Step 1. □

When $f \in W^{1/p,p}$ with $p > 2$, or more generally when $f \in VMO$, Proposition 12.4 cannot be applied since there need not exist $U \in H^1$ such that $\operatorname{tr} U = f$. In this case, it is still possible to give a meaning to (12.29), but this requires a new idea.

Theorem 12.2. *Given $f \in VMO(\mathbb{S}^1; \mathbb{S}^1)$, there exists some $U : \mathbb{D} \to \mathbb{R}^2$ such that*

$$U \in \operatorname{Lip}_{loc}(\mathbb{D}; \mathbb{R}^2); \tag{12.36}$$

$$|U(x)| = 1 \text{ near } \mathbb{S}^1 \\ (\text{and thus } \operatorname{Jac} U = 0 \text{ near } \mathbb{S}^1, \text{ so that } \operatorname{Jac} U \in L^\infty(\mathbb{D})); \tag{12.37}$$

$$U(rx) \to f(x) \text{ in } BMO(\mathbb{S}^1) \text{ as } r \nearrow 1. \tag{12.38}$$

Moreover, for any such U we have

$$\deg f = \frac{1}{\pi} \int_{\mathbb{D}} \operatorname{Jac} U. \tag{12.39}$$

Proof. Step 1. Construction of U satisfying (12.36)–(12.38)
For $x \in \mathbb{S}^1$ and $0 < \rho < 1/2$, let $J_\rho(x)$ denote the arc of $\mathbb{S}^1$ of length $4\pi\rho$ centered at x. We also let $J_\rho(x) = \mathbb{S}^1$ when $1/2 \le \rho \le 1$.

Let $0 < \alpha < 1$, and consider some

$$\Pi \in C^\infty(\mathbb{R}^2; \mathbb{R}^2) \text{ such that } \Pi(z) = \frac{z}{|z|} \text{ when } |z| \ge \alpha; \tag{12.40}$$

take, for example, $\Pi(z) := \zeta(|z|) \dfrac{z}{|z|}$, where $\zeta \in C^\infty([0, \infty); [0, 1])$ satisfies $\zeta(t) = 0$ for t near 0, and $\zeta(t) = 1$ for $t \ge \alpha$.

Set

$$w(rx) := \fint_{J_{1-r}(x)} f, \ \forall x \in \mathbb{S}^1, \ \forall r \in [0, 1), \text{ and } U := \Pi \circ w. \tag{12.41}$$

Clearly, U satisfies (12.36). Property (12.37) is a consequence of the existence of some $r_0 \in (0, 1)$ such that

$$|w(rx)| > \alpha, \ \forall\, r \in (r_0, 1), \ \forall\, x \in \mathbb{S}^1, \tag{12.42}$$

which in turn follows from the proof of Lemma 5.6.

We next note that, by a variant of item 6 in Section 5.2.1 (see Brezis and Nirenberg [104, Corollary 1]), we have

$$w(r\cdot) \to f \text{ in } BMO\,(\mathbb{S}^1) \text{ as } r \nearrow 1. \tag{12.43}$$

Combining (12.42), (12.43), and Theorem 15.9, we obtain (12.38).

Step 2. Proof of (12.39)

We apply the Kronecker formula in the disk $\mathbb{D}_r$, with $0 < r < 1$ sufficiently close to 1, and obtain that

$$\deg U(r\cdot) = \frac{1}{\pi} \int_{\mathbb{D}_r} \operatorname{Jac} U. \tag{12.44}$$

We next let $r \nearrow 1$ in (12.44) and use:

1. The continuity of the degree with respect to BMO convergence and (12.38).
2. The fact that the right-hand side of (12.44) does not depend on r sufficiently close to 1 (by (12.37)). □

Remark 12.6. With more work, it is possible to construct some $U \in C^\infty(\mathbb{D}; \mathbb{R}^2)$ satisfying (12.37)–(12.39). For example, one may let w be the harmonic extension of f, which satisfies the crucial property (12.42) (see Brezis and Nirenberg [105, Theorem A3.2]), and set as above $U := \Pi \circ w$.

Remark 12.7. The proof of Theorem 12.2 sheds some light on a "mysterious discrepancy." On the one hand, we know (see Theorem 12.1) that the degree is stable under uniform (and even VMO) convergence. On the other hand, the Cauchy formula ((12.27) and (12.28)) provides stability under a stronger mode of convergence ($W^{1,1}$ or $H^{1/2}$). Similarly, the Kronecker formula (12.29) seems to require convergence of U in $H^1(\mathbb{D})$, and thus convergence of f in $H^{1/2}(\mathbb{S}^1)$.

The heart of the matter is the following. Assume that

$$f_n, f \in VMO\,(\mathbb{S}^1; \mathbb{S}^1) \text{ and } |f_n - f|_{BMO} \to 0 \text{ as } n \to \infty. \tag{12.45}$$

Then there exist $U_n, U \in \operatorname{Lip}_{loc}(\mathbb{D}; \mathbb{R}^2)$ such that

$$\operatorname{Jac} U_n, \ \operatorname{Jac} U \in L^\infty(\mathbb{D}), \tag{12.46}$$

$$\operatorname{Jac} U_n \to \operatorname{Jac} U \text{ in } L^1(\mathbb{D}), \tag{12.47}$$

and

$$\deg f_n = \frac{1}{\pi}\int_{\mathbb{D}} \operatorname{Jac} U_n \to \frac{1}{\pi}\int_{\mathbb{D}} \operatorname{Jac} U = \deg f. \tag{12.48}$$

Indeed, we know, by (12.16), that there exists $r_0 \in (0, 1)$ such that (12.42) holds uniformly in n. Therefore,

$$\operatorname{Jac} U_n = \operatorname{Jac} U = 0 \text{ on } \mathbb{D}\setminus\mathbb{D}_{r_0},\ \forall n,$$

and

$$\nabla U_n \to \nabla U \text{ in } L^p(\mathbb{D}_r),\ \forall p<\infty,\ \forall r<1,\ \text{as } n\to\infty.$$

12.4 A "distributional degree"

We next define a distribution closely connected to the degree: the "distributional degree." Given $f \in VMO\,(\mathbb{S}^1;\mathbb{S}^1)$, fix any $g_0 \in C^\infty(\mathbb{S}^1;\mathbb{S}^1)$ such that $\deg g_0 = \deg f$. Write $f = e^{\imath\varphi} g_0$ for some $\varphi \in VMO\,(\mathbb{S}^1;\mathbb{R})$ and set

$$j(f) := \dot\varphi + g_0 \wedge \dot g_0 \in \mathscr{D}'(\mathbb{S}^1;\mathbb{R}). \tag{12.49}$$

When f is smooth, so is φ, and then the identity $f = e^{\imath\varphi} g_0$ leads to

$$j(f) = f\wedge\dot f,\ \forall f\in C^\infty(\mathbb{S}^1;\mathbb{S}^1). \tag{12.50}$$

Clearly, $j(f)$ is independent of the choice of g_0 and (using (12.27))

$$\langle j(f), 1\rangle = \langle j(g_0), 1\rangle = 2\pi\ \deg g_0 = 2\pi\ \deg f. \tag{12.51}$$

Other straightforward properties are

$$j(f\,g) = j(f) + j(g),\ \forall f, g \in VMO\,(\mathbb{S}^1;\mathbb{S}^1), \tag{12.52}$$

$$j(f\,\overline{g}) = j(f) - j(g),\ \forall f, g \in VMO\,(\mathbb{S}^1;\mathbb{S}^1) \tag{12.53}$$

and

$$j(\alpha f) = j(f),\ \forall f\in VMO\,(\mathbb{S}^1;\mathbb{S}^1),\ \forall \alpha\in\mathbb{S}^1. \tag{12.54}$$

A basic property of the distribution $j(f)$ is its continuity with respect to f.

Proposition 12.5. *Let $f_n, f \in VMO\,(\mathbb{S}^1;\mathbb{S}^1)$ be such that $|f_n - f|_{BMO} \to 0$. Then, $j(f_n)\to j(f)$ in the sense of distributions.*

Proof. By Lemma 12.1, it suffices to establish the conclusion in
Case 1. $f_n \to f$ in BMO.
Case 2. $f \equiv C_1$ and $f_n \to C_2$ in BMO.

We consider only Case 1. Case 2 is similar and is left to the reader.

Step 1. Proof in Case 1 when $f \equiv 1$
For large n, we have $\deg f_n = 0$. For such n, we let $g_0 \equiv 1$. By the proof of Theorem 5.8, the *VMO* phase φ_n of f_n satisfies $|\varphi_n|_{BMO} \to 0$ as $n \to \infty$. Using Poincaré's inequality (5.26), we may assume that $\varphi_n \to \varphi$ in *BMO*, and then φ is a *VMO* phase of f (thus a constant multiple of 2π), and $j(f_n) = \dot{\varphi}_n \to \dot{\varphi} = j(f)$ in the sense of distributions.

Step 2. Proof in Case 1 for an arbitrary f
By (12.52), we have $j(f_n) = j(f) + j(f_n\overline{f})$. It then suffices to note that $f_n\overline{f} \to 1$ in *BMO* (by Corollary 15.4) and to use Step 1. □

When $f \in H^{1/2}(\mathbb{S}^1;\mathbb{S}^1)$, there are two convenient formulas for $j(f)$, inspired respectively by (12.28) and (12.29).

Proposition 12.6. *Let $f \in H^{1/2}(\mathbb{S}^1;\mathbb{S}^1)$. Then,*

$$\langle j(f), \zeta\rangle = [\langle \dot{f}_2, \zeta f_1\rangle - \langle \dot{f}_1, \zeta f_2\rangle], \ \forall \zeta \in C^\infty(\mathbb{S}^1;\mathbb{R}). \tag{12.55}$$

Proof. When f is smooth, the conclusion follows from (12.50). The general case follows by density (via Corollary 15.8), using the continuity of the quantities in (12.55) with respect to strong $H^{1/2}$ convergence. [For the right-hand side, this is clear. For the left-hand side, we rely on Proposition 12.5 and on the embedding $H^{1/2} \hookrightarrow VMO$.] □

Proposition 12.7. *Let $f \in H^{1/2}(\mathbb{S}^1;\mathbb{S}^1)$ and $\zeta \in C^\infty(\mathbb{S}^1;\mathbb{R})$. Let $U \in H^1(\mathbb{D};\mathbb{R}^2)$ (resp. $Z \in C^\infty(\overline{\mathbb{D}};\mathbb{R})$) be an extension of f (resp. ζ). Then,*

$$\langle j(f), \zeta\rangle = 2\int_{\mathbb{D}} (\operatorname{Jac} U)\, Z - \int_{\mathbb{D}} ((U \wedge U_x)\, Z_y - (U \wedge U_y)\, Z_x). \tag{12.56}$$

Proof. We note that (12.32) leads to the identity

$$\int_{\mathbb{D}} \operatorname{Jac} V = -\langle \dot{g}_1, g_2\rangle, \ \forall\, V \in H^1(\mathbb{D};\mathbb{R}^2) \text{ with } \operatorname{tr} V = (g_1, g_2). \tag{12.57}$$

If we sum (12.57) applied to $V := (Z\, U_1, U_2)$ and $g := (\zeta f_1, f_2)$ to (12.57) applied to $V := (U_1, Z\, U_2)$ and $g := (f_1, \zeta f_2)$, and use (12.55), we find that

$$2\int_{\mathbb{D}} (\operatorname{Jac} U)\, Z + \int_{\mathbb{D}} (U_1 \operatorname{Jac}(Z, U_2) - U_2 \operatorname{Jac}(Z, U_1)) = \langle j(f), \zeta\rangle. \tag{12.58}$$

It then suffices to note the identity

$$U_1 \operatorname{Jac}(Z, U_2) - U_2 \operatorname{Jac}(Z, U_1) = (U \wedge U_y)\, Z_x - (U \wedge U_x)\, Z_y.$$ □

12.5 Estimates for the degree

Recall that $\deg f$ is well-defined for every $f \in VMO(\mathbb{S}^1; \mathbb{S}^1)$. However, it is impossible to bound $|\deg f|$ in terms of $|f|_{BMO}$, since $|f|_{BMO} \le 2$, whereas $|\deg f|$ can be arbitrarily large. On the other hand, $|\deg f| \le (1/2\pi)\,|f|_{W^{1,1}}$, by the Cauchy formula (12.27). Since $\deg f$ is well-defined for $f \in W^{1/p,p}(\mathbb{S}^1; \mathbb{S}^1)$, it is natural to ask whether $|\deg f|$ can be controlled in terms of $|f|_{W^{1/p,p}}$, with $1 < p < \infty$. Our goal in this section is to prove that this is indeed true, and in fact sharper versions of such estimates hold.

Here is a first result.

Theorem 12.3. *Given any $1 < p < \infty$, there exists a constant C_p (depending only on p) such that*

$$|\deg f| \le C_p\, |f|^p_{W^{1/p,p}}, \ \forall f \in W^{1/p,p}(\mathbb{S}^1; \mathbb{S}^1). \tag{12.59}$$

Proof. Set $u(t) := f(e^{\imath t})$, $0 < t < 4\pi$. Clearly, $u \in W^{1/p,p}((0,4\pi); \mathbb{S}^1)$, with

$$|u|_{W^{1/p,p}} \le C\, |f|_{W^{1/p,p}}, \tag{12.60}$$

and we may write $u = e^{\imath\varphi}$ with $\varphi \in W^{1/p,p}((0,4\pi); \mathbb{R})$, by Theorem 5.1. We *cannot* control $|\varphi|_{W^{1/p,p}}$ in terms of $|u|_{W^{1/p,p}}$ (see Proposition 9.2). However, we can bound $|\varphi|_{BMO}$, by Theorem 9.3. More precisely, we have

$$|\varphi|_{BMO} \le C_p(|u|_{W^{1/p,p}} + |u|^p_{W^{1/p,p}}) \le C_p(|f|_{W^{1/p,p}} + |f|^p_{W^{1/p,p}}) \tag{12.61}$$

(by (12.60)).

Recall (see (12.23)) that

$$\varphi(t+2\pi) - \varphi(t) = 2\pi\ \deg f \text{ for a.e. } t \in (0, 2\pi). \tag{12.62}$$

Note that

$$\begin{aligned}\left|\fint_{(0,2\pi)} \varphi - \fint_{(0,4\pi)} \varphi\right| &\le \fint_{(0,2\pi)} \left|\varphi - \fint_{(0,4\pi)} \varphi\right| \\ &\le 2\fint_{(0,4\pi)} \left|\varphi - \fint_{(0,4\pi)} \varphi\right| \le 2|\varphi|_{BMO},\end{aligned} \tag{12.63}$$

and similarly

$$\left|\fint_{(2\pi,4\pi)} \varphi - \fint_{(0,4\pi)} \varphi\right| \le 2|\varphi|_{BMO}. \tag{12.64}$$

Combining (12.63) and (12.64) yields

$$\left| \fint_{(0,2\pi)} \varphi - \fint_{(2\pi,4\pi)} \varphi \right| \leq 4|\varphi|_{BMO}$$

and therefore, by (12.62),

$$|\deg f| \leq C\,|\varphi|_{BMO}. \tag{12.65}$$

From (12.61) and (12.65), we obtain

$$|\deg f| \leq C_p\,(|f|_{W^{1/p,p}} + |f|^p_{W^{1/p,p}}). \tag{12.66}$$

Applying (12.66) with $|f|_{W^{1/p,p}} \leq \beta$, where $C_p(\beta + \beta^p) < 1$, we find that $\deg f = 0$. On the other hand, if $|f|_{W^{1/p,p}} > \beta$, then we have $|f|_{W^{1/p,p}} \leq (1/\beta^{p-1})\,|f|^p_{W^{1/p,p}}$ and therefore $|\deg f| \leq C'_p\,|f|^p_{W^{1/p,p}}$ in all cases. □

Remark 12.8. Unfortunately, Theorem 12.3 has a mild defect: it does not provide the optimal estimate for $\deg f$ when $f \in C^{0,\alpha}$. Indeed, it is easy to see (counting how many times f can cover $\mathbb{S}^1$) that

$$|\deg f| \leq C|f|^{1/\alpha}_{C^{0,\alpha}}. \tag{12.67}$$

However, such an estimate does not follow from Theorem 12.3; using the embedding $C^{0,\alpha} \hookrightarrow W^{1/p,p}$ with $p > 1/\alpha$, we merely obtain $|\deg f| \lesssim |f|^p_{C^{0,\alpha}}$, $\forall p > 1/\alpha$.

Note that, for every $\delta > 0$,

$$|f|^p_{W^{1/p,p}} = \int_{\mathbb{S}^1}\int_{\mathbb{S}^1} \frac{|f(x) - f(y)|^p}{|x-y|^2}\,dxdy \geq \delta^p \iint\limits_{\substack{\mathbb{S}^1\times\mathbb{S}^1\\ |f(x)-f(y)|\geq\delta}} \frac{1}{|x-y|^2}\,dxdy. \tag{12.68}$$

Similarly,

$$|f|^{1/\alpha}_{C^{0,\alpha}} \gtrsim \delta^{1/\alpha} \iint\limits_{\substack{\mathbb{S}^1\times\mathbb{S}^1\\ |f(x)-f(y)|\geq\delta}} \frac{1}{|x-y|^2}\,dxdy. \tag{12.69}$$

In view of (12.68) and (12.69), the next result is a substantial improvement of Theorem 12.3, which in addition implies (12.67).

Theorem 12.4. *For every $0 < \delta < \sqrt{2}$, there exists a constant $C = C_\delta$ such that, for every $f \in VMO(\mathbb{S}^1;\mathbb{S}^1)$,*

$$|\deg f| \leq C \iint\limits_{\substack{\mathbb{S}^1\times\mathbb{S}^1\\ |f(x)-f(y)|\geq\delta}} \frac{1}{|x-y|^2}\,dxdy := C\,I_\delta(f). \tag{12.70}$$

Remark 12.9. If $f \in C(\mathbb{S}^1; \mathbb{S}^1)$, then the integral $I_\delta(f)$ is finite for every $\delta > 0$. However, if $f \in VMO\,(\mathbb{S}^1; \mathbb{S}^1)$ it may happen that $I_\delta(f) = \infty, \forall\, 0 < \delta < 2$ (see Brezis and Nguyen [97, Theorem 4]). On the other hand, if $f \in L^1(\mathbb{S}^1; \mathbb{R})$ is such that $I_\delta(f) < \infty$, $\forall\, \delta > 0$, then $f \in VMO$ [97, Theorem 1].

Proof of Theorem 12.4. Let w, α, Π, and U be as in the proof of Theorem 12.2, with $0 < \alpha < 1$ to be chosen later (depending on $0 < \delta < \sqrt{2}$). In view of (12.39), it suffices to prove that

$$\int_{\mathbb{D}} |\operatorname{Jac} U| \leq C\, I_\delta(f). \tag{12.71}$$

On the other hand, we clearly have the pointwise estimate

$$|\operatorname{Jac} U(y)| \leq C\, |\nabla w(y)|^2\, \chi_A(y), \text{ with } A := \{y \in \mathbb{D};\ |w(y)| \leq \alpha\}. \tag{12.72}$$

By the definition of w and the fact that $|f| \equiv 1$, we find that

$$|\nabla w(r x)| \leq \frac{C}{1-r},\ \forall f,\ \forall x \in \mathbb{S}^1,\ \forall r \in (0, 1). \tag{12.73}$$

[In estimate (12.73), the *VMO* assumption on f plays no role.]

If we let, for $x \in \mathbb{S}^1$,

$$d(x) := \begin{cases} \infty, & \text{if } |w(r x)| > \alpha,\ \forall r \in [0, 1) \\ 1 - \sup\{r \in [0, 1);\ |w(r x)| \leq \alpha\}, & \text{otherwise} \end{cases}, \tag{12.74}$$

it follows from (12.72) and (12.73) that

$$\int_{\mathbb{D}} |\operatorname{Jac} U| \leq C \int_{\mathbb{S}^1} \frac{1}{d(x)}\, dx. \tag{12.75}$$

In view of (12.71) and (12.75), it suffices to prove that

$$\int_{\mathbb{S}^1} \frac{1}{d(x)}\, dx \leq C\, I_\delta(f). \tag{12.76}$$

In turn, estimate (12.76) follows from Lemma 12.2 below, which provides a pointwise estimate for $1/d(x)$. □

Remark 12.10. Let us note that, in the above proof, the assumption $f \in VMO\,(\mathbb{S}^1; \mathbb{S}^1)$ appears only (via (12.39)) through the estimate

$$|\deg f| \leq C \int_{\mathbb{S}^1} \frac{1}{d(x)}\, dx.$$

The heart of the matter is the estimate (12.76), which is valid for every *measurable* $f : \mathbb{S}^1 \to \mathbb{S}^1$. A similar remark holds for Theorem 12.5 below.

Lemma 12.2. *Let $0 < \delta < \sqrt{2}$. Let $f : \mathbb{S}^1 \to \mathbb{S}^1$ be a measurable map. Then,*

$$\frac{1}{d(x)} \leq \frac{C}{(2-\delta^2)} \int\limits_{\substack{y \in \mathbb{S}^1 \\ |f(x)-f(y)| \geq \delta}} \frac{1}{|x-y|^2}\, dy, \text{ for a.e. } x \in \mathbb{S}^1, \tag{12.77}$$

where C does not depend on f or x.

Proof. We may assume that $d(x) < \infty$ (otherwise (12.77) is trivial). We claim that

$$0 \leq d(x) \leq \frac{1}{2}. \tag{12.78}$$

Indeed, from the definition (12.41) of w, we have

$$w(rx) = w(0),\ \forall\, r \in [0, 1/2],\ \forall\, x \in \mathbb{S}^1. \tag{12.79}$$

Moreover, the function $r \mapsto w(rx)$ is continuous on $[0, 1)$. Assume by contradiction that $d(x) > 1/2$. Then, by (12.74),

$$\theta := \sup\{r \in [0, 1);\ |w(rx)| \leq \alpha\} < \frac{1}{2}.$$

Hence, $|w(rx)| > \alpha$, $\forall\, r \in [\theta, 1)$ and, by (12.79), we have $|w(rx)| > \alpha$, $\forall\, r \in [0, 1)$. Consequently, $d(x) = \infty$, which contradicts the assumption that $d(x) < \infty$.

Let us note that, for a.e. $x \in \mathbb{S}^1$, we have $w(rx) \to f(x)$ as $r \nearrow 1$ (by Lebesgue's differentiation theorem), and thus we may also assume that $d(x) > 0$.

With $x \in \mathbb{S}^1$ such that $0 < d(x) \leq 1/2$, let $X := (1 - d(x))x$. Then we have, by (12.74),

$$|w(X)| = \alpha. \tag{12.80}$$

Therefore, by (12.41) and (12.80), we have for a.e. $x \in \mathbb{S}^1$,

$$\begin{aligned} 2(1-\alpha) \leq 2\big(1 - f(x) \cdot w(X)\big) &= \fint_{y \in J_{d(x)}(x)} |f(x) - f(y)|^2\, dy \\ &\leq \delta^2 + \frac{4}{|J_{d(x)}(x)|} \int\limits_{\substack{y \in J_{d(x)}(x) \\ |f(x)-f(y)| \geq \delta}} dy \\ &= \delta^2 + \frac{4}{|J_{d(x)}(x)|} \big|\{y \in J_{d(x)}(x);\ |f(x) - f(y)| \geq \delta\}\big|. \end{aligned} \tag{12.81}$$

Choosing $\alpha := \dfrac{1}{4}(2 + \delta^2)$, we see that

$$\frac{1}{2}(2-\delta^2) \le \frac{4}{|J_{d(x)}(x)|}\left|\left\{y \in J_{d(x)}(x);\ |f(x)-f(y)| \ge \delta\right\}\right|, \tag{12.82}$$

and thus

$$\int\limits_{\substack{y\in J_{d(x)}(x)\\ |f(x)-f(y)|\ge\delta}} \frac{1}{|x-y|^2}\,dy \ge \frac{C}{(d(x))^2}\left|\left\{y \in J_{d(x)}(x);\ |f(x)-f(y)| \ge \delta\right\}\right|$$
$$\ge \frac{C(2-\delta^2)}{d(x)},$$

which implies (12.77). □

With more work, it is possible to improve Theorem 12.4 as follows.

Theorem 12.5. *There exists a constant C such that, for every $f \in VMO\,(\mathbb{S}^1; \mathbb{S}^1)$,*

$$|\deg f| \le C \iint\limits_{\substack{\mathbb{S}^1\times\mathbb{S}^1\\ |f(x)-f(y)|\ge\sqrt{3}}} \frac{1}{|x-y|^2}\,dxdy = C\,I_{\sqrt{3}}(f). \tag{12.83}$$

Proof. Again, the strategy is to obtain (12.76), with $\delta = \sqrt{3}$, for some appropriate choice of α.

We rely on the following version of Lemma 12.2, whose proof is postponed to Appendix 15.5.

Lemma 12.3. *There exist constants $\alpha, \varepsilon > 0$ such that, for any measurable map $g : (0, 1) \to \mathbb{S}^1$, we have*

$$\left|\int_0^1 g\right| \le \alpha \implies |A| \ge \varepsilon, \tag{12.84}$$

where

$$A := \left\{(z, t) \in (0, 1)^2;\ |z-t| \ge \varepsilon, |g(z)-g(t)| \ge \sqrt{3}\right\}. \tag{12.85}$$

Granted Lemma 12.3, we proceed as follows. Consider α and ε as above. Let $x \in \mathbb{S}^1$ be such that $d(x) < \infty$, and thus $|w((1-d(x))\,x)| = \alpha$. By scaling and (12.84), we have

$$d(x) < \infty \implies |A_x| \ge C_2\,\varepsilon\,d^2(x), \tag{12.86}$$

where

$$A_x := \left\{(z, t) \in (J_{d(x)}(x))^2;\ |z-t| \ge C_1\,\varepsilon\,d(x),\ |f(z)-f(t)| \ge \sqrt{3}\right\}. \tag{12.87}$$

By (12.86), we have

$$\frac{1}{d(x)} \lesssim \frac{1}{d^3(x)} \iint_{A_x} dzdt, \ \forall x \in \mathbb{S}^1 \text{ such that } d(x) < \infty. \tag{12.88}$$

Therefore,

$$\int_{\mathbb{S}^1} \frac{1}{d(x)} dx \lesssim \int_{\mathbb{S}^1} \frac{1}{d^3(x)} \iint_{A_x} dzdtdx. \tag{12.89}$$

We next note that

$$(z, t) \in A_x \implies d(x) \sim |z - t| \text{ and } |z - x| \lesssim |z - t|. \tag{12.90}$$

By (12.90), we find that

$$\int\limits_{\substack{x \in \mathbb{S}^1 \\ (z,t)\in A_x}} \frac{1}{d^3(x)} dx \lesssim \frac{1}{|z-t|^3} |\{x \in \mathbb{S}^1;\ (z, t) \in A_x\}| \lesssim \frac{1}{|z-t|^2}, \ \forall z, t \in \mathbb{S}^1. \tag{12.91}$$

Estimate (12.83) follows from (12.89) and (12.91). □

Remark 12.11. As noticed by Nguyen [279], the constant $\sqrt{3}$ is optimal in the following sense: there exists a sequence (f_n) in $C^\infty(\mathbb{S}^1; \mathbb{S}^1)$ such that $\deg f_n = 1$, $\forall n$, whereas

$$\lim_{n\to\infty} \iint\limits_{\substack{\mathbb{S}^1 \times \mathbb{S}^1 \\ |f_n(x) - f_n(y)| > \sqrt{3}}} \frac{1}{|x-y|^2} dxdy = 0. \tag{12.92}$$

Indeed, consider $f_n(e^{\imath\theta}) := e^{\imath\varphi_n(\theta)}$, with $\varphi_n : [0, 2\pi] \to [0, 2\pi]$ smooth and non-decreasing such that

$$\varphi_n(\theta) = \begin{cases} 0, & \text{if } 0 \le \theta \le 2\pi/3 - 1/n \\ 2\pi/3, & \text{if } 2\pi/3 \le \theta \le 4\pi/3 - 1/n \\ 4\pi/3, & \text{if } 4\pi/3 \le \theta \le 2\pi - 1/n \\ 2\pi, & \text{if } \theta = 2\pi \end{cases}.$$

It is straightforward that the set

$$A_n := \left\{(x, y) \in \mathbb{S}^1 \times \mathbb{S}^1;\ |f_n(x) - f_n(y)| > \sqrt{3}\right\}$$

satisfies

$$A_n \subset \left\{(x, y) \in \mathbb{S}^1 \times \mathbb{S}^1;\ |x - y| \ge \sqrt{3} - 2/n\right\} \text{ and } |A_n| \to 0,$$

from which (12.92) follows easily.

12.6 Homotopy classes

Recall a standard fact about continuous $\mathbb{S}^1$-valued functions: if f, g are connected by a path, then $\deg f = \deg g$, and conversely. [For the converse, see, for example, the proof of Proposition 12.8 *1.*]

As a consequence, the path-connected components of $C(\mathbb{S}^1; \mathbb{S}^1)$ are classified by the degree. We prove below that the same holds in the spaces $VMO\,(\mathbb{S}^1; \mathbb{S}^1)$ and $W^{s,p}(\mathbb{S}^1; \mathbb{S}^1)$ with $sp \geq 1$.

Proposition 12.8. *1. The path-connected components of* $VMO\,(\mathbb{S}^1; \mathbb{S}^1)$ *are given by* $\{f \in VMO\,(\mathbb{S}^1; \mathbb{S}^1);\ \deg f = d\}$, *with* $d \in \mathbb{Z}$.

2. *Let* $0 < s < \infty$ *and* $1 \leq p < \infty$ *be such that* $sp \geq 1$. *Then the path-connected components of* $W^{s,p}(\mathbb{S}^1; \mathbb{S}^1)$ *are given by* $\{f \in W^{s,p}(\mathbb{S}^1; \mathbb{S}^1);\ \deg f = d\}$, *with* $d \in \mathbb{Z}$.

Proof. Let X be one of the spaces VMO or $W^{s,p}$ with $sp \geq 1$. Let $f, g \in X(\mathbb{S}^1; \mathbb{S}^1)$ be connected by a path. Since $X \hookrightarrow VMO$, we find that the map deg is continuous on $X(\mathbb{S}^1; \mathbb{S}^1)$, and thus $\deg f = \deg g$.

Conversely, assuming that $f, g \in X(\mathbb{S}^1; \mathbb{S}^1)$ satisfy $\deg f = \deg g$, we connect f to g. This is done as follows. By the results in Section 12.1, we may write $f = e^{\imath\varphi} g$, with $\varphi \in X(\mathbb{S}^1; \mathbb{R})$. We then consider the mapping $F(t, \cdot) := e^{\imath(1-t)\varphi} g$, which satisfies $F(0, \cdot) = f$ and $F(1, \cdot) = g$.

It remains to prove that $F \in C([0, 1]; X(\mathbb{S}^1; \mathbb{S}^1))$. When $X = W^{s,p}$, this follows from Theorems 15.6, 15.7, and Lemma 15.41. When $X = VMO$, we rely on Theorem 15.9 and Corollary 15.4. □

Remark 12.12. By contrast, when $0 < s < 1$ and $1 \leq p < \infty$ are such that $sp < 1$, the space $W^{s,p}(\mathbb{S}^1; \mathbb{S}^1)$ is contractible (see Mironescu [259], and also Section 14.6.3).

Further results concerning the homotopy classes are presented in Section 12.8.3.

12.7 Degree and Fourier coefficients

12.7.1 Some preliminaries

Given a function $f \in L^2(\mathbb{S}^1; \mathbb{R}^2) \simeq L^2(\mathbb{S}^1; \mathbb{C})$, consider its Fourier coefficients

$$a_n = a_n(f) := \frac{1}{2\pi}\int_0^{2\pi} f(e^{\imath\theta})\, e^{-\imath n\theta}\, d\theta = \frac{1}{2\pi}\int_{\mathbb{S}^1} f(z)\, z^{-n}\, d\ell, \tag{12.93}$$

so that

$$f(e^{\imath\theta}) = \sum_{n\in\mathbb{Z}} a_n\, e^{\imath n\theta} \text{ in } L^2(\mathbb{S}^1; \mathbb{C}). \tag{12.94}$$

Since we are concerned with maps taking their values in $\mathbb{S}^1$, it is useful to characterize their Fourier coefficients.

Lemma 12.4. *Let $f \in L^2(\mathbb{S}^1; \mathbb{C})$. Then, $f \in L^2(\mathbb{S}^1; \mathbb{S}^1)$ if and only if*

$$\sum_{n\in\mathbb{Z}} a_n \overline{a_{n+k}} = 0, \ \forall k \in \mathbb{Z} \setminus \{0\}, \tag{12.95}$$

and

$$\sum_{n\in\mathbb{Z}} |a_n|^2 = 1. \tag{12.96}$$

Proof. Recall that, if $f, g \in L^2(\mathbb{S}^1; \mathbb{C})$, then, $\forall k \in \mathbb{Z}$,

$$a_{-k}(fg) = \sum_{n\in\mathbb{Z}} a_n(f)\, a_{-n-k}(g). \tag{12.97}$$

Applying this with $g := \overline{f}$, we find that

$$a_{-k}(|f|^2) = \sum_{n\in\mathbb{Z}} a_n \overline{a_{n+k}}, \ \forall k \in \mathbb{Z}. \tag{12.98}$$

From (12.98), we obtain, by identification of the Fourier coefficients, that

$$|f|^2 = 1 \iff [(12.95) \text{ and } (12.96) \text{ hold}].$$ □

Remark 12.13. As a consequence of Lemma 12.96, we see that a *finite* trigonometric series, i.e., $f(e^{\imath\theta}) = \sum_{n\in I} a_n e^{\imath n\theta}$ with I finite, cannot belong to $L^2(\mathbb{S}^1; \mathbb{S}^1)$, unless $f(e^{\imath\theta}) = e^{\imath n\theta}$ for some $n \in \mathbb{Z}$.

We next consider maps $f \in H^{1/2}(\mathbb{S}^1; \mathbb{C})$. It is standard that such functions can be characterized in terms of their Fourier coefficients. More precisely,

$$|f|_{H^{1/2}(\mathbb{S}^1)} \sim \left(\sum_{n\in\mathbb{Z}} |n|\, |a_n|^2\right)^{1/2}, \tag{12.99}$$

where $|f|_{H^{1/2}}$ denotes any seminorm on $H^{1/2}$ (defined via trace theory, interpolation, etc.). In fact, there is a simple formula expressing the Gagliardo $H^{1/2}(\mathbb{S}^1)$ seminorm in terms of the Fourier coefficients.

Lemma 12.5. *We have, for every $f \in L^2(\mathbb{S}^1; \mathbb{C})$,*

$$|f|^2_{H^{1/2}(\mathbb{S}^1)} := \int_{\mathbb{S}^1}\int_{\mathbb{S}^1} \frac{|f(x) - f(y)|^2}{|x-y|^2}\, dxdy = 4\pi^2 \sum_{n\in\mathbb{Z}} |n|\, |a_n(f)|^2, \tag{12.100}$$

where $|\ |$ denotes the Euclidean distance in $\mathbb{R}^2$.

Proof. Let us first note that

$$\int_0^{2\pi} \frac{|e^{\imath n\xi}-1|^2}{|e^{\imath\xi}-1|^2}\,d\xi = \int_0^{2\pi} \left|\sum_{k=0}^{n-1} e^{\imath k\xi}\right|^2 d\xi = 2\pi\,|n|,\ \forall n \in \mathbb{Z}. \tag{12.101}$$

Let $f \in H^{1/2}(\mathbb{S}^1;\mathbb{S}^1)$, and write $f = \sum_{n\in\mathbb{Z}} a_n\, e^{\imath n\theta}$. Using (12.101) and Parseval's identity, we find that

$$\begin{aligned}|f|^2_{H^{1/2}(\mathbb{S}^1)} =& \int_0^{2\pi}\int_0^{2\pi} \frac{\left|\sum a_n\, e^{\imath n(\theta+\xi)} - \sum a_n\, e^{\imath n\theta}\right|^2}{|e^{\imath(\theta+\xi)}-e^{\imath\theta}|^2}\, d\theta d\xi \\ =& 2\pi \sum |a_n|^2 \int_0^{2\pi} \frac{|e^{\imath n\xi}-1|^2}{|e^{\imath\xi}-1|^2}\, d\xi \\ =& 4\pi^2 \sum |n|\,|a_n|^2.\end{aligned} \qquad \square \tag{12.102}$$

Remark 12.14. Warning. Given $f \in L^2(\mathbb{S}^1;\mathbb{C})$, set

$$H(f)(\theta) := f(e^{\imath\theta}),\ 0 < \theta < 2\pi.$$

Clearly, H is an isomorphism from $L^2(\mathbb{S}^1)$ to $L^2((0,2\pi))$. Note that if $f \in H^{1/2}(\mathbb{S}^1)$, then $H(f) \in H^{1/2}((0,2\pi))$. However, H is not surjective onto $H^{1/2}((0,2\pi))$. The correct space on $(0,2\pi)$ is

$$H^{1/2}_p((0,2\pi);\mathbb{C}) := \left\{u_{|(0,2\pi)};\ u \in H^{1/2}_{loc}(\mathbb{R};\mathbb{C}) \text{ and } u \text{ is } 2\pi - \text{periodic}\right\},$$

and H is an isomorphism from $H^{1/2}(\mathbb{S}^1;\mathbb{C})$ onto $H^{1/2}_p((0,2\pi);\mathbb{C})$.

12.7.2 Computing the degree using Fourier coefficients

Recall that maps $f \in H^{1/2}(\mathbb{S}^1;\mathbb{S}^1)$ admit a well-defined degree. The next result provides a remarkable formula relating the degree of f and the Fourier coefficients of f.

Theorem 12.6. *For every $f \in H^{1/2}(\mathbb{S}^1;\mathbb{S}^1)$, we have*

$$\deg f = \sum_{n\in\mathbb{Z}} n\,|a_n|^2, \tag{12.103}$$

where the a_n's are the Fourier coefficients of f, given by (12.93).

Note that, by (12.100), the right-hand side of (12.103) is *absolutely convergent*, since $f \in H^{1/2}$.

Proof. We start with $f \in C^1(\mathbb{S}^1;\mathbb{S}^1)$. From the Cauchy formula (12.27), we have

$$\deg f = \frac{1}{2\imath\pi}\int_{\mathbb{S}^1} \overline{f}\dot{f} = \frac{1}{\imath}\, a_0(\overline{f}\dot{f}). \tag{12.104}$$

Combining (12.104) with (12.97) and the fact that $a_n(\dot{f}) = \imath\, n\, a_n(f), \forall\, n$, we obtain (12.103) for $f \in C^1$.

We now turn to the general case $f \in H^{1/2}(\mathbb{S}^1; \mathbb{S}^1)$. Note that both sides in (12.103) are continuous with respect to strong $H^{1/2}$ convergence. The continuity of $\deg f$ follows from Theorem 12.1 and the embedding $H^{1/2} \hookrightarrow VMO$; for the right-hand side of (12.103), one uses the continuity of the function $(a_n) \mapsto \sum_{n\in\mathbb{Z}} n\, |a_n|^2$ from $\ell^2_\#$ to $\mathbb{R}$, where

$$\ell^2_\# := \left\{ (a_n);\ \sum_{n\in\mathbb{Z}} |n|\, |a_n|^2 < \infty \right\},$$

equipped with its natural seminorm. Equation (12.103) holds for every $f \in H^{1/2}(\mathbb{S}^1; \mathbb{S}^1)$ since $C^1(\mathbb{S}^1; \mathbb{S}^1)$ is dense in $H^{1/2}(\mathbb{S}^1; \mathbb{S}^1)$ (Corollary 15.8 in Appendix 15.5). □

Theorem 12.6 raises some intriguing questions. But we first present an immediate consequence of the theorem, which is seemingly unrelated to the degree.

Corollary 12.2. *Let (a_n) be a sequence of complex numbers satisfying*

$$\sum_{n\in\mathbb{Z}} a_n\, \overline{a_{n+k}} = 0,\ \forall\, k \in \mathbb{Z} \setminus \{0\}, \tag{12.105}$$

$$\sum_{n\in\mathbb{Z}} |a_n|^2 = 1, \tag{12.106}$$

and

$$\sum_{n\in\mathbb{Z}} |n|\, |a_n|^2 < \infty. \tag{12.107}$$

Then,

$$\sum_{n\in\mathbb{Z}} n\, |a_n|^2 \in \mathbb{Z}. \tag{12.108}$$

Proof. Set $f(e^{\imath\theta}) := \sum_{n\in\mathbb{Z}} a_n\, e^{\imath n\theta}$. By (12.107), we have $f \in H^{1/2}$, and, by (12.105) and (12.106), we have $f : \mathbb{S}^1 \to \mathbb{S}^1$. It then suffices to invoke Theorem 12.6. □

Suppose now $f \in C(\mathbb{S}^1; \mathbb{S}^1)$ (or more generally $f \in VMO\,(\mathbb{S}^1; \mathbb{S}^1)$) and $f \notin H^{1/2}$. Then the series $\sum_{n\in\mathbb{Z}} |n|\, |a_n|^2$ is divergent. Thus, the left-hand side of (12.103) is well-defined, but the right-hand side is not. It is natural to ask whether $\deg f$ may still be computed using one of the standard summation processes for the series $\sum_{n\in\mathbb{Z}} n\, |a_n|^2$ (which is not absolutely convergent). Consider, for example,

$$\sigma_N := \sum_{n=-N}^{N} n\, |a_n|^2,\ N \in \mathbb{N},$$

or

$$P_r := \sum_{n\in Z} n\,|a_n|^2\, r^{|n|},\ 0 < r < 1.$$

Is it true that, for $f \in VMO\,(\mathbb{S}^1; \mathbb{S}^1)$,

$$\deg f = \lim_{N\to\infty} \sigma_N \text{ or } \deg f = \lim_{r\nearrow 1} P_r? \tag{12.109}$$

Korevaar [214] proved that the answer is positive, for example, if $f \in (BV \cap C)(\mathbb{S}^1; \mathbb{S}^1)$, but – surprisingly – negative in general. He has constructed interesting examples of maps $f \in C(\mathbb{S}^1; \mathbb{S}^1)$ of degree 0 such that σ_N (resp. P_r) need not have a limit when $N \to \infty$ (resp. when $r \nearrow 1$) or may converge to any given number $\lambda \neq 0$ (including $\pm\infty$).

Of course, the answer to question (12.109) is positive if $f \in H^{1/2}(\mathbb{S}^1; \mathbb{S}^1)$ (by Theorem 12.6). The next result exhibits a summation process that is valid beyond $H^{1/2}$.

Theorem 12.7. *For every $f \in W^{1/3,3}(\mathbb{S}^1; \mathbb{S}^1)$, we have*

$$\deg f = \lim_{\varepsilon\searrow 0} \sum_{n\in\mathbb{Z}\setminus\{0\}} n\,|a_n|^2\, \frac{\sin n\varepsilon}{n\varepsilon}. \tag{12.110}$$

Recall that, on $\mathbb{S}^1$, we have $H^{1/2} \hookrightarrow W^{1/3,3}$, with strict inclusion. More generally, the inclusion $W^{1/p,p} \hookrightarrow W^{1/3,3}$ is strict when $1 < p < 3$. Indeed, the spaces $W^{1/p,p}(\mathbb{R})$ and $W^{1/3,3}(\mathbb{R})$ do not coincide (see, for example, Triebel [337, Section 2.12, Theorem (a), p. 242]), and the argument in [337] shows that the same holds in $(0, 1)$. By localization, the same is true on $\mathbb{S}^1$. Alternatively, one could argue as in Mironescu and Sickel [268, Section 3] and prove that the map $z \mapsto \sum_{j=1}^{\infty} j^{-1/p}\, z^{2^j}$ belongs to $W^{1/3,3}(\mathbb{S}^1; \mathbb{S}^1)$, but not to $W^{1/p,p}(\mathbb{S}^1; \mathbb{S}^1)$.

The case $p = 1$ is a little special, since $W^{1,1}$ is not contained in any of the spaces $W^{1/p,p}$ with $p > 1$ (see Lemma 15.4). However, the analogue of Theorem 12.7 holds for $p = 1$.

Theorem 12.8. *Let $f \in W^{1,1}(\mathbb{S}^1; \mathbb{S}^1)$. Then,* (12.110) *holds.*
More generally, (12.110) *holds for $f \in (BV \cap C)(\mathbb{S}^1; \mathbb{S}^1)$.*

Proof of Theorem 12.7. Let $k := \deg f$. By Proposition 12.1, we may write $f(e^{\imath\theta}) = e^{\imath(k\theta+\varphi(\theta))}$, for some periodic function $\varphi \in W^{1/3,3}_{loc}(\mathbb{R}; \mathbb{R})$. A simple calculation gives

$$\begin{aligned}\int_0^{2\pi} e^{\imath k\varepsilon}\, e^{\imath(\varphi(\theta+\varepsilon)-\varphi(\theta))}\, d\theta &= \int_0^{2\pi} f(e^{\imath(\theta+\varepsilon)})\overline{f}(e^{\imath\theta})\, d\theta \\ &= 2\pi \sum_{n\in\mathbb{Z}} |a_n|^2\, e^{\imath n\varepsilon},\end{aligned} \tag{12.111}$$

$$e^{\imath k\varepsilon} = 1 + \imath k\varepsilon + O(\varepsilon^2), \tag{12.112}$$

and

$$e^{\imath(\varphi(\theta+\varepsilon)-\varphi(\theta))} = 1 + \imath(\varphi(\theta+\varepsilon)-\varphi(\theta)) - \frac{1}{2}(\varphi(\theta+\varepsilon)-\varphi(\theta))^2 + O(|\varphi(\theta+\varepsilon)-\varphi(\theta)|^3). \tag{12.113}$$

Thus,

$$\begin{aligned} \operatorname{Im}[e^{\imath k\varepsilon} e^{\imath(\varphi(\theta+\varepsilon)-\varphi(\theta))}] &= \operatorname{Im}[(1+\imath k\varepsilon)\, e^{\imath(\varphi(\theta+\varepsilon)-\varphi(\theta))}] + O(\varepsilon^2) \\ &= (\varphi(\theta+\varepsilon)-\varphi(\theta)) + k\varepsilon \\ &\quad + O(\varepsilon^2) + O(\varepsilon\, |\varphi(\theta+\varepsilon)-\varphi(\theta)|^2) \\ &\quad + O(|\varphi(\theta+\varepsilon)-\varphi(\theta)|^3). \end{aligned} \tag{12.114}$$

Integrating (12.114) with respect to θ yields (via (12.111))

$$\begin{aligned} \left| \sum_{n\in\mathbb{Z}} |a_n|^2 \sin n\varepsilon - k\varepsilon \right| &\le C\varepsilon^2 + C\varepsilon \int_0^{2\pi} |\varphi(\theta+\varepsilon)-\varphi(\theta)|^2\, d\theta \\ &\quad + C\int_0^{2\pi} |\varphi(\theta+\varepsilon)-\varphi(\theta)|^3\, d\theta. \end{aligned} \tag{12.115}$$

We find that

$$\begin{aligned} \left| \sum_{n\in\mathbb{Z}} |a_n|^2 \frac{\sin n\varepsilon}{\varepsilon} - k \right| &\le C\varepsilon + C\int_0^{2\pi} |\varphi(\theta+\varepsilon)-\varphi(\theta)|^2\, d\theta \\ &\quad + \frac{C}{\varepsilon}\int_0^{2\pi} |\varphi(\theta+\varepsilon)-\varphi(\theta)|^3\, d\theta \\ &= o(1) + \frac{C}{\varepsilon}\int_0^{2\pi} |\varphi(\theta+\varepsilon)-\varphi(\theta)|^3\, d\theta \end{aligned} \tag{12.116}$$

as $\varepsilon \to 0$.

We complete the proof of the theorem via Lemma 15.3 (in Appendix 15.2) applied with $N = 1$, $s = 1/3$ and $p = 3$. [Strictly speaking, we apply the lemma not to φ but to a function $\zeta \in W^{1/3,3}(\mathbb{R})$ with compact support such that $\zeta = \varphi$ on $(-\pi, 3\pi)$.] □

Proof of Theorem 12.8. We proceed as in the previous proof. Since $f \in BV \cap C$, the corresponding φ satisfies $\varphi \in BV \cap C$. We rely again on (12.116).

We note that

$$\int_0^{2\pi} |\varphi(\theta+\varepsilon)-\varphi(\theta)| d\theta \le \varepsilon |\varphi|_{BV}. \tag{12.117}$$

Inserting (12.117) in (12.116) gives

$$\left| \frac{1}{\varepsilon} \sum_{n\in\mathbb{Z}} |a_n|^2 \sin n\varepsilon - k \right| \le o(1) + C \sup_\theta |\varphi(\theta+\varepsilon)-\varphi(\theta)|^2 \tag{12.118}$$

and the conclusion follows, since $\varphi \in C$. □

The previous results imply the following

Corollary 12.3. *Let $f, g \in W^{1/3,3}(\mathbb{S}^1; \mathbb{S}^1)$ (resp. $f, g \in (BV \cap C)(\mathbb{S}^1; \mathbb{S}^1)$). Assume that $|a_n(f)| = |a_n(g)|$, $\forall n \in \mathbb{Z}$. Then,* $\deg f = \deg g$.

In particular, the conclusion holds for $f, g \in C^{0,\alpha}(\mathbb{S}^1; \mathbb{S}^1)$ with $\alpha > 1/3$.

One may wonder whether the same conclusion holds in a more general setting, e.g. $f, g \in VMO$. This question was raised by Brezis [74] under the name "Can one hear the degree?" It turns out that the answer is negative. A very elaborate construction by Bourgain and Kozma [59] provides an example of two functions $f, g \in C(\mathbb{S}^1; \mathbb{S}^1)$ such that $|a_n(f)| = |a_n(g)|$, $\forall n \in \mathbb{Z}$, but $\deg f \neq \deg g$. Note, however, that there is a wide "gap" between the positive result (Corollary 12.3) and the counterexample by Bourgain and Kozma. It is not known whether $W^{1/3,3}$ is the exact borderline (see Open Problem 22 for related questions).

12.8 Complements and open problems

12.8.1 Least energy with prescribed degree

In this section, we let $1 \leq p < \infty$, and investigate the least $W^{1/p,p}$ energy required to "produce" a map of degree $d \in \mathbb{Z}$, i.e.,

$$m_{p,d} := \inf\{|f|^p_{W^{1/p,p}};\ f \in W^{1/p,p}(\mathbb{S}^1; \mathbb{S}^1),\ \deg f = d\}, \tag{12.119}$$

where $|\ |_{W^{1/p,p}}$ denotes the Gagliardo seminorm.

This quantity is the 1D counterpart of $S_{1/p,p}(u)$ introduced in (8.27). The main result is the following

Theorem 12.9. *Let $1 \leq p < \infty$. Then we have*

$$|d|\, m_p \leq m_{p,d} \leq |d|\, m_{p,1},\ \forall d \in \mathbb{Z}, \tag{12.120}$$

for some constant $m_p > 0$.

Moreover,

$$m_{1,d} = 2\pi\, |d|,\ \forall d \in \mathbb{Z} \tag{12.121}$$

and

$$m_{2,d} = 4\pi^2\, |d|,\ \forall d \in \mathbb{Z}. \tag{12.122}$$

In order to prove Theorem 12.9, we start with a useful

Lemma 12.6. *We have*

$$m_{p,d_1+d_2} \le m_{p,d_1} + m_{p,d_2}, \ \forall\, d_1, d_2 \in \mathbb{Z}. \tag{12.123}$$

In turn, Lemma 12.6 follows from the following result.

Lemma 12.7. *Let* $1 < p < \infty$, $I \subset \mathbb{S}^1$ *be an arc and* $h \in W^{1/p,p}(\mathbb{S}^1; \mathbb{S}^1)$. *Then there exists a sequence* $(h_n) \subset C^\infty(\mathbb{S}^1; \mathbb{S}^1)$ *such that*

a) $h_n \equiv 1$ *on* $\mathbb{S}^1 \setminus I$
b) $h_n \to 1$ *a.e. on* $\mathbb{S}^1$ *as* $n \to \infty$
c) $\deg h_n \equiv \deg h$
d) $\limsup_{n\to\infty} |h_n|_{W^{1/p,p}} \le |h|_{W^{1/p,p}}$.

Proof of Lemma 12.7. By density (see Brezis, Mironescu, and Shafrir [93, Lemma 2.2]), it suffices to consider the case of a map h constant near some point of $\mathbb{S}^1$, say, $h \equiv 1$ on some arc J centered at 1. We consider a sequence $(a_n) \subset \mathbb{D}$ such that $a_n \to 1$ and define $h_n := h \circ M_{a_n}$, where $M_a(z) := \dfrac{a - z}{1 - \overline{a}\, z}$. As explained in Section 12.8.2, we have $|h_n|_{W^{1/p,p}} = |h|_{W^{1/p,p}}$, $\forall\, n$. Moreover, given any arc K centered at 1, we have, for sufficiently large n (depending on K), $h_n \equiv 1$ on $\mathbb{S}^1 \setminus K$. We find that, up to a rotation, (h_n) has all the required properties. □

Proof of Lemma 12.6. Let $f \in \mathscr{E}_{d_1}$ be such that $\deg f = d_1$ and f is constant, say $f \equiv 1$, on an arc I. Such f's are dense in $W^{1/p,p}(\mathbb{S}^1; \mathbb{S}^1)$ [93, Lemma 2.2].

Let $h \in \mathscr{E}_{d_2}$, and consider a sequence (h_n) as in Lemma 12.7. As in (8.31), we have

$$\begin{aligned} |f\, h_n|^p_{W^{1/p,p}} &= |f|^p_{W^{1/p,p}} + |h_n|^p_{W^{1/p,p}} + o(1) \\ &\le |f|^p_{W^{1/p,p}} + |h|^p_{W^{1/p,p}} + o(1) \text{ as } n \to \infty. \end{aligned}$$

Since h is arbitrary, we find that

$$m_{p,d_1+d_2} \le |f|^p_{W^{1/p,p}} + m_{p,d_2}, \ \forall f \in \mathscr{E}_{d_1} \text{ with } f \text{ constant on an arc.} \tag{12.124}$$

By density, (12.124) holds for any $f \in \mathscr{E}_{d_1}$, whence (12.123). □

Proof of Theorem 12.9. *Step 1.* Proof of (12.120) when $1 < p < \infty$
The inequality $m_{p,d} \le |d|\, m_{p,1}$ follows from Lemma 12.6 and the fact that

$$m_{p,-d} = m_{p,d}. \tag{12.125}$$

On the other hand, the inequality $|d|\, m_p \le m_{p,d}$ corresponds precisely to Theorem 12.3.

Step 2. Proof of (12.121)
We may assume that $d > 0$. By the Cauchy formula (12.27), we have $2\pi\, d \le m_{1,d}$. On the other hand, if we choose $f_0(e^{\imath\theta}) := e^{\imath d\theta}$ we see that $m_{1,d} \le \int_{\mathbb{S}^1} |\dot{f}| = 2\pi\, d$.

Step 3. Proof of (12.122)
We may again assume that $d > 0$. By (12.100) and (12.103), we have

$$4\pi^2 d \leq |f|^2_{H^{1/2}}, \ \forall f \in H^{1/2}(\mathbb{S}^1; \mathbb{S}^1) \text{ such that } \deg f = d,$$

so that $4\pi^2 d \leq m_{2,d}$.

On the other hand, if we choose $f_0(e^{\imath\theta}) := e^{\imath d\theta}$ we see that $m_{2,d} \leq |f_0|^2_{H^{1/2}}$, and $|f_0|^2_{H^{1/2}} = 4\pi^2 d$, by (12.100). Thus, $m_{2,d} \leq 4\pi^2 d$. □

12.8.2 Existence of least energy maps

The question whether the infimum in the definition of $m_{p,d}$ (see (12.119)) is achieved when $d \neq 0$, and the characterization of minimizers (if they exist), are extremely delicate and widely open. We start with the only two cases, $p = 1$ and $p = 2$, where the answer is complete.

Let us first recall some standard maps used in complex analysis. For $a \in \mathbb{D}$, let $M_a : \mathbb{S}^1 \to \mathbb{S}^1$ be the restriction to $\mathbb{S}^1$ of the Moebius transform $M_a(z) := \dfrac{a - z}{1 - \overline{a}\,z}$, $z \in \overline{\mathbb{D}}$. [Recall that, up to a multiplicative factor $\alpha \in \mathbb{S}^1$, the Moebius transforms give all the conformal representations $u : \mathbb{D} \to \mathbb{D}$. In particular, $M_a : \mathbb{S}^1 \to \mathbb{S}^1$ is an orientation-preserving diffeomorphism, and thus $\deg M_a = 1$.] We will still refer to M_a as a Moebius transform.

A Blaschke product of degree $d > 0$ is a map of the form

$$B(z) = \alpha\, M_{a_1}(z) \cdots M_{a_d}(z), \ \alpha \in \mathbb{S}^1, \ a_1, \ldots, a_d \in \mathbb{D}.$$

Theorem 12.10. *1. $m_{1,d}$ is achieved precisely by the maps of the form $f(e^{\imath\theta}) = e^{\imath\varphi(\theta)}$, where $\varphi \in W^{1,1}((0, 2\pi); \mathbb{R})$ is monotone and satisfies $\varphi(2\pi) - \varphi(0) = 2\pi d$.*
2. If $d > 0$, $m_{2,d}$ is achieved precisely by the Blaschke products of degree d. If $d < 0$, $m_{2,d}$ is achieved precisely by the conjugates of Blaschke products of degree $|d|$.

Proof of Theorem 12.10. *Proof of item 1*
If φ is as in item *1*, then $|f|_{W^{1,1}} = |\varphi|_{W^{1,1}} = 2\pi\,|d| = m_{1,d}$, by (12.121). Therefore, f is a minimizer of $m_{1,d}$.

Conversely, if f achieves $m_{1,d}$ and we write $f(e^{\imath\theta}) = e^{\imath\varphi(\theta)}$ with $\varphi \in W^{1,1}$, then $|\dot{f}| = |\varphi'|$ and thus

$$\int_0^{2\pi} |\varphi'| = 2\pi\,|d| = |\varphi(2\pi) - \varphi(0)| = \left|\int_0^{2\pi} \varphi'\right|$$

An inspection of the case of equality shows that φ is as in item *1*.

Step 2. Proof of item 2
It suffices to consider the case $d > 0$. From (12.122), we know that

$$|f|^2_{H^{1/2}} \geq 4\pi^2 d = m_{2,d}, \ \forall f \in H^{1/2}(\mathbb{S}^1; \mathbb{S}^1) \text{ such that } \deg f = d. \tag{12.126}$$

Write $f(e^{\imath\theta}) = \sum_{n\in\mathbb{Z}} a_n e^{\imath n\theta}$. Equality occurs in (12.126) if and only if $a_n = 0$ for $n < 0$. This is equivalent to the fact that the harmonic extension u of f, which is given by $u(z) = \sum_{n\geq 0} a_n z^n + \sum_{n<0} a_n \overline{z}^{-n}$, is holomorphic. In order to complete the proof, it suffices to prove that the latter property of u is equivalent to u being a Blaschke product. One implication is clear. Conversely, assume that $u \in H^1(\mathbb{D})$ is holomorphic and that $f := \operatorname{tr} u$ satisfies $|f| = 1$ a.e. on $\mathbb{S}^1$. Since $f \in H^{1/2} \hookrightarrow VMO$ and since u is the harmonic extension of f, we have $\lim_{|z|\to 1} |u(z)| = 1$ uniformly on $\mathbb{D}$. [This version of Lemma 5.6 is established in Brezis and Nirenberg [105, Theorem A3.2].] Therefore, u has a finite number of zeros in $\mathbb{D}$. We denote them $a_1, \ldots, a_n$, counted with their multiplicities. Let $v := M_{a_1} \cdots M_{a_n}$ and set $w := u/v$. Then, w is holomorphic, $w \neq 0$ in $\mathbb{D}$ and $\lim\limits_{|z|\to 1} |w(z)| = 1$ uniformly. Therefore, we may write $w = e^h$, where h is holomorphic and satisfies $\lim\limits_{|z|\to 1} \operatorname{Re} h(z) = 0$. By the maximum principle, we have $\operatorname{Re} h \equiv 0$, and thus $\operatorname{Im} h \equiv const$. Thus, $u = \alpha v$ for some $\alpha \in \mathbb{S}^1$. Finally, $d = \deg u = \deg v =$ the number of zeros of v, and thus $n = d$. □

We now return to the question of the existence of minimizers for $m_{p,d}$, $d \neq 0$, when $p \neq 1$, $p \neq 2$. The main difficulty comes from a *lack of compactness*. Indeed, let (f_n) be a minimizing sequence; one may assume, up to a subsequence, that $f_n \rightharpoonup f$. Clearly, $|f|_{W^{1/p,p}} \leq \liminf |f_n|_{W^{1/p,p}}$ and thus f is a candidate for a minimizer. However, it *may happen* that $\deg f \neq d$.

This lack of compactness is even more "dramatic." If the set of minimizers is non-empty, then it is never weakly sequentially compact. This is well illustrated by the following

Basic example. Let $1 \leq p < \infty$. For $f \in W^{1/p,p}(\mathbb{S}^1; \mathbb{S}^1)$, set $f_a := f \circ M_a$. Then,

1. $\deg f_a = \deg f$.
2. $|f_a|_{W^{1/p,p}} = |f|_{W^{1/p,p}}$.
3. If $|a_n| \to 1$, then, up to a subsequence, $f_{a_n} \to \beta$ a.e. for some constant $\beta \in \mathbb{S}^1$, $f_{a_n} \rightharpoonup \beta$ weakly in $W^{1/p,p}$ for $1 < p < \infty$, and $f_{a_n} \to \beta$ strongly in $W^{s,p}$ for every $s > 0$ and every $p \geq 1$ with $sp < 1$.

In particular, the last property implies that, if $\deg f \neq 0$, then, in the limit as $|a| \to 1$, the degree changes.

Proof of Properties 1–3. Since $\deg M_a = 1$, we have $\deg f_a = \deg f$ when f is smooth. [This is an immediate consequence of the definition (1.19).] By density of $C^\infty(\mathbb{S}^1; \mathbb{S}^1)$ in $W^{1/p,p}(\mathbb{S}^1, \mathbb{S}^1)$, we have Property *1*.

We next establish Property 2. We start with $p = 1$. The key fact is that $M_a(e^{\imath\theta}) = e^{\imath\varphi(\theta)}$, $0 \leq \theta \leq 2\pi$, with φ *smooth* and *increasing*. Write $f(e^{\imath\theta}) = e^{\imath\psi(\theta)}$, with $\psi \in W^{1,1}_{loc}(\mathbb{R})$. Then,

$$
\begin{aligned}
|f_a|_{W^{1,1}} &= \int_0^{2\pi} |(\psi \circ \varphi)'(\theta)|\, d\theta = \int_{\varphi^{-1}(0)}^{\varphi^{-1}(2\pi)} |\psi'(\theta)|\, d\theta \\
&= \int_0^{2\pi} |\psi'(\theta)|\, d\theta = |f|_{W^{1,1}}.
\end{aligned}
$$

Let now $1 < p < \infty$. Then, Property *2* follows from the following identity, valid for every measurable function $F : \mathbb{S}^1 \times \mathbb{S}^1 \to [0, \infty]$:

$$
\int_{\mathbb{S}^1} \int_{\mathbb{S}^1} \frac{F(x, y)}{|x - y|^2}\, dxdy = \int_{\mathbb{S}^1} \int_{\mathbb{S}^1} \frac{F(M_a(x), M_a(y))}{|x - y|^2}\, dxdy. \tag{12.127}
$$

Identity (12.127) is obtained as follows. We have $[M_a]^{-1} = M_a$ and thus, after change of variables, (12.127) amounts to

$$
|x - y|^2\, |\dot{M}_a(x)|\, |\dot{M}_a(y)| = |M_a(x) - M_a(y)|^2, \ \forall x, y \in \mathbb{S}^1. \tag{12.128}
$$

In turn, (12.128) follows immediately from the straightforward equality $|\dot{M}_a(x)| = \dfrac{1 - |a|^2}{|1 - \overline{a}x|^2}$.

We now turn to Property *3*. Assume, for example, that $a_n \to 1$. By the proof of Property 2, for every arc $I \subset \mathbb{S}^1$ we have

$$
|f_{a_n}|^p_{W^{1/p,p}(I)} = |f|^p_{W^{1/p,p}(M_{a_n}(I))}. \tag{12.129}
$$

Fix now any closed arc I not containing 1. Then, $J_n := M_{a_n}(I)$ is an arc shrinking to $\{1\}$ as $n \to \infty$. Combining this with (12.129), we obtain that $|f_{a_n}|^p_{W^{1/p,p}(I)} \to 0$ as $n \to \infty$, and thus (I being arbitrary) $f_{a_n} - \fint f_{a_n} \to 0$ in $W^{1/p,p}_{loc}(\mathbb{S}^1 \setminus \{1\})$ as $n \to \infty$. This easily implies Property *3*. □

We conclude this section with a very partial existence result due to Mironescu [258].

Proposition 12.9. *There exists some* $\varepsilon > 0$ *such that* $m_{p,1}$ *is achieved for* $2 - \varepsilon < p < 2$.

We do not know what happens in the remaining cases (see Open Problem 24).

Remark 12.15. The existence of a minimizer of $m_{p,d}$ is very sensitive to the choice of the seminorm in $W^{1/p,p}$. For example, let $p = 2$ and let A be a symmetric uniformly elliptic matrix in $\mathbb{D}$. Let, for $f \in H^{1/2}(\mathbb{S}^1)$, $u_A \in H^1_f(\mathbb{D})$ denote the solution of $-\operatorname{div}(A\nabla u) = 0$ in $\mathbb{D}$, and define $|f|^2_A := \int_{\mathbb{D}} (A\nabla u_A) \cdot \nabla u_A$. The seminorms $|\ |_A$ are equivalent to the standard seminorm $|\ |_{H^{1/2}}$, but the attainability of $m_{2,d}$ for such seminorms depends heavily on A (see Mironescu and Pisante [264]).

12.8.3 Distances between homotopy classes

We present here some properties of the usual distance and of the Hausdorff distance between homotopy classes in $W^{1,1}(\mathbb{S}^1;\mathbb{S}^1)$, in $W^{s,p}(\mathbb{S}^1;\mathbb{S}^1)$ with $sp > 1$, and in $W^{1/p,p}(\mathbb{S}^1;\mathbb{S}^1)$ with $1 < p < \infty$. These classes were characterized in Proposition 12.8.

We follow mainly Brezis, Mironescu, and Shafrir [93], but some results are from Brezis and Nirenberg [104], Rubinstein and Shafrir [308], and Shafrir [322].

12.8.3.1 Distances in $W^{1,1}$

In this subsection, we denote

$$\mathscr{E}_d := \{f \in W^{1,1}(\mathbb{S}^1;\mathbb{S}^1);\ \deg f = d\} = \{e^{\imath\varphi(z)} z^d;\ \varphi \in W^{1,1}(\mathbb{S}^1;\mathbb{R})\}. \quad (12.130)$$

The quantities of interest are the $W^{1,1}$ distance, denoted by $\operatorname{dist}(\mathscr{E}_{d_1}, \mathscr{E}_{d_2})$, and the Hausdorff $W^{1,1}$ distance, denoted by $H - \operatorname{dist}(\mathscr{E}_{d_1}, \mathscr{E}_{d_2})$:

$$\operatorname{dist}(\mathscr{E}_{d_1}, \mathscr{E}_{d_2}) := \inf_{f\in\mathscr{E}_{d_1}} \inf_{g\in\mathscr{E}_{d_2}} \int_{\mathbb{S}^1} |\dot f - \dot g|, \quad (12.131)$$

$$H - \operatorname{dist}(\mathscr{E}_{d_1}, \mathscr{E}_{d_2}) := \max\left\{\operatorname{Dist}(\mathscr{E}_{d_1}, \mathscr{E}_{d_2}),\ \operatorname{Dist}(\mathscr{E}_{d_2}, \mathscr{E}_{d_1})\right\}, \quad (12.132)$$

where

$$\operatorname{Dist}(\mathscr{E}_{d_1}, \mathscr{E}_{d_2}) := \sup_{f\in\mathscr{E}_{d_1}} \inf_{g\in\mathscr{E}_{d_2}} \int_{\mathbb{S}^1} |\dot f - \dot g| = \sup_{f\in\mathscr{E}_{d_1}} \operatorname{dist}(f, \mathscr{E}_{d_2}). \quad (12.133)$$

Proposition 12.10. ([308]) *We have*

$$\operatorname{dist}(\mathscr{E}_{d_1}, \mathscr{E}_{d_2}) = 4|d_1 - d_2|,\ \forall\, d_1,\, d_2 \in \mathbb{Z}. \quad (12.134)$$

In particular, we have

$$|\deg f - \deg g| \le \frac{1}{4}\int_{\mathbb{S}^1} |\dot f - \dot g|,\ \forall f, g \in W^{1,1}(\mathbb{S}^1;\mathbb{S}^1). \quad (12.135)$$

12.8.3.2 Distances in $W^{s,p}$, $sp > 1$

Let $0 < s < \infty$ and $1 \le p < \infty$ be such that $sp > 1$. In this functional setting, we consider the appropriate classes

$$\mathscr{E}_d := \{f \in W^{s,p}(\mathbb{S}^1;\mathbb{S}^1);\ \deg f = d\} = \{e^{\imath\varphi(z)} z^d;\ \varphi \in W^{s,p}(\mathbb{S}^1;\mathbb{R})\}$$

and distances

$$\operatorname{dist}(\mathscr{E}_{d_1}, \mathscr{E}_{d_2}) := \inf\{|f - g|_{W^{s,p}};\ f \in \mathscr{E}_{d_1},\ g \in \mathscr{E}_{d_2}\}, \tag{12.136}$$

$$\operatorname{Dist}(\mathscr{E}_{d_1}, \mathscr{E}_{d_2}) := \sup_{f\in\mathscr{E}_{d_1}} \inf_{g\in\mathscr{E}_{d_2}} |f - g|_{W^{s,p}} = \sup_{f\in\mathscr{E}_{d_1}} \operatorname{dist}(f, \mathscr{E}_{d_2}), \tag{12.137}$$

and

$$H - \operatorname{dist}(\mathscr{E}_{d_1}, \mathscr{E}_{d_2}) := \max\left\{\operatorname{Dist}(\mathscr{E}_{d_1}, \mathscr{E}_{d_2}),\ \operatorname{Dist}(\mathscr{E}_{d_2}, \mathscr{E}_{d_1})\right\}. \tag{12.138}$$

Proposition 12.11. ([93]) *Let* $0 < s < \infty$ *and* $1 \le p < \infty$ *be such that* $sp > 1$. *Then,*

$$\operatorname{dist}(\mathscr{E}_{d_1}, \mathscr{E}_{d_2}) \sim |d_1 - d_2|^s,\ \forall\, d_1,\ d_2 \in \mathbb{Z}. \tag{12.139}$$

Remark 12.16. When $s = 1$, it is possible to improve the conclusion of Proposition 12.11, and to obtain the following counterpart of Proposition 12.10:

$$\inf_{f\in\mathscr{E}_{d_1}} \inf_{g\in\mathscr{E}_{d_2}} |f - g|_{W^{1,p}} = \frac{2^{1+1/p}}{\pi^{1-1/p}} |d_1 - d_2|,\ \forall\, 1 < p < \infty,\ \forall\, d_1,\ d_2 \in \mathbb{Z}; \tag{12.140}$$

see [308, Remark 2.1].

Proposition 12.12. ([93]) *If* $s > 0$ *and* $1 \le p < \infty$ *are such that* $sp > 1$, *then*

$$\operatorname{Dist}(\mathscr{E}_{d_1}, \mathscr{E}_{d_2}) = \infty,\ \forall\, d_1, d_2 \in \mathbb{Z} \text{ such that } d_1 \ne d_2. \tag{12.141}$$

In particular,

$$H - \operatorname{dist}(\mathscr{E}_{d_1}, \mathscr{E}_{d_2}) = \infty,\ \forall\, d_1, d_2 \in \mathbb{Z} \text{ such that } d_1 \ne d_2.$$

12.8.3.3 Distances in $W^{1/p,p}(\mathbb{S}^1;\mathbb{S}^1)$, $1 < p < \infty$

We define dist, Dist and $H -$ dist as in the previous section, with $s = 1/p$ and $1 < p < \infty$.

When $p = 2$, the next result appears in [104, Lemma 6]. For the general case, see [93].

Proposition 12.13. *Let* $1 < p < \infty$. *Then we have*

$$\operatorname{dist}(\mathscr{E}_{d_1}, \mathscr{E}_{d_2}) = 0,\ \forall\, d_1, d_2 \in \mathbb{Z}. \tag{12.142}$$

We call the attention of the reader to the following beautiful result of Shafrir [322].

Theorem 12.11. *Let $1 < p < \infty$. Then we have*

$$\operatorname{Dist}(\mathscr{E}_{d_1}, \mathscr{E}_{d_2}) = \inf\{|f|_{W^{1/p,p}};\ f \in \mathscr{E}_{d_1-d_2}\},\ \forall\, d_1, d_2 \in \mathbb{Z}. \tag{12.143}$$

In particular (see Theorem 12.9), there exist $0 < C'_p \le C_p < \infty$ such that

$$C'_p|d_1 - d_2|^{1/p} \le \operatorname{Dist}(\mathscr{E}_{d_1}, \mathscr{E}_{d_2}) \le C_p|d_1 - d_2|^{1/p},\ \forall\, d_1, d_2 \in \mathbb{Z}. \tag{12.144}$$

When $p = 2$, we have the exact formula

$$\operatorname{Dist}(\mathscr{E}_{d_1}, \mathscr{E}_{d_2}) = (2\pi|d_1 - d_2|)^{1/2},\ \forall\, d_1, d_2 \in \mathbb{Z}. \tag{12.145}$$

12.8.4 Degree of maps $u : \Omega \times \mathbb{S}^1 \to \mathbb{S}^1$

Let $\Omega \subset \mathbb{R}^N$ be a domain (i.e., open and connected set). If $u \in C(\Omega \times \mathbb{S}^1; \mathbb{S}^1)$, then $\deg u(\lambda, \cdot)$ is independent of $\lambda \in \Omega$. This follows from the fact that deg is locally constant in $C(\mathbb{S}^1; \mathbb{S}^1)$. We investigate here the analogue of this property when $u \in W^{s,p}(\Omega \times \mathbb{S}^1; \mathbb{S}^1)$. For such u, we have $u(\lambda, \cdot) \in W^{s,p}(\mathbb{S}^1; \mathbb{S}^1)$ for a.e. $\lambda \in \Omega$, and thus $\deg u(\lambda, \cdot)$ is well-defined for a.e. $\lambda \in \Omega$, provided $sp \ge 1$. The natural question is whether this degree is constant a.e.

Here is our main result.

Proposition 12.14. *Let $\Omega \subset \mathbb{R}^N$ be a domain. Let $1 < p < \infty$ and let $u \in W^{2/p,p}(\Omega \times \mathbb{S}^1; \mathbb{S}^1)$. Then, $\Omega \ni \lambda \mapsto \deg u(\lambda, \cdot)$ is constant a.e. on Ω.*

In this case, we set

$$\deg u = \deg u(\lambda, \cdot),\ \text{ for a.e. } \lambda \in \Omega. \tag{12.146}$$

Remark 12.17. The regularity assumption on u in Proposition 12.14 is optimal. Indeed, let $u(\lambda, x) := \dfrac{\lambda x - 2}{|\lambda x - 2|}$, with $\lambda \in \Omega := (1, 3)$ and $x \in \mathbb{S}^1$. Then, u may be identified with the map $v(z) = \dfrac{z - 2}{|z - 2|}$, $1 < |z| < 3$. By Lemma 15.12, we have $u \in W^{s,p}$ for every s and p such that $sp < 2$. However, we have $\deg u(\lambda, \cdot) = \begin{cases} 0, & \text{if } 1 < \lambda < 2 \\ 1, & \text{if } 2 < \lambda < 3 \end{cases}$, and thus $\lambda \mapsto \deg u(\cdot, \lambda)$ is not constant a.e.

Proof of Proposition 12.14. We may assume that Ω is a ball. By the proof of Theorem 10.2' in Chap. 14 (see also item 4 in Section 14.6.2), there exists a sequence $(u_n) \subset C^\infty(\overline{\Omega} \times \mathbb{S}^1; \mathbb{S}^1)$ such that $u_n \to u$ in $W^{2/p,p}$. [In order to be in a position to apply this result, we identify u with a map on $\Omega \times \mathbb{R}$.] By Corollary 15.2, possibly after passing to a subsequence, still denoted by (u_n), we may assume that for a.e. $\lambda \in \Omega$ we have $u_n(\lambda, \cdot) \to u(\lambda, \cdot)$ in $W^{2/p,p}$ as $n \to \infty$. For such λ, we have $\deg u_n(\lambda, \cdot) \to \deg u(\lambda, \cdot)$. Let $d_n := \deg u_n(\lambda, \cdot)$, which is independent of $\lambda \in \Omega$. Since, for a.e. λ, we have $d_n \to \deg u(\lambda, \cdot)$, we find that $d_n \to d \in \mathbb{Z}$, and that $\deg u(\lambda, \cdot) = d$ for a.e. λ. □

Remark 12.18. The above proof shows that a stronger variant of Proposition 12.14 holds. Assume that $u \in W^{1/p,p}(\Omega \times \mathbb{S}^1; \mathbb{S}^1)$, $1 < p < \infty$, satisfies $Ju = 0$. Then, $\Omega \ni \lambda \mapsto \deg u(\lambda, \cdot)$ is constant a.e. [Recall that $Ju = 0$ if $u \in W^{2/p,p}(\Omega \times \mathbb{S}^1; \mathbb{S}^1)$ (see (8.12)).]

Indeed, under the assumption $Ju = 0$ we may argue as in the proof of Proposition 12.14, using Theorem 10.4' instead of Theorem 10.2'.

We end this section with an unexpected stability result.

Proposition 12.15. *Let* $1 < p < \infty$ *and let* $(u_n) \subset W^{2/p,p}(\Omega \times \mathbb{S}^1; \mathbb{S}^1)$ *be a sequence such that* $u_n \to u$ *strongly in* $W^{1/p,p}(\Omega \times \mathbb{S}^1; \mathbb{S}^1)$ – *this holds, for example, if* $u_n \rightharpoonup u$ *weakly in* $W^{2/p,p}$. *Then,* $\deg u_n \to \deg u$.

In particular, for every $d \in \mathbb{Z}$, *the problem*

$$\inf \left\{ \int_{\Omega \times \mathbb{S}^1} |\nabla u|^2;\ u \in H^1(\Omega \times \mathbb{S}^1; \mathbb{S}^1),\ \deg u = d \right\}$$

admits a minimizer.

Proof. We argue as in the proof of Proposition 12.14. Possibly after passing to a subsequence, we have $u_n(\lambda, \cdot) \to u(\lambda, \cdot)$ in $W^{1/p,p}$ for a.e. λ as $n \to \infty$. This implies that $\deg u_n \to \deg u$. □

Remark 12.19. Using Remark 12.18, we see that the conclusion Proposition 12.15 holds under the weaker assumptions $(u_n) \subset W^{1/p,p}(\Omega \times \mathbb{S}^1; \mathbb{S}^1)$ and $Ju_n = 0$, $\forall\, n$.

Remark 12.20. We emphasize that the conclusion of Proposition 12.15 is somewhat counterintuitive. Take, for example, $p = 2$ and $\Omega = (1/2, 1)$, so that

$$\Omega \times \mathbb{S}^1 \simeq \omega := \{x \in \mathbb{R}^2;\ 1/2 < |x| < 1\}.$$

Clearly, if $u \in H^1(\omega; \mathbb{S}^1)$, then $\deg u = \deg(u_{|\mathbb{S}^1})$. Assume that $u_n \rightharpoonup u$ weakly in $H^1(\Omega; \mathbb{S}^1)$. Then, Proposition 12.15 asserts that $\deg(u_{n|\mathbb{S}^1}) \to \deg(u_{|\mathbb{S}^1})$.

On the other hand, the assumption $u_n \rightharpoonup u$ only implies that $u_{n|\mathbb{S}^1} \rightharpoonup u_{|\mathbb{S}^1}$ *weakly* in $H^{1/2}(\mathbb{S}^1; \mathbb{S}^1)$, and we know that the degree is *not* stable under weak convergence in $H^{1/2}(\mathbb{S}^1)$.

Remark 12.21. Here is a simple inequality closely related to Proposition 12.15. We have, for every $u, v \in H^1(\Omega \times \mathbb{S}^1; \mathbb{S}^1)$,

$$|\deg u - \deg v| \le \frac{1}{2\pi\, |\Omega|} \|u - v\|_{L^2} \left(|u|_{H^1} + |v|_{H^1}\right). \tag{12.147}$$

Indeed, for a.e. $\lambda \in \Omega$ we have, by the Cauchy formula (12.27),

$$\begin{aligned}\deg u(\lambda, \cdot) - \deg v(\lambda, \cdot) &= \frac{1}{2\pi} \int_{\mathbb{S}^1} [(u(\lambda, \cdot) \wedge u(\lambda, \cdot)_\tau) - (v(\lambda, \cdot) \wedge v(\lambda, \cdot)_\tau)] \\ &= \frac{1}{2\pi} \int_{\mathbb{S}^1} (u(\lambda, \cdot) - v(\lambda, \cdot)) \wedge (u(\lambda, \cdot)_\tau + v(\lambda, \cdot)_\tau),\end{aligned}$$

and thus

$$|\deg u(\lambda,\cdot)-\deg v(\lambda,\cdot)|\le\frac{1}{2\pi}\,\|u(\lambda,\cdot)-v(\lambda,\cdot)\|_{L^2(\mathbb{S}^1)}$$
$$\times\left(\|u(\lambda,\cdot)_\tau\|_{L^2(\mathbb{S}^1)}+\|v(\lambda,\cdot)_\tau\|_{L^2(\mathbb{S}^1)}\right).$$

Integrating in λ and using Proposition 12.14, we obtain (12.147).

12.8.5 Trace theory revisited

We return to the *global* aspect of trace theory, already discussed in Section 12.2 for a simple geometry ($\Omega=\mathbb{D}$). More precisely, let $\Omega\subset\mathbb{R}^2$ be smooth bounded, $0<s<\infty$ be a non-integer, and $1\le p<\infty$. We do not assume Ω simply connected, and we denote Γ_j the components of $\partial\Omega$. Given $f\in W^{s,p}(\partial\Omega;\mathbb{S}^1)$, we ask whether $f=\operatorname{tr} u$ for some $u\in W^{s+1/p,p}(\Omega;\mathbb{S}^1)$. In view of the proof of Theorem 11.1, the answer is positive when $sp<1$. We thus assume that $sp\ge 1$; in this case, Theorem 11.1 implies the existence of an extension of f near $\partial\Omega$, but not on all of Ω. In the range $sp\ge 1$, f has a degree on each component Γ_j of $\partial\Omega$. This degree is defined as follows. Consider on Γ_j the orientation induced by the outward normal n to Ω (i.e., (n,τ) is a direct basis). Let $\Phi_j:\mathbb{S}^1\to\Gamma_j$ be an orientation-preserving parametrization of Γ_j. Then we set $\deg(f,\Gamma_j):=\deg f\circ\Phi_j$. It is easy to see that this degree does not depend on the choice of Φ_j, and it has all the properties of the degree established in the previous sections. Moreover, when $f\in W^{1,1}$ we have

$$\deg(f,\Gamma_j)=\frac{1}{2\pi}\int_{\Gamma_j}f\wedge f_\tau. \tag{12.148}$$

The main result here is the following

Proposition 12.16. *Let $\Omega\subset\mathbb{R}^2$ be a smooth bounded domain, whose boundary has components $\Gamma_1,\ldots,\Gamma_k$. Let $0<s<\infty$ be non-integer, and $1\le p<\infty$ be such that $sp\ge 1$. Let $f\in W^{s,p}(\partial\Omega;\mathbb{S}^1)$. Then, f has an extension $u\in W^{s+1/p,p}(\Omega;\mathbb{S}^1)$ if and only if*

$$\sum_{j=1}^{k}\deg(f,\Gamma_j)=0. \tag{12.149}$$

In particular, a map $f\in H^{1/2}(\partial\Omega;\mathbb{S}^1)$ has an extension $u\in H^1(\Omega;\mathbb{S}^1)$ if and only if (12.149) *holds.*

Proof. Step 1. Proof of "$\Longleftarrow$"
We assume that $k\ge 2$. The case $k=1$ is easier and is left to the reader.

Assume that Γ_1 encloses Ω; therefore, Γ_1 is oriented counterclockwise, whereas Γ_j, $j\ge 2$, is oriented clockwise. Let $d_1:=\deg(f,\Gamma_1)$ and $d_j:=-\deg(f,\Gamma_j)$, $j=$

$2, \ldots, k$. For each $j \geq 2$, let a_j be a point in the interior of Γ_j, and set $v_j(z) := \left(\dfrac{z - a_j}{|z - a_j|}\right)^{d_j}$, $z \neq a_j$. Then,

$$\deg(v_j, \Gamma_\ell) = \begin{cases} \deg(f, \Gamma_j), & \text{if } \ell = j \\ d_j, & \text{if } \ell = 1 \\ 0, & \text{otherwise} \end{cases}.$$

Set $v := \prod_{j=2}^k v_j \in C^\infty(\overline{\Omega}; \mathbb{S}^1)$ and $w := v_{|\partial\Omega}$. By the above, we have $\deg(w, \Gamma_j) = \deg(f, \Gamma_j)$, for each j. Let $g := f\,\overline{w}$. Then, $\deg(g, \Gamma_j) = 0$ for each j. We may thus find some $\varphi \in W^{s,p}(\partial\Omega; \mathbb{R})$ such that $g = e^{\imath\varphi}$ (see Proposition 12.1, item 2). Let $\psi \in W^{s+1/p,p}(\Omega; \mathbb{R})$ be an extension of φ. By Corollary 15.3, $u := e^{\imath\psi}\, v$ is a $W^{s+1/p,p}(\Omega; \mathbb{S}^1)$ extension of f.

Step 2. Proof of " $\Longrightarrow$ "

Since $(s + 1/p)\,p \geq 2$, we may approximate u in $W^{s+1/p,p}$ by a sequence $(u_n) \subset C^\infty(\overline{\Omega}; \mathbb{S}^1)$ (see Theorem 10.2' in Chapter 14). Then, $\operatorname{tr} u_n \to \operatorname{tr} u$ in $W^{s,p} \hookrightarrow VMO$, and therefore $\deg(u_n, \Gamma_j) \to \deg(f, \Gamma_j)$ as $n \to \infty$. By the above, it suffices to establish " $\Longrightarrow$ " for $v \in C^\infty(\overline{\Omega}; \mathbb{S}^1)$. When v is smooth and $|v| = 1$ on $\partial\Omega$, we have, via integration by parts (as in (12.30)) and (12.148):

$$\frac{1}{\pi}\int_\Omega \operatorname{Jac} v = \frac{1}{2\pi}\int_{\partial\Omega} v \wedge v_\tau = \frac{1}{2\pi}\sum_{j=1}^k \int_{\Gamma_j} v \wedge v_\tau = \sum_{j=1}^k \deg(u, \Gamma_j).$$

In our case, we have $\operatorname{Jac} v = 0$ on Ω (since $|v| = 1$), whence the conclusion. □

Corollary 12.4. *Let Ω be simply connected. Then a map $f \in H^{1/2}(\partial\Omega; \mathbb{S}^1)$ has an extension $u \in H^1(\Omega; \mathbb{S}^1)$ if and only if $\deg f = 0$.*

Moreover, if $\deg f = 0$, and we write $f = e^{\imath\varphi_0}$, with $\varphi_0 \in H^{1/2}(\partial\Omega; \mathbb{R})$, then

$$\begin{aligned} H^1_f(\Omega; \mathbb{S}^1) :=& \{u \in H^1(\Omega; \mathbb{S}^1);\ \operatorname{tr} u = f \text{ on } \partial\Omega\} \\ =& \{e^{\imath\varphi};\ \varphi \in H^1(\Omega; \mathbb{R}),\ \operatorname{tr}\varphi = \varphi_0 \text{ on } \partial\Omega\}. \end{aligned} \tag{12.150}$$

In particular, for such f the minimum

$$\min\left\{\int_\Omega |\nabla u|^2;\ u \in H^1(\Omega; \mathbb{S}^1) \text{ and } u \in H^1_f(\Omega; \mathbb{S}^1)\right\} \tag{12.151}$$

is achieved by $e^{\imath\varphi}$, where φ is the harmonic extension of φ_0.

Proof. Inclusion "$\supset$" in (12.150) follows from the straightforward equality

$$\operatorname{tr} e^{\imath\varphi} = e^{\imath \operatorname{tr}\varphi}, \ \forall \varphi \in H^1(\Omega; \mathbb{R}). \tag{12.152}$$

Inclusion "$\subset$" is obtained as follows. If $u \in H^1(\Omega; \mathbb{S}^1)$, then we may write $u = e^{\imath\varphi}$ with $\varphi \in H^1(\Omega; \mathbb{R})$ (Theorem 1.1). By (12.152) and Corollary 6.2, we have (mod 2π) $\operatorname{tr}\varphi = \varphi_0$, whence "$\subset$."

The last part of the corollary follows from (12.150) and the equality $|\nabla(e^{\imath\varphi})| = |\nabla\varphi|$. □

We conclude this section with a stability result related to Proposition 12.15.

Proposition 12.17. *Let $\Omega \subset \mathbb{R}^2$ be a smooth bounded domain, whose boundary has components $\Gamma_1, \ldots, \Gamma_k$. Let $1 < p < \infty$. Let $(u_n) \subset W^{2/p,p}(\Omega; \mathbb{S}^1)$ be such that $u_n \rightharpoonup u$ weakly in $W^{2/p,p}$. Then, $\deg(u_n, \Gamma_j) \to \deg(u, \Gamma_j)$ as $n \to \infty$, for every j.*

In particular, given integers $d_1, \ldots, d_k$ satisfying the compatibility condition $\sum_{j=1}^k d_j = 0$, the non-empty class

$$\{u \in H^1(\Omega; \mathbb{S}^1);\ \deg(u, \Gamma_j) = d_j,\ j = 1, \ldots, k\}$$

is weakly sequentially closed in H^1.

Proof. We reduce this result to Proposition 12.15. Assume, for example, that Γ_1 encloses Ω. We present the proof for $j = 1$; the other cases are similar. Let $\Phi_1 : \mathbb{S}^1 \to \Gamma_1$ be a smooth orientation-preserving parametrization of Γ_1. For small ε, this Φ_1 induces a natural parametrization $\Psi : [0, \varepsilon] \times \mathbb{S}^1 \to \{x \in \overline{\Omega};\ \operatorname{dist}(x, \Gamma_1) \le \varepsilon\}$ via $\Psi(t, z) := \Phi_1(z) - t\,\nu_{\Phi_1(z)}$. Here, ν_y is the outer normal to $\partial\Omega$ at y. Set $v_n := u_n \circ \Psi$ and $u := u \circ \Psi$, so that (v_n) is bounded in $W^{2/p,p}((0, \varepsilon) \times \mathbb{S}^1; \mathbb{S}^1)$ and $v_n \rightharpoonup v$ weakly in $W^{2/p,p}$.

We clearly have $\deg v_n = \deg(u_n, \Gamma_1)$ and $\deg v = \deg(u, \Gamma_1)$. The conclusion of Proposition 12.17 follows from Proposition 12.15. □

Remark 12.22. As a consequence of Proposition 12.17, the following problem

$$\inf\left\{\int_\Omega |\nabla u|^2;\ u \in H^1(\Omega; \mathbb{S}^1),\ \deg(u, \Gamma_j) = d_j,\ j = 1, \ldots, k\right\} \text{ with } \sum_{j=1}^k d_j = 0,$$

admits a minimizer. This problem has been extensively studied in Bethuel, Brezis, and Hélein [31, Chapter I] via a direct approach.

12.8.6 Maps from $\mathbb{S}^N$ to $\mathbb{S}^N$

Many results from this chapter admit extensions to maps from $\mathbb{S}^N$ to $\mathbb{S}^N$, $N \ge 2$. Here are some of them.

1. A degree theory for $VMO\,(\mathbb{S}^N;\mathbb{S}^N)$ has been developed in Brezis and Nirenberg [104].
2. An integral representation for the degree. There are two aspects.

   a) The first Kronecker formula

   $$\deg f = \frac{1}{|\mathbb{S}^N|}\int_{\mathbb{S}^N} \mathrm{Jac} f \tag{12.153}$$

   makes trivially sense for $f \in W^{1,N}(\mathbb{S}^N;\mathbb{S}^N)$; it can be extended (via the second Kronecker formula below) to $f \in W^{1-1/(N+1),N+1}(\mathbb{S}^N;\mathbb{S}^N)$.

   b) The second Kronecker formula

   $$\deg f = \frac{1}{|\mathbb{B}^{N+1}|}\int_{\mathbb{B}^{N+1}} \mathrm{Jac}\, U, \tag{12.154}$$

   where U is an appropriate extension of f to $\mathbb{B}^{N+1}$ and with values in $\mathbb{R}^{N+1}$. This makes trivially sense for any $f \in W^{1-1/(N+1),N+1}(\mathbb{S}^N;\mathbb{S}^N)$. When $f \in VMO\,(\mathbb{S}^N;\mathbb{S}^N)$, one can still give a meaning to (12.154) using the same strategy as in the proof of (12.39).

3. The analogue of Section 12.4 is considered in Brezis and Nguyen [98], where a distributional degree is defined for $f \in (VMO \cap W^{1-1/N,N})(\mathbb{S}^N;\mathbb{S}^N)$.
4. The estimate

   $$|\deg f| \le C_p \int_{\mathbb{S}^N}\int_{\mathbb{S}^N} \frac{|f(x)-f(y)|^p}{|x-y|^{2N}}\,dxdy,\ \forall N<p<\infty, \tag{12.155}$$

   is proved in Bourgain, Brezis, and Mironescu [56, Theorem 0.6].
5. The estimate

   $$|\deg f| \le C_N \iint\limits_{\substack{\mathbb{S}^N\,\mathbb{S}^N\\ |f(x)-f(y)|\ge \ell_N}} \frac{1}{|x-y|^{2N}}\,dxdy$$

   where $\ell_N := \left(2+\dfrac{2}{N+1}\right)^{1/2}$ and $f \in C(\mathbb{S}^N;\mathbb{S}^N)$ is due to Nguyen [279]; the value of ℓ_N is optimal (as in Theorem 12.5).
   This was raised as an open problem in [56, Open Problem 2] with a small constant δ in place of ℓ_N and was proved with $\delta<\sqrt{2}$ (instead of ℓ_N) in Bourgain, Brezis, and Nguyen [58].
   A refined estimate

   $$|\deg f| \le C_N\delta^N \iint\limits_{\substack{\mathbb{S}^N\,\mathbb{S}^N\\ |f(x)-f(y)|\ge \delta}} \frac{1}{|x-y|^{2N}}\,dxdy,\ \forall\,\delta\le \ell_N,\ \forall f \in C(\mathbb{S}^N;\mathbb{S}^N),$$

was established by Nguyen [282] when $N \geq 2$. The case $N = 1$ is not settled; see below, Open Problem 17.

6. All the results in Section 12.6 were extended to higher dimensions in Brezis, Mironescu, and Shafrir [93].
7. Extensions of some of the results in Section 12.8.4 can be found in Brezis, Li, Mironescu, and Nirenberg [81].

12.8.7 Open problems

The first question goes somewhat in the "opposite direction" of Theorems 12.4 and 12.5. Our goal is to bound $|\deg f|$ in terms of $I_\delta(f)$ for *small* δ. This seems absurd because $I_\delta(f) \nearrow \infty$ as $\delta \searrow 0$ (unless f is constant). However, we may attempt to bound $|\deg f|$ by $c_\delta I_\delta(f)$ with $c_\delta \to 0$ as $\delta \to 0$. Here is a question raised by Brezis [74, Open Problem 3].

Open Problem 17. Is it true that for some C independent of small δ we have

$$|\deg f| \leq C\,\delta \iint_{\substack{\mathbb{S}^1\times\mathbb{S}^1\\ |f(x)-f(y)|\geq\delta}} \frac{1}{|x-y|^2}\,dxdy = C\,\delta\, I_\delta(f),\ \forall f \in VMO\,(\mathbb{S}^1;\mathbb{S}^1)?$$

Such an estimate does hold in the limit as $\delta \to 0$ when f is smooth. More precisely, one has (see Brezis [76, Theorem 3.1] and Brezis and Nguyen [102, Proposition 1])

$$\lim_{\delta\to 0} \delta \iint_{\substack{\mathbb{S}^1\times\mathbb{S}^1\\ |f(x)-f(y)|\geq\delta}} \frac{1}{|x-y|^2}\,dxdy = C\int_{\mathbb{S}^1} |\dot{f}|.$$

Open Problem 18. Is there a simpler, more geometric, proof of Theorem 12.5? What is the best constant C in (12.83)? Is it achieved?

From Theorem 12.3, we know that, for every $1 < p < \infty$,

$$|\deg f| \leq C_p\, |f|^p_{W^{1/p,p}} = C_p\,(A_{p,\delta} + B_{p,\delta}), \tag{12.156}$$

where

$$A_{p,\delta} := \iint_{\substack{\mathbb{S}^1\times\mathbb{S}^1\\ |f(x)-f(y)|\geq\delta}} \frac{|f(x)-f(y)|^p}{|x-y|^2}\,dxdy$$

and

$$B_{p,\delta} := \iint\limits_{\substack{\mathbb{S}^1\times\mathbb{S}^1\\ |f(x)-f(y)|<\delta}} \frac{|f(x)-f(y)|^p}{|x-y|^2}\,dxdy.$$

Note that, with $I_\delta(f)$ defined in (12.70), we have

$$\delta^p\, I_\delta(f) \le A_{p,\delta} \le 2^p\, I_\delta(f)$$

and thus Theorem 12.4 implies

$$|\deg f| \le \frac{C}{\delta^p} A_{p,\delta}. \tag{12.157}$$

In view of (12.156) and (12.157), one may wonder whether $|\deg f|$ can also be bounded in terms of $B_{p,\delta}$. More precisely, we raise

Open Problem 19. Is it true that $\forall\, 1 < p < \infty$, $\forall\, \delta > 0$,

$$|\deg f| \le C_{p,\delta} \iint\limits_{\substack{\mathbb{S}^1\times\mathbb{S}^1\\ |f(x)-f(y)|<\delta}} \frac{|f(x)-f(y)|^p}{|x-y|^2}\,dxdy,\ \ \forall f \in VMO\,(\mathbb{S}^1;\mathbb{S}^1),$$

for some constant $C_{p,\delta}$? Can one take $C_{p,\delta} = \dfrac{C\,(p-1)}{\delta^{p-1}}$ with C independent of p and δ?

This question is suggested by the fact that

$$\lim_{\delta\to 0} \frac{1}{\delta^{p-1}} \iint\limits_{\substack{\mathbb{S}^1\times\mathbb{S}^1\\ |f(x)-f(y)|<\delta}} \frac{|f(x)-f(y)|^p}{|x-y|^2}\,dxdy = \frac{2}{p-1}\int\limits_{\mathbb{S}^1} |\dot{f}(x)|\,dx$$

for every $f \in C^1(\mathbb{S}^1;\mathbb{S}^1)$ (see Brezis [76, Theorem 3.1] and Brezis and Nguyen [102, Proposition 1 and Remark 6]).

Open Problem 20. Let $N \ge 1$ and $1 < p < \infty$. Is it true that $|\deg f_n - \deg g_n| \to 0$ for any sequences (f_n), (g_n) in $W^{N/p,p}(\mathbb{S}^N;\mathbb{S}^N)$ such that

$$|f_n - g_n|_{W^{N/p,p}} \to 0 \text{ as } n \to \infty$$

and

$$|f_n|_{W^{N/p,p}} + |g_n|_{W^{N/p,p}} \text{ remains bounded as } n \to \infty?$$

The answer is positive when $N = 1$ (see Mironescu and Shafrir [267]) and when $N \ge 2$ and $p \le N+1$ (see Brezis, Mironescu, and Shafrir [93, Proposition 7.9]).

Open Problem 21. This is a pedagogical question. Is there a proof of Corollary 12.2 that does not rely on Theorem 12.6?

Open Problem 22. What happens to Corollary 12.3 when $f, g \in W^{1/p,p}(\mathbb{S}^1; \mathbb{S}^1)$ with $p > 3$ (resp. $f, g \in C^{0,\alpha}(\mathbb{S}^1; \mathbb{S}^1)$ with $\alpha \leq 1/3$)? The problem is open even under the stronger assumption that $f \in W^{1/p,p}(\mathbb{S}^1; \mathbb{S}^1)$ with $2 < p \leq 3$ and $g \in W^{1/q,q}(\mathbb{S}^1; \mathbb{S}^1)$ with $q > 3$ (resp. $f \in C^{0,\alpha}(\mathbb{S}^1; \mathbb{S}^1)$ with $\alpha > 1/3$ and $g \in C^{0,\beta}(\mathbb{S}^1; \mathbb{S}^1)$ with $\beta \leq 1/3$).

Open Problem 23. Let $1 < p < \infty$. Is it true that

$$m_{p,d} = |d|\, m_{p,1}, \ \forall d \in \mathbb{Z}? \tag{12.158}$$

The answer is positive when $p = 2$.

Assuming that (12.158) does not hold, is there an explicit formula for the limit

$$\lim_{d\to\infty} \frac{m_{p,d}}{d} = \inf_{d>0} \frac{m_{p,d}}{d}?$$

[The equality lim = inf follows from Lemma 12.6 and Fekete's lemma.]

Open Problem 24. Let $1 < p < \infty$ and $d > 0$. Is

$$m_{p,d} := \inf\{|f|^p_{W^{1/p,p}};\ f \in W^{1/p,p}(\mathbb{S}^1; \mathbb{S}^1),\ \deg f = d\}$$

achieved?

Recall that the answer is positive for $p = 2$, with any $d > 0$, and for $2 - \varepsilon < p < 2$ with $d = 1$ (Theorem 12.10 item 2 and Proposition 12.9).

12.9 Comments

1. The extension of degree theory to *VMO* $(\mathbb{S}^N; \mathbb{S}^N)$ is due to Brezis and Nirenberg [104], and the proof of Theorem 12.1 follows [104]. There were two predecessors:
   (a) In connection with their proof of existence of "large harmonic maps," Brezis and Coron [78] (see also Brezis [68, 70]) were led to consider degree for H^1 maps from $\mathbb{S}^2$ to $\mathbb{S}^2$. The degree is defined by an integral via the Kronecker formula (see (12.153)). To prove that this integral is an integer one relies on the fact that smooth maps from $\mathbb{S}^2$ to $\mathbb{S}^2$ are dense in $H^1(\mathbb{S}^2; \mathbb{S}^2)$ (see Schoen and Uhlenbeck [318] and Comment 1 in Chapter 10).
   (b) Motivated by a question concerning the Ginzburg–Landau equation (see [67]), L. Boutet de Monvel and Gabber introduced the degree for maps $u \in H^{1/2}(\mathbb{S}^1; \mathbb{S}^1)$, using the approach described in Proposition 12.3, via the Cauchy formula and the $H^{1/2}, H^{-1/2}$ duality. They proved that the expression in (12.28) is an integer, by approximation, as in (a).

2. The Kronecker formula for maps in $H^{1/2}(\mathbb{S}^1; \mathbb{S}^1)$ (see Proposition 12.4) has been known for a while. It is implicit in Bethuel, Brezis, and Hélein [31, Introduction]. The proof of a cousin result appears in Bourgain, Brezis, and Mironescu [54, proof of Lemma 3]. Theorem 12.2 is from Mironescu [252, Section 2.2] (with roots in Bourgain, Brezis, and Mironescu [56, proof of Theorem 1.1]).
3. The distributional degree was introduced by Brezis and Nguyen [98] for Sobolev maps. Proposition 12.5 is a special case of [98, Theorem 1]. Propositions 12.6 and 12.7 are strictly speaking new, but reminiscent of Propositions 12.3 and 12.4, which were known before.
4. Theorem 12.3 comes from Bourgain, Brezis, and Mironescu [56, Corollary 0.5]. A weak version of Theorem 12.4 (for very small δ) was obtained in Bourgain, Brezis, and Mironescu [55, Theorem 4]. Theorem 12.4 is due to Bourgain, Brezis, and Nguyen [58]. The optimal result Theorem 12.5 was established by Nguyen [279] for continuous f. The version stated here (with $f \in VMO$) is new, but the proof borrows some ideas from [279].
5. In Proposition 12.8, the path-connectedness of $\{f \in VMO(\mathbb{S}^1; \mathbb{S}^1);\ \deg f = d\}$ is due to Brezis and Nirenberg [104, Property 2, Section I.4]. The second part of the proposition is new.
6. The striking formula (12.103), relating the degree and the Fourier coefficients of f, was brought to light by Brezis [74], in the wake of a question raised by Gelfand in 1995 at a seminar by Brezis at Rutgers.
7. Theorem 12.7 is due to Kahane [208] (see also Kahane [209, 210]) under the stronger assumption $f \in C^{0,\alpha}$, $\alpha > 1/3$. The proof we present borrows some ideas from [208]. Korevaar [214] established a result similar to Theorem 12.8 for the summation process σ_N that appears in (12.109).
8. Theorem 12.10 item *2* appears in Berlyand, Mironescu, Rybalko, and Sandier [24, Lemma 3.1]. Theorem 12.9 is new.
9. The basic example in Section 12.8 is reminiscent of bubbling phenomena in non-compact problems, for which there exists a vast body of literature. In the context of the minimization of $m_{2,d}$, it appeared first in Berlyand and Mironescu [23]. Proposition 12.9 is from Mironescu [258], and its proof relies heavily on Berlyand, Mironescu, Rybalko, and Sandier [24, Theorem 4.4].
10. Propositions 12.14 and 12.15 are from Brezis, Li, Mironescu, and Nirenberg [81]. They were motivated by an earlier striking result of Rubinstein and Sternberg [309] dealing with the case $p = 2$ (with roots in White [345, Theorem 2.1]).
11. Proposition 12.16 is new, but has been known for a while when $s = 1/2$ and $p = 2$ (see, for example, Bethuel, Brezis and Hélein [31, Chapter I]). Proposition 12.17 is new, but the special case $p = 2$ was obtained (with a different argument) by Lassoued [218, 219].

Chapter 13
Dirichlet problems. Gaps. Infinite energies

13.0 Introduction

In this chapter, we investigate minimization problems of the form

$$A_g := \inf\left\{\int_\Omega |\nabla u|^p;\ u \in W^{1,p}_g(\Omega;\mathbb{S}^1)\right\}, \tag{13.1}$$

where $\Omega \subset \mathbb{R}^N$, $N \geq 2$, is a smooth bounded domain, $g \in C^\infty(\partial\Omega;\mathbb{S}^1)$ is a given boundary condition, and $1 \leq p < \infty$. For simplicity, we assume throughout this chapter (except in Theorem 13.5) that Ω is *diffeomorphic to a ball*.

When $N \geq 3$, we always have $W^{1,p}_g(\Omega;\mathbb{S}^1) \neq \emptyset$. Indeed, we may write $g = e^{\imath\psi_0}$ for some $\psi_0 \in C^\infty(\partial\Omega;\mathbb{R})$ (since $\partial\Omega$ is simply connected). Let $\varphi_0 \in C^\infty_{\psi_0}(\overline{\Omega};\mathbb{R})$. Then, $u_0 := e^{\imath\varphi_0} \in C^\infty_g(\overline{\Omega};\mathbb{S}^1) \subset W^{1,p}_g(\Omega;\mathbb{S}^1)$.

When $N = 2$ and $p \geq 2$, we have $W^{1,p}_g(\Omega;\mathbb{S}^1) \neq \emptyset$ if and only if $\deg g = 0$, by Proposition 12.16 applied with $s = 1 - 1/p$, so that $sp = p - 1 \geq 1$. When $N = 2$ and $1 \leq p < 2$, we always have $W^{1,p}_g(\Omega;\mathbb{S}^1) \neq \emptyset$. Indeed, assume, for example, that $\Omega = \mathbb{D}$ and note that $u(x) := g(x/|x|) \in W^{1,p}_g(\mathbb{D};\mathbb{S}^1)$, $\forall\, p \in [1,2)$.

Section 13.1 is devoted to the case $p \geq 2$. Theorem 13.1 asserts that the infimum in (13.1) is uniquely achieved and minimizers belong to $C^{1,\alpha}(\overline{\Omega})$ for some $\alpha \in (0,1)$. This stands in sharp contrast to general $\mathbb{S}^k$-valued variational problems for $k \geq 2$, where minimizers are usually discontinuous and one can only assert their partial regularity (in the spirit of Schoen and Uhlenbeck [317, 318] and Hardt and Lin [183]). The key ingredient is the basic lifting Theorem 1.1, which reduces problem (13.1) to a scalar situation.

The rest of this chapter concerns the case $1 < p < 2$, which is much more delicate, since we do not have lifting at our disposal. Here, new phenomena occur and the class of competitors plays a subtle role.

In Section 13.2, we present a partial regularity result for minimizers of (13.1): in general, they admit a compact singular set $K \subset \Omega$, of finite $(N-2)$-Hausdorff

H. Brezis and P. Mironescu, *Sobolev Maps to the Circle*, Progress in Nonlinear Differential Equations and Their Applications 96,
https://doi.org/10.1007/978-1-0716-1512-6_13

measure when $N \geq 3$, resp. consisting of a finite number of points when $N = 2$ (see Theorem 13.2).

We may also restrict ourselves to continuous competitors. More precisely, set

$$B_g := \inf\left\{\int_\Omega |\nabla u|^p;\ u \in W_g^{1,p}(\Omega;\mathbb{S}^1) \cap C(\overline{\Omega};\mathbb{S}^1)\right\}. \tag{13.2}$$

Then, $B_g < \infty$ if and only if either $N \geq 3$, or $N = 2$ and $\deg g = 0$. In this case, (surprisingly,) minimizers still exist at a (possibly) higher energy level B_g (see Theorem 13.4).

Clearly, $A_g \leq B_g$. In Section 13.3, we show that a gap – which is reminiscent of a Lavrentiev phenomenon – can occur, i.e., it may happen that $A_g < B_g < \infty$ (see Theorem 13.5).

We also introduce a third, intermediate, minimization problem,

$$C_g := \inf\left\{\int_\Omega |\nabla u|^p;\ u \in W_g^{1,p}(\Omega;\mathbb{S}^1) \text{ and } Ju = 0\right\}, \tag{13.3}$$

so that $A_g \leq C_g \leq B_g$. It is easy to see that $C_g < \infty$ and that C_g is achieved, $\forall\, g \in C^\infty(\partial\Omega;\mathbb{S}^1)$. In Theorem 13.5, we produce, for every $1 < p < 2$ and Ω, boundary conditions $g \in C^\infty(\partial\Omega;\mathbb{S}^1)$ such that $A_g < C_g < B_g < \infty$. For such g we have the coexistence of three minimizers with distinct features. At the lowest energy level A_g, the minimizers are singular inside Ω. At the intermediate energy level C_g, minimizers are regular inside Ω, but not up to the boundary. At the highest energy level B_g, minimizers are regular in $\overline{\Omega}$.

Section 13.4 deals with refined properties of minimizers of (13.1) when $N = 2$ and $1 < p < 2$. More specifically, we prove that, if $\Omega = \mathbb{D}$ and $g(x) = x$, then $x/|x|$ is the unique minimizer of A_g (see Theorem 13.6). In a related direction, we prove that, if $g \in C^\infty(\mathbb{S}^1;\mathbb{S}^1)$, then $g(x/|x|)$ is a minimizer of A_g if and only if g is either a constant or an isometry (see Theorem 13.7). This analysis, combined with additional ingredients, implies that, if Ω is a smooth bounded domain in $\mathbb{R}^2$, $g \in C^\infty(\partial\Omega;\mathbb{S}^1)$, and u is a minimizer of A_g, then, near each singularity $a \in \Omega$ of u, we have $u(x) \sim \mathscr{R}(x-a)/|x-a|$ for some linear isometry $\mathscr{R}$ of the plane (see Theorem 13.3).

In Section 13.5, we establish that, when $N = 2$, $g \in C^\infty(\partial\Omega;\mathbb{S}^1)$, and $\deg g = 0$, minimizers of (13.1) are regular in $\overline{\Omega}$ provided $p_0(g) < p < 2$, so that no gap occurs in this range (see Theorem 13.8).

13.1 Minimizing the $W^{1,p}$ energy when $p \geq 2$

The main result in this section is

Theorem 13.1. *Let $N \geq 2$, $p \geq 2$ and $g \in C^\infty(\partial\Omega;\mathbb{S}^1)$. When $N = 2$, we also assume that* $\deg g = 0$. *Then, problem*

$$A_g := \inf\left\{\int_\Omega |\nabla u|^p;\ u \in W_g^{1,p}(\Omega;\mathbb{S}^1)\right\} \tag{13.4}$$

admits a unique minimizer U. Moreover, $U \in C^{1,\alpha}(\overline{\Omega}; \mathbb{S}^1)$ for some $\alpha \in (0, 1)$.

Proof. In view of Theorem 1.1 and Corollary 6.2 (applied with $s = 1 - 1/p$, so that $sp = p - 1 \geq 1$), we know that

$$W^{1,p}_g(\Omega; \mathbb{S}^1) = \{e^{\imath\varphi};\ \varphi \in W^{1,p}_{\psi_0}(\Omega; \mathbb{R})\}, \tag{13.5}$$

where $\psi_0 \in C^\infty(\partial\Omega; \mathbb{R})$ is such that $g = e^{\imath\psi_0}$.

Therefore, (13.4) corresponds to the standard *scalar* minimization problem

$$\inf \left\{ \int_\Omega |\nabla\varphi|^p;\ \varphi \in W^{1,p}_{\psi_0}(\Omega; \mathbb{R}) \right\}, \tag{13.6}$$

which admits a unique minimizer Φ, and then $U := e^{\imath\Phi}$ is the unique minimizer in (13.4).

Classical elliptic regularity yields $\Phi \in C^{1,\alpha}(\overline{\Omega})$ for some $\alpha \in (0, 1)$ (see Lieberman [226] and also Lin and Li [230]). Therefore, $U \in C^{1,\alpha}(\overline{\Omega})$. □

Remark 13.1. From Theorem 13.1, we deduce that

$$A_g = \inf \left\{ \int_\Omega |\nabla u|^p;\ u \in C^1_g(\overline{\Omega}; \mathbb{S}^1) \right\}.$$

In fact, we also have

$$A_g = \inf \left\{ \int_\Omega |\nabla u|^p;\ u \in C^\infty_g(\overline{\Omega}; \mathbb{S}^1) \right\}.$$

This is a direct consequence of the density Theorem 10.6 *1*.

13.2 Minimizing the $W^{1,p}$ energy when $1 < p < 2$

This case is much more subtle and new phenomena appear. As above, fix $g \in C^\infty(\partial\Omega; \mathbb{S}^1)$. Here, we *always* have $W^{1,p}_g(\Omega; \mathbb{S}^1) \neq \emptyset$ (even if $N = 2$ and $\deg g \neq 0$).

The first result in this section concerns the *partial regularity* of minimizers.

Theorem 13.2. *Let $N \geq 2$ and $1 < p < 2$. Then, problem* (13.1) *admits at least one minimizer.*

In general, the minimizers of (13.1) *need not be continuous in Ω. However, they belong to $C^{1,\alpha}(\overline{\Omega} \setminus K)$ for some $\alpha \in (0, 1)$, where K is a compact subset of Ω, of finite $(N - 2)$-Hausdorff measure when $N \geq 3$, resp. K consists of a finite number of points in Ω when $N = 2$.*

Proof. Here, we do not have (13.5) at our disposal. The existence of minimizers of (13.1) is clear. The partial regularity result was proved by Hardt and Lin [183], with roots in Schoen and Uhlenbeck [317, 318] (see also Luckhaus [233] and Hardt and Lin [184]). □

Remark 13.2. There are presumably cases where the minimizers of (13.1) are not unique (see Open Problem 25).

Remark 13.3. When $N = 2$ and $\deg g \neq 0$, any $u \in W^{1,p}_g(\Omega; \mathbb{S}^1)$ has to be discontinuous, and in particular any minimizer in (13.1) must be discontinuous. However, when $N = 2$ and $\deg g = 0$, or when $N \geq 3$, we have $C^\infty_g(\overline{\Omega}; \mathbb{S}^1) \neq \emptyset$, and one may be tempted to believe that minimizers of (13.1) are continuous because "discontinuities tend to increase the energy." This intuition is misleading. In fact, singularities may sometimes "help" to decrease the energy. This is an immediate consequence of the gap phenomenon described in the next section. We present there, for any $1 < p < 2$, an example of $g \in C^\infty(\partial\Omega; \mathbb{S}^1)$ such that $C^\infty_g(\overline{\Omega}; \mathbb{S}^1) \neq \emptyset$ and any minimizer of (13.1) is discontinuous.

When $N = 2$ and $1 < p < 2$, we have much more information about the nature of the singularities of minimizers. Here is a typical result due to Lin [229], with roots in Hardt and Lin [183] and Hardt, Lin, and Wang [185, 186].

Theorem 13.3. *Assume $N = 2$ and $1 < p < 2$. Let u be a minimizer in* (13.1) *and let $a \in \Omega$ be a singularity of u. Then there exists a linear isometry $\mathscr{R}$ of $\mathbb{R}^2$ such that*

$$\lim_{x \to a} [u(x) - \mathscr{R}(x - a)/|x - a|] = 0. \tag{13.7}$$

Remark 13.4. The corresponding result for minimizing harmonic maps $u : \Omega \to \mathbb{S}^2$ with $\Omega \subset \mathbb{R}^3$ (i.e., minimizers of $\int_\Omega |\nabla u|^2$ in $H^1_g(\Omega; \mathbb{S}^2)$) is known to be true (see Brezis, Coron, and Lieb [79, Section VII]). The proof in [79] involves two ingredients:
a) The existence of a map $g : \mathbb{S}^2 \to \mathbb{S}^2$ such that $u(x) - g((x - a)/|x - a|) \to 0$ as $x \to a$ (this relies on fundamental works by Schoen and Uhlenbeck [317] and Simon [323, Section 8]).
b) The identification of g as an isometry ([79, Section VII]).

This program can be adapted to the context of Theorem 13.3. The analogue of b) is Theorem 13.7 below.

We now turn to a more surprising effect: the existence of minimizers within the class of *continuous competitors*.

Theorem 13.4. *Let $N \geq 2$ and $1 < p < 2$. When $N = 2$, we also assume that* $\deg g = 0$. *Then, problem* (13.2) *admits a unique minimizer U. Moreover, $U \in C^{1,\alpha}(\overline{\Omega})$ for some $\alpha \in (0, 1)$.*

Proof. Let $\psi_0 \in C^\infty(\partial\Omega; \mathbb{R})$ be such that $g = e^{\imath\psi_0}$. From Corollary 1.1, we know that

$$W^{1,p}_g(\Omega; \mathbb{S}^1) \cap C(\overline{\Omega}; \mathbb{S}^1) = \{e^{\imath\varphi};\ \varphi \in W^{1,p}_{\psi_0}(\Omega; \mathbb{R}) \cap C(\overline{\Omega}; \mathbb{R})\},$$

and thus (13.2) is equivalent to

$$B_g = \inf\left\{\int_\Omega |\nabla\varphi|^p;\ \varphi \in W^{1,p}_{\psi_0}(\Omega; \mathbb{R}) \cap C(\overline{\Omega}; \mathbb{R})\right\}. \tag{13.8}$$

Consider the minimization problem

$$\widetilde{B}_g := \inf\left\{\int_\Omega |\nabla\varphi|^p;\ \varphi \in W^{1,p}_{\psi_0}(\Omega; \mathbb{R})\right\}. \tag{13.9}$$

Clearly, $\widetilde{B}_g \le B_g$ and (13.9) admits a unique minimizer Φ. Moreover, as in Section 13.1, we have $\Phi \in C^{1,\alpha}(\overline{\Omega}; \mathbb{R})$ for some $\alpha \in (0, 1)$. Thus, Φ is also a competitor in (13.8). We find that $U := e^{\imath\Phi} \in C^{1,\alpha}(\overline{\Omega}; \mathbb{S}^1)$ is the unique minimizer of (13.2). □

Remark 13.5. We may also introduce variants of (13.2), e.g.,

$$B'_g := \inf\left\{\int_\Omega |\nabla u|^p;\ u \in C^1_g(\overline{\Omega}; \mathbb{S}^1)\right\} \tag{13.10}$$

or

$$B''_g := \inf\left\{\int_\Omega |\nabla u|^p;\ u \in C^\infty_g(\overline{\Omega}; \mathbb{S}^1)\right\}. \tag{13.11}$$

Equality $B_g = B'_g$ is an immediate consequence of Theorem 13.4. Equality $B_g = B''_g$ comes from the density of $C^\infty_g(\overline{\Omega}; \mathbb{S}^1)$ in $W^{1,p}_g(\Omega; \mathbb{S}^1) \cap C(\overline{\Omega}; \mathbb{S}^1)$ (which is obtained via lifting, as in the proof of Theorem 10.6 *1*).

We now investigate the "intermediate" minimization problem (13.3), for a general $g \in C^\infty(\partial\Omega; \mathbb{S}^1)$ (no further assumption when $N = 2$). Clearly,

$$A_g \le C_g \le B_g \le \infty. \tag{13.12}$$

We always have $C_g < \infty$. Indeed, when $N = 3$ or $N = 2$ and $\deg g = 0$, this is clear, since $B_g < \infty$. When $N = 2$ and $\deg g \ne 0$, consider some BV lifting ψ_0 of g. Since $BV(\partial\Omega) \hookrightarrow W^{1-1/p,p}(\partial\Omega)$ (here, we use $1 < p < 2$ and the one-dimensional embedding $BV \hookrightarrow W^{s,q}$ with $sq < 1$), there exists some $\varphi \in W^{1,p}(\Omega) \cap C^\infty(\Omega)$ such that $\operatorname{tr}_{|\partial\Omega} \varphi = \psi_0$, and then $u := e^{\imath\varphi}$ is a competitor in (13.3) (by Corollary 1.1 applied in any $\omega \Subset \Omega$).

Moreover, the infimum in (13.3) is achieved, as a consequence of the continuity of J under weak $W^{1,p}$ convergence (see (1.155)). In addition,

$$\text{any minimizer of (13.3) belongs to } C^{1,\alpha}(\Omega) \text{ for some } \alpha \in (0, 1). \tag{13.13}$$

Indeed, (13.3) corresponds via lifting to

$$C_g = \inf\left\{\int_\Omega |\nabla\varphi|^p;\ \varphi \in W^{1,p}(\Omega;\mathbb{R}) \text{ and } \operatorname{tr}_{|\partial\Omega}\varphi - \psi_0 \in 2\pi\,\mathbb{Z} \text{ a.e. on } \partial\Omega\right\} \tag{13.14}$$

(where ψ_0 is a BV lifting of g) and we can apply local regularity theory (see, for example, DiBenedetto [140], Lewis [224], and Tolksdorf [336]) to (13.14).

Warning. Here, we cannot assert (as in (13.6)) that $\operatorname{tr}_{|\partial\Omega}\varphi - \psi_0 = 2\pi\, k$ for some $k \in \mathbb{Z}$, since $W^{1-1/p,p}(\partial\Omega)$ is not a uniqueness class in the sense of Chapter 6, because $(1-1/p)p = p-1 < 1$. Hence, we cannot apply regularity theory up to the boundary of Ω. In fact, a minimizer U in (13.3) is not, in general, continuous up to $\partial\Omega$ (even when $N = 2$ and $\deg g = 0$); see Theorem 13.5.

13.3 Gaps

As in Section 13.2, we let $1 < p < 2$. Given $g \in C^\infty(\partial\Omega;\mathbb{S}^1)$ with $C^1_g(\overline{\Omega};\mathbb{S}^1) \neq \emptyset$, we set, as above,

$$A_g := \inf\left\{\int_\Omega |\nabla u|^p;\ u \in W^{1,p}_g(\Omega;\mathbb{S}^1)\right\},$$

$$C_g := \inf\left\{\int_\Omega |\nabla u|^p;\ u \in W^{1,p}_g(\Omega;\mathbb{S}^1) \text{ and } Ju = 0\right\},$$

and

$$B_g := \inf\left\{\int_\Omega |\nabla u|^p;\ u \in W^{1,p}_g(\Omega;\mathbb{S}^1) \cap C(\overline{\Omega};\mathbb{S}^1)\right\}$$
$$= \inf\left\{\int_\Omega |\nabla u|^p;\ u \in C^1_g(\overline{\Omega};\mathbb{S}^1)\right\}.$$

The main result of this section is the following

Theorem 13.5. *Let $N \geq 2$ and $1 < p < 2$. Let $\Omega \subset \mathbb{R}^N$ be a smooth bounded domain. Then, there exists some $g \in C^\infty(\partial\Omega;\mathbb{S}^1)$ such that $C^1_g(\overline{\Omega};\mathbb{S}^1) \neq \emptyset$ and*

$$A_g < C_g < B_g < \infty. \tag{13.15}$$

As a consequence, if U is a minimizer in (13.3) *for this g, then U is continuous in Ω but not up to $\partial\Omega$.*

Proof. Step 1. Construction of g such that $C_g < B_g$
Let B be a small geodesic ball on $\partial\Omega$. Let U be a smooth, non-empty, open subset of $\partial\Omega$ such that $U \subsetneqq B$, and set $\zeta := 2\pi\,\chi_U$. Then, $\zeta \in W^{1-1/p,p}(\partial\Omega;\mathbb{R})$ (see Lemma

15.9) and $e^{\imath\zeta} = 1$. If (ψ_k) is a sequence of smooth functions $\psi_k : \partial\Omega \to \mathbb{R}$ such that $\psi_k \to \zeta$ in $W^{1-1/p,p}(\partial\Omega)$ as $k \to \infty$, then

$$|\psi_k|_{W^{1-1/p,p}(B)} \to |\zeta|_{W^{1-1/p,p}(B)} > 0 \text{ as } k \to \infty \tag{13.16}$$

and

$$|e^{\imath\psi_k}|_{W^{1-1/p,p}(\partial\Omega)} = |e^{\imath\psi_k} - 1|_{W^{1-1/p,p}(\partial\Omega)} \to 0 \text{ as } k \to \infty. \tag{13.17}$$

Set $g_k := e^{\imath\psi_k}$. Then, $C^1_{g_k}(\overline{\Omega}; \mathbb{S}^1) \neq \emptyset$ (since $C^1_{g_k}(\overline{\Omega}; \mathbb{S}^1)$ contains $e^{\imath\varphi}$, where φ is any smooth extension of ψ_k).

On the one hand, we have

$$\begin{aligned} C_{g_k} &\le \inf\left\{\int_\Omega |\nabla(e^{\imath\varphi})|^p;\ \varphi \in W^{1,p}_{\psi_k-\zeta}(\Omega; \mathbb{R})\right\} \\ &= \inf\left\{\int_\Omega |\nabla\varphi|^p;\ \varphi \in W^{1,p}_{\psi_k-\zeta}(\Omega; \mathbb{R})\right\} \\ &\le C|\psi_k - \zeta|^p_{W^{1-1/p,p}(\partial\Omega)} \to 0 \text{ as } k \to \infty. \end{aligned} \tag{13.18}$$

On the other hand, let $\varepsilon > 0$ be sufficiently small such that the nearest point projection Π on $\partial\Omega$ is smooth on the set $\{x \in \mathbb{R}^N;\ \operatorname{dist}(x, \partial\Omega) < 2\varepsilon\}$. Consider the open set

$$\omega := \{x \in \Omega;\ \operatorname{dist}(x, \partial\Omega) < \varepsilon \text{ and } \Pi(x) \in B\}.$$

Then,

$$\overline{\omega} \text{ is bi-Lipschitz homeomorphic with } \overline{\mathbb{B}^{N-1}} \times [0, 1], \tag{13.19}$$

and in particular is simply connected. It follows that, if $u \in C^1_{g_k}(\overline{\Omega}; \mathbb{S}^1)$, then we may write $u = e^{\imath\varphi}$ for some $\varphi \in C^1(\overline{\omega}; \mathbb{R})$ such that $\varphi = \psi_k$ on B. By standard trace theory, (13.16) and (13.19), we have

$$\begin{aligned} |u|_{W^{1,p}(\Omega)} &\ge |u|_{W^{1,p}(\omega)} = |\varphi|_{W^{1,p}(\omega)} \\ &\gtrsim |\psi_k|_{W^{1-1/p,p}(B)} \to |\zeta|_{W^{1-1/p,p}(B)} > 0. \end{aligned} \tag{13.20}$$

Combining (13.18) and (13.20), we obtain, for large k, $C_{g_k} < B_{g_k}$.

Step 2. Proof of $A_g < C_g$ under the assumption $C_g < B_g$
We claim the following. If $A_g = C_g$, then $A_g = C_g = B_g$ (whence the conclusion). Indeed, let U be a minimizer of C_g. Then, $U \in C^{1,\alpha}(\Omega)$ for some $\alpha \in (0, 1)$ (see (13.13)). If $A_g = C_g$, then U is also a minimizer of A_g, and thus U is in $C^{1,\alpha}$ near $\partial\Omega$ (Theorem 13.2). We find that $U \in C^{1,\alpha}(\overline{\Omega})$, and thus U is a competitor for B_g. This implies $B_g \le A_g = C_g \le B_g$ and thus $A_g = C_g = B_g$. □

Remark 13.6. Part of Theorem 13.15, namely, the existence of g such that $C_g < B_g$ (and thus $A_g < B_g$), still holds for $p = 1$. Indeed, it suffices to consider, with $\zeta := 2\pi\, \chi_U$, a sequence (ψ_k) of smooth functions such that $\psi_k \to \zeta$ in $L^1(\partial\Omega)$, and to replace, in the last line of (13.18) and in (13.20), $W^{1-1/p,p}$ by L^1.

13.4 More about minimizers when $\Omega = \mathbb{D}$ and $1 < p < 2$

In this section, we let $N = 2$, $1 < p < 2$, and consider the minimization problem (13.1).

A first result is the following

Theorem 13.6. *Let* $1 < p < 2$. *Then,* $u_0(x) := x/|x|$ *is the unique minimizer in*

$$\inf \left\{ \int_{\mathbb{D}} |\nabla u|^p;\ u \in W^{1,p}(\mathbb{D}; \mathbb{S}^1),\ u(x) = x \text{ on } \mathbb{S}^1 \right\}. \tag{13.21}$$

The next result supplements Theorem 13.6.

Theorem 13.7. *Let* $g \in C^1(\mathbb{S}^1; \mathbb{S}^1)$ *and set* $U(x) := g(x/|x|)$, $\forall\, x \in \mathbb{D}$.
Consider, for $1 < p < 2$, *the minimization problem*

$$\inf \left\{ \int_{\mathbb{D}} |\nabla u|^p;\ u \in W^{1,p}_g(\mathbb{D}; \mathbb{S}^1) \right\}. \tag{13.22}$$

Then, U *is a minimizer in* (13.22) *if and only if either* g *is a constant, or* g *is an isometry.*

Sketch of the proof of Theorem 13.7. If U is a minimizer in (13.22), then g is a p-harmonic map, i.e., a critical point of

$$g \mapsto \int_{\mathbb{S}^1} \left| \frac{\partial g}{\partial \tau} \right|^p.$$

On the other hand, all p-harmonic maps $g : \mathbb{S}^1 \to \mathbb{S}^1$ are of the form $g(e^{\imath\theta}) = e^{\imath(d\theta+\beta)}$ for some $d \in \mathbb{Z}$ and $\beta \in \mathbb{R}$ (see Hardt and Chen [179, Section 4]). Clearly, when $d = 0$, g is a constant and U is a minimizer of (13.22). When $|d| \geq 2$, U is not a minimizer of (13.22) [179, Lemma 7.1].

It remains to establish that, if $g : \mathbb{S}^1 \to \mathbb{S}^1$ is an isometry, then U is a minimizer for (13.22). Clearly, if *one* isometry has this property, then *all* the isometries have it (by the invariance of the energy under isometries), so that it suffices to prove this property for *some* isometry.

Let u be a minimizer in (13.21). Let $a \in \mathbb{D}$. Then we have the following monotonicity formula (see, for example, Hardt and Lin [183, Section 4]):

$$r \mapsto f(r) := \frac{1}{r^{2-p}} \int_{B_r(a)} |\nabla u|^p,\ 0 < r < 1 - |a|, \text{ is non-decreasing.} \tag{13.23}$$

On the other hand, there exists some $g \in C^{1,\alpha}(\mathbb{S}^1; \mathbb{S}^1)$ (possibly depending on a) satisfying the following two properties:

$$\lim_{r\to 0} \frac{1}{r^{2-p}} \int_{B_r(a)} |\nabla u|^p = \frac{1}{2-p} \int_{\mathbb{S}^1} \left|\frac{\partial g}{\partial \tau}\right|^p, \tag{13.24}$$

$$g(x/|x|) \text{ is a minimizer in (13.22)} \tag{13.25}$$

(see [183, Corollary 4.4]).

We now specify our choice of a: we let $a \in \mathbb{D}$ be a singularity of u. [Such an a has to exist, since the boundary condition $u(x) = x$ on $\mathbb{S}^1$ prevents u from being globally smooth.] By the first part of the proof of the theorem, the above g is either a constant, or an isometry. Actually, g cannot be a constant, for otherwise

$$\lim_{r\to 0} \frac{1}{r^{2-p}} \int_{B_r(a)} |\nabla v|^p = 0,$$

and then u is $C^{1,\alpha}$ in a neighborhood of a ([183, Corollary 2.6]). Thus, g is an isometry. From (13.25), it follows that there exists *some* isometry $g : \mathbb{S}^1 \to \mathbb{S}^1$ such that the corresponding U is a minimizer for (13.22). This completes the proof of Theorem 13.7. □

Proof of Theorem 13.6. It suffices to prove the uniqueness. Let u be any minimizer in (13.21), and consider the map $v : \mathbb{R}^2 \to \mathbb{S}^1$,

$$v(x) := \begin{cases} u(x), & \text{if } |x| \le 1 \\ x/|x|, & \text{if } |x| > 1 \end{cases}.$$

Using the minimality of u_0 (which follows from Theorem 13.7), we see that, for any $R > 1$,

$$v \text{ is a minimizer of } \inf \left\{ \int_{B_R(0)} |\nabla w|^p;\ w(x) = x/|x| \text{ on } C(0, R) \right\}. \tag{13.26}$$

Let $f(r)$, with $r > 0$, be as in (13.23). By a direct calculation, we have

$$\lim_{r\to\infty} f(r) = \frac{2\pi}{2-p},\quad \forall a \in \mathbb{R}^2. \tag{13.27}$$

On the other hand,

$$f(r) := \frac{1}{r^{2-p}} \int_{B_r(a)} |\nabla v|^p \text{ is non-increasing for } r > 0 \tag{13.28}$$

(the argument is the same as for (13.23)), and (13.24) and (13.25) hold for v. Choosing, as in the proof of Theorem 13.7, $a \in \mathbb{D}$ to be a singularity of u, we find, using (13.24) and the fact that g in (13.24) is an isometry (see the proof of Theorem 13.7), that

$$\lim_{r\to 0}\frac{1}{r^{2-p}}\int_{B_r(a)}|\nabla v|^p=\frac{2\pi}{2-p}. \tag{13.29}$$

Combining (13.27)–(13.29), we find that

$$r\mapsto\frac{1}{r^{2-p}}\int_{B_r(a)}|\nabla v|^p \text{ is constant,}$$

and thus v is constant on each ray emanating from a [183, Corollary 4.2].

Noting that $v(x)=x/|x|$ for $|x|>1$, we find that a has to be the origin, and then $u=u_0$. □

13.5 Further regularity and no gap for $p<2$, near $p=2$, when $N=2$

We return to the setting of Theorem 13.2 when $N=2$ and establish a regularity result for $p<2$, p near 2.

Theorem 13.8. *Let $N=2$ and $g\in C^\infty(\partial\Omega;\mathbb{S}^1)$ be such that* $\deg g=0$. *Then there exists some $p_0=p_0(g)<2$ such that, $\forall\, p_0<p<2$, problem* (13.1) *admits a unique minimizer U_p. Moreover, $U_p\in C^{1,\alpha}(\overline{\Omega};\mathbb{S}^1)$ for some $\alpha\in(0,1)$.*

Remark 13.7. Theorem 13.8 implies in particular that $A_g=B_g$, i.e., no gap when $N=2$ and $p_0<p<2$.

Proof. Let $U_p\in W^{1,p}_g(\Omega;\mathbb{S}^1)$ minimize (13.1).
Step 1. We have

$$\int_\Omega|\nabla U_p|^p\le C,\ \forall\,1\le p<2. \tag{13.30}$$

Indeed, if $v\in C^1_g(\overline{\Omega};\mathbb{S}^1)$, then

$$\int_\Omega|\nabla U_p|^p\le\int_\Omega|\nabla v|^p\le|\Omega|\,(\|\nabla v\|_{L^\infty}+\|\nabla v\|^2_{L^\infty}),\ \forall\,1\le p<2.$$

We now fix a small number $r_0>0$, and let Π be the nearest point projection on $\partial\Omega$. For $x_0\in\overline{\Omega}$ and $0<r<r_0$, set

$$Q_r(x_0):=\begin{cases}B_r(x_0), & \text{if } \operatorname{dist}(x_0,\partial\Omega)>r\\ B_{2r}(\Pi(x_0))\cap\Omega, & \text{if } \operatorname{dist}(x_0,\partial\Omega)\le r\end{cases}.$$

Step 2. Let $\delta > 0$. Then there exists some $p_1 = p_1(g, \delta) < 2$ such that:

$$\forall x_0 \in \overline{\Omega},\ \forall p \in (p_1, 2),\ \exists r \in (0, r_0) \text{ such that } \operatorname*{Var}_{\partial Q_r(x_0)} U_p < \delta. \tag{13.31}$$

[Here, $\operatorname{Var}_{\partial Q_r(x_0)} U_p$ stands for the total variation of U_p on $\partial Q_r(x_0)$, i.e.,

$$\operatorname*{Var}_{\partial Q_r(x_0)} U_p := \int_{\partial Q_r(x_0)} |\partial U_p/\partial \tau|.$$

In particular, the radius r has the property that $U_p \in W^{1,1}(\partial Q_r(x_0))$.]

Indeed, by taking a smaller r_0 if necessary, we may assume that

$$\operatorname*{Var}_{\partial Q_r(x_0)\cap\partial\Omega} g < \frac{\delta}{2},\ \forall r < r_0,\ \forall x_0 \in \overline{\Omega}. \tag{13.32}$$

It thus suffices to find some $0 < r < r_0$ such that $\operatorname{Var}_{\partial Q_r(x_0)\cap\Omega} U_p < \delta/2$.

Set

$$d = d(x_0) := \min\{\operatorname{dist}(x_0, \partial\Omega), r_0\}.$$

We have

$$\begin{aligned} C &\geq \int_{B_d(x_0)} |\nabla U_p|^p = \int_0^d \left(\int_{C_r(x_0)} |\nabla U_p|^p \, d\ell \right) dr \\ &\geq \int_0^d \left[\operatorname*{Var}_{C_r(x_0)} U_p \right]^p \frac{dr}{(2\pi r)^{p-1}} \\ &\geq \frac{1}{2\pi(2-p)} d^{2-p} \operatorname*{essinf}_{0<r<d} \left[\operatorname*{Var}_{C_r(x_0)} U_p \right]^p, \end{aligned} \tag{13.33}$$

where we have used (13.30).

We obtain the existence of some $r \in (0, d)$ such that $U_p \in W^{1,p}(C_r(x_0))$ and

$$\left[\operatorname*{Var}_{C_r(x_0)} U_p \right]^p \leq \frac{C'(2-p)}{d^{2-p}}. \tag{13.34}$$

Similarly, assuming $d \leq r_0/4$ and letting $y_0 := \Pi(x_0)$, we have

$$\begin{aligned} C &\geq \int_{(B_{r_0}(y_0)\setminus B_{2d}(y_0))\cap\Omega} |\nabla U_p|^p \\ &\geq \frac{1}{(2-p)\,2\pi} [r_0^{2-p} - (2d)^{2-p}] \operatorname*{essinf}_{2d<r<r_0} \left[\operatorname*{Var}_{C_r(y_0)\cap\Omega} U_p \right]^p. \end{aligned} \tag{13.35}$$

We find that there exists some $r \in (2d, r_0)$ such that $U_p \in W^{1,p}(\partial Q_r(x_0))$ and

$$\left[\operatorname*{Var}_{C_r(y_0)\cap\Omega} U_p\right]^p \leq \frac{C'(2-p)}{r_0^{2-p}-(2d)^{2-p}}. \tag{13.36}$$

With no loss of generality, we may assume that $r_0 < 1$. If $d > r_0/4$, then (13.34) yields the existence of some $r \in (0, d)$ such that $U_p \in W^{1,p}(C_r(x_0))$ and

$$\left[\operatorname*{Var}_{C_r(x_0)} U_p\right]^p \leq \frac{2\,C'(2-p)}{d^{2-p}} \leq \frac{32\,C'(2-p)}{r_0}. \tag{13.37}$$

If $d \leq r_0/4$, then either $(2d)^{2-p} > (1/2)r_0^{2-p}$, and then (by (13.34)) we obtain some $r \in (0, d)$ such that $U_p \in W^{1,p}(C_r(x_0))$ and

$$\left[\operatorname*{Var}_{C_r(x_0)} U_p\right]^p \leq \frac{2^{4-p}\,C'(2-p)}{r_0^{2-p}} \leq \frac{8\,C'(2-p)}{r_0}, \tag{13.38}$$

or $(2d)^{2-p} \leq (1/2)r_0^{2-p}$, and then (13.36) leads to the existence of some $r \in (2d, r_0)$ satisfying $U_p \in W^{1,p}(\partial Q_r(x_0))$ and

$$\left[\operatorname*{Var}_{C_r(y_0)\cap\Omega} U_p\right]^p \leq \frac{4\,C'(2-p)}{r_0^{2-p}} \leq \frac{4\,C'(2-p)}{r_0}. \tag{13.39}$$

Since the cases considered above exhaust all possibilities, we obtain the existence of p_1 such that (13.31) holds. Step 2 is complete.

Step 3. Let $p_0 := p_1(g, \pi/2)$. Then, for $p_0 < p < 2$, we may locally write $U_p = e^{\imath\eta}$, with $\eta \in W^{1,p}$

Let $x_0 \in \overline{\Omega}$ and let r be as in (13.31). Then, $U_p(\partial Q_r(x_0))$ is strictly contained in a half circle. With no loss of generality, we assume that there exists some $\varepsilon > 0$ such that

$$U_p(\partial Q_r(x_0)) \subset \{e^{\imath\theta};\ |\theta| \leq \pi/2 - \varepsilon\}. \tag{13.40}$$

We consider the competitor

$$v_p(x) := \begin{cases} U_p(x), & \text{if } x \in \Omega \setminus Q_p(x_0) \\ U_p(x), & \text{if } x \in Q_p(x_0) \text{ and } \operatorname{Re} U_p(x) \geq 0\,. \\ -\operatorname{Re} U_p(x) + \imath \operatorname{Im} U_p(x), & \text{if } x \in Q_p(x_0) \text{ and } \operatorname{Re} U_p(x) < 0 \end{cases}$$

Clearly, $v_p \in W^{1,p}_g(\Omega; \mathbb{S}^1)$,

$$v_p(\partial Q_r(x_0)) = U_p(\partial Q_r(x_0)) \subset \{e^{\imath\theta};\ |\theta| \leq \pi/2 - \varepsilon\} \tag{13.41}$$

and (by the de la Vallée Poussin chain rule in $W^{1,p}$ (see, for example, Serrin and Varberg [320]), $|\nabla v_p| = |\nabla U_p|$ a.e. in Ω, so that v_p is a minimizer of the left-hand side of (13.1).

Consider now

$$w_p(x) := \begin{cases} v_p(x), & \text{if } x \in \Omega \setminus Q_p(x_0) \\ v_p(x), & \text{if } x \in Q_p(x_0) \text{ and } \operatorname{Re} v_p(x) \geq \sin \varepsilon \\ e^{\imath(\pi/2-\varepsilon)}, & \text{if } x \in Q_p(x_0),\ \operatorname{Re} v_p(x) < \sin \varepsilon \text{ and } \operatorname{Im} v_p(x) > 0 \\ e^{-\imath(\pi/2-\varepsilon)}, & \text{if } x \in Q_p(x_0),\ \operatorname{Re} v_p(x) < \sin \varepsilon \text{ and } \operatorname{Im} v_p(x) < 0 \end{cases}.$$

By (13.41), we have $w_p \in W_g^{1,p}(\Omega; \mathbb{S}^1)$.

Since $\int_\Omega |\nabla v_p|^p \leq \int_\Omega |\nabla w_p|^p$, we find that

$$\int_{Q_r(x_0)\cap[\operatorname{Re} v_p<\sin\varepsilon]} |\nabla v_p|^p = 0,$$

and in particular

$$\int_{Q_r(x_0)\cap[\operatorname{Re} v_p<\sin\varepsilon]} |\nabla \operatorname{Re} v_p|^p = 0,$$

which implies (via the chain rule) that

$$\int_{Q_r(x_0)\cap[\operatorname{Re} v_p<\sin\varepsilon]} |\nabla(\operatorname{Re} v_p - \sin\varepsilon)_-|^p = 0.$$

Combining this with the fact that $\operatorname{Re} v_p \geq \sin\varepsilon$ on $\partial Q_r(x_0)$, we find that $\operatorname{Re} v_p \geq \sin\varepsilon$ a.e. in $Q_r(x_0)$. Taking the definition of v_p into account, we obtain

$$\operatorname{Re} U_p(x) \in [-1, -\sin\varepsilon] \cup [\sin\varepsilon, 1] \text{ for a.e. } x \in Q_r(x_0).$$

Since the essential range of $\operatorname{Re} U_p$ on $Q_r(x_0)$ is connected (see Theorem 6.5 *c)*), we find that $\operatorname{Re} U_p(x) \in [\sin\varepsilon, 1]$ for a.e. $x \in Q_r(x_0)$. We therefore have $U_p = e^{\imath\eta}$ in $Q_r(x_0)$, where $\eta := (1/\imath)\ln u_p$, and ln is the principal branch of the logarithm. By Theorem 15.6, we have $\eta \in W^{1,p}(Q_r(x_0))$.

For further use, we note that there exist some $t \in \mathbb{R}$ and $\varepsilon > 0$ such that

$$\eta(x) \in (t + \varepsilon, t + \pi - \varepsilon) \text{ for a.e. } x \in Q_r(x_0). \tag{13.42}$$

Step 4. Conclusion

By Step 3 and Theorem 1.2, for $p_0 < p < 2$ we know that $JU_p = 0$, locally, and therefore globally, in Ω. We thus have $U_p = e^{\imath\varphi}$, with $\varphi \in W^{1,p}(\Omega; \mathbb{R})$.

As in Section 13.1, write, on $\partial\Omega$, $g = e^{\imath\psi_0}$, with $\psi_0 \in C^\infty(\partial\Omega; \mathbb{R})$. Then, $\operatorname{tr}_{|\partial\Omega} \varphi - \psi_0 \in 2\pi\mathbb{Z}$ a.e. on $\partial\Omega$. Fix $x_0 \in \partial\Omega$. Using (13.42), we deduce that, near x_0, $\operatorname{tr}_{|\partial\Omega} \varphi - \psi_0 = 2\pi k$ for some $k \in \mathbb{Z}$. Thus, $\operatorname{tr}_{|\partial\Omega} \varphi - \psi_0 = 2\pi k$ *globally* on $\partial\Omega$. Adding $2\pi k$ to ψ_0, we may always assume that $\varphi \in W^{1,p}_{\psi_0}(\Omega; \mathbb{R})$. Therefore, $U_p = e^{\imath\varphi}$ is a competitor for $\widetilde{B}_g$ in (13.9). As in the proof of Theorem 13.4, we find that U_p is the unique minimizer of B_g, and in particular $U_p \in C^{1,\alpha}(\overline{\Omega}; \mathbb{S}^1)$. □

13.6 Complements and open problems

13.6.1 A generalization of Theorem 13.6 to dimension $N \geq 2$

Theorem 13.6 can be extended to all dimensions $N \geq 2$.

Theorem 13.9. *Let $N \geq 2$ and $1 < p < N$. Then, $u_0(x) := x/|x|$ is the unique minimizer in*

$$\inf \left\{ \int_{\mathbb{B}^N} |\nabla u|^p;\ u \in W^{1,p}(\mathbb{B}^N; \mathbb{S}^{N-1}),\ u(x) = x \text{ on } \mathbb{S}^{N-1} \right\}.$$

Theorem 13.9 is due to Hardt, Lin, and Wang [186] when $N - 1 \leq p < N$ and to Hong [190] when $1 < p < N - 1$. Earlier partial results on this problem include Jäger and Kaul [203], Brezis, Coron, and Lieb [79], Lin [227], Coron and Gulliver [120], and Avellaneda and Lin [17].

13.6.2 Minimizing the $W^{1,p}$ energy with prescribed Jacobian

In Section 2.11.1, we already investigated minimization problems in $W^{1,1}$ with prescribed Jacobian. We consider here similar questions in $W^{1,p}$, with $1 < p < 2$. For simplicity, we discuss only an elementary example, but it is easy to imagine more complex situations. Let $N = 2$, $g \in C^\infty(\partial\Omega; \mathbb{S}^1)$, and $1 < p < 2$. Fix two distinct points $a, b \in \Omega$. Note that

$$\{u \in W^{1,p}_g(\Omega; \mathbb{S}^1);\ Ju = \pi\,(\delta_a - \delta_b)\} \neq \emptyset$$

(even when $\deg g \neq 0$). In the spirit of the minimization problem (13.3), we may consider

$$\inf \left\{ \int_{\Omega} |\nabla u|^p;\ u \in W^{1,p}_g(\Omega; \mathbb{S}^1) \text{ and } Ju = \pi\,(\delta_a - \delta_b) \right\}. \tag{13.43}$$

Clearly, the infimum in (13.43) is achieved by some U (see (1.155)). Using lifting and local regularity theory, we know that $U \in C^{1,\alpha}(\Omega \setminus \{a, b\})$ for some $\alpha \in (0, 1)$. It would be interesting to analyze the precise behavior of U near $\partial\Omega$ and near a and b. Minimization problems with prescribed singularities have also been studied by Hardt and Chen [179] from a different perspective.

13.6.3 Minimizing infinite energies when $N = 2$ and $p \geq 2$. The BBH theory of Ginzburg–Landau vortices

Throughout this subsection, we assume that $\Omega \subset \mathbb{R}^2$ and that $p \geq 2$. In Theorem 13.1, we assumed that $g \in C^\infty(\partial\Omega; \mathbb{S}^1)$ and $\deg g = 0$. By contrast, we deal here with the case where $\deg g \neq 0$, so that $W^{1,p}_g(\Omega; \mathbb{S}^1) = \emptyset$. In other words, $\int_\Omega |\nabla u|^p = \infty$ for every $u \in W^{1,1}_g(\Omega; \mathbb{S}^1)$, and thus problem (13.1) is meaningless. The strategy is to "bypass" this phenomenon of "infinite energies" via a family of approximate problems involving no topological obstruction. There are two natural approaches.

Approach 1. Change the topology of the target space

Here, we relax the constraint $u : \Omega \to \mathbb{S}^1$ and replace it by $u : \Omega \to \mathbb{R}^2$. On the other hand, we introduce a stiff "penalty" in the energy when a competitor takes its values outside $\mathbb{S}^1$. More precisely, define the p-Ginzburg–Landau energy

$$E_{p,\varepsilon}(u) := \int_\Omega |\nabla u|^p + \frac{1}{\varepsilon^2}\int_\Omega (|u|^2 - 1)^2,$$

where $\varepsilon > 0$ is small, replace problem (13.1) by

$$m_{p,\varepsilon} := \inf\left\{E_{p,\varepsilon}(u);\ u \in W^{1,p}_g(\Omega; \mathbb{R}^2)\right\}, \tag{13.44}$$

and analyze the asymptotic behavior of its minimizers u_ε as $\varepsilon \to 0$. This program was successfully carried out by Bethuel, Brezis, and Hélein when $p = 2$. The whole monograph [31] is dedicated to this situation, which is inspired by the Ginzburg–Landau theory of superfluids and superconductors. A central result from [31] is the following

Theorem 13.10. *Assume $p = 2$ and $D := \deg g > 0$. Then,*

$$m_{2,\varepsilon} = 2\pi D|\ln\varepsilon| + O(1) \textit{ as } \varepsilon \to 0.$$

Moreover, up to a subsequence, u_ε converges to a function u_* of the form

$$u_*(z) = e^{\imath\varphi(z)}\frac{z - a_1}{|z - a_1|}\cdots\frac{z - a_D}{|z - a_D|}, \tag{13.45}$$

where $a_1, \ldots, a_D$ are distinct points in Ω, and $\varphi \in C^\infty(\overline{\Omega}; \mathbb{R})$ is the harmonic function in Ω, uniquely determined (up to an integer multiple of 2π) by the boundary condition $u_* = g$ on $\partial\Omega$.

It would be very interesting to try a similar approach when $p > 2$ (see Open Problem 28).

Approach 2. Change the topology of the domain

Let $k \in \mathbb{N}^*$, and assume, for example, that $D = \deg g > 0$; the case where $D < 0$ is similar. Consider a collection $a = (a_1, \dots, a_k)$ of k distinct points $a_1, \dots, a_k \in \Omega$ and a collection $d = (d_1, \dots, d_k)$ of k integers $d_1, \dots, d_k \in \mathbb{Z}$ such that $\sum_{j=1}^k d_j = D$. Drill k small holes $\overline{B}_r(a_j)$, $r > 0$, $1 \le j \le k$, in Ω, and set $\Omega_r := \Omega \setminus \cup_j \overline{B}_r(a_j)$. Consider the minimization problem

$$E_{p,r}(a,d) := \inf \left\{ \int_{\Omega_r} |\nabla u|^p;\ u \in W^{1,p}(\Omega_r; \mathbb{S}^1),\ \mathrm{tr}_{|\partial\Omega}\, u = g \text{ and } \deg(u, C(a_j, r)) = d_j, \forall j = 1, \dots, k \right\}. \tag{13.46}$$

We know (see Proposition 12.17) that the class of competitors in (13.46) is nonempty, and that the infimum in (13.46) is achieved. When $p = 2$, the asymptotic behavior of the minimal energy $E_{2,r}(a, d)$ as $r \to 0$ has been carefully studied in [31]. In particular, it has been proved that $E_{2,r}(a, d)$ blows up with leading term $2\pi \left(\sum_{j=1}^k d_j^2\right) |\ln r|$. In order to minimize the blow-up rate of this energy as $r \to 0$, one is led to choose $k = D$ and $d_j = 1$, $\forall j$. Returning to $E_{2,r}(a, d)$, with this choice of d, we may then compute the next term in the expansion of $E_{2,r}(a, d)$ as $r \to 0$.

The first two terms in the asymptotic expansion are

$$E_{2,r}(a,d) = 2\pi D\, |\ln r| + W(a, g) + O(r) \text{ as } r \to 0,$$

where $W(a, g)$ is called the *renormalized energy*. In other words, $W(a, g)$ is what remains after the singular core energy has been removed.

A basic result, due to Bethuel, Brezis, and Hélein [31], relates this approach to the Ginzburg–Landau-type strategy discussed above:

Theorem 13.11. *The configuration $a = (a_1, \dots, a_D)$ given in Theorem 13.10 is a minimizer of the renormalized energy $W(a, (1, \dots, 1))$ among all configurations $a \subset \Omega^D$ such that $a_j \ne a_\ell$ when $j \ne \ell$.*

It would be very interesting to adapt this methodology to the case $p > 2$ (see Open Problem 29).

Remark 13.8. In the analysis of various Ginzburg–Landau models, it is sometimes useful to consider a more general version of Approach 2, in which holes are of different sizes and varying locations. In this setting, lower bounds for the analogue of $E_{2,r}(a, d)$ were obtained by Han and Shafrir [174] and by Sandier [313, 314].

Remark 13.9. When $p = 2$, there is a further natural way of bypassing the topological obstruction via approximation. Given $1 < q < 2$, consider, as in Section 13.2,

$$\inf \left\{ \int_\Omega |\nabla u|^q;\ u \in W^{1,q}_g(\Omega; \mathbb{S}^1) \right\}, \tag{13.47}$$

and analyze the asymptotic behavior of minimizers u_q as $q \nearrow 2$. [Recall that $W^{1,q}_g(\Omega; \mathbb{S}^1) \ne \emptyset$ even if $\deg g \ne 0$.] This strategy, proposed by Bethuel, Brezis, and Hélein [31], has been successfully carried out by Hardt and Lin [184], who established that it is consistent with those presented above; namely, $u_{q_n} \to u_*$ given by (13.45), where $a = (a_1, \ldots, a_D)$ is a minimizer of the renormalized energy.

13.6.4 Minimizing infinite energies when $N \geq 3$ and $2 \leq p < N$

Throughout this subsection, we assume that $\Omega \subset \mathbb{R}^N$, with $N \geq 3$.

In Theorem 13.1, we assumed that $g \in C^\infty(\partial\Omega; \mathbb{S}^1)$, so that $W^{1,p}_g(\Omega; \mathbb{S}^1) \neq \emptyset$. By contrast, here we deal with the case where g is *not* smooth. More precisely, we assume that singularities of g cause $W^{1,p}_g(\Omega; \mathbb{S}^1)$ to be *empty*, whereas $W^{1,p}_g(\Omega; \mathbb{R}^2)$ is *non-empty*. This can happen when $2 \leq p < N$. It suffices to take some $g \in W^{1-1/p,p}(\partial\Omega; \mathbb{S}^1)$ that admits no lifting in $W^{1-1/p,p}(\partial\Omega; \mathbb{R})$ (by Theorem 5.2, this is possible when $1 \leq (1 - 1/p)p < N - 1$, i.e., when $2 \leq p < N$; see also Theorem 11.1 in Chapter 11).

For such g, the minimization problem

$$\inf \left\{ \int_\Omega |\nabla u|^p;\ u \in W^{1,p}_g(\Omega; \mathbb{S}^1) \right\} \tag{13.48}$$

makes no sense, and one can approach it via a Ginzburg–Landau-type approximation,

$$m_{p,\varepsilon} := \inf \left\{ \int_\Omega |\nabla u|^p + \frac{1}{\varepsilon^2} \int_\Omega (|u|^2 - 1)^2;\ u \in W^{1,p}_g(\Omega; \mathbb{R}^2) \right\}. \tag{13.49}$$

Questions concerning the asymptotic behavior as $\varepsilon \to 0$ of $m_{p,\varepsilon}$ and of minimizers $u_{p,\varepsilon}$ in (13.49) have been investigated when $p = 2$. Here are some results; for simplicity, we restrict ourselves to $N = 3$.

1. Assume that g is smooth except at a finite number of singularities $a_j \in \partial\Omega$, and that, near each a_j,

$$g(z) \sim \frac{(z - a_j)^{d_j}}{|z - a_j|^{d_j}},$$

where we identify a neighborhood of a_j in $\partial\Omega$ with a disk in $\mathbb{C}$ centered at a_j.

Clearly, $\sum d_j = 0$, and thus one can list the points a_j with multiplicities $|d_j|$ as collections (p_ℓ), (n_ℓ) as in Section 2.0. Note that $g \in H^{1/2}(\partial\Omega; \mathbb{S}^1)$ and that g has no lifting in $H^{1/2}(\partial\Omega; \mathbb{R})$ (see Section 5.1.2). Therefore, there exists no $u \in H^1_g(\Omega; \mathbb{S}^1)$.

Theorem 13.12. *(Lin and Rivière [231, 232], with roots in Rivière [304]) Up to a subsequence, $u_{2,\varepsilon_n} \to u_*$ pointwise in Ω, except along a minimal connection Σ connecting in $\overline{\Omega}$ the points p_ℓ to the points n_ℓ. Moreover, u_* is smooth in $\overline{\Omega} \setminus \Sigma$, and locally of the form $u_* = e^{\imath\varphi_*}$, with φ_* harmonic.*

2. The case of a general boundary condition $g \in H^{1/2}(\partial\Omega; \mathbb{S}^1)$ was investigated by Bourgain, Brezis, and Mironescu [54]. One knows (see Theorem 11.3 with $s = 1/2$, $p = 2$) that there are two distinguished subclasses in $H^{1/2}(\partial\Omega; \mathbb{S}^1)$,

$$X \subsetneqq Y \subsetneqq H^{1/2}(\partial\Omega; \mathbb{S}^1),$$

where

$$\begin{aligned} X :=& \{g = e^{\imath\psi_0} \text{ with } \psi_0 \in H^{1/2}(\partial\Omega; \mathbb{R})\}, \\ Y :=& \{g \in H^{1/2}(\partial\Omega; \mathbb{S}^1);\ Jg = 0\} = \overline{C^\infty(\partial\Omega; \mathbb{S}^1)}^{H^{1/2}}. \end{aligned}$$

When $g \in X$, it is easy to see that (13.48) has a unique solution, $u_* := e^{\imath\varphi}$, with $\varphi \in C^\infty(\Omega) \cap H^1(\Omega)$ being the harmonic extension of ψ_0, and that $u_{2,\varepsilon} \to u_*$ strongly in H^1, with $m_{2,\varepsilon} \to \int_\Omega |\nabla u_*|^2$ as $\varepsilon \to 0$.

In fact, the condition $\liminf_{\varepsilon\to 0} m_{2,\varepsilon} < \infty$ implies that $g \in X$.

When $g \in Y$, we may write, by Theorem 8.8, $g = e^{\imath\psi_0}$, where $\psi_0 \in H^{1/2} + W^{1,1}(\partial\Omega; \mathbb{R})$. In fact, we know that Y coincides precisely with

$$\{g \in H^{1/2}(\partial\Omega; \mathbb{S}^1);\ g = e^{\imath\psi} \text{ for some } \psi \in H^{1/2} + W^{1,1}(\partial\Omega; \mathbb{R})\}$$

(see also the proof of Theorem 10.4 in Case 3). Moreover, the lifting ψ_0 of g is unique (modulo an integer multiple of 2π) in $H^{1/2} + W^{1,1}$, by Theorem 6.6.

Let φ be the harmonic extension of ψ_0 in Ω. We claim that $\varphi \in W^{1,3/2}(\Omega)$. This is an immediate consequence of the fractional Sobolev embeddings $H^{1/2}(\partial\Omega) \hookrightarrow W^{1/3,3/2}(\partial\Omega)$ and $W^{1,1}(\partial\Omega) \hookrightarrow W^{1/3,3/2}(\partial\Omega)$ (see Remark 15.2 and Theorem 15.4), and of standard trace theory.

It follows that

$$u_* := e^{\imath\varphi} \in W^{1,3/2}(\Omega; \mathbb{S}^1) \cap C^\infty(\Omega),$$

but we do not have $u_* \in H^1(\Omega)$ – unless $g \in X$, a case that has already been dealt with above.

A delicate result of Bourgain, Brezis, and Mironescu [51, 54] (as refined by Bethuel, Orlandi, and Smets [35, 36]) asserts that

$$u_{2,\varepsilon} \to u_* \text{ as } \varepsilon \to 0, \text{ in } W^{1,q}(\Omega) \cap C^\infty(\Omega),\ \forall\, q < 3/2, \tag{13.50}$$

$$\|u_{2,\varepsilon}\|_{W^{1,3/2}(\Omega)} \le C, \tag{13.51}$$

$$m_{2,\varepsilon} = o(|\ln \varepsilon|) \text{ as } \varepsilon \to 0. \tag{13.52}$$

The outcome of this analysis is that u_* can be viewed as the (unique) generalized "minimizer" of the impossible problem (13.48), even though u_* has infinite energy!

Finally, for a general $g \in H^{1/2}(\partial\Omega; \mathbb{S}^1)$, it is established in [35, 36, 49, 54] that

$$\|u_{2,\varepsilon}\|_{W^{1,3/2}(\Omega)} \le C, \tag{13.53}$$

$$m_{2,\varepsilon} = 2\pi\, \Sigma(g)\, |\ln \varepsilon| + o(|\ln \varepsilon|) \text{ as } \varepsilon \to 0, \tag{13.54}$$

$$u_{2,\varepsilon} \text{ converges (modulo a subsequence) in } W^{1,q}(\Omega),\ \forall\, q < 3/2$$
$$\text{and in } W^{1,q}_{loc}(\Omega),\ \forall\, q < 2, \text{ to some limit } u_*. \tag{13.55}$$

Here, $\Sigma(g)$ is computed via duality (as in Theorem 2.5), using the metric induced on $\partial\Omega$ by the geodesic distance in $\overline{\Omega}$ (so that (13.54) is consistent with Theorem 13.11).

13.6.5 Open problems

Open Problem 25. Let $1 < p < 2$. Can one construct examples of boundary data $g \in C^\infty(\mathbb{S}^1; \mathbb{S}^1)$ such that $\deg g = 0$ and (13.1) admits multiple minimizers in $W^{1,p}_g(\mathbb{D}; \mathbb{S}^1)$?

Open Problem 26. Assume $1 < p < 2$, $g \in C^\infty(\partial\Omega; \mathbb{S}^1)$ and, if $N = 2$, $\deg g = 0$. Suppose that $A_g = B_g$, i.e., no gap. Let $U \in C^{1,\alpha}(\overline{\Omega}; \mathbb{S}^1)$ be the unique minimizer in (13.2) (see Theorem 13.4). Is U also the *unique* minimizer in (13.1)? Alternatively, could there be additional minimizers in (13.1) that are discontinuous?

Open Problem 27. Let Ω be the unit ball in $\mathbb{R}^3$ and let $g \in C^\infty(\mathbb{S}^2; \mathbb{S}^2)$ be such that $\deg g = 0$. By analogy with Theorem 13.4, it is tempting to consider the minimization problem

$$\inf \left\{ \int_\Omega |\nabla u|^2;\ u \in H^1_g(\Omega; \mathbb{S}^2) \cap C(\overline{\Omega}; \mathbb{S}^2) \right\}, \tag{13.56}$$

and to ask whether (13.56) admits minimizers. [By contrast with the situation of Theorem 13.4, we do not have lifting at our disposal.] This is a long-standing open question (see Hardt and Lin [182, Question at the end of the paper], Hardt [178, Section 6], and Brezis [73, Open problem 3]). A natural approach, involving an appropriate relaxed energy, was attempted in Bethuel, Brezis, and Coron [30], but it did not yield any conclusive evidence.

Open Problem 28. Assume $N = 2$, $p > 2$, and $\deg g \neq 0$. What happens to $m_{p,\varepsilon}$ as $\varepsilon \to 0$? Let u_ε be a minimizer in (13.44). What happens to u_ε as $\varepsilon \to 0$? Is there an analogue of Theorem 13.10 (dealing with the case $p = 2$)?

Similar questions when $N \ge 3$, with $u : \Omega \subset \mathbb{R}^N \to \mathbb{R}^N$ and $p \ge N$.

Open Problem 29. Assume $N = 2$, $p > 2$, and $\deg g \neq 0$. What happens to Approach 2 described in Section 13.6.3 for $p = 2$?

Open Problem 30. Let $p > 2$ and let u_ε be a minimizer of the Ginzburg–Landau energy $E_{p,\varepsilon}$ in $W_g^{1,p}(\Omega; \mathbb{R}^2)$, where $\Omega := \mathbb{D}$ and $g(x) := x$. Does $u_\varepsilon(x) \to x/|x|$ as $\varepsilon \to 0$?

This question, when $p = 2$, was originally raised by Matano [237] and led to the theory developed in Bethuel, Brezis, and Hélein [31]. The answer is known to be positive when $p = 2$ (see [31, Theorem VIII.6]), but the proof is quite intricate, and relies on a careful analysis of the renormalized energy. [Incidentally, we do not know any elementary proof of this result.]

More generally, let $N \geq 2$, $p > 1$, and let u_ε be a minimizer of the Ginzburg-Landau energy $E_{p,\varepsilon}$ in $W_g^{1,p}(\Omega; \mathbb{R}^N)$, where $\Omega := \mathbb{B}^N$ and $g(x) := x$. One may ask whether $u_\varepsilon(x) \to x/|x|$ as $\varepsilon \to 0$. The answer is known to be positive if $N \geq 2$ and $1 < p < N$ (as a consequence of Theorems 13.6 and 13.9), but it is unknown if $N \geq 2$ and $p \geq N$, except when $N = 2$ and $p = 2$ (see the comment after Open Problem 30).

In a closely related direction, one may raise the question of uniqueness of the solutions of the Ginzburg-Landau equation,

$$\begin{cases} -\Delta u = \dfrac{1}{\varepsilon^2} u(1 - |u|^2) & \text{in } \Omega = \mathbb{B}^N \\ u(x) = x & \text{on } \partial\Omega = \mathbb{S}^{N-1} \end{cases} \tag{13.57}$$

(where $N \geq 2$ and $u : \Omega \to \mathbb{R}^N$), i.e., of critical points of $E_{2,\varepsilon}$ in $W_g^{1,2}(\Omega; \mathbb{R}^N)$.

Clearly, there exists only one solution of (13.57) when $\varepsilon \geq 1/\sqrt{\lambda_1}$, where λ_1 is the first eigenvalue of $-\Delta$ (under zero Dirichlet boundary condition), since for such ε the functional $E_{2,\varepsilon}$ is strictly convex on $W_g^{1,2}(\Omega; \mathbb{R}^N)$. The problem of the uniqueness of the solutions of (13.57) when $\varepsilon < 1/\sqrt{\lambda_1}$, originally raised by Bethuel, Brezis, and Hélein [31, Problem 10, p. 139] when $N = 2$ (see also Brezis [71] when $N \geq 2$), turns out to be difficult. [Note that uniqueness for small ε implies that any solution u_ε of (13.57) is of the form $u_\varepsilon(x) = f_\varepsilon(|x|)x/|x|$, and then it is easy to show that $u_\varepsilon(x) \to x/|x|$ as $\varepsilon \to 0$.]

Pacard and Rivière [284] established that, when $N = 2$ and $0 < \varepsilon < \varepsilon_0$, with ε_0 sufficiently small, (13.57) admits a unique solution. The proof is highly non-trivial, and fills a significant part of the monograph [284]; it would be interesting to find a simpler proof. We also point out that uniqueness is still open when $N = 2$ and $\varepsilon_0 \leq \varepsilon < 1/\sqrt{\lambda_1}$.

One may raise an array of related questions, for example, replacing critical points of $E_{2,\varepsilon}$ by minimizers of $E_{2,\varepsilon}$, replacing $\mathbb{B}^N$ by $\mathbb{R}^N$ (with an appropriate "boundary condition" at infinity), replacing $E_{2,\varepsilon}$ by $E_{p,\varepsilon}$, etc. We refer the interested reader to Ignat, Nguyen, Slastikov, and Zarnescu [198] and the references therein for a glimpse of the recent progress on these topics.

13.7 Comments

1. Strictly speaking, Theorems 13.1 and 13.4 are new, but they rely heavily, via lifting, on the classical regularity theory for the p-Laplacian.
2. Theorem 13.2 is a very special case of a remarkable result of Hardt and Lin [183], who extended to minimizing p-harmonic maps, with arbitrary p, the analysis of Schoen and Uhlenbeck for minimizing 2-harmonic maps [317, 318]. In their work, boundary regularity is only $C^{0,\beta}$. In our context, the improved $C^{1,\alpha}$ regularity near the boundary comes from an adaptation of the proof of Theorem 13.4.
3. As mentioned in Comment 15 d) in Chapter 1, a Lavrentiev-type phenomenon in the spirit of Theorem 13.5 was first discovered by Hardt and Lin [182], who constructed a smooth map $g : \mathbb{S}^2 \to \mathbb{S}^2$ such that $C^1_g(\overline{\mathbb{B}^3}; \mathbb{S}^2) \neq \emptyset$, but
$$\inf\left\{\int_{\mathbb{B}^3} |\nabla u|^2;\ u \in W^{1,2}_g(\mathbb{B}^3; \mathbb{S}^2)\right\} < \inf\left\{\int_{\mathbb{B}^3} |\nabla u|^2;\ u \in C^1_g(\overline{\mathbb{B}^3}; \mathbb{S}^2)\right\}$$
(see also Brezis [69] for a simpler construction).
Demengel and Hadiji [133] proved that the same phenomenon occurs for the $W^{1,1}$ energy and maps $g : \mathbb{S}^1 \to \mathbb{S}^1$; this corresponds to $A_g < B_g$ in the special case where $p = 1$ and $\Omega = \mathbb{D}$ (see Remark 13.6).
These results were extended to more general domains and target manifolds by Giaquinta, Modica, and Souček [158]. Inequality $A_g < B_g$ (for arbitrary N and Ω) is a special case of the main result in [158], but the approach we present, via Theorem 13.5, is new. The "double gap" phenomenon presented in Theorem 13.5 is new.
4. The core of Theorems 13.6 and 13.7 goes back to Chen [117] and Hardt and Chen [179, Section 4 and Lemma 7.1]. Their results were subsequently refined and generalized by Hardt, Lin, and Wang [185, 186]. The uniqueness assertion in Theorem 13.6 is not explicitly stated in these works, but it is implicit in [186, proof of the main theorem]. The elegant argument we present in Section 13.4 was communicated to us by Lin [229]; it relies on ideas that are underlying in the papers quoted above.
As already mentioned in Remark 13.4, the main substance of Theorems 13.6 and 13.7 originates in the work of Brezis, Coron, and Lieb [79, Section VII].
5. Theorem 13.8 is closely related to results by Hardt and Lin [184] and Hardt and Chen [179, Theorem E].

Chapter 14
Domains with topology

14.0 Introduction

In the previous chapters, we discussed several properties of maps $u \in W^{s,p}(\Omega; \mathbb{S}^1)$ (existence of lifting, relaxed energy, density of smooth maps, etc.), with Ω *simply connected* or even contractible. We assume *throughout this chapter* that $\Omega \subset \mathbb{R}^N$ is smooth, bounded, and connected (but not simply connected) and we investigate the effect of the topology of Ω on these properties. This is relevant only when the space dimension is $N \geq 2$.

In order to keep this chapter of reasonable length, we do not revisit all the main theorems, but only some typical ones. In particular, we investigate in detail $W^{1,p}(\Omega; \mathbb{S}^1)$, which presents new features when Ω is not simply connected.

In Sections 14.1 and 14.2, we revisit Chapters 1, 2, and 3. In Section 14.3, we investigate the analogues of the main results in Chapters 5 and 8. The basic results from Chapter 10 are extended in Section 14.4. Finally, in Section 14.5 we describe the path-connected components, i.e., the homotopy classes of $W^{s,p}(\Omega; \mathbb{S}^1)$. The main results (Theorems 14.6 and 14.7) assert that, if $sp < 2$, $W^{s,p}(\Omega; \mathbb{S}^1)$ is *always* path-connected, whereas if $sp \geq 2$, the homotopy classes of $W^{s,p}(\Omega; \mathbb{S}^1)$ are similar to the standard homotopy classes of $C(\overline{\Omega}; \mathbb{S}^1)$.

14.1 Lifting in $W^{1,p}(\Omega; \mathbb{S}^1)$ revisited

The first question we consider is whether $W^{1,p}(\Omega; \mathbb{S}^1)$ has the lifting property; recall that in a ball in $\mathbb{R}^N$ with $N \geq 2$, this occurs only when $p \geq 2$ (Theorem 1.1 and Remark 1.6). In general, the lifting property fails for every $p \geq 1$.

Proposition 14.1. *Let* $\Omega := \{x \in \mathbb{R}^2;\ 1 < |x| < 2\}$. *Then, for any* $p \geq 1$, $W^{1,p}(\Omega; \mathbb{S}^1)$ *does not have the lifting property.*

H. Brezis and P. Mironescu, *Sobolev Maps to the Circle*, Progress in Nonlinear Differential Equations and Their Applications 96,
https://doi.org/10.1007/978-1-0716-1512-6_14

Proof. Consider $u(x) := x/|x|$, which is smooth in $\overline{\Omega}$. Then, u has no lifting in $W^{1,p}$ (see the proof of (1.62) in Remark 1.6). □

In particular, the above example shows that the implication

$$Ju = 0 \implies u \text{ has a lifting within } W^{1,p}, \tag{14.1}$$

valid in simply connected domains (Theorem 1.2), does not hold in arbitrary domains. However, the converse implication is true (see (1.58)). We also point out that property (1.61), i.e.,

$$Ju = 0, \ \forall\, u \in W^{1,p}(\Omega; \mathbb{S}^1), \ p \geq 2, \tag{14.2}$$

is still valid in arbitrary domains. In view of the above example, we see that the property "$Ju = 0$" no longer governs the existence of a lifting in $W^{1,p}$, $p \geq 1$. One possible "substitute" is the following factorization result (which is a special case of Corollary 14.3 below).

Proposition 14.2. *Given* $u \in W^{1,p}(\Omega; \mathbb{S}^1)$, $p \geq 1$, *the following assertions are equivalent:*

a) u *can be written as* $u = e^{\imath\varphi} v$ *for some* $\varphi \in W^{1,p}(\Omega; \mathbb{R})$ *and some* $v \in C^\infty(\overline{\Omega}; \mathbb{S}^1)$

b) $Ju = 0$.

In particular, a) holds for every $u \in W^{1,p}(\Omega; \mathbb{S}^1)$ *when* $p \geq 2$.

In another direction, we note that a smallness condition in $W^{1,p}$, $p \geq 2$, implies the existence of a lifting in arbitrary domains.

Proposition 14.3. *Given* $p \geq 2$, *there exists some* $\delta = \delta(\Omega, p) > 0$ *such that:*

$$[u \in W^{1,p}(\Omega; \mathbb{S}^1) \text{ and } \|\nabla u\|_{L^p} < \delta] \implies [\exists\, \varphi \in W^{1,p}(\Omega; \mathbb{R}) \text{ such that } u = e^{\imath\varphi}].$$

See Theorem 14.5 for a more general result.

Remark 14.1. The condition $p \geq 2$ is optimal, since for $1 \leq p < 2$ we may find $(u_n) \subset W^{1,p}(\Omega; \mathbb{S}^1)$ such that $\|\nabla u_n\|_{L^p} \to 0$ and $Ju_n \neq 0$, $\forall\, n$. Here is an example when $N = 2$. Fix some point $a \in \Omega$ and consider a sequence $(a_n) \subset \Omega \setminus \{a\}$ such that $a_n \to a$. Let $u_n(z) := \dfrac{z-a}{|z-a|}\dfrac{\overline{z-a_n}}{|z-a_n|}$. Then, $Ju_n = \pi(\delta_a - \delta_{a_n})$, whereas, clearly, $u_n \to 1$ in $W^{1,p}$. [This is a consequence of Lemma 15.41 and the fact that $F(\cdot + h) \to F(\cdot)$ in $W^{1,p}$ as $h \to 0$ when $F \in W^{1,p}$.]

However, under the additional assumption $Ju = 0$, we have the following variant of Proposition 14.3.

Proposition 14.4. *Given* $1 \leq p < 2$, *there exists some* $\delta = \delta(\Omega, p) > 0$ *such that:*

$$[u \in W^{1,p}(\Omega; \mathbb{S}^1),\ Ju = 0 \text{ and } \|\nabla u\|_{L^p} < \delta] \implies$$
$$[\exists\, \varphi \in W^{1,p}(\Omega; \mathbb{R}) \text{ such that } u = e^{\imath\varphi}].$$

Adopting the same program as in Section 1.4, we work with maps $u \in W^{1,1}(\Omega;\mathbb{S}^1)$ and investigate their lifting in BV. We first point out that Theorem 1.4 holds in every domain, and in particular every $u \in W^{1,1}(\Omega;\mathbb{S}^1)$ may be written as $u = e^{\imath\varphi}$ for some $\varphi \in BV(\Omega;\mathbb{R})$. As in Section 1.4 we set, for every $u \in W^{1,1}(\Omega;\mathbb{S}^1)$,

$$E(u) := \inf\{|\varphi|_{BV};\ \varphi \in BV(\Omega;\mathbb{R}) \text{ is such that } u = e^{\imath\varphi} \text{ in } \Omega\}.$$

We also define $\Sigma(u)$ as in (1.80), i.e.,

$$\Sigma(u) := \inf_{\eta \in W^{1,1}(\Omega;\mathbb{R})} \|u \wedge \nabla u - \nabla\eta\|_{L^1},$$

and prove that (1.82) still holds, i.e.,

$$\Sigma(u) = \min_{\eta \in BV(\Omega;\mathbb{R})} \|u \wedge \nabla u - D\eta\|_{\mathscr{M}}.$$

[However, (1.85) fails for general domains Ω.]

In particular we have

$$[\Sigma(u) = 0] \iff [u \text{ admits a lifting in } W^{1,1}(\Omega;\mathbb{R})].$$

With the same proof as in Section 1.4.2 we obtain the following

Theorem 1.5′. *For every $u \in W^{1,1}(\Omega;\mathbb{S}^1)$ we have*

$$E(u) = \int_\Omega |\nabla u| + \Sigma(u). \tag{14.3}$$

Moreover, every minimal lifting $\varphi \in BV(\Omega;\mathbb{R})$ of u satisfies

$$\Sigma(u) = \|u \wedge \nabla u - D\varphi\|_{\mathscr{M}}, \tag{14.4}$$

i.e., φ is a minimizer in (1.82).

We now address the question of how to compute $\Sigma(u)$ by duality (in the spirit of the results in Section 1.5). The next result is inspired by Corollary 1.6.

Corollary 1.6′. *Let $u \in W^{1,1}(\Omega;\mathbb{S}^1)$. Then,*

$$\Sigma(u) = \max\left\{\int_\Omega (u \wedge \nabla u)\cdot\xi;\ \xi \in L^\infty(\Omega;\mathbb{R}^N),\ \|\xi\|_{L^\infty} \le 1,\right.$$
$$\left.\operatorname{div}\xi = 0 \textit{ in } \Omega \textit{ and } \xi\cdot\nu = 0 \textit{ on } \partial\Omega\right\}.$$

Here, the condition [$\operatorname{div}\xi = 0$ in Ω and $\xi\cdot\nu = 0$ on $\partial\Omega$] is understood in the weak sense $\left[\int_\Omega \xi\cdot\nabla\varphi = 0,\ \forall\,\varphi \in W^{1,1}(\Omega;\mathbb{R})\right]$, or equivalently $\left[\int_\Omega \xi\cdot\nabla\varphi = 0,\ \forall\,\varphi \in C^\infty(\overline{\Omega};\mathbb{R})\right]$.

Proof. Apply Hahn–Banach as in Comment 12 of Chapter 1. □

We now present a geometric interpretation of $\Sigma(u)$ in the special case $N = 2$. [We have not studied the higher dimensional analogue; see Open Problem 32.]

For this purpose, we start by establishing a 2D version of Corollary 1.6. Set

$$W^{1,\infty}_{const}(\Omega; \mathbb{R}) := \{\zeta \in W^{1,\infty}(\Omega; \mathbb{R});\ \zeta \text{ is constant on each } \Gamma_j\}.$$

Corollary 1.6″. *Assume $N = 2$. Let $u \in W^{1,1}(\Omega; \mathbb{S}^1)$. Then,*

$$\Sigma(u) = \max\left\{\int_\Omega (u \wedge \nabla u)\cdot \nabla^\perp \zeta;\ \zeta \in W^{1,\infty}_{const}(\Omega; \mathbb{R}) \text{ and } \|\nabla \zeta\|_{L^\infty} \le 1\right\}.$$

Proof. It suffices to show that

$$\begin{aligned}&\{\xi \in L^\infty(\Omega; \mathbb{R}^2);\ \operatorname{div}\xi = 0 \text{ in } \Omega \text{ and } \xi\cdot\nu = 0 \text{ on } \partial\Omega\}\\ &= \{\nabla^\perp\zeta;\ \zeta \in W^{1,\infty}_{const}(\Omega; \mathbb{R})\}.\end{aligned} \tag{14.5}$$

For the inclusion "$\supset$," we verify that

$$\int_\Omega \nabla^\perp\zeta\cdot\nabla\varphi = 0,\ \forall\varphi \in C^\infty(\overline{\Omega}; \mathbb{R}),\ \forall\zeta \in W^{1,\infty}_{const}(\Omega; \mathbb{R});$$

this is clear since $\operatorname{curl}(\nabla\varphi) = 0$ and, $\forall j$,

$$\int_{\Gamma_j} \zeta\,\nabla\varphi\cdot\tau = \zeta(\Gamma_j)\int_{\Gamma_j}\frac{\partial\varphi}{\partial\tau} = 0.$$

Here, $\zeta(\Gamma_j)$ is the (constant) value assumed by ζ on Γ_j.

For the inclusion "$\subset$," we start with some $\xi \in L^\infty(\Omega; \mathbb{R}^2)$ such that

$$\int_\Omega \xi\cdot\nabla\varphi = 0,\ \forall\varphi \in W^{1,1}(\Omega; \mathbb{R}). \tag{14.6}$$

Set $\overline{\xi} := \begin{cases}\xi, & \text{in } \Omega\\ 0, & \text{in } \mathbb{R}^2\setminus\Omega\end{cases}$.

Then, by (14.6),

$$\int_{\mathbb{R}^2} \overline{\xi}\cdot\nabla\Phi = \int_\Omega \xi\cdot\nabla\left(\Phi_{|\Omega}\right) = 0,\ \forall\Phi \in C^1_c(\mathbb{R}^2; \mathbb{R}). \tag{14.7}$$

Thus, we may invoke the generalized Poincaré lemma (Lemma 1.9) in $\mathbb{R}^2$ and conclude that $\overline{\xi} = \nabla^\perp\overline{\zeta}$ for some $\overline{\zeta} \in W^{1,\infty}(\mathbb{R}^2; \mathbb{R})$. Clearly, $\zeta := \overline{\zeta}_{|\Omega} \in W^{1,\infty}_{const}(\Omega; \mathbb{R})$ and $\nabla^\perp\zeta = \xi$. □

We next assume that

$$u \in W^{1,1}(\Omega; \mathbb{S}^1) \cap C(\Omega\setminus\cup_{i=1}^{\ell}\{a_i\}; \mathbb{S}^1), \tag{14.8}$$

and present an explicit formula for $\Sigma(u)$, in the spirit of Theorem 2.5 in Chapter 2.

Denote by $\Gamma_j, j = 1, \ldots, k$, the components of $\partial\Omega$, oriented such that (τ, n) is a direct basis of $\mathbb{R}^2$, with n the outward normal to Ω. [Warning: this is the opposite of the usual orientation, which we have used so far, in particular in Section 12.8.5.] We define $d_i := \deg(u, a_i)$ as usual. Note that the standard (counterclockwise) orientation on $\partial\mathbb{D}(a_i, \varepsilon)$ is consistent with the orientation we have adopted on $\partial(\Omega \setminus \overline{\mathbb{D}}(a_i, \varepsilon))$. We also define $D_j := \deg(u, \Gamma_j)$ by considering the restriction of u to

$$\Gamma_{j,\varepsilon} := \{x \in \Omega;\ \operatorname{dist}(x, \Gamma_j) = \varepsilon\}, \tag{14.9}$$

with $\varepsilon > 0$ small.

Let us note the compatibility condition

$$\sum_{j=1}^{k} D_j + \sum_{i=1}^{\ell} d_i = 0, \tag{14.10}$$

which is obtained via the excision and additivity properties of the degree (see, for example, Dinca and Mawhin [141, Theorem 1.2.1, Theorem 1.2.3]), using the fact that u is continuous and $\mathbb{S}^1$-valued on

$$\Omega_\varepsilon := \Omega \setminus \Big(\{x \in \Omega;\ \operatorname{dist}(x, \partial\Omega) \le \varepsilon\} \cup \cup_{i=1}^{\ell} \overline{\mathbb{D}}(a_i, \varepsilon)\Big). \tag{14.11}$$

The next result plays an important role.

Lemma 14.1. *Assume that u satisfies* (14.8). *Then,* $\forall\, \zeta \in W^{1,\infty}_{const}(\Omega; \mathbb{R})$,

$$-\frac{1}{2\pi}\int_\Omega (u \wedge \nabla u) \cdot \nabla^\perp \zeta = \sum_{j=1}^{k} D_j\, \zeta(\Gamma_j) + \sum_{i=1}^{\ell} d_i\, \zeta(a_i). \tag{14.12}$$

Proof. It suffices to prove (14.12) when $\zeta \in W^{1,\infty}_{const}(\Omega; \mathbb{R})$ and, in addition,

$$\zeta \text{ is constant near each } \Gamma_j \text{ and each } a_i. \tag{14.13}$$

Indeed, given any $\zeta \in W^{1,\infty}_{const}(\Omega; \mathbb{R})$, there exists a sequence $(\zeta_n) \subset W^{1,\infty}_{const}(\Omega; \mathbb{R})$ such that each ζ_n satisfies (14.13) and, in addition,

$$|\nabla\zeta_n| \le C \text{ in } \Omega,\ \forall n,$$
$$\nabla\zeta_n \to \nabla\zeta \text{ a.e. in } \Omega \text{ as } n \to \infty,$$
$$\zeta_n(\Gamma_j) = \zeta(\Gamma_j),\ \zeta_n(a_i) = \zeta(a_i),\ \forall n, \forall i, \forall j.$$

To construct ζ_n, we may, for example, modify ζ near Γ_j by choosing

$$\zeta_n(x) := (\zeta(x) - \zeta(\Gamma_j))\, H(n \operatorname{dist}(x, \Gamma_j)) + \zeta(\Gamma_j),$$

where H is a smooth function on $[0, \infty)$ such that $H(t) = 0$ for $t \leq 1$ and $H(t) = 1$ for $t \geq 2$.

Next, we assume that ζ satisfies (14.13). Then, for sufficiently small ε, we have

$$\int_{\Omega} (u \wedge \nabla u) \cdot \nabla^{\perp} \zeta = \int_{\Omega_\varepsilon} (u \wedge \nabla u) \cdot \nabla^{\perp} \zeta, \tag{14.14}$$

with Ω_ε as in (14.11). On the other hand, since $u \in C(\overline{\Omega_\varepsilon})$, we may find a sequence $(u_m) \subset C^{\infty}(\overline{\Omega_\varepsilon}; \mathbb{S}^1)$ such that $u_m \to u$ in $W^{1,1}(\Omega_\varepsilon)$ and in $L^{\infty}(\Omega_\varepsilon)$.

An easy calculation yields (with $\Gamma_{j,\varepsilon}$ as in (14.9))

$$\begin{aligned} -\frac{1}{2\pi} \int_{\Omega_\varepsilon} (u_m \wedge \nabla u_m) \cdot \nabla^{\perp} \zeta =& \sum_{j=1}^{k} \deg(u_m, \Gamma_{j,\varepsilon})\, \zeta(\Gamma_j) \\ &+ \sum_{i=1}^{\ell} \deg(u_m, \partial \mathbb{D}(a_i, \varepsilon))\, \zeta(a_i). \end{aligned} \tag{14.15}$$

Passing to the limit in (14.15) as $m \to \infty$ (for fixed ε), we see that

$$\begin{aligned} -\frac{1}{2\pi} \int_{\Omega_\varepsilon} (u \wedge \nabla u) \cdot \nabla^{\perp} \zeta =& \sum_{j=1}^{k} \deg(u, \Gamma_{j,\varepsilon})\, \zeta(\Gamma_j) \\ &+ \sum_{i=1}^{\ell} \deg(u, \partial \mathbb{D}(a_i, \varepsilon))\, \zeta(a_i). \end{aligned} \tag{14.16}$$

Finally, we combine (14.14) and (14.16) and obtain (14.12). □

Next, we define on $\overline{\Omega}$ a pseudometric as in (2.40)–(2.42). When $\overline{\Omega}$ is multiply connected, there are two distinct definitions. Given $P, N \in \overline{\Omega}$, set

$$\begin{aligned} d(P, N) := \inf\{&\text{length}\, (\gamma \cap \Omega);\ \gamma \subset \overline{\Omega} \text{ is a rectifiable curve} \\ &\text{from } P \text{ to } N\}, \end{aligned} \tag{14.17}$$

and

$$\delta(P, N) := \min\{|P - N|,\ \text{dist}(P, \partial\Omega) + \text{dist}(N, \partial\Omega)\}. \tag{14.18}$$

Clearly, $\delta(P, N) \leq d(P, N)$, and it may happen that $\delta(P, N) < d(P, N)$.

Note that the inf in (14.17) is achieved by a curve γ that consists of a finite union of line segments in Ω and of pieces of curves in $\partial\Omega$. Clearly,

$$d(P, N) = 0 \iff [\text{either } P = N \text{ or } P,\ N \in \Gamma_j \text{ for some } j]. \tag{14.19}$$

By analogy with Lemma 2.3 (and with a similar proof) we have the following

Lemma 2.3′. *Let $\zeta : \overline{\Omega} \to \mathbb{R}$. Then, the following properties are equivalent:*

$$\zeta \in W^{1,\infty}_{const}(\Omega; \mathbb{R}) \text{ and } \|\nabla \zeta\|_{L^\infty} \leq 1. \tag{14.20}$$

$$|\zeta(x) - \zeta(y)| \leq d(x, y), \ \forall x, y \in \overline{\Omega}. \tag{14.21}$$

In view of (14.19), we may identify each Γ_j with a single point in $\overline{\Omega}$, still denoted by Γ_j, and associated with the integer D_j. We then repeat the a_i's and the Γ_j's according to their multiplicities and we rewrite them as a collection of m positive points (P_l) and m negative points (N_l), $1 \leq l \leq m$. [This is possible, by (14.10).] Finally, we set

$$L(a, d, \Gamma, D) := \min_{\sigma} \sum_{l=1}^{m} d(P_l, N_{\sigma(l)}), \tag{14.22}$$

where the minimum is taken over all permutations σ of the integers $\{1, \ldots, m\}$.

The main result is the following

Theorem 14.1. *Assume* (14.8). *Then,* $\Sigma(u) = 2\pi\, L(a, d, \Gamma, D)$.

Proof. Apply Corollary 1.6″, (14.12), Lemma 2.3′, and Theorem 2.4. □

Example 14.1. Assume that Ω has a single hole, so that $\partial\Omega = \Gamma_1 \cup \Gamma_2$, and that u is smooth in Ω. Then, $\deg(u, \Gamma_1) = -\deg(u, \Gamma_2)$ and

$$\Sigma(u) = 2\pi\, |\deg(u, \Gamma_1)|\ \mathrm{dist}(\Gamma_1, \Gamma_2) = 2\pi\, |\deg(u, \Gamma_2)|\ \mathrm{dist}(\Gamma_1, \Gamma_2).$$

14.2 Relaxed energy revisited

Define the relaxed energy $R_1(u)$ as in (1.183), i.e.,

$$R_1(u) := \inf \left\{ \liminf_{n\to\infty} \int_\Omega |\nabla u_n|;\ u_n \in C^\infty(\overline{\Omega}; \mathbb{S}^1),\ u_n \to u \text{ a.e.} \right\}. \tag{14.23}$$

Here, we need not have $R_1(u) = \int_\Omega |\nabla u| + \Sigma(u)$, as we had in Theorem 1.11, as can be seen from the following example.

Proposition 14.5. *Let* $\Omega := \{x \in \mathbb{R}^2;\ 1 < |x| < 2\}$ *and* $u(x) := x/|x|$. *Then,* $R_1(u) = \int_\Omega |\nabla u| = 2\pi$ *and* $\Sigma(u) = 2\pi$.

Proof. First equality is clear, since u is smooth. Equality $\Sigma(u) = 2\pi$ follows from Example 14.1. □

An explicit formula for $R_1(u)$ involves the quantity

$$\begin{aligned} \Sigma'(u) :&= \inf_{\substack{\eta \in BV(\Omega;\mathbb{R}) \\ v \in C^\infty(\overline{\Omega};\mathbb{S}^1)}} \|u \wedge \nabla u - v \wedge \nabla v - D\eta\|_{\mathcal{M}} \\ &= \inf_{v \in C^\infty(\overline{\Omega};\mathbb{S}^1)} \Sigma(u\overline{v}). \end{aligned} \tag{14.24}$$

Remark 14.2. As we see in Theorem 14.2, the first inf in (14.24) is achieved (and thus so is the second).

Clearly, we have $\Sigma'(u) \leq \Sigma(u), \forall\, u \in W^{1,1}(\Omega; \mathbb{S}^1)$. To see that the two quantities need not be equal, let u be as in Proposition 14.5. Then, u is smooth and thus $\Sigma'(u) = 0$. On the other hand, Example 14.1 shows that $\Sigma(u) = 2\pi$. Thus, we may have $\Sigma'(u) < \Sigma(u)$.

However, if Ω is simply connected, then $\Sigma'(u) = \Sigma(u)$. Indeed, if $v \in C^\infty(\overline{\Omega}; \mathbb{S}^1)$, then $v = e^{\imath\psi}$ for some $\psi \in C^\infty(\overline{\Omega}; \mathbb{R})$, and thus

$$u \wedge \nabla u - v \wedge \nabla v - D\eta = u \wedge \nabla u - D(\eta + \psi).$$

Theorem 14.2. *For $u \in W^{1,1}(\Omega; \mathbb{S}^1)$, we have*

$$R_1(u) = \int_\Omega |\nabla u| + \Sigma'(u). \tag{14.25}$$

Moreover, let $(u_n) \subset C^\infty(\overline{\Omega}; \mathbb{S}^1)$ be such that $u_n \to u$ a.e. and $\int_\Omega |\nabla u_n| \to R_1(u)$. Then, possibly up to a subsequence, we may write $u_n = e^{\imath\varphi_n} v$, with $\varphi_n \in C^\infty(\overline{\Omega}; \mathbb{R})$, $v \in C^\infty(\overline{\Omega}; \mathbb{S}^1)$, $\varphi_n \to \eta$ a.e., for some $\eta \in BV(\Omega; \mathbb{R})$, and the pair (η, v) is a minimizer in (14.24).

The proof of Theorem 14.2 relies on the following lemma, whose proof is postponed to Section 14.5.

Lemma 14.2. *Let $(u_n) \subset C^\infty(\overline{\Omega}; \mathbb{S}^1)$ be a sequence bounded in $W^{1,1}$. Then, possibly up to a subsequence, we may write $u_n = e^{\imath\varphi_n} v$, with $\varphi_n \in C^\infty(\overline{\Omega}; \mathbb{R})$ and $v \in C^\infty(\overline{\Omega}; \mathbb{S}^1)$.*

Proof of Theorem 14.2. Let $(u_n) \subset C^\infty(\overline{\Omega}; \mathbb{S}^1)$ be such that $u_n \to u$ a.e. and $\int_\Omega |\nabla u_n| \to R_1(u)$. Possibly up to a subsequence, we write u_n as in Lemma 14.2. Then, (φ_n) is bounded in $W^{1,1}$, and, again up to a subsequence, we may find some $\eta \in BV(\Omega; \mathbb{R})$ such that $\varphi_n \to \eta$ a.e. Note that $u = e^{\imath\eta} v$, and that

$$\|D\eta + v \wedge \nabla v\|_{\mathscr{M}} \leq \liminf_{n\to\infty} \int_\Omega |\nabla\varphi_n + v \wedge \nabla v| = \liminf_{n\to\infty} \int_\Omega |\nabla u_n| = R_1(u). \tag{14.26}$$

On the other hand, it is clear that, whenever $\eta \in BV(\Omega; \mathbb{R})$ and $v \in C^\infty(\overline{\Omega}; \mathbb{R})$ are such that $u = e^{\imath\eta} v$, we have

$$\|D\eta + v \wedge \nabla v\|_{\mathscr{M}} \geq R_1(u). \tag{14.27}$$

Indeed, it suffices to consider in (14.23) the competitors $u_n := e^{\imath\eta_n} v$, with η_n as in formula (1.83) (also valid in multiply connected domains) applied to $F := -v \wedge \nabla v$.

Thus,

$$R_1(u) = \min_{\substack{\eta \in BV(\Omega;\mathbb{R}) \\ v \in C^\infty(\overline{\Omega};\mathbb{S}^1) \\ u = e^{\imath\eta} v}} \|D\eta + v \wedge \nabla v\|_{\mathscr{M}}. \tag{14.28}$$

In view of (14.24) and (14.28), the conclusions of the theorem follow from the equalities

$$\min_{\substack{\eta \in BV(\Omega;\mathbb{R}) \\ v \in C^\infty(\overline{\Omega};\mathbb{S}^1) \\ u = e^{\imath\eta} v}} \|D\eta + v \wedge \nabla v\|_{\mathscr{M}} = \int_\Omega |\nabla u| + \inf_{\substack{\eta \in BV(\Omega;\mathbb{R}) \\ v \in C^\infty(\overline{\Omega};\mathbb{S}^1)}} \|u \wedge \nabla u - v \wedge \nabla v - D\eta\|_{\mathscr{M}} \tag{14.29}$$

and

$$|D\eta + v \wedge \nabla v| = |\nabla u| + |u \wedge \nabla u - v \wedge \nabla v - D\eta|, \tag{14.30}$$

for $\eta \in BV(\Omega;\mathbb{R})$ and $v \in C^\infty(\overline{\Omega};\mathbb{R})$ such that $u = e^{\imath\eta} v$. We now prove the above assertions.

Equality (14.30) follows from (1.108) and the equality $e^{\imath\eta} = u\overline{v}$, which imply that $D\eta = u \wedge \nabla u - v \wedge \nabla v + D^j\eta$, and thus

$$|D\eta + v \wedge \nabla v| = |u \wedge \nabla u| + |D^j\eta| = |\nabla u| + |u \wedge \nabla u - v \wedge \nabla v - D\eta|.$$

It remains to prove the equality (14.29). Inequality "$\geq$" is clear, by (14.30). On the other hand, as in Lemma 1.13, we have

$$\Sigma'(u) = \inf_{\substack{\psi \in W^{1,1}(\Omega;\mathbb{R}) \\ v \in C^\infty(\overline{\Omega};\mathbb{S}^1)}} \int_\Omega |u \wedge \nabla u - v \wedge \nabla v - \nabla\psi| \tag{14.31}$$

(the proof of this equality does not rely on the simple connectedness of Ω).

In view of (14.31), it suffices to find, for every $v \in C^\infty(\overline{\Omega};\mathbb{S}^1)$ and every $\psi \in W^{1,1}(\Omega;\mathbb{R})$, a sequence $(u_n) \subset C^\infty(\overline{\Omega};\mathbb{S}^1)$ such that $u_n \to u$ a.e. and

$$\limsup_{n\to\infty} \int_\Omega |\nabla u_n| \leq \int_\Omega |\nabla u| + \int_\Omega |u \wedge \nabla u - v \wedge \nabla v - \nabla\psi|. \tag{14.32}$$

Set $w := u\overline{v}e^{-\imath\psi}$. By the proof of Lemma 1.15 (valid in any domain), there exist Lipschitz functions $\Phi_n : \mathbb{S}^1 \to \mathbb{S}^1$, of degree 1, such that $w_n := \Phi_n(w)$ satisfies $w_n \to 1$ a.e. and

$$\limsup_{n\to\infty} \int_\Omega |\nabla w_n| \leq \int_\Omega |\nabla w| = \int_\Omega |u \wedge \nabla u - v \wedge \nabla v - \nabla\psi|. \tag{14.33}$$

By (1.71), we have $Jw_n = Jw = Ju$, and thus $v_n := u\,\overline{w_n}$ satisfies $v_n \to u$ a.e. and $Jv_n = 0$. It follows that $v_n \in \overline{C^\infty(\overline{\Omega}; \mathbb{S}^1)}^{W^{1,1}}$ (Theorem 10.4′ below). By approximating each v_n and a diagonal process, we obtain a sequence $(u_n) \subset C^\infty(\overline{\Omega}; \mathbb{S}^1)$ such that $u_n \to u$ a.e. and

$$\begin{aligned}\limsup_{n\to\infty} \int_\Omega |\nabla u_n| &= \limsup_{n\to\infty} \int_\Omega |\nabla v_n| = \limsup_{n\to\infty} \int_\Omega |\nabla (u\,\overline{w_n})| \\ &\le \int_\Omega |\nabla u| + \limsup_{n\to\infty} \int_\Omega |\nabla w_n| \\ &\le \int_\Omega |\nabla u| + \int_\Omega |u \wedge \nabla u - v \wedge \nabla v - \nabla \psi|\end{aligned}$$

(by (14.33)). Thus, (14.32) holds. □

We continue with an alternative formula for $\Sigma'(u)$, in the spirit of (1.88).

Corollary 14.1. *For $u \in W^{1,1}(\Omega; \mathbb{S}^1)$, we have*

$$\begin{aligned}\Sigma'(u) = \min\{\|u \wedge \nabla u - v \wedge \nabla v - D\eta\|_{\mathcal{M}};\ &v \in C^\infty(\overline{\Omega}; \mathbb{S}^1), \\ &\eta \in BV(\Omega; \mathbb{R}) \text{ is an optimal lifting of } u\,\overline{v}\}.\end{aligned} \tag{14.34}$$

Proof. Let $(v, \eta_0) \in C^\infty(\overline{\Omega}; \mathbb{S}^1) \times BV(\Omega; \mathbb{R})$ be such that

$$\Sigma'(u) = \|u \wedge \nabla u - v \wedge \nabla v - D\eta_0\|_{\mathcal{M}}.$$

[Such a couple exists, by Theorem 14.2.] If $\eta \in BV(\Omega; \mathbb{R})$ is an optimal lifting of $u\,\overline{v}$ then, by Theorem 1.5, we have

$$\|u \wedge \nabla u - v \wedge \nabla v - D\eta\|_{\mathcal{M}} \le \|u \wedge \nabla u - v \wedge \nabla v - D\eta_0\|_{\mathcal{M}} = \Sigma'(u). \tag{14.35}$$

We conclude by combining (14.24) and (14.35). □

Consider next, for $u \in W^{1,1}(\Omega; \mathbb{S}^1)$, $S_1(u)$ as in (1.185) with $p = 1$, i.e.,

$$S_1(u) := \inf\left\{\int_\Omega |\nabla v|;\ v \in W^{1,1}(\Omega; \mathbb{S}^1),\ Jv = Ju \text{ in } \mathscr{D}'\right\}.$$

The analogue of (1.84) is the following

Proposition 14.6. *For $u \in W^{1,1}(\Omega; \mathbb{S}^1)$, we have*

$$S_1(u) = \Sigma'(u). \tag{14.36}$$

Proof. By (1.57) and Proposition 14.2, for $u, w \in W^{1,1}(\Omega; \mathbb{S}^1)$ we have

$$Ju = Jw \iff J(u\,\overline{w}) = 0 \iff u\,\overline{w} = e^{\imath\eta}\, v \text{ for some } \eta \in W^{1,1} \text{ and } v \in C^\infty.$$

We find that

$$S_1(u) = \inf_{\substack{w\in W^{1,1}(\Omega;\mathbb{S}^1)\\ Jw=Ju}} \int_\Omega |\nabla w| = \inf_{\substack{\eta\in W^{1,1}(\Omega;\mathbb{R})\\ v\in C^\infty(\overline{\Omega};\mathbb{S}^1)}} \int_\Omega |u\wedge\nabla u - v\wedge\nabla v - \nabla\eta|,$$

and we conclude using (14.31). □

We next address the question of how to compute $\Sigma'(u)$ by duality and geometrically, in line with Sections 1.5, 2.3, and 3.6. It turns out that $\Sigma'(u)$ plays a role very similar to that played by $\Sigma(u)$ in simply connected domains.

Here is a first example. It involves maps $u \in W^{1,1}(\Omega;\mathbb{S}^1)$, where Ω is a two-dimensional domain, and u has a finite number of singularities, a_i, $i = 1,\ldots,\ell$, with corresponding degrees d_i, $i = 1,\ldots,\ell$. For such a collection (a, d), we consider an integer m and points $P_l, N_l \in \overline{\Omega}$, $l = 1,\ldots,m$, such that

$$\sum_{l=1}^m (\delta_{P_l} - \delta_{N_l}) = \sum_{i=1}^\ell d_i\delta_{a_i} \text{ in } \mathscr{D}'(\Omega),$$

and set

$$L(a,d) := \min_{\sigma\in S_m} \sum_{l=1}^m \delta(P_l, N_{\sigma(l)}), \tag{14.37}$$

where δ is the distance defined in (14.18).

Theorem 14.3. *Assume that $N = 2$, $u \in W^{1,1}(\Omega;\mathbb{S}^1)\cap C(\Omega\setminus\cup_{i=1}^\ell\{a_i\};\mathbb{S}^1)$, and set $d_i := \deg(u, a_i)$. Then,*

$$\Sigma'(u) = 2\pi\, L(a,d), \tag{14.38}$$

where $L(a, d)$ is the minimal length in (14.37).

Proof. Let $\Gamma_1,\ldots,\Gamma_k$ denote the connected components of $\partial\Omega$. We associate with each Γ_j an integer $E_j \in \mathbb{Z}$ such that

$$\sum_{j=1}^k E_j + \sum_{i=1}^\ell d_i = 0. \tag{14.39}$$

Step 1. We have

$$\Sigma'(u) = \inf\{L(a,d,\Gamma,E);\ E\in\mathbb{Z}^k \text{ satisfies (14.39)}\}. \tag{14.40}$$

Indeed, "$\geq$" follows from (14.24), Theorem 14.1, and the fact that, if $v \in C^\infty(\overline{\Omega};\mathbb{S}^1)$, then (by the proof of (14.10)) the integers $E_j := \deg(u\overline{v}, \Gamma_j)$, $j = 1,\ldots,k$, satisfy (14.39).

For the opposite inequality, let E satisfy (14.39). By the proof of Proposition 12.16 and the fact that

$$\sum_{j=1}^{k}(-E_j + \deg(u, \Gamma_j)) = 0$$

(which follows from (14.10) and (14.39)), we find that there exists some $v \in C^\infty(\overline{\Omega}; \mathbb{S}^1)$ such that

$$\deg(v, \Gamma_j) = -E_j + \deg(u, \Gamma_j),\ j = 1, \ldots, k. \tag{14.41}$$

By Theorem 14.1 (applied to $u\,\overline{v}$) and (14.24), we obtain "$\leq$."

Step 2. $L(a, d)$ computed by duality

Repeating the proof of Lemma 2.3 we obtain, via Theorem 2.4, Corollary 1.6, and Lemma 14.12, that

$$\begin{aligned}
2\pi\, L(a, d) &= 2\pi \max\left\{\sum_{i=1}^{\ell} d_i\zeta(a_i);\ \zeta \in W_0^{1,\infty}(\Omega),\ \|\nabla\zeta\|_{L^\infty} \leq 1\right\} \\
&= 2\pi \sup\left\{\sum_{i=1}^{\ell} d_i\zeta(a_i);\ \zeta \in C_c^\infty(\Omega),\ \|\nabla\zeta\|_{L^\infty} \leq 1\right\} \\
&= \sup\left\{\int_\Omega (u \wedge \nabla u)\cdot\nabla^\perp\zeta;\ \zeta \in C_c^\infty(\Omega),\ \|\nabla\zeta\|_{L^\infty} \leq 1\right\} \\
&:= \Xi(u).
\end{aligned} \tag{14.42}$$

Step 3. We have $\Xi(u) \leq \Sigma'(u)$

If $v \in C^\infty(\overline{\Omega}; \mathbb{S}^1)$, $\eta \in W^{1,1}(\Omega; \mathbb{R})$ and $\zeta \in C_c^\infty(\Omega)$ satisfies $\|\nabla\zeta\|_{L^\infty} \leq 1$, then

$$\begin{aligned}
\int_\Omega (u \wedge \nabla u)\cdot\nabla^\perp\zeta &= \int_\Omega (u \wedge \nabla u - v \wedge \nabla v - \nabla\eta)\cdot\nabla^\perp\zeta \\
&\leq \int_\Omega |u \wedge \nabla u - v \wedge \nabla v - \nabla\eta|.
\end{aligned} \tag{14.43}$$

We conclude via (14.24) and (14.43).

Step 4. We have $\Xi(u) \geq \Sigma'(u)$

Using (14.42) and repeating the proof of Lemma 2.4, there exist points $P_l, N_l \in \{a_1, \ldots, a_\ell\} \cup \partial\Omega$, $l = 1, \ldots, n$, such that (2.81)–(2.83) hold and, in addition,

$$\Xi(u) = \sum_{l=1}^{n} |P_l - N_l|. \tag{14.44}$$

Set, for $i = 1, \ldots, k$,

$$E_i := \#\{l;\ P_l \in \Gamma_i\} - \#\{l;\ N_l \in \Gamma_i\}.$$

It follows from (2.83) that (14.39) holds.

On the other hand, the pseudometric d introduced in (14.17) clearly satisfies

$$d(P_l, N_l) \le |P_l - N_l|, \ l = 1, \dots, n. \tag{14.45}$$

Thus, with $E := (E_1, \dots, E_k)$, we have

$$\Xi(u) = \sum_{l=1}^{n} |P_l - N_l| \ge \sum_{l=1}^{n} d(P_l, N_l) \ge L(a, d, \Gamma, E); \tag{14.46}$$

here, we use (14.45) and (14.22). [Note that, strictly speaking, there may be more than $|E_i|$ "positive" points P_l and "negative" points N_l on Γ_i. However, this amounts to adding, in (14.22), artificial couples of Γ_i's with opposite degrees, and does not change the value of $L(a, d, \Gamma, E)$ in (14.22).] We conclude via (14.40) and (14.46).

The proof of Theorem 14.3 is complete. □

In two dimensions and for a general $u \in W^{1,1}(\Omega; \mathbb{S}^1)$, we have the following analogue of Theorem 1.9.

Theorem 1.9′. *Assume* $N = 2$. *Let* $u \in W^{1,1}(\Omega; \mathbb{S}^1)$. *We have*

$$\Sigma'(u) = \sup\left\{ \int_\Omega (u \wedge \nabla u) \cdot \nabla^\perp \zeta; \ \zeta \in C_c^\infty(\Omega; \mathbb{R}), \ \|\nabla \zeta\|_{L^\infty} \le 1 \right\}. \tag{14.47}$$

Granted Theorem 1.9′, whose proof is presented below, we deduce the following

Corollary 14.2. *Assume* $N = 2$. *Let* $u \in W^{1,1}(\Omega; \mathbb{S}^1)$. *We have*

$$\Sigma'(u) = \inf\{\|u \wedge \nabla u - \mu\|_{\mathcal{M}}; \ \mu \in \mathcal{M}(\Omega; \mathbb{R}^2), \ \operatorname{curl} \mu = 0\}. \tag{14.48}$$

Proof of Corollary 14.2. Combining (14.24) with the fact that, for every $\eta \in BV(\Omega; \mathbb{R})$ and $v \in C^\infty(\Omega; \mathbb{S}^1)$, we have $\operatorname{curl}(v \wedge \nabla v + D\eta) = 0$, we obtain "$\ge$" in (14.48).

On the other hand, if $\mu \in \mathcal{M}(\Omega; \mathbb{R}^2)$ satisfies $\operatorname{curl} \mu = 0$, then, $\forall \, \zeta \in C_c^\infty(\Omega; \mathbb{R})$ such that $\|\nabla \zeta\|_{L^\infty} \le 1$, we have

$$\int_\Omega (u \wedge \nabla u) \cdot \nabla^\perp \zeta = \int_\Omega (u \wedge \nabla u - \mu) \cdot \nabla^\perp \zeta \le \|u \wedge \nabla u - \mu\|_{\mathcal{M}}. \tag{14.49}$$

Taking, in (14.49), sup in ζ and inf in μ, we obtain "$\le$" in (14.48). □

Proof of Theorem 1.9′. "$\ge$" is clear. Let $\Xi(u)$ denote the right-hand side of (14.47). If $u, w \in W^{1,1}(\Omega; \mathbb{S}^1)$, we clearly have (by (14.24) and the definition of $\Xi(u)$)

$$|\Sigma'(u) - \Sigma'(w)| \le \|u \wedge \nabla u - w \wedge \nabla w\|_{L^1} \text{ and } |\Xi(u) - \Xi(w)| \le \|u \wedge \nabla u - w \wedge \nabla w\|_{L^1}.$$

It follows that both Σ' and Ξ depend continuously on u, and thus it suffices to establish "$\leq$" when u belongs to a dense class of $W^{1,1}(\Omega;\mathbb{S}^1)$. By Theorem 10.3′ below, we may thus assume that u satisfies (14.8), and then we are done, by the proof of Theorem 14.1. □

Remark 14.3. We do not know whether there exists a formula analogous to (14.47) when $N \geq 3$ (see Open Problem 33).

We next present the analogue of (2.190) for general two-dimensional domains.

Theorem 14.4. *Assume $N = 2$. Let $u \in W^{1,1}(\Omega;\mathbb{S}^1)$. We have*

$$\Sigma'(u) = 2\pi \min\{|\mathscr{C}|;\ \mathscr{C} \subset \Omega \text{ is a 1-rectifiable current such that } \pi\, \partial\mathscr{C} = Ju\}. \tag{14.50}$$

Proof. Step 1. Proof of "$\leq$" in (14.50)
Let $\mathscr{C}$ be a competitor in (14.50). If $\zeta \in C_c^\infty(\Omega;\mathbb{R})$ satisfies $|\nabla\zeta| \leq 1$, then

$$\int_\Omega (u \wedge \nabla u)\cdot \nabla^\perp \zeta = 2\langle Ju, \zeta\rangle = 2\pi\langle \partial\mathscr{C}, \zeta\rangle = 2\pi\langle \mathscr{C}, d\zeta\rangle \leq 2\pi\,|\mathscr{C}|. \tag{14.51}$$

Taking in (14.51) sup in ζ and inf in $\mathscr{C}$ and using Theorem 1.9′, we obtain "$\leq$."
Step 2. Proof of "$\geq$" in (14.50)
We first discuss the multiply connected version of Theorem 2.10. Properties (2.154) and (2.155) still hold, but (in view of Theorem 1.9′) (2.156) becomes

$$\begin{aligned}\Sigma'(u) &= 2\|Ju\|_W \\ &= 2\pi \inf\Big\{ \sum |\widetilde{P}_l - \widetilde{N}_l|;\ \widetilde{P}_l, \widetilde{N}_l \in \overline{\Omega} \\ &\qquad \text{and } \sum\big(\delta_{\widetilde{P}_l} - \delta_{\widetilde{N}_l}\big) = \sum\big(\delta_{P_l} - \delta_{N_l}\big) \text{ in } \mathscr{D}'(\Omega)\Big\}.\end{aligned} \tag{14.52}$$

Let $\widetilde{P}_l, \widetilde{N}_l$ be competitors in (14.52). Then,

$$\mathscr{C} := \sum_l \tau_l \mathscr{H}^1 \llcorner ([\widetilde{N}_l, \widetilde{P}_l] \cap \Omega)$$

is a competitor in (14.50), of mass

$$|\mathscr{C}| \leq \sum |\widetilde{P}_l - \widetilde{N}_l|; \tag{14.53}$$

here, $[\widetilde{N}_l, \widetilde{P}_l]$ denotes the oriented line segment from N_l to P_l.

By (14.52) and (14.53), we obtain "$\geq$" in (14.50). □

Remark 14.4. Theorem 14.4 can be extended to any dimension $N \geq 2$. More specifically, we have

$$\Sigma'(u) = 2\pi \inf\{|M|;\ M \subset \Omega \text{ is an } (N-1)\text{-rectifiable current such that } \pi\,\partial M = *Ju\}. \tag{14.54}$$

Here is a sketch of proof of (14.54) for $u \in \mathscr{R}$; the general case is obtained by density via the N-dimensional version of Theorem 3.10. Inequality "$\geq$" follows from the coarea formula, as in Section 3.3. The opposite inequality, "$\leq$," follows from the equalities $\Sigma'(u) = S_1(u)$ (see Proposition 14.6), $A_0(\Gamma) = A(\Gamma)$, where $\Gamma := (1/\pi) * Ju$ and $A_0(\Gamma), A(\Gamma)$ are as in Section 3.8.2, and a dipole construction in the spirit of Section 3.4.

14.3 Lifting in $W^{s,p}(\Omega; \mathbb{S}^1)$ revisited

We start with two straightforward results concerning lifting, in which the topology of Ω plays no role.

Proposition 14.7. *Let $0 < s < 1$ and $1 \leq p < \infty$ be such that $sp < 1$. Then, for every Ω, $W^{s,p}(\Omega; \mathbb{S}^1)$ has the lifting property.*

Proof. It suffices to repeat the proof of Theorem 5.2 *c)* in Section 5.1.2. □

Proposition 14.8. *Assume $N \geq 2$. Let $s > 0$ and $1 \leq p < \infty$ be such that:*

1. *Either $0 < s < 1$ and $1 \leq sp < N$*
2. *Or $s \geq 1$ and $1 \leq sp < 2$.*

Then, for every Ω, $W^{s,p}(\Omega; \mathbb{S}^1)$ does not have the lifting property.

Proof. We consider the same examples of maps without lifting as in the proof of Theorem 5.2 *b)* in Section 5.1.2, resp. Theorem 5.3 *b)* in Section 5.1.3. □

However, in general the topology of Ω does play a role, as shown by the following variant of Proposition 14.1.

Proposition 14.9. *Let $\Omega := \{x \in \mathbb{R}^2;\ 1 < |x| < 2\}$. Let $s > 0$ and $1 \leq p < \infty$ be such that $sp \geq 1$. Then $W^{s,p}(\Omega; \mathbb{S}^1)$ does not have the lifting property.*

Proof. Let $u(x) := x/|x|$, $\forall\, x \in \Omega$. Then, u is smooth in Ω and thus admits locally a smooth lifting. By uniqueness (Corollary 6.2), if u has a lifting φ in $W^{s,p}$, then φ has to be smooth. However, u does not have a global continuous lifting. □

We now present $W^{s,p}$ versions of Propositions 14.2, 14.3, and 14.4. As we have seen in Section 8.2, when $s > 0$ and $1 \leq p < \infty$ are such that $sp \geq 1$, a natural lifting space for $W^{s,p}(\Omega; \mathbb{S}^1)$ is:

$$X_{s,p} = X_{s,p}(\Omega) := \begin{cases} W^{s,p}(\Omega; \mathbb{R}), & \text{if } sp \geq N \\ (W^{s,p} + W^{1,sp})(\Omega; \mathbb{R}), & \text{if } 0 < s < 1,\ 1 \leq sp < N \\ (W^{s,p} \cap W^{1,sp})(\Omega; \mathbb{R}), & \text{if } s \geq 1,\ 1 \leq sp < N \end{cases}. \tag{14.55}$$

Let us recall that, for any Ω, and $1 < p < \infty$, it is possible to define Ju when $u \in W^{1/p,p}(\Omega; \mathbb{S}^1)$ (see Theorem 8.1).

A basic result is the following

Theorem 14.5. *Let $N \geq 2$, $s > 0$, and $1 \leq p < \infty$ be such that $sp \geq 1$. Then there exists some $\delta = \delta(\Omega, s, p) > 0$ such that every map $u \in W^{s,p}(\Omega; \mathbb{S}^1)$ satisfying $|u|_{W^{s,p}} < \delta$ has a lifting $\varphi \in X_{s,p}$, in the following cases:*

1. $1 \leq sp < 2$ *and* $Ju = 0$
2. $sp \geq 2$.

Proof. We cover Ω with a finite number of open sets $\Omega_i \subset \Omega$, $i = 1, \ldots, n$, such that:

a) For every i, Ω_i is a Lipschitz simply connected domain.
b) For every i and j, $\Omega_i \cap \Omega_j$ is either connected or empty.

We set $\Omega_0 := \Omega$.
Recall Poincaré's inequality

$$\left\| f - \fint_{\Omega_i} f \right\|_{L^1(\Omega_i)} \leq C_1 \, |f|_{W^{s,p}(\Omega_i)}, \ \forall i, \ \forall f \in W^{s,p}(\Omega_i; \mathbb{R}). \tag{14.56}$$

We leave to the reader the following easy consequence of (14.56). There exists some $C_2 > 0$ such that for every i we have

$$[u \in W^{s,p}(\Omega_i; \mathbb{S}^1)] \Rightarrow [\exists \, \alpha_i \in \mathbb{S}^1 \text{ such that } \|u - \alpha_i\|_{L^1(\Omega_i)} \leq C_2 \, |u|_{W^{s,p}(\Omega_i)}]. \tag{14.57}$$

[The key point in (14.57) is that $\alpha_i = \alpha_i(u) \in \mathbb{S}^1$.]

For further use, we also note the inequality

$$\left\| f - \fint_{\Omega_i} f \right\|_{L^1(\Omega_i)} \leq C_3 \, |f|_{X_{s,p}(\Omega_i)}, \ \forall i, \ \forall f \in X_{s,p}(\Omega_i), \tag{14.58}$$

with $| \ |_{X_{s,p}(\Omega_i)}$ the natural seminorm associated with the space $X_{s,p}(\Omega_i)$.

Let $u \in W^{s,p}(\Omega; \mathbb{S}^1)$. With no loss of generality, we may assume that $\alpha_0 = \alpha_0(u) = 1$ (where α_0 is as in (14.57)), and then we have

$$\|u - 1\|_{L^1(\Omega)} \leq C_2 \, |u|_{W^{s,p}(\Omega)}. \tag{14.59}$$

Using Theorems 9.1, 9.6, and 9.7, on each Ω_i with $i \geq 1$ we may write $u = e^{\imath \varphi_i}$, with $\varphi_i \in X_{s,p}(\Omega_i)$ and

$$|\varphi_i|_{X_{s,p}(\Omega_i)} \leq C_4 \, (|u|_{W^{s,p}(\Omega_i)} + |u|^{1/s}_{W^{s,p}(\Omega_i)}) \leq C_4 \, (|u|_{W^{s,p}(\Omega)} + |u|^{1/s}_{W^{s,p}(\Omega)}). \tag{14.60}$$

By (14.58) and (14.60), for every $i \geq 1$ we have

$$\left\|\varphi_i - \fint_{\Omega_i}\varphi_i\right\|_{L^1(\Omega_i)} \le C_3\, C_4\,(|u|_{W^{s,p}(\Omega)} + |u|^{1/s}_{W^{s,p}(\Omega)}). \tag{14.61}$$

If we set $t_i := \fint_{\Omega_i}\varphi_i$, (14.61) leads to

$$\left\|u - e^{\imath t_i}\right\|_{L^1(\Omega_i)} \le C_3\, C_4\,(|u|_{W^{s,p}(\Omega)} + |u|^{1/s}_{W^{s,p}(\Omega)}). \tag{14.62}$$

By subtracting a suitable multiple of 2π, we may assume that $t_i \in [-\pi,\pi)$. Then, (14.59) and (14.62) imply

$$|t_i| \le C_5\,(|u|_{W^{s,p}(\Omega)} + |u|^{1/s}_{W^{s,p}(\Omega)}). \tag{14.63}$$

Combining (14.61) and (14.63), we find

$$\|\varphi_i\|_{L^1(\Omega_i)} \le C_6\,(|u|_{W^{s,p}(\Omega)} + |u|^{1/s}_{W^{s,p}(\Omega)}),\ \forall\, i \ge 1. \tag{14.64}$$

To summarize, on each Ω_i with $i \ge 1$ we may write $u = e^{\imath\varphi_i}$, with φ_i satisfying (14.64).

Consider now $i, j \ge 1$ such that $\Omega_i \cap \Omega_j \ne \emptyset$. By Corollary 6.2 and property b) of the Ω_i's, on $\Omega_i \cap \Omega_j$ we have $\varphi_j - \varphi_i = 2\pi\, k_{i,j}$ for some integer $k_{i,j}$. Thus,

$$\|\varphi_j - \varphi_i\|_{L^1(\Omega_i\cap\Omega_j)} = 2\pi\,|k_{i,j}|\,|\Omega_i \cap \Omega_j|. \tag{14.65}$$

By (14.64) and (14.65), when $|u|_{W^{s,p}(\Omega)}$ is sufficiently small we have $\varphi_i = \varphi_j$ whenever $\Omega_i \cap \Omega_j \ne \emptyset$. We find that the function $\varphi = \varphi_i$ in Ω_i, $i \ge 1$, is a well-defined global $X_{s,p}$ phase of u. □

In the spirit of Theorems 5.2 *a)*, 5.3 *a)*, 8.5, and 8.8, we have the following

Corollary 14.3. *Let $N \ge 2$, $s > 0$, and $1 \le p < \infty$ be such that $sp \ge 1$. Let $u \in W^{s,p}(\Omega;\mathbb{S}^1)$. Then, u may be written $u = e^{\imath\varphi}\, v$, with $\varphi \in X_{s,p}$ and $v \in C^\infty(\overline{\Omega};\mathbb{S}^1)$, in the following cases*

1. *$1 \le sp < 2$ and $Ju = 0$.*
2. *$sp \ge 2$.*

Proof. Let δ be as in Theorem 14.5. Consider some $v \in C^\infty(\overline{\Omega};\mathbb{S}^1)$ such that $|u\,\overline{v}|_{W^{s,p}} < \delta$. [Existence of such v follows from Theorems 10.2′ and 10.4′ below, and Lemma 15.41.] By the above and the fact that $J(u\,\overline{v}) = Ju = 0$, the map $u\,\overline{v}$ has a global lifting $\varphi \in X_{s,p}$, whence the conclusion. □

We next derive a necessary and sufficient condition for the existence of a global lifting in $X_{s,p}$. This requires some preliminaries.

Let us recall the well-known fact that $\pi_1(\Omega)$, the fundamental group of Ω, is finitely generated. We fix some point x_0 in Ω and a finite number of curves $\gamma_1,\ldots,\gamma_n$ such that

$$\gamma_1, \ldots, \gamma_n \text{ are smooth simple (oriented) loops generating } \pi_1(\Omega) \tag{14.66}$$

(with base point x_0).

Then we have, $\forall\, u \in C^\infty(\overline{\Omega}; \mathbb{S}^1)$,

$$[\exists\, \varphi \in C^\infty(\overline{\Omega}; \mathbb{R}) \text{ such that } u = e^{\imath\varphi}] \iff [\deg(u, \gamma_j) = 0, \ \forall\, j], \tag{14.67}$$

and a similar statement holds for $u \in C(\overline{\Omega}; \mathbb{S}^1)$. Equivalence (14.67) is obtained as follows. Recall that $u \in C^\infty(\overline{\Omega}; \mathbb{S}^1)$ has a smooth phase if and only if on every loop $\gamma \subset \Omega$ the winding number of $\deg(u, \gamma)$ is zero. We obtain (14.67) using the invariance by homotopy of the winding number and the fact that the γ_j's generate $\pi_1(\Omega)$. Our next goal is to present a $W^{s,p}$ version of (14.67) when $sp \geq 1$.

If B is the unit ball in $\mathbb{R}^{N-1}$, then for sufficiently small $\delta > 0$, the sets

$$U_j := \{x \in \Omega;\ \text{dist}(x, \gamma_j) < \delta\}, j = 1, \ldots, n, \tag{14.68}$$

are diffeomorphic to $B \times \mathbb{S}^1$ through diffeomorphisms

$$\Phi_j : B \times \mathbb{S}^1 \to U_j \text{ such that } \Phi_j(0, \cdot)(\mathbb{S}^1) = \gamma_j,\ j = 1, \ldots, n. \tag{14.69}$$

We may assume that $\Phi_j(0, \cdot)$ is orientation-preserving.

Let $s > 0$ and $1 \leq p < \infty$ be such that $sp \geq 2$. Consider some $u \in W^{s,p}(\Omega; \mathbb{S}^1)$ and the maps

$$u_j : B \times \mathbb{S}^1 \to \mathbb{S}^1,\ u_j := u \circ \Phi_j, j = 1, \ldots, n. \tag{14.70}$$

Clearly, $u_j \in W^{s,p}(B \times \mathbb{S}^1; \mathbb{S}^1)$, and thus u_j has a degree on $B \times \mathbb{S}^1$ (Proposition 12.14). We denote

$$\deg(u, U_j) := \deg u_j = \deg u_j(\lambda, \cdot) \text{ for a.e. } \lambda \in B. \tag{14.71}$$

For further use, let us note the following.

$$\forall\, u \in C^\infty(\overline{\Omega}; \mathbb{S}^1),\ \deg(u, U_j) = \deg(u, \gamma_j). \tag{14.72}$$

It is still possible to define $\deg(u, U_j)$ when $1 \leq sp < 2$, under the additional assumption $Ju = 0$. In order to establish this assertion, we first note that the condition $Ju = 0$ is invariant under diffeomorphisms. Indeed, in view of (8.1) and (8.2), it suffices to prove the invariance when $u \in W^{1,1}$. In that case, the conclusion is clear, since (at least locally) $Ju = 0 \iff u = e^{\imath\varphi}$ for some $\varphi \in W^{1,1}$ (Lemma 1.8). Consider now some $u \in W^{1/p,p}(\Omega; \mathbb{S}^1)$ such that $Ju = 0$. Let u_j be as in (14.70). Then, $Ju_j = 0$ and thus we may set (see Remark 12.17)

$$\deg(u, U_j) := \deg u_j = \deg u_j(\lambda, \cdot) \text{ for a.e. } \lambda \in B. \tag{14.73}$$

To summarize, let $u \in W^{s,p}(\Omega; \mathbb{S}^1)$. If either $sp \geq 2$ (and then $Ju = 0$) or $1 \leq sp < 2$ and we assume that $Ju = 0$, then we may define $\deg(u, U_j)$ as in (14.73).

Corollary 14.4. *Let $N \geq 2$, $s > 0$ and $1 \leq p < \infty$ be such that $sp \geq 1$. Let $u \in W^{s,p}(\Omega; \mathbb{S}^1)$. Then, u may be written $u = e^{\imath\varphi}$, with $\varphi \in X_{s,p}$, in the following cases*

1. $1 \leq sp < 2$, $Ju = 0$, *and* $\deg(u, U_j) = 0$, $\forall j$.
2. $sp \geq 2$ *and* $\deg(u, U_j) = 0$, $\forall j$.

And conversely.

Proof. Write $u = e^{\imath\varphi} v$, with φ and v as in Corollary 14.3. Using the homotopy $t \mapsto e^{\imath\, t\varphi} v$, Proposition 12.15, Remark 12.19, and the results in Appendix 15.3, we have $\deg(v, U_j) = \deg(u, U_j)$, $\forall j$.

We find that $\deg(v, \gamma_j) = \deg(v, U_j) = \deg(u, U_j) = 0$, and thus $v = e^{\imath\psi}$, with $\psi \in C^\infty(\overline{\Omega}; \mathbb{S}^1)$ (by (14.67)). We find that $u = e^{\imath(\varphi+\psi)}$, with $\varphi + \psi \in X_{s,p}$.

Conversely, let $\varphi \in X_{s,p}$ be a lifting of u. By Theorem 8.8, we have $Ju = 0$ (locally, thus globally). In view of the embedding $X_{s,p}(\mathbb{S}^1) \hookrightarrow VMO$, for a.e. $\lambda \in B$, the map $u_j(\lambda, \cdot)$ (defined in (14.70)) has a VMO phase, and thus $\deg u_j(\lambda, \cdot) = 0$ (see (12.14)). We find that $\deg(u, U_j) = 0$ for each j. □

14.4 Density in $W^{s,p}(\Omega; \mathbb{S}^1)$ revisited

In this section, we return to density questions (discussed in Chapter 10) within the framework of multiply connected domains. As we are going to see, the statements of the basic results are unchanged, but some proofs require more work. We start with

Theorem 10.2′. *Assume that $N \geq 2$, $s > 0$, $1 \leq p < \infty$. Then, $C^\infty(\overline{\Omega}; \mathbb{S}^1)$ is dense in $W^{s,p}(\Omega; \mathbb{S}^1)$ if and only if $sp < 1$ or $sp \geq 2$.*

Proof. Density when $sp < 1$ and non-density when $1 \leq sp < 2$ are obtained as in the proof of Theorem 10.2 (Cases 1 and 2).

Assume next that $sp \geq 2$, and let $u \in W^{s,p}(\Omega; \mathbb{S}^1)$.

We first assume, in addition, that u is defined in a neighborhood ω of $\overline{\Omega}$. [We will remove this assumption later.] Cover, for sufficiently small $\varepsilon > 0$, $\overline{\Omega}$ with a finite family of balls $B(x_i, \varepsilon)$ such that $B(x_i, 2\varepsilon) \subset \omega$. On each $B(x_i, 2\varepsilon)$, we may write $u = e^{\imath\varphi_i}$, with $\varphi_i \in X_{s,p}(B(x_i, 2\varepsilon))$. Here, $X_{s,p}$ is as in (14.55), and the existence of φ_i follows from several results of Section 8.2. Let (ρ_ε) be a family of standard mollifiers. For $n > 1/\varepsilon$, set $\varphi_{i,n}(x) := \varphi_i * \rho_{1/n}(x)$, $x \in B(x_i, \varepsilon)$, and $u_{i,n} := e^{\imath\varphi_{i,n}}$.

We claim that $u_{i,n} = u_{j,n}$ in $B(x_i, \varepsilon) \cap B(x_j, \varepsilon)$. [We do not claim that $\varphi_{i,n} = \varphi_{j,n}$ in $B(x_i, \varepsilon) \cap B(x_j, \varepsilon)$.] Indeed, by Corollary 6.2, there exists some integer $k_{i,j}$ such that $\varphi_j - \varphi_i = 2\pi\, k_{i,j}$ in $B(x_i, 2\varepsilon) \cap B(x_j, 2\varepsilon)$. Thus, $\varphi_{j,n} - \varphi_{i,n} = 2\pi\, k_{i,j}$ in $B(x_i, \varepsilon) \cap B(x_j, \varepsilon)$, and therefore $u_{i,n} = u_{j,n}$ in $B(x_i, \varepsilon) \cap B(x_j, \varepsilon)$, as claimed.

We now define $u_n(x) := u_{i,n}(x)$ if $x \in B(x_i, \varepsilon)$. By the above, the definition is consistent, and clearly $u_n \in C^\infty(\overline{\Omega}; \mathbb{S}^1)$.

On the other hand, on each $B(x_i, \varepsilon)$ we have $\varphi_{i,n} \to \varphi_i$ in $X_{s,p}$, and thus $u_n \to u$ in $W^{s,p}(B(x_i, \varepsilon))$. [We apply here the continuity of the map $\varphi \mapsto e^{\imath\varphi}$ from $X_{s,p}$ to $W^{s,p}$; see Appendix 15.3.] In conclusion, $u_n \to u$ in $W^{s,p}(\Omega; \mathbb{S}^1)$.

We finally remove the assumption that u is defined in a neighborhood of $\overline{\Omega}$. By Lemma 15.26 in Appendix 15.6, u is a strong limit in $W^{s,p}(\Omega; \mathbb{S}^1)$ of a family (u^δ) of $\mathbb{S}^1$-valued maps, each one defined in a neighborhood of $\overline{\Omega}$ (depending on δ). Thus, it suffices to replace in the above u by u^δ and then let $\delta \to 0$. □

In the spirit of Sections 10.2 and 10.3, we have the following

Theorem 10.3′. *Assume that* $N \geq 2$, $s > 0$, $1 \leq p < \infty$, *and* $1 \leq sp < 2$. *Then, the class* $\mathscr{R}$ *given by* (10.1) *is dense in* $W^{s,p}(\Omega; \mathbb{S}^1)$.

Theorem 10.4′. *Assume that* $N \geq 2$, $s > 0$, $1 \leq p < \infty$, *and* $1 \leq sp < 2$. *Let* $u \in W^{s,p}(\Omega; \mathbb{S}^1)$. *Then, the following assertions are equivalent:*
a) $Ju = 0$
b) $u \in \overline{C^\infty(\overline{\Omega}; \mathbb{S}^1)}^{W^{s,p}}$.

Recall (see Theorem 8.1) that, for any Ω, Ju is well-defined for $u \in W^{1/p,p}(\Omega; \mathbb{S}^1)$.

The proof of Theorem 10.3′ is identical to that of Theorem 10.3 and is omitted.

Proof of Theorem 10.4′. "*b)* $\implies$ *a)*" is clear, since $Ju = 0$ when $u \in C^\infty(\overline{\Omega}; \mathbb{S}^1)$, and the map $u \mapsto Ju$ is continuous, by Corollary 8.1.

For the implication "*a)* $\implies$ *b)*," when $sp > 1$ or [$s = 1$ and $p = 1$] we follow exactly the same argument as in the proof of Theorem 10.2′. However, this strategy cannot be used when $1 < p < \infty$ and $s = 1/p$, since the mapping $\varphi \mapsto e^{\imath\varphi}$ does not map $X_{1/p,p}$ in $W^{1/p,p}$. In this case, we adapt the argument of Case 3 in the proof of Theorem 10.4.

Let $1 < p < \infty$ and $u \in W^{1/p,p}(\Omega; \mathbb{S}^1)$ be such that $Ju = 0$. Write, as in Theorem 7.1, $u = e^{\imath\varphi} v$, with $\varphi \in W^{1/p,p}(\Omega; \mathbb{R})$ and $v \in (W^{1,1} \cap W^{1/p,p})(\Omega; \mathbb{S}^1)$. Since $v \in W^{1,1}(\Omega; \mathbb{S}^1)$ and $Jv = Ju = 0$, we have $v \in \overline{C^\infty(\overline{\Omega}; \mathbb{S}^1)}^{W^{1,1}}$ (by the implication "*a)* $\implies$ *b)*" when $s = 1$ and $p = 1$). We consider a sequence $(w_k) \subset C^\infty(\overline{\Omega}; \mathbb{S}^1)$ such that $w_k \to v$ in $W^{1,1}$.

By Corollary 10.1 (which holds in any smooth domain), there exists a sequence $(v_k) \subset \mathscr{R}$ such that $v_k \to v$ in $W^{1/p,p}$ and in $W^{1,1}$. Then, $y_k := v_k \overline{w_k}$ clearly satisfies $Jy_k = Jv_k$ and $y_k \to 1$ strongly in $W^{1,1}$.

We next invoke Lemma 15.17 (which does not rely on the simple connectedness of Ω). For each $\lambda > 0$ there exists some $z_{k,\lambda} \in \mathscr{R}$ such that $Jz_{k,\lambda} = Jy_k$, $|z_{k,\lambda}|^p_{W^{1/p,p}} \leq C_p |y_k|_{W^{1,1}}$ and $|[z_{k,\lambda} \neq 1]| \leq \lambda$. As in the proof of Theorem 10.4, we are led to

$$|v_k - v_k \overline{z_{k,\lambda}}|^p_{W^{1/p,p}} \leq |z_{k,\lambda}|^p_{W^{1/p,p}} + o(1) \leq C_p |y_k|_{W^{1,1}} + o(1) \text{ as } \lambda \to 0. \tag{14.74}$$

On the other hand, we have $J(v_k\overline{z_{k,\lambda}}) = 0$ and $v_k\overline{z_{k,\lambda}} \in \widetilde{\mathscr{R}}$ (where $\widetilde{\mathscr{R}}$ is as in (7.12)). By Remark 15.11, we have $v_k\overline{z_{k,\lambda}} \in W^{s,p}$ when s is such that $1 < sp < 2$. By the first part of the proof of "$a) \implies b)$," we have $v_k\overline{z_{k,\lambda}} \in \overline{C^\infty(\overline{\Omega};\mathbb{S}^1)}^{W^{s,p}}$, and thus $v_k\overline{z_{k,\lambda}} \in \overline{C^\infty(\overline{\Omega};\mathbb{S}^1)}^{W^{1/p,p}}$. Combining this with (14.74), we find that

$$\operatorname{dist}_{W^{1/p,p}}\left(v_k, \overline{C^\infty(\overline{\Omega};\mathbb{S}^1)}^{W^{1/p,p}}\right) \le C_p|y_k|_{W^{1,1}}. \tag{14.75}$$

We obtain the desired conclusion by letting $k \to \infty$ in (14.75). □

14.5 Homotopy classes

Throughout this section, we assume that Ω is a general smooth bounded domain in $\mathbb{R}^N$, $N \ge 2$. Given $s > 0$ and $1 \le p < \infty$, we address the question whether $W^{s,p}(\Omega;\mathbb{S}^1)$ is path-connected and, if not, we describe its path-connected components. The starting point of the results in this section is the following

Proposition 14.10. *Let Ω be a smooth bounded simply connected domain in $\mathbb{R}^N$, with $N \ge 1$. Then, $W^{s,p}(\Omega;\mathbb{S}^1)$ is path-connected for every $s > 0$ and every $1 \le p < \infty$.*

The conclusion of Proposition 14.10 may fail when Ω is not simply connected. Here is a typical example.

Proposition 14.11. *Let $\Omega := \{x \in \mathbb{R}^2;\ 1 < |x| < 2\}$. Then,*

1. *$W^{1,p}(\Omega;\mathbb{S}^1)$ is path-connected when $1 \le p < 2$.*
2. *$W^{1,p}(\Omega;\mathbb{S}^1)$ is not path-connected when $p \ge 2$.*

The proof of Proposition 14.10 is postponed to the end of this section. Proposition 14.11 is a direct consequence of the two main results of this section, which we state now.

Theorem 14.6. *Let $s > 0$ and $1 \le p < \infty$ be such that $sp < 2$. Then, $W^{s,p}(\Omega;\mathbb{S}^1)$ is path-connected.*

Moreover, if $sp < 1$, then $W^{s,p}(\Omega;\mathbb{S}^1)$ is contractible.

In connection with the above result, we mention Open Problem 35.

Theorem 14.7. *Let $s > 0$ and $1 \le p < \infty$ be such that $sp \ge 2$. Then there is a natural bijection between the path-connected components of $W^{s,p}(\Omega;\mathbb{S}^1)$ and of $C(\overline{\Omega};\mathbb{S}^1)$. More precisely:*

1. *Each $u \in W^{s,p}(\Omega;\mathbb{S}^1)$ is homotopic in $W^{s,p}$ to some $v \in C^\infty(\overline{\Omega};\mathbb{S}^1)$.*
2. *Two maps $u, v \in C^\infty(\overline{\Omega};\mathbb{S}^1)$ are homotopic in $C(\overline{\Omega};\mathbb{S}^1)$ if and only if they are homotopic in $W^{s,p}(\Omega;\mathbb{S}^1)$.*

The proof of Theorem 14.7 relies on the degree of maps as defined in Proposition 12.14; in some sense this section is a follow-up of Sections 12.8.4 and 14.3.

Proof of Theorem 14.6. *Step 1.* $W^{s,p}(\Omega;\mathbb{S}^1)$ is path-connected when $sp<2$
The main ingredient in the proof is the following

Lemma 14.3. *Let $s>0$ and $1\le p<\infty$ be such that $sp<2$. Then, every map $u\in W^{s,p}(\Omega;\mathbb{S}^1)$ has an extension $w\in W^{s,p}_{loc}(\mathbb{R}^N;\mathbb{S}^1)$. In addition, we may assume that w is constant outside some large ball B.*

Granted Lemma 14.3, whose proof is presented in Appendix 15.6 (see Lemma 15.29), we argue as follows. Let $u\in W^{s,p}(\Omega;\mathbb{S}^1)$ and w as above. Fix a non-zero vector $e\in\mathbb{R}^N$ and set $F(t)(x):=w(x-t\,e)$, $\forall x\in\Omega$. Then, $F\in C(\mathbb{R};W^{s,p}(\Omega;\mathbb{S}^1))$, $F(0)=u$ and, for sufficiently large t, $F(t)$ is a constant map. Consequently, every map in $W^{s,p}(\Omega;\mathbb{S}^1)$ can be connected to a constant map. We conclude using the fact that two constants $\alpha,\beta\in\mathbb{S}^1$ can clearly be connected.

Step 2. $W^{s,p}(\Omega;\mathbb{S}^1)$ is contractible when $sp<1$
By Lemma 5.5, the operator $W^{s,p}(\Omega;\mathbb{S}^1)\ni u\mapsto w\in W^{s,p}_{loc}(\mathbb{R}^N;\mathbb{S}^1)$, $w=w(u):=\begin{cases}u, & \text{in } \Omega\\ 1, & \text{in } \mathbb{R}^N\setminus\Omega\end{cases}$, is continuous. If we consider, in the same vein as Step 1, $G(t,u)(x):=w(x-t\,e)$, $\forall x\in\Omega$, then $G\in C(\mathbb{R}\times W^{s,p}(\Omega;\mathbb{S}^1);W^{s,p}(\Omega;\mathbb{S}^1))$, $G(0,\cdot)=\mathrm{Id}$ and $G(t,\cdot)\equiv 1$ for sufficiently large t. Thus, $W^{s,p}(\Omega;\mathbb{S}^1)$ is contractible.
□

Proof of Theorem 14.7. Let $u\in W^{s,p}(\Omega;\mathbb{S}^1)$. Write $u=e^{\imath\varphi}v$, with $v\in C^\infty(\overline{\Omega};\mathbb{S}^1)$ and φ as in Corollary 14.3. By the results in Appendix 15.3, the map $[0,1]\ni t\mapsto e^{\imath(1-t)\varphi}v\in W^{s,p}(\Omega;\mathbb{S}^1)$ connects u to v continuously in $W^{s,p}$. Thus, every $u\in W^{s,p}(\Omega;\mathbb{S}^1)$ is homotopic in $W^{s,p}$ to some $v\in C^\infty(\overline{\Omega};\mathbb{S}^1)$.

It remains to prove that two maps $v,w\in C^\infty(\overline{\Omega};\mathbb{S}^1)$ are homotopic as continuous maps if and only if they are homotopic as $W^{s,p}$ maps.

Assume first that v and w are homotopic as continuous maps, via some homotopy map $F\in C([0,1]\times\overline{\Omega};\mathbb{S}^1)$. Then, $w=e^{\imath\varphi}v$ for some $\varphi\in C(\overline{\Omega};\mathbb{R})$. [This can be obtained directly as follows: for every $x\in\Omega$, let $\eta_x\in C([0,1];\mathbb{R})$ satisfy $\eta_x(0)=0$ and $e^{\imath\eta_x(t)}=F(t,x)/v(x)$, $\forall t\in[0,1]$. Set $\varphi(x):=\eta_x(1)$, which is clearly a lifting of w/v. Then, one may prove that φ is continuous. Alternatively, we may obtain the existence of φ by invoking Proposition 5.3 in Hu [191].] Since v and w are smooth, we have $\varphi\in C^\infty(\overline{\Omega};\mathbb{R})$. Clearly, $t\mapsto e^{\imath t\varphi}v$ connects v to w continuously in $W^{s,p}$.

Assume next that v and w are homotopic as $W^{s,p}$ maps. We may assume that $w\equiv 1$. Proposition 12.15 and (14.72) imply that $\deg(v,\gamma_j)=\deg(w,\gamma_j)=0$, where γ_j are as in (14.66). By (14.67), v has a global smooth phase φ, and we may connect v to $w\equiv 1$ continuously in $C(\overline{\Omega};\mathbb{S}^1)$ via $t\mapsto e^{\imath(1-t)\varphi}$. □

Incidentally, the arguments developed above are useful in the

Proof of Lemma 14.2. Let $(u_n)\subset C^\infty(\overline{\Omega};\mathbb{S}^1)$ be bounded in $W^{1,1}$. With the notation in (14.68)–(14.70), we claim that the sequences $(\deg(u_n,U_j))_n$ are bounded. Indeed, the sequences $(u_n \circ \Phi_j)_n$ are bounded in $W^{1,1}(B \times \mathbb{S}^1)$, and thus we may find $\lambda_{j,n} \in B$ such that

$$\int_{\mathbb{S}^1} \left|\partial_\tau \left[u_n \circ \Phi_j(\lambda_{j,n}, \cdot)\right]\right| \le C, \text{ with } C \text{ independent of } n \text{ and } j. \tag{14.76}$$

The claim follows from (14.76) and the Cauchy formula (1.23).

Possibly after extracting a subsequence, we may assume that $\deg(u_n, U_j)$ is independent of n, and thus $\deg(u_n, \gamma_j)$ is independent of $n \ge 0$. This implies that $u_n \overline{u_0}$ has winding number zero on each γ_j. Using (14.67), we may write $u_n = e^{\imath\varphi_n} u_0$ for some smooth φ_n. □

Corollary 14.5. *Let $s > 0$ and $1 \le p < \infty$ be such that $sp \ge 2$. Let $u \in W^{s,p}(\Omega; \mathbb{S}^1)$. Then, the homotopy class of u in $W^{s,p}(\Omega; \mathbb{S}^1)$ is $\{e^{\imath\varphi} u;\ \varphi \in X_{s,p}\}$.*

Proof. As in the proof of Theorem 14.7, if $y = e^{\imath\varphi} u$ with $\varphi \in X_{s,p}$, then y is homotopic to u in $W^{s,p}$.

Conversely, assume that $y \in W^{s,p}(\Omega; \mathbb{S}^1)$ is homotopic in $W^{s,p}$ to u. Write $y = e^{\imath\psi} v$ and $u = e^{\imath\eta} w$, with $v, w \in C^\infty(\overline{\Omega}; \mathbb{S}^1)$ and $\psi, \eta \in X_{s,p}$ (Corollary 14.3). By the above, y and v, respectively u and w, are homotopic in $W^{s,p}$. Thus, v and w are homotopic in $W^{s,p}$. By the proof of Theorem 14.7, there exists some $\lambda \in C^\infty(\overline{\Omega}; \mathbb{R})$ such that $v = e^{\imath\lambda} w$. Finally, $y = e^{\imath\varphi} u$, where $\varphi := \psi + \lambda - \eta \in X_{s,p}$. □

Proof of Proposition 14.10. *Step 1.* Proof when $N = 1$
Write $u \in W^{s,p}(\Omega; \mathbb{S}^1)$ as $u = e^{\imath\varphi}$, with $\varphi \in W^{s,p}$ (Theorem 5.1). Then, $t \mapsto e^{\imath(1-t)\varphi}$ connects u to 1 continuously in $W^{s,p}$ (Corollary 15.3).
Step 2. Proof when $N \ge 2$
The case $sp < 2$ follows from Theorem 14.6. When $sp \ge 2$, we rely on Theorem 14.7 and on the fact that $C^\infty(\overline{\Omega}; \mathbb{S}^1)$ is path-connected. □

14.6 Complements and open problems

14.6.1 Domains that admit global liftings

In this section, we discuss a condition on Ω that generalizes simple connectedness, and that is relevant to the existence of global liftings. More specifically, we consider Ω satisfying

$$\text{(P) every map } u \in C^\infty(\overline{\Omega}; \mathbb{S}^1) \text{ has a phase } \varphi \in C^\infty(\overline{\Omega}; \mathbb{R}). \tag{14.77}$$

Clearly, (P) holds in smooth bounded simply connected domains. However, there exist domains that satisfy (P), but are not simply connected (see Remark 14.6 item 4 below). The interest of such domains is that, for the matters discussed in the previous sections, they behave as simply connected domains. Here are two typical examples.

Corollary 14.3′. *Let $N \geq 2$ and Ω satisfy* (P). *Let $N \geq 2$, $s > 0$, and $1 \leq p < \infty$ be such that $sp \geq 1$. Let $u \in W^{s,p}(\Omega; \mathbb{S}^1)$. Then, u may be written $u = e^{\imath\varphi}\, v$, with $\varphi \in X_{s,p}$, in the following cases:*

1. $1 \leq sp < 2$ *and* $Ju = 0$.
2. $sp \geq 2$.

Proof. Combine Corollary 14.3 with (P). □

Theorem 14.8. *Let $N \geq 2$ and Ω satisfy* (P). *Then, $W^{s,p}(\Omega; \mathbb{S}^1)$ is connected, $\forall\, s > 0$, $\forall\, 1 \leq p < \infty$.*

Proof. Combine Theorems 14.6 and 14.7 with the fact that $C(\overline{\Omega}; \mathbb{S}^1)$ is connected (which follows easily from Remark 14.6 item 1 below). □

Remark 14.5. Using Corollary 14.3, it is easy to see that property (P) is equivalent, for every $p \geq 2$, to

$$(\mathrm{P})_{1,p}\ \forall u \in W^{1,p}(\Omega; \mathbb{S}^1),\ u \text{ has a phase } \varphi \in W^{1,p}(\Omega; \mathbb{R}), \tag{14.78}$$

and to an analogous "$(\mathrm{P})_{s,p}$" condition for $s > 0$ and $1 \leq p < \infty$ such that $sp \geq 2$.

Remark 14.6. The characterization of domains Ω satisfying (P) is a classical subject in homotopy theory. We briefly sketch a few facts related to (P).

1. It is straightforward that (P) is equivalent to

   (P′) every map $u \in C(\overline{\Omega}; \mathbb{S}^1)$ has a phase $\varphi \in C(\overline{\Omega}; \mathbb{R})$.

2. A sufficient condition for having (P′) is the following: the fundamental group $\pi_1(\overline{\Omega})$ of $\overline{\Omega}$ is perfect, i.e.,

   $$\pi_1(\overline{\Omega}) = G(\overline{\Omega}), \tag{14.79}$$

   where $G(\overline{\Omega})$ is the commutator of $\pi_1(\overline{\Omega})$ (see Hu [191, Theorem 6.1, p. 45, Theorem 7.1, p. 49]; see also Hatcher [188, Proposition 2.7, p. 109, Theorem 3.2, p. 195]).
3. The condition $\pi_1(M) = G(M)$ is satisfied, for example, by the Poincaré homology sphere M, which is connected but not simply connected (see Poincaré [292]).
4. This M is a compact connected orientable three-dimensional manifold. It can thus be embedded in $\mathbb{R}^5$, and we consider it as a closed submanifold of $\mathbb{R}^5$. If we let, for sufficiently small $\varepsilon > 0$, $\Omega := \{x \in \mathbb{R}^5;\ \mathrm{dist}(x, M) < \varepsilon\}$, then $\pi_1\left(\overline{\Omega}\right) \sim \pi_1(M)$.
   Thus, Ω provides an example of a smooth bounded domain that is not simply connected, but satisfies (P′) (and thus (P)).

14.6.2 What happens when Ω is a manifold?

In this section, we assume that Ω is an N-dimensional compact manifold with or without boundary, not necessarily simply connected. We revisit briefly some of the above results concerning $W^{s,p}(\Omega; \mathbb{S}^1)$.

1. Lifting. Propositions 14.2, 14.3, and 14.4 remain unchanged. Similarly for the results in Section 14.3. A systematic approach allowing lifting and density results to be transferred from open sets to manifolds is presented in Mironescu and Van Schaftingen [269, proof of Theorem 1]. This is also relevant for item 3 below.

Turning to $u \in W^{1,1}(\Omega; \mathbb{S}^1)$ and its lifting in $BV(\Omega; \mathbb{R})$, Theorem 1.5′ and Corollary 1.6′ still hold. However, the geometric interpretation of $\Sigma(u)$ (in the spirit of Theorem 14.1) is still missing even in the simple case where Ω is a closed surface (see Open Problem 31).

2. Relaxed energy. Theorem 14.2 still holds, with $\Sigma'(u)$ defined as in (14.24). The reason is that the main ingredient of the proof, Lemma 14.2, is valid on manifolds (with the same proof), and so is the implication

$$[v \in W^{1,1}(\Omega; \mathbb{S}^1),\ Jv = 0] \implies v \in \overline{C^\infty(\overline{\Omega}; \mathbb{S}^1)}^{W^{1,1}}$$

(see the next item). However, there is no known explicit formula for $\Sigma'(u)$, even when Ω is a closed surface (see Open Problem 31).

3. Density. All the results in Section 14.4 are unchanged.

4. Homotopy classes. The conclusions of Theorem 14.6 still hold, but the proofs need to be modified (see Bousquet [62], with roots in Brezis and Li [80], and Mironescu [259]). Theorem 14.7 holds with the same proof.

14.6.3 Replacing $\mathbb{S}^1$ by a general target manifold $\mathcal{N}$

In this section, we assume that Ω is an N-dimensional compact manifold, $N \geq 2$, and that $\mathcal{N}$ is a compact manifold without boundary contained in some $\mathbb{R}^\ell$. We focus on the homotopy classes of $W^{s,p}(\Omega; \mathcal{N})$. [The study of homotopy classes in $W^{1,p}(\Omega; \mathcal{N})$ was initiated by White [345]. His definition of $W^{1,p}(\Omega; \mathcal{N})$ is different from ours (see Section 5.4.1); in [345], $W^{1,p}(\Omega; \mathcal{N})$ denotes the closure of $C^\infty(\Omega; \mathcal{N})$ in $W^{1,p}$. Of course, the two definitions coincide when $C^\infty(\Omega; \mathcal{N})$ is dense in "our" $W^{1,p}(\Omega; \mathcal{N})$.]

When $sp < 1$, the proof of Theorem 14.6 leads to the following

Theorem 14.9. *Let Ω be a smooth bounded domain in $\mathbb{R}^N$, and let $\mathcal{N} \subset \mathbb{R}^\ell$ be an arbitrary set.*

Let $0 < s < 1$ and $1 \leq p < \infty$ be such that $sp < 1$. Then, $W^{s,p}(\Omega; \mathcal{N})$ is contractible.

On the other hand, when $s = 1$ we have the following result, due to Brezis and Li [80].

Theorem 14.10. *For* $1 \le p < 2$, $W^{1,p}(\Omega; \mathscr{N})$ *is path-connected.*

The conclusion of Theorem 14.10 also holds in $W^{s,p}(\Omega; \mathscr{N})$ when $sp < 2$ (Bousquet [62, Theorem 1]).

Brezis and Li [80] introduced a concept that plays a very important role in describing the possible changes in homotopy classes for $W^{1,p}$ as p varies. More specifically, let $p > 1$ and $0 < \varepsilon < p - 1$. Clearly,

$$W^{1,p+\varepsilon}(\Omega; \mathscr{N}) \subset W^{1,p-\varepsilon}(\Omega; \mathscr{N}).$$

Given $u \in W^{1,p}(\Omega; \mathscr{N})$, we denote by $[u]_p$ its homotopy class in $W^{1,p}$. We denote by $W^{1,p}/_\sim$ the quotient of $W^{1,p}$ by the equivalence relation $f \sim g$ in $W^{1,p}$.

Obviously, if $u, v \in W^{1,p+\varepsilon}$ and $u \sim v$ in $W^{1,p+\varepsilon}$, then $u \sim v$ in $W^{1,p-\varepsilon}$. As a result, we have a canonical map

$$i_{p+\varepsilon,p-\varepsilon} : W^{1,p+\varepsilon}/_\sim \to W^{1,p-\varepsilon}/_\sim.$$

Definition 14.1. We say that a *change of topology* occurs at p if, for every $\varepsilon \in (0, p-1)$, the map $i_{p+\varepsilon,p-\varepsilon}$ is *not* bijective. We denote by $CT(\Omega, \mathscr{N})$ the set of p's for which a change of topology occurs at p.

If a change of topology occurs at p, two things may happen, $i_{p+\varepsilon,p-\varepsilon}$ is not injective or $i_{p+\varepsilon,p-\varepsilon}$ is not surjective (or both!). Roughly speaking, *not injective* means that there are two maps $u, v \in W^{1,p+\varepsilon}$ such that u and v are *not* homotopic in $W^{1,p+\varepsilon}$, whereas $u \sim v$ in $W^{1,p-\varepsilon}$.

In other words, (at least) one component of $W^{1,q}(\Omega, \mathscr{N})$ *splits* into two distinct components as q *increases* from $p - \varepsilon$ to $p + \varepsilon$. Another view point is to say that two distinct components of $W^{1,q}$ have *merged* as q decreases from $p + \varepsilon$ to $p - \varepsilon$. This is a common situation, which we have already encountered above. For example, if $\Omega := \{x \in \mathbb{R}^2; 1 \le |x| \le 2\}$, a change of topology for $W^{1,p}(\Omega; \mathbb{S}^1)$ occurs at $p = 2$, because $i_{2+\varepsilon,2-\varepsilon}$ is *not injective*: any two maps $u, v \in W^{1,2+\varepsilon}$ are homotopic in $W^{1,2-\varepsilon}$, whereas they need not be in $W^{1,2+\varepsilon}$ (unless their degree is the same).

On the other hand, the fact that $i_{p+\varepsilon,p-\varepsilon}$ is *not surjective* means that a new component of $W^{1,q}$ appears (out of nowhere!) as q decreases from $p + \varepsilon$ to $p - \varepsilon$. This situation is more unusual and it does not occur in the example $W^{1,p}(\mathbb{S}^k; \mathbb{S}^k)$. It means that some map $u \in W^{1,p-\varepsilon}$ *cannot be homotopied* in $W^{1,p-\varepsilon}$ to any $v \in W^{1,p+\varepsilon}$. In particular, such a map u cannot be smooth and it cannot be homotopied in $W^{1,p-\varepsilon}$ to any smooth map – so it must have a rather "robust" singularity!

In the case of spheres, the following holds.

Theorem 14.11. *(Brezis and Li [80]) We have* $CT(\mathbb{S}^2, \mathbb{S}^2) = \{2\}$ *or, more generally,* $CT(\mathbb{S}^k, \mathbb{S}^k) = \{k\}$ *for any* $k \ge 2$.

Theorem 14.12. *([80]) We have $CT(\mathbb{S}^k \times \Lambda, \mathbb{S}^k) = \{k+1\}$, where Λ is the unit ball in $\mathbb{R}^m$ (any $m \geq 1$).*

It was also observed in [80] that a change in topology may occur at *several* values of p – not just one, as in the above examples. Here is such a situation, with a "cascade of mergings."

Theorem 14.13. *([80]) We have $CT(\mathbb{S}^1 \times \mathbb{S}^2, \mathbb{S}^1 \times \mathbb{S}^2) = \{2, 3\}$.*

This is somewhat natural because we have here (at least) two invariants: write $u = (u_1(x, y), u_2(x, y))$ and set $d_1 := \deg u_1(\cdot, y)$ and $d_2 := \deg u_2(x, \cdot)$. Note that d_1 is well-defined and independent of $y \in \mathbb{S}^2$ when $u \in W^{1,p}$ and $p \geq 2$ (Proposition 12.14); similarly, d_2 is well-defined and independent of $x \in \mathbb{S}^1$, but only when $p \geq 3$. Some new invariants appear, as p increases from 1 to ∞, when crossing the values $p = 2$ and $p = 3$.

Brezis and Li [80] formulated two conjectures.

Conjecture 1. $CT(\Omega, \mathcal{N})$ consists only of integers, i.e., the change of topology for $W^{1,p}$ occurs only when p is an integer.

Conjecture 2. Any map $f \in W^{1,p}(\Omega; \mathcal{N})$ can always be connected in $W^{1,p}$ to a smooth map (any p, any $\Omega, \mathcal{N}$).

They were both solved in a beautiful piece of work by Hang and Lin [177]. The answer to Conjecture 1 is positive.

Theorem 14.14. *$CT(\Omega, \mathcal{N})$ consists only of integers.*

Concerning Conjecture 2, it holds, for example, when $\dim \Omega = 2$ (any $\mathcal{N}$), or $\dim \Omega = 3$ and $\partial\Omega \neq \emptyset$ (any $\mathcal{N}$), or $\mathcal{N} = \mathbb{S}^1$ (any Ω) [80]. However, Hang and Lin found some situations where Conjecture 2 fails. Here is such an example.

Theorem 14.15. *Let $\Omega = \mathbb{RP}^3$ and $\mathcal{N} = \mathbb{RP}^2$. Then, $CT(\Omega, \mathcal{N}) = \{2, 3\}$. Moreover, there are maps $u \in W^{1,p}(\Omega; \mathcal{N})$, with $p \in (2, 3)$, such that f cannot be connected in $W^{1,p}$ to any smooth map.*

Some of the above results have been extended to $W^{s,p}(\Omega; \mathcal{N})$, under additional assumptions, by Bousquet [62].

14.6.4 Open problems

Open Problem 31. Assume Ω is a closed surface, e.g., $\Omega = \mathbb{S}^1 \times \mathbb{S}^1$. Let $u \in W^{1,1}(\Omega; \mathbb{S}^1) \cap C(\Omega \setminus \cup_{i=1}^{\ell}\{a_i\}; \mathbb{S}^1)$. Find explicit formulas for $\Sigma(u)$ and $\Sigma'(u)$ in the spirit of Theorems 14.1 and 14.3.

Open Problem 32. Assume $N \geq 3$. Let $u \in W^{1,1}(\Omega; \mathbb{S}^1)$. Give a geometric interpretation of $\Sigma(u)$, in the spirit of Sections 3.2 and 3.8.2.

[For $N = 2$ and nice u, see Theorem 14.1.]

Open Problem 33. Assume $N \geq 3$. Let $u \in W^{1,1}(\Omega; \mathbb{S}^1)$. Find a duality formula for the quantity $\Sigma'(u)$ defined by (14.24).

[For $N = 2$, see Theorem 1.9′.]

Open Problem 34. Assume $N \geq 3$. Let $u \in W^{1,1}(\Omega; \mathbb{S}^1)$. By (14.24), we have

$$\Sigma'(u) \geq \inf\{\|u \wedge \nabla u - \mu\|_{\mathcal{M}};\ \mu \in \mathcal{M}(\Omega; \mathbb{R}^N),\ \operatorname{curl} \mu = 0\}. \tag{14.80}$$

Do we have equality in (14.80)?

[For $N = 2$, see Corollary 14.2.]

The following basic question was raised by Bahri [19].

Open Problem 35. Let B be a ball in $\mathbb{R}^N$, with $N \geq 2$. Is $H^{1/2}(B; \mathbb{S}^1)$ simply connected?

More generally, let $s > 0$ and $1 \leq p < \infty$ be such that $1 \leq sp < 2$. Is $W^{s,p}(B; \mathbb{S}^1)$ simply connected?

Same questions where simply connected is replaced by contractible.

The answers are not known even when $N = 2$, $s = 1$, and $1 \leq p < 2$.

[When $N = 2$ and $sp \geq 2$, $W^{s,p}(\Omega; \mathbb{S}^1)$ is not simply connected. This follows by combining the embedding $W^{s,p} \hookrightarrow VMO$ with the fact that $VMO(\Omega; \mathbb{S}^1)$ is not simply connected. The latter assertion is a consequence of the fact that the loop $z \mapsto u_z \in VMO(\Omega; \mathbb{S}^1)$, $z \in \mathbb{S}^1$, where $u_z : \Omega \to \mathbb{S}^1$ is the constant map $x \mapsto z$, $\forall x \in \Omega$, is not homotopic to a constant in $VMO(\Omega; \mathbb{S}^1)$.]

14.7 Comments

1. Strictly speaking, the results in Sections 14.1–14.4 are new. Several proofs are straightforward adaptations of the analogous results in simply connected domains, but sometimes they require much more work.
   We call the attention of the reader to the results involving smallness conditions (Propositions 14.3 and 14.4, Theorem 14.5), which have no counterpart in simply connected domains. They are efficient tools in this chapter.
2. Theorems 14.6 and 14.7 are due to Brezis and Li [80] when $s = 1$. The extension to general s appears in Brezis and Mironescu [84]. The argument leading to contractibility is from Mironescu [259].
3. In full generality, Corollary 14.5 is new, but special cases appeared in [84].

Chapter 15
Appendices

15.1 Sobolev spaces

In this appendix, we present characterizations of Sobolev spaces via differences or Littlewood–Paley decompositions. We follow Triebel [338, 339] and Runst and Sickel [312]. Other useful references for fractional Sobolev spaces are Maz'ya [240], [139], Leoni [220], and Mironescu [261].

Definition 15.1. A *standard domain* Ω in $\mathbb{R}^N$ is: either $\mathbb{R}^N$, or a half space, or a Lipschitz bounded domain.

We start by recalling the definition of the spaces $W^{s,p}(\Omega)$, when Ω is a standard domain and s is not an integer.

Let $0 < s < 1$ and $1 \le p < \infty$. Then we let

$$|u|_{W^{s,p}}^p := \int_\Omega \int_\Omega \frac{|u(x) - u(y)|^p}{|x - y|^{N+sp}} \, dxdy, \quad \|u\|_{W^{s,p}} := \|u\|_{L^p} + |u|_{W^{s,p}}, \tag{15.1}$$

$$W^{s,p}(\Omega) := \{u : \Omega \to \mathbb{R};\ \|u\|_{W^{s,p}} < \infty\}.$$

When $p = \infty$, we let $W^{s,\infty}(\Omega)$ be the Hölder space $C^s(\Omega)$, and set

$$|u|_{W^{s,\infty}} := |u|_{C^s} = \sup_{x \neq y} \frac{|u(x) - u(y)|}{|x - y|^s}, \quad \|u\|_{W^{s,\infty}} := \|u\|_{C^s} = \|u\|_{L^\infty} + |u|_{C^s}.$$

When $s > 1$ is not an integer, we write $s = m + \sigma$, with $m \in \mathbb{N}$ and $\sigma \in (0, 1)$, and then we let

$$W^{s,p}(\Omega) := \{u \in W^{m,p}(\Omega);\ D^m u \in W^{\sigma,p}(\Omega)\},$$

with norm

$$\|u\|_{W^{s,p}} := \|u\|_{W^{m,p}} + \|D^m u\|_{W^{\sigma,p}}.$$

H. Brezis and P. Mironescu, *Sobolev Maps to the Circle*, Progress in Nonlinear Differential Equations and Their Applications 96,
https://doi.org/10.1007/978-1-0716-1512-6_15

The above spaces are fractional order Sobolev spaces (also known as Slobodeskii spaces). When $s \geq 1$, we let $|\ |_{W^{s,p}}$ denote the following seminorm on $W^{s,p}$:

$$|u|_{W^{s,p}} = \|Du\|_{W^{s-1,p}},$$

so that

$$\|u\|_{W^{s,p}} \sim \|u\|_{L^p} + |u|_{W^{s,p}}$$

(see Theorem 15.2 below).

Remark 15.1. In Chapter 12, we work with $W^{s,p}$ maps defined on $\mathbb{S}^1$. In order to obtain "cheap" versions of the estimates presented in the appendices in the case where $\Omega = \mathbb{S}^1$, it is convenient to identify $W^{s,p}(\mathbb{S}^1)$ with

$$\{u : \mathbb{R} \to \mathbb{R};\ u \in W^{s,p}_{loc}(\mathbb{R}),\ u \text{ is } 2\pi - \text{periodic},\ \|u\|_{W^{s,p}} < \infty\},$$

where the norm is given by

$$u \mapsto \|u\|_{L^p((-\pi,3\pi))} + |u|_{W^{s,p}((-\pi,3\pi))}.$$

For the convenience of the reader, whenever a result is used in the main text with $\Omega = \mathbb{S}^1$, we state its variant for $\Omega = \mathbb{S}^1$.

We start by noting the following refinement of the embedding $W^{s_1,p} \hookrightarrow W^{s_2,p}$, $s_1 > s_2$.

Lemma 15.1. *Let Ω be a Lipschitz bounded domain and $0 < s_2 < s_1 < \infty$, $1 \leq p \leq \infty$. Then,*

$$|u|_{W^{s_2,p}} \lesssim |u|_{W^{s_1,p}}. \tag{15.2}$$

Proof. We sketch the argument when $1 \leq p < \infty$. If we have, for example, $\ell < s_2 < s_1 < \ell + 1$ for some integer ℓ, then the conclusion is clear by inspection of the definition. The remaining cases easily reduce to the following special assertions:

1. If $0 < s_1 \leq 1$ and $\int_\Omega u = 0$, then $\|u\|_{L^p} \lesssim |u|_{W^{s_1,p}}$.
2. (15.2) holds when $0 < s_2 < s_1 = 1$.

In the first case, the inequality to prove is nothing else than the well-known Poincaré inequality in $W^{s_1,p}(\Omega)$.

We next assume that $0 < s_2 < s_1 = 1$. We may assume that $\int_\Omega u = 0$. Then, Poincaré's inequality in $W^{1,p}(\Omega)$ yields

$$|u|_{W^{s_2,p}} \lesssim \|u\|_{W^{1,p}} \lesssim |u|_{W^{1,p}}.$$ □

It is possible to characterize the spaces $W^{s,p}$ using differences, as in the case where $s \in (0, 1)$ [339, Section 3.5.3], [312, Section 2.3.1].

Theorem 15.1. *Let $s > 0$ be non-integer and $1 \le p < \infty$. Let Ω be a standard domain. Then, $\|u\|^p_{W^{s,p}}$ is equivalent to the quantity*

$$\|u\|^p_{L^p} + \sum_{j=1}^{N} \int_{-1}^{1} \left(\int \frac{|\delta^M_{te_j} u(x)|^p}{|t|^{1+sp}}\, dx \right) dt. \tag{15.3}$$

Here:

1. *t is a one-dimensional variable.*
2. *M is any integer satisfying $M > s$.*
3. *The integral in x is computed over the set $\{x \in \Omega;\ [x, x + Mte_j] \subset \Omega\}$.*
4. *$\delta_h u(x) := u(x+h) - u(x)$, and $\delta^M_h := \underbrace{\delta_h \circ \ldots \circ \delta_h}_{M \text{ times}}$.*

We next present several results in the same vein as Theorem 15.1 when $\Omega = \mathbb{R}^N$. [We often work in $\mathbb{R}^N$, by extending functions initially defined on a standard domain Ω to the whole of $\mathbb{R}^N$, via an appropriate extension operator.] First, an immediate consequence of (15.3).

Corollary 15.1. *Let $s > 0$ be non-integer and $1 \le p < \infty$. If $u : \mathbb{R}^N \to \mathbb{R}$, then we have*

$$\|u\|^p_{W^{s,p}(\mathbb{R}^N)} \sim \sum_{j=1}^{N} \int \|u(x_1, \ldots, x_{j-1}, \cdot, x_{j+1}, \ldots, x_N)\|^p_{W^{s,p}(\mathbb{R})}\, d\widehat{x_j}. \tag{15.4}$$

Here, $d\widehat{x_j} := dx_1 \ldots dx_{j-1} dx_{j+1} \ldots dx_N$.

A straightforward consequence of (15.4) and of the "converse" to the dominated convergence theorem is the following Fubini type convergence result.

Corollary 15.2. *Let $s > 0$ be a non-integer and $1 \le p < \infty$. Assume that $u_n \to u$ in $W^{s,p}(\mathbb{R}^N)$. Then, possibly up to a subsequence, still denoted by (u_n), we have $u_n(x_1, \ldots, x_{N-1}, \cdot) \to u(x_1, \ldots, x_{N-1}, \cdot)$ in $W^{s,p}(\mathbb{R})$ for a.e. $(x_1, \ldots, x_{N-1}) \in \mathbb{R}^{N-1}$.*

Let us also note that a variant of (15.4) holds for both fractional and integer Sobolev spaces. Let $u : \mathbb{R}^N \to \mathbb{R}$. If $\omega \in \mathbb{S}^{N-1}$, let $\omega^\perp$ denote the hyperplane orthogonal to ω, and consider the partial functions

$$\omega^\perp \ni x' \mapsto u^{x'}_\omega, \text{ with } u^{x'}_\omega(t) := u(x' + t\omega),\ \forall t \in \mathbb{R}.$$

Then we have the following

Proposition 15.1. *([261, Lemma 22]) Let $s \ge 0$ and $1 \le p < \infty$. Let $u : \mathbb{R}^N \to \mathbb{R}$. Then,*

$$\|u\|^p_{W^{s,p}} \sim \int_{\mathbb{S}^{N-1}} \int_{\omega^\perp} \left\| u^{x'}_\omega \right\|^p_{W^{s,p}} dx'\, d\omega. \tag{15.5}$$

As in the case of the standard Sobolev spaces $W^{s,p}$ with integer s, it is possible to define the fractional Sobolev spaces by induction [338, Section 2.3.8], [312, Section 2.1.4].

Theorem 15.2. *Let $s > m$, with m integer, and $1 \le p < \infty$. Let Ω be a standard domain and $u \in L^p(\Omega)$. Then, $u \in W^{s,p}(\Omega)$ if and only if $D^m u \in W^{s-m,p}(\Omega)$, and the quantity $\|u\|_{L^p} + \|D^m u\|_{W^{s-m,p}}$ is equivalent to $\|u\|_{W^{s,p}}$.*

We next recall the (most commonly used) Littlewood–Paley decomposition of a tempered distribution.

Definition 15.2. Let $\psi \in C_c^\infty(\mathbb{R}^N)$ be such that $\psi = 1$ in $B_1(0)$ and $\operatorname{supp}\psi \subset B_2(0)$. Define $\psi_0 = \psi$ and, for $j \ge 1$, $\psi_j(x) := \psi(x/2^j) - \psi(x/2^{j-1})$. Set $\varphi_j := \mathscr{F}^{-1}\psi_j \in \mathscr{S}$. [Equivalently, we have $\varphi_0 = \mathscr{F}^{-1}\psi$ and

$$\varphi_j(x) = 2^{Nj}\varphi_0(2^j x) - 2^{N(j-1)}\varphi_0(2^{j-1}x),\ \forall j \ge 1.]$$

Then, for each tempered distribution u, we have

$$u = \sum_j u_j \text{ in } \mathscr{S}', \quad \text{with } u_j := u * \varphi_j. \tag{15.6}$$

$u = \sum u_j$ is *"the" Littlewood–Paley decomposition* of $u \in \mathscr{S}'$.

Note that $\mathscr{F} u_j = \psi_j \mathscr{F} u$ is compactly supported, and therefore for each j we have $u_j \in C^\infty$.

Definition 15.3. Starting from the Littlewood–Paley decomposition, we define the *Triebel–Lizorkin spaces* $F^s_{p,q} = F^s_{p,q}(\mathbb{R}^N)$ as follows: for $s \in \mathbb{R}$, $0 < p < \infty$, and $0 < q \le \infty$, we let

$$\|u\|_{F^s_{p,q}} := \left\| \left\| \left(2^{sj}u_j(x)\right)_{j\ge 0} \right\|_{l^q(\mathbb{N})} \right\|_{L^p(\mathbb{R}^N)}, \quad F^s_{p,q} := \{u \in \mathscr{S}';\ \|u\|_{F^s_{p,q}} < \infty\}.$$

Same definition when $p = q = \infty$.

Most of the Sobolev spaces can be identified with Triebel–Lizorkin spaces [338, Section 2.3.5], [312, Section 2.1.2].

Theorem 15.3. *The following equalities of spaces hold, with equivalence of norms:*

1. *If $s > 0$ is not an integer and $1 \le p \le \infty$, then $W^{s,p}(\mathbb{R}^N) = F^s_{p,p}(\mathbb{R}^N)$*
2. *If $s \ge 0$ is an integer and $1 < p < \infty$, then $W^{s,p}(\mathbb{R}^N) = F^s_{p,2}(\mathbb{R}^N)$.*

When $s \ge 0$ is an integer and either $p = 1$ or $p = \infty$, the Sobolev space $W^{s,p}$ cannot be identified with a Triebel–Lizorkin space.

Theorem 15.3 is usually used in conjunction with Lemma 15.2 below. The reason is that, in practice, we do not know the Littlewood–Paley decomposition of u, but only a *generalized Nikol'skij decomposition* of u. Such a representation is of the form $u = \sum u^j$, with $\operatorname{supp}\mathscr{F}u^j \subset B_{2^{j+3}}(0)$. For the next result, see Yamazaki [346]; for special cases, see also [338, Section 2.5.2] and [312, Section 2.3.2].

Lemma 15.2. *Let $0 < s < \infty$, $1 \le p < \infty$, and $0 < q \le \infty$. Let $u = \sum u^j$ be a generalized Nikol'skij decomposition of $u \in \mathscr{S}'$. Then,*

$$\|u\|_{F^s_{p,q}(\mathbb{R}^N)} \lesssim \left\| \left\| \left(2^{sj} u^j(x)\right)_{j\ge 0} \right\|_{l^q(\mathbb{N})} \right\|_{L^p(\mathbb{R}^N)}. \tag{15.7}$$

Same conclusion when $p = q = \infty$.

15.2 Sobolev embeddings and Gagliardo–Nirenberg inequalities

Throughout this section, we let Ω be a standard domain (as in Definition 15.1).

We first investigate the validity of optimal and suboptimal Sobolev embeddings and Gagliardo–Nirenberg inequalities in the full scale of Sobolev spaces.

An optimal Sobolev embedding, if true, is of the form $W^{s,p} \hookrightarrow W^{r,q}$, where s, r, p, q satisfy

$$s > r \ge 0,\ 1 \le p < q \le \infty,\ s - \frac{N}{p} = r - \frac{N}{q}. \tag{15.8}$$

Theorem 15.4. *Let s, r, p, q, N satisfy* (15.8). *Let Ω be a standard domain in $\mathbb{R}^N$. Then we have*

$$W^{s,p}(\Omega) \hookrightarrow W^{r,q}(\Omega) \tag{15.9}$$

with the following exceptions, when (15.9) *does not hold.*

1. *When*

$$N = 1,\ s \text{ is an integer } \ge 1,\ p = 1,\ 1 < q < \infty,\ \text{and } r = s - 1 + 1/q, \tag{15.10}$$

we have

$$W^{s,1}(\Omega) \not\hookrightarrow W^{s-1+1/q,q}(\Omega). \tag{15.11}$$

In particular, we have

$$W^{1,1}((0,1)) \not\hookrightarrow W^{1/q,q}((0,1)),\ 1 < q < \infty \tag{15.12}$$

2. *When*

$$N \ge 1,\ 1 < p < \infty,\ q = \infty,\ \text{and } s - \frac{N}{p} = r \text{ is an integer}, \tag{15.13}$$

we have

$$W^{s,p}(\Omega) \not\hookrightarrow W^{r,\infty}(\Omega). \tag{15.14}$$

The above result is well-known for most of values of s, r, p, q. For a full proof, see, for example, Brezis and Mironescu [89, Appendix].

Remark 15.2. Suboptimal embeddings $W^{s,p}(\Omega) \hookrightarrow W^{r,q}(\Omega)$, characterized by

$$s > r \geq 0,\ 1 \leq p \leq q \leq \infty,\ s - \frac{N}{p} > r - \frac{N}{q}, \tag{15.15}$$

are always true. The proof of this assertion is omitted.

Remark 15.3. We call the attention of the reader to the following non-embedding. Let Ω be bounded and $s > 0$ be non-integer. Assume that $1 \leq q < p \leq \infty$. Then, $W^{s,p}(\Omega) \not\hookrightarrow W^{s,q}(\Omega)$ (see Mironescu and Sickel [268]). This contrasts with the embedding $W^{s,p}(\Omega) \hookrightarrow W^{s,q}(\Omega)$, valid when $s \in \mathbb{N}$ and $1 \leq q < p \leq \infty$.

However, some additional embeddings are true in bounded domains. Indeed, assume that Ω is Lipschitz and bounded and that the following suboptimal condition holds

$$s > r \geq 0,\ 1 \leq p, q < \infty,\ s - \frac{N}{p} > r - \frac{N}{q}. \tag{15.16}$$

[We do not assume $p \leq q$.] Then we have $W^{s,p}(\Omega) \hookrightarrow W^{r,q}(\Omega)$. Here is a proof of this fact. In view of Remark 15.2, we have $W^{s,p} \hookrightarrow W^{s-\varepsilon,p}$ and $W^{r+\varepsilon,q} \hookrightarrow W^{r,q}$. By considering an appropriate small ε, we may assume that both s and r are non-integers. Again by Remark 15.2, we may assume that $q < p$, and thus $L^p(\Omega) \hookrightarrow L^q(\Omega)$. In view of Theorem 15.1, it remains to prove that

$$\iint \frac{|\delta^M_{te_j} u(x)|^q}{|t|^{1+rq}}\, dxdt < \infty, \text{ provided } \iint \frac{|\delta^M_{te_j} u(x)|^p}{|t|^{1+sp}}\, dxdt < \infty.$$

This follows from

$$\begin{aligned}\iint \frac{|\delta^M_{te_j} u(x)|^q}{|t|^{1+rq}}\, dxdt &\leq \left(\iint \frac{|\delta^M_{te_j} u(x)|^p}{|t|^{1+sp}}\, dxdt\right)^{q/p} \times \\ &\quad \left(\iint \frac{1}{|t|^{1-(s-r)pq/(p-q)}}\, dxdt\right)^{(p-q)/p} \\ &\lesssim \left(\iint \frac{|\delta^M_{te_j} u(x)|^p}{|t|^{1+sp}}\, dxdt\right)^{q/p}. \end{aligned}$$

□

Embeddings can be considered in other functional settings, for example, in the scale of Besov or Triebel–Lizorkin spaces. We do not discuss them here, except for a simple example equivalent to the embedding $W^{s,p} \hookrightarrow B^s_{p,\infty}$, which is used in Section 12.7.2.

Lemma 15.3. *Let $0 < s < 1$ and $1 \le p < \infty$. If $u : \mathbb{R}^N \to \mathbb{R}$, then for every $h \in \mathbb{R}^N$,*

$$\|u(x+h) - u(x)\|_{L^p} \le C\,|h|^s \left(\iint_{|x-y|\le|h|} \frac{|u(x)-u(y)|^p}{|x-y|^{N+sp}}\,dxdy\right)^{1/p}. \tag{15.17}$$

In particular, if $u \in W^{s,p}$, then

$$\|u(x+h) - u(x)\|_{L^p} = o(|h|^s) \text{ as } h \to 0. \tag{15.18}$$

Proof. Set, for $h \in \mathbb{R}^N$, $I(h) := \|u(\cdot + h) - u\|_{L^p}$. By the triangle inequality we have $I(h_1 + h_2) \le I(h_1) + I(h_2)$, $\forall\, h_1, h_2 \in \mathbb{R}^N$, and thus

$$I(h) \le I(\xi) + I(h-\xi),\ \forall\, \xi \in \mathbb{R}^N. \tag{15.19}$$

Set $B_h := B(h/2, |h|/2)$. Taking the L^p norm over $\xi \in B_h$ in (15.19) and using once more the triangle inequality gives

$$I(h)\,|B_h|^{1/p} \le \|I(\xi)\|_{L^p(B_h)} + \|I(h-\xi)\|_{L^p(B_h)}. \tag{15.20}$$

On the other hand, noting that, if $\xi \in B_h$, then $|\xi| \le |h|$, we find that

$$\|I(\xi)\|^p_{L^p(B_h)} \le \iint_{|x-y|\le|h|} |u(x)-u(y)|^p\,dxdy,$$

and the same inequality holds for $\|I(h-\xi)\|_{L^p(B_h)} = \|I(\xi)\|_{L^p(B_h)}$ (since $B_h = h - B_h$).

Going back to (15.20) yields

$$I(h)\,|B_h|^{1/p} \le 2\left(\iint_{|x-y|\le|h|} |u(x)-u(y)|^p\,dxdy\right)^{1/p}.$$

Since $|B_h|^{1/p} = c|h|^{N/p}$, we conclude that

$$\begin{aligned} I(h) &\le \frac{C}{|h|^{N/p}}\left(\iint_{|x-y|\le|h|} |u(x)-u(y)|^p\,dxdy\right)^{1/p} \\ &\le C|h|^s\left(\iint_{|x-y|\le|h|} \frac{|u(x)-u(y)|^p}{|x-y|^{N+sp}}\,dxdy\right)^{1/p}. \end{aligned}$$

□

We continue by investigating, in the full scale of Sobolev spaces, the validity of the "pure" Gagliardo–Nirenberg-type estimates

$$\|u\|_{W^{s,p}} \lesssim \|u\|^{\theta}_{W^{s_1,p_1}} \|u\|^{1-\theta}_{W^{s_2,p_2}},\ \forall\, u \in W^{s_1,p_1} \cap W^{s_2,p_2}(\Omega), \tag{15.21}$$

where $s_1, s_2, s \ge 0$ and $1 \le p_1, p_2, p \le \infty$ are related by

$$\begin{cases} s = \theta s_1 + (1-\theta)s_2 \\ \dfrac{1}{p} = \dfrac{\theta}{p_1} + \dfrac{1-\theta}{p_2} \end{cases} \quad \text{for some } \theta \in (0,1). \tag{15.22}$$

With no loss of generality, we may assume that

$$s_1 \leq s_2. \tag{15.23}$$

The following condition plays an essential role in the validity of (15.21):

$$s_2 \text{ is an integer } \geq 1, \ p_2 = 1 \text{ and } s_2 - s_1 \leq 1 - \frac{1}{p_1}. \tag{15.24}$$

[The latter condition can also be written in the more symmetric form $s_1 - 1/p_1 \geq s_2 - 1/p_2$.]

We have the following result (see Brezis and Mironescu [86]).

Theorem 15.5. *Let s, s_1, s_2, p, p_1, and p_2 satisfy* (15.22) *and* (15.23). *Then,* (15.21) *holds if and only if* (15.24) *fails.*

More precisely, we have

1. *If* (15.24) *fails, then, for every $\theta \in (0,1)$, there exists a constant C depending on s_1, s_2, p_1, p_2, θ, and Ω such that*

$$\|u\|_{W^{s,p}(\Omega)} \lesssim \|u\|^{\theta}_{W^{s_1,p_1}(\Omega)} \|u\|^{1-\theta}_{W^{s_2,p_2}(\Omega)}, \ \forall u \in W^{s_1,p_1} \cap W^{s_2,p_2}(\Omega) \tag{15.25}$$

2. *If* (15.24) *holds, there exists some $u \in W^{s_1,p_1} \cap W^{s_2,p_2}(\Omega)$ such that $u \notin W^{s,p}(\Omega)$, $\forall \theta \in (0,1)$.*

Remark 15.4. In particular, the validity of (15.21) does not depend on N, Ω, or θ, but only on s_1, s_2, p_1, and p_2.

Remark 15.5. Let us emphasize that the embedding

$$W^{s,p} \cap L^{\infty} \hookrightarrow W^{\theta s, p/\theta}, \ \forall 0 < \theta < 1,$$

and the corresponding estimate (15.21) hold, $\forall s > 0$, $\forall 1 \leq p \leq \infty$, except when $s = p = 1$. [The behavior of the best constant in the corresponding inequality when $s \nearrow 1$ and $1 < p < \infty$ was established by Maz'ya and Shaposhnikova [244] (see also Maz'ya [240, Section 10.3]), solving a conjecture raised in Brezis and Mironescu [83].] In particular, we have

$$W^{1,1} \cap L^{\infty} \not\hookrightarrow W^{1/p,p}, \ \forall 1 < p < \infty. \tag{15.26}$$

The failure of Gagliardo–Nirenberg inequality in this case is usually established by passing to BV. The next result provides an explicit example illustrating (15.26).

Lemma 15.4. *There exists a compactly supported function $u : (0,1) \to \mathbb{R}$ such that:*

1. $u \in W^{1,1}((0,1))$
2. *For every* $p \in (1,\infty)$, $u \notin W^{1/p,p}((0,1))$.

Proof. For $0 < \varepsilon < 1/2$, $a > 0$, $b > 0$, we consider the function $v = v_{\varepsilon,a,b} : (-2a, 2a) \to \mathbb{R}$ given by:

$$v(x) := \begin{cases} 0, & \text{if } |x| > (1+\varepsilon)a \\ b, & \text{if } |x| < a \\ \text{affine}, & \text{in } (a,(1+\varepsilon)a) \text{ and in } (-(1+\varepsilon)a,-a) \end{cases}.$$

Then,

$$|v|^p_{W^{1/p,p}} \ge C_p b^p |\ln\varepsilon|, \forall\, 1 < p < \infty, \text{ and } \|v\|_{W^{1,1}} \le C_1(a+1)b.$$

Consider a sequence $u_j := v_{\varepsilon_j,a_j,b_j}(\cdot - d_j)$, where $b_j := 1/j^2$, $\varepsilon_j := e^{-e^j}$, and a_j and d_j are chosen such that the intervals $I_j := (d_j - 2a_j, d_j + 2a_j)$ have mutually disjoint supports contained in $(0,1)$. Let $u := \sum u_j$. Then, clearly, $u \in W^{1,1}$, whereas for $1 < p < \infty$ we have

$$|u|^p_{W^{1/p,p}((0,1))} \ge \sum_j |u_j|^p_{W^{1/p,p}(I_j)} \ge C_p \sum_j \frac{1}{j^{2p}} e^j = \infty. \qquad \square$$

Remark 15.6. Suboptimal Gagliardo–Nirenberg-type estimates, corresponding to

$$\begin{cases} s < \theta s_1 + (1-\theta)s_2 \\ \dfrac{1}{p} = \dfrac{\theta}{p_1} + \dfrac{1-\theta}{p_2} \end{cases} \quad \text{for some } \theta \in (0,1)$$

are always valid. The proof of this assertion is omitted.

We continue this appendix by discussing a variant of the Gagliardo–Nirenberg estimates in Lipschitz bounded domains, in which full norms are replaced by seminorms. The general rule is that whenever such an estimate holds for norms, it also hold for seminorms.

Lemma 15.5. *Let Ω be a Lipschitz bounded domain. Let $s_1, s_2, s > 0$, $1 \le p_1, p_2, p \le \infty$ be such that* (15.21) *holds. Then we have*

$$|u|_{W^{s,p}} \lesssim |u|^\theta_{W^{s_1,p_1}} |u|^{1-\theta}_{W^{s_2,p_2}}, \ \forall u \in W^{s_1,p_1} \cap W^{s_2,p_2}(\Omega). \tag{15.27}$$

A similar result holds when at least one of s, s_1, s_2 is zero. We mention here only one case, of interest for us.

Lemma 15.6. *Let Ω be a Lipschitz bounded domain. Let $s > 0$, $1 \le p < \infty$ and $\theta \in (0,1)$ be such that $(s,p) \ne (1,1)$. Then,*

$$|u|_{W^{\theta s,p/\theta}} \lesssim |u|^\theta_{W^{s,p}} \|u\|^{1-\theta}_{L^\infty}, \ \forall u \in W^{s,p} \cap L^\infty(\Omega). \tag{15.28}$$

Proof. The argument is the same for (15.27) and (15.28): we apply (15.21) to $u - \fint u$ and then we note that, whenever $\sigma > 0$, we have $|u - \fint u|_{W^{\sigma,q}} = |u|_{W^{\sigma,q}}$ and (by Poincaré's inequality) $\|u - \fint u\|_{W^{\sigma,q}} \lesssim |u|_{W^{\sigma,q}}$. Details are left to the reader. □

In Chapter 5, we have used Lemma 5.9, a refinement of (15.28), that we now recall and prove.

Lemma 15.7. *Let $0 < s < 1$ and $1 \le p < \infty$. Let $\Omega \subset \mathbb{R}^N$ be a bounded Lipschitz domain. Then there exists some constant $C' = C'(s, p, \Omega)$ such that for every $f \in W^{s,p}(\Omega) \cap BMO$ we have*

$$|f|_{W^{s/2,2p}} \le C' |f|_{BMO}^{1/2} |f|_{W^{s,p}}^{1/2}. \tag{15.29}$$

More generally, for $0 < \theta < 1$, there exists some $C'' = C''(s, p, \theta, \Omega)$ such that

$$|f|_{W^{\theta s,p/\theta}} \le C'' |f|_{BMO}^{1-\theta} |f|_{W^{s,p}}^{\theta}. \tag{15.30}$$

Proof. For simplicity, we work in the unit ball B and establish only (15.29). It suffices to consider maps f satisfying the additional assumption $\int_B f = 0$. As we will see below, we have

$$\|g\|_{W^{s/2,2p}(\mathbb{R}^N)}^{2p} \le C\|g\|_{BMO(\mathbb{R}^N)}^{p} \|g\|_{W^{s,p}(\mathbb{R}^N)}^{p}, \ \forall g \in W^{s,p} \cap BMO(\mathbb{R}^N). \tag{15.31}$$

In turn, this estimate implies (15.29). Indeed, let T be the extension operator defined in Corollary 15.9 in Appendix 15.6. If we apply (15.31) to $g := Tf$, and use Corollary 15.9, we find that

$$|f|_{W^{s/2,2p}}^{2p} \le \|g\|_{W^{s/2,2p}(\mathbb{R}^N)}^{2p} \le C\|f\|_{BMO}^{p} \|f\|_{W^{s,p}}^{p} \le C|f|_{BMO}^{p} |f|_{W^{s,p}}^{p},$$

the latter inequality following from the assumption $\int_B f = 0$ and the Poincaré inequalities in $W^{s,p}$ and in BMO (see (5.26)).

It therefore remains to prove (15.31). This inequality is true for every non-integer $s > 0$ and for every $1 \le p < \infty$. In order to prove (15.31), we rely on Littlewood–Paley decompositions as in (15.6). For $W^{s,p}$, we use Theorem 15.3, and for BMO we rely on the corresponding Littlewood–Paley theory (see, for example, Triebel [339, Section 1.7.1]). [Alternatively, estimate (15.33) below could have been derived directly from the definition of BMO.] More specifically, if $f \in BMO$ and φ_j are as in (15.6), then there exists a sequence $(f^j) \subset L^\infty$ such that we have $f = \sum_{j\ge 0} f^j * \varphi_j$ in $\mathscr{S}'$ and [339, Theorem, p. 47]

$$\|f^j\|_{L^\infty} \le C\|f\|_{BMO}, \text{ for some } C \text{ independent of } j \text{ and } f. \tag{15.32}$$

We claim that the Littlewood–Paley decomposition $f = \sum_{j\ge 0} f_j$ of $f \in BMO$ satisfies

$$\|f_j\|_{L^\infty} \le C\|f\|_{BMO}, \text{ for some } C \text{ independent of } j \text{ and } f. \tag{15.33}$$

Indeed, this follows from the fact that $\varphi_j * \varphi_k = 0$ if $|k-j| \geq 2$, and thus,

$$f_j = f * \varphi_j = \sum_{|k-j|\leq 1} f^k * \varphi_k * \varphi_j;$$

this leads to

$$\begin{aligned}\|f_j\|_{L^\infty} &\leq \sum_{|k-j|\leq 1} \|f^k * \varphi_k * \varphi_j\|_{L^\infty} \leq \sum_{|k-j|\leq 1} \|f^k\|_{L^\infty} \|\varphi_k\|_{L^1} \|\varphi_j\|_{L^1} \\ &\leq C \sum_{|k-j|\leq 1} \|f^k\|_{L^\infty},\end{aligned}$$

and thus to (15.33).

Finally, we combine Theorem 15.3 with (15.33), and find that

$$\|f\|_{W^{s/2,2p}}^{2p} \lesssim \sum 2^{sjp} \|f_j\|_{L^{2p}}^{2p} \leq \sum 2^{sjp} \|f_j\|_{L^\infty}^p \|f_j\|_{L^p}^p \lesssim \|f\|_{BMO}^p \|f\|_{W^{s,p}}^p. \qquad \square$$

Remark 15.7. The conclusion of Lemma 15.7 holds whenever $s > 0$ and θs are not integers. This is obtained by combining (15.31) with Remark 15.13.

Remark 15.8. Define, for $f \in L^1_{loc}(\mathbb{R}^N)$,

$$\|f\|_{bmo} := \sup_{|B|\leq 1} \int_B |f| + \sup_{|B|\leq 1} \fint_B \fint_B |f(x) - f(y)|\, dxdy,$$

the sup being taken over the balls of volume ≤ 1. Thus,

$$bmo := \{f \in L^1_{loc}(\mathbb{R}^n);\ \|f\|_{bmo} < \infty\}$$

is a "local BMO" space, with a natural norm. Clearly, $BMO \hookrightarrow bmo$.

Given $f \in bmo$, there exists a sequence $(f^j) \subset L^\infty$ such that (with φ_j as in (15.6)) we have $f = \sum_{j\geq 0} f^j * \varphi_j$ in $\mathscr{S}'$ and [339, Theorem, p. 47]

$$\operatorname*{esssup}_{x\in\mathbb{R}^N} \sum_j |f^j(x)|^2 \leq C\|f\|_{bmo}^2 \text{ for some } C \text{ independent of } f. \tag{15.34}$$

This leads to the following improvement of (15.31). If $0 < s < 1$, $1 \leq p < \infty$, and $0 < \theta < 1$, then we have

$$\|g\|_{W^{s/2,2p}(\mathbb{R}^N)}^{2p} \leq C\|g\|_{bmo(\mathbb{R}^N)}^p \|g\|_{W^{s,p}(\mathbb{R}^N)}^p,\ \forall g \in W^{s,p} \cap bmo(\mathbb{R}^N). \tag{15.35}$$

15.3 Composition in Sobolev spaces

Given $\Phi : \mathbb{R} \to \mathbb{R}$ and $u : \Omega \to \mathbb{R}$, set

$$F(u) := \Phi \circ u. \tag{15.36}$$

[In applications, we may also consider $\Phi : \mathbb{R}^M \to \mathbb{R}^K$ and $u : \Omega \to \mathbb{R}^M$.]

In this appendix, we recall some basic results concerning the following question: find reasonable conditions on Φ and $u \in W^{s,p}$ in order to have $F(u) \in W^{s,p}$. We start with the best known result in this direction.

Theorem 15.6. *Let Ω be a standard domain. Assume that Φ is Lipschitz. If Ω is unbounded, assume also that $\Phi(0) = 0$. Let $0 < s \leq 1$, $1 \leq p < \infty$. Then, F is continuous from $W^{s,p}$ in itself.*

Same if $\Omega = \mathbb{S}^1$.

The fact that F maps $W^{s,p}$ into itself is straightforward. More delicate is the continuity. When $s = 1$, this is due to Marcus and Mizel [236]. When $0 < s < 1$, the proof of the continuity is easier (see, for example, Bourgain, Brezis, and Mironescu [54, proof of (5.43)]).

When $s > 1$, things are more complicated. A celebrated result of Dahlberg [127] asserts that, if F maps $W^{s,p}$ into itself, with $s \in \mathbb{N}$, $s \geq 2$, $1 < p < \infty$, and $sp < N$, then Φ has to be affine. The same holds if $s > 1 + 1/p$ is not an integer (see Bourdaud [42]). However, when s is sufficiently close to 1, there are non-trivial Φ's such that F maps $W^{s,p}$ into itself. For example, Bourdaud and Meyer [44] proved that the function $\Phi(t) = |t|$ maps $W^{s,p}$ in $W^{s,p}$ provided $1 < p < \infty$ and $1 < s < 1 + 1/p$. For a thorough discussion of this subject in the setting of $W^{s,p}$ spaces (with integer or non-integer s), we refer the reader to the survey of Bourdaud and Sickel [45].

On the other hand, extra properties of u can lead to the conclusion that $\Phi(u) \in W^{s,p}$. A first result in this direction is the following

Theorem 15.7. *Let Ω be a standard domain. Let $s > 1$ and $1 \leq p < \infty$. Let m denote the first integer $\geq s$. Let $\Phi \in C^m$. If Ω is unbounded, assume also that $\Phi^{(j)}(0) = 0$, $0 \leq j \leq m - 1$.*

Then, F maps $W^{s,p} \cap L^\infty$ into $W^{s,p}$ and is continuous in the following sense:

if $u_n \to u$ in $W^{s,p}$ and $\|u_n\|_{L^\infty} \leq C$, then $\Phi(u_n) \to \Phi(u)$ in $W^{s,p}$.

Same if $\Omega = \mathbb{S}^1$.

For an elementary proof of the above result for arbitrary s, see Escobedo [145].

A far-reaching improvement of Theorem 15.7 is the following

Theorem 15.8. *Let $s > 1$ and let $1 \leq p < \infty$. Let m be the least integer $\geq s$. Let $\Phi \in C^m(\mathbb{R}^M; \mathbb{R}^K)$ have bounded derivatives of order $\leq m$. If Ω is unbounded, assume also that $\Phi^{(j)}(0) = 0$, $0 \leq j \leq m - 1$.*

Then, for every $u \in W^{s,p} \cap W^{1,sp}(\Omega; \mathbb{R}^M)$, we have $F(u) \in W^{s,p}(\Omega; \mathbb{R}^K)$. In addition, F is continuous from $W^{s,p} \cap W^{1,sp}$ to $W^{s,p}$.

A result that has a "family resemblance" to Theorem 15.8 appeared originally in Adams and Frazier [3]. The version stated above is due to Brezis and Mironescu [83], under the slightly stronger assumption that $1 < p < \infty$; however, the argument there (which relies on a paraproduct technique) also applies to the case where $p = 1$. For a more elementary proof, see Maz'ya and Shaposhnikova [242] (see also Maz'ya [240, Section 12.4]). The fact that, Theorem 15.8 implies Theorem 15.7 is a consequence of the Gagliardo–Nirenberg embedding $W^{s,p} \cap L^\infty \hookrightarrow W^{1,sp}$.

Corollary 15.3. *Let $0 < s < \infty$, $1 \le p < \infty$, and let Ω be a Lipschitz bounded domain. If either $s \le 1$ or $sp \ge N$, then $\varphi \mapsto e^{\imath\varphi}$ acts continuously from $W^{s,p}(\Omega;\mathbb{R})$ to $W^{s,p}(\Omega;\mathbb{S}^1)$.*

Same if $\Omega = \mathbb{S}^1$.

Proof. When $s \le 1$, this is a special case of Theorem 15.6. When $s > 1$, by assumption we have $sp \ge N$, and in this case we rely on Theorem 15.8 and on the Sobolev embedding $W^{s,p} \hookrightarrow W^{1,sp}$. □

Remark 15.9. Condition $u \in W^{1,sp}$ in Theorem 15.8 is necessary in the following sense. Assume that $u \in W^{s,p}$ (with $s \ge 1$ and $1 \le p < \infty$) and that $U := e^{\imath u}$ in $W^{s,p}$. Then, $u \in W^{1,sp}$. Indeed, we have

$$U \in W^{s,p} \implies U \in W^{1,sp} \implies \nabla u = \frac{1}{\imath U}\nabla U \in L^{sp} \implies u \in W^{1,sp}.$$

Here, we used the Gagliardo–Nirenberg embedding $W^{s,p} \cap L^\infty \hookrightarrow W^{1,sp}$.

15.4 Standard (and non-standard) examples of maps in Sobolev spaces

In the first three lemmas, we present examples of functions in $W^{s,p}$, with $sp < 1$.

Lemma 15.8. *Let $0 < s < 1$ and $1 \le p < \infty$ be such that $sp < 1$. Let $Q := (0,1)^N$ and $\Omega := (0,1)^{N-1} \times (-1,1)$. Then, $\varphi_N := \chi_Q$ belongs to $W^{s,p}(\Omega)$.*

Proof. In view of Theorem 15.1 (applied with $M = 1$), the conclusion of the lemma is equivalent to the straightforward fact that

$$\int_{-1}^{0}\int_{0}^{1} \frac{1}{|x-y|^{1+\sigma}}\,dxdy < \infty, \quad \forall\, 0 < \sigma < 1. \qquad \square$$

Lemma 15.8 leads to our next example (see Runst and Sickel [312, Example 2, p. 48]).

Lemma 15.9. *Let $s > 0$ and $1 \le p < \infty$. Let ω be a non-empty smooth relatively compact domain such that $\overline{\omega} \subset \Omega$. Then,*

$$\varphi := \chi_\omega \in W^{s,p}(\Omega) \iff sp < 1.$$

Proof. The function φ is integer-valued and is not constant a.e. Thus, $\varphi \notin W^{s,p}$ when $sp \geq 1$, by Corollary 6.2.

Assume next that $sp < 1$. Let (U_j) be a finite covering of $\overline{\Omega}$ such that for each j we have either $\varphi = 0$ in U_j, or there exists a bi-Lipschitz diffeomorphism Φ_j of $(0, 1)^{N-1} \times (-1, 1)$ onto U_j such that $\varphi \circ \Phi_j = \varphi_N$ (with φ_N as in the previous lemma). For any such j, we have $\varphi \circ \Phi_j \in W^{s,p}((0, 1)^{N-1} \times (-1, 1))$ (and thus $\varphi \in W^{s,p}$ in U_j). Finally, we have $\varphi \in W^{s,p}(U_j)$ for each j, which implies that $\varphi \in W^{s,p}(\Omega)$. □

Another procedure producing $W^{s,p}$ functions, with $sp < 1$, consists of gluing together $W^{s,p}$ functions. This is explained below.

Lemma 15.10. *Let $0 < s < 1$ and $1 \leq p < \infty$ be such that $sp < 1$.*

Let Ω_j, $j = 1, 2$, be two disjoint Lipschitz domains and let $\Gamma \subset \overline{\Omega_1} \cap \overline{\Omega_2}$. Assume that $\Omega := \Omega_1 \cup \Omega_2 \cup \Gamma$ is a Lipschitz domain.

Let $f_j \in W^{s,p}(\Omega_j)$, $j = 1, 2$, and set $f := \begin{cases} f_1, & \text{in } \Omega_1 \\ f_2, & \text{in } \Omega_2 \end{cases}$. Then, $f \in W^{s,p}(\Omega)$.

Same if $\Omega = \mathbb{S}^1$ and Ω_1, Ω_2 are arcs on $\mathbb{S}^1$ with $\Omega_1 \cap \Omega_2 = \emptyset$ and $\mathbb{S}^1 = \overline{\Omega_1} \cup \Omega_2$.

Proof. Let $\widetilde{f_j}$ denote the extension of f_j to $\mathbb{R}^N$ with the value 0 outside Ω_j. Then, $\widetilde{f_j} \in W^{s,p}(\mathbb{R}^N)$ (see Lemma 5.5). We next note that f is the restriction to Ω of $\widetilde{f_1} + \widetilde{f_2}$, and thus $f \in W^{s,p}(\Omega)$. □

We next investigate power-type functions, following Runst and Sickel [312, Example 1, p. 43].

Lemma 15.11. *Let $\Omega := B_1(0) \subset \mathbb{R}^N$. Let $\alpha > 0$ and set $\varphi(x) := \dfrac{1}{|x|^\alpha}$, $\forall x \in \Omega \setminus \{0\}$. Then,*

$$\varphi \in W^{s,p}(\Omega) \iff \alpha < \frac{N}{p} - s. \tag{15.37}$$

Proof. If s is an integer, the conclusion is clear. We may therefore assume that s is not an integer and that $\alpha < N$ (the latter condition ensures that $u \in L^1(\Omega)$). We rely on Theorem 15.1. Before going further, let us note that, since the specific φ we consider is smooth outside the origin, we have $\varphi \in W^{s,p}$ if and only if the double integral in (15.3) considered over the larger set $[-1, 1] \times \Omega$ is convergent.

We use (15.3) with an integer M satisfying, in addition to $M > s$, the condition

$$(\alpha + M)p > N. \tag{15.38}$$

For simplicity, we drop the subscript j in e_j.

Using the homogeneity of φ, we find that $\delta^M_{te}\varphi(x) = \dfrac{1}{|t|^\alpha}\delta^M_e \varphi(x/t)$. Hence,

$$\begin{aligned} I :=& \int_{-1}^{1} \int_{B_1(0)} \frac{|\delta_{te}^M \varphi(x)|^p}{t^{1+sp}} \, dxdt \\ =& 2 \int_0^1 \int_{B_{1/t}(0)} t^{N-(\alpha+s)p-1} |\delta_e^M \varphi(y)|^p \, dydt. \end{aligned} \tag{15.39}$$

" $\Longleftarrow$ " Assume that $\alpha < \dfrac{N}{p} - s$. In this case, we show that $I < \infty$. Indeed, we have

$$I \leq 2 \int_0^1 t^{N-(\alpha+s)p-1} \, dt \int_{\mathbb{R}^N} |\delta_e^M \varphi(y)|^p \, dy.$$

Note that the assumption on α implies that $\int_0^1 t^{N-(\alpha+s)p-1} \, dt < \infty$. On the other hand, we have $\int_{\mathbb{R}^N} |\delta_e^M \varphi(y)|^p \, dy < \infty$, since $|\delta_e^M \varphi(y)| \sim \dfrac{1}{|y|^{\alpha+M}}$ at infinity and $\left|\delta_e^M \varphi\right|^p$ has integrable singularities. [These singularities, located at $y = 0$, $-e$, ..., $-Me$, behave like $|x|^{-\alpha p}$.] This proves " $\Longleftarrow$."

" $\Longrightarrow$ " We note that (15.39) implies that

$$I \geq 2 \int_0^1 \int_{B_1(0)} t^{N-(\alpha+s)p-1} |\delta_e^M \varphi(y)|^p \, dydt = \infty.$$

Here, we use the fact that $\displaystyle\int_{B_1(0)} |\delta_e^M \varphi(y)|^p \, dy > 0$ (since φ is not a polynomial in the e direction). □

The same argument leads to our third example.

Lemma 15.12. *Let $u(x) := \dfrac{x}{|x|}$, $\forall x \in \mathbb{R}^2 \setminus \{0\}$. Let $s > 0$, $1 \leq p < \infty$. Then, $u \in W^{s,p}(B_1(0))$ if and only if $sp < 2$.*

More generally, let $k \in \{1, \ldots, N\}$ and let

$$u : \mathbb{R}^N \to \mathbb{R}^k, \; u(x_1, \ldots, x_N) := \frac{(x_1, \ldots, x_k)}{|(x_1, \ldots, x_k)|}, \; \forall x \in \mathbb{R}^N \setminus (\{0\} \times \mathbb{R}^{N-k}).$$

Then, $u \in W^{s,p}(B_1(0))$ if and only if $sp < k$.

A more general result is the following

Lemma 15.13. *Let U be a neighborhood of $\overline{\Omega}$ and let S be a d-dimensional submanifold of U, with $d \leq N-1$. Let $u \in C^\infty(\overline{U} \setminus S)$ satisfy*

$$|D^\ell u(x)| \leq C_\ell [\, dist\,(x, S)]^{-\ell}, \; \ell = 0, \ldots, N-d, \; \forall x \in \overline{U} \setminus S. \tag{15.40}$$

Then, $u_{|\Omega} \in W^{s,p}(\Omega)$ provided $sp < N-d$.

Proof. We consider the more complicated case where $d \leq N-2$. The case where $d = N-1$ is left to the reader.

Without loss of generality, we may assume that $sp > N - d - 1$. Furthermore, we can also suppose that $s > N - d - 1$, and thus $N - d - 1 < s < N - d$ (since $sp < N - d$). Indeed, if $s \leq N - d - 1$, fix any $\theta \in (0, 1)$ such that

$$\frac{1}{p} < \theta < \frac{s}{N-d-1} \leq 1,$$

and let $r := s/\theta > N - d - 1$, and $q := p\theta > 1$. From the Gagliardo–Nirenberg embedding $W^{r,q} \cap L^\infty \hookrightarrow W^{s,p}$ (see Remark 15.5), and the fact that $u_{|\Omega} \in L^\infty$ (by (15.40) with $\ell = 0$), it suffices to prove that $u_{|\Omega} \in W^{r,q}$. Thus, as claimed, we may add into the lemma the extra assumption $s > N - d - 1$.

Let $m := N - d - 1$ and write $s = m + \sigma$, with $0 < \sigma < 1$. The assumption $sp < N - d$ reads $(m + \sigma)p < N - d$.

We set $d(x) := \operatorname{dist}(x, S)$. Since S is d-dimensional, we find that

$$\int_\Omega \frac{1}{d^q(x)}\,dx < \infty, \ \forall\, 1 \leq q < N - d. \tag{15.41}$$

We next invoke the following well-known consequence of Lemma 1.10.

Lemma 15.14. *Let S be a d-submanifold of the open set $\Omega \subset \mathbb{R}^N$, with $d \leq N - 2$. Let $u \in C^1(\Omega \setminus S)$ be such that $\nabla u \in L^1_{loc}(\Omega)$. Then, $u \in W^{1,1}_{loc}(\Omega)$.*

Proof of Lemma 15.13 *continued.* Combining (15.40) with (15.41) and Lemma 15.14, we find that $D^j u \in L^p(\Omega), j = 0, \ldots, m$. It remains to check that

$$\int_\Omega \int_\Omega \frac{|D^m u(x) - D^m u(y)|^p}{|x-y|^{N+\sigma p}}\,dxdy < \infty. \tag{15.42}$$

For this purpose, we note that, with constants depending on u but not on x or y, we have:

$$|D^m u(x) - D^m u(y)| \leq C \max\left\{\frac{|x-y|}{d^{m+1}(x)}, \frac{|x-y|}{d^{m+1}(y)}\right\}, \tag{15.43}$$

$$|D^m u(x) - D^m u(y)| \leq C \max\left\{\frac{1}{d^m(x)}, \frac{1}{d^m(y)}\right\}. \tag{15.44}$$

We split

$$\iint \frac{|D^m u(x) - D^m u(y)|^p}{|x-y|^{N+\sigma p}}\,dx\,dy = 2(I + J),$$

where

$$I := \iint_{|x-y|<d(y)\leq d(x)} \ldots, \ J := \iint_{d(y)\leq \min\{d(x),\,|x-y|\}} \ldots$$

Using (15.43) (resp. (15.44)), in order to estimate I (resp. J), we find that

$$\begin{aligned}\iint \frac{|D^m u(x) - D^m u(y)|^p}{|x-y|^{N+\sigma p}}\,dx\,dy &\lesssim \iint_{|x-y|<d(y)\le d(x)} \frac{|x-y|^{p-N-\sigma p}}{d^{(m+1)p}(y)}\,dx\,dy \\ &+ \iint_{d(y)\le \min\{d(x),\,|x-y|\}} \frac{dx\,dy}{|x-y|^{N+\sigma p} d^{mp}(y)} \\ &\lesssim \int \frac{1}{d^{mp+\sigma p}(y)}\,dy < \infty,\end{aligned}$$

by (15.41) (since $mp + \sigma p < N - d$). □

Remark 15.10. In Lemma 15.13, when $d \le N-2$, the assumptions on u can be weakened to $u \in C^{N-d}(\Omega \setminus S)$ and $|D^{N-d}u(x)| \le C[\mathrm{dist}\,(x, S)]^{d-N}$. This is left as an exercise.

Remark 15.11. The conclusion of Lemma 15.13 is still valid under the weaker assumption that S is a finite union of d-dimensional submanifolds. The following application of this fact (with $d = N-2$) is of special interest. Recall (see (7.12)) the definition of the class $\widetilde{\mathscr{R}}$. We let U be a smooth bounded open neighborhood of $\overline{\Omega}$ and set

$$\widetilde{\mathscr{R}} := \Big\{ u_{|\Omega};\ u \in C^\infty(\overline{U} \setminus S; \mathbb{S}^1) \text{ for some finite union } S \text{ of } (N-2) - \text{submanifolds of } U, \text{ and } |D^\ell u(x)| \le C(\ell)[\mathrm{dist}(x,S)]^{-\ell}, \forall\, \ell \ge 1,\ \forall x \in \overline{U} \setminus S \Big\}.$$

By the above, if $u \in \widetilde{\mathscr{R}}$ and $sp < 2$, then $u_{|\Omega} \in W^{s,p}(\Omega)$. Moreover, we have

$$\int_\Omega \frac{1}{[\mathrm{dist}(x,S)]^q}\,dx < \infty,\ \forall q < 2. \tag{15.45}$$

We next turn to the proof of Lemma 7.1, restated below for the convenience of the reader.

Lemma 15.15. *Let $u \in \widetilde{\mathscr{R}}$, $1 < p < \infty$ and $\delta > 0$. Then there exist universal constants C_1 and C_p and some $v : \Omega \to \mathbb{S}^1$ such that*

a) $v \in W^{1/r,r}(\Omega; \mathbb{S}^1)$, $\forall\, r \ge 1$
b) $|v|_{W^{1,1}} \le C_1 |u|_{W^{1,1}}$
c) $|v|^p_{W^{1/p,p}} \le C_p |u|_{W^{1,1}}$
d) $|[v \ne 1]| \le \delta$
e) $Jv = Ju$.

Proof. As in the proof of Lemma 1.15, we take v of the form $v = \Phi_{\varepsilon,\theta} \circ u$, where

$$\Phi_{\varepsilon,\theta}(z) := \begin{cases} \exp(\imath\pi(2\delta/\varepsilon + 1)), & \text{if } z = e^{\imath(\theta+\delta)} \text{ with } |\delta| < \varepsilon/2 \\ 1, & \text{otherwise} \end{cases}.$$

Since $\Phi_{\varepsilon,\theta}$ is Lipschitz and $u \in \widetilde{\mathscr{R}} \subset W^{1/r,r}$, $\forall\, 1 \le r < \infty$, item *a)* holds. As noted in the proof of Lemma 1.15, for each ε, θ we have $J\,(\Phi_{\varepsilon,\theta} \circ u) = Ju$, and also item *d)* holds for small ε. In order to complete the proof of the lemma, it suffices to prove the existence of $\varepsilon_0 = \varepsilon_0(u, q)$ (depending on u and $1 \le q < \infty$) and of a constant C_q (independent of u or ε) such that

$$\int_0^{2\pi} |\Phi_{\varepsilon,\theta} \circ u|^q_{W^{1/q,q}}\, d\theta \le C_q |u|_{W^{1,1}}, \ \forall\, \varepsilon < \varepsilon_0. \tag{15.46}$$

[Granted (15.46), in order to obtain items *b)* and *c)*, it suffices to let $v := \Phi_{\varepsilon,\theta} \circ u$, with $\varepsilon < \min(\varepsilon_0(u, 1), \varepsilon_0(u, p))$.]

By (1.100), estimate (15.46) is valid for $q = 1$ (with $C_1 = 2\pi$). We next turn to the case $1 < q < \infty$.

In this case, we rely on the following straightforward estimate, whose proof is left to the reader.

Lemma 15.16. *Let* $1 < q < \infty$. *For* $\omega, \eta \in \mathbb{S}^1$, *let*

$$\Psi_\varepsilon(\omega, \eta) := \int_0^{2\pi} |\Phi_{\varepsilon,\theta}(\omega) - \Phi_{\varepsilon,\theta}(\eta)|^q\, d\theta.$$

Then,

$$\Psi_\varepsilon(\omega, \eta) \le C_q \min\left\{\varepsilon, \frac{|\omega - \eta|^q}{\varepsilon^{q-1}}\right\}. \tag{15.47}$$

Proof of Lemma 15.15 *continued.* Estimate (15.46) is equivalent to

$$I_\varepsilon := \iint_{\Omega^2} \frac{\Psi_\varepsilon(u(x), u(y))}{|x - y|^{N+1}}\, dxdy \le C_q |u|_{W^{1,1}}, \tag{15.48}$$

that we prove below.

In what follows, C_q denotes a constant depending only on q (and possibly N) that may change from one line to another.

Let, for $x \in \Omega$, $d(x) := \operatorname{dist}(x, S)$. [Here, $S = S_u$ is as in the definition of the class $\widetilde{\mathscr{R}}$.] The starting point in the proof of (15.48) is the following straightforward property of maps in $\widetilde{\mathscr{R}}$: if $x, y \in \Omega$ are such that $|x - y| \le \dfrac{1}{2} d(x)$, then (by the second-order Taylor formula)

$$|u(x) - u(y)| \le |x - y||\nabla u(x)| + \frac{C(u)}{d^2(x)}|x - y|^2. \tag{15.49}$$

We decompose $\Omega^2 = A_1 \cup A_2 \cup A_3$, where

$$A_1 := \{(x, y) \in (\Omega \setminus S)^2;\ |x - y| > \sqrt{\varepsilon} d(x)\},$$
$$A_2 := \left\{(x, y) \in (\Omega \setminus S)^2;\ |x - y| > \frac{\varepsilon}{|\nabla u(x)|}\right\},$$
$$A_3 := (\Omega \setminus S)^2 \setminus (A_1 \cup A_2).$$

We take $\varepsilon < \dfrac{1}{4}$. We have $I_\varepsilon \le I_{\varepsilon,1} + I_{\varepsilon,2} + I_{\varepsilon,3}$, where $I_{\varepsilon,j} = \iint_{A_j} \cdots$

We start by estimating $I_{\varepsilon,1}$. Since (15.45) holds and (by Lemma 15.16) we have $\Psi_\varepsilon \le C_q \varepsilon$, we find that

$$\begin{aligned} I_{\varepsilon,1} &\le C_q \varepsilon \iint_{|x-y| > \sqrt{\varepsilon} d(x)} \frac{1}{|x-y|^{N+1}} \\ &\le C_q \sqrt{\varepsilon} \int \frac{1}{d(x)} \le K(u, q)\sqrt{\varepsilon}. \end{aligned} \tag{15.50}$$

The integral $I_{\varepsilon,2}$ is estimated in a similar way:

$$\begin{aligned} I_{\varepsilon,2} &\le C_q \varepsilon \iint_{|x-y| > \varepsilon/|\nabla u(x)|} \frac{1}{|x-y|^{N+1}} \\ &\le C_q \int |\nabla u| = C_q |u|_{W^{1,1}}. \end{aligned} \tag{15.51}$$

In order to estimate $I_{\varepsilon,3}$, we rely on the validity of (15.49) in A_3 and on the bound

$$\Psi_\varepsilon(u(x), u(y)) \le C_q \frac{|u(x) - u(y)|^q}{\varepsilon^{q-1}}$$

(see Lemma 15.16). We find that

$$\begin{aligned} I_{\varepsilon,3} \le & C_q \iint_{A_3} \left(\frac{|x-y|^{q-N-1}}{\varepsilon^{q-1}} |\nabla u(x)|^q + \frac{C^q(u)|x-y|^{2q-N-1}}{\varepsilon^{q-1} d^{2q}(x)} \right) \\ \le & C_q \iint_{|x-y| \le \varepsilon/|\nabla u(x)|} \frac{|x-y|^{q-N-1}}{\varepsilon^{q-1}} |\nabla u(x)|^q \\ & + K(u, q) \iint_{|x-y| \le \sqrt{\varepsilon} d(x)} \frac{|x-y|^{2q-N-1}}{\varepsilon^{q-1} d^{2q}(x)} \\ = & C_q \int |\nabla u| + K(u, q)\sqrt{\varepsilon} \int \frac{1}{d(x)} \le C_q |u|_{W^{1,1}} + K'(u, q)\sqrt{\varepsilon}. \end{aligned} \tag{15.52}$$

Combining (15.50)–(15.52), we obtain

$$I_\varepsilon \le C_q |u|_{W^{1,1}} + K(u, q)\sqrt{\varepsilon},\ \forall\, \varepsilon < \frac{1}{4}.$$

If u is non-constant, we find that (15.46) holds for small ε. When u is constant, we may take $v := u$. □

Here is a variant of Lemma 15.15 used in Chapter 10.

Lemma 15.17. *Let $u \in \widetilde{\mathscr{R}}$. Let $\lambda > 0$ and $1 \le p < \infty$. Then there exists some $v \in \widetilde{\mathscr{R}}$ such that*

a) $|v|^p_{W^{1/p,p}} \le C_p |u|_{W^{1,1}}$
b) $Jv = Ju$
c) $|[v \neq 1]| \le \lambda$.

Moreover, if $u \in \mathscr{R}$, then we may take $v \in \mathscr{R}$.

Proof. We repeat the proof of Lemma 15.15, but we change the definition of $\Phi_{\varepsilon,\theta}$. Fix some smooth map $\Phi : \mathbb{S}^1 \to \mathbb{S}^1$ such that $\deg \Phi = 1$ and

$$\Phi(e^{\imath\varphi}) = 1 \text{ if } \varphi \in (-\pi, \pi] \text{ and } |\varphi| \ge 1.$$

We set

$$\Phi_{\varepsilon,\theta}(z) := \begin{cases} \Phi\left(e^{\imath\delta/\varepsilon}\right), & \text{if } z = e^{\imath(\theta+\delta)}, \text{ with } |\delta| < \varepsilon \\ 1, & \text{otherwise} \end{cases}.$$

Then, for appropriate ε and θ, $v := \Phi_{\varepsilon,\theta} \circ u$ has all the required properties. The details are left to the reader. □

By copying the proof of Proposition 1.1 and by using Lemma 15.17 instead of Lemma 1.15, we obtain the following

Proposition 15.2. *Let $1 < p < \infty$. Let $u \in \mathscr{R}$. Then there exists a sequence $(u_n) \subset \mathscr{R}$ such that*

$$Ju_n = Ju, \ \forall n, \tag{15.53}$$

$$u_n \to 1 \text{ a.e.} \tag{15.54}$$

and

$$\limsup_{n\to\infty} |u_n|^p_{W^{1/p,p}} \lesssim \Sigma(u). \tag{15.55}$$

The next result is an analogue of Proposition 15.2 for maps f from $\mathbb{S}^1$ to $\mathbb{S}^1$ (recall that in that context the analogue of $\Sigma(u)$ is $\deg f$).

Lemma 15.18. *Let $1 \le p < \infty$. Fix an arc $I \subset \mathbb{S}^1$ and an integer d. Then there exists a smooth map $h : \mathbb{S}^1 \to \mathbb{S}^1$ (depending on I and d) such that*

a) $\deg h = d$
b) $h = 1$ *outside* I
c) $|h|_{W^{1/p,p}} \le C_p |d|^{1/p}$.

Proof. We may as well work on an interval I in $\mathbb{R}$. We may assume that $d > 0$. Fix d points $P_1, \ldots, P_d$ in I. Consider a smooth map $T : \mathbb{R} \to \mathbb{S}^1$ such that $T(x) = 1$ when $|x| \geq 1$ and $\deg T = 1$. For large n, let

$$h(x) := \begin{cases} T(n(x - P_j)), & \text{if } |x - P_j| < 1/n \text{ for some } j \\ 1, & \text{otherwise} \end{cases}.$$

Clearly, h satisfies properties *a)* and *b)*. We claim that h also satisfies *c)*. This is clear when $p = 1$. Let $1 < p < \infty$ and fix some small $\delta > 0$. Consider the intervals $I_1, \ldots, I_d$ of length δ centered at $P_1, \ldots, P_d$ and set $I_{d+1} := I \setminus (I_1 \cup \cdots \cup I_d)$. By straightforward calculations, we have, as $n \to \infty$:

$$\int_{I_j} \int_{I_k} \frac{|h(x) - h(y)|^p}{|y - x|^{1+(1/p)p}} \, dxdy = \begin{cases} C_p + o(1), & \text{if } 1 \leq j = k \leq d \\ o(1), & \text{otherwise} \end{cases}; \tag{15.56}$$

this implies that $|h|^p_{W^{1/p,p}} = C_p \, d + o(1)$ and completes the proof of the lemma. □

15.5 Further results on *BMO* and *VMO*

We recall here a few results concerning the superposition operators in *BMO*, the embedding of $W^{s,p}$ in *VMO*, and the effect of these properties on the extensions "by smoothing and projection." Throughout this section, we let $\Omega \subset \mathbb{R}^N$ be a Lipschitz open set.

We start by mentioning the following version of Theorem 15.6 in the *BMO* context (see Brezis and Nirenberg [104, Lemma A.8]).

Theorem 15.9. *Let $\Phi : \mathbb{R}^\ell \to \mathbb{R}^k$ be uniformly continuous. Let $(\varphi_n) \subset BMO\,(\Omega; \mathbb{R}^\ell)$ be such that $\varphi_n \to \varphi$ in BMO for some $\varphi \in VMO\,(\Omega; \mathbb{R}^\ell)$. Then, $\Phi \circ \varphi_n \to \Phi \circ \varphi$ in BMO.*

Corollary 15.4. *If $(u_n), (v_n) \subset BMO\,(\Omega; \mathbb{C})$, $u, v \in VMO\,(\Omega; \mathbb{C})$ are such that*

$$\|u_n - u\|_{BMO} \to 0, \ \|v_n - v\|_{BMO} \to 0, \tag{15.57}$$

$$\|u_n\|_{L^\infty} \leq C, \ \|v_n\|_{L^\infty} \leq C, \tag{15.58}$$

then $\|u_n v_n - uv\|_{BMO} \to 0$.

Same if $\Omega = \mathbb{S}^1$.

In particular, $VMO\,(\Omega) \cap L^\infty$ is an algebra.

Proof. Apply Theorem 15.9 with $\varphi_n := (u_n, v_n)$, $\varphi := (u, v)$ and a uniformly continuous function $\Phi : \mathbb{C}^2 \to \mathbb{C}$ such that $\Phi(z, w) = zw$, $\forall\, z, w \in \mathbb{C}$ with $|z|, |w| \leq C$. □

Let ρ be an integrable function supported in $\overline{B}_1(0)$. We introduce the following notation.

$$\Omega_\varepsilon := \{x \in \Omega;\ \text{dist}\,(x, \partial\Omega) > \varepsilon\}, \tag{15.59}$$

$$\rho_\varepsilon(x) := \frac{1}{\varepsilon^N}\rho(x/\varepsilon),\ \forall\, x \in \mathbb{R}^N,\ \forall\, \varepsilon > 0, \tag{15.60}$$

$$u_\varepsilon(x) := u * \rho_\varepsilon(x),\ \forall\, \varepsilon > 0,\ \forall\, x \in \Omega_\varepsilon. \tag{15.61}$$

The next result is a slight generalization of Lemma 5.6.

Lemma 15.19. *Let $u \in VMO\ (\Omega; F)$, where $F \subset \mathbb{R}^d$. Let ρ be a bounded function supported in $\overline{B}_1(0)$ and such that $\int \rho = 1$. Then,*

$$\lim_{\varepsilon\to 0}\sup_{x\in\Omega_\varepsilon} \mathit{dist}\,(u_\varepsilon(x), F) = 0. \tag{15.62}$$

Proof. For $x \in \Omega_\varepsilon$ we have

$$\begin{aligned}\fint_{B_\varepsilon(x)} |u(y) - u_\varepsilon(x)|\, dy &\le \int_{B_\varepsilon(x)} \fint_{B_\varepsilon(x)} |u(y) - u(z)|\ |\rho_\varepsilon(x-z)|\, dydz \\ &\le \omega_N \|\rho\|_{L^\infty} \fint_{B_\varepsilon(x)} \fint_{B_\varepsilon(x)} |u(y) - u(z)|\ dydz;\end{aligned}$$

here, ω_N is the volume of the Euclidean unit ball of $\mathbb{R}^N$.

We find that there exists some $y_0 \in B_\varepsilon(x)$ such that

$$|u(y_0) - u_\varepsilon(x)| \le \omega_N \|\rho\|_{L^\infty} \fint_{B_\varepsilon(x)} \fint_{B_\varepsilon(x)} |u(y) - u(z)|\ dydz.$$

For such a y_0, we have

$$\begin{aligned}\text{dist}(u_\varepsilon(x), F) &\le |u_\varepsilon(x) - u(y_0)| \\ &\le \omega_N \|\rho\|_{L^\infty} \fint_{B_\varepsilon(x)} \fint_{B_\varepsilon(x)} |u(y) - u(z)|\ dydz.\end{aligned}$$

We conclude using the definition of *VMO*. □

Remark 15.12. If F is bounded, then we may remove the assumption that ρ is bounded. This is left as an exercise.

Corollary 15.5. *Let $u \in VMO\,(\Omega; \mathbb{S}^1)$ and set $u_\varepsilon(x) := \fint_{B_\varepsilon(x)} u$, $\forall\, x \in \Omega_\varepsilon$. Then,*

a) $u_\varepsilon \to u$ *in* $BMO_{\,loc}$ *as* $\varepsilon \to 0$
b) $|u_\varepsilon| \to 1$ *uniformly on compact subsets of* Ω *as* $\varepsilon \to 0$
c) $u_\varepsilon/|u_\varepsilon| \to u$ *in* $BMO_{\,loc}$ *as* $\varepsilon \to 0$.

When $\Omega = \mathbb{S}^1$, set $u_\varepsilon(x) := \fint_{I_\varepsilon(x)} u(t)\, dt$, $\forall\, x \in \mathbb{S}^1$, where $I_\varepsilon(x)$ is the arc on $\mathbb{S}^1$ of length 2ε centered at x. Then,

a) $u_\varepsilon \to u$ *in* BMO *as* $\varepsilon \to 0$
b) $|u_\varepsilon| \to 1$ *uniformly on* $\mathbb{S}^1$ *as* $\varepsilon \to 0$
c) $u_\varepsilon/|u_\varepsilon| \to u$ *in* BMO *as* $\varepsilon \to 0$.

Proof. We combine item 6 in Section 5.2.1, Lemma 5.6 and Theorem 15.6 (applied to a uniformly continuous function $\Phi : \mathbb{C} \to \mathbb{C}$ such that $\Phi(z) = z/|z|$ when $|z| \geq 1/2$) . □

For the next result, we refer the reader to Brezis and Nirenberg [104] (but special cases were known before; see, for example, Boutet de Monvel-Berthier, Georgescu, and Purice [67]).

Lemma 15.20. *Assume that* $0 < s < \infty$, $1 \leq p < \infty$, *and* $sp \geq N$. *Then,* $W^{s,p}(\Omega) \hookrightarrow VMO(\Omega)$.

Same if $\Omega = \mathbb{S}^1$.

Proof. When $sp > N$, we have $W^{s,p} \hookrightarrow C(\overline{\Omega}) \hookrightarrow VMO$. The same holds when $s = N$ and $p = 1$. When $sp = N$ and $s \neq N$, for each $0 < \sigma < 1$ we have $W^{s,p} \hookrightarrow W^{\sigma,q}$, with $q = \dfrac{N}{\sigma}$, by the Sobolev embeddings. It thus suffices to prove that $W^{s,p} \hookrightarrow VMO$ when $0 < s < 1$ and $sp = N$. For such s and p, let $u \in W^{s,p}$. Since $u \in L^1(\Omega)$, we have $u \in VMO$ provided we prove that $\lim_{|B|\to 0} \fint_B \fint_B |u(x) - u(y)|\, dxdy = 0$. This is a consequence of the fact that

$$|u|^p_{W^{s,p}} = \int_\Omega \int_\Omega \frac{|u(x) - u(y)|^p}{|x - y|^{2N}}\, dxdy < \infty. \tag{15.63}$$

Indeed, Hölder's inequality implies that

$$\fint_B \fint_B |u(x) - u(y)|\, dxdy \leq C \left(\int_B \int_B \frac{|u(x) - u(y)|^p}{|x - y|^{2N}}\, dxdy \right)^{1/p},$$

and the last integral tends to 0 as $|B| \to 0$, in view of (15.63).

The case where $\Omega = \mathbb{S}^1$ is similar. □

We next combine $W^{s,p}$ theory with the properties of *VMO* maps and obtain the existence of "good" extensions of $W^{s,p} \cap VMO$ maps.

Let Ω_ε and u_ε be as in (15.59)–(15.61), and set

$$U := \{(x, \varepsilon);\ \varepsilon > 0,\ x \in \Omega_\varepsilon\}, \tag{15.64}$$

$$v(x, \varepsilon) := u_\varepsilon(x) = u * \rho_\varepsilon(x),\ \forall\, (x, \varepsilon) \in U. \tag{15.65}$$

Corollary 15.6. *Assume that* $s > 0$ *is non-integer,* $1 \leq p < \infty$, $sp \geq N$. *Let* $F \subset \mathbb{R}^d$ *and* $u \in W^{s,p}(\Omega; F)$. *Let* ρ *be a normalized bump function. Then,*

a) $v \in W^{s+1/p,p}(U)$ *and (for appropriate seminorms)* $|v|_{W^{s+1/p,p}(U)} \leq C|u|_{W^{s,p}(\Omega)}$
b) $\|v\|_{W^{s+1/p,p}(U)} \leq C\|u\|_{W^{s,p}(\Omega)}$
c) $u_\varepsilon \to u$ *in* $W^{s,p}_{loc}(\Omega)$ *as* $\varepsilon \to 0$
d) $tr\, v_{|\Omega} = u$
e) $\lim_{\varepsilon\to 0} \sup_{x\in\Omega_\varepsilon} dist\,(u_\varepsilon(x), F) = 0$.

In the special case $F = \mathbb{S}^1$, *we have*

$$\lim_{\varepsilon\to 0} \sup_{x\in\Omega_\varepsilon} ||u_\varepsilon(x)| - 1| = 0.$$

When $\Omega = \mathbb{R}^N$, *the above properties hold with* $U := \mathbb{R}^N \times (0, T)$, *where* $T < \infty$. *When* $\Omega = \mathbb{S}^1$, *we have*

$$\lim_{\varepsilon\to 0} \sup_{x\in\mathbb{S}^1} \mathit{dist}\,(u_\varepsilon(x), F) = 0,$$

and if, in addition, $F = \mathbb{S}^1$,

$$\lim_{\varepsilon\to 0} \sup_{x\in\mathbb{S}^1} ||u_\varepsilon(x)| - 1| = 0.$$

Proof. Properties *a)–d)* are not specific to the range $sp \geq N$. They hold in all fractional Sobolev spaces (see Lemmas 15.46 and 15.47 in Appendix 15.14). Property *e)* follows from Lemmas 15.20 and 15.19. □

Corollary 15.7. *Let* $0 < s < 1$ *and* $1 \leq p < \infty$. *Let* $u \in W^{s,p} \cap \mathit{VMO}\,(\Omega; \mathbb{S}^1)$ *and set* $u_\varepsilon(x) := \fint_{B_\varepsilon(x)} u$, $\forall x \in \Omega_\varepsilon$. *Then,*

a) $u_\varepsilon \to u$ *in* $W^{s,p}_{loc}$ *and in* $\mathit{BMO}_{\,loc}$ *as* $\varepsilon \to 0$
b) $|u_\varepsilon| \to 1$ *uniformly on compacts of* Ω *as* $\varepsilon \to 0$
c) $u_\varepsilon/|u_\varepsilon| \to u$ *in* $W^{s,p}_{loc}$ *and* $\mathit{BMO}_{\,loc}$ *as* $\varepsilon \to 0$.

When $\Omega = \mathbb{S}^1$, *set* $u_\varepsilon(x) := \fint_{I_\varepsilon(x)} u(t)\,dt$, $\forall x \in \mathbb{S}^1$, *where* $I_\varepsilon(x)$ *is the arc on* $\mathbb{S}^1$ *of length* 2ε *centered at* x. *Then,*

a) $u_\varepsilon \to u$ *in* $W^{s,p}$ *and in* BMO *as* $\varepsilon \to 0$
b) $|u_\varepsilon| \to 1$ *uniformly on* $\mathbb{S}^1$ *as* $\varepsilon \to 0$
c) $u_\varepsilon/|u_\varepsilon| \to u$ *in* $W^{s,p}$ *and* BMO *as* $\varepsilon \to 0$.

Proof. The convergence $u_\varepsilon \to u$ in $W^{s,p}_{loc}$ holds for each ρ such that $\operatorname{supp}\rho \subset \overline{B}_1(0)$ and $\int \rho = 1$ (see Lemma 15.46). In particular, we may take $\rho := \dfrac{1}{|B_1(0)|}\chi_{B_1(0)}$, and then $u_\varepsilon(x) = \fint_{B_\varepsilon(x)} u$ satisfies $u_\varepsilon \to u$ in $W^{s,p}_{loc}$. Combining this with Corollaries 15.5 and 15.6, we obtain the conclusions of the corollary. □

Corollary 15.8. *Let* $1 < p < \infty$. *Then,* $C^\infty(\mathbb{S}^1; \mathbb{S}^1)$ *is dense in* $W^{1/p,p}(\mathbb{S}^1; \mathbb{S}^1)$.

Proof. The maps $u_\varepsilon/|u_\varepsilon|$ in the previous corollary are Lipschitz. Thus, $\mathrm{Lip}(\mathbb{S}^1; \mathbb{S}^1)$ is dense in $W^{1/p,p}(\mathbb{S}^1; \mathbb{S}^1)$. The conclusion follows by smoothing. □

We continue this section with the proofs of the embedding $\mathit{BMO}\,(\Omega) \hookrightarrow L^1(\Omega)$ and of the related estimates (5.25) and (5.26), mentioned in Section 5.2.1, which we recall here for the convenience of the reader.

Lemma 15.21. *Let $\Omega \subset \mathbb{R}^N$ be a smooth bounded domain. Then, $BMO(\Omega) \hookrightarrow L^1(\Omega)$ and, for every ball $\overline{B} \subset \Omega$, we have*

$$\|f\|_{L^1(\Omega)} \leq C_B \left(|f|_{BMO(\Omega)} + \left| \int_B f \right| \right). \tag{15.66}$$

Moreover, we have the following Poincaré-type inequality

$$\left\| f - \fint_\Omega f \right\|_{L^1(\Omega)} \leq C |f|_{BMO(\Omega)}. \tag{15.67}$$

Same if $\Omega = \mathbb{S}^1$.

Lemma 15.21 is an easy consequence of the next two results.

Lemma 15.22. *Let $\Phi : V \to U$ be a bi-Lipschitz homeomorphism, with U, V open sets in $\mathbb{R}^N$. Then, $u \mapsto u \circ \Phi$ maps $BMO(U)$ in $BMO(V)$ (resp. $VMO(U)$ in $VMO(V)$).*

Same if $\Omega = \mathbb{S}^1$.

Lemma 15.23. *Let $\Omega \subset \mathbb{R}^N$ be smooth and bounded. Let $K_1, K_2 \subset \Omega$ be compact sets of positive measure. Then, $f \mapsto |f|_{BMO} + \left| \int_{K_j} f \right|$, $j = 1, 2$, are equivalent norms on $BMO(\Omega)$.*

Proof of Lemma 15.21. *Step 1.* Let $\omega := (-1, 1)^N$, and let $K \subset \omega$ be a compact set of positive measure. We claim that

$$\|f\|_{L^1(\omega)} \leq C_K |f|_{BMO(\omega)}, \ \forall f \in BMO(\omega) \text{ such that } \int_K f = 0. \tag{15.68}$$

In order to prove (15.68), we rely on the fact that the geometry of the balls does not change the *BMO* spaces and yields equivalent seminorms (see Jones [207]; see also [105]). It is convenient to work here with balls for the sup norm, i.e., cubes of the form $a + (-r, r)^N$, with $a \in \mathbb{R}^N$. If we let $B_r := (-r, r)^N$ then, by monotone convergence,

$$\fint_\omega \fint_\omega |f(x) - f(y)| \, dxdy = \lim_{r \nearrow 1} \fint_{B_r} \fint_{B_r} |f(x) - f(y)| \, dxdy \leq |f|_{BMO(\omega)},$$

and thus $f \in L^1(\omega)$ and

$$\int_\omega \left| f - \fint_\omega f \right| \leq \int_\omega \fint_\omega |f(x) - f(y)| \, dxdy \leq 2^N |f|_{BMO(\omega)}. \tag{15.69}$$

Combining (15.69) and the assumption $\int_K f = 0$, we find that

$$|K| \left| \fint_\omega f \right| = |K| \left| \fint_K \left(f - \fint_\omega f \right) \right| \leq \int_K \left| f - \fint_\omega f \right| \leq 2^N |f|_{BMO(\omega)}, \tag{15.70}$$

and therefore

$$\|f\|_{L^1(\omega)} \leq \int_\omega \left| f - \fint_\omega f \right| + 2^N \left| \fint_\omega f \right| \leq 2^N (1 + 1/|K|) |f|_{BMO(\omega)}.$$

Step 2. Proof of (15.66)
We cover Ω with a finite number of open subsets Ω_j, $j = 1, \ldots, k$, each one bi-Lipschitz diffeomorphic to ω. Let, for each j, $K_j \subset \Omega_j$ be a compact set of positive measure. By Lemma 15.22 and Step 1, we have

$$\|f\|_{L^1(\Omega_j)} \leq D_j \left(|f|_{BMO(\Omega_j)} + \left| \int_{K_j} f \right| \right), \quad \forall j, \ \forall f \in BMO(\Omega). \tag{15.71}$$

for some appropriate finite constants D_j.

By Lemma 15.23 and (15.71), we find that (15.66) holds (and it actually holds when B is replaced by any compact subset of Ω, of positive measure).
Step 3. Proof of (15.67)
This follows from (15.66) and the inequality

$$\int_\Omega \left| g - \fint_\Omega g \right| \leq 2 \int_\Omega |g|. \tag{15.72}$$

Combining (15.72) (with $g := f - \fint_B f$) with (15.66) (applied to the function $f - \fint_B f$), we obtain (15.67). □
Proof of Lemma 15.22. For $\varepsilon \leq 1$, we set

$$M_{t,1,\varepsilon}(u) := \sup \left\{ \fint_B \fint_B |u(x) - u(y)| \, dxdy; \ B = B_r(z), \ z \in U, \ 0 < r \leq t, \right.$$
$$\left. \operatorname{dist}(z, \partial U) \geq r/\varepsilon \right\},$$

so that $M_{t,1,1}(u) = M_{t,1}(u) = M_t(u)$ (see (5.27) and (5.28)).

We have $M_{t,1,\varepsilon}(u) \sim M_t(u)$. When $\varepsilon = 1/2$, this is proved in Brezis and Nirenberg [105, Theorem A1.1], but the argument there applies to any ε. Therefore, it suffices to prove that, for sufficiently small $\varepsilon > 0$ and sufficiently large $C, K > 0$ we have $M_{t,1,\varepsilon}(u \circ \Phi) \leq C M_{Kt}(u)$. In turn, this inequality follows easily from

1. The double inclusion $B_{r/K}(\Phi(w)) \subset \Phi(B_r(w)) \subset B_{Kr}(\Phi(w))$ (valid for large K);
2. The fact that, for fixed K as above, we have $B_{Kr}(\Phi(w)) \subset U$ provided $\operatorname{dist}(w, \partial V) \geq r/\varepsilon$ for sufficiently small ε (depending on K). □

Proof of Lemma 15.23. For $K \subset \Omega$ a compact set of positive measure, set $f_K := \fint_K f$. In order to prove the lemma, it suffices to establish the inequality

$$|f_{K_1} - f_{K_2}| \leq C(K_1, K_2) |f|_{BMO}, \ \forall f \in BMO(\Omega). \tag{15.73}$$

For this purpose, we start from the following inequality, stated in [105, Lemma 1],

$$\|f - f_K\|_{L^1(K)} \leq C_K |f|_{BMO(\Omega)}, \ \forall f \in BMO(\Omega). \tag{15.74}$$

Applying (15.74) to $K := K_1 \cup K_2$, we find that

$$|K_1|\,|f_{K_1} - f_K| = \left|\int_{K_1} (f - f_K)\right| \leq \int_{K_1} |f - f_K| \leq C_K |f|_{BMO(\Omega)},$$

and thus

$$|f_{K_1} - f_K| \leq \frac{C_K}{|K_1|} |f|_{BMO(\Omega)}. \tag{15.75}$$

Estimate (15.73) follows from (15.75) and the similar estimate for $|f_{K_2} - f_K|$. □

For the convenience of the reader, we sketch a proof of inequality (15.74).

We note the following clear consequence of (15.72):

$$\int_K |f - f_K| \leq 2 \int_K |f - c|, \ \forall c \in \mathbb{R}. \tag{15.76}$$

In view of (15.76), it suffices to prove inequality (15.74) in the following seemingly weaker form

$$\int_K |f - c(f, K)| \leq C_K |f|_{BMO(\Omega)}, \ \forall f. \tag{15.77}$$

Let

$$\mathscr{K} := \{K \subset \Omega;\ K \text{ is a compact set of positive measure and (15.77) holds, } \forall f\}.$$

We have the following straightforward properties:

$$\overline{B} \in \mathscr{K}, \ \forall \overline{B} \text{ closed ball in } \Omega, \tag{15.78}$$

$$[K_1 \subset K_2, |K_1| > 0 \text{ and } K_2 \in \mathscr{K}] \implies K_1 \in \mathscr{K}, \tag{15.79}$$

$$[K_1, K_2 \in \mathscr{K}, |K_1 \cap K_2| > 0] \implies K_1 \cup K_2 \in \mathscr{K}. \tag{15.80}$$

For the latter property, we proceed as in the proof of Lemma 15.23 and estimate

$$\begin{aligned} |c(f, K_1) - c(f, K_2)| &\leq \fint_{K_1 \cap K_2} [|f - c(f, K_1)| + |f - c(f, K_2)|] \\ &\leq \frac{C_{K_1} + C_{K_2}}{|K_1 \cap K_2|} |f|_{BMO(\Omega)}. \end{aligned}$$

Since Ω is connected, for each compact set $K \subset \Omega$ there exists a chain of balls, $B_1, \ldots, B_n$, such that $\overline{B_j} \subset \Omega, \forall j, B_{\ell+1} \cap \cup_{j=1}^{\ell} B_j \neq \emptyset, \forall \ell = 1, \ldots, n-1$, and $K \subset \cup_{j=1}^{n} B_j$. By (15.78) and (15.80), we obtain, by induction on ℓ, that $\cup_{j=1}^{\ell} \overline{B_j} \in \mathscr{K}$, $\forall \ell = 1, \ldots, n$, and in particular $\cup_{j=1}^{n} \overline{B_j} \in \mathscr{K}$. If, in addition, $|K| > 0$, then (15.79) implies that $K \in \mathscr{K}$. In conclusion, each compact set $K \subset \Omega$ such that $|K| > 0$ belongs to $\mathscr{K}$, and thus (15.77) (and, consequently, (15.74)) holds for any such K.

We conclude this section with the

Proof of Lemma 12.3. The heart of the proof consists of proving the existence of $\alpha > 0$ and $\varepsilon > 0$ such that

$$\left|\int_0^1 g\right| \leq \alpha \implies \left|\left\{(z, t) \in (0, 1)^2;\ |g(z) - g(t)| \geq \sqrt{3}\right\}\right| \geq 3\varepsilon. \tag{15.81}$$

Assuming temporarily that (15.81) holds, we conclude as follows. Set

$$A := \left\{(z, t) \in (0, 1)^2;\ |z - t| \geq \varepsilon, |g(z) - g(t)| \geq \sqrt{3}\right\}$$

and

$$B := \left\{(z, t) \in (0, 1)^2;\ |g(z) - g(t)| \geq \sqrt{3}\right\}. \tag{15.82}$$

Since

$$\left|\{(z, t) \in (0, 1)^2;\ |z - t| \leq \varepsilon\}\right| \leq 2\varepsilon,$$

we find that $|A| \geq |B| - 2\varepsilon \geq \varepsilon$, and therefore (12.84) holds.

It thus remains to prove (15.81). With $\alpha > 0$ and $\varepsilon > 0$ to be fixed later, we will obtain a contradiction by assuming that for some g we have

$$\left|\int_0^1 g\right| \leq \alpha \text{ and } |B| = \left|\left\{(z, t) \in (0, 1)^2;\ |g(z) - g(t)| \geq \sqrt{3}\right\}\right| < 3\varepsilon. \tag{15.83}$$

To start with, if (15.83) holds, then there exists some $x_0 \in (0, 1)$ such that

$$|\{y \in (0, 1);\ |g(y) - g(x_0)| \geq \sqrt{3}\}| < 3\varepsilon.$$

With no loss of generality, we may assume that $g(x_0) = 1$, and thus, with $I := \{e^{\imath\theta};\ -2\pi/3 < \theta < 2\pi/3\}$, we have

$$|E| > 1 - 3\varepsilon, \text{ where } E := \{x \in (0, 1);\ g(x) \in I\}. \tag{15.84}$$

Define

$$J_+ := \{e^{\imath\theta};\ \pi/3 \leq \theta < 2\pi/3\} \text{ and } J_- := \{e^{\imath\theta};\ -2\pi/3 < \theta \leq -\pi/3\}.$$

Note that

$$[\zeta_+ \in J_+ \text{ and } \zeta_- \in J_-] \implies |\zeta_+ - \zeta_-| \geq \sqrt{3}.$$

With B as in (15.82), we thus have

$$\begin{aligned} &[g(z_+) \in J_+ \text{ and } g(z_-) \in J_-] \implies (z_+, z_-) \in B, \\ &\text{i.e., } (g^{-1}(J_+)) \times (g^{-1}(J_-)) \subset B. \end{aligned} \tag{15.85}$$

Since $|B| < 3\varepsilon$ (by assumption (15.83)), we have (by (15.85)) either $|g^{-1}(J_+)| < \sqrt{3\varepsilon}$ or $|g^{-1}(J_-)| < \sqrt{3\varepsilon}$. With no loss of generality, we may assume that

$$|g^{-1}(J_-)| < \sqrt{3\varepsilon}. \tag{15.86}$$

Set $K := \{e^{\imath\theta};\ -\pi/3 < \theta < 2\pi/3\}$ and set $F := g^{-1}(K)$. Note that K is a half unit circle. By (15.84) and (15.86), we have

$$g : F \to K \text{ and } |F| > 1 - 3\varepsilon - \sqrt{3\varepsilon}. \tag{15.87}$$

On the other hand, we have

$$\left|\int_F g\right| \leq \left|\int_0^1 g\right| + |(0, 1) \setminus F| \leq \alpha + 3\varepsilon + \sqrt{3\varepsilon}. \tag{15.88}$$

We next invoke the following lemma, whose proof is postponed.

Lemma 15.24. *Let Σ be a half unit circle. Let $0 < \lambda < 2$ and $0 < \gamma < 1$. Let $X \subset \mathbb{R}^N$ be any measurable set such that $0 < |X| < \infty$. Then there exists $\beta > 0$ such that*

$$\left[h : X \to \Sigma,\ \left|\fint_X h\right| \leq \beta\right] \implies |\{(x, y) \in X^2;\ |h(x) - h(y)| > \lambda\}| > \gamma\, |X|^2/4.$$

[Similar statement in a measured space $(X, \mathscr{T}, \mu)$, with $0 < \mu(X) < \infty$.]

Proof of Lemma 12.3 *continued.* We apply Lemma 15.24 with $\lambda := \sqrt{3}$, $\gamma := 1/2$, $\Sigma := K, N = 1, X := F$ and $h := g_{|F}$. When α and ε are sufficiently small we have (by (15.87) and (15.88)) $\left|\fint_F g\right| \leq \beta$, and thus

$$|\{(x, y) \in F^2;\ |g(x) - g(y)| > \sqrt{3}\}| \geq 1/9.$$

This contradicts (15.83) when $\varepsilon < 1/27$. $\square$

Proof of Lemma 15.24. With no loss of generality, we may assume that $K = \{z \in \mathbb{S}^1;\ \operatorname{Re} z \geq 0\}$ and that $|X| = 1$.

Let $\beta > 0$ be a small constant to be fixed later. Let also μ be such that

$$2\sqrt{\gamma} < \mu < 2. \tag{15.89}$$

Consider the following sets:

$$X_1 := \{x \in X;\ |\mathrm{Im}\, h(x)| \le \mu/2\} = \left\{x \in X;\ \mathrm{Re}\, h(x) \ge (1-\mu^2/4)^{1/2}\right\},$$
$$X_2 := \{x \in X;\ \mathrm{Im}\, h(x) > \mu/2\},$$
$$X_3 := \{x \in X;\ \mathrm{Im}\, h(x) < -\mu/2\}.$$

Note that

$$[x \in X_2,\ y \in X_3] \implies |h(x) - h(y)| > \mu > \lambda,$$
$$\text{so that } X_2 \times X_3 \subset \{(x, y) \in X^2;\ |h(x) - h(y)| > \lambda\},$$

and thus it suffices to prove that, for appropriate β and μ, we have

$$|X_2| > \frac{\sqrt{\gamma}}{2} \text{ and } |X_3| > \frac{\sqrt{\gamma}}{2}. \tag{15.90}$$

In order to establish (15.90), we rely on the following simple inequalities:

$$|X_1|\,(1-\mu^2/4)^{1/2} \le \int_{X_1} \mathrm{Re}\, h \le \int_X \mathrm{Re}\, h \le \beta, \tag{15.91}$$

$$(\mu/2)\,|X_2| - |X_3| \le \int_{X_2 \cup X_3} \mathrm{Im}\, h \le \int_X \mathrm{Im}\, h + |X_1| \le \beta + |X_1|, \tag{15.92}$$

and, similarly,

$$|X_2| - (\mu/2)\,|X_3| \ge \int_{X_2 \cup X_3} \mathrm{Im}\, h \ge -\beta - |X_1|. \tag{15.93}$$

We conclude as follows. First, from (15.91), (15.92), and $|X_2| \le 1$, we find that

$$\begin{aligned} |X_2| - |X_3| &= (\mu/2)\,|X_2| - |X_3| + (1-\mu/2)\,|X_2| \\ &\le \beta + |X_1| + (1-\mu/2) \\ &\le \beta\,\left(1 + 1/(1-\mu^2/4)^{1/2}\right) + (1-\mu/2), \end{aligned} \tag{15.94}$$

and, similarly,

$$|X_3| - |X_2| \le \beta\,\left(1 + 1/(1-\mu^2/4)^{1/2}\right) + (1-\mu/2). \tag{15.95}$$

On the other hand, we have

$$|X_2| + |X_3| = 1 - |X_1| \ge 1 - \beta/(1-\mu^2/4)^{1/2}. \tag{15.96}$$

Subtracting (15.95) from (15.96) (resp. (15.94) from (15.96)), we obtain

$$|X_j| \ge \frac{\mu}{4} - \beta/(1-\mu^2/4)^{1/2},\ j = 1, 2. \tag{15.97}$$

Estimate (15.90) follows from (15.97) by taking β sufficiently small (depending on γ and μ) such that the right-hand side of (15.97) is $> \dfrac{\sqrt{\gamma}}{2}$; this is possible, by (15.89). □

15.6 Enlarging Ω

The common theme of the results in this section is that, when dealing with $\mathbb{S}^1$-valued maps in some function space in Ω, we can assume that these maps are defined in a neighborhood of $\overline{\Omega}$. The two basic techniques for "enlarging" the domain that we present are:

1. The deformation of Ω. This works in $W^{s,p}$ for any s and p (and with more work we could prove that it also works in *VMO*). The price to pay is that, in Ω, we replace a given map u with an approximation of it (see Lemma 15.26).
2. The extension by reflections across $\partial\Omega$. This works in *BMO* and in *VMO*. The limitation of this method is that, in order to make it work in $W^{s,p}$, s has to be small. More specifically, the method is limited to the range $1 \le p < \infty$, $0 < s < 1 + 1/p$ (see Lemma 15.27).

Let $\Omega \subset \mathbb{R}^N$ be a smooth bounded open set. For sufficiently small $\delta > 0$, let $\Omega^\delta := \{x \in \mathbb{R}^N;\ \operatorname{dist}(x, \Omega) < \delta\}$.

Lemma 15.25. *There exists a family of diffeomorphisms $\Phi_\delta : \Omega^\delta \to \Omega$ such that:*

a) Φ_δ admits a continuous extension to $\overline{\Omega^\delta}$, and this extension, still denoted by Φ_δ, is a diffeomorphism of $\overline{\Omega^\delta}$ onto $\overline{\Omega}$
b) $\|(\Phi_\delta)^{-1} - Id\|_{C^k(\overline{\Omega})} = O(\delta)$ and $\|\Phi_\delta - Id\|_{C^k(\overline{\Omega^\delta})} = O(\delta)$ for every $k \in \mathbb{N}$.

Proof. For $x \in \mathbb{R}^N$ sufficiently close to $\partial\Omega$, let $\Psi(x)$ denote the "outer normal to $\partial\Omega$" at x. More specifically, if y is the nearest point projection of x onto $\partial\Omega$, then $\Psi(x)$ is the outer normal $\nu(y)$ to $\partial\Omega$ at y. For sufficiently small $\delta_0 > 0$, we thus have

$$\Psi(y + t\nu(y)) = \nu(y),\ \forall\, y \in \partial\Omega,\ \forall\, |t| < \delta_0,$$

and Ψ is smooth in $V := \{x \in \mathbb{R}^N;\ \operatorname{dist}(x, \partial\Omega) < \delta_0\}$. Let $\zeta \in C_c^\infty(V; [0, 1])$ be such that $\zeta = 1$ on $\partial\Omega$. For $\delta > 0$, the map $\overline{\Omega} \ni x \mapsto \Psi_\delta(x) := x + \delta\zeta(x)\Psi(x)$ is smooth. For sufficiently small δ, Ψ_δ is a diffeomorphism, whose image is $\overline{\Omega^\delta}$. [To see the latter property, it suffices to note that $\Phi_\delta(x) = x$ if $x \in \Omega$ is such that $\operatorname{dist}(x, \partial\Omega) \ge \delta_0$, and to note that, when $y \in \partial\Omega$, the mapping

$$y + t\nu(y) \mapsto y + t\nu(y) + \delta\zeta(y + t\nu(y))\,\nu(y)$$

maps the segment $(y - \delta_0\nu(y), y]$ onto the segment $(y - \delta_0\nu(y), y + \delta\nu(y)]$.]

If we set $\Phi_\delta := (\Phi_{\delta|\Omega^\delta})^{-1}$, Φ_δ clearly has all the required properties. □

Lemma 15.26. *Let $s > 0$ and $1 \le p < \infty$. With Φ_δ as above and $u \in W^{s,p}(\Omega)$, we have $u \circ \Phi_\delta \in W^{s,p}(\Omega^\delta)$, and $u \circ \Phi_\delta \to u$ in $W^{s,p}(\Omega)$, as $\delta \to 0$.*

[Here, u can be scalar or $\mathbb{R}^M$-valued.]

Proof. Let us first note that there exist constants $C > 0$ and $\delta_1 > 0$ such that

$$\|u \circ \Phi_\delta\|_{W^{s,p}} \le C\|u\|_{W^{s,p}}, \ \forall u \in W^{s,p}, \ \forall 0 < \delta < \delta_1. \tag{15.98}$$

On the other hand, if $u \in C^\infty(\overline{\Omega})$, we clearly have

$$\begin{aligned} &u \circ \Phi_\delta \to u \text{ in } C^k(\overline{\Omega}) \text{ as } \delta \to 0, \ \forall k \in \mathbb{N}^*, \\ &\text{and thus } u \circ \Phi_\delta \to u \text{ in } W^{s,p}(\Omega) \text{ as } \delta \to 0. \end{aligned} \tag{15.99}$$

The conclusion of the lemma is an immediate consequence of (15.98), (15.99), and of the density of $C^\infty(\overline{\Omega})$ in $W^{s,p}(\Omega)$. □

Lemma 15.27. *If $s < 1 + 1/p$ and $u \in W^{s,p}(\Omega; \mathbb{R}^M)$, then the extension of u by reflection across $\partial\Omega$ belongs to $W^{s,p}$ (in a neighborhood of $\overline{\Omega}$).*

Proof. The case $s = 1$ being classical, we assume that $s \ne 1$.

Using a partition of unity and composing with appropriate diffeomorphisms near $\partial\Omega$, we may assume that $\Omega = \mathbb{R}^N_+$. Let $u : \mathbb{R}^N_+ \to \mathbb{R}^M$ and let $\widetilde{u}$ be its extension by reflection to $\mathbb{R}^N$. We will prove that

$$|\widetilde{u}|_{W^{s,p}} \le C|u|_{W^{s,p}}. \tag{15.100}$$

Step 1. Proof of (15.100) when $0 < s < 1$

In this case, it suffices to estimate the cross term $\int_{\mathbb{R}^N_+}\int_{\mathbb{R}^N_-}$ in the Gagliardo seminorm (15.1) as follows:

$$\begin{aligned} \int_{\mathbb{R}^N_+ \times \mathbb{R}^N_-} \frac{|\widetilde{u}(x) - \widetilde{u}(y)|^p}{|x-y|^{N+sp}}\,dxdy &= \int_{\mathbb{R}^N_+ \times \mathbb{R}^N_+} \frac{|u(x) - u(y)|^p}{|x-\widetilde{y}|^{N+sp}}\,dxdy \\ &\le \int_{\mathbb{R}^N_+ \times \mathbb{R}^N_+} \frac{|u(x) - u(y)|^p}{|x-y|^{N+sp}}\,dxdy = |u|^p_{W^{s,p}}, \end{aligned}$$

where $\widetilde{y} := (y_1, \ldots, y_{N-1}, -y_N)$.

Step 2. Proof of (15.100) when $1 < s < 1 + 1/p$

Write $s = 1 + \sigma$, with $0 < \sigma < 1/p$. Then we have to prove that

$$|\partial_j \widetilde{u}|_{W^{\sigma,p}} \le C|u|_{W^{s,p}}, \ \forall j = 1, \ldots, N. \tag{15.101}$$

When $j \le N - 1$, we have $\partial_j \widetilde{u} = \widetilde{\partial_j u}$, and then (15.101) follows from Step 1 applied with σ and $\partial_j u$ instead of s and u.

When $j = N$, we have $\partial_N \widetilde{u}(x', x_N) = \operatorname{sgn}(x_N)\partial_N u(x', |x_N|)$, and then (15.101) is obtained from the following calculation:

$$\begin{aligned}
&\int_{\mathbb{R}^N_+\times\mathbb{R}^N_-} \frac{|\partial_N \widetilde{u}(x) - \partial_N \widetilde{u}(y)|^p}{|x-y|^{N+\sigma p}}\, dxdy \\
&= \int_{\mathbb{R}^N_+\times\mathbb{R}^N_-} \frac{|\partial_N u(x, x_N) + \partial_N u(y', -y_N)|^p}{|x-y|^{N+\sigma p}}\, dxdy \\
&\le C \int_{\mathbb{R}^N_+\times\mathbb{R}^N_-} \frac{|\partial_N u(x, x_N)|^p}{|x-y|^{N+\sigma p}}\, dxdy \le C \int_{\mathbb{R}^N_+} \frac{|\partial_N u(x)|^p}{[\operatorname{dist}(x, \partial\mathbb{R}^N_+)]^{\sigma p}}\, dx \\
&= C \int_{\mathbb{R}^{N-1}} \int_0^\infty \frac{|\partial_N u(x', x_N)|^p}{(x_N)^{\sigma p}}\, dx_N dx' \\
&\le C |\partial_N u|^p_{W^{\sigma,p}}.
\end{aligned} \tag{15.102}$$

The latter inequality (which completes the proof of Lemma 15.27) follows from Fubini's theorem combined with:

1. The norm equivalence (15.4) (in its version for half-spaces).
2. The density of $C^\infty_c((0,\infty))$ in $W^{\sigma,p}((0,\infty))$ (which holds, since $\sigma p < 1$).
3. The following well-known Hardy-type inequality (see, for example, Mironescu [261, Lemma 2]): if $v \in C^\infty_c((0,\infty), \mathbb{R})$ and if $0 < t < 1$, $1 \le p < \infty$ are such that $tp \ne 1$, then

$$\int_0^\infty \frac{|v(x)|^p}{x^{tp}}\, dx \le C \int_0^\infty \int_0^\infty \frac{|v(x) - v(y)|^p}{|x-y|^{1+tp}}\, dxdy. \qquad \square \tag{15.103}$$

We now turn to extensions of *BMO* and *VMO* maps. Here is a first result, in the spirit of Jones [207].

Lemma 15.28. *Let Ω be a smooth bounded domain and let $\widetilde{\Omega}$ be a small neighborhood of $\overline{\Omega}$. Then the extension $u \mapsto \widetilde{u}$ by reflection near $\partial\Omega$ maps $BMO(\Omega)$ into $BMO(\widetilde{\Omega})$ and $VMO(\Omega; \mathbb{R}^M)$ into $VMO(\widetilde{\Omega})$.*

Moreover, there exists a constant $C > 0$ such that

$$M_t(\widetilde{u}, \widetilde{\Omega}) \le C M_t(u, \Omega),\ \forall\, t > 0, \ \text{and } |\widetilde{u}|_{BMO(\widetilde{\Omega})} \le C|u|_{BMO(\Omega)}.$$

[Here, u can be scalar or $\mathbb{R}^M$-valued.]

Proof. We first note that it suffices to establish the seemingly weaker estimate $M_t(\widetilde{u}, \widetilde{\Omega}) \le C M_{Kt}(u, \Omega)$ for some K, since we always have

$$M_{Kt}(u, \Omega) \le C_K M_t(u, \Omega)$$

(see, for example, Brezis and Nirenberg [104]).

When $\widetilde{\Omega}$ is a small neighborhood of $\overline{\Omega}$, the reflection across $\partial\Omega$ is a smooth diffeomorphism of $\overline{\widetilde{\Omega}} \setminus \Omega$ onto its image (contained in $\overline{\Omega}$). Fix a ball B_0 such that $\overline{B_0} \subset \Omega$. Since the extension of a constant C is C, it suffices to consider only maps u such that $\int_{B_0} u = 0$, and in that case we have $\|u\|_{L^1(\Omega)} \le C|u|_{BMO(\Omega)}$ (see (15.66)). Clearly, for balls $B \subset \widetilde{\Omega}$ of sufficiently large radius, $r \ge r_0$, we have, by the above

$$\fint_B \fint_B |\widetilde{u}(x) - \widetilde{u}(y)| \, dxdy \le C(r_0)\|u\|_{L^1(\Omega)} \le C(r_0)|u|_{BMO(\Omega)}.$$

Therefore, with the notation in (5.27), it suffices to prove that there exist constants $K, C > 0$ such that $M_t(\widetilde{u}, \widetilde{\Omega}) \le CM_{Kt}(u, \Omega)$ provided t is sufficiently small, $t < r_0$. Let $B := B_r(z)$, with $x \in \widetilde{\Omega}$. The inequality

$$\fint_B \fint_B |\widetilde{u}(x) - \widetilde{u}(y)| \, dxdy \le CM_{Kt}(u) \tag{15.104}$$

holds clearly (using a change of variables) provided

1. $r \le t$.
2. C and K are sufficiently large.
3. $\operatorname{dist}(z, \partial\Omega) \ge r/\varepsilon$ for some sufficiently small ε (independent of z, r and t).

It therefore remains to investigate the validity of (15.104) when $r \le t$, t is sufficiently small, and when, in addition, $\operatorname{dist}(z, \partial\Omega) \le r/\varepsilon$. In that case, we may flatten $\partial\Omega$ (see Lemma 15.22 above) and replace the Euclidean norm with the sup norm (see the proof of Lemma 15.21). After all these reductions, it suffices to prove the following. If $u : \mathbb{R}^N_+ \to \mathbb{R}^M$, and if $\widetilde{u}$ is its extension by reflection to $\mathbb{R}^N$, then (for balls corresponding to the sup norm) we have

$$M_t(\widetilde{u}, \mathbb{R}^N) \le CM_t(u, \mathbb{R}^N_+). \tag{15.105}$$

Let $\widetilde{Q}$ be a ball in $\mathbb{R}^N$. In order to prove (15.105) (with $C = 4$), it suffices to find another ball $Q \subset \mathbb{R}^N_+$, of the same size as $\widetilde{Q}$, such that

$$\fint_{\widetilde{Q}} \fint_{\widetilde{Q}} |\widetilde{u}(x) - \widetilde{u}(y)| \, dxdy \le 4 \fint_Q \fint_Q |u(x) - u(y)| \, dxdy. \tag{15.106}$$

If $\widetilde{Q} \subset \mathbb{R}^N_+$, we may take $Q := \widetilde{Q}$. If $\widetilde{Q} \subset \mathbb{R}^N_-$, we may take $Q := -\widetilde{Q}$. Finally, if $\widetilde{Q} = (a, b) \times Q'$ intersects $\partial\mathbb{R}^N_-$, let $Q := (0, b - a) \times Q'$. Then, clearly, (15.106) holds for this Q. □

The next result in this section is a consequence of both the statement and the proof of Lemma 15.28.

Corollary 15.9. *Let $0 < s < 1$ and $1 \le p < \infty$. Let B be a ball in $\mathbb{R}^N$ and let B^* be the double of B. Then there exists a linear extension operator $T = T_B$ that maps $W^{s,p}(B)$ into $W^{s,p}_c(B^*)$, $BMO\,(B)$ into $BMO_c(B^*)$, and $VMO\,(B)$ into $VMO_c(B^*)$, and such that $\|Tu\|_{W^{s,p}(\mathbb{R}^N)} \le C\|u\|_{W^{s,p}(B)}$, $\|Tu\|_{BMO\,(\mathbb{R}^N)} \le C\|u\|_{BMO\,(B)}$ for some C independent of B.*

Proof. By scaling, it suffices to let B be the unit ball. Let $\zeta \in C^\infty_c(B_{4/3}(0); [0, 1])$ satisfy $\zeta = 1$ in $\overline{B}$. Let $\widetilde{u}$ be the extension of u by reflection and set $Tu := \zeta\,\widetilde{u}$. The $W^{s,p}$ part of the statement being clear (via Lemma 15.27), we investigate the *BMO*/*VMO* part.

If $C = B_r(x)$ is a ball of radius $r \leq 1/6$, then either $|x| \geq 3/2$ and then $Tu = 0$ on C, or $|x| < 3/2$ and then

$$\fint_C \fint_C |Tu(x) - Tu(y)|\, dxdy \leq \fint_C \fint_C |\widetilde{u}(x) - \widetilde{u}(y)|\, dxdy + \fint_C \fint_C |\widetilde{u}(x)||\zeta(x) - \zeta(y)|\, dxdy$$
$$\lesssim M_r(\widetilde{u}, B_{5/3}(0)) + r \fint_C |\widetilde{u}|$$
$$\lesssim M_r(\widetilde{u}, B_{5/3}(0)) + \|\widetilde{u}\|_{L^N(C)},$$

and thus

$$M_r(Tu, \mathbb{R}^N) \lesssim M_r(\widetilde{u}, B_{5/3}(0)) + \sup_{\substack{C \text{ ball of radius } \leq r \\ \overline{C} \subset B_{5/3}(0)}} \|\widetilde{u}\|_{L^N(C)}. \tag{15.107}$$

The conclusion of Corollary 15.9 follows by combining (15.107) with Lemma 15.28 and with the embedding $BMO \hookrightarrow L^N$ (see John and Nirenberg [206]). □

Remark 15.13. With more work, it is possible to remove the restriction $s < 1$ and to obtain the conclusion of Lemma 15.9 for any $s > 0$. We sketch the argument when $\Omega = \mathbb{R}^N_+$. In that case, we define a Luxemburg-type extension

$$\widetilde{u}(x', x_N) := \begin{cases} u(x', x_N), & \text{if } x_N > 0 \\ \sum_{j=1}^k \alpha_j u(x', -jx_N), & \text{if } x_N > 0 \end{cases},$$

where $k > s$ and the α_j's satisfy the moment conditions $\sum_{j=1}^k (-j)^\ell \alpha_j = 1$, $\ell = 0, \ldots, k-1$. It is well-known that this extension operator maps $W^{s,p}$ into $W^{s,p}$ (see, for example, Adams [4]), and, by the proof of (15.105), it also maps BMO into BMO and VMO into VMO.

It is likely (but we did not investigate this) that, in a Lipschitz domain, the universal extension operator T devised by Stein [330, Section VI.3] maps $BMO\,(\Omega)$ into $BMO\,(\mathbb{R}^N)$. Recall that this operator does not depend on s and p and maps $W^{s,p}(\Omega)$ into $W^{s,p}(\mathbb{R}^N)$ for every s and p.

We end this section with some extension results used in Section 14.5.

Lemma 15.29. *Let $s > 0$ and $1 \leq p < \infty$ be such that $sp < 2$. Then, every map $u \in W^{s,p}(\Omega; \mathbb{S}^1)$ has an extension $w \in W^{s,p}_{loc}(\mathbb{R}^N; \mathbb{S}^1)$. In addition, we may assume that w is constant outside some large ball B.*

Proof. By Lemma 5.5, it suffices to consider the case where $1 \leq sp < 2$.

Step 1. Proof when $1 < sp < 2$ or [$s = 1$ and $p = 1$]
In this case, standard trace theory applies: u has a trace $v \in W^{\sigma,p}(\partial\Omega; \mathbb{S}^1)$, with $\sigma := s - 1/p$. Moreover, we can glue two $W^{s,p}$ maps as explained by the following standard result, whose proof is omitted.

Lemma 15.30. *Let $s > 0$ and $1 \le p < \infty$ be such that $1/p < s < 1 + 1/p$. Let Ω_j, $j = 1, 2$, be two disjoint Lipschitz domains and let $\Gamma \subset \overline{\Omega_1} \cap \overline{\Omega_2}$ be a hypersurface. Assume that $\Omega := \Omega_1 \cup \Omega_2 \cup \Gamma$ is a Lipschitz domain.*

Let $f_j \in W^{s,p}(\Omega_j)$, $j = 1, 2$, and set $f := \begin{cases} f_1, & \text{in } \Omega_1 \\ f_2, & \text{in } \Omega_2 \end{cases}$. If $\operatorname{tr} f_1 = \operatorname{tr} f_2$ on Γ, then $f \in W^{s,p}(\Omega)$.

Same conclusion when $s = 1$ and $p = 1$.

Step 1 continued. Assume first that $(s, p) \neq (1, 1)$. Consider some ψ such that $v = e^{\imath\psi}$ and $\psi \in \bigcap_{0<t\le 1} W^{t\sigma,p/t}(\partial\Omega; \mathbb{R})$. [Existence of ψ follows from Theorem 5.10, whose proof does not require the simple connectedness.] Extend ψ by some map $\varphi \in \bigcap_{0<t\le 1} W^{t(\sigma+1/p),p/t}(\mathbb{R}^N \setminus \overline{\Omega}; \mathbb{R})$ such that $\varphi \equiv 0$ outside some large ball B. Then the map $y := e^{\imath\varphi}$ belongs to $W^{s,p}_{loc}(\mathbb{R}^N \setminus \overline{\Omega}; \mathbb{S}^1)$ (see the proof of Theorem 11.1) and, clearly, $\operatorname{tr} y = v$ on $\partial\Omega$. By Lemma 15.30, the map $w := \begin{cases} u, & \text{in } \Omega \\ y, & \text{in } \mathbb{R}^N \setminus \overline{\Omega} \end{cases}$ has all the required properties.

Assume next that $s = 1$ and $p = 1$. We argue as above, with $\psi \in L^1$ and with φ a $W^{1,1}$ extension of ψ vanishing outside B.

Step 2. Proof when $1 < p < \infty$ and $s = 1/p$
In this case, trace theory fails, and we have to devise a separate argument. We first extend u by reflections across $\partial\Omega$ and obtain a map, still denoted by u, in $W^{1/p,p}(U; \mathbb{S}^1)$; here, U is a smooth domain such that $\overline{\Omega} \subset U$ (Lemma 15.27). Fix some small $\delta > 0$ and let $0 < \varepsilon < \delta$. Let

$$V_\varepsilon := \{x \in \mathbb{R}^N \setminus \overline{\Omega};\ \operatorname{dist}(x, \partial\Omega) < \varepsilon\} \text{ and } \Sigma_\varepsilon := \{x \in \mathbb{R}^N \setminus \overline{\Omega};\ \operatorname{dist}(x, \partial\Omega) = \varepsilon\}.$$

Let $\Pi_\varepsilon(x)$ denote the nearest point projection on Σ_ε of a point $x \in V_\delta$. Then, $V_\varepsilon \subset U$ and Π_ε is smooth.

We next use the following properties of $W^{1/p,p}$ (we refer the reader to Brezis and Mironescu [84, Appendix B] for their proofs).

First, for a.e. $0 < \varepsilon < \delta$, we have $u_{|\Sigma_\varepsilon} \in W^{1/p,p}(\Sigma_\varepsilon)$. In addition, if we set

$$u_\varepsilon(x) := \begin{cases} u(x), & \text{if } x \in \overline{\Omega} \cup V_\varepsilon \\ u(\Pi(x)), & \text{if } x \in V_\delta \setminus V_\varepsilon \end{cases},$$

then for a.e. $0 < \varepsilon < \delta$ we have $u_\varepsilon \in W^{1/p,p}(\overline{\Omega} \cup V_\delta; \mathbb{S}^1)$ [84, Lemma B5]. We refer to any ε having these two properties as a "good ε."

Next, consider a good ε. Let $W_\varepsilon := \mathbb{R}^N \setminus (\Omega \cup \overline{V_\varepsilon})$, which is a smooth domain with boundary Σ_ε. Let $1/p < t < 1$. If $y \in W^{t,p}_{loc}(W_\varepsilon)$ is such that $\operatorname{tr}_{\Sigma_\varepsilon} y = u_{|\Sigma_\varepsilon}$, then the glued map $w := \begin{cases} u_\varepsilon, & \text{in } \overline{\Omega} \cup \overline{V_\varepsilon} \\ y, & \text{in } W_\varepsilon \end{cases}$ belongs to $W^{1/p,p}_{loc}(\mathbb{R}^N)$ [84, Lemma 2].

In order to complete Step 2, it remains to construct some $y \in W^{t,p}_{loc}(W_\varepsilon; \mathbb{S}^1)$ such that $y \equiv 1$ outside some large ball B, for some t such that $1/p < t < 1$.

Let t satisfy $1/p < t < 1$ and, in addition, $t < 2/p$. Since $u_{|\Sigma_\varepsilon} \in W^{1/p,p}$, we may write $u_{|\Sigma_\varepsilon} = e^{\imath\psi}$, with $\psi \in W^{t-1/p,p}$. [Here, we use the fact that $(t - 1/p)\,p < 1$, and thus $W^{t-1/p,p}(\omega; \mathbb{S}^1)$ has the lifting property, even when ω is not simply connected.] We extend ψ by some function $\varphi \in W^{t,p}$ such that $\varphi \equiv 0$ outside some large ball B. Then, $y := e^{\imath\varphi}$ has all the required properties. □

15.7 Fine theory of BV maps

We briefly sketch here some aspects the theory of BV maps, as developed by De Giorgi, Federer, and Volpert from 1954 to 1967. We follow essentially Ambrosio, Fusco, and Pallara [13, Chapter 3]; the interested reader will find there proofs of the assertions presented in this appendix.

We start with interior regularity properties, valid without any smoothness assumption on Ω. Let $\varphi \in BV(\Omega; \mathbb{R})$. A point $x \in \Omega$ is a point of approximate continuity if there exists a number $a \in \mathbb{R}$ such that

$$\lim_{r\to 0} \fint_{B_r(x)} |\varphi(y) - a|\, dy = 0.$$

This (unique) a is the approximate limit of φ at x and is denoted by $\widetilde{\varphi}(x)$. Let C_φ be the set of points of approximate continuity of φ. By the Lebesgue differentiability theorem, if φ is merely in L^1_{loc}, then C_φ is of full measure: the complement S_φ of C_φ (the set of points of approximate discontinuity) has zero Lebesgue measure. When φ belongs to BV, this conclusion can be strengthened to: S_φ is countably $\mathscr{H}^{N-1}$-rectifiable. That is, we may write, modulo an $\mathscr{H}^{N-1}$-null set, $S_\varphi = \cup_{i\in\mathbb{N}} S_i$; here, the S_i's are disjoint Borel subsets of C^1 hypersurfaces Σ_i (see Federer [147, 3.1.16]).

For further use, we note that $\mathscr{H}^{N-1}$-null sets are irrelevant in the description of $D\varphi$ (with $\varphi \in BV$), since $D\varphi$ does not charge $\mathscr{H}^{N-1}$-null sets: $|D\varphi|(B) = 0$ if B is an $\mathscr{H}^{N-1}$-null Borel set.

At each $x \in S_i$, let $\nu_i = \nu_i(x)$ be a normal vector at Σ_i. We further assume that $x \mapsto \nu_i(x)$ is Borel measurable. For $\mathscr{H}^{N-1}$-a.e. $x \in S_\varphi$, there are numbers $b, c \in \mathbb{R}$ such that

$$\lim_{r\to 0} \fint_{B_r(x)\cap\{(y-x)\cdot\nu_i(x)>0\}} |\varphi(y) - b|\, dy = 0$$

and

$$\lim_{r\to 0} \fint_{B_r(x)\cap\{(y-x)\cdot\nu_i(x)<0\}} |\varphi(y) - c|\, dy = 0.$$

We let $\varphi^+(x) := b$ and $\varphi^-(x) := c$; these are the one-sided approximate limits of φ at x and are uniquely determined by the choice of $\nu_i(x)$.

We define

$J_\varphi := \{x \in \cup_i S_i;\ \varphi \text{ has one-sided approximate limits at } x\}.$

[Our definition is slightly different from that in [13], but the two definitions coincide up to an $\mathcal{H}^{N-1}$-null Borel set.]

According to the Lebesgue decomposition theorem, we may write $D\varphi = D^a\varphi + D^s\varphi$, where $D^a\varphi$ is absolutely continuous with respect to the Lebesgue measure $\mathcal{L}^N$, and $D^s\varphi$ and $\mathcal{L}^N$ are mutually orthogonal. We next let $D^c\varphi := D^s\varphi \llcorner C_\varphi$ and $D^j\varphi := D^s\varphi \llcorner S_\varphi$. We have $D^j\varphi = D^s\varphi \llcorner J_\varphi$, since $S_\varphi = J_\varphi$ modulo an $\mathcal{H}^{N-1}$-null Borel set. $D^c\varphi$ and $D^j\varphi$ are respectively the Cantor and the jump part of $D\varphi$.

We next describe in more detail the absolutely continuous and the jump part of $D\varphi$.

According to the Radon–Nikodym theorem, we may write $D^a\varphi = \nabla\varphi\mathcal{L}^N$, an equality that we will write in the simpler form $D^a\varphi = \nabla\varphi$. Here, $\nabla\varphi$ is the density of $D^a\varphi$. When $\varphi \in W^{1,1}$, $\nabla\varphi$ is precisely the distributional gradient of φ. For a general $\varphi \in BV$, $\nabla\varphi$ is , for $\mathcal{H}^{N-1}$-a.e. $x \in C_\varphi$, the approximate gradient of φ at x, i.e.,

$$\lim_{r\to 0} \fint_{B_r(x)} \frac{|\varphi(y) - \tilde{\varphi}(x) - \nabla\varphi(x)\cdot(y-x)|}{r}\, dy = 0.$$

On the other hand, the jump part $D^j\varphi$ is given by

$$D^j\varphi = \sum_i (\varphi^+ - \varphi^-)\nu_i \mathcal{H}^{N-1} \llcorner S_i.$$

We now explain how to read the jumps of φ in the case where $D^j\varphi$ is not represented via a disjoint union of S_i's. More specifically, assume that $\varphi \in BV$ is such that

$$D^j\varphi = \sum_i \alpha_i \nu_i \mathcal{H}^{N-1} \llcorner S_i. \tag{15.108}$$

Here, each S_i is a Borel subset of a C^1 hypersurface Σ_i, ν_i is a Borel measurable normal vector to Σ_i, and α_i is a Borel measurable function on S_i. We assume satisfied the finite mass condition

$$\sum_i \int_{S_i} |\alpha_i| < \infty. \tag{15.109}$$

However, we do not assume that the S_i's are mutually disjoint. If $x \in S_i \cap S_\ell$ and the vectors $\nu_i(x)$ and $\nu_\ell(x)$ are not collinear, then Σ_i and Σ_ℓ are transversal at x, and thus the intersection of Σ_i and Σ_ℓ is locally $(N-2)$-dimensional, and hence $\mathcal{H}^{N-1}$-negligible; therefore, if $x \in S_i \cap S_\ell$, then $\nu_i(x) = \pm\nu_\ell(x)$ "most of the time." This leads immediately to the following result (see, for example, Giaquinta, Modica, and Souček [156, Section 2.1.4] for more details).

Lemma 15.31. *Assume that* (15.108) *and* (15.109) *hold. Let* $S' := \cup_i S_i$ *and set, for* $x \in S'$,

$$\nu(x) := \nu_i(x) \quad \text{if } x \in S_i \setminus (S_1 \cup \ldots \cup S_{i-1}),$$

$$\alpha(x) := \sum_j \chi_{S_j}(x)\alpha_j(x)\nu_j(x) \cdot \nu(x) = \sum_j \chi_{S_j}(x)\alpha_j(x)\nu_j(x) \cdot \nu_i(x).$$

Define $S := \{x; \alpha(x) \neq 0\}$. *Then:*

1. ν *defines a Borel orientation of the rectifiable set* S
2. *We have* $S = S_\varphi$ *up to a* $\mathscr{H}^{N-1}$*-negligible set*
3. *We have,* $\mathscr{H}^{N-1}$*-a.e. on* S, $\varphi^+ - \varphi^- = \alpha$.

In the special case where the α_i*'s are integer-valued, the jumps of* φ *are integer-valued* $\mathscr{H}^{N-1}$*-a.e. on* S_φ.

We next present the

Theorem 15.10. (Chain rule in BV). *Let* $f \in C^1(\mathbb{R})$ *be a Lipschitz map. If* Ω *is unbounded, then we also assume that* $f(0) = 0$. *If* $\varphi \in BV(\Omega)$, *then* $f \circ \varphi \in BV$. *In addition, we have*

$$D^a(f \circ \varphi) = f'(\widetilde{\varphi})\nabla\varphi = f'(\varphi)\nabla\varphi, \; D^c(f \circ \varphi) = f'(\widetilde{\varphi})D^c\varphi$$

and

$$D^j(f \circ \varphi) = \sum_i (f(\varphi^+) - f(\varphi^-))\nu_i \mathscr{H}^{N-1} \llcorner S_i.$$

We continue this section with the proof of a standard result used in Section 1.4, and with some of its variants.

Lemma 15.32. *Let* $F \in L^1(\Omega; \mathbb{R}^N)$ *and* $\varphi \in BV(\Omega)$. *Then there exists a sequence* $(\varphi_n) \subset C^\infty(\Omega) \cap W^{1,1}(\Omega)$ *such that*

$$\varphi_n \to \varphi \text{ in } L^1(\Omega) \text{ and } \lim_{n\to\infty} \int_\Omega |F - \nabla\varphi_n| = \|F - D\varphi\|_{\mathscr{M}}. \tag{15.110}$$

If, in addition, Ω *is smooth and bounded, we may choose* $(\varphi_n) \subset C^\infty(\overline{\Omega})$.

Remark 15.14. We warn the reader that the conclusion of the lemma fails when F is a measure. Here is an example. Let φ be any function in $BV \setminus W^{1,1}$, for example, the characteristic function of a ball $B \subset \Omega$. Let $F := D\varphi$. Then there exists no sequence $(\varphi_n) \subset W^{1,1}(\Omega)$ such that (15.110) holds. For otherwise we would have $\varphi_n \to \varphi$ in BV, and thus $\varphi \in W^{1,1}$, a contradiction.

Proof of Lemma 15.32. We first recall a classical fact. If $(\varphi_n) \subset BV$, $\varphi \in BV$ and $\varphi_n \to \varphi$ in L^1, then

$$\liminf_{n\to\infty} \|F - D\varphi_n\|_{\mathscr{M}} \geq \|F - D\varphi\|_{\mathscr{M}}, \; \forall F \in L^1(\Omega; \mathbb{R}^N). \tag{15.111}$$

Therefore, in order to complete the proof of the lemma, it suffices to construct, for each $\delta > 0$, a map $\psi = \psi_\delta \in C^\infty(\Omega) \cap W^{1,1}(\Omega)$ such that

$$\|\varphi - \psi\|_{L^1(\Omega)} < \delta \text{ and } \int_\Omega |F - \nabla\psi| \leq \|F - D\varphi\|_{\mathscr{M}} + \delta.$$

For this purpose, we consider an exhaustion $\Omega = \cup_{j\geq 0}\Omega_j$ of Ω, with each Ω_j open and $\overline{\Omega_j} \Subset \Omega_{j+1}$, $\forall j \geq 0$. Let $U_0 := \Omega_1$ and, for $j \geq 1$, $U_j := \Omega_{j+1} \setminus \overline{\Omega_{j-1}}$. Thus, $\Omega = \cup_{j\geq 0} U_j$ and each $x \in \Omega$ belongs to at most two U_j's. Let (ξ_j) be a partition of unity subordinate to the covering (U_j) of Ω. Let (ρ_ε) be a family of standard mollifiers. With the sequence (ε_j) to be fixed later, we let $\psi := \sum_{j\geq 0} \psi_j$, where $\psi_j := (\varphi\xi_j) * \rho_{\varepsilon_j}$. We claim that, for an appropriate choice of ε_j, ψ has the required properties.

To start with we choose, for every $j \geq 0$, an $\overline{\varepsilon}_j > 0$ such that

$$\overline{\varepsilon}_j < \operatorname{dist}(\operatorname{supp}\xi_j, \partial U_j), \ \forall j \geq 0. \tag{15.112}$$

Next we choose, for every $j \geq 0$, $\widehat{\varepsilon}_j > 0$ such that

$$\|(\varphi\xi_j) * \rho_\varepsilon - \varphi\xi_j\|_{L^1} < \frac{\delta}{2^{j+1}}, \ \forall j \geq 0, \ \forall \varepsilon < \widehat{\varepsilon}_j, \tag{15.113}$$

$$\|(\varphi\nabla\xi_j) * \rho_\varepsilon - \varphi\nabla\xi_j\|_{L^1} < \frac{\delta}{2^{j+2}}, \ \forall j \geq 0, \ \forall \varepsilon < \widehat{\varepsilon}_j, \tag{15.114}$$

and

$$\|(\xi_j F) * \rho_\varepsilon - \xi_j F\|_{L^1} < \frac{\delta}{2^{j+2}}, \ \forall j \geq 0, \ \forall \varepsilon < \widehat{\varepsilon}_j. \tag{15.115}$$

Finally, we choose $\varepsilon_j < \min\{\overline{\varepsilon}_j, \widehat{\varepsilon}_j\}$. From (15.113), we have

$$\|\psi - \varphi\|_{L^1} \leq \sum_{j\geq 0} \|\psi_j - \varphi\xi_j\|_{L^1} = \sum_{j\geq 0} \|(\varphi\xi_j) * \rho_{\varepsilon_j} - \varphi\xi_j\|_{L^1} < \delta.$$

We now prove that ψ belongs to $C^\infty(\Omega) \cap W^{1,1}(\Omega)$ and satisfies

$$\int_\Omega |F - \nabla\psi| \leq \|F - D\varphi\|_{\mathscr{M}} + \delta. \tag{15.116}$$

Fix an open set ω such that $\overline{\omega} \Subset \Omega$. Then, $\omega \subset \Omega_{k_0}$ for some k_0, and thus $\omega \cap U_j = \emptyset$, $\forall j \geq k_0 + 1$. From (15.112), we have $\psi_j \in C_c^\infty(U_j)$, $\forall j \geq 0$, and thus $\psi_j = 0$ on ω, $\forall j \geq k_0 + 1$. Consequently, $\psi = \sum_{j=0}^{k_0} \psi_j$ on ω and $\psi \in C^\infty(\Omega)$.

On ω, we write

$$
\begin{aligned}
\nabla\psi &= \sum_{j=0}^{k_0} \nabla\psi_j = \sum_{j=0}^{k_0} \nabla((\varphi\xi_j) * \rho_{\varepsilon_j}) = \sum_{j=0}^{k_0} (D(\varphi\xi_j)) * \rho_{\varepsilon_j} \\
&= \sum_{j=0}^{k_0} (\varphi\nabla\xi_j) * \rho_{\varepsilon_j} + \sum_{j=0}^{k_0} (\xi_j D\varphi) * \rho_{\varepsilon_j} \\
&= \sum_{j=0}^{k_0} [(\varphi\nabla\xi_j) * \rho_{\varepsilon_j} - \varphi\nabla\xi_j] + \sum_{j=0}^{k_0} (\xi_j D\varphi) * \rho_{\varepsilon_j},
\end{aligned}
$$

since $\sum_{j=0}^{k_0} \xi_j = 1$ on ω, so that $\sum_{j=0}^{k_0} \nabla\xi_j = 0$ on ω.

Clearly, on ω we have

$$
F = -\sum_{j=0}^{k_0} [(\xi_j F) * \rho_{\varepsilon_j} - \xi_j F] + \sum_{j=0}^{k_0} (\xi_j F) * \rho_{\varepsilon_j}.
$$

It follows from (15.114) and (15.115) that

$$
\begin{aligned}
\int_\omega |F - \nabla\psi| &\le \sum_{j=0}^{k_0} \int_\omega |(\xi_j (F - D\varphi)) * \rho_{\varepsilon_j}| + \delta \\
&\le \sum_{j=0}^{k_0} \|\xi_j (F - D\varphi)\|_{\mathcal{M}(\omega)} + \delta \\
&= \|F - D\varphi\|_{\mathcal{M}(\omega)} + \delta \le \|F - D\varphi\|_{\mathcal{M}(\Omega)} + \delta.
\end{aligned}
\tag{15.117}
$$

Here, we have used the fact that, if ν is an $\mathbb{R}^k$-valued (finite) Borel measure on an open set $\omega \subset \mathbb{R}^N$ and if ξ_j are continuous *non-negative* functions on ω, then $|\xi_j \nu| = \xi_j\,|\nu|$ in the sense of measures, so that

$$
\sum_j \|\xi_j \nu\|_{\mathcal{M}(\omega)} = \sum_j \|\xi_j |\nu|\|_{\mathcal{M}(\omega)} = \left\| \left(\sum_j \xi_j\right) |\nu| \right\|_{\mathcal{M}(\omega)} = \left\| \left(\sum_j \xi_j\right) \nu \right\|_{\mathcal{M}(\omega)}.
$$

Since (15.117) holds for every $\omega \Subset \Omega$, we conclude that $\psi \in W^{1,1}(\Omega)$ and that (15.116) holds.

The last assertion of the lemma follows easily from the fact that $C^\infty(\overline{\Omega})$ is dense in $W^{1,1}(\Omega)$ when Ω is smooth. □

We next present a variant of Lemma 15.32 involving forms. This result was used in Section 1.9.2.

Lemma 15.33. *Let $\Omega \subset \mathbb{R}^N$ be a smooth bounded contractible domain. Let $1 \le k \le N-1$. Let $F \in L^1(\Omega; \Lambda^k)$, and let $\nu \in \mathcal{M}(\Omega; \Lambda^k)$ be such that $d\nu = 0$. Then there exists a sequence $(\varphi_j) \subset C^\infty(\overline{\Omega}; \Lambda^{k-1})$ such that*

$$\lim_{j\to\infty}\int_\Omega |F - d\varphi_j| = \|F - \nu\|_{\mathscr{M}}. \tag{15.118}$$

Proof. Define, for $\varepsilon > 0$,

$$\Omega_\varepsilon := \{x \in \Omega;\ \mathrm{dist}(x, \partial\Omega) > \varepsilon\},\ \ \Omega_{-\varepsilon} := \{x \in \mathbb{R}^N;\ \mathrm{dist}(x, \Omega) < \varepsilon\}.$$

Let (ρ_ε) be a family of standard mollifiers, and set $\nu_\varepsilon := \nu * \rho_\varepsilon$, defined in $\overline{\Omega_\varepsilon}$. Then,

$$d\nu_\varepsilon = (d\nu) * \rho_\varepsilon = 0 \text{ in } \Omega_\varepsilon. \tag{15.119}$$

For small ε, Ω_ε is diffeomorphic to Ω, and thus contractible, as Ω. This fact combined with (15.119) allows us to apply Poincaré's lemma and obtain the existence of some $\varphi_\varepsilon \in C^\infty(\Omega_\varepsilon;\ \Lambda^{k-1})$ such that $d\varphi_\varepsilon = \nu_\varepsilon$ in Ω_ε. Consider now (in the spirit of Appendix 15.6), for small $\delta > 0$ and for $0 < \varepsilon < \delta/2$, a map $\Phi_{\delta,\varepsilon} : \Omega_{-\delta} \to \Omega_\varepsilon$ such that: $\Phi_{\delta,\varepsilon}$ is a C^∞-diffeomorphism, $\Phi_{\delta,\varepsilon} = \mathrm{Id}$ on Ω_δ, and

$$\|\Phi_{\delta,\varepsilon}\|_{C^1(\Omega_{-\delta})} \le C,\ \ \|\Phi_{\delta,\varepsilon}^{-1}\|_{C^1(\Omega_\varepsilon)} \le C, \tag{15.120}$$

with C independent of δ and ε.

Consider the form $\Phi_{\delta,\varepsilon}^\sharp\varphi_\varepsilon \in C^\infty(\Omega_{-\delta};\ \Lambda^{k-1})$, and let $\varphi_{\delta,\varepsilon} \in C^\infty(\overline{\Omega};\ \Lambda^{k-1})$ be its restriction to $\overline{\Omega}$. [Here, $f^\sharp\omega$ stands for the pullback of the form ω by the map f.] We will prove that, for every sequence $\delta_j \to 0$ and an appropriate sequence $\varepsilon_j \to 0$, the forms $\varphi_j := \varphi_{\delta_j,\varepsilon_j}$ satisfy (15.118). We start by collecting some straightforward properties of $\varphi_{\delta,\varepsilon}$. On the one hand, we have

$$d\varphi_{\delta,\varepsilon} = \Phi_{\delta,\varepsilon}^\sharp\nu_\varepsilon \text{ in } \Omega. \tag{15.121}$$

In particular, (15.121) implies that

$$d\varphi_{\delta,\varepsilon} = \nu_\varepsilon = \nu * \rho_\varepsilon \text{ in } \Omega_\delta \tag{15.122}$$

and (in conjunction with (15.120)) that

$$\|d\varphi_{\delta,\varepsilon}\|_{L^1(\Omega\setminus\Omega_\delta)} \le C\|\nu_\varepsilon\|_{L^1(\Omega_\varepsilon\setminus\Omega_\delta)} \le C|\nu|(\Omega \setminus \overline{\Omega_{\delta+\varepsilon}}). \tag{15.123}$$

The last inequality in (15.123) follows from the straightforward fact that, for every open set $\omega \subset \Omega_\varepsilon$, we have

$$\|\nu_\varepsilon\|_{L^1(\omega)} \le |\nu|(\{x;\ \mathrm{dist}(x, \omega) < \varepsilon\}).$$

Let

$$A(\delta, \varepsilon) := \int_{\Omega_\delta} |F - F * \rho_\varepsilon| + \int_{\Omega\setminus\Omega_\delta} |F| + \int_{\Omega\setminus\Omega_\delta} |d\varphi_{\delta,\varepsilon}|.$$

By (15.123), we have

$$\lim_{\delta\to 0}\limsup_{\varepsilon\to 0} A(\delta,\varepsilon)=0.$$

On the other hand, by the triangle inequality and (15.122) we clearly have

$$\int_{\Omega_\delta}|(F-\nu)*\rho_\varepsilon|-A(\delta,\varepsilon)\le\int_\Omega|F-d\varphi_{\delta,\varepsilon}|\le\int_{\Omega_\delta}|(F-\nu)*\rho_\varepsilon|+A(\delta,\varepsilon).$$

Therefore, it suffices to prove that, for every measure $\mu\in\mathscr{M}(\Omega;\Lambda^k)$, we have

$$\lim_{\delta\to 0}\liminf_{\varepsilon\to 0}\int_{\Omega_\delta}|\mu*\rho_\varepsilon|=\lim_{\delta\to 0}\limsup_{\varepsilon\to 0}\int_{\Omega_\delta}|\mu*\rho_\varepsilon|=|\mu|(\Omega). \tag{15.124}$$

In turn, (15.124) is a consequence of the following straightforward estimate (applied with $\omega:=\Omega_\delta$): if $\omega\subset\Omega$ is open and $\overline{\omega}\subset\Omega$, then

$$|\mu|(\omega)\le\liminf_{\varepsilon\to 0}\int_\omega|\mu*\rho_\varepsilon|\le\limsup_{\varepsilon\to 0}\int_\omega|\mu*\rho_\varepsilon|\le|\mu|(\overline{\omega}).\qquad\square$$

We now present a proof of the N-dimensional version of Lemma 3.5 used in Section 3.6.

Lemma 15.34. *Let M be an $(N-1)$-rectifiable current in $\Omega\subset\mathbb{R}^N$. Then there exists some $\psi\in BV(\Omega;\mathbb{R})$ such that $D^j\psi=-2\pi*M$ and $D^c\psi=0$.*

[Equivalently, there exists some $\psi\in SBV(\Omega;\mathbb{R})$ such that the jump part of $D\psi$ is $-2\pi*M$.]

Proof. We write $M=\sum_i S_i$, where each S_i is a Borel subset of an oriented C^1 hypersurface Σ_i in Ω and $\sum_i|S_i|<\infty$. Possibly after using a suitable partition of each S_i, we may assume that S_i is relatively compact in Σ_i and that Σ_i is the graph of a C^1 function. It suffices to construct, for each i, an L^1 function f_i and a BV function ψ_i such that $D\psi_i=f_i-2\pi*S_i$ and $\|f_i\|_{L^1}\le C\,|S_i|$. Assuming that ψ_i is as above, the function $\psi:=\sum_i\left(\psi_i-⨍_\Omega\psi_i\right)$ belongs to $BV(\Omega;\mathbb{R})$ and satisfies $D\psi=\sum_i f_i-2\pi*M$.

We have thus reduced Lemma 15.34 to the following. Let $\Sigma\subset\Omega$ be an oriented C^1 hypersurface that is a graph, and let $S\subset\Sigma$ be a relatively compact Borel set of positive $\mathscr{H}^{N-1}$ measure. Then there exist some $f\in L^1(\Omega)$ and $\varphi\in BV(\Omega;\mathbb{R})$ such that $D\varphi=f-2\pi*S=f-2\pi\nu\mathscr{H}^{N-1}\llcorner S$ and $\|f\|_{L^1}\le C\,|S|$ (with C independent of Σ and S). With no loss of generality, we may write $\Sigma=\{(x',\xi(x'));\ x'\in\omega\}$, with $\omega\subset\mathbb{R}^{N-1}$ an open set and $\xi\in C^1$. Set $U^+=\{(x',t);\ x'\in\omega,\ t>\xi(x')\}$ and $U^-=\{(x',t);\ x'\in\omega,\ t<\xi(x')\}$. Again, with no loss of generality, we may assume that Σ is oriented by the outward normal at U^+ on Σ.

In order to construct f, we rely on the following quantitative form of trace theory in $W^{1,1}$ (see, for example, Giusti [167, proof of Proposition 2.15, Theorem 2.16, Remark 2.17]; see also Mironescu [257] in the case of a half-space, and Ponce

[299, Section 15.3]). Given $g \in L^1_c(\Sigma)$ and $\varepsilon > 0$, there exist some compact $K \subset \Sigma \cup (U^+ \cap \Omega)$ and some $h \in W^{1,1}(U^+)$ such that $\mathrm{tr}_{|\Sigma}\, h = g$, $h = 0$ in $U^+ \setminus K$ and $\|\nabla h\|_{L^1(U^+)} \le \|g\|_{L^1(\Sigma)} + \varepsilon$. Applying this with $g := \chi_S$ and $\varepsilon := \mathscr{H}^{N-1}(S)$, we find some $h \in W^{1,1}(U^+)$ such that $\mathrm{tr}_{|\Sigma}\, h = \chi_S$ and $\|\nabla h\|_{L^1(U^+)} \le 2\mathscr{H}^{N-1}(S)$.

Now set $\psi := \begin{cases} 2\pi h, & \text{in } \Omega \cap U^+ \\ 0, & \text{in } \Omega \setminus U^+ \end{cases}$. Then (see [13, Corollary 3.89]), we have $\psi \in BV(\Omega; \mathbb{R})$, with

$$\nabla\psi = \begin{cases} \nabla h, & \text{in } \Omega \cap U^+ \\ 0, & \text{in } \Omega \setminus U^+ \end{cases}, \; D^c\psi = 0, \; D^j\psi = -2\pi\nu\mathscr{H}^{N-1} \llcorner S.$$

We find that ψ has all the required properties. □

We next present a proof of Lemma 8.1 used in the proof of Theorem 8.13 in Section 8.6.

Proof of Lemma 8.1. It clearly suffices to prove the result in the model case where $\Omega = (0, 1)^N$.

Step 1. Proof when $N = 1$

In this case, the conclusions of the lemma amount to $f \in BV \cap C$. We prove this property under the weaker assumption that $f \in VMO$. We may assume that g is given by its precise representative. Consider the finite set $A := \{x \in (0, 1);\ |g(x+) - g(x-)| \ge 1/2\}$.

We first prove that $g - f$ is locally constant on $(0, 1) \setminus A$. Let $x \in (0, 1) \setminus A$ and consider some interval $J = [x - 2\eta, x + 2\eta]$ such that $|g(y) - g(x)| \le 1/2, \forall y \in J$. Thus, $f(y) \in F_x$, for a.e. $y \in J$, where F_x is the closed set $2\pi Z + [g(x) - 1/2, g(x) + 1/2]$.

Let $f_\varepsilon(y) := \fint_{y-\varepsilon}^{y+\varepsilon} f$. By Lemma 15.19, we have

$$\lim_{\varepsilon\to 0}\ \sup_{x-\eta\le y\le x+\eta} \mathrm{dist}(f_\varepsilon(y), F_x) = 0.$$

Since the range of f_ε is connected, we find that, for each small ε, there exists some $k_\varepsilon \in \mathbb{Z}$ such that $f_\varepsilon(y) \in 2\pi k_\varepsilon + [g(x) - 1/2, g(x) + 1/2]$, $\forall y \in [x - \eta, x + \eta]$. Letting $\varepsilon \to 0$, we obtain the existence of some $k \in \mathbb{Z}$ such that $f(y) \in 2\pi k + [g(x) - 1/2, g(x) + 1/2]$, a.e. on $[x - \eta, x + \eta]$. We find that $g(y) - f(y) \in -2\pi k + [g(x - 1), g(x) + 1]$, a.e. on $[x - \eta, x + \eta]$, and thus $g - f = -2\pi k$ a.e. on $[x - \eta, x + \eta]$.

We next claim that g (and thus f) is continuous on $(0, 1) \setminus A$. Indeed, argue by contradiction and consider some discontinuity point x_0 of g in $(0, 1) \setminus A$. Then, g (and thus f) has different side limits $g(x_0\pm)$. This clearly implies that

$$\lim_{\varepsilon\to 0} \fint_{x-\varepsilon}^{x+\varepsilon} \fint_{x-\varepsilon}^{x+\varepsilon} |f(y) - f(z)|\, dydz = \frac{|g(x_0+) - g(x_0-)|}{2} > 0, \tag{15.125}$$

and thus $f \notin VMO$. In particular, we have $D^j f = 0$.

A similar argument shows that actually f is also continuous at the points of A. For further use, let us note that the continuity of f combined with the fact that $g - f$ is constant outside the jump set of g implies that

$$|f|_{BV} = \|f' + D^c f\|_{\mathscr{M}} = \|g' + D^c g\|_{\mathscr{M}} \le |g|_{BV}. \tag{15.126}$$

Step 2. Proof when $N \ge 2$

Set $f^1_{x'}(t) := f(t, x')$, with $x' = (x_2, \ldots, x_N) \in (0,1)^{N-1}$; $g^1_{x'}$ is defined similarly. By the analysis of the case $N = 1$, for a.e. $x' \in (0,1)^{N-1}$ we have $f^1_{x'} \in BV \cap C$ and (15.126) holds for $f^1_{x'}$ and $g^1_{x'}$. The same holds for all the partial functions $f^i_{x'} = f(x_1, \ldots, x_{i-1}, \cdot, x_{i+1}, \ldots, x_N)$, $1 \le i \le N-1$. Integrating (15.126) over x' and summing over i, we find (see [13, Theorem 3.103, Remark 3.104]) that

$$|f|_{BV} \simeq \sum_{i=1}^{N} \int_{(0,1)^{N-1}} |f^i_{x'}|_{BV}\, dx' \le \sum_{i=1}^{N} \int_{(0,1)^{N-1}} |g^i_{x'}|_{BV}\, dx' \simeq |g|_{BV},$$

and thus $f \in BV$.

Finally, we prove that, if $f \in W^{1/p,p} \cap BV$, then $D^j f = 0$. We thus have to prove that J_f is an $\mathscr{H}^{N-1}$-null set. After localization and local straightening of coordinates, this amounts to the following. Let S be a Borel set in $(0,1)^{N-1}$. Consider two measurable functions f^+, f^- on S and a function $f \in W^{1/p,p}((0,1)^{N-1} \times (-1,1))$ such that, for a.e. $x' \in S$, we have $\lim_{t \searrow 0} f(x', t) = f^+(x')$ and $\lim_{t \nearrow 0} f(x', t) = f^-(x')$. Then, $f^+ = f^-$ a.e. on S. This follows from the fact that, for every x' such that $f^+(x') \ne f^-(x')$ we have

$$\int_{-1}^{1} \int_{-1}^{1} \frac{|f(x', t) - f(x', \tau)|^p}{(t-\tau)^2}\, dt d\tau = \infty,$$

combined with the assumption $f \in W^{1/p,p}$ and Corollary 15.1. □

15.8 Description of (minimal) connections associated with (a, d)

In this appendix, we establish Theorems 2.7 and 2.8, stated without proof in Chapter 2.

15.8.1 Structure of general connections associated with (a, d)

In this section, we prove Theorem 2.7, restated below.

Theorem 15.11. *Every connection μ associated with (a, d) is a sum (possibly infinite) of the form*

$$\mu = \sum_i \nu_i \mathscr{H}^1 \llcorner \gamma_i, \tag{15.127}$$

where every γ_i is an oriented rectifiable Jordan arc or curve (in the sense of Section 2.7) contained in $\overline{\Omega}$ and

$$\sum_i \mathscr{H}^1(\gamma_i) < \infty.$$

Moreover, for each i:

1. *Either $\gamma_i \subset \Omega$ and γ_i connects some a_j to some a_m*
2. *Or $\gamma_i \subset \Omega \cup \{b\}$, with $b \in \partial\Omega$, and γ_i connects some a_j to b*
3. *Or $\gamma_i \subset \Omega \cup \{b, c\}$, with $b, c \in \partial\Omega$, $b \neq c$, and γ_i connects b to c*
4. *Or γ_i is a closed curve in $\overline{\Omega}$ such that $\gamma_i \cap \partial\Omega$ consists of at most one point.*

Remark 15.15. It will be clear from the proof that the decomposition (15.127) is valid for a connection associated with an arbitrary $T = \sum_j (\delta_{P_j} - \delta_{N_j}) \in \mathscr{E}$, provided we replace the collection (a_i) with the collection of points (P_j, N_j).

In the proof of Theorem 15.11, we use the following result, which we temporarily take for granted.

Theorem 15.12. *Let $E \subset \mathbb{R}^2$ be a bounded set of finite perimeter. Then there exists a (possibly infinite) sequence (γ_i) of rectifiable Jordan curves such that:*

1. $\sum_i \mathscr{H}^1(\gamma_i) = \operatorname{Per} E < \infty$
2. *We have*

$$D\chi_E = \sum_i \nu_i \mathscr{H}^1 \llcorner \gamma_i. \tag{15.128}$$

Here, Per E stands for the (measure theoretic) perimeter of the set E, i.e., $\operatorname{Per} E := \|D\chi_E\|_{\mathscr{M}}$.

In item *2* above, ν_i is the normal to γ_i, defined as in (2.134).

Proof of Theorem 15.11. *Step 1.* Reduction to the case where a is empty
We fix a connection μ_0 associated with (a, d), for example

$$\mu_0 = \sum_j d_j\, \nu_j \mathscr{H}^1 \llcorner [Q_j, a_j], \tag{15.129}$$

where $Q_j \in \partial\Omega$ satisfies $|Q_j - a_j| = \operatorname{dist}(a_j, \partial\Omega)$ and $[Q_j, a_j]$ is the oriented segment from Q_j to a_j (with normal ν_j). Set $\widetilde{\mu} := \mu - \mu_0$, which is a connection associated with the empty collections $a = \emptyset$ and $d = \emptyset$. Equivalently, $\widetilde{\mu}$ is a connection such that $\operatorname{curl} \widetilde{\mu} = 0$. We will prove that one may write $\widetilde{\mu}$ as in (15.127), where the arcs γ_i satisfy either *3* or *4*.

Step 2. From connections to integer-valued BV functions
Since $\operatorname{curl} \widetilde{\mu} = 0$, there exists some $\widetilde{\psi} \in BV(\Omega; \mathbb{R})$ such that $D\widetilde{\psi} = \widetilde{\mu}$. We write $\widetilde{\mu}$ as in Definition 2.1, i.e., $\widetilde{\mu} = \sum_i \nu_i \mathscr{H}^1 \llcorner (S_i \cap \Omega)$. We first note that $\nabla\widetilde{\psi} = 0$ and

$D^c\widetilde{\psi} = 0$. By Lemma 15.31, the jump set of $\widetilde{\psi}$ is $\mathscr{H}^1$-a.e. contained in $\cup_i(S_i \cap \Omega)$, and the jumps $\widetilde{\psi}^+ - \widetilde{\psi}^-$ of $\widetilde{\psi}$ are $\mathscr{H}^1$-a.e. integer-valued. Using Volpert's chain rule (see Appendix 15.7) and the specific form of $\widetilde{\psi}$ and $D\widetilde{\psi}$, we find that, with $\widetilde{S}_i := (S_i \setminus \cup_{j<i}S_j) \cap J_\psi$, we have

$$D[\cos(2\pi\widetilde{\psi})] = -2\pi\,\sin(2\pi\widetilde{\psi})\,\nabla\widetilde{\psi} - 2\pi\,\sin(2\pi\widetilde{\psi})\,D^c\widetilde{\psi} + \sum_i \frac{\cos(2\pi\widetilde{\psi}^+) - \cos(2\pi\widetilde{\psi}^-)}{\widetilde{\psi}^+ - \widetilde{\psi}^-}\nu_i\mathscr{H}^1 \llcorner (\widetilde{S}_i) = 0.$$

Similarly, we have $D[\sin(2\pi\widetilde{\psi})] = 0$. Therefore, $\widetilde{\psi}$ takes, a.e., values in $\mathbb{Z} + c$, for some constant $c \in \mathbb{R}$. We may thus assume that $\widetilde{\psi} \in BV(\Omega; \mathbb{Z})$ and $D\widetilde{\psi} = \widetilde{\mu}$. We next extend $\widetilde{\psi}$ to $\mathbb{R}^2$ by setting

$$\psi := \begin{cases} \widetilde{\psi}, & \text{in } \Omega \\ 0, & \text{in } \mathbb{R}^2 \setminus \Omega. \end{cases}.$$

Then, $\psi \in BV(\mathbb{R}^2; \mathbb{Z})$ (see, for example, Ambrosio, Fusco, and Pallara [13, Corollary 3.89]), and clearly

$$\widetilde{\mu} = D\widetilde{\psi} = D\psi_{|\Omega}. \tag{15.130}$$

Step 3. From integer-valued *BV* functions to finite perimeter sets
By the Fleming–Rishel coarea formula for *BV* functions (see, for example, [13, Theorem 3.40]), for a.e. $t \in \mathbb{R}$, the set $[\psi > t]$ has finite perimeter and we have

$$\|D\psi\|_{\mathscr{M}} = \int_{-\infty}^{\infty} \operatorname{Per}\,[\psi > t]\,dt. \tag{15.131}$$

Equivalently, for a.e. t we have $\chi_{[\psi>t]} \in BV(\mathbb{R}^2)$, and

$$\|D\psi\|_{\mathscr{M}} = \int_{-\infty}^{\infty} \|D\chi_{[\psi>t]}\|_{\mathscr{M}}\,dt.$$

Let $j \in \mathbb{Z}$ and set $F_j := [\psi \geq j]$. Since, for $t \in (j-1, j)$, we have $[\psi > t] = F_j$, we find that F_j has finite perimeter for every j, and that

$$\|D\psi\|_{\mathscr{M}} = \sum_{j=-\infty}^{\infty} \operatorname{Per} F_j. \tag{15.132}$$

Let now, for $j \in \mathbb{Z}$,

$$E_j := \begin{cases} F_j, & \text{if } j \geq 0 \\ (F_{j+1})^c = \mathbb{R}^2 \setminus F_{j+1}, & \text{if } j < 0 \end{cases}. \tag{15.133}$$

By (15.132), we have

$$\|D\psi\|_{\mathscr{M}} = \sum_{j=-\infty}^{\infty} \operatorname{Per} E_j. \tag{15.134}$$

We claim that

$$\psi = \sum_{j=-\infty}^{\infty} (\operatorname{sgn} j)\, \chi_{E_j} \text{ in } BV(\mathbb{R}^2). \tag{15.135}$$

The claim is proved as follows. Let $m \geq 1$ be an integer, and set

$$\psi_m := \begin{cases} \psi, & \text{if } |\psi| \leq m \\ m, & \text{if } \psi \geq m \\ -m, & \text{if } \psi \leq -m \end{cases}.$$

Clearly, we have

$$\psi_m \to \psi \text{ in } L^1(\mathbb{R}^2) \text{ as } m \to \infty. \tag{15.136}$$

Applying the coarea formula as above, we have

$$\|D(\psi - \psi_m)\|_{\mathscr{M}} = \sum_{|j|>m} \operatorname{Per} F_j \to 0 \text{ as } m \to \infty. \tag{15.137}$$

On the other hand, we clearly have

$$\psi_m = \sum_{j=-m}^{m} (\operatorname{sgn} j)\, \chi_{E_j} \text{ in } BV(\mathbb{R}^2). \tag{15.138}$$

We obtain (15.135) from (15.136)–(15.138).

Step 4. Conclusion

We apply Theorem 15.12 to each E_j as in (15.133). Using the equalities (15.134) and (15.135), we find that

$$D\psi = \sum_i \overline{\nu}_i \mathscr{H}^1 \llcorner \overline{\gamma}_i, \tag{15.139}$$

with $\overline{\gamma}_i$ rectifiable Jordan curves such that $\sum_i \mathscr{H}^1(\overline{\gamma}_i) < \infty$. For any such curve, we have:

1. Either $\overline{\gamma}_i \cap \partial\Omega = \emptyset$, and then $\overline{\gamma}_i \subset \Omega$ or $\overline{\gamma}_i \subset \mathbb{R}^2 \setminus \overline{\Omega}$
2. Or $\#(\overline{\gamma}_i \cap \partial\Omega) \geq 2$, and then $\overline{\gamma}_i \cap \Omega$ consists of a disjoint and at most countable union of rectifiable Jordan arcs $\gamma_{i,j}$, each one connecting two points of $\partial\Omega$ and such that $\sum_j \mathscr{H}^1(\gamma_{i,j}) \leq \mathscr{H}^1(\overline{\gamma}_i)$
3. Or $\overline{\gamma}_i \cap \partial\Omega = \{Q\}$, and then $\overline{\gamma}_i \setminus \{Q\}$ is contained in either Ω or $\mathbb{R}^2 \setminus \overline{\Omega}$, and "connects" $Q \in \partial\Omega$ to itself.

Using this, (15.130) and the specific form (15.129) of μ_0, we find that μ can be written as in (15.127). □

We now turn to Theorem 15.12. Under an additional assumption on E, this result was obtained by Fleming [151, Theorem 4]. [One can prove that the assumption in [151, Theorem 4] is equivalent to the fact that E is simply connected.] The generalization of this result to arbitrary sets is stated in Federer [147, Section 4.2.25], with a sketch of proof. For the convenience of the reader, we present below a full proof. It relies on a version of Theorem 15.12, involving reduced boundaries of sets (thus $\mathscr{H}^1$-rectifiable sets) instead of distributional derivatives (i.e., vector-valued measures), and proved in Ambrosio, Caselles, Masnou, and Morel [10, Corollary 1]. We explain below how to obtain Theorem 15.12 using only the arguments presented in [10].

Proof of Theorem 15.12. If E is a set of finite perimeter, there are (at least) two natural notions of "measure theoretic" boundary:

$$\begin{aligned}\partial^* E := \Big\{x \in \mathbb{R}^2;\ &\limsup_{r\to 0} \frac{|E \cap B(x, r)|}{|B(x, r)|} > 0 \text{ and}\\ &\limsup_{r\to 0} \frac{|E^c \cap B(x, r)|}{|B(x, r)|} > 0\Big\},\end{aligned} \tag{15.140}$$

respectively

$$\partial^D E := \operatorname{supp} D\chi_E. \tag{15.141}$$

The two clearly coincide (as sets) when E is smooth. When E is rectifiable, the theorems of De Giorgi and Federer (see, for example, Ambrosio, Fusco, and Pallara [13, Theorems 3.59 and 3.61]) assert that

$$\partial^* E = \partial^D E \text{ up to an } \mathscr{H}^1\text{-null set,} \tag{15.142}$$

and that

$$\operatorname{Per} E = \|D\chi_E\|_{\mathscr{M}} = \mathscr{H}^1(\operatorname{supp} D\chi_E) = \mathscr{H}^1(\partial^* E). \tag{15.143}$$

For any Jordan curve (not necessarily rectifiable) γ, the set $\mathbb{R}^2 \setminus \gamma$ has precisely two connected components (Jordan curve theorem). We denote by $\operatorname{int}(\gamma)$ the *interior of* γ, i.e., the one of the two components which is bounded. If, in addition, γ is rectifiable, then $\operatorname{int}(\gamma)$ is a set of finite perimeter [10, second part of Theorem 7]. More specifically, for every rectifiable Jordan curve γ we have [10, Lemma 4]

$$\operatorname{Per} \operatorname{int}(\gamma) = \mathscr{H}^1(\gamma). \tag{15.144}$$

On the other hand, for any open set E we clearly have $\partial^* E \subset \partial E$ (where the latter set is the topological boundary of E). In particular, we have $\partial^* \operatorname{int}(\gamma) \subset \gamma$. Combining this with (15.143) and (15.144), we find that, for every rectifiable Jordan curve γ,

$$\partial^* \operatorname{int}(\gamma) = \gamma\ \mathscr{H}^1\text{-a.e.} \tag{15.145}$$

Given a bounded set $E \subset \mathbb{R}^2$ of finite perimeter, we may write [10, Corollary 1]

$$E = \bigcup_j E_j \text{ a.e.,} \tag{15.146}$$

where the sets E_j of the (possibly countable) family (E_j) satisfy:

1. $E_j \cap E_\ell = \emptyset$ if $j \neq \ell$.
2. $E_j = \operatorname{int}(\gamma_j) \setminus \cup_{i \in I_j} \operatorname{int}(\gamma_{j,i})$ (with I_j at most countable) for some rectifiable Jordan curves γ_j and $\gamma_{j,i}$.
3. $\operatorname{int}(\gamma_{j,i}) \subset \operatorname{int}(\gamma_j)$, $\forall j, \forall i \in I_j$.
4. $\operatorname{int}(\gamma_{j,i}) \cap \operatorname{int}(\gamma_{j,k}) = \emptyset$, $\forall j, \forall i \neq k$.
5. We have

$$\sum_j \left[\mathscr{H}^1(\gamma_j) + \sum_{i \in I_j} \mathscr{H}^1(\gamma_{j,i}) \right] = \mathscr{H}^1(\partial^* E) = \|D\chi_E\|_{\mathscr{M}}. \tag{15.147}$$

A final ingredient of the proof of Theorem 15.12 is the following result, that we temporarily take for granted.

Lemma 15.35. *Let γ be a rectifiable Jordan curve, and define $\mu := \nu \mathscr{H}^1 \llcorner \gamma$ as in* (2.135). *Then we have*

$$D\chi_{\operatorname{int}(\gamma)} = \pm \mu. \tag{15.148}$$

Proof of Theorem 15.12 *completed.* For each j, we have (using (15.144) and (15.147))

$$\chi_{E_j} = \chi_{\operatorname{int}(\gamma_j)} - \sum_{i \in I_j} \chi_{\operatorname{int}(\gamma_{j,i})} \text{ in } BV(\mathbb{R}^2). \tag{15.149}$$

Using (15.149) and Lemma 15.35 we find that, with appropriate orientations on γ_j and $\gamma_{j,i}$, we have

$$D\chi_{E_j} = \nu_j \mathscr{H}^1 \llcorner \gamma_j + \sum_{i \in I_j} \nu_{j,i} \mathscr{H}^1 \llcorner \gamma_{j,i} \text{ in } \mathscr{M}(\mathbb{R}^2; \mathbb{R}^2). \tag{15.150}$$

By (15.146), (15.147), and (15.150), we have

$$\chi_E = \sum_j \chi_{E_j} \text{ in } BV(\mathbb{R}^2) \tag{15.151}$$

and

$$D\chi_E = \sum_j \left(\nu_j \mathscr{H}^1 \llcorner \gamma_j + \sum_{i \in I_j} \nu_{j,i} \mathscr{H}^1 \llcorner \gamma_{j,i} \right) \text{ in } \mathscr{M}(\mathbb{R}^2; \mathbb{R}^2). \tag{15.152}$$

Equality (15.128) follows from (15.152). □

Proof of Lemma 15.35. Set $E := \operatorname{int}(\gamma)$. By [10, second part of Theorem 7], E is a "simple set" in the sense of [10, Definition 3]. As a consequence, the proof of

Theorem 7 in [10] applies to E. More specifically, there exists a sequence of smooth uniformly bounded sets (G_n) such that:

1. $\chi_{G_n} \to \chi_E$ in $L^1(\mathbb{R}^2)$, $D\chi_{G_n} \rightharpoonup D\chi_E$ in $\mathscr{M}(\mathbb{R}^2; \mathbb{R}^2)$, and $\mathrm{Per}\,(G_n) \to \mathrm{Per}\, E$ as $n \to \infty$.
2. For each n, we have $\partial G_n = \Gamma_n \cup \cup_{i \in J_n} \Gamma_{n,i}$, with Γ_n and $\Gamma_{n,i}$ mutually disjoint smooth connected closed curves.
3. For appropriate Lipschitz parametrizations $f_n : [0, 1] \to \Gamma_n$ and $f_{n,i} : [0, 1] \to \Gamma_{n,i}$, we have $|f_n'| \le C$ and $|f_{n,i}'| \le C$.
4. In addition, these parametrizations satisfy $f_n \to f$ uniformly and $f_{n,i} \to C_i$ uniformly as $n \to \infty$, for some constants $C_i \in \mathbb{R}^2$.

With no loss of generality, we may assume that the normal vectors ν_n and $\nu_{n,i}$ induced by f_n and $f_{n,i}$ satisfy

$$D\chi_{G_n} = \nu_n \mathscr{H}^1 \llcorner \Gamma_n + \sum_{i \in J_n} \nu_{n,i} \mathscr{H}^1 \llcorner \Gamma_{n,i}. \tag{15.153}$$

Let us note that items 1 and 2 imply that

$$\int_0^1 |f_n'(t)|\, dt + \sum_{i \in J_n} \int_0^1 |f_{n,i}'(t)|\, dt \le C < \infty. \tag{15.154}$$

By (15.153), (15.154), and items 1, 3, and 4, for every $\xi \in C_c^\infty(\mathbb{R}^2; \mathbb{R}^2)$ we have

$$\begin{aligned} \langle D\chi_{G_n}, \xi \rangle &= -\int_0^1 \left[\xi(f_n(t)) \cdot (f_n')^\perp(t) + \sum_{i \in J_n} \xi(f_{n,i}(t)) \cdot (f_{n,i}')^\perp(t) \right] dt \\ &\to -\int_0^1 \xi(f(t)) \cdot (f')^\perp(t)\, dt \text{ as } n \to \infty. \end{aligned} \tag{15.155}$$

By (15.155) and item 1, we thus have

$$\begin{aligned} \langle D\chi_E, \xi \rangle &= -\int_0^1 \xi(f(t)) \cdot (f')^\perp(t)\, dt \\ &= \int_0^1 [\xi_1(f(t)) f_2'(t) - \xi_2(f(t)) f_1'(t)]\, dt. \end{aligned} \tag{15.156}$$

Since E is not a null set, we have $L := \int_0^1 |f'(t)|\, dt > 0$. Therefore, we may consider a "reparametrization" of $f([0, 1])$ by the arc length, i.e., some $\widetilde{f} : [0, L] \to f([0, 1])$ such that $|\widetilde{f}'(t)| = 1$ a.e. and

$$\langle D\chi_E, \xi \rangle = \int_0^1 [\xi_1(\widetilde{f}(t)) \widetilde{f}_2'(t) - \xi_2(\widetilde{f}(t)) \widetilde{f}_1'(t)]\, dt. \tag{15.157}$$

[At this stage, we do not claim that $\widetilde{f}$ is one-to-one on $[0, L)$.] Since $\int_0^L |\widetilde{f}'(t)|\, dt > 0$, the $\mathscr{H}^1$-rectifiable set $\widetilde{f}([0, L])$ contains a rectifiable Jordan curve Γ [10, Lemma 5]. By [10, proof of Theorem 7, end of Step 2], we have

$$L = \int_0^L |\widetilde{f}'(t)|\, dt = \mathscr{H}^1(\Gamma).$$

Since, on the other hand, we have

$$\mathscr{H}^1(\Gamma) \le \mathscr{H}^1(\widetilde{f}([0, L]) \le \int_0^L |\widetilde{f}'(t)|\, dt = L,$$

we find first that $\Gamma = \widetilde{f}([0, L])$, next that $\widetilde{f}$ is a Lipschitz parametrization of Γ.

Therefore, (15.157) reads

$$D\chi_E = \nu \mathscr{H}^1 \llcorner \Gamma$$

(for an appropriate orientation on Γ).

To summarize, we know that for every rectifiable Jordan curve γ there exists some rectifiable Jordan curve Γ such that

$$D\chi_{\text{int}\,(\gamma)} = \nu \mathscr{H}^1 \llcorner \Gamma. \tag{15.158}$$

By (15.141), (15.142), (15.145), and (15.158) we have $\gamma = \Gamma$, and thus (15.148) holds. □

Lemma 2.10 in Section 2.10 can be obtained as a byproduct of the above analysis.

Proof of Lemma 2.10. Assertion b) follows from the coarea formula (see, for example, Evans and Gariepy [146, Remark (ii), p. 112]).

Let $t \in \mathbb{R}$ and $x \in \Omega$ be such that $\zeta(x) = t$ and $\nabla\zeta(x) \neq 0$. Set $E := [\zeta > t]$. Then, clearly, $x \in \partial^* E$ (see (15.140)). On the other hand, if y is such that $\zeta(y) \neq t$, then again, clearly, $y \notin \partial^* E$. From b), it follows that for a.e. $t \in \mathbb{R}$ we have $\partial^* E = [\zeta = t]$ $\mathscr{H}^1$-a.e. We complete the proof of the lemma using the construction of curves γ_i in the proof of Theorem 15.12. □

15.8.2 Structure of minimal connections associated with (a, d)

We prove Theorem 2.8, which we restate here for the convenience of the reader.

Theorem 15.13. *Let μ_0 be a minimal connection associated with (a, d). Then, $\mu_0 = \sum_{\ell=1}^n \nu_\ell \mathscr{H}^1 \llcorner \gamma_\ell$, where γ_ℓ is the oriented line segment from N_ℓ to P_ℓ and the collection $(P_\ell, N_\ell)_{\ell=1}^n$ of points in $\overline{\Omega}$ satisfies*

$$L(a,d) = \sum_{\ell=1}^{n} |P_\ell - N_\ell|, \tag{15.159}$$

$$\sum_{\ell=1}^{n} \delta_{P_\ell} = \sum_{j=1}^{k} d_j^+ \, \delta_{a_j} \ \textit{in}\ \mathscr{D}'(\Omega), \tag{15.160}$$

$$\sum_{\ell=1}^{n} \delta_{N_\ell} = \sum_{j=1}^{k} d_j^- \, \delta_{a_j} \ \textit{in}\ \mathscr{D}'(\Omega); \tag{15.161}$$

in particular,

$$\sum_{\ell=1}^{n} \left(\delta_{P_\ell} - \delta_{N_\ell}\right) = \sum_{j=1}^{k} d_j \, \delta_{a_j} \ \textit{in}\ \mathscr{D}'(\Omega). \tag{15.162}$$

Moreover,

$$\textit{for every } \ell = 1, \dots, n,\ P_\ell \neq N_\ell \textit{ and the open interval } (P_\ell, N_\ell) \textit{ is contained in } \Omega. \tag{15.163}$$

The conclusion of Theorem 15.13 is reminiscent of the well-known structure of 1-rectifiable currents that are mass-minimizing in all of $\mathbb{R}^2$ (see, for example, Morgan [273, Theorem 10.1]) and in fact this ingredient plays an essential role in our proof, but the localization to a domain Ω requires some additional work.

Proof. It will be convenient to work with currents instead of connections, using the dictionary (2.224)–(2.225). In this setting, we have to prove the following. Let $\mathscr{C}$ be a 1-rectifiable current in Ω such that

$$\partial \mathscr{C} = \sum_{j=1}^{k} d_j \, \delta_{a_j} \ \text{in}\ \mathscr{D}'(\Omega) \tag{15.164}$$

and

$$\|\mathscr{C}\|_{\mathscr{M}} \le \|\widetilde{\mathscr{C}}\|_{\mathscr{M}}, \ \forall \widetilde{\mathscr{C}} \text{ satisfying (15.164).} \tag{15.165}$$

Then,

$$\mathscr{C} = \sum_{\ell=1}^{n} \tau_\ell \mathscr{H}^1 \llcorner \gamma_\ell, \tag{15.166}$$

where $\gamma_\ell := [N_\ell, P_\ell]$, τ_ℓ is the unit tangent vector to the oriented segment from N_ℓ to P_ℓ, and the collection $(P_\ell, N_\ell)_{\ell=1}^n$ satisfies (15.159)–(15.163).

[Here and in what follows, $\| \ \|_{\mathscr{M}}$ is the mass of measures on Ω.]

Recall that (15.164) means that

$$\langle \mathscr{C}, \nabla \zeta \rangle = \sum_{j=1}^{k} d_j \, \zeta(a_j), \ \forall \zeta \in C_c^\infty(\Omega; \mathbb{R}).$$

Let $\mathscr{C}$ satisfy (15.164) and (15.165). We write

$$\mathscr{C} = \sum_i \tau_i \mathscr{H}^1 \llcorner \gamma_i, \tag{15.167}$$

with

$$\sum_i \mathscr{H}^1(\gamma_i) < \infty \tag{15.168}$$

and γ_i as in Theorem 15.11.

Set $\omega := \Omega \setminus \{a_1, \ldots, a_k\}$. The proof of the theorem has three main parts. We first derive (using slicing and the local structure of mass-minimizing 1-rectifiable currents) that, locally in ω, $\mathscr{C}$ consists of oriented line segments with constant multiplicities. We next prove that $\mathscr{C}$ is a finite sum of such segments. Finally, we "recombine" this sum in order to obtain segments with endpoints satisfying (15.159)–(15.163).

Step 1. Local slicing in Ω

This step can be treated using the general theory of slicing, discussed in Federer [147, Section 4.2.1], but we take advantage of (15.167) and present a self-contained argument.

Let $x \in \mathbb{R}^2$. If γ is a Lipschitz Jordan arc or curve parametrized by $f : [a, b] \to \gamma$, then the area formula (see, for example, Evans and Gariepy [146, Section 3.3, Theorem 1]) yields

$$\begin{aligned} \int_0^\infty \#\{y \in \gamma;\ |y - x| = r\}\, dr &= \int_0^\infty \#\{t \in [a, b];\ |f(t) - x| = r\}\, dr \\ &= \int_a^b ||f(t) - x|'|\, dt \\ &\leq \int_a^b |f'(t)|\, dt = \mathscr{H}^1(\gamma). \end{aligned} \tag{15.169}$$

By (15.168) and (15.169), for every $x \in \mathbb{R}^2$

$$\text{the set } \{y \in C(x, \varepsilon);\ y \in \gamma_i \text{ for some } i\} \text{ is finite} \tag{15.170}$$

for a.e. $\varepsilon > 0$.

Fix some $x \in \omega$, and let $\varepsilon > 0$ such that (15.170) holds and, in addition, $\overline{B}(x, \varepsilon) \subset \omega$. Then, for every i, we have:

1. Either $\gamma_i \subset \overline{B}(x, \varepsilon)$ and $\gamma_i \cap C(x, \varepsilon)$ is finite
2. Or $\gamma_i \subset \Omega \setminus B(x, \varepsilon)$ and $\gamma_i \cap C(x, \varepsilon)$ is finite
3. Or $\gamma_i \cap B(x, \varepsilon)$ consists of a finite number of arcs connecting points of $C(x, \varepsilon)$.

The latter possibility occurs only for a finite number of i's.

Taking into account the structure of γ_i (described in Theorem 15.11) and the fact that $\overline{B}(x, \varepsilon)$ contains no singularity a_j, we find that $\mathscr{C}_1 := \mathscr{C}_{|B(x,\varepsilon)}$ is of the form

$$\mathscr{C}_1 = \sum_{i=1}^{m} \overline{\tau}_i \, \mathscr{H}^1 \llcorner \overline{\gamma}_i + \sum_{\ell} \tau_\ell^* \mathscr{H}^1 \llcorner \gamma_\ell^*,$$

where

1. $\overline{\gamma}_i$ connects $M_i \in C(x, \varepsilon)$ to $Q_i \in C(x, \varepsilon)$, $\forall\, i$.
2. γ_ℓ^* is a rectifiable Jordan curve, $\forall\, \ell$.
3. $\overline{\tau}_i$ and τ_ℓ^* are corresponding unit tangent vectors.

Moreover, we have $\mathscr{C}_{|C(x,\varepsilon)} = 0$ (equality of measures) and thus

$$\mathscr{C} = \mathscr{C}_1 + \mathscr{C}_{|\Omega \setminus \overline{B}(x,\varepsilon)} := \mathscr{C}_1 + \mathscr{C}_2. \tag{15.171}$$

By the above and (15.164), we have

$$\partial \mathscr{C}_1 = \sum_{i=1}^{m} \left(\delta_{Q_i} - \delta_{M_i}\right) \text{ in } \mathscr{D}'(\mathbb{R}^2), \tag{15.172}$$

$$\partial \mathscr{C}_2 = \sum_{j=1}^{k} d_j \, \delta_{a_j} - \sum_{i=1}^{m} \left(\delta_{Q_i} - \delta_{M_i}\right) \text{ in } \mathscr{D}'(\Omega), \tag{15.173}$$

and

$$\|\mathscr{C}_1\|_{\mathscr{M}} + \|\mathscr{C}_2\|_{\mathscr{M}} = \|\mathscr{C}\|_{\mathscr{M}} = L(a, d). \tag{15.174}$$

We claim that, for every 1-rectifiable current $\widetilde{\mathscr{C}_1}$ in $\mathbb{R}^2$ such that

$$\partial \widetilde{\mathscr{C}_1} = \sum_{i=1}^{m} \left(\delta_{Q_i} - \delta_{M_i}\right) \text{ in } \mathscr{D}'(\mathbb{R}^2), \tag{15.175}$$

we have

$$\|\mathscr{C}_1\|_{\mathscr{M}(\mathbb{R}^2)} = \|\mathscr{C}_1\|_{\mathscr{M}} \le \|(\widetilde{\mathscr{C}_1})_{|\Omega}\|_{\mathscr{M}} \le \|\widetilde{\mathscr{C}_1}\|_{\mathscr{M}(\mathbb{R}^2)}. \tag{15.176}$$

Equality in (15.176) is clear. On the other hand, for any $\widetilde{\mathscr{C}_1}$ as above, the 1-rectifiable current $(\widetilde{\mathscr{C}_1})_{|\Omega}$ satisfies (15.164), and thus (by (15.165) and (15.174))

$$\|\mathscr{C}_1\|_{\mathscr{M}} + \|\mathscr{C}_2\|_{\mathscr{M}} = \|\mathscr{C}\|_{\mathscr{M}} \le \|(\widetilde{\mathscr{C}_1})_{|\Omega} + \mathscr{C}_2\|_{\mathscr{M}} \le \|(\widetilde{\mathscr{C}_1})_{|\Omega}\|_{\mathscr{M}} + \|\mathscr{C}_2\|_{\mathscr{M}},$$

whence (15.176).

Step 2. Local structure of $\mathscr{C}$ in ω
We invoke the fact that, for any $\mathscr{C}_1$ satisfying (15.176), $\operatorname{supp}\mathscr{C}_1 \setminus \operatorname{supp}\partial\mathscr{C}_1$ consists of a (locally finite) union of disjoint line segments (see, for example, Morgan [273, Theorem 10.1]).

Let x, ε and $\mathscr{C}_1$ be as in Step 1. Since $\mathscr{C}_{|B(x,\varepsilon)} = \mathscr{C}_1$, we find that $(\operatorname{supp}\mathscr{C}) \cap B(x, \varepsilon)$ consists of a locally finite union of disjoint line segments, and thus $(\operatorname{supp}\mathscr{C}) \cap \omega$ consists of a locally finite union of disjoint line segments.

Step 3. "Extraction" of a line segment from $\mathscr{C}$
Let $x \in \operatorname{supp}\mathscr{C}$. Let D be the straight line passing through x such that, in an open neighborhood V of x, we have $\operatorname{supp}\mathscr{C} \cap V = D \cap V$. Let S be the connected component of $D \cap \omega$ containing x. Thus, S is an open line segment of the form $S = (M, Q)$, with $M, Q \in \partial\omega \cup \{a_1, \ldots, a_k\}$. We claim that there exist a neighborhood U of S in ω and an integer $\ell \in \mathbb{Z} \setminus \{0\}$ such that

$$\mathscr{C}_{|U} = \ell\, \tau\, \mathscr{H}^1 \llcorner (M, Q). \tag{15.177}$$

Moreover, we claim that

$$\text{either } M \in \{a_1, \ldots, a_k\} \text{ or } Q \in \{a_1, \ldots, a_k\}. \tag{15.178}$$

Indeed, the set

$$\{y \in S;\ \exists\, \delta > 0 \text{ such that } (\operatorname{supp}\mathscr{C}) \cap B(y, \delta) = D \cap B(y, \delta)\}$$

is non-empty, open in S and, by [273, Theorem 10.1], closed in S. This implies that the above set coincides with S and that there exists some open set $U \subset \omega$ containing S and such that

$$(\operatorname{supp}\mathscr{C}) \cap U = S. \tag{15.179}$$

Using the structure of 1-rectifiable currents (see Federer [147, Section 4.1.28]), we have

$$\mathscr{C}_{|U} = \ell(x)\, \tau\, \mathscr{H}^1 \llcorner (M, Q),$$

with $\ell(x) : S \to \mathbb{Z} \setminus \{0\}$ an $\mathscr{H}^1$-integrable function on S.

We next use the fact that we have (by construction of S), $\partial\mathscr{C}_{|U} = 0$ in $\mathscr{D}'(U)$. To fix the ideas, assume, for example, that $S = ((b, 0), (c, 0))$, with $b, c \in \mathbb{R}$, $b < c$. Then, $\tau = (1, 0)$ and thus

$$0 = \langle \partial\mathscr{C}, \zeta \rangle = \int_b^c \ell(t, 0) \frac{\partial \zeta}{\partial x_1}(t, 0)\, dt,\ \ \forall\, \zeta \in C_c^\infty(U). \tag{15.180}$$

By testing (15.180) on functions $\zeta \in C_c^\infty(U)$ of the form

$$\zeta(x_1, x_2) := \psi(x_1)\, \eta(x_2),\ \text{ with } \psi \in C_c^\infty((b, c)),\ \eta \in C_c^\infty(\mathbb{R}),\ \eta(0) = 1,$$

we find that $\ell' = 0$ in $\mathscr{D}'((b, c))$, and thus ℓ =const. This implies (15.177).

Assume next, by contradiction, that (15.178) does not hold. Define $\mathscr{C}_3 := \mathscr{C}_{|U}$ and $\mathscr{C}_4 := \mathscr{C} - \mathscr{C}_3$. Then we have, in the sense of $\mathscr{D}'(\Omega)$, $\partial \mathscr{C}_3 = 0$,

$$\partial \mathscr{C}_4 = \partial \mathscr{C}, \tag{15.181}$$

and

$$\|\mathscr{C}_4\|_{\mathscr{M}} = \|\mathscr{C}\|_{\mathscr{M}} - \|\mathscr{C}_3\|_{\mathscr{M}} < \|\mathscr{C}\|_{\mathscr{M}} = L(a, d). \tag{15.182}$$

We obtain a contradiction from (15.181), (15.182), and Proposition 2.2.

Step 3. Global structure of $\mathscr{C}$

We fix $a_1, \ldots, a_k$, and prove that, for every collection $d = (d_1, \ldots, d_k) \in \mathbb{Z}^k$, we may write every 1-rectifiable current $\mathscr{C}$ satisfying (15.174) and (15.175) in the form

$$\mathscr{C} = \sum_{\ell=1}^{\overline{n}} \overline{\tau}_\ell \, \mathscr{H}^1 \llcorner \overline{\gamma}_\ell, \tag{15.183}$$

where the collection $(\overline{P}_\ell, \overline{N}_\ell)_{\ell=1}^{\overline{n}}$ satisfies

$$\sum_{\ell=1}^{\overline{n}} \left(\delta_{\overline{P}_\ell} - \delta_{\overline{N}_\ell}\right) = \sum_{j=1}^{k} d_j \, \delta_{a_j} \text{ in } \mathscr{D}'(\Omega), \tag{15.184}$$

$$\overline{N}_\ell \neq \overline{P}_\ell, \ \forall \, \ell = 1, \ldots, \overline{n}, \tag{15.185}$$

$$(\overline{N}_\ell, \overline{P}_\ell) \subset \omega, \ \forall \, \ell = 1, \ldots, \overline{n}, \tag{15.186}$$

and

$$\sum_{\ell=1}^{\overline{n}} |\overline{P}_\ell - \overline{N}_\ell| = L(a, d). \tag{15.187}$$

It is convenient to allow, in the above, some of the d_j's to be zero.

Set

$$c := \min\left\{|a_j - a_l|, \ 1 \leq j < l \leq k, \ \operatorname{dist}(a_j, \partial\Omega), \ 1 \leq j \leq k\right\} > 0$$

and $M := [L(a, d)/c]$ (the integer part of $L(a, d)/c$). We will prove (15.183) by induction on M.

When $M = 0$, we have $L(a, d) = 0$ (since every segment in Lemma 2.4 has length $\geq c$), and then the conclusion is clear.

Assume next that the conclusion has been obtained for every $\overline{d} \in \mathbb{Z}^k$ such that $L(a, \overline{d}) < M\,c$, and let $d \in \mathbb{Z}^k$ be such that $M\,c \leq L(a, d) < (M + 1)\,c$. By Step 2, there exists an oriented line segment S and an integer $\ell \in \mathbb{Z} \setminus \{0\}$ such that the 1-rectifiable currents $\mathscr{C}_5 := \ell\,\tau\,\mathscr{H}^1 \llcorner S$ and $\mathscr{C}_6 := \mathscr{C} - \mathscr{C}_5$ satisfy

$$\|\mathscr{C}_5\|_{\mathscr{M}} + \|\mathscr{C}_6\|_{\mathscr{M}} = \|\mathscr{C}\|_{\mathscr{M}} = L(a, d), \tag{15.188}$$

$$\partial \mathscr{C}_5 = \sum_{j=1}^{k} d_j^* \, \delta_{a_j} \text{ in } \mathscr{D}'(\Omega), \tag{15.189}$$

$$\partial \mathscr{C}_6 = \sum_{j=1}^{k} \overline{d}_j \, \delta_{a_j} \text{ in } \mathscr{D}'(\Omega), \tag{15.190}$$

for appropriate collections $d^*, \overline{d}$ of integers. By Proposition 2.2 and the dictionary (2.224)–(2.225), we have

$$\|\mathscr{C}_5\|_{\mathscr{M}} \geq L(a, d^*), \tag{15.191}$$

$$\|\mathscr{C}_6\|_{\mathscr{M}} \geq L(a, \overline{d}). \tag{15.192}$$

On the other hand, by Theorem 2.3 we clearly have

$$L(a, d^*) + L(a, \overline{d}) \leq L(a, d^* + \overline{d}) = L(a, d). \tag{15.193}$$

We find that $\mathscr{C}_5$ and $\mathscr{C}_6$ are mass-minimizing 1-rectifiable currents associated with (a, d^*), resp. $(a, \overline{d})$, and that $L(a, \overline{d}) = L(a, d) - \|\mathscr{C}_5\|_{\mathscr{M}}$. Since, by construction, we have $\|\mathscr{C}_5\|_{\mathscr{M}} \geq c$, we obtain that $L(a, \overline{d}) < M\, c$. We conclude with the help of the induction hypothesis.

Step 4. "Recombining" the sum $\sum_{\ell=1}^{\overline{n}} \overline{\tau}_\ell \, \mathscr{H}^1 \llcorner \overline{\gamma}_\ell$

Set $|d| := \sum_{j=1}^{k} |d_j|$. We prove by induction on $|d| + \overline{n}$ that the sum in (15.183) can be "recombined" as $\sum_{\ell=1}^{n} \tau_\ell \, \mathscr{H}^1 \llcorner \gamma_\ell$, with $\gamma_\ell := [N_\ell, P_\ell]$ and N_ℓ, P_ℓ as in (15.159)–(15.163).

The starting point in Step 4 is the fact that, by the inductive construction,

$$\text{two oriented segments } (\overline{P}_\ell, \overline{N}_\ell),\ (\overline{P}_i, \overline{N}_i) \text{ are either disjoint or equal.} \tag{15.194}$$

The conclusion of Step 4 is clear when $d = 0$. Assume that $d \neq 0$ and consider some a_j such that $d_j \neq 0$, say $d_j > 0$. Since clearly

$$d_j = \#\{\ell;\ a_j = \overline{P}_\ell\} - \#\{\ell;\ a_j = \overline{N}_\ell\}, \tag{15.195}$$

there exists some ℓ such that $a_j = \overline{P}_\ell$. For this ℓ, one of the following occurs:

1. Either $\overline{N}_\ell \in \partial\Omega$
2. Or $\overline{N}_\ell = a_l$, with $d_l < 0$
3. Or $\overline{N}_\ell = a_l$, with $d_l \geq 0$.

In the first case, we let

$$\mathscr{C}_7 := \tau_\ell \, \mathscr{H}^1 \llcorner (\overline{N}_\ell, \overline{P}_\ell),\ \mathscr{C}_8 := \mathscr{C} - \mathscr{C}_7,\ d_l^* := \begin{cases} 1, & \text{if } l = j \\ 0, & \text{if } l \neq j \end{cases},\ \overline{d} := d - d^*.$$

Since $\|\mathscr{C}_7\|_{\mathscr{M}} + \|\mathscr{C}_8\|_{\mathscr{M}} = \|\mathscr{C}\|_{\mathscr{M}}$ (here, we use (15.194)), we find, as in Step 2, that $\mathscr{C}_8$ is a mass-minimizing 1-rectifiable current associated with $(a, \overline{d})$. We then conclude via the induction hypothesis applied to $\mathscr{C}_8$.

The second case, where $d_l < 0$, is similar. Assume next that we are in the third case, where $d_l \geq 0$. By (15.195) applied to d_l, there exists some index p such that $a_l = \overline{P}_p$. Consider now in (15.165) the "competitor"

$$\widetilde{\mathscr{C}} := \mathscr{C} - \tau_\ell\, \mathscr{H}^1 \llcorner (\overline{N}_\ell, \overline{P}_\ell) - \tau_p\, \mathscr{H}^1 \llcorner (\overline{N}_p, \overline{P}_p) + \tau\, \mathscr{H}^1 \llcorner (\overline{N}_p, \overline{P}_\ell).$$

By (15.194), we have

$$\|\widetilde{\mathscr{C}}\|_{\mathscr{M}} \leq \|\mathscr{C}\|_{\mathscr{M}} - |\overline{P}_\ell - \overline{N}_\ell| - |\overline{P}_p - \overline{N}_p| + |\overline{P}_\ell - \overline{N}_p|,$$

so that $\|\widetilde{\mathscr{C}}\|_{\mathscr{M}} < \|\mathscr{C}\|_{\mathscr{M}}$ unless a_l is a point of the segment $(\overline{N}_p, \overline{P}_\ell)$. Therefore, a_l has to be on this segment, and thus we have

$$\mathscr{C} = \sum_{q \neq \ell, p} \tau_q\, \mathscr{H}^1 \llcorner (\overline{N}_q, \overline{P}_q) + \tau\, \mathscr{H}^1 \llcorner (\overline{N}_p, \overline{P}_\ell).$$

We conclude again using the induction hypothesis.

The proof of Theorem 15.13 is complete. □

15.9 A pathological case

This appendix is mainly devoted to the proof of Proposition 2.4. As a byproduct of our arguments, we derive the characterization of boundaries of 1-rectifiable currents in $\mathbb{R}^2$, stated in Remark 2.13 (see Proposition 15.4).

In what follows, we identify $\mathbb{R}$ (endowed with the natural orientation) with $\mathbb{R} \times \{0\}$. We assume that

$$\overline{\mathbb{D}}(0, 8) \subset \Omega. \tag{15.196}$$

Proposition 15.3. *Given a Borel subset A of $[0, 1]$, we define $\mu_A := \nu \mathscr{H}^1 \llcorner A$ and $T_A :=$ curl $\mu_A \in \mathscr{D}'(\Omega)$. Then:*

1. *$T_A \in \mathscr{E}$*
2. *The following are equivalent:*

   a) There exist (P_j, N_j) such that

$$T_A = \sum_j (\delta_{P_j} - \delta_{N_j}) \tag{15.197}$$

   and

$$\sum_j |P_j - N_j| = \|T_A\|_W \tag{15.198}$$

b) A is an open set in the measure theoretic sense, i.e., there exists an open set $U \subset \mathbb{R}$ such that $|A \bigtriangleup U| = 0$.

Proof of Proposition 2.4. Take, in Proposition 15.3, a Cantor set A of positive measure, or any other Borel set that is not open in the measure theoretical sense. □

Proof of Proposition 15.3. We have

$$\langle T_A, \zeta \rangle = \int_A \frac{\partial \zeta}{\partial x_1}(x_1, 0)\, dx_1, \ \forall \zeta \in C_c^\infty(\Omega). \tag{15.199}$$

By an approximation argument in the spirit of Section 1.9.2, (15.199) still holds for $\zeta \in W_0^{1,\infty}(\Omega)$, provided we interpret $\dfrac{\partial \zeta}{\partial x_1}$ as the derivative of $x_1 \mapsto \zeta(x_1, 0)$.

We claim that

$$\|T_A\|_W = |A|. \tag{15.200}$$

Inequality "$\le$" is clear, from (15.199). For the reverse inequality, consider a map $\zeta \in W_0^{1,\infty}(\Omega)$ such that $\|\nabla \zeta\|_{L^\infty} \le 1$ and $\zeta(x_1, 0) = x_1$ for $x_1 \in [0, 1]$. Such a ζ clearly exists. Then, $\|T_A\|_W \ge \langle T_A, \zeta \rangle = |A|$.

We next prove item *1*, which actually holds for any Borel subset of $\mathbb{R} \cap \Omega$. Assume first that $A \subset \mathbb{R} \cap \Omega$ is open. Then we may write A as a disjoint union $A = \cup(N_j, P_j)$, and it is clear that

$$T_A = \sum (\delta_{P_j} - \delta_{N_j}), \text{ with } \sum |P_j - N_j| = |A| = \|T_A\|_W. \tag{15.201}$$

In particular, we have "*b*) $\Longrightarrow$ *a*)" in item 2. Assume next that $A = U \setminus V$, with U, V open subsets of $\mathbb{R} \cap \Omega$ and $V \subset U$. Then, $T_A = T_U - T_V$, and thus $T_A \in \mathscr{E}$, by the previous step. In addition, by (15.200) and Proposition 2.3, we may write

$$T_{U \setminus V} = \sum (\delta_{P_j} - \delta_{N_j}), \text{ with } \sum |P_j - N_j| \le 2|U \setminus V|. \tag{15.202}$$

Next let A be a general Borel subset of $\mathbb{R} \cap \Omega$. Consider a non-increasing sequence $(U_k)_{k \ge 0}$ of open subsets of $\mathbb{R} \cap \Omega$ such that $U_k \supset A$, $|U_k| \to |A|$ and

$$|U_k \setminus U_{k+1}| \le 2^{-k}. \tag{15.203}$$

Then,

$$T_A = T_{U_0} - \sum_{k \ge 0} T_{U_k \setminus U_{k+1}}. \tag{15.204}$$

We complete the proof of item *1* by combining (15.201)–(15.204).

We next turn to the proof of "*a)* $\implies$ *b)*" in item 2. Here, we do use the fact that $A \subset [0, 1]$. We assume that *a)* holds and that, in addition, we have $P_j \neq N_j$ for each j. The key ingredient is the following

Lemma 15.36. *Let* $\zeta_0(x_1, 0) := |A \cap [0, x_1]|$, $\forall x_1 \in [0, 1]$. *Consider the set*

$$X_A := \left\{\zeta \in W_0^{1,\infty}(\Omega);\ \|\nabla\zeta\|_{L^\infty} \leq 1,\ \zeta = \zeta_0 \text{ in } [0, 1] \times \{0\}\right\}.$$

Let $P, N \in \overline{\Omega}$ *be such that* $P \neq N$ *and at least one of the points* P *or* N *does not belong to* $[0, 1] \times \{0\}$. *Then there exists some* $\zeta \in X_A$ *such that* $|\zeta(P) - \zeta(N)| < |P - N|$.

Proof of Proposition 15.3 *completed.* If $\zeta \in X_A$, then $\dfrac{\partial}{\partial x_1}\zeta(\cdot, 0) = \chi_A$ in $[0, 1]$, and thus $\langle T_A, \zeta\rangle = \|T_A\|_W$. By *a)*, we have

$$\sum |P_j - N_j| = \|T_A\|_W = \sum(\zeta(P_j) - \zeta(N_j)) \leq \sum |P_j - N_j|;$$

this leads to $\zeta(P_j) - \zeta(N_j) = |P_j - N_j|$ for each j and for each $\zeta \in X_A$. By Lemma 15.36, we must have $P_j, N_j \in [0, 1]$ for each j, and thus $\zeta_0(P_j) - \zeta_0(N_j) = |P_j - N_j|$. Since ζ_0 is non-decreasing in $[0, 1]$, we find that $N_j < P_j$. Therefore, we have

$$P_j - N_j = \zeta_0(P_j) - \zeta_0(N_j) = |A \cap [P_j, N_j]|,$$

and thus $(N_j, P_j) \subset A$ a.e. Set $U := \cup(N_j, P_j)$, so that $U \subset A$ a.e. Let $B := A \setminus U$. We complete the proof by showing that $|B| = 0$. Indeed, consider a 1-Lipschitz map ξ in $W_0^{1,\infty}(\Omega)$ such that $\xi(x_1) = |B \cap [0, x_1]|$, $\forall x_1 \in [0, 1]$. Then, on the one hand we have $\langle T_A, \xi\rangle = |B|$, and on the other hand $\xi(P_j) - \xi(N_j) = 0$ for each j. Thus, $|B| = 0$. □

Proof of Lemma 15.36. We start with a preliminary fact, whose easy proof is left to the reader. Let $\mathscr{A} := [0, 1] \times \{0\}$ and let $P \notin \mathscr{A}$. Define

$$l_1 := \max\{\zeta_0(R) - |P - R|;\ R \in \mathscr{A}\},\ l_2 := \min\{\zeta_0(R) + |P - R|;\ R \in \mathscr{A}\}.$$

Then, $l_1 < l_2$, and the function $g : \mathscr{A} \cup \{P\} \to \mathbb{R}$, $g(R) := \begin{cases}\zeta_0(R), & \text{if } R \in \mathscr{A}\\ l, & \text{if } R = P\end{cases}$, is 1-Lipschitz if and only if $l \in [l_1, l_2]$. In particular, for sufficiently small $\varepsilon > 0$, the function $g(R) := \begin{cases}\zeta_0(R), & \text{if } R \in \mathscr{A}\\ l_2 - \varepsilon, & \text{if } R = P\end{cases}$, is 1-Lipschitz.

We next note that the function $\zeta_1 := \begin{cases}\zeta_0, & \text{in } \mathscr{A}\\ 0, & \text{in } \overline{\Omega} \setminus \mathbb{D}(0, 2)\end{cases}$ is 1-Lipschitz, and thus has a 1-Lipschitz extension $\zeta_2 \in W_0^{1,\infty}(\Omega; [0, 2])$, given by (2.52). We may take $\zeta = \zeta_2$ in the following cases: either $|P - N| > 2$, or $|P| \geq 2$ and $|N| \geq 2$. Therefore, we may assume that we are in the remaining case, and thus $|P| \leq 4$ and $|N| \leq 4$. We will then construct a 1-Lipschitz extension $\zeta_3 : \mathscr{A} \cup \{P, N\} \to [0, 4]$ of ζ_0 satisfying $|\zeta_3(P) - \zeta_3(N)| < |P - N|$. Once ζ_3 is constructed, we extend it to $\overline{\Omega}$ as above (using (15.196)).

Set

$$\zeta_4(R) := \min\{\zeta_0(S) + |R - S|;\ S \in \mathscr{A}\}.$$

If $|\zeta_4(P) - \zeta_4(N)| < |P - N|$, then we let $\zeta_3 = \zeta_4$ and we are done. It remains to investigate the case where $|\zeta_4(P) - \zeta_4(N)| = |P - N|$; we may assume that $\zeta_4(P) > \zeta_4(N)$, and thus $\zeta_4(P) - \zeta_4(N) = |P - N|$.

Assume first that $P \notin \mathscr{A}$. We then let

$$\zeta_3(x) := \begin{cases} \zeta_4(x), & \text{if } x \in \mathscr{A} \cup \{N\} \\ \zeta_4(P) - \varepsilon, & \text{if } x = P \end{cases}. \tag{15.205}$$

Using the preliminary fact and (15.205), it is easy to see that ζ_3 has all the required properties when $\varepsilon > 0$ is sufficiently small.

We complete the proof by showing that the remaining case, i.e., $P \in \mathscr{A}$, $N \notin \mathscr{A}$, $\zeta_4(P) - \zeta_4(N) = |P - N|$, cannot occur. Argue by contradiction. Let $R \in \mathscr{A}$ be such that $\zeta_4(N) = \zeta_0(R) + |N - R|$. Since $\zeta_4(P) = \zeta_0(P)$, we find that

$$|N - R| + |P - N| = \zeta_0(P) - \zeta_0(R) \le |P - R| \le |N - R| + |P - N|,$$

and thus N is on the segment $[P, R]$. This contradicts the assumption that $N \notin \mathscr{A}$. □

Remark 15.16. The proof of Lemma 15.36 leads to the following more general fact: let C be a compact convex set in a vector space E equipped with a strictly convex norm $|\ |$. Let $f : C \to \mathbb{R}$ be 1-Lipschitz and let $x, y \in E$ be such that $x \notin C$ and $x \ne y$. Then, f has a 1-Lipschitz extension $g : E \to \mathbb{R}$ such that $|g(x) - g(y)| < |x - y|$.

Remark 15.17. We mention here without proof that a sufficient condition for having $\inf = \min$ in (2.156) is that we may write

$$T = \sum (\delta_{P_j} - \delta_{N_j}),\ \text{with} \sum |P_j - N_j| < \infty \text{ and } \mathscr{H}^1\left(\overline{\cup\{P_j, N_j\}}\right) = 0$$

(see Ponce [298, Theorem 3]).

Proposition 15.4. *Let $\mathscr{C}$ be a 1-rectifiable current in $\Omega \subset \mathbb{R}^2$. Then there exist sequences $P_j, N_j \in \overline{\Omega}$ such that $\sum_j |P_j - N_j| < \infty$ and $\partial\mathscr{C} = \sum_j (\delta_{P_j} - \delta_{N_j})$.*

Proof. We write, as in (2.175), $\mathscr{C} = \sum_i \mathscr{C}_i$, with $\mathscr{C}_i = \tau_i \mathscr{H}^1 \llcorner (S_i \cap \Omega)$. It suffices to find, for each i, sequences $P_j = P_j^i, N_j = N_j^i \in \overline{\Omega}$ such that

$$\partial\mathscr{C}_i = \sum_j (\delta_{P_j} - \delta_{N_j}) \text{ and } \sum_j |P_j - N_j| \le 2\mathscr{H}^1(S_i \cap \Omega). \tag{15.206}$$

Let C_i be a C^1 oriented curve such that $S_i \subset C_i$. Property (15.206) is obtained by adapting the proof of item *1* in Proposition 15.3 (see, more specifically, (15.201)–(15.204)); it suffices to replace the line segment $[0, 1]$ with C_i. □

15.10 Jacobians of $W^{1,p}(\Omega;\mathbb{S}^1)$ maps

This appendix is a follow-up of Section 2.11.2. More specifically, given $1 < p < 2$, we discuss the possibility of characterizing distributions of the form $T = \pi\sum(\delta_{P_j} - \delta_{N_j})$ arising as Jacobians of some $u \in W^{1,p}(\Omega;\mathbb{S}^1)$. More generally, we discuss (in any dimension $N \geq 2$) properties of the set

$$\mathscr{A}_p := \left\{\frac{1}{\pi} * Ju;\ u \in W^{1,p}(\Omega;\mathbb{S}^1)\right\},\ 1 \leq p < 2.$$

As usual, when $N = 2$ we omit the $*$ operator.

Proposition 15.5. *Let $N = 2$ and $1 < p < 2$. Let $P_j, N_j \in \overline{\Omega}$ be such that $\sum |P_j - N_j|^{2/p-1} < \infty$ and set $T := \sum(\delta_{P_j} - \delta_{N_j})$. Then, $T \in \mathscr{A}_p$, i.e., there exists some $u \in W^{1,p}(\Omega;\mathbb{S}^1)$ such that $Ju = \pi T$.*

Proof. This is an immediate consequence of Theorem 2.20 and of the Morrey embedding $W^{1,q}(\Omega) \hookrightarrow C^\alpha(\overline{\Omega})$, with $\alpha := 2/p - 1$. Indeed, we have

$$|\langle T, \zeta\rangle| \leq \sum |\zeta(P_j) - \zeta(N_j)| \leq |\zeta|_{C^\alpha} \sum |P_j - N_j|^\alpha \leq C|\zeta|_{W^{1,q}},\ \forall\, \zeta \in C_c^\infty(\Omega).$$

We find that T belongs to $[W_0^{1,q}(\Omega)]^*$. We conclude via Theorem 2.20. □

Proposition 15.6. *Let $1 < p < 2$. Then there exists some $u \in W^{1,p}(\Omega;\mathbb{S}^1)$ such that $T := \dfrac{1}{\pi}Ju$ cannot be written as*

$$T = \sum(\delta_{P_j} - \delta_{N_j}) \text{ with } P_j, N_j \in \overline{\Omega} \text{ and } \sum |P_j - N_j|^{2/p-1} < \infty. \tag{15.207}$$

Proof. Let $\alpha := 2/p - 1$ and $\beta := 2 - p$, so that $0 < \alpha < \beta < 1$.

We start by presenting a dipole construction adapted to our context. Given two points P, N, we denote by $S_{P,N}$ the square having P and N as opposite vertices. We claim that there exists some $u = u_{P,N} \in W^{1,p}_{loc}(\mathbb{R}^2;\mathbb{S}^1)$ such that:

$$u = 1 \text{ outside } S_{P,N},\ Ju = \pi(\delta_P - \delta_N), \text{ and } \|\nabla u\|^p_{L^p} \leq C|P - N|^\beta.$$

Here, C depends only on p. Indeed, when $P = (0,0)$ and $N = (0,1)$, this follows from the dipole construction in Section 2.5 (with parameters $L := 1/2$ and $n := 1$). The general case is obtained by scaling. As a consequence, if we consider a sequence of disjoint intervals (M_k, R_k) on the real line (identified with $\mathbb{R}\times\{0\}$) such that $\sum(R_k - M_k)^\beta < \infty$, then there exists some $u \in W^{1,p}_{loc}(\mathbb{R}^2;\mathbb{S}^1)$ such that $Ju = \pi\sum(\delta_{R_k} - \delta_{M_k})$. Indeed, it suffices to let $u := \prod u_{R_k,M_k}$.

We will prove that u as above satisfies the conclusion of the proposition for an appropriate choice of Ω, M_k, and R_k. We let Ω contain $(-1,2)^2$ and choose M_k, and R_k as follows:

$$M_1 := 0,\ R_1 := l_1,\ M_{k+1} := R_k + l_k,\ R_{k+1:} = M_{k+1} + l_{k+1},\ \forall\, k \geq 1, \tag{15.208}$$

where

$$l_k := \frac{K}{j^{1/\alpha}},\ K > 0 \text{ the constant such that } \sum l_k = \frac{1}{2}.$$

[Here, we identify $M_k \in \mathbb{R}$ with $(M_k, 0) \in \mathbb{R}^2$. Similarly for R_k.]

By construction, all the points M_k, R_k are in Ω and we have

$$\sum (R_k - M_k)^\beta < \infty \tag{15.209}$$

and

$$\sum (l_k)^\alpha = \infty. \tag{15.210}$$

Let $T := \sum(\delta_{R_k} - \delta_{M_k})$. By (15.209) and the above dipole construction, we have $T \in \mathscr{A}_p$. We claim that there are no points $P_j, N_j \in \mathbb{R}^2$ (and thus no points in $\overline{\Omega}$) such that

$$T = \sum(\delta_{P_j} - \delta_{N_j}) \text{ and } \sum |P_j - N_j|^\alpha < \infty. \tag{15.211}$$

The proof of this fact relies on the following result.

Lemma 15.37. *Let $t \in \mathbb{N}$. Consider the function $\xi_t : [-1, 2] \to \mathbb{R}$ defined as follows:*

a) $\xi_t(x_1) = 0$ for $x_1 \le 0$ or $x_1 \ge M_{t+1}$
b) ξ_t is affine on $[M_k, R_k]$ and on $[R_k, M_{k+1}]$, $\forall\, k \le t$
c) $\xi_t(R_k) = (l_k)^\alpha$ and $\xi_t(M_{k+1}) = 0$, $\forall\, k \le t$.

Then, $|\xi_t|_{C^\alpha} \le C$, with C independent of t.

Proof of Lemma 15.37. Let $-1 \le x < y \le 2$. We have to obtain a uniform upper bound for the quotient $\dfrac{|\zeta_t(x) - \zeta_t(y)|}{(y-x)^\alpha}$. We consider only the case where $0 \le x < y \le M_{t+1}$; the remaining cases are easier. Let k and s be such that $M_k \le x < M_{k+1}$ and $M_s < y \le M_{s+1}$; thus, $0 \le k \le s \le t$.

Consider first the case where $s \ge k+2$. Then,

$$\frac{|\zeta_t(x) - \zeta_t(y)|}{(y-x)^\alpha} \le \frac{(l_k)^\alpha}{(y-x)^\alpha} \le \frac{(l_k)^\alpha}{(l_{k+1})^\alpha} \le C.$$

When $s \in \{k, k+1\}$, we have

$$\frac{|\zeta_t(x) - \zeta_t(y)|}{(y-x)^\alpha} \le (y-x)^{1-\alpha} \sup_{[M_k, M_{k+2}]} |\zeta_t'| \le (2(l_k + l_{k+1}))^{1-\alpha}(l_{k+1})^{\alpha-1} \le C. \quad \square$$

Proof of Proposition 15.6 *completed.* Argue by contradiction and assume that (15.211) does hold. Then we have

$$|\langle T, \zeta\rangle| \le C|\zeta|_{C^\alpha},\ \forall\, \zeta \in W_0^{1,\infty}(\Omega). \tag{15.212}$$

We now let $\zeta_t(x_1, x_2) := \xi_t(x_1)\varphi(x_2)$, where $\varphi \in C_c^\infty((-1, 2))$ is fixed and satisfies $\varphi(0) = 1$. By (15.212) and Lemma 15.37, the sequence $(\langle T, \zeta_t\rangle)$ is bounded. However, we have

$$\langle T,\zeta_t\rangle=\sum_{k=1}^{t}(l_k)^\alpha\to\infty\quad\text{as } t\to\infty.$$ □

We next prove that the sets $\mathscr{A}_p$, $1\le p<2$ (defined in Sections 2.11.2 and 3.8.3), are (strictly) decreasing with p. For simplicity, we treat in detail only the case $N=2$ (see, however, Remark 15.18 below when $N\ge 3$).

Lemma 15.38. *Let $N=2$ and $1\le p<q<2$. Then $\mathscr{A}_q$ is strictly contained in $\mathscr{A}_p$.*

Proof. Since $\mathscr{A}_p$ is non-increasing with p, we may assume that $1<p<q<2$. We adjust the construction in Proposition 15.6. We still define R_k, M_k as in (15.208) and set $T:=\sum(\delta_{R_k}-\delta_{M_k})$, but we adjust differently the numbers l_k and K. First, we require (15.209) to hold, so that $T\in\mathscr{A}_p$, by the proof of Proposition 15.6. We next require that

$$\sum_k k^{-q/(q-1)}\,(l_k)^{2-q/(q-1)}<\infty. \tag{15.213}$$

In view of the assumption $1<p<q<2$, requirements (15.209) and (15.213) are satisfied, for example, with the choice $l_k:=1/k^{(1+\varepsilon)/(2-p)}$ for sufficiently small ε.

We claim that

$$T\notin[W_0^{1,q/(q-1)}(\Omega)]^*, \tag{15.214}$$

and thus, by Theorem 2.20, $T\notin\mathscr{A}_q$. The proof of (15.214) relies on an adapted "dipole style" construction. First, consider some smooth function ξ such that $\operatorname{supp}\xi\subset(0,2)\times(-1,1)$ and $\xi(1,0)=1$. In view of condition (15.213), the series

$$\zeta(x):=\sum_k\frac{1}{k}\,\xi((x-M_k)/l_k)$$

converges in $W_0^{1,q/(q-1)}(\Omega)$. However, we have

$$T(\zeta)=\sum_k\frac{1}{k}\xi(R_k)=\sum_k\frac{1}{k}=\infty,$$

and thus (15.214) holds, as claimed. □

Remark 15.18. The above construction can be adapted to $N\ge 3$. Indeed, assume, for example, that $(-1,2)\times B_2(0)\subset\Omega$ (the ball $B_2(0)$ is $(N-1)$-dimensional). Let γ be the unit sphere of $\mathbb{R}^{N-1}$, with a fixed orientation. When $P\in\mathbb{R}$, let $\delta_{P\times\gamma}$ denote the integration over $P\times\gamma$ (with the orientation inherited from that of γ). Consider M_j, R_j as in the proof of Lemma 15.38. By adapting the proof of Lemma 15.38, we may prove that, if we set $T:=\sum(\delta_{R_j\times\gamma}-\delta_{M_j\times\gamma})$, then $T\in\mathscr{A}_p$ but $T\notin\mathscr{A}_q$.

15.11 When *Ju* is a measure

In this section, we establish the version of the Federer–Fleming boundary rectifiability theorem (see Federer [147, Theorem 4.2.16 (2), p. 413]) required in the proof of Theorem 3.11, and prove the assertion stated in Remark 3.9 in Section 3.7.

Lemma 15.39. *Let $1 \leq \ell \leq N$. Let T be an ℓ-rectifiable current on Ω such that ∂T has finite mass. Then, ∂T is rectifiable.*

Proof. Let us note that, if U is an m-current that has finite mass and is locally rectifiable (i.e., if each point $x_0 \in \Omega$ has an open neighborhood V such that $U \llcorner V$ is rectifiable), then U is rectifiable. It therefore suffices to prove that ∂T is locally rectifiable.

Let $x_0 \in \Omega$ and $0 < R < \operatorname{dist}(x_0, \partial\Omega)$. Since T has finite mass, for a.e. $0 < r < R$ the current $T(r) := \partial(T \llcorner B(x_0, r)) - (\partial T) \llcorner B(x_0, r)$ has finite mass [147, Section 4.2.1]. For any such r, $T \llcorner B(x_0, r)$ is a compactly supported ℓ-rectifiable current on Ω, whose boundary $T(r) + (\partial T) \llcorner B(x_0, r)$ has finite mass (since, by assumption, ∂T has finite mass). By the standard form of the boundary rectifiability theorem [147, Theorem 4.2.16 (2), p. 413], $T(r) + (\partial T) \llcorner B(x_0, r)$ is an $(\ell - 1)$-rectifiable current. Since, on the other hand, $T(r)$ is supported on the sphere $S(x_0, r)$ [147, Section 4.2.1], we find that $(\partial T) \llcorner B(x_0, r)$ is an $(\ell - 1)$-rectifiable current. □

Lemma 15.40. *Let $N \geq 3$ and $\Omega \subset \mathbb{R}^N$ be a smooth bounded simply connected domain. There exists some $u \in W^{1,1}(\Omega; \mathbb{S}^1)$ such that $Ju \in \mathcal{M}$ and, for every $p > N/(N-1)$, there exists no $\varphi \in W^{1,1}(\Omega; \mathbb{S}^1)$ such that $u \wedge \nabla u - \nabla\varphi \in L^p$.*

Proof. We let $N = 3$. [The argument in other dimensions is similar.] Assume that Ω contains the closed unit ball, denoted by B.

For $0 < r \leq 1$, let $C_r \subset \Omega$ denote the circle $\{(x_1, x_2, 0);\ |(x_1, x_2)| = r\}$, with the orientation inherited from $\mathbb{R}^2$; we identify C_r with a 1-current. Let, for $j \geq 4$, $r_j := 1/(j \ln^2 j)$.

We set $T := \sum_j C_{r_j}$. Clearly, $T = \partial M$, where $M := \sum_j \mathbb{D}_{r_j}$ and (with the appropriate orientation) $\mathbb{D}_r$ is the disk bounded by C_r. Since $\sum_j (r_j)^2 < \infty$, M is 2-rectifiable, which implies that $T \in \mathcal{F}$, and thus $\pi\, T = Ju$ for some $u \in W^{1,1}(\Omega; \mathbb{S}^1)$ (see Theorem 3.8). On the other hand, we have $\sum_j |C_{r_j}| < \infty$, and thus Ju is a measure.

For further use, let us note the following estimate, obtained by a straightforward application of the Maclaurin–Cauchy test:

$$\sum_{r_j \geq 2A} r_j \ln(1/r_j) \geq K \ln\ln(1/A), \tag{15.215}$$

with $K > 0$ independent of small $A > 0$.

We claim that, for every $p > 3/2$, there exists no $\varphi \in W^{1,1}(\Omega; \mathbb{R})$ such that $u \wedge \nabla u - \nabla\varphi \in L^p$. By duality (see Theorem 1.16), this amounts to

$$\sup\{\langle Ju, *\zeta\rangle;\ \zeta \in C_c^\infty(\Omega; \Lambda^1),\ \|d\zeta\|_{L^q} \leq 1\} = \infty \tag{15.216}$$

for any $q < 3$.

In turn, (15.216) is obtained as follows. For $f \in C_c^\infty((0, 1]; \mathbb{R})$, set

$$\zeta_f(x) := f(|x|)\,(x_1\,dx_2 - x_2\,dx_1), \;\; \forall x = (x_1, x_2, x_3) \in \Omega. \tag{15.217}$$

Clearly, we have $\zeta_f \in C_c^\infty(\Omega; \Lambda^1)$ and, for any $q \geq 1$,

$$\|d\zeta_f\|_{L^q(\Omega)} \leq C(\|f(|x|)\|_{L^q(B)} + \||x|f'(|x|)\|_{L^q(B)}).$$

On the other hand, we have $\langle \pi\, T, \zeta_f\rangle = 2\pi^2 \sum_j (r_j)^2 f(r_j)$. In order to obtain (15.216), it thus suffices to construct a sequence (f_k) such that

$$\|f_k(|x|)\|_{L^q(B)} + \||x|(f_k)'(|x|)\|_{L^q(B)} \leq C, \text{ with } C \text{ independent of } k, \tag{15.218}$$

and

$$\sum_j (r_j)^2 f_k(r_j) \to \infty \text{ as } k \to \infty. \tag{15.219}$$

Let $\psi \in C^\infty([0, \infty); [0, 1])$ be such that $\psi(r) = 1$ if $r \leq 1/2$ and $\psi(r) = 0$ if $r \geq 1$. Let $g(r) := -\ln r/r, 0 < r < 1$. We set $f_k(|x|) := \psi(|x|)\,(1 - \psi(|x|/(2r_k)))\,g(|x|)$, $\forall x \in \Omega$, and claim that f_k has the desired properties (15.218) and (15.219).

Indeed, on the one hand we have $\|g(|x|)\|_{L^q(B)} + \||x|g'(|x|)\|_{L^q(B)} < \infty$ (since $q < 3$) and $\psi(|x|/(2r_k)) \to 0$ in $W^{1,q}(\Omega)$ (since $q < 3$ and $r_k \to 0$). This implies (15.218).

On the other hand, for large k we have (using the properties of ψ and (15.215))

$$\sum_j (r_j)^2 f_k(r_j) \geq \sum_{r_j \geq 2r_k} r_j \ln(1/r_j) \geq K \ln\ln(1/r_k) \to \infty \text{ as } k \to \infty. \qquad \square$$

15.12 Products in fractional Sobolev spaces

We start by recalling the following fact (see, for example, Runst and Sickel [312] for the corresponding result in $\mathbb{R}^N$).

Lemma 15.41. *Let $s > 0$, $1 \leq p < \infty$. If $u, v \in W^{s,p}(\mathbb{R}^N) \cap L^\infty(\mathbb{R}^N)$, then $uv \in W^{s,p}$ and*

$$\|uv\|_{W^{s,p}} \leq C(\|u\|_{L^\infty}\|v\|_{W^{s,p}} + \|v\|_{L^\infty}\|u\|_{W^{s,p}}). \tag{15.220}$$

In addition, the map $(u, v) \mapsto uv$ is continuous in the following sense: if $\|u_n - u\|_{W^{s,p}} \to 0$, $\|v_n - v\|_{W^{s,p}} \to 0$ and if $\|u_n\|_{L^\infty}, \|v_n\|_{L^\infty} \leq C$, then $u_n v_n \to uv$ in $W^{s,p}$.

Same in $W^{s,p}(\mathbb{S}^1) \cap L^\infty(\mathbb{S}^1)$.

Proof of Lemma 15.41 *on* $\mathbb{S}^1$. We establish, for example, (15.220), the remaining part of the proof being similar.

We identify maps on $\mathbb{S}^1$ with 2π-periodic maps, as explained in Remark 15.1. Let $\zeta \in C_c^\infty(\mathbb{R}; [0, 1])$ be such that $\zeta(x) = 1$ when $|x| \leq 4\pi$. Let us note the straightforward estimate

$$\|\zeta u\|_{W^{s,p}} \lesssim \|u\|_{W^{s,p}}. \tag{15.221}$$

We apply Lemma 15.41 (on $\mathbb{R}$) to $\widetilde{u} := \zeta\, u$ and $\widetilde{v} := \zeta\, v$. We find that

$$\begin{aligned}\|uv\|_{W^{s,p}} &\leq \|\widetilde{u}\widetilde{v}\|_{W^{s,p}} \lesssim \|\widetilde{u}\|_{L^\infty}\|\widetilde{v}\|_{W^{s,p}} + \|\widetilde{v}\|_{L^\infty}\|\widetilde{u}\|_{W^{s,p}}\\ &\leq \|u\|_{L^\infty}\|\widetilde{v}\|_{W^{s,p}} + \|v\|_{L^\infty}\|\widetilde{u}\|_{W^{s,p}} \lesssim \|u\|_{L^\infty}\|v\|_{W^{s,p}} + \|v\|_{L^\infty}\|u\|_{W^{s,p}},\end{aligned}$$

the last inequality being a consequence of (15.221). □

We next prove a regularity result for products in Sobolev spaces.

Lemma 15.42. *Let* $N \geq 1, 1 < s < \infty, 1 \leq p < \infty$. *Let* $u, v \in W^{s,p}(\mathbb{R}^N) \cap L^\infty(\mathbb{R}^N)$. *Then,* $uDv \in W^{s-1,p}(\mathbb{R}^N)$.

Proof. If s is an integer, the conclusion follows easily from the Gagliardo–Nirenberg inequality. We henceforth assume that s is not an integer. In this case, we use a paraproduct technique (see, for example, Bony [40], Alinhac and Gérard [7], or Chemin [116]). Given $f \in \mathscr{S}'$, let $f = \sum_{j\geq 0} f_j$ be "its" Littlewood–Paley decomposition, as in (15.6). Thus, $u = \sum u_j$ and $Dv = \sum Dv_k$ in $\mathscr{S}'$.

We have $uDv = \sum(r_j + s_j)$, where

$$r_j := u_j \sum_{k\leq j-1} Dv_k \text{ and } s_j := Dv_j \sum_{k\leq j} u_k.$$

Since, clearly, the φ_j's in (15.6) satisfy

$$\Big\|\sum_{k\leq j}\varphi_k\Big\|_{L^1} \leq C, \ \Big\|\sum_{k\leq j} D\varphi_k\Big\|_{L^1} \leq C2^j, \ \forall j \geq 0,$$

we obtain

$$\Big\|\sum_{k\leq j} v_k\Big\|_{L^q} \leq C\|v\|_{L^q}, \ \forall q, \tag{15.222}$$

$$\Big\|\sum_{k\leq j} Dv_k\Big\|_{L^q} \leq C2^j\|v\|_{L^q}, \ \forall q, \tag{15.223}$$

and the same inequalities hold for u. Therefore,

$$\|r_j\|_{L^p}^p \leq \|u_j\|_{L^p}^p \Big\|\sum_{k\leq j-1} Dv_k\Big\|_{L^\infty}^p \leq C2^{jp}\|v\|_{L^\infty}^p\|u_j\|_{L^p}^p. \tag{15.224}$$

On the other hand, we have $v_j = \sum_{k\leq j+2}(v_j)_k$, since, for $k \geq j+3$,

$$\mathscr{F}((v_j)_k) = \psi_j \psi_k \mathscr{F}(v) = 0.$$

Therefore,

$$\|Dv_j\|_{L^q} = \left\| \sum_{k \le j+2} D(v_j)_k \right\|_{L^q} \le C2^j \|v_j\|_{L^q}, \ \forall q,$$

by (15.223) applied to v_j. Consequently,

$$\|s_j\|_{L^p}^p \le C\|u\|_{L^\infty}^p \|Dv_j\|_{L^p}^p \le C2^{jp} \|u\|_{L^\infty}^p \|v_j\|_{L^p}^p. \tag{15.225}$$

Using (15.224), (15.225), and Theorem 15.3, we find

$$\sum_{j \ge 0} 2^{(s-1)jp} \|r_j + s_j\|_{L^p}^p \le C(\|u\|_{L^\infty}^p \|v\|_{W^{s,p}}^p + \|v\|_{L^\infty}^p \|u\|_{W^{s,p}}^p). \tag{15.226}$$

Since $\operatorname{supp} \mathscr{F}(r_j + s_j) \subset B_{2^{j+3}}$, Lemma 15.2 (applied with $u^j := r_j + s_j$) combined with (15.226) yields that $uDv \in W^{s-1,p}$ and that

$$\|uDv\|_{W^{s-1,p}} \le C(\|u\|_{L^\infty} \|v\|_{W^{s,p}} + \|v\|_{L^\infty} \|u\|_{W^{s,p}}). \qquad \square$$

15.13 Maximal estimates

This short section is essentially devoted to the proof of Lemma 15.45, originally devised in Bousquet [61, Section 5], and used in the proof of Theorem 10.3.

To start with, we recall the following elementary estimate (see Stein [331, eq (17), p. 57]). If $\zeta \in L^1(\mathbb{R}^N; [0, \infty])$ is a radial non-increasing function, then

$$\int_{\mathbb{R}^N} \zeta(y - x)|g(y)|\, dy \le \left(\int_{\mathbb{R}^N} \zeta(y)\, dy \right) \mathscr{M} g(x), \ \forall x \in \mathbb{R}^N. \tag{15.227}$$

Here, $\mathscr{M} g$ stands for the standard maximal function of a measurable function $g : \mathbb{R}^N \to \mathbb{R}$.

Lemma 15.43. *Let $f : \mathbb{R}^N \to \mathbb{R}$ be a Lipschitz function, $r > 0$ and $0 < \beta \le 1$. Then,*

$$\int_{S(x,r)} |f(y) - f(x)|^\beta\, dy \lesssim r^{N+\beta-1} [\mathscr{M} |\nabla f|(x)]^\beta. \tag{15.228}$$

Proof. We may assume that $x = 0$. Then,

$$
\begin{aligned}
\int_{S(0,r)} |f(y)-f(0)|^\beta\,dy &= r^{N-1}\int_{S(0,1)} |f(r\omega)-f(0)|^\beta\,d\omega \\
&\le r^{N+\beta-1}\int_{S(0,1)}\left(\int_0^1 |(\nabla f)(tr\omega)|\,dt\right)^\beta d\omega \\
&\lesssim r^{N+\beta-1}\left(\int_{S(0,1)}\int_0^1 |(\nabla f)(tr\omega)|\,dtd\omega\right)^\beta \\
&= r^{N-1}\left(\int_{B_r(0)} \frac{1}{|y|^{N-1}}|(\nabla f)(y)|\,dy\right)^\beta \\
&\lesssim r^{N+\beta-1}[\mathcal{M}|\nabla f|(0)]^\beta,
\end{aligned}
$$

the last inequality following from (15.227). □

Multiplying (15.228) by $r^{-N-\gamma}$ and integrating over r, we obtain the following

Lemma 15.44. *Let* $0<\gamma<\beta\le 1$, *and let* f *be a Lipschitz function. Then,*

$$\int_{B_\delta(x)} \frac{|f(y)-f(x)|^\beta}{|y-x|^{N+\gamma}}\,dy \lesssim \delta^{\beta-\gamma}[\mathcal{M}|\nabla f|(x)]^\beta. \tag{15.229}$$

Lemma 15.45. *Let* f *be a Lipschitz function such that* $\|f\|_{L^\infty}\le 1$. *Let* $0<\gamma<\beta\le 1$. *Then,*

$$\int_{\mathbb{R}^N} \frac{|f(y)-f(x)|^\beta}{|y-x|^{N+\gamma}}\,dy \lesssim [\mathcal{M}|\nabla f|(x)]^\gamma. \tag{15.230}$$

Proof. Let $\delta>0$. Then,

$$\int_{|y-x|>\delta} \frac{|f(y)-f(x)|^\beta}{|y-x|^{N+\gamma}}\,dy \lesssim \delta^{-\gamma}. \tag{15.231}$$

By (15.229) and (15.231), we have

$$\int_{\mathbb{R}^N} \frac{|f(y)-f(x)|^\beta}{|y-x|^{N+\gamma}}\,dy \lesssim \delta^{-\gamma}+\delta^{\beta-\gamma}[\mathcal{M}|\nabla f|(x)]^\beta. \tag{15.232}$$

We obtain (15.230) by taking $\delta := [\mathcal{M}|\nabla f|(x)]^{-1}$ in (15.232). □

15.14 Smoothing

In this section, we let, as in (15.59)–(15.61), (15.64), and (15.65) in Appendix 15.5,

$$\Omega_\varepsilon := \{x \in \Omega;\ \mathrm{dist}\,(x, \partial\Omega) > \varepsilon\},$$
$$\rho : \mathbb{R}^N \to \mathbb{R} \text{ is an integrable function such that } \mathrm{supp}\,\rho \subset \overline{B}_1(0),$$
$$\rho_\varepsilon(x) := \frac{1}{\varepsilon^N}\rho(x/\varepsilon),\ \forall x \in \mathbb{R}^N,\ \forall \varepsilon > 0,$$
$$u_\varepsilon(x) := u * \rho_\varepsilon(x),\ \forall \varepsilon > 0,\ \forall x \in \Omega_\varepsilon,$$
$$U := \{(x, \varepsilon);\ \varepsilon > 0,\ x \in \Omega_\varepsilon\},$$
$$v(x, \varepsilon) := u_\varepsilon(x) = u * \rho_\varepsilon(x), \forall\,(x, \varepsilon) \in U.$$

The general theme here is the following. Assuming that u has "some smoothness," what can be said about v and u_ε?

As the next result shows, in the Sobolev setting the convergence $u_\varepsilon \to u$ holds under weak assumptions on ρ.

Lemma 15.46. *Let $\Omega \subset \mathbb{R}^N$ be an open set. Let ρ be a in integrable function supported in $\overline{B}_1(0)$ and such that $\int \rho = 1$. Let $s > 0$, $1 \le p < \infty$, and $u \in W^{s,p}(\Omega)$. Then, $u_\varepsilon \to u$ in $W^{s,p}_{loc}(\Omega)$ as $\varepsilon \to 0$.*

Proof. It clearly suffices to prove that

$$\|u_\varepsilon - u\|_{W^{s,p}(\mathbb{R}^N)} \to 0 \text{ as } \varepsilon \to 0,\ \forall u \in W^{s,p}(\mathbb{R}^N). \tag{15.233}$$

In turn, when s is not an integer, (15.233) is obtained by repeating the proof in Leoni [220, Proposition 17.12]. There, ρ is supposed to be a normalized bump function, but the proof uses only the fact that ρ is integrable, with compact support and with integral 1. The key ingredient is the fact that, for such ρ and $1 \le p < \infty$, we have

$$\|f * \rho_\varepsilon - f\|_{L^p(\mathbb{R}^N)} \to 0 \text{ as } \varepsilon \to 0,\ \forall f \in L^p(\mathbb{R}^N). \tag{15.234}$$

When s is an integer, we obtain (15.233) by applying (15.234) with $f = \partial^\alpha u$, $\forall\,\alpha \in \mathbb{N}^N$ with $|\alpha| \le s$. □

The following easy consequence of (15.233) was used in the proof of Theorem 10.2.

Corollary 15.10. *Let $\Omega \subset \mathbb{R}^N$ be a standard domain (in the sense of Definition 15.1), $s > 0$, and $1 \le p < \infty$. Then, $C^\infty(\overline{\Omega}; \mathbb{R}) \cap W^{s,p}(\Omega; \mathbb{R})$ is dense in $W^{s,p}(\Omega; \mathbb{R})$.*

Proof. Given $u \in W^{s,p}(\Omega)$, let $w \in W^{s,p}(\mathbb{R}^N)$ be an extension of u (which exists, since Ω is a standard domain). Let (ρ_ε) be a family of standard mollifiers and let u^ε denote the restriction of $w * \rho_\varepsilon$ to $\overline{\Omega}$. Then, $u^\varepsilon \in C^\infty(\overline{\Omega}; \mathbb{R}) \cap W^{s,p}(\Omega; \mathbb{R})$ and, by (15.233), we have $u^\varepsilon \to u$ in $W^{s,p}(\Omega)$ as $\varepsilon \to 0$. □

We next recall a standard result in trace theory.

Lemma 15.47. *Let $\Omega \subset \mathbb{R}^N$ be a Lipschitz bounded domain. Let ρ be a normalized bump function. Let $s > 0$ be non-integer and $1 \le p < \infty$. For $u \in W^{s,p}(\Omega)$, we have:*

a) $v \in W^{s+1/p,p}(U)$ *and (for appropriate seminorms satisfying* (15.235) *and* (15.236) *below)* $|v|_{W^{s+1/p,p}(U)} \le C|u|_{W^{s,p}(\Omega)}$

b) $\|v\|_{W^{s+1/p,p}(U)} \le C\|u\|_{W^{s,p}(\Omega)}$

c) $tr\, v_{|\Omega} = u$.

The same conclusions hold with $\Omega = \mathbb{R}^N$ *and* $U := \mathbb{R}^N \times (0, T)$, *with any* $T < \infty$.

Proof. Item *c)* follows from item *a)*, Lemma 15.46, and trace theory. Although items *a)* and *b)* are well known to experts, we were unable to find in the literature a proof in the full range $s > 0$, $1 \le p < \infty$. For the convenience of the reader, we present here an argument.

First, let us note that, clearly, $\|v\|_{L^p(U)} \lesssim \|u\|_{L^p(\Omega)}$. Therefore, in order to obtain item *b)* it suffices to prove item *a)* for seminorms such that

$$u \mapsto \|u\|_{L^p(\Omega)} + |u|_{W^{s,p}(\Omega)} \text{ is equivalent to } \|u\|_{W^{s,p}(\Omega)}, \tag{15.235}$$

$$v \mapsto \|v\|_{L^p(U)} + |v|_{W^{s+1/p,p}(U)} \text{ is equivalent to } \|v\|_{W^{s+1/p,p}(U)}. \tag{15.236}$$

When we work on $\Omega = \mathbb{R}^N$, it will be convenient to define, for *every* $\varepsilon > 0$ (and not only for $0 < \varepsilon < T$), $v(x, \varepsilon) := u * \rho_\varepsilon(x)$, $\forall x \in \mathbb{R}^N$.

Step 1. Proof of item *a)* when $\Omega = \mathbb{R}^N$ and $s + 1/p$ is an integer

Set $m := s + 1/p \in \mathbb{N}$. In this case, we let $|v|_{W^{m,p}(U)} := \|D^m v\|_{L^p(U)}$ (where D stands for the full differential, in x and ε) and

$$|u|^p_{W^{s,p}(\mathbb{R}^N)} := \iint_{\mathbb{R}^N \times \mathbb{R}^N} \frac{|\delta_h^m u(x)|^p}{|h|^{N+sp}}\, dxdh. \tag{15.237}$$

[Here, δ_h^m is defined as in Theorem 15.1.]

We have

$$|v|_{W^{m,p}(U)} \lesssim |u|_{W^{s,p}}, \ \forall u \in W^{s,p}(\mathbb{R}^N) \tag{15.238}$$

(see, for example, Mironescu and Russ [265, Theorem 5.4]). In this case, (15.236) is clear. For (15.235), see, for example, Triebel [339, Section 2.6.1, Theorem, p. 140].

Step 2. Proof of item *a)* when $\Omega = \mathbb{R}^N$ and $s + 1/p$ is not an integer

In this case, we fix m an integer such that $m > s + 1/p$. With $|u|_{W^{s,p}(\Omega)}$ as in (15.237) (with the m we have fixed), we have [265, Theorem 5.4]

$$\begin{aligned} I_m(v) = I_{m,s,p}(v) := \iint_{\mathbb{R}^N \times (0,\infty)} & \varepsilon^{p(m-s)-1} |D^m v(x, \varepsilon)|^p \, dx d\varepsilon \\ & \lesssim |u|^p_{W^{s,p}(\mathbb{R}^N)}, \ \forall u \in W^{s,p}(\mathbb{R}^N). \end{aligned} \tag{15.239}$$

In order to complete this step, we rely on the estimate

$$|v|^p_{W^{s+1/p,p}(U)} \lesssim I_m(v), \ \forall v \in C^\infty(\mathbb{R}^N \times (0, \infty)). \tag{15.240}$$

By contrast with (15.239), where m could be *any* integer $> s + 1/p$, in (15.240) m is *the smallest integer* $> s + 1/p$. Also, in (15.240), $|\ |_{W^{s+1/p,p}(U)}$ is defined as in Theorem 15.1 (with $M := m$ and with s replaced by $s + 1/p$). The validity of (15.240) for this value of m and for this seminorm is established in Lemma 15.48 below.

We complete this step by noting that Theorem 15.1 implies (15.235) and (15.236).

Step 3. Proof of item *a)* when Ω is a bounded Lipschitz domain
We still use the seminorms considered in Steps 1 and 2. In order to emphasize the dependence on m of the seminorm considered on Ω, we write $|u|_{W^{s,p}_m(\Omega)}$ instead of $|u|_{W^{s,p}(\Omega)}$. We rely on the following well-known version of Poincaré's inequality. There exists some polynomial P (depending on u) of degree $\le m - 1$ such that

$$\|u - P\|_{L^p(\Omega)} \lesssim |u|_{W^{s,p}_m(\Omega)} = |u - P|_{W^{s,p}_m(\Omega)}, \ \forall\, u \in W^{s,p}(\Omega) \tag{15.241}$$

(the equality on the right-hand side of (15.241) follows from the fact that $|P|_{W^{s,p}_m(\Omega)} = 0$ for every polynomial of degree $\le m - 1$).

Let us briefly recall the proof of (15.241). Argue by contradiction. If (15.241) does not hold, then there exists a sequence $(u_j) \subset W^{s,p}(\Omega)$ such that

$$\begin{aligned} \|u_j\|_{L^p(\Omega)} = 1 \text{ and } \|u_j - P\|_{L^p(\Omega)} \ge 1,\ \forall j \ge 1, \\ \forall P \text{ of degree } \le m-1, \end{aligned} \tag{15.242}$$

$$|u_j|_{W^{s,p}_m(\Omega)} \le 1/j,\ \forall j \ge 1. \tag{15.243}$$

Possibly up to a subsequence, we have $u_j \to u$ in $L^p(\Omega)$, for some $u \in W^{s,p}(\Omega)$ such that $|u|_{W^{s,p}_m(\Omega)} = 0$, and thus u is a polynomial P_0 of degree $\le m - 1$ (since Ω is connected). This contradicts the fact that, by (15.242), we have $\|u - P_0\|_{L^p(\Omega)} \ge 1$.

We now complete Step 3. Let $u \in W^{s,p}(\Omega)$ and let P be as in (15.241). Let $\widetilde{u} := u - P$. We extend $\widetilde{u}$ to $\mathbb{R}^N$; the extension is still denoted by $\widetilde{u}$. By (15.241), we have

$$\|\widetilde{u}\|_{W^{s,p}(\mathbb{R}^N)} \lesssim \|\widetilde{u}\|_{W^{s,p}(\Omega)} \lesssim |u|_{W^{s,p}_m(\Omega)}. \tag{15.244}$$

On the other hand, since P is a polynomial of degree $\le m - 1$, $P * \rho_\varepsilon$ is also a polynomial of degree $\le m - 1$ in the variables (x, ε). Thus, if we set $\widetilde{v}(x, \varepsilon) := \widetilde{u} * \rho_\varepsilon(x)$, then we have $|v|_{W^{s+1/p,p}(U)} = |\widetilde{v}|_{W^{s+1/p,p}(U)}$ (since $s + 1/p > m - 1$, and therefore $|\ |_{W^{s+1/p,p}(U)}$ "does not see" the polynomials of degree $\le m - 1$). By Steps 1 and 2, and (15.244),

$$|v|^p_{W^{s+1/p,p}(U)} = |\widetilde{v}|^p_{W^{s+1/p,p}(U)} \lesssim |\widetilde{u}|^p_{W^{s,p}_m(\mathbb{R}^N)} \lesssim |u|^p_{W^{s,p}_m(\Omega)}. \tag{15.245}$$

Estimate (15.245) completes the proof of Lemma 15.47, granted the inequality (15.240), whose proof we present now. □

Lemma 15.48. *Let $s > 0$ and $1 \le p < \infty$ be such that $s + 1/p$ is not an integer. Let m be the smallest integer $> s + 1/p$. Then we have*

$$|v|^p_{W^{s+1/p,p}(U)} \lesssim I_m(v) = I_{m,s,p}(v),\ \forall v \in C^\infty(\mathbb{R}^N \times (0,\infty)),$$

for the seminorm (15.3) *(with $M := m$ and s replaced by $s + 1/p$). More precisely, we have*

$$\iint_{\mathbb{R}^N \times (0,\infty)} \left(\int_0^\infty \sum_{j=1}^N \frac{|\delta^m_{he_j} v(x,\varepsilon)|^p}{h^{2+sp}} dh \right) dx d\varepsilon \lesssim I_m(v) \tag{15.246}$$

and

$$\int_{\mathbb{R}^N} \left(\iint_{(0,\infty)\times(0,\infty)} \frac{|\delta^m_{he_{N+1}} v(x,\varepsilon)|^p}{h^{2+sp}} dh d\varepsilon \right) dx \lesssim I_m(v). \tag{15.247}$$

Proof. Step 1. Dimensional reduction
We claim that it suffices to establish (15.246) when $N = 1$, i.e.,

$$J(v) := \int_0^\infty \left(\iint_{\mathbb{R}\times(0,\infty)} \frac{|\delta^m_{he_1} v(x,\varepsilon)|^p}{h^{2+sp}} dx dh \right) d\varepsilon \lesssim I_m(v), \tag{15.248}$$

(with $v \in C^\infty(\mathbb{R} \times (0,\infty))$) and the one-dimensional estimate

$$\begin{aligned} K(w) := &\iint_{(0,\infty)\times(0,\infty)} \frac{|\delta^m_h w(\varepsilon)|^p}{h^{2+sp}} d\varepsilon dh \\ &\lesssim \int_0^\infty \varepsilon^{p(m-s)-1} |w^{(m)}(\varepsilon)|^p d\varepsilon \end{aligned} \tag{15.249}$$

(with $w \in C^\infty((0,\infty))$).

Indeed, (15.246) (resp. (15.247)) follows from (15.248) (resp. (15.249)) and Fubini.

Step 2. Proof of (15.248)
We rely on the following homogeneous version of Lemma 7.3. If $V \in C^\infty(\mathbb{R}^N \times [0,\infty))$, $r > 0$ is not an integer, and $m > r$ is an integer, then (see, for example, Maz'ya [240, Section 10.1.1, Theorem 1, p. 512], or [265, Theorem 1.1]),

$$|V(\cdot,0)|^p_{W^{r,p}(\mathbb{R}^N)} \lesssim \int_{\mathbb{R}^N\times(0,\infty)} \varepsilon^{p(m-r)-1} |D^m V(x,\varepsilon)|^p dx d\varepsilon,\ \forall V. \tag{15.250}$$

Here, D stands for the full differential, in x and ε, and $|\ |_{W^{r,p}}$ is as in (15.3), with $M := m$ and $s := r$.

Applying (15.250) with $N := 1$ and $r := s + 1/p$ to $V(x,\varepsilon) = \tau_t v(x,\varepsilon) := v(x, t + \varepsilon)$, $t > 0$, we find that

$$
\begin{aligned}
J(v) &= \int_0^\infty |\tau_t v|^p_{W^{s+1/p,p}(\mathbb{R})}\, dt \\
&\lesssim \int_0^\infty \left(\iint_{\mathbb{R}\times(0,\infty)} \varepsilon^{p(m-s)-2} |D^m v(x, t+\varepsilon)|^p \, dx d\varepsilon \right) dt \\
&= \int_0^\infty \left(\iint_{\mathbb{R}\times(t,\infty)} (\varepsilon - t)^{p(m-s)-2} |D^m v(x, \varepsilon)|^p \, dx d\varepsilon \right) dt \\
&= \iint_{\mathbb{R}^N\times(0,\infty))} \left(\int_0^\varepsilon (\varepsilon - t)^{p(m-s)-2}\, dt \right) |D^m v(x,\varepsilon)|^p \, dx d\varepsilon \\
&= C \iint_{\mathbb{R}^N\times(0,\infty))} \varepsilon^{p(m-s)-1} |D^m v(x,\varepsilon)|^p \, dx d\varepsilon = C\, I_m(v),
\end{aligned}
$$

so that (15.248) holds. Here, we have used the fact that $m > s + 1/p$, and thus $p(m - s) - 2 > -1$. [Let us note that this part of the proof does not use the fact that m is the smallest integer $> s + 1/p$.]

Step 3. Proof of (15.249)

By Taylor's formula in the integral form, we have

$$
|\delta_h^m w(\varepsilon)| \le C h^{m-1} \int_\varepsilon^{\varepsilon+mh} |w^{(m)}(t)|\, dt, \ \forall\, \varepsilon > 0,\ \forall\, t > 0 \tag{15.251}
$$

(see, for example, [265, Lemma 3.1]).

We next note that, since m is the smallest integer $> s + 1/p$ and since $s + 1/p$ is not an integer, we have $m < s + 1/p + 1$, and thus $2 + sp - (m-1)p > 1$. Combining this with (15.251) and with the weighted Hardy inequality (see, for example, [265, (2.2)]), we find that

$$
\begin{aligned}
\int_0^\infty \frac{|\delta_h^m w(\varepsilon)|^p}{h^{2+sp}}\, dh &\lesssim \int_0^\infty \frac{1}{h^{2+sp-(m-1)p}} \left(\int_\varepsilon^{\varepsilon+mh} |w^{(m)}(t)|\, dt \right)^p dh \\
&\lesssim \int_\varepsilon^\infty (t-\varepsilon)^{p-[2+sp-(m-1)p]} |w^{(m)}(t)|^p\, dt \\
&= \int_\varepsilon^\infty (t-\varepsilon)^{mp-sp-2} |w^{(m)}(t)|^p\, dt.
\end{aligned}
\tag{15.252}
$$

Integrating (15.252) over $\varepsilon > 0$ and using the fact that $mp - sp - 2 > -1$, we obtain

$$
\begin{aligned}
K(w) &\lesssim \int_0^\infty \left(\int_0^t (t-\varepsilon)^{mp-sp-2}\, d\varepsilon \right) |w^{(m)}(t)|^p\, dt \\
&= C \int_0^\infty t^{mp-sp-1} |w^{(m)}(t)|^p\, dt.
\end{aligned}
$$

This implies (15.249) and completes the proof of Lemma 15.48. □

Remark 15.19. Trace theory consists of "direct" theorems (estimates of the semi-norm of the trace $u := v(\cdot, 0)$ of a function v defined on $\mathbb{R}^N \times [0, \infty)$) and "inverse" theorems (construction of a function v on $\mathbb{R}^N \times [0, \infty)$ with prescribed regularity in $\mathbb{R}^N \times (0, \infty)$ and prescribed trace u on $\mathbb{R}^N$). When u belongs to a fractional Sobolev space $W^{s,p}$, u can be obtained either as the trace of a function v in $W^{s+1/p,p}$, or as the trace of a function w in a suitable weighted Sobolev space. From this perspective, Lemma 15.48 asserts that a suitable "weighted regularity" of w implies that $w \in W^{s+1/p,p}$, and thus "inverse trace theory in weighted (integer order) Sobolev spaces implies inverse trace theory in (fractional) Sobolev spaces."

We continue with the proof of two estimates used in the proof of Theorem 10.3.

Lemma 15.49. *Let* $0 < s < 1$, $1 \le p < \infty$, *and* $u \in W^{s,p}(\mathbb{R}^N)$. *Let* $\rho \in C_c^\infty(B_1(0))$. *We have*

$$\begin{aligned} \|\nabla u_\varepsilon\|_{L^p(\mathbb{R}^N)}^p &\lesssim \varepsilon^{(s-1)p} \iint_{|y-x|<\varepsilon} \frac{|u(y)-u(x)|^p}{|y-x|^{N+sp}}\, dxdy \\ &= o(\varepsilon^{(s-1)p}) \text{ as } \varepsilon \to 0. \end{aligned} \tag{15.253}$$

Proof. We note that

$$\begin{aligned} |\partial_j u_\varepsilon(x)| &= \frac{1}{\varepsilon^{N+1}} \left| \int (u(x+z)-u(x)) \partial_j \rho(z/\varepsilon)\, dz \right| \\ &\lesssim \frac{1}{\varepsilon^{N+1}} \int_{|z|<\varepsilon} |u(x+z)-u(x)|\, dz. \end{aligned}$$

It then suffices to apply Hölder's inequality and find that

$$\begin{aligned} \int |\nabla u_\varepsilon|^p &\lesssim \frac{1}{\varepsilon^{N+p}} \iint_{|y-x|<\varepsilon} |u(y)-u(x)|^p\, dxdy \\ &\le \varepsilon^{(s-1)p} \iint_{|y-x|<\varepsilon} \frac{|u(y)-u(x)|^p}{|y-x|^{N+sp}}\, dxdy. \end{aligned}$$
□

Lemma 15.50. *Let* $0 < s < 1$, $1 \le p < \infty$, *and* $u \in W^{s,p}(\mathbb{R}^N)$. *Let* $\rho \in C_c^\infty(B_1(0))$. *We have*

$$\|u_\varepsilon - u\|_{L^p(\mathbb{R}^N)} = o(\varepsilon^s) \text{ as } \varepsilon \to 0. \tag{15.254}$$

Proof. We start from

$$|u_\varepsilon(x) - u(x)| \lesssim \frac{1}{\varepsilon^N} \int_{|y-x|<\varepsilon} |u(y)-u(x)|\, dy$$

and proceed as in the proof of Lemma 15.49. □

References

1. E. Acerbi, G. Bouchitté, and I. Fonseca. Relaxation of convex functionals: the gap problem. *Ann. Inst. H. Poincaré Anal. Non Linéaire*, 20(3):359–390, 2003.
2. C.C. Adams. *The knot book: An elementary introduction to the mathematical theory of knots*. American Mathematical Society, Providence, RI, 2004. Revised reprint of the 1994 original.
3. D.R. Adams and M. Frazier. Composition operators on potential spaces. *Proc. Amer. Math. Soc.*, 114(1):155–165, 1992.
4. R.A. Adams, *Sobolev Spaces*, volume 65 of *Pure and Applied Mathematics*. Academic Press, New York–London, 1975.
5. S.N. Afriat. The system of inequalities $a_{rs} \geq X_r - X_s$. *Mathematical Proceedings of the Cambridge Philosophical Society*, 59(1):125–133, 1963.
6. G. Alberti, S. Baldo, and G. Orlandi. Functions with prescribed singularities. *J. Eur. Math. Soc. (JEMS)*, 5(3):275–311, 2003.
7. S. Alinhac and P. Gérard. *Pseudo-differential operators and the Nash–Moser theorem*, volume 82 of *Graduate Studies in Mathematics*. American Mathematical Society, 2007.
8. F. Almgren, W. Browder, and E.H. Lieb. Co-area, liquid crystals, and minimal surfaces. In *Partial differential equations (Tianjin, 1986)*, volume 1306 of *Lecture Notes in Math.*, pages 1–22. Springer, Berlin, 1988.
9. L. Ambrosio, J. Bourgain, H. Brezis, and A. Figalli. BMO-type norms related to the perimeter of sets. *Comm. Pure Appl. Math.*, 69(6):1062–1086, 2016.
10. L. Ambrosio, V. Caselles, S. Masnou, and J.-M. Morel. Connected components of sets of finite perimeter and applications to image processing. *J. Eur. Math. Soc. (JEMS)*, 3(1):39–92, 2001.
11. L. Ambrosio, C. De Lellis, and C. Mantegazza. Line energies for gradient vector fields in the plane. *Calc. Var. Partial Differential Equations*, 9(4):327–255, 1999.
12. L. Ambrosio, G. De Philippis, and L. Martinazzi. Gamma-convergence of nonlocal perimeter functionals. *Manuscripta Math.*, 134(3–4):377–403, 2011.
13. L. Ambrosio, N. Fusco, and D. Pallara. *Functions of bounded variation and free discontinuity problems*. Oxford Mathematical Monographs. The Clarendon Press Oxford University Press, New York, 2000.
14. G. Angelsberg and D. Pumberger. A regularity result for polyharmonic maps with higher integrability. *Ann. Global Anal. Geom.*, 35(1):63–81, 2009.
15. C. Antonucci, M. Gobbino, and N. Picenni. On the gap between the Gamma-limit and the pointwise limit for a nonlocal approximation of the total variation. *Anal. PDE*, 13(3):627–649, 2020.
16. G. Aubert and P. Kornprobst. Can the nonlocal characterization of Sobolev spaces by Bourgain et al. be useful for solving variational problems? *SIAM J. Numer. Anal.*, 47(2):844–860, 2009.

H. Brezis and P. Mironescu, *Sobolev Maps to the Circle*, Progress in Nonlinear Differential Equations and Their Applications 96,
https://doi.org/10.1007/978-1-0716-1512-6

17. M. Avellaneda and F.H. Lin. Fonctions quasi affines et minimisation de $\int |\nabla \mathbf{u}|^p$. *C. R. Acad. Sci. Paris Sér. I Math.*, 306(8):355–358, 1988.
18. P. Aviles and Y. Giga. On lower semicontinuity of a defect energy obtained by a singular limit of the Ginzburg–Landau type energy for gradient fields. *Proc. Roy. Soc. Edinburgh Sect. A*, 129(1):1–17, 1999.
19. A. Bahri. Personal communication to the authors, 2014.
20. J.M. Ball. Convexity conditions and existence theorems in nonlinear elasticity. *Arch. Rational Mech. Anal.*, 63(4):337–403, 1976/1977.
21. J.M. Ball and A. Zarnescu. Orientability and energy minimization in liquid crystal models. *Arch. Ration. Mech. Anal.*, 202(2):493–535, 2011.
22. D. Barbieri. Approximations of Sobolev norms in Carnot groups. *Commun. Contemp. Math.*, 13(5):765–794, 2011.
23. L.V. Berlyand and P. Mironescu. Ginzburg–Landau minimizers with prescribed degrees. Capacity of the domain and emergence of vortices. *J. Funct. Anal.*, 239(1):76–99, 2006.
24. L.V. Berlyand, P. Mironescu, V. Rybalko, and E. Sandier. Minimax critical points in Ginzburg–Landau problems with semi-stiff boundary conditions: existence and bubbling. *Comm. Partial Differential Equations*, 39(5):946–1005, 2014.
25. F. Bethuel. A characterization of maps in $H^1(B^3, S^2)$ which can be approximated by smooth maps. *Ann. Inst. H. Poincaré Anal. Non Linéaire*, 7(4):269–286, 1990.
26. F. Bethuel. The approximation problem for Sobolev maps between two manifolds. *Acta Math.*, 167(3–4):153–206, 1991.
27. F. Bethuel. Approximations in trace spaces defined between manifolds. *Nonlinear Anal.*, 24(1):121–130, 1995.
28. F. Bethuel. A new obstruction to the extension problem for Sobolev maps between manifolds. *J. Fixed Point Theory Appl.*, 15(1):155–183, 2014.
29. F. Bethuel. A counterexample to the weak density of smooth maps between manifolds in Sobolev spaces. *Invent. Math.*, 219(2):507–651, 2020.
30. F. Bethuel, H. Brezis, and J.-M. Coron. Relaxed energies for harmonic maps. In *Variational methods* (Paris, 1988), volume 4 of *Progress in Nonlinear Differential Equations and their Applications*, pages 37–52. Birkhäuser, Boston, MA, 1990.
31. F. Bethuel, H. Brezis, and F. Hélein. *Ginzburg-Landau vortices*, volume 13 of *Progress in Nonlinear Differential Equations and their Applications*. Birkhäuser, Boston, MA, 1994.
32. F. Bethuel and D. Chiron. Some questions related to the lifting problem in Sobolev spaces. In *Perspectives in nonlinear partial differential equations*, volume 446 of *Contemp. Math.*, pages 125–152. Amer. Math. Soc., Providence, RI, 2007.
33. F. Bethuel, J.-M. Coron, F. Demengel, and F. Hélein. A cohomological criterion for density of smooth maps in Sobolev spaces between two manifolds. In *Nematics (Orsay, 1990)*, volume 332 of *NATO Adv. Sci. Inst. Ser. C Math. Phys. Sci.*, pages 15–23. Kluwer Acad. Publ., Dordrecht, 1991.
34. F. Bethuel and F. Demengel. Extensions for Sobolev mappings between manifolds. *Calc. Var. Partial Differential Equations*, 3(4):475–491, 1995.
35. F. Bethuel, G. Orlandi, and D. Smets. On an open problem for Jacobians raised by Bourgain, Brezis and Mironescu. *C. R. Math. Acad. Sci. Paris*, 337(6):381–385, 2003.
36. F. Bethuel, G. Orlandi, and D. Smets. Approximations with vorticity bounds for the Ginzburg-Landau functional. *Commun. Contemp. Math.*, 6(5):803–832, 2004.
37. F. Bethuel and X.M. Zheng. Density of smooth functions between two manifolds in Sobolev spaces. *J. Funct. Anal.*, 80(1):60–75, 1988.
38. B. Bojarski. Remarks on the Bourgain–Brezis–Mironescu approach to Sobolev spaces. *Bull. Pol. Acad. Sci. Math.*, 59(1):65–75, 2011.
39. B. Bojarski, L. Ihnatsyeva, and J. Kinnunen. How to recognize polynomials in higher order Sobolev spaces. *Math. Scand.*, 112(2):161–181, 2013.
40. J.-M. Bony. Calcul symbolique et propagation des singularités pour les équations aux dérivées partielles non linéaires. *Ann. Sci. École Norm. Sup. (4)*, 14(2):209–246, 1981.
41. R. Borghol. Some properties of Sobolev spaces. *Asymptot. Anal.*, 51(3–4):303–318, 2007.

42. G. Bourdaud. Fonctions qui opèrent sur les espaces de Sobolev. In *Seminar on Harmonic Analysis, 1980–1981*, volume 2 of *Publ. Math. Orsay 82*, pages 6–17. Univ. Paris XI, Orsay, 1982.
43. G. Bourdaud. Ondelettes et espaces de Besov. *Rev. Mat. Iberoamericana*, 11(3):477–512, 1995.
44. G. Bourdaud and Y. Meyer. Fonctions qui opèrent sur les espaces de Sobolev. *J. Funct. Anal.*, 97(2):351–360, 1991.
45. G. Bourdaud and W. Sickel. Composition operators on function spaces with fractional order of smoothness. In *Harmonic analysis and nonlinear partial differential equations*, RIMS Kôkyûroku Bessatsu, B26, pages 93–132. Res. Inst. Math. Sci. (RIMS), Kyoto, 2011.
46. M. Bourdon and H. Pajot. Cohomologie l_p et espaces de Besov. *J. Reine Angew. Math.*, 558:85–108, 2003.
47. J. Bourgain. Personal communication to the authors, 2015.
48. J. Bourgain and H. Brezis. On the equation $\operatorname{div} Y = f$ and application to control of phases. *J. Amer. Math. Soc.*, 16(2):393–426 (electronic), 2003.
49. J. Bourgain and H. Brezis. New estimates for elliptic equations and Hodge type systems. *J. Eur. Math. Soc. (JEMS)*, 9(2):277–315, 2007.
50. J. Bourgain, H. Brezis, and P. Mironescu. Lifting in Sobolev spaces. *J. Anal. Math.*, 80:37–86, 2000.
51. J. Bourgain, H. Brezis, and P. Mironescu. On the structure of the Sobolev space $H^{1/2}$ with values into the circle. *C. R. Acad. Sci. Paris Sér. I Math.*, 331(2):119–124, 2000.
52. J. Bourgain, H. Brezis, and P. Mironescu. Another look at Sobolev spaces. In *Optimal Control and Partial Differential Equations – Innovations and Applications*, pages 439–455, Amsterdam, 2001. IOS Press.
53. J. Bourgain, H. Brezis, and P. Mironescu. Limiting embedding theorems for $W^{s,p}$ when $s \uparrow 1$ and applications. *J. Anal. Math.*, 87:77–101, 2002. Dedicated to the memory of Thomas H. Wolff.
54. J. Bourgain, H. Brezis, and P. Mironescu. $H^{1/2}$ maps with values into the circle: minimal connections, lifting, and the Ginzburg–Landau equation. *Publ. Math. Inst. Hautes Études Sci.*, 99:1–115, 2004.
55. J. Bourgain, H. Brezis, and P. Mironescu. Complements to the paper "Lifting, degree, and the distributional Jacobian revisited", 2005, hal-00747668.
56. J. Bourgain, H. Brezis, and P. Mironescu. Lifting, degree, and distributional Jacobian revisited. *Comm. Pure Appl. Math.*, 58(4):529–551, 2005.
57. J. Bourgain, H. Brezis, and P. Mironescu. A new function space and applications. *J. Eur. Math. Soc. (JEMS)*, 17(9):2083–2101, 2015.
58. J. Bourgain, H. Brezis, and H.-M. Nguyen. A new estimate for the topological degree. *C. R. Math. Acad. Sci. Paris*, 340(11):787–791, 2005.
59. J. Bourgain and G. Kozma. One cannot hear the winding number. *J. Eur. Math. Soc. (JEMS)*, 9(4):637–658, 2007.
60. J. Bourgain and H.M. Nguyen. A new characterization of Sobolev spaces. *C. R. Math. Acad. Sci. Paris*, 343(2):75–80, 2006.
61. P. Bousquet. Topological singularities in $W^{s,p}(S^N, S^1)$. *J. Anal. Math.*, 102:311–346, 2007.
62. P. Bousquet. Fractional Sobolev spaces and topology. *Nonlinear Anal.*, 68(4):804–827, 2008.
63. P. Bousquet and P. Mironescu. Prescribing the Jacobian in critical Sobolev spaces. *J. Anal. Math.*, 122:317–373, 2014.
64. P. Bousquet, A.C. Ponce, and J. Van Schaftingen. A case of density in $W^{2,p}(M; N)$. *C. R. Math. Acad. Sci. Paris*, 346(13–14):735–740, 2008.
65. P. Bousquet, A.C. Ponce, and J. Van Schaftingen. Strong approximation of fractional Sobolev maps. *J. Fixed Point Theory Appl.*, 15(1):133–153, 2014.
66. P. Bousquet, A.C. Ponce, and J. Van Schaftingen. Strong density for higher order Sobolev spaces into compact manifolds. *J. Eur. Math. Soc. (JEMS)*, 17(4):763–817, 2015.
67. A. Boutet de Monvel-Berthier, V. Georgescu, and R. Purice. A boundary value problem related to the Ginzburg-Landau model. *Comm. Math. Phys.*, 142(1):1–23, 1991.

68. H. Brezis. Large harmonic maps in two dimensions. In *Nonlinear variational problems (Isola d'Elba, 1983)*, volume 127 of *Res. Notes in Math.*, pages 33–46. Pitman, Boston, MA, 1985.
69. H. Brezis. Liquid crystals and energy estimates for S^2-valued maps. In *Theory and applications of liquid crystals (Minneapolis, Minn., 1985)*, volume 5 of *IMA Vol. Math. Appl.*, pages 31–52. Springer, New York, 1987.
70. H. Brezis. Metastable harmonic maps. In *Metastability and incompletely posed problems (Minneapolis, Minn., 1985)*, volume 3 of *IMA Vol. Math. Appl.*, pages 33–42. Springer, New York, 1987.
71. H. Brezis. Symmetry in nonlinear PDE's. In *Differential equations: La Pietra 1996 (Florence)*, volume 65 of *Proc. Sympos. Pure Math.*, pages 1–12. Amer. Math. Soc., Providence, RI, 1999.
72. H. Brezis, How to recognize constant functions. A connection with Sobolev spaces. *Uspekhi Mat. Nauk*, 57(4(346)):59–74, 2002 (in Russian). English translation in *Russian Math. Surveys* 57:693–708, 2002.
73. H. Brezis. The interplay between analysis and topology in some nonlinear PDE problems. *Bull. Amer. Math. Soc. (N.S.)*, 40(2):179–201 (electronic), 2003.
74. H. Brezis. New questions related to the topological degree. In *The unity of mathematics*, volume 244 of *Progress in Mathematics*, pages 137–154. Birkhäuser Boston, Boston, MA, 2006.
75. H. Brezis. *Functional analysis, Sobolev spaces and partial differential equations*. Universitext. Springer, New York, 2011.
76. H. Brezis. New approximations of the total variation and filters in imaging. *Atti Accad. Naz. Lincei Rend. Lincei Mat. Appl.*, 26(2):223–240, 2015.
77. H. Brezis. Remarks on the Monge-Kantorovich problem in the discrete setting. *C. R. Math. Acad. Sci. Paris*, 356(2):207–213, 2018.
78. H. Brezis and J.-M. Coron. Large solutions for harmonic maps in two dimensions. *Comm. Math. Phys.*, 92(2):203–215, 1983.
79. H. Brezis, J.-M. Coron, and E.H. Lieb. Harmonic maps with defects. *Comm. Math. Phys.*, 107(4):649–705, 1986.
80. H. Brezis and Y. Li. Topology and Sobolev spaces. *J. Funct. Anal.*, 183(2):321–369, 2001.
81. H. Brezis, Y. Li, P. Mironescu, and L. Nirenberg. Degree and Sobolev spaces. *Topol. Methods Nonlinear Anal.*, 13(2):181–190, 1999.
82. H. Brezis and E.H. Lieb. A relation between pointwise convergence of functions and convergence of functionals. *Proc. Amer. Math. Soc.*, 88(3):486–490, 1983.
83. H. Brezis and P. Mironescu. Gagliardo–Nirenberg, composition and products in fractional Sobolev spaces. *J. Evol. Equ.*, 1(4):387–404, 2001. Dedicated to the memory of Tosio Kato.
84. H. Brezis and P. Mironescu. On some questions of topology for S^1-valued fractional Sobolev spaces. *RACSAM Rev. R. Acad. Cienc. Exactas Fís. Nat. Ser. A Mat.*, 95(1):121–143, 2001.
85. H. Brezis and P. Mironescu. Density in $W^{s,p}(\Omega; N)$. *J. Funct. Anal.*, 269:2045–2109, 2015.
86. H. Brezis and P. Mironescu. Gagliardo–Nirenberg inequalities and non-inequalities: the full story. *Ann. Inst. H. Poincaré Anal. Non Linéaire*, 35(5):1355–1376, 2018.
87. H. Brezis and P. Mironescu. Minimizers of the $W^{1,1}$-energy of $\mathbb{S}^1$-valued maps with prescribed singularities. Do they exist? *Nonlinear Anal.*, 177, part A:105–134, 2018.
88. H. Brezis and P. Mironescu. The Plateau problem from the perspective of optimal transport. *C. R. Math. Acad. Sci. Paris*, 357(7):597–612, 2019.
89. H. Brezis and P. Mironescu. Where Sobolev interacts with Gagliardo–Nirenberg. *J. Funct. Anal.*, 277(8):2839–2864, 2019.
90. H. Brezis and P. Mironescu. Relaxed energies involving BV norms and boundary effects, 2021. In preparation
91. H. Brezis, P. Mironescu, and A. Ponce. Complements to the paper "$W^{1,1}$-maps with values into S^1", 2005, hal-00747667.
92. H. Brezis, P. Mironescu, and A. Ponce. $W^{1,1}$-maps with values into S^1. In *Geometric analysis of PDE and several complex variables*, volume 368 of *Contemp. Math.*, pages 69–100. Amer. Math. Soc., Providence, RI, 2005.

93. H. Brezis, P. Mironescu, and I. Shafrir. Distances between homotopy classes of $W^{s,p}(\mathbb{S}^N; \mathbb{S}^N)$. *ESAIM COCV*, 22:1204–1235, 2016.
94. H. Brezis, P. Mironescu, and I. Shafrir. Distances between classes in $W^{1,1}(\Omega; \mathbb{S}^1)$. *Calc. Var. Partial Differential Equations*, 57(1):Art. 14, 32, 2018.
95. H. Brezis, P. Mironescu, and I. Shafrir. Radial extensions in fractional Sobolev spaces. *Rev. R. Acad. Cienc. Exactas Fís. Nat. Ser. A Mat. RACSAM*, 113(2):707–714, 2019.
96. H. Brezis and H.-M. Nguyen. The Jacobian determinant revisited. *Invent. Math.*, 185(1):17–54, 2011.
97. H. Brezis and H.-M. Nguyen. On a new class of functions related to VMO. *C. R. Math. Acad. Sci. Paris*, 349(3–4):157–160, 2011.
98. H. Brezis and H.-M. Nguyen. On the distributional Jacobian of maps from $\mathbb{S}^N$ into $\mathbb{S}^N$ in fractional Sobolev and Hölder spaces. *Ann. of Math. (2)*, 173(2):1141–1183, 2011
99. H. Brezis and H.-M. Nguyen. The BBM formula revisited. *Atti Accad. Naz. Lincei Rend. Lincei Mat. Appl.*, 27(4):515–533, 2016.
100. H. Brezis and H.-M. Nguyen. Two subtle convex nonlocal approximations of the BV-norm. *Nonlinear Anal.*, 137:222–245, 2016.
101. H. Brezis and H.-M. Nguyen. Non-convex, non-local functionals converging to the total variation. *C. R. Math. Acad. Sci. Paris*, 355(1):24–27, 2017.
102. H. Brezis and H.-M. Nguyen. Non-local functionals related to the total variation and connections with image processing. *Ann. PDE*, 4(1):Art. 9, 77, 2018.
103. H. Brezis and H.-M. Nguyen. Non-local functionals related to the Sobolev norms. *Nonlinear Anal.*, 191:111626, 9, 2020.
104. H. Brezis and L. Nirenberg. Degree theory and BMO. I. Compact manifolds without boundaries. *Selecta Math. (N.S.)*, 1(2):197–263, 1995.
105. H. Brezis and L. Nirenberg. Degree Theory and BMO. II. Compact manifolds with boundaries. *Selecta Math.*, 2:309–368, 1996.
106. H. Brezis, J. Van Schaftingen, and P.-L. Yung. Going to Lorentz when fractional Sobolev, Gagliardo and Nirenberg estimates fail, *Calc. Var. Partial Differential Equations*, 60(4):129, 2021.
107. A. Buades, B. Coll, and J. M. Morel. A review of image denoising algorithms, with a new one. *Multiscale Model. Simul.*, 4(2):490–530, 2005.
108. A. Buades, B. Coll, and J. M. Morel. Image denoising methods. A new nonlocal principle. *SIAM Rev.*, 52(1):113–147, 2010. Reprint of "A review of image denoising algorithms, with a new one".
109. C. Bucur and E. Valdinoci. *Nonlocal diffusion and applications*, volume 20 of *Lecture Notes of the Unione Matematica Italiana*. Springer, [Cham]; Unione Matematica Italiana, Bologna, 2016.
110. G. Buttazzo. *Semicontinuity, relaxation and integral representation in the calculus of variations*, volume 207 of *Pitman Research Notes in Mathematics Series*. Longman Scientific & Technical, Harlow; copublished in the United States with John Wiley & Sons, Inc., New York, 1989.
111. G. Buttazzo and M. Belloni. A survey on old and recent results about the gap phenomenon in the calculus of variations. In *Recent developments in well-posed variational problems*, volume 331 of *Math. Appl.*, pages 1–27. Kluwer Acad. Publ., Dordrecht, 1995.
112. G. Buttazzo and V.J. Mizel. Interpretation of the Lavrentiev phenomenon by relaxation. *J. Funct. Anal.*, 110(2):434–460, 1992.
113. L. Caffarelli, J.-M. Roquejoffre, and O. Savin. Nonlocal minimal surfaces. *Comm. Pure Appl. Math.*, 63(9):1111–1144, 2010.
114. L. Caffarelli and E. Valdinoci. Uniform estimates and limiting arguments for nonlocal minimal surfaces. *Calc. Var. Partial Differential Equations*, 41(1–2):203–240, 2011.
115. G. Carbou. Applications harmoniques à valeurs dans un cercle. *C. R. Acad. Sci. Paris Sér. I Math.*, 314:359–362, 1992.
116. J.-Y. Chemin. *Perfect incompressible fluids*, volume 14 of *Oxford Lecture Series in Mathematics and its Applications*. The Clarendon Press Oxford University Press, New York, 1998. Translated from the 1995 French original by Isabelle Gallagher and Dragoş Iftimie.

117. B. Chen. Singularities of p-harmonic mappings, 1989. Doctoral Dissertation, the Univ. of Minnesota.
118. D. Chiron. On the definitions of Sobolev and BV spaces into singular spaces and the trace problem. *Commun. Contemp. Math.*, 9(4):473–513, 2007.
119. R. Coifman and Y. Meyer. Une généralisation du théorème de Calderón sur l'intégrale de Cauchy. In *Fourier analysis (Proc. Sem., El Escorial, 1979)*, volume 1 of *Asoc. Mat. Española*, pages 87–116. Asoc. Mat. Española, 1980.
120. J.-M. Coron and R. Gulliver. Minimizing p-harmonic maps into spheres. *J. Reine Angew. Math.*, 401:82–100, 1989.
121. X. Cui, N. Lam, and G. Lu. New characterizations of Sobolev spaces on the Heisenberg group. *J. Funct. Anal.*, 267(8):2962–2994, 2014.
122. X. Cui, N. Lam, and G. Lu. Characterizations of second order Sobolev spaces. *Nonlinear Anal.*, 121:241–261, 2015.
123. E. Curcă. On the representation as exterior differentials of closed forms with L^1-coefficients. *C. R. Math. Acad. Sci. Paris*, 357(4):355–359, 2019.
124. E. Curcă. Minimal BV-liftings of $W^{1,1}(\Omega, \mathbb{S}^1)$ maps in 2D are "often" unique. *Nonlinear Anal.*, 199:111943, 14, 2020
125. F. Da Lio and T. Rivière. Critical Chirality in Elliptic Systems, 2019, arXiv.1907.10520.
126. F. Da Lio, T. Rivière, and Wettstein J. Bergman–Bourgain–Brezis-type Inequality, 2020, arXiv.2011.03950.
127. B.E.J. Dahlberg. A note on Sobolev spaces. In *Harmonic analysis in Euclidean spaces (Proc. Sympos. Pure Math., Williams Coll., Williamstown, Mass., 1978), Part 1*, volume XXXV of *Proc. Sympos. Pure Math.*, pages 183–185. Amer. Math. Soc., Providence, R.I., 1979.
128. J. Dávila. On an open question about functions of bounded variation. *Calc. Var. Partial Differential Equations*, 15(4):519–527, 2002.
129. J. Dávila and R. Ignat. Lifting of BV functions with values in S^1. *C. R. Math. Acad. Sci. Paris*, 337(3):159–164, 2003.
130. C. De Lellis and R. Ignat. A regularizing property of the $2D$-eikonal equation. *Comm. Partial Differential Equations*, 40(8):1543–1557, 2015.
131. G. De Marco, C. Mariconda, and S. Solimini. An elementary proof of a characterization of constant functions. *Adv. Nonlinear Stud.*, 8(3):597–602, 2008.
132. F. Demengel. Une caractérisation des applications de $W^{1,p}(B^N, S^1)$ qui peuvent être approchées par des fonctions régulières. *C. R. Acad. Sci. Paris Sér. I Math.*, 310(7):553–557, 1990.
133. F. Demengel and R. Hadiji. Relaxed energies for functionals on $W^{1,1}(B^2, S^1)$. *Nonlinear Anal.*, 19(7):625–641, 1992.
134. A. DeSimone, R.V. Kohn, S. Müller, and F. Otto. Magnetic microstructures–a paradigm of multiscale problems. In *ICIAM 99 (Edinburgh)*, pages 175–190. Oxford Univ. Press, Oxford, 2000.
135. A. Desimone, R.V. Kohn, S. Müller, and F. Otto. A reduced theory for thin-film micromagnetics. *Comm. Pure Appl. Math.*, 55(11):1408–1460, 2002.
136. A. DeSimone, S. Müller, R.V. Kohn, and F. Otto. A compactness result in the gradient theory of phase transitions. *Proc. Roy. Soc. Edinburgh Sect. A*, 131(4):833–844, 2001.
137. R.A. DeVore and V.A. Popov. Interpolation of Besov spaces. *Trans. Amer. Math. Soc.*, 305(1):397–414, 1988.
138. S. Di Marino and M. Squassina. New characterizations of Sobolev metric spaces. *J. Funct. Anal.*, 276(6):1853–1874, 2019.
139. E. Di Nezza, G. Palatucci, and E. Valdinoci. Hitchhiker's guide to the fractional Sobolev spaces. *Bull. Sci. Math.*, 136(5):521–573, 2012.
140. E. DiBenedetto. $C^{1+\alpha}$ local regularity of weak solutions of degenerate elliptic equations. *Nonlinear Anal.*, 7(8):827–850, 1983.
141. G. Dinca and J. Mawhin. *Brouwer degree. The core of nonlinear analysis*, volume 95 of *Progress in Nonlinear Differential Equations and their Applications*. Birkhäuser, Boston, MA, 2021.

142. L. Dupaigne and A.C. Ponce. Singularities of positive supersolutions in elliptic PDEs. *Selecta Math. (N.S.)*, 10(3):341–358, 2004.
143. J. Ericksen. Personal communication to H. Brezis, 1984.
144. J. L. Ericksen and D. Kinderlehrer, editors. *Theory and applications of liquid crystals*, volume 5 of *The IMA Volumes in Mathematics and its Applications*. Springer-Verlag, New York, 1987. Papers from the IMA workshop held in Minneapolis, Minn., 21–25 January, 1985.
145. M. Escobedo. Some remarks on the density of regular mappings in Sobolev classes of S^M-valued functions. *Rev. Mat. Univ. Complut. Madrid*, 1(1–3):127–144, 1988.
146. L.C. Evans and R.F. Gariepy. *Measure theory and fine properties of functions*. Studies in Advanced Mathematics. CRC Press, Boca Raton, FL, 1992.
147. H. Federer. *Geometric measure theory*. Die Grundlehren der mathematischen Wissenschaften, Band 153. Springer-Verlag, New York, 1969.
148. H. Federer. Real flat chains, cochains and variational problems. *Indiana Univ. Math. J.*, 24:351–407, 1974/1975.
149. H. Federer and W.H. Fleming. Normal and integral currents. *Ann. of Math. (2)*, 72:458–520, 1960.
150. A. Figalli, N. Fusco, F. Maggi, V. Millot, and M. Morini. Isoperimetry and stability properties of balls with respect to nonlocal energies. *Comm. Math. Phys.*, 336(1):441–507, 2015.
151. W.H. Fleming. Functions with generalized gradient and generalized surfaces. *Ann. Mat. Pura Appl. (4)*, 44:92, 93–103, 1957.
152. F. Frankl and L. Pontrjagin. Ein Knotensatz mit Anwendung auf die Dimensionstheorie. *Math. Ann.*, 102(1):785–789, 1930.
153. M. Frazier. Personal communication to the authors, 2015.
154. A.M. Garsia and E. Rodemich. Monotonicity of certain functionals under rearrangement. *Ann. Inst. Fourier (Grenoble)*, 24(2):vi, 67–116, 1974. Colloque International sur les Processus Gaussiens et les Distributions Aléatoires (Colloque Internat. du CNRS, No. 222, Strasbourg, 1973).
155. M. Giaquinta, G. Modica, and J. Souček. Connectivity properties of the range of a weak diffeomorphism. *Ann. Inst. H. Poincaré Anal. Non Linéaire*, 12(1):61–73, 1995.
156. M. Giaquinta, G. Modica, and J. Souček. *Cartesian currents in the calculus of variations. I. Cartesian currents*, volume 37 of *Ergebnisse der Mathematik und ihrer Grenzgebiete. 3. Folge. A Series of Modern Surveys in Mathematics [Results in Mathematics and Related Areas. 3rd Series. A Series of Modern Surveys in Mathematics]*. Springer-Verlag, Berlin, 1998.
157. M. Giaquinta, G. Modica, and J. Souček. *Cartesian currents in the calculus of variations. II. Variational integrals*, volume 38 of *Ergebnisse der Mathematik und ihrer Grenzgebiete. 3. Folge. A Series of Modern Surveys in Mathematics [Results in Mathematics and Related Areas. 3rd Series. A Series of Modern Surveys in Mathematics]*. Springer-Verlag, Berlin, 1998.
158. M. Giaquinta, G. Modica, and J. Souček. The gap phenomenon for variational integrals in Sobolev spaces. *Proc. Roy. Soc. Edinburgh Sect. A*, 120(1–2):93–98, 1992.
159. M. Giaquinta and D. Mucci. The relaxed Dirichlet energy of mappings into a manifold. *Calc. Var. Partial Differential Equations*, 24(2):155–166, 2005.
160. M. Giaquinta and D. Mucci. The BV-energy of maps into a manifold: relaxation and density results. *Ann. Sc. Norm. Super. Pisa Cl. Sci. (5)*, 5(4):483–548, 2006.
161. M. Giaquinta and D. Mucci. *Maps into manifolds and currents: area and $W^{1,2}$-, $W^{1/2}$-, BV-energies*, volume 3 of *Centro di Ricerca Matematica Ennio De Giorgi (CRM) Series*. Edizioni della Normale, Pisa, 2006.
162. M. Giaquinta and D. Mucci. Erratum and addendum to: "The BV-energy of maps into a manifold: relaxation and density results" [Ann. Sc. Norm. Super. Pisa Cl. Sci. (5) **5** (2006), no. 4, 483–548; mr2297721]. *Ann. Sc. Norm. Super. Pisa Cl. Sci. (5)*, 6(1):185–194, 2007.
163. M. Giaquinta and D. Mucci. Graphs of $W^{1,1}$-maps with values in S^1: relaxed energies, minimal connections and lifting. *Proc. Roy. Soc. Edinburgh Sect. A*, 137(5):937–962, 2007.
164. D. Gilbarg and N. Trudinger. *Elliptic partial differential equations of second order*. Classics in Mathematics. Springer-Verlag, Berlin, 2001.

165. G. Gilboa and S. Osher. Nonlocal linear image regularization and supervised segmentation. *Multiscale Model. Simul.*, 6(2):595–630, 2007.
166. G. Gilboa and S. Osher. Nonlocal operators with applications to image processing. *Multiscale Model. Simul.*, 7(3):1005–1028, 2008.
167. E. Giusti, *Minimal surfaces and functions of bounded variation*, volume 80 of *Monographs in Mathematics*. Birkhäuser Verlag, Basel, 1984.
168. M. Gobbino and M.G. Mora. Finite-difference approximation of free-discontinuity problems. *Proc. Roy. Soc. Edinburgh Sect. A*, 131(3):567–595, 2001.
169. C. Goffman and J. Serrin. Sublinear functions of measures and variational integrals. *Duke Math. J.*, 31:159–178, 1964.
170. A. Grigor'yan, J. Hu, and K.-S. Lau. Heat kernels on metric measure spaces and an application to semilinear elliptic equations. *Trans. Amer. Math. Soc.*, 355(5):2065–2095, 2003.
171. Q. Gu and K.-S. Lau. Dirichlet forms and convergence of Besov norms on self-similar sets. *Ann. Acad. Sci. Fenn. Math.*, 45(2):625–646, 2020.
172. P. Hajłasz. Approximation of Sobolev mappings. *Nonlinear Anal.*, 22(12):1579–1591, 1994.
173. P Hajłasz. Sobolev mappings, co-area formula and related topics. In *Proceedings on Analysis and Geometry (Russian) (Novosibirsk Akademgorodok, 1999)*, pages 227–254. Izdat. Ross. Akad. Nauk Sib. Otd. Inst. Mat., Novosibirsk, 2000.
174. Z.-C. Han and I. Shafrir. Lower bounds for the energy of S^1-valued maps in perforated domains. *J. Anal. Math.*, 66:295–305, 1995.
175. F. Hang. Density problems for $W^{1,1}(M, N)$. *Comm. Pure Appl. Math.*, 55(7):937–947, 2002.
176. F.B. Hang and F.H. Lin. A remark on the Jacobians. *Commun. Contemp. Math.*, 2(1):35–46, 2000.
177. F.B. Hang and F.H. Lin. Topology of Sobolev mappings. II. *Acta Math.*, 191(1):55–107, 2003.
178. R. Hardt. Singularities of harmonic maps. *Bull. Amer. Math. Soc. (N.S.)*, 34(1):15–34, 1997.
179. R. Hardt and B. Chen. Prescribing singularities for p-harmonic maps. *Indiana Univ. Math. J.*, 44(2):575–602, 1995.
180. R. Hardt, D. Kinderlehrer, and F.H. Lin. Stable defects of minimizers of constrained variational principles. *Ann. Inst. H. Poincaré Anal. Non Linéaire*, 5(4):297–322, 1988.
181. R. Hardt, D. Kinderlehrer, and F.H. Lin. The variety of configurations of static liquid crystals. In *Variational methods (Paris.* 1988), volume 4 of *Progress in Nonlinear Differential Equations and their Applications*, pages 115–131. Birkhäuser Boston, Boston, MA, 1990.
182. R. Hardt and F.H. Lin. A remark on H^1 mappings. *Manuscripta Math.*, 56(1):1–10, 1986.
183. R. Hardt and F.H. Lin. Mappings minimizing the L^p norm of the gradient. *Comm. Pure Appl. Math.*, 40(5):555–588, 1987.
184. R. Hardt and F.H. Lin. Singularities for p-energy minimizing unit vectorfields on planar domains. *Calc. Var. Partial Differential Equations*, 3(3):311–341, 1995.
185. R. Hardt, F.H. Lin, and C.Y. Wang. Singularities of p-energy minimizing maps. *Comm. Pure Appl. Math.*, 50(5):399–447, 1997.
186. R. Hardt, F.H. Lin, and C.Y. Wang. The p-energy minimality of $x/|x|$. *Comm. Anal. Geom.*, 6(1):141–152, 1998.
187. R.M. Hardt and J.T. Pitts. Solving Plateau's problem for hypersurfaces without the compactness theorem for integral currents. In *Geometric measure theory and the calculus of variations (Arcata, Calif., 1984)*, volume 44 of *Proc. Sympos. Pure Math.*, pages 255–259. Amer. Math. Soc., Providence, RI, 1986.
188. A. Hatcher. *Algebraic topology*. Cambridge University Press, Cambridge, 2002.
189. J. Heinonen, P. Koskela, N. Shanmugalingam, and J.T. Tyson. *Sobolev spaces on metric measure spaces. An approach based on upper gradients*, volume 27 of *New Mathematical Monographs*. Cambridge University Press, Cambridge, 2015.
190. M.C. Hong. On the minimality of the p-harmonic map $\frac{x}{|x|} : B^n \to S^{n-1}$. *Calc. Var. Partial Differential Equations*, 13(4):459–468, 2001.
191. S. Hu. *Homotopy theory*. Pure and Applied Mathematics, Vol. VIII. Academic Press, New York/London, 1959.

192. R. Ignat. On an open problem about how to recognize constant functions. *Houston J. Math.*, 31(1):285–304, 2005.
193. R. Ignat. The space $\mathrm{BV}(S^2, S^1)$: minimal connection and optimal lifting. *Ann. Inst. H. Poincaré Anal. Non Linéaire*, 22(3):283–302, 2005.
194. R. Ignat. Singularities of divergence-free vector fields with values into S^1 or S^2. Applications to micromagnetics. *Confluentes Math.*, 4(3):1230001, 80, 2012.
195. R. Ignat. Two-dimensional unit-length vector fields of vanishing divergence. *J. Funct. Anal.*, 262(8):3465–3494, 2012.
196. R. Ignat and X. Lamy. Lifting of $\mathbb{RP}^{d-1}$-valued maps in BV and applications to uniaxial Q-tensors. With an appendix on an intrinsic BV-energy for manifold-valued maps. *Calc. Var. Partial Differential Equations*, 58(2):Paper No. 68, 26, 2019
197. R. Ignat and B. Merlet. Entropy method for line-energies. *Calc. Var. Partial Differential Equations*, 44(3–4):375–418, 2012.
198. R. Ignat, L. Nguyen, V. Slastikov, and A. Zarnescu. Uniqueness of degree-one Ginzburg-Landau vortex in the unit ball in dimensions $N \geq 7$. *C. R. Math. Acad. Sci. Paris*, 356(9):922–926, 2018.
199. R. Ignat and A. Poliakovsky. On the relation between minimizers of a Γ-limit energy and optimal lifting in BV-space. *Commun. Contemp. Math.*, 9(4):447–472, 2007.
200. T. Isobe. Obstructions to the extension problem of Sobolev mappings. *Topol. Methods Nonlinear Anal.*, 21(2):345–368, 2003.
201. T. Iwaniec. *Integrability theory of the Jacobians: Rudolph-Lipschitz-Vorlesung*. Vorlesungsreihe. Sonderforschungsbereich 256, 1995.
202. P.-E. Jabin, F. Otto, and B. Perthame. Line-energy Ginzburg–Landau models: zero-energy states. *Ann. Sc. Norm. Super. Pisa Cl. Sci. (5)*, 1(1):187–202, 2002.
203. W. Jäger and H. Kaul. Rotationally symmetric harmonic maps from a ball into a sphere and the regularity problem for weak solutions of elliptic systems. *J. Reine Angew. Math.*, 343:146–161, 1983.
204. R.L. Jerrard and H.M. Soner. Rectifiability of the distributional Jacobian for a class of functions. *C. R. Acad. Sci. Paris Sér. I Math.*, 329(8):683–688, 1999.
205. R.L. Jerrard and H.M. Soner. Functions of bounded higher variation. *Indiana Univ. Math. J.*, 51(3):645–677, 2002.
206. F. John and L. Nirenberg. On functions of bounded mean oscillation. *Comm. Pure Appl. Math.*, 14:415–426, 1961.
207. P. Jones. Extension theorems for BMO. *Indiana Univ. Math. J.*, 29(1):41–66, 1980.
208. J.-P. Kahane. Sur l'équation fonctionnelle $\int_{\mathbb{T}}(\psi(t+s)-\psi(s))^3 ds = \sin t$. *C. R. Math. Acad. Sci. Paris*, 341(3):141–145, 2005.
209. J.-P. Kahane. Winding numbers and summation processes. *Complex Var. Elliptic Equ.*, 55(8–10):911–922, 2010.
210. J.-P. Kahane. Summation processes viewed from the Fourier properties of continuous unimodular functions on the circle. In *Marcinkiewicz centenary volume*, volume 95 of *Banach Center Publ.*, pages 75–87. Polish Acad. Sci. Inst. Math., Warsaw, 2011.
211. L. Kantorovitch, On the translocation of masses. *C. R. (Doklady) Acad. Sci. URSS (N.S.)*, 37:199–201, 1942. English translation: *J. Math. Sc.*, 133:1381–1382, 2006.
212. G.E. Karadzhov, M. Milman, and J. Xiao. Limits of higher-order Besov spaces and sharp reiteration theorems. *J. Funct. Anal.*, 221(2):323–339, 2005.
213. R.C. Kirby, *The Topology of 4-Manifolds*, volume 1374 of *Lecture Notes in Mathematics*. Springer-Verlag, Berlin, 1989
214. J. Korevaar. On a question of Brezis and Nirenberg concerning the degree of circle maps. *Selecta Math. (N.S.)*, 5(1):107–122, 1999.
215. P. Koskela. Removable sets for Sobolev spaces. *Ark. Mat.*, 37(2):291–304, 1999.
216. H. Kozono and H. Wadade. Remarks on Gagliardo-Nirenberg type inequality with critical Sobolev space and BMO. *Math. Z.*, 259(4):935–950, 2008.
217. S. Kuksin and A. Shirikyan. *Mathematics of two-dimensional turbulence*, volume 194 of *Cambridge Tracts in Mathematics*. Cambridge University Press, Cambridge, 2012.

218. L. Lassoued. Sur quelques équations aux dérivées partielles non linéaires issues de la géométrie et de la physique, PhD thesis, Paris 6, 1996.
219. L. Lassoued. A remark on the weak topology and the degree. *Northeast. Math. J.*, 14(2): 171–176, 1998.
220. G. Leoni. *A first course in Sobolev spaces*, volume 181 of *Graduate Studies in Mathematics*. American Mathematical Society, Providence, RI, second edition, 2017.
221. G. Leoni and D. Spector. Characterization of Sobolev and BV spaces. *J. Funct. Anal.*, 261(10):2926–2958, 2011.
222. G. Leoni and D. Spector. Corrigendum to "Characterization of Sobolev and BV spaces" [J. Funct. Anal. 261 (10) (2011) 2926–2958]. *J. Funct. Anal.*, 266(2):1106–1114, 2014.
223. S. Levi and I. Shafrir. On the distance between homotopy classes of maps between spheres. *J. Fixed Point Theory Appl.*, 15(2):501–518, 2014.
224. J.L. Lewis. Regularity of the derivatives of solutions to certain degenerate elliptic equations. *Indiana Univ. Math. J.*, 32(6):849–858, 1983.
225. E.H. Lieb and M. Loss. *Analysis*, volume 14 of *Graduate Studies in Mathematics*. American Mathematical Society, Providence, RI, second edition, 2001.
226. G.M. Lieberman. Boundary regularity for solutions of degenerate elliptic equations. *Nonlinear Anal.*, 12(11):1203–1219, 1988.
227. F.H. Lin. A remark on the map $x/|x|$. *C. R. Acad. Sci. Paris Sér. I Math.*, 305(12):529–531, 1987.
228. F.H. Lin. Personal communication to the authors, July 2019.
229. F.H. Lin. Personal communication to the authors, January 2020.
230. F.H. Lin and Y. Li. Boundary $C^{1,\alpha}$ regularity for variational inequalities. *Comm. Pure Appl. Math.*, 44(6):715–732, 1991.
231. F.H. Lin and T. Rivière. Complex Ginzburg–Landau equations in high dimensions and codimension two area minimizing currents. *J. Eur. Math. Soc. (JEMS)*, 1(3):237–311, 1999.
232. F.H. Lin and T. Rivière. Erratum: "Complex Ginzburg–Landau equations in high dimensions and codimension two area minimizing currents" [J. Eur. Math. Soc. (JEMS) **1** (1999), no. 3, 237–311; MR1714735 (2000g:49048)]. *J. Eur. Math. Soc. (JEMS)*, 2(1):87–91, 2000.
233. S. Luckhaus. Partial Hölder continuity for minima of certain energies among maps into a Riemannian manifold. *Indiana Univ. Math. J.*, 37(2):349–367, 1988.
234. J. Malý, D. Swanson, and W.P. Ziemer. The co-area formula for Sobolev mappings. *Trans. Amer. Math. Soc.*, 355(2):477–492, 2003.
235. M. Marcus and V. J. Mizel. Absolute continuity on tracks and mappings of Sobolev spaces. *Arch. Rational Mech. Anal.*, 45:294–320, 1972.
236. M. Marcus and V.J. Mizel. Every superposition operator mapping one Sobolev space into another is continuous. *J. Funct. Anal.*, 33(2):217–229, 1979.
237. H. Matano. Personal communication to H. Brezis, 1992.
238. J.M. Mazón, J.D. Rossi, and J. Toledo. Nonlocal perimeter, curvature and minimal surfaces for measurable sets. *J. Anal. Math.*, 138(1):235–279, 2019.
239. V. Maz'ya. Bourgain–Brezis type inequality with explicit constants. In *Interpolation theory and applications*, volume 445 of *Contemp. Math.*, pages 247–252. Amer. Math. Soc., Providence, RI, 2007.
240. V. Maz'ya. *Sobolev spaces with applications to elliptic partial differential equations*, volume 342 of *Grundlehren der Mathematischen Wissenschaften [Fundamental Principles of Mathematical Sciences]*. Springer, Heidelberg, augmented edition, 2011.
241. V. Maz'ya and J. Nagel. Über äquivalente Normierung der anisotropen Funktionalräume $H^{\mu}(R^n)$. *Beiträge Anal.*, 12:7–17, 1978.
242. V. Maz'ya and T. Shaposhnikova. An elementary proof of the Brezis and Mironescu theorem on the composition operator in fractional Sobolev spaces. *J. Evol. Equ.*, 2(1):113–125, 2002.
243. V. Maz'ya and T. Shaposhnikova. On the Bourgain, Brezis, and Mironescu theorem concerning limiting embeddings of fractional Sobolev spaces. *J. Funct. Anal.*, 195(2):230–238, 2002.
244. V. Maz'ya and T. Shaposhnikova. On the Brezis and Mironescu conjecture concerning a Gagliardo–Nirenberg inequality for fractional Sobolev norms. *J. Math. Pures Appl. (9)*, 81(9):877–884, 2002.

245. V. Maz'ya and T. Shaposhnikova. Erratum to: "On the Bourgain, Brezis and Mironescu theorem concerning limiting embeddings of fractional Sobolev spaces" [J. Funct. Anal. **195** (2002), no. 2, 230–238; MR1940355 (2003j:46051)]. *J. Funct. Anal.*, 201(1):298–300, 2003.
246. T. Mengesha. Nonlocal Korn-type characterization of Sobolev vector fields. *Commun. Contemp. Math.*, 14(4):1250028, 28, 2012.
247. B. Merlet. Two remarks on liftings of maps with values into S^1. *C. R. Math. Acad. Sci. Paris*, 343(7):467–472, 2006.
248. Y. Meyer. Régularité des solutions des équations aux dérivées partielles non linéaires (d'après J.-M. Bony). In *Bourbaki Seminar, Vol. 1979/80*, volume 842 of *Lecture Notes in Math.*, pages 293–302. Springer, Berlin, 1981.
249. Y. Meyer and T. Rivière. A partial regularity result for a class of stationary Yang-Mills fields in high dimension. *Rev. Mat. Iberoamericana*, 19(1):195–219, 2003.
250. M. Milman. Notes on limits of Sobolev spaces and the continuity of interpolation scales. *Trans. Amer. Math. Soc.*, 357(9):3425–3442 (electronic), 2005.
251. G. Mingione and D. Mucci. Integral functionals and the gap problem: sharp bounds for relaxation and energy concentration. *SIAM J. Math. Anal.*, 36(5):1540–1579 (electronic), 2005.
252. P. Mironescu. Sobolev maps on manifolds: degree, approximation, lifting. In *Perspectives in nonlinear partial differential equations*, volume 446 of *Contemp. Math.*, pages 413–436. Amer. Math. Soc., Providence, RI, 2007.
253. P. Mironescu. Lifting default for $\mathbb{S}^1$-valued maps. *C. R. Math. Acad. Sci. Paris*, 346(19–20):1039–1044, 2008.
254. P. Mironescu. Decomposition of $\mathbb{S}^1$-valued maps in Sobolev spaces. *C. R. Math. Acad. Sci. Paris*, 348(13–14):743–746, 2010.
255. P. Mironescu. On some inequalities of Bourgain, Brezis, Maz'ya, and Shaposhnikova related to L^1 vector fields. *C. R. Math. Acad. Sci. Paris*, 348(9–10):513–515, 2010.
256. P. Mironescu. $\mathbb{S}^1$-valued Sobolev mappings. *Sovrem. Mat. Fundam. Napravl.*, 35:86–100, 2010. Translation in *J. Math. Sci. (N. Y.)* 170(3):340–355, 2010.
257. P. Mironescu. Note on Gagliardo's theorem "tr $W^{1,1} = L^1$". *Ann. Univ. Buchar. Math. Ser.*, 6(LXIV)(1):99–103, 2015.
258. P. Mironescu. Profile decomposition and phase control for circle-valued maps in one dimension. *C. R. Math. Acad. Sci. Paris*, 353(12):1087–1092, 2015.
259. P. Mironescu. Low regularity function spaces of N-valued maps are contractible. *Math. Scand.*, 121(1):144–150, 2017.
260. P. Mironescu. Sum-intersection property of Sobolev spaces. In T.M. Rassias, editor, *Current Research in Nonlinear Analysis*, volume 135 of *Springer Optimization and Its Applications*, pages 203–228. Springer, 2018, hal-01574338.
261. P. Mironescu. The role of the Hardy type inequalities in the theory of function spaces. *Rev. Roumaine Math. Pures Appl.*, 63(4):447–525, 2018.
262. P. Mironescu. Sobolev spaces of circle-valued maps, 2021. In preparation.
263. P. Mironescu and I. Molnar. Phases of unimodular complex valued maps: optimal estimates, the factorization method, and the sum-intersection property of Sobolev spaces. *Ann. Inst. H. Poincaré Anal. Non Linéaire*, 32(5):965–1013, 2015.
264. P. Mironescu and A. Pisante. A variational problem with lack of compactness for $H^{1/2}(S^1; S^1)$ maps of prescribed degree. *J. Funct. Anal.*, 217(2):249–279, 2004.
265. P. Mironescu and E. Russ. Traces of weighted Sobolev spaces. Old and new. *Nonlinear Anal.*, 119:354–381, 2015.
266. P. Mironescu, E. Russ, and Y. Sire. Lifting in Besov spaces. *Nonlinear Anal.*, 193:111489, 2020.
267. P. Mironescu and I. Shafrir. A uniform continuity property of the winding number of self-mappings of the circle. *Pure Appl. Funct. Anal.*, 5(5):1199–1204, 2020, hal-02151571.
268. P. Mironescu and W. Sickel. A Sobolev non embedding. *Rend. Lincei Mat. Appl.*, 26:1–8, 2015.

269. P. Mironescu and J. Van Schaftingen. Lifting in compact covering spaces, 2019, hal-02171533. To appear in *Analysis and PDE*.
270. P. Mironescu and J. Van Schaftingen. Trace theory for Sobolev mappings into a manifold, 2020, hal-02431628. To appear in *Ann. Fac. Sci, Toulouse*.
271. I. Molnar. Prescribed singularities with weights. *Adv. Nonlinear Anal.*, 1(4):355–381, 2012.
272. F. Morgan. Area-minimizing currents bounded by higher multiples of curves. *Rend. Circ. Mat. Palermo (2)*, 33(1):37–46, 1984.
273. F. Morgan. *Geometric measure theory. A beginner's guide*. Elsevier/Academic Press, Amsterdam, fourth edition, 2009.
274. D. Mucci. BV-maps with values into S^1: graphs, minimal connections and optimal lifting. *Forum Math.*, 20(5):859–880, 2008.
275. D. Mucci. Strong density results in trace spaces of maps between manifolds. *Manuscripta Math.*, 128(4):421–441, 2009.
276. V. Munnier. Integral energy characterization of Hajłasz–Sobolev spaces. *J. Math. Anal. Appl.*, 425(1):381–406, 2015.
277. H.-M. Nguyen. Some new characterizations of Sobolev spaces. *J. Funct. Anal.*, 237(2):689–720, 2006.
278. H.-M. Nguyen. Γ-convergence and Sobolev norms. *C. R. Math. Acad. Sci. Paris*, 345(12):679–684, 2007.
279. H.-M. Nguyen. Optimal constant in a new estimate for the degree. *J. Anal. Math.*, 101:367–395, 2007.
280. H.-M. Nguyen. Inequalities related to liftings and applications. *C. R. Math. Acad. Sci. Paris*, 346:957–962, 2008.
281. H.-M. Nguyen, Γ-convergence, Sobolev norms, and BV functions. *Duke Math. J.*, 157(3):495–533, 2011.
282. H.-M. Nguyen. A refined estimate for the topological degree. *C. R. Math. Acad. Sci. Paris*, 355(10):1046–1049, 2017.
283. L. Nirenberg. On elliptic partial differential equations. *Ann. Scuola Norm. Sup. Pisa (3)*, 13:115–162, 1959.
284. F. Pacard and T. Rivière. *Linear and nonlinear aspects of vortices. The Ginzburg-Landau model*, volume 39 of *Progress in Nonlinear Differential Equations and their Applications*. Birkhäuser Boston, Inc., Boston, MA, 2000.
285. M.R. Pakzad. Relaxing the Dirichlet energy for maps into S^2 in high dimensions. *Commun. Contemp. Math.*, 4(3):513–528, 2002.
286. M.R. Pakzad. Weak density of smooth maps in $W^{1,1}(M, N)$ for non-abelian $\pi_1(N)$. *Ann. Global Anal. Geom.*, 23(1):1–12, 2003.
287. M.R. Pakzad and T. Rivière. Weak density of smooth maps for the Dirichlet energy between manifolds. *Geom. Funct. Anal.*, 13(1):223–257, 2003.
288. M. A. Peletier, R. Planqué, and M. Röger. Sobolev regularity via the convergence rate of convolutions and Jensen's inequality. *Ann. Sc. Norm. Super. Pisa Cl. Sci. (5)*, 6(4):499–510, 2007.
289. K. Pietruska-Pałuba. On function spaces related to fractional diffusions on d-sets. *Stochastics Stochastics Rep.*, 70(3–4):153–164, 2000.
290. K. Pietruska-Pałuba. Heat kernels on metric spaces and a characterisation of constant functions. *Manuscripta Math.*, 115(3):389–399, 2004.
291. A. Pinamonti, M. Squassina, and E. Vecchi. Magnetic BV-functions and the Bourgain–Brezis–Mironescu formula. *Adv. Calc. Var.*, 12(3):225–252, 2019.
292. H. Poincaré. Cinquième complément à l'analysis situs. *Rend. Circ. Mat. Palermo*, 18:45–110, 1904.
293. A. Poliakovsky. On a minimization problem related to lifting of BV functions with values in S^1. *C. R. Math. Acad. Sci. Paris*, 339(12):855–860, 2004.
294. A. Poliakovsky. On a singular perturbation problem related to optimal lifting in BV-space. *Calc. Var. Partial Differential Equations*, 28(4):411–426, 2007.

295. A.C. Ponce. On the distributions of the form $\sum_i(\delta_{p_i} - \delta_{n_i})$. *C. R. Math. Acad. Sci. Paris*, 336(7):571–576, 2003.
296. A.C. Ponce. An estimate in the spirit of Poincaré's inequality. *J. Eur. Math. Soc. (JEMS)*, 6(1):1–15, 2004.
297. A.C. Ponce. A new approach to Sobolev spaces and connections to Γ-convergence. *Calc. Var. Partial Differential Equations*, 19(3):229–255, 2004.
298. A.C. Ponce. On the distributions of the form $\sum_i(\delta_{p_i} - \delta_{n_i})$. *J. Funct. Anal.*, 210(2):391–435, 2004.
299. A.C. Ponce. *Elliptic PDEs, measures and capacities. From the Poisson equations to nonlinear Thomas–Fermi problems*, volume 23 of *EMS Tracts in Mathematics*. European Mathematical Society (EMS), Zürich, 2016.
300. A.C. Ponce and D. Spector. On formulae decoupling the total variation of BV functions. *Nonlinear Anal.*, 154:241–257, 2017.
301. A.C. Ponce and J. Van Schaftingen. Closure of smooth maps in $W^{1,p}(B^3; S^2)$. *Differential Integral Equations*, 22(9–10):881–900, 2009.
302. A. Ranjbar-Motlagh. An integral type characterization of constant functions on metric-measure spaces. *J. Math. Anal. Appl.*, 385(1):194–201, 2012.
303. A. Ranjbar-Motlagh. A remark on the Bourgain–Brezis–Mironescu characterization of constant functions. *Houston J. Math.*, 46(1):113–115, 2020.
304. T. Rivière. Line vortices in the U(1)-Higgs model. *ESAIM Contrôle Optim. Calc. Var.*, 1:77–167, 1995/1996.
305. T. Rivière. Dense subsets of $H^{1/2}(S^2, S^1)$. *Ann. Global Anal. Geom.*, 18(5):517–528, 2000.
306. T. Rivière and S. Serfaty. Limiting domain wall energy for a problem related to micromagnetics. *Comm. Pure Appl. Math.*, 54(3):294–338, 2001.
307. R. T. Rockafellar. Characterization of the subdifferentials of convex functions. *Pacific J. Math.*, 17:497–510, 1966.
308. J. Rubinstein and I. Shafrir. The distance between homotopy classes of S^1-valued maps in multiply connected domains. *Israel J. Math.*, 160:41–59, 2007.
309. J. Rubinstein and P. Sternberg. Homotopy classification of minimizers of the Ginzburg-Landau energy and the existence of permanent currents. *Comm. Math. Phys.*, 179(1):257–263, 1996.
310. L.I. Rudin, S. Osher, and E. Fatemi. Nonlinear total variation based noise removal algorithms. *Physica D*, 60:259–268, 1992.
311. W. Rudin. *Real and complex analysis*. McGraw-Hill Book Co., New York, third edition, 1987.
312. T. Runst and W. Sickel. *Sobolev spaces of fractional order, Nemytskij operators, and nonlinear partial differential equations*, volume 3 of *de Gruyter Series in Nonlinear Analysis and Applications*. Walter de Gruyter & Co., Berlin, 1996.
313. É. Sandier. Lower bounds for the energy of unit vector fields and applications. *J. Funct. Anal.*, 152(2):379–403, 1998.
314. É. Sandier. Erratum: "Lower bounds for the energy of unit vector fields and applications" [J. Funct. Anal. 152 (1998), no. 2, 379–403]. *J. Funct. Anal.*, 171(1):233, 2000.
315. É. Sandier and S. Serfaty. *Vortices in the magnetic Ginzburg–Landau model*, volume 70 of *Progress in Nonlinear Differential Equations and their Applications*. Birkhäuser, Boston, MA, 2007.
316. D. Sarason. Functions of vanishing mean oscillation. *Trans. Amer. Math. Soc.*, 207:391–405, 1975.
317. R. Schoen and K. Uhlenbeck. A regularity theory for harmonic maps. *J. Differential Geom.*, 17(2):307–335, 1982.
318. R. Schoen and K. Uhlenbeck. Boundary regularity and the Dirichlet problem for harmonic maps. *J. Differential Geom.*, 18(2):253–268, 1983.
319. H. Seifert. Über das Geschlecht von Knoten. *Math. Ann.*, 110(1):571–592, 1935.
320. J. Serrin and D.E. Varberg. A general chain rule for derivatives and the change of variables formula for the Lebesgue integral. *Amer. Math. Monthly*, 76:514–520, 1969.
321. I. Shafrir. Personal communication to the authors, 2010.

322. I. Shafrir. On the distance between homotopy classes in $W^{1/p,p}(\mathbb{S}^1;\mathbb{S}^1)$. *Confluentes Math.*, 10(1):125–136, 2018.

323. L. Simon. Asymptotics for a class of nonlinear evolution equations, with applications to geometric problems. *Ann. of Math. (2)*, 118(3):525–571, 1983.

324. L. Simon. *Lectures on geometric measure theory*, volume 3 of *Proceedings of the Centre for Mathematical Analysis, Australian National University*. Australian National University, Centre for Mathematical Analysis, Canberra, 1983.

325. D. Smets. On some infinite sums of integer valued Dirac's masses. *C. R. Math. Acad. Sci. Paris*, 334(5):371–374, 2002.

326. D. Spector. L^p-Taylor approximations characterize the Sobolev space $W^{1,p}$. *C. R. Math. Acad. Sci. Paris*, 353(4):327–332, 2015.

327. D. Spector. On a generalization of L^p-differentiability. *Calc. Var. Partial Differential Equations*, 55(3):Art. 62, 21, 2016.

328. M. Squassina and B. Volzone. Bourgain–Brezis–Mironescu formula for magnetic operators. *C. R. Math. Acad. Sci. Paris*, 354(8):825–831, 2016.

329. G. Stampacchia, *Équations elliptiques du second ordre à coefficients discontinus*. Séminaire de Mathématiques Supérieures, No. 16 (Été, 1965). Les Presses de l'Université de Montréal, Montreal, Que., 1966.

330. E.M. Stein. *Singular integrals and differentiability properties of functions*. Princeton Mathematical Series, No. 30. Princeton University Press, Princeton, NJ, 1970.

331. E.M. Stein. *Harmonic analysis: real-variable methods, orthogonality, and oscillatory integrals*, volume 43 of *Princeton Mathematical Series*. Princeton University Press, Princeton, NJ, 1993. With the assistance of Timothy S. Murphy, Monographs in Harmonic Analysis, III.

332. E.M. Stein. Personal communication to H. Brezis, 2001.

333. P. Strzelecki. Gagliardo–Nirenberg inequalities with a BMO term. *Bull. London Math. Soc.*, 38(2):294–300, 2006.

334. U. Tarp-Ficenc. On the minimizers of the relaxed energy functional of mappings from higher dimensional balls into S^2. *Calc. Var. Partial Differential Equations*, 23(4):451–467, 2005.

335. X. Tian and Q. Du. A class of high order nonlocal operators. *Arch. Ration. Mech. Anal.*, 222(3):1521–1553, 2016.

336. P. Tolksdorf. Regularity for a more general class of quasilinear elliptic equations. *J. Differential Equations*, 51(1):126–150, 1984.

337. H. Triebel. *Interpolation theory, function spaces, differential operators*, volume 18 of *North-Holland Mathematical Library*. North-Holland Publishing Co., Amsterdam, 1978.

338. H. Triebel. *Theory of function spaces*, volume 78 of *Monographs in Mathematics*. Birkhäuser Verlag, Basel, 1983.

339. H. Triebel. *Theory of function spaces. II*, volume 84 of *Monographs in Mathematics*. Birkhäuser Verlag, Basel, 1992.

340. S.V. Uspenskiĭ. Imbedding theorems for classes with weights. *Trudy Mat. Inst. Steklov.*, 60:282–303, 1961. English translation: *Am. Math. Soc. Transl.* 87:121–145, 1970.

341. J. Van Schaftingen and M. Willem. Set transformations, symmetrizations and isoperimetric inequalities. In *Nonlinear analysis and applications to physical sciences*, pages 135–152. Springer Italia, Milan, 2004.

342. C. Villani. *Optimal transport. Old and new*, volume 338 of *Grundlehren der Mathematischen Wissenschaften [Fundamental Principles of Mathematical Sciences]*. Springer-Verlag, Berlin, 2009.

343. B. White. The least area bounded by multiples of a curve. *Proc. Amer. Math. Soc.*, 90(2):230–232, 1984.

344. B. White. Homotopy classes in Sobolev spaces and energy minimizing maps. *Bull. Amer. Math. Soc. (N.S.)*, 13(2):166–168, 1985.

345. B. White. Homotopy classes in Sobolev spaces and the existence of energy minimizing maps. *Acta Math.*, 160(1–2):1–17, 1988.

346. M. Yamazaki. A quasihomogeneous version of paradifferential operators. II. A symbol calculus. *J. Fac. Sci. Univ. Tokyo Sect. IA Math.*, 33(2):311–345, 1986.

347. L.C. Young. Generalized curves and the existence of an attained absolute minimum in the calculus of variations. *C. R. Soc. Sci. Lett. Varsovie, Classe III*, 30:212–234, 1937.
348. L.C. Young. Some extremal questions for simplicial complexes. V. The relative area of a Klein bottle. *Rend. Circ. Mat. Palermo (2)*, 12:257–274, 1963.
349. W.P. Ziemer. *Weakly differentiable functions. Sobolev spaces and functions of bounded variation*, volume 120 of *Graduate Texts in Mathematics*. Springer-Verlag, New York, 1989.

Symbol Index

Symbols

$\#A$, Notation N. 42, xxxiv
|A|, Notation N.9, xxxi
$A_0(\Gamma)$, 143
$a \wedge b$, Notation N.28, xxxiii
A^c, Notation N.14, xxxi
A_g, 381
$A^\ell(T)$, 183
$A(T)$, 153
$A(u)$, 44
$B(\Gamma)$, 149
B_g, 382
$B^\ell(T)$, 183
BMO, Notation N.25, xxxiii
$|\mathbb{B}^N|$, Notation N.10, xxxi
$\mathbb{B}^N$, Notation N.8, xxxi
$B_R(x)$, Notation N.4, xxxi
$B(T)$, 153
$B(u)$, 44
BV, Notation N.23, xxxii
$B(x, R)$, Notation N.4, xxxi
$C(\Gamma)$, 144
C_g, 382
$C^\ell(T)$, 183
conv, Notation N.35, xxxiv
$C(T)$, 153
$C(u)$, 44
$C(x, R)$, Notation N.6, xxxi
$D\varphi$, 30
$D^a\varphi$, 30, 468
$D^c\varphi$, 30, 468
$D^j\varphi$, 30, 468
$D^s\varphi$, 468
$\mathscr{D}'(\Omega; \Lambda^k)$, Notation N. 41, xxxiv
$\deg u$, 7
d^*, 38
$\mathbb{D}$, Notation N.7, xxxi
$\mathbb{D}_r$, Notation N.7, xxxi
$D(u)$, 173, 174
$\mathscr{E}$, 108
$E(u)$, 25
Ext, Notation N.35, xxxiv
$\mathscr{F}$, 150
$\mathscr{F}^\ell$, 183
$f^\sharp\omega$, Notation N.37, xxxiv
$\mathscr{H}^{N-1} \llcorner S$, 31, 464
$\mathscr{H}^\ell$, Notation N.34, xxxiv
$\mathscr{H}^\ell \llcorner S$, Notation N.34, xxxiv
J_φ, 31
Ju, 12, 174
$\widetilde{J}u$, 14
$L^1(\Omega; \Lambda^k)$, Notation N. 41, xxxiv
$L(a, d)$, 66
$\mathscr{M}$, Notation N.21, xxxii
$\mathscr{M}(\Omega; \Lambda^k)$, Notation N. 41, xxxiv
$M_t(f)$, 217
$M_{t,p}(f, \Omega)$, 217
$\mathbb{N}$, Notation N.1, xxxi
$\mathbb{N}^*$, Notation N.1, xxxi
$\mathscr{P}(X)$, Notation N.36, xxxiv
$\mathscr{Q}(T)$, 153
$\mathscr{Q}(u)$, 46
$\mathscr{R}$, 255
$\widetilde{\mathscr{R}}$, 256
$R_k(u)$, 176
$R_p(u)$, 49
$\mathscr{R}(u)$, 46
SBV, 31

H. Brezis and P. Mironescu, *Sobolev Maps to the Circle*, Progress in Nonlinear Differential Equations and Their Applications 96,
https://doi.org/10.1007/978-1-0716-1512-6

$\mathbb{S}^1$, Notation N.8, xxxi
$S_k(u)$, 175
$\mathbb{S}^k$, Notation N.8, xxxi
$S_p(u)$, 51
$S(x, R)$, Notation N.5, xxxi
$u \wedge \nabla v$, Notation N.31, xxxiii
VMO, Notation N.25, xxxiii
VMO_{loc}, Notation N.27, xxxiii
$W^{s,p}$, Notation N.17–N.18, xxxii
$W^{s,p}(\Omega; \mathbb{S}^1)$, Notation N.26, xxxiii
$W^{s,p}_{loc}$, Notation N.27, xxxiii
$W^{s,p}_{loc}(\Omega; \mathbb{S}^1)$, Notation N.27, xxxiii
$X_{s,p}$, 417
χ_A, Notation N.12, xxxi
Λ^k, Notation N. 41, xxxiv
Ω, Notation N.13, xxxi
ω_N, Notation N.10, xxxi
$\Sigma(u)$, 25
$\Sigma'(u)$, 409
$\Sigma^*(u)$, 189, 192
$*$ (Hodge operator), Notation N.32, xxxiv
$| \, |_{\mathrm{Lip}}$, Notation N.16, xxxi
$| \, |_{W^{s,p}}$, Notation N.19–N.20, xxxii
$\| \, \|_{\mathscr{M}}$, Notation N.22, xxxii
$\| \, \|_W$, 41
$\| \, \|_{W^{s,p}}$, Notation N.17–N.18, xxxii
$⨍$, Notation N.11, xxxi
$\nabla^{\perp}$, Notation N.33, xxxiv
$\perp$, Notation N.33, xxxiv
$\varphi^{\pm}$, 31

Subject Index

A
Approximate projection, 258

B
Bourgain–Brezis–Mironescu (BBM) formula, 231
BV functions
approximate limit of, 31
jump set of, 31
one-sided limits of, 31
precise representative of, 31

C
Cauchy formula, 8, 339
Chain rule
in $W^{1,p}$, 11
Volpert's chain rule in BV, 31
Coarea formula, 92, 145
Connection, 101, 112
minimal connection, 104, 113
Current
area-minimizing 2-rectifiable current, 151
ℓ-rectifiable current, 115
mass-minimizing 1-rectifiable current, 117
1-rectifiable current, 115

D
Degree
and Fourier coefficients, 358
classical, 7, 65, 339
estimates for the, 352
VMO, 341
Density
of smooth maps, 312
of the class $\mathscr{R}$, 313
Dipole
construction, 28, 94, 147
removing, 183
Distributional Jacobian, 12
Duality, 37

F
Factorization, 253
Family
of mollifiers, Notation N.40, xxxiv
of standard mollifiers, Notation N.39, xxxiv
of standard radial mollifiers, Notation N.39, xxxiv
Frankl–Pontryagin theorem, 141

G
Gagliardo–Nirenberg inequality, 437
Gap phenomenon, 386
Geometric measure theory (GMT), 115
Ginzburg–Landau vortices, 394

H
Homotopy classes, 358, 369

H. Brezis and P. Mironescu, *Sobolev Maps to the Circle*, Progress in Nonlinear Differential Equations and Their Applications 96,
https://doi.org/10.1007/978-1-0716-1512-6

I
Infinite energies, 396
Integral representation for the degree, 376
Integral representation of Ju, 119

J
Jacobian of u, 12, 174

K
Kronecker formula, 339, 376

L
Least area spanned by a contour, 142, 149, 189
Least energy with prescribed degree, 364
Least energy with prescribed Jacobian, 51, 175
Lifting, 1
 in fractional Sobolev spaces, 206
 minimal, 8, 25
 uniqueness of, 227
 in *VMO*, 216
 in $W^{1,p}$, 11
Littlewood–Paley decomposition, Definition 15.2, 434

M
Minimal configuration, 67, 85
Monge–Kantorovich (MK), 44, 134

N
Nice map, 65
Non-local
 approximation, 247
 energy, 242
 functional, 251
 perimeter, 242
Normalized bump function, Notation N.38, xxxiv

O
Optimal transport (OT), 76

P
Phase, 1
 estimates of, 299
Profile decomposition, 307

R
Rectifiable Jordan arc, 101
Rectifiable Jordan curve, 101
Relaxed energy, 4, 49, 176, 289, 409
Renormalized energy, 396

S
Seifert hypersurface, 165
Singularities
 line, 133
 point, 65
 topological, 65
Sobolev spaces
 composition in, 441
 definition, norms, seminorms, Notation N.17–N.20, xxxi
 embeddings of, 435
 equivalent (spatial) norms, Theorems 15.1–15.2, 434
 products in fractional, 497
 weighted, 259
Sphere-valued maps, 173
Square root, 22
Standard domain, Notation N.13, xxxi
Surface
 minimal, 134
 nice, 160
 Seifert, 141

T
Traces in Sobolev spaces, 331
Triebel–Lizorkin spaces, Definition 15.3, 434

Author Index

A
Acerbi, 62
Adams, C.C., 141
Adams, D.R., 226, 442
Adams, R.A., 14, 230, 465
Afriat, 128
Alberti, 57, 124, 130, 170, 171, 179, 187–189, 191, 193, 203
Alinhac, 498
Almgren, 91, 124, 129, 145, 170, 171, 189, 193, 203
Ambrosio, ix, 24, 26, 30, 101, 102, 130, 238, 252, 467, 477, 479
Angelsberg, 226
Antonucci, 248
Aubert, 252
Avellaneda, 394

B
Bahri, 430
Baldo, 57, 124, 130, 170, 171, 179, 187–189, 191, 193, 203
Ball, ix, 12, 56, 57
Barbieri, 252
Belloni, 62
Berlyand, 380
Bethuel, vii, viii, 11, 34, 49, 56, 57, 63, 111, 130, 179, 183, 196–199, 203, 224, 250, 255, 312, 313, 320, 326–329, 336, 375, 380, 395–400
Bojarski, 250, 252
Bony, 498
Borghol, 250
Bouchitté, 62
Bourdaud, 210, 442
Bourdon, 251
Bourgain, viii, 34, 56, 60, 130, 132, 158, 171, 201, 207, 225, 230, 238, 249–251, 276, 277, 283, 284, 298, 301, 307–309, 328, 337, 364, 376, 380, 398, 442
Bousquet, ix, 57, 126, 169, 170, 258, 278, 283, 295, 298, 317, 326, 328, 427, 429, 499
Boutet de Monvel-Berthier, 453
Boutet de Monvel, L., 379
Brezis, vii, 3, 11, 12, 14, 34, 47, 49, 51, 56–61, 63, 73, 75, 90, 113, 123, 125, 127, 128, 130, 132, 158, 170, 171, 190, 194, 199, 201, 203, 207–209, 217, 219, 225, 228, 233, 238, 246–252, 276, 277, 283, 284, 296, 298, 301, 306–309, 313, 321, 322, 326, 328, 337, 339, 341, 344, 349, 354, 364, 365, 367, 369, 375–380, 384, 394–401, 427–430, 436, 438, 442, 451, 453, 456, 463, 466
Browder, 91, 124, 129, 145, 170, 171, 189, 193, 203
Buades, 251
Buttazzo, 61, 62

C
Carbou, 56
Caselles, 101, 102, 130, 479
Cheeger, 252

H. Brezis and P. Mironescu, *Sobolev Maps to the Circle*, Progress in Nonlinear Differential Equations and Their Applications 96,
https://doi.org/10.1007/978-1-0716-1512-6

Chemin, 498
Chen, 388, 394, 401
Chiron, 57, 224, 252
Coifman, 225
Coll, 251
Coron, viii, 34, 49, 57, 60, 61, 63, 125, 128, 130, 131, 170, 190, 194, 197, 201, 203, 326, 329, 379, 384, 394, 399, 401
Cui, 250, 252
Curcă, ix, 42, 56

D
Da Lio, 276
Dahlberg, 442
David, 219
Dávila, ix, 24, 56, 57, 240, 250
De Giorgi, 467
De Marco, 251
Demengel, 56, 130, 197, 250, 326, 328, 329, 336, 401
DeVore, 210
DiBenedetto, 386
Di Marino, 252
Dinca, 19, 178, 339, 407
Di Nezza, 431
Du, 250
Dupaigne, 18

E
Ericksen, viii, 201
Escobedo, 57, 181, 208, 328, 442
Evans, ix, 92, 120, 122, 195, 482, 484

F
Fatemi, 251
Federer, 13, 102, 124, 130, 131, 145, 160, 170, 187, 193, 328, 467, 479, 484, 486, 496
Figalli, ix, 238
Fleming, 102, 130, 328, 479
Fonseca, 62
Franchi, 252
Frankl, 141
Frazier, ix, 226, 442
Fusco, 24, 26, 30, 467, 477, 479

G
Gérard, 498
Gabber, 379
Gariepy, 92, 120, 122, 195, 482, 484
Garsia, 250, 251
Georgescu, 453
Giaquinta, ix, 13, 24, 27, 38, 41, 49, 52, 53, 58, 63, 131, 190, 195, 201, 204, 250, 328, 401, 468
Gilbarg, 11, 229
Gilboa, 245, 252
Giusti, 23, 26, 473
Gobbino, 248, 250
Goffman, 62
Grigor'yan, 251
Gromov, 252
Gu, 251
Gulliver, 394

H
Hadiji, 130, 401
Hajłasz, 124, 252, 327
Han, 396
Hang, 130, 327, 328, 429
Hardt, ix, 62, 145, 171, 250, 328, 336, 381, 384, 388, 394, 397, 399, 401
Hatcher, 426
Heinonen, 252
Hélein, vii, ix, 197, 326, 329, 375, 380, 395–397, 400
Hong, 394
Hu, 251, 424, 426

I
Ignat, ix, 24, 27, 52, 53, 56, 57, 60, 127, 251, 400
Ihnatsyeva, 251
Isobe, 337
Iwaniec, 41

J
Jabin, 127
Jäger, 394
Jerrard, 131, 171, 187, 297
John, 216, 465
Jones, 455, 463
Jost, 252

K
Kahane, 380
Kantorovich, 128
Karadzhov, 250
Kaul, 394

Kinderlehrer, viii, 201, 250, 328, 336
Kinnunen, 251, 252
Kirby, 165
Korevaar, J., 362, 380
Korevaar, N., 252
Kornprobst, 252
Koskela, 18, 252
Kozma, 364
Kozono, 226
Kuksin, 250

L
Lam, 250, 252
Lamy, ix, 53
Lassoued, 380
Lau, ix, 251
Leoni, 230, 245, 246, 249, 251, 431, 501
Levi, 328
Lewis, 386
Lieb, viii, 21, 34, 57, 60, 61, 63, 91, 124, 125, 128, 129, 131, 145, 170, 171, 189, 190, 193, 194, 201, 203, 384, 394, 401
Lieberman, 383
Lin, ix, 62, 130, 185, 250, 327, 328, 336, 381, 383, 384, 388, 394, 397, 399, 401, 429
Li, Yanyan, ix, 250, 377, 380, 427–430
Li, Yi, 383
Loss, 21
Lu, 250, 252
Luckhaus, 384

M
Malý, 124
Marcus, 21, 442
Mariconda, 251
Masnou, 101, 102, 130, 479
Matano, 400
Mawhin, ix, 19, 178, 339, 407
Maz'ya, ix, 16, 208, 250, 251, 276, 431, 438, 443
Mengesha, 251
Merlet, ix, 24, 57, 309
Meyer, 208, 225, 442
Millot, ix
Milman, 250
Mingione, 62
Mironescu, 3, 51, 56–60, 63, 75, 90, 123, 127, 128, 130, 132, 170, 171, 201, 207, 208, 213, 224, 226, 238, 250, 251, 258, 259, 274, 276–278, 283, 284, 287, 292, 295, 296, 298, 300, 301, 307–309, 313, 321, 322, 326, 328, 337, 344, 358, 362, 365, 368, 369, 376–378, 380, 398, 427, 430, 431, 436, 438, 442, 463, 466, 473, 502
Mizel, 21, 62, 442
Modica, 13, 24, 27, 38, 41, 49, 53, 131, 204, 250, 401, 468
Molnar, ix, 170, 189, 194, 203, 226, 278, 298, 300, 301, 308, 309
Mora, 250
Morel, 101, 102, 130, 251, 479
Morgan, 130, 131, 160, 184, 483, 486
Mucci, 52, 58, 62, 63, 190, 195, 201, 204, 328
Munnier, 252

N
Nagel, 250
Nguyen, Hoai-Minh, ix, 12, 61, 199, 246–248, 252, 278, 298, 306, 307, 309, 340, 354, 357, 376–378, 380
Nguyen, Luc, 400
Nirenberg, ix, 21, 208, 209, 216, 219, 225, 228, 250, 339, 341, 344, 349, 367, 369, 376, 377, 379, 380, 451, 453, 456, 463, 465

O
Orlandi, 57, 124, 130, 170, 171, 179, 187–189, 191, 193, 203, 398
Osher, 245, 251, 252
Otto, 127

P
Pacard, 400
Pajot, 251
Pakzad, 190, 201, 203, 328
Palatucci, 431
Pallara, 24, 26, 30, 467, 477, 479
Peletier, 252
Perthame, 127
Picenni, 248
Pietruska-Pałuba, 251
Pisante, ix, 368
Pitts, 145, 171
Planqué, 252
Poincaré, 426
Poliakovsky, ix, 57, 58, 60, 233

Ponce, ix, 18, 56–58, 60, 63, 127, 130, 131, 197, 244, 326, 328, 473, 492
Pontryagin, 141
Popov, 210
Pumberger, 226
Purice, 453

R
Ranjbar-Motlagh, 233, 251
Reshetnyak, 252
Rivière, ix, 226, 276, 328, 397, 400
Rockafellar, 128
Rodemich, 250, 251
Röger, 252
Rubinstein, ix, 369, 380
Rudin, 251, 295
Runst, 431, 443, 444, 497
Russ, ix, 259, 274, 278, 502
Rybalko, 380

S
Sandier, vii, ix, 380, 396
Sarason, 216, 217, 237
Savin, ix
Schoen, 196, 252, 327, 328, 379, 381, 384, 401
Seifert, 141
Serfaty, vii
Serrin, 21, 62, 392
Shafrir, ix, 3, 57, 60, 321, 322, 328, 344, 365, 369, 370, 377, 378, 396
Shanmugalingam, 252
Shaposhnikova, 208, 250, 251, 438, 443
Shirikyan, 250
Sickel, 362, 431, 436, 442–444, 497
Simon, 130, 384
Sire, ix, 259, 278
Slastikov, 400
Smets, 131, 398
Solimini, 251
Soner, 131, 171, 187, 297
Souček, 13, 24, 27, 38, 41, 49, 53, 58, 131, 204, 250, 401, 468
Spector, 245, 246, 249, 251
Squassina, 252
Stampacchia, 229
Stein, 250, 261, 271, 465, 499
Sternberg, ix, 380
Strzelecki, 226
Swanson, 124

T
Tarp-Ficenc, 190, 204
Tian, 250
Tolksdorff, 386
Triebel, 362, 440, 502
Trudinger, 11, 229

U
Uhlenbeck, vii, 196, 327, 328, 379, 381, 384, 401
Uspenskiĭ, 258, 259

V
Valdinoci, 431
Van Schaftingen, ix, 197, 224, 233, 250, 292, 298, 326, 328, 337, 427
Varberg, 21, 392
Villani, ix, 128
Volpert, 467

W
Wadade, 226
Wang, 384, 394, 401
Wettstein, 276
White, 184, 328, 427
Willem, 250

X
Xiao, 250

Y
Yamazaki, 434
Young, 101, 184
Yung, 233

Z
Zarnescu, 400
Zheng, 11, 56, 111, 179, 183, 196, 197, 203, 328, 329
Ziemer, 124

CPSIA information can be obtained
at www.ICGtesting.com
Printed in the USA
BVHW010222130122
626134BV00002B/6

9 781071 615102